Communications
in Computer and Information Science   1440

More information about this series at http://www.springer.com/series/7899

Mayank Singh · Vipin Tyagi ·
P. K. Gupta · Jan Flusser ·
Tuncer Ören · V. R. Sonawane (Eds.)

Advances in Computing and Data Sciences

5th International Conference, ICACDS 2021
Nashik, India, April 23–24, 2021
Revised Selected Papers, Part I

 Springer

Editors
Mayank Singh
Consilio Research Lab
Tallinn, Estonia

P. K. Gupta
Jaypee University of Information
Technology
Waknaghat, Himachal Pradesh, India

Tuncer Ören
University of Ottawa
Ottawa, ON, Canada

Vipin Tyagi
Jaypee University of Engineering
and Technology
Guna, Madhya Pradesh, India

Jan Flusser
Institute of Information Theory
and Automation
Prague, Czech Republic

V. R. Sonawane
MVPS's Karmaveer Adv. Baburao
Ganpatrao Thakare College of Engineering
Nashik, Maharashtra, India

ISSN 1865-0929   ISSN 1865-0937 (electronic)
Communications in Computer and Information Science
ISBN 978-3-030-81461-8   ISBN 978-3-030-81462-5 (eBook)
https://doi.org/10.1007/978-3-030-81462-5

This Springer imprint is published by the registered company Springer Nature Switzerland AG
The registered company address is: Gewerbestrasse 11, 6330 Cham, Switzerland

Preface

Computing techniques like big data, cloud computing, machine learning, and the Internet of Things (IoT) etc. are playing a key role in the processing of data and retrieving of advanced information. Several state-of-art techniques and computing paradigms have been proposed based on these techniques. This volume contains papers presented at 5th International Conference on Advances in Computing and Data Sciences (ICACDS 2021) held during April 23–24, 2021, by MVPS'S Karmaveer Adv. Baburao Ganpatrao Thakare College of Engineering, Nashik, Maharashtra, India. Due to the COVID-19 pandemic, ICACDS 2021 was organized virtually. The conference was organized specifically to help bring together researchers, academicians, scientists, and industry experts and to derive benefits from the advances of next generation computing technologies in the areas of advanced computing and data sciences.

The Program Committee of ICACDS 2021 is extremely grateful to the authors who showed an overwhelming response to the call for papers, with over 781 papers submitted in the two tracks of Advanced Computing and Data Sciences. All submitted papers went through a double-blind peer-review process, and finally 103 papers were accepted for publication in the Springer CCIS series. We are thankful to the reviewers for their efforts in finalizing the high-quality papers.

The conference featured many distinguished personalities like Vineet Kansal, Pro-Vice Chancellor, A. P. J. Abdul Kalam Technical University, India; Komal Bhatia, J. C. Bose University of Science and Technology, India; Tuncer Ören, University of Ottawa, Canada; Aparna Pandey, Gulf Medical University, UAE; Shailendra Mishra, Majmaah University, Saudi Arabia; Robert Siegfried, Aditerna GmbH, Germany; Anup Girdhar, Sedulity Group, India; Arun Sharma, Indira Gandhi Delhi Technical University for Women, India; Sarika Sharma, Symbiosis International (Deemed University), India; Shashi Kant Dargar, University of KwaZulu-Natal, South Africa; and Prathmesh Churi, NMIMS University, India, among many others. We are very grateful for the participation of all speakers in making this conference a memorable event.

The Organizing Committee of ICACDS 2021 is indebted to Smt. Neelimatai V. Pawar, Sarchitnis, MVPS, India, for the confidence that she gave to us during organization of this international conference, and all faculty members and staff of MVPS'S Karmaveer Adv. Baburao Ganpatrao Thakare College of Engineering (KBTCOE), India, for their support in organizing the conference and for making it a grand success.

We would also like to thank N. S. Patil, N. B. Desale, and S. P. Jadhav, KBTCoE, India; Sameer Kumar Jasra, University of Malta, Malta; Hemant Gupta, Carleton University, Canada; Nishant Gupta and Archana Sar, MGM CoET, India; Arun Agarwal, University of Delhi, India; Kunj Bihari Meena, Neelesh Jain, Nilesh Patel, and Kriti Tyagi, JUET Guna, India; Vibhash Yadav, REC Banda, India; Sandhya Tarar, Gautam Buddha University, India; Vimal Dwivedi, Abhishek Dixit, and Vipin

Deval, Tallinn University of Technology, Estonia; Sumit Chaudhary, Indrashil University, India; Supraja P., SRM Institute of Science and Technology, India; Lavanya Sharma, Amity University, Noida, India; Sheshang Degadwala, Sigma Institute of Engineering, India; Poonam Tanwar and Rashmi Agarwal, MRIIS, India; Rohit Kapoor, SK Info Techies, Noida, India; and Akshay Chaudhary and Tarun Pathak, Consilio Intelligence Research Lab, India, for their support.

Our sincere thanks to Consilio Intelligence Research Lab, India; the GISR Foundation, India; SK Info Techies, India; and Print Canvas, India, for sponsoring the event.

August 2021

Mayank Singh
Vipin Tyagi
P. K. Gupta
Jan Flusser
Tuncer Ören
V. R. Sonawane

Organization

Steering Committee

Alexandre Carlos Brandão Ramos	UNIFEI, Brazil
Mohit Singh	Georgia Institute of Technology, USA
H. M. Pandey	Edge Hill University, UK
M. N. Hooda	BVICAM, India
S. K. Singh	IIT BHU, India
Jyotsna Kumar Mandal	University of Kalyani, India
Ram Bilas Pachori	IIT Indore, India
Alex Norta	Tallinn University of Technology, Estonia

Chief Patron

Neelimatai V. Pawar	Sarchitnis, Maratha Vidya Prasarak Samaj, India

Patrons

Tushar R. Shewale (President)	Maratha Vidya Prasarak Samaj, India
Manikrao M. Boraste Sabhapati	Maratha Vidya Prasarak Samaj, India

Honorary Chairs

N. S. Patil (Education Officer and Principal)	MVPS's KBTCOE, India
N. B. Desale (Vice-principal)	MVPS's KBTCOE, India

General Chairs

Jan Flusser	Institute of Information Theory and Automation, Czech Republic
Mayank Singh	Consilio Research Lab, Estonia

Advisory Board Chairs

Shailendra Mishra	Majmaah University, Saudi Arabia
P. K. Gupta	JUIT Solan, India
Vipin Tyagi	JUET Guna, India

Technical Program Committee Chairs

Tuncer Ören	University of Ottawa, Canada
Viranjay M. Srivastava	University of KwaZulu-Natal, South Africa
Ling Tok Wang	National University of Singapore, Singapore
Ulrich Klauck	Aalen University, Germany
Anup Girdhar	Sedulity Group, Delhi, India
Arun Sharma	Indira Gandhi Delhi Technical University for Women, India

Conference Chair

V. R. Sonawane	MVPS's KBTCOE, India

Conference Co-chair

S. P. Jadhav	MVPS's KBTCOE, India

Conveners

Sameer Kumar Jasra	University of Malta, Malta
Hemant Gupta	Carleton University, Canada

Co-conveners

V. C. Shewale	MVPS's KBTCOE, India
Ghanshyam Raghuwanshi	Manipal University, India
Prathamesh Churi	NMIMS, Mumbai, India
Lavanya Sharma	Amity University, Noida, India

Organizing Chairs

Shashi Kant Dargar	University of KwaZulu-Natal, South Africa
V. S. Pawar	MVPS's KBTCOE, India

Organizing Co-chairs

Abhishek Dixit	Tallinn University of Technology, Estonia
Vibhash Yadav	REC Banda, India
Nishant Gupta	MGM CoET, India

Organizing Secretaries

Akshay Kumar	CIRL, India
Rohit Kapoor	SKIT, India
M. P. Kadam	MVPS's KBTCOE, India

Creative Head

Tarun Pathak   Consilio Intelligence Research Lab, India

Program Committee

A. K. Nayak   Computer Society of India, India
A. J. Nor'aini   Universiti Teknologi MARA, Malaysia
Aaradhana Deshmukh   Alaborg University, Denmark
Abdel Badeeh Salem   Ain Shams University, Egypt
Abdelhalim Zekry   Ain Shams University, Egypt
Abdul Jalil Manshad Khalaf   University of Kufa, Iraq
Abhhishek Verma   Indian Institute of Information Technology and
                                    Management, Gwalior, India
Abhinav Vishnu   Pacific Northwest National Laboratory, USA
Abhishek Gangwar   Center for Development of Advanced Computing,
                                    India
Aditi Gangopadhyay   IIT Roorkee, India
Adrian Munguia   AI MEXICO, USA
Amit K. Awasthi   Gautam Buddha University, India
Antonina Dattolo   University of Udine, Italy
Arshin Rezazadeh   University of Western Ontario, Canada
Arun Chandrasekaran   National Institute of Technology Karnataka, India
Arun Kumar Yadav   National Institute of Technology Hamirpur, India
Asma H. Sbeih   Palestine Ahliya University, Palestine
Brahim Lejdel   University of El-Oued, Algeria
Chandrabhan Sharma   University of the West Indies, West Indies
Ching-Min Lee   I-Shou University, Taiwan
Deepanwita Das   National Institute of Technology Durgapur, India
Devpriya Soni   Jaypee Institute of Information Technology, India
Donghyun Kim   Georgia State University, Georgia
Eloi Pereira   University of California, Berkeley, USA
Felix J. Garcia Clemente   Universidad de Murcia, Spain
Gangadhar Reddy   RajaRajeswari College of Engineering, India
    Ramireddy
Hadi Erfani   Islamic Azad University, Iran
Harpreet Singh   Alberta Emergency Management Agency, Canada
Hussain Saleem   University of Karachi, Pakistan
Jai Gopal Pandey   CSIR-Central Electronics Engineering Research
                                    Institute, Pilani, India
Joshua Booth   University of Alabama in Huntsville, Alabama
Khattab Ali   University of Anbar, Iraq
Lokesh Jain   Delhi Technological University, India
Manuel Filipe Santos   University of Minho, Portugal
Mario José Diván   National University of La Pampa, Argentina
Megat Farez Azril Zuhairi   Universiti Kuala Lumpur, Malaysia

Mitsunori Makino	Chuo University, Japan
Moulay Akhloufi	Université de Moncton, Canada
Naveen Aggarwal	Panjab University, India
Nawaz Mohamudally	University of Technology, Mauritius
Nileshkumar R. Patel	Jaypee University of Engineering and Technology, India
Nirmalya Kar	National Institute of Technology Agartala, India
Nitish Kumar Ojha	Indian Institute of Technology Allahabad, India
Paolo Crippa	Università Politecnica delle Marche, Italy
Parameshachari B. D.	GSSS Institute of Engineering and Technology for Women, India
Patrick Perrot	Gendarmerie Nationale, France
Prathamesh Chur	SVKM's NMIMS Mukesh Patel School of Technology Management and Engineering, India
Pritee Khanna	Indian Institute of Information Technology, Design and Manufacturing Jabalpur, India
Purnendu Shekhar Pandey	Indian Institute of Technology (Indian School of Mines) Dhanbad, India
Quoc-Tuan Vien	Middlesex University, UK
Rubina Parveen	Canadian All Care College, Canada
Saber Abd-Allah	Beni-Suef University, Egypt
Sahadeo Padhye	Motilal Nehru National Institute of Technology, India
Sarhan M. Musa	Prairie View A&M University, Texas, USA
Shamimul Qamar	King Khalid University, Saudi Arabia
Shashi Poddar	University at Buffalo, USA
Shefali Singhal	Madhuben & Bhanubhai Patel Institute of Technology, India
Siddeeq Ameen	University of Mosul, Iraq
Sotiris Kotsiantis	University of Patras, Greece
Subhasish Mazumdar	New Mexico Tech, New Mexico
Sudhanshu Gonge	Symbiosis International University, India
Tomasz Rak	Rzeszow University of Technology, Poland
Vigneshwar Manoharan	Bharath Corporate, India
Xiangguo Li	Henan University of Technology, China
Youssef Ouassit	Hassan II University, Morocco

Sponsor

Consilio Intelligence Research Lab, India

Co-sponsors

GISR Foundation, India
Print Canvas, India
SK Info Techies, India

Contents – Part I

Contents – Part II

An Energy-Efficient Hybrid Hierarchical Clustering Algorithm for Wireless Sensor Devices in IoT

Nitesh Chouhan[1(✉)] and S. C. Jain[2]

[1] Department of IT, MLV Textile and Engineering College, Bhilwara 311001, India
[2] Computer Science and Engineering Department, Rajasthan Technical University, Kota 324010, India

Abstract. An advancement made in wireless technologies has developed a greater impact over the Internet of Things (IoT) systems. Clustering is one of the efficient approaches that connects and organizes the sensor nodes by balancing the loads and maximizing the lifespan of the network. This paper presents an Energy-Efficient Hybrid Hierarchical Clustering Algorithm that performs the characteristics of static and dynamic clustering formation. The proposed algorithm performs by two processes, viz, a) cluster head selection using fuzzy C-Mean (FCM) approach and shortest route path finding using Reliable Cluster-based Energy-aware Routing protocol. The main characteristics is the improvement done in the cluster formation and selection. The designed protocol is simulated using a programming language, NS2 and the performance measures studied are packet delivery ratio, end-to-end delay, and energy utilization. The results have stated that the developed HHCA approach outperforms better than the AODV protocol.

Keywords: Clustering · IoT · WSNs · Hierarchical routing and Fuzzy C-mean approach

1 Introduction

Advances made in the wireless sensor networks have attracted and gained momentum in the various fields of real-time applications like medicine, military, monitoring systems, tracking systems and so forth. The different aspects and requirements of the applications have led to the development of low-cost and power consumption wireless devices. Decision making process is one of the supporting tools which helps to take action by analyzing several parameters. The incorporation of different devices exposes a different form of information and thus, the decision making process becomes quite complex tasks [1]. In specific, information obtained from the WSNs has the potentiality of associating towards different devices which can lead to delayed information transfer processes. In Spite of those unique features rendered by WSNs, despite its application in real-world is limited. In general, the radio range is used to connect the different devices. Relied upon the requirement of application, the sensor nodes are deployed and communicated. Nodes form networks by organizing among themselves to reachable and great value

© Springer Nature Switzerland AG 2021
M. Singh et al. (Eds.): ICACDS 2021, CCIS 1440, pp. 1–14, 2021.
https://doi.org/10.1007/978-3-030-81462-5_1

information from the physical environment [2].The nodes are managed by the clustering approach. It is performed in two ways, namely, centralized and distributed [3]. In the viewpoint of centralized clustering approach, the sink node takes charge of collecting the information from wireless networks. Each sensor node is provided with the global knowledge since the sink node is limited from the aspects like energy constraints, and storage constraints. Finally, the sink node estimates the CHs and also its members. However, it is not suitable for the optimal based large-scale environment [4]. The distributed clustering approach makes use of local knowledge wherein each sensor node is capable of electing the CHs on the basis of requirements. In the case of heterogeneous sensor networks, the deployment of static and dynamic clustering approach has brought the challenges like network congestion, heavy traffic rate, under sampling and oversampling of the cluster centers. In order to eradicate (or) minimize the effects of energy and storage constraints, a hybrid clustering approach [5] is innovated in this study. The rest of the work is divided as follows: Sect. 2 gives the related work; Sect. 3 presents the scope of clustering technologies in WSNs; Research methodology is given in Sect. 4; Experimental Results and Analysis in Sect. 5 and finally, concludes the work in Sect. 6.

2 Related Study

This segment discusses the reviews of existing techniques. Different sensor nodes are placed on the wireless environment under multi-hop communications [6]. Sink nodes have consumed an additional energy to transmit/ receive the data packets. Energy hole issue is one of the vital concepts which was evaluated under AODV, DSR and TORA protocols. The results have stated that the AODV and DSR protocols performed better than the TORA from the aspects of packet delivery ratio, throughput and overheads. Though it has improved the network lifetime, the increased sensors node in the topology has lowered the performance. Reaching the base station has become quite complex in large-scale networks. This leads to the routing problem which was resolved by the protocol, named, On-Hole Children Reconnection (OHCR) and On-Hole Alert (OHA) [7]. The connectivity factor between the sensors nodes were efficiently handled under energy metrics. Compared to Shortest Path Tree (SPT) and Degree Constrained Tree (DCT), the suggested technique has achieved 75% increased network lifetime. Further, an energy-efficient LEACH protocol [8] was studied to improve the residual energy. Each deployed sensor node was projected into a clustering process. Depending on the instruction given by cluster heads, the resources are optimized. Though it has reduced the energy-consumption rate, the traffic flow between sensors under the clustering process is not explored.

Quality of Service (QoS) based routing protocols [9] were introduced under multi-objective functions. Here, a heuristic based neighbor selection models were formulated under geographic based routing models. It has followed the distance, delay and path based metrics to obtain optimal routing path. It has significantly reduced the network consumption, yet the network congestion rate becomes increased. Several offloading computational algorithms [10] were introduced to effectively utilize the route-discovery mechanisms for building topologies, cluster head formation and cluster head selection. The time taken for cluster head selection is higher while discovering the routes. It develops an overhead over the protocol. Owing to it, several clustering based routing protocols were developed using non-deterministic approaches [11]. PSO protocol for Hierarchical Clustering (PSO-HC) was designed to improve the lifetime of the CHs as well as network scalability. System has reduced the cluster head, and the link quality of networks can't be studied.

With the baseline of PSO, Multiple-sink Placement algorithm [12] was suggested for encoding the particles and evaluating the fitness. Depending on the hop count, the multiple hop count was employed for energy-efficient systems. The position of the sink has significantly depleted the energy and thus, heuristics models were used for finding the minimum sink utilization. Then, Enhanced PSO-Based Clustering Energy Optimization (EPSO-CEO) algorithm [13] was suggested to improve the local searching algorithms for cluster head selection. With the help of multihop routing protocols, the network lifetime and the consumed energy were enhanced. The data collection and aggregation methods have to be enhanced. Delay-sensitive based multi-hop routing protocols were established by [14]. Here, an end-to-end delay was improvised using probability blocking mechanisms. Different numerical simulations were formulated for relay nodes and thus, reduced the end-to-end delay. Though QoS models are improved, the probability of the hop count was increased in small-scale networks.

Different optimization techniques were studied by different researchers. Since the energy transmission/consumption [15] remains to be challenged, a different numerical solver was suggested. Cluster head selection was done by the clusters rate, hop-count rate and the relay nodes. Cluster based aggregation mechanisms were designed for inter-cluster and intra-cluster communication. Network overloads [16] were re-formulated to reduce the congestion near cluster nodes. However, extracting the cluster head nodes are not properly defined. Energy consumption rate was improvised the position of the sink nodes. Ring routing protocols were introduced to reduce the overhead of the mobile sinks. System has increased the packet delivery rate on small-scale networks. Along with the similar objectives, Genetic algorithm [17] was studied for network lifetime enhancement and the cost minimization.

In the case of virtual networks, wireless connectivity plays a key role in cluster head formation. The deployment of Weakly Connected Dominating Set (WCDS) [18] has explored the proper utilization of the cluster head selection. The network edge and balancing of loads were improved in this study. The system has obtained better utilization of cluster heads, irrespective of the node size. Similar approaches were explored in the game theory applications [19]. Again, the clustering algorithm has been stimulated in the IoT applications. It was explored in the LEACH protocol [20] and obtained 60% development in throughput, network span and residual energy. Similarly, Game theory based Energy Efficient Clustering routing protocol (GEEC) [21] was developed to balance the energy efficiency and the lifespan without compromising the QoS of wireless networks.

Owing to it, an Energy Efficient Data Aggregation scheme for Clustered Wireless Sensor Network (EEDAC-WSN) [22] was established for small-sized control frames of cluster member nodes. By monitoring the stability of the node, the member nodes are efficiently communicated in the networks. Compared to the LEACH protocol, the suggested protocols have lowered the delay rate even for small-scale networks as well as large scale networks. Induction trees in hierarchical based clustering nodes [23] have been studied to resolve the inference problems during induced tree formation. Induced tree of the crossed cube (ITCC) was designed on the basis of the degree of the graph nodes. During clustering based communication process, the sensor node takes maximized energy which was resolved by WEMER protocol [24]. Though it was concentrated on improvising the gateway nodes, the congestion between those nodes are not concentrated.

3   Clustering in WSN

This Clustering technologies play a key role in the WSNs which have assisted in improving the performance of the network. The WSNs is classified into two networks, viz, flat networks and clustered networks.

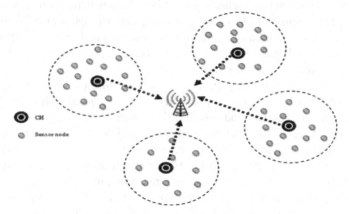

Fig. 1. Clustered WSNs

The Fig. 1 presents the clustered wireless sensor networks [25]. Initially, the sensor nodes are distributed arbitrarily in the wireless environment. Cluster Head (CH) is selected by the deployed sensor nodes based on the energy parameter. It helps to collect, aggregate and transmit the data to the sink nodes. This takes the hierarchy structure of the sensor nodes. It also diversifies the member nodes and the cluster heads. The other parameters in clustering are the cluster number/size, complexity of algorithms, overlapping, relay-overhead in inter-cluster and routing policies. The cluster heads are elected by its co-sensor nodes of assigned probabilities known as probabilistic algorithms. Whereas, cluster heads are selected under a certain criteria as well as proximity of sensor nodes known as non-probabilistic algorithms.

4   Research Methodology

This section presents the methodology of the research study. A novel routing protocol, named Reliable Cluster-based Energy-aware Routing (RCER) is designed with the objectives of enhancement of network lifetime and reduced usage of cluster number/size. The proposed phases are presented as follows:

4.1 Creating Wireless Sensor Based IoT Environment:

Consider a set of sensor nodes, as $S = \{N_1, N_2....N_n\}$ deployed in the IoT environment which are randomly associated with each other. Initially, the sensor nodes and the sink nodes are discrete in nature. All sensor nodes are equipped with a similar amount of energy. By the use of signal strength obtained value, the distance between the nodes are computed.

4.2 Hybrid Hierarchical Clustering Approach (HHCA)

This phase aim is to increase the network lifetime of the deployed sensor nodes. It follows the static as well as dynamic wireless environment. Here, the clustering has been done on demand basis i.e. whenever a specific sensor area initiates for transmission purpose, at that time, the cluster head has elected. Let us assume, network lifetime of a sensor node is denoted as, $N_{Lifetime(t)}$. The elapsed time taken by nodes until the first node depletes its energy is given as $N_{ElapTime}$. Collectively, it is represented as,

$$N_{ElapTime} = min \qquad (1)$$

The hybrid HCA eliminates the demerits of static and dynamic clustering approaches, by not performing clustering at each round. Based on the sync_pulse window, the clustering process is initiated for the particular sensor area. The below pseudo-code depicts the working of HHCA.

Algorithm 1: Hybrid Hierarchical Clustering Approach

Setup stage:
         The residual energy of a cluster head node $\rightarrow \Re_{CH}$
         $\beta: constant\,value\,ranges\,0 \leq \beta \leq 1$
         $S = \{N \in S_{CH} \vee \Re_N > 0\}$
$\forall N \in S: If \Re_N < \beta \Re_{CH}(N)$ Then
         The node N with data packets allocates the time bits.
         Collects the data packets
         Transmits to the Sink node in a multi-hop manner.
If the sync_ pulse has obtained then
         The node N gets ready to do the clustering process.
Once the sink node obtains data packet:
         If time_bits (T) = true   then
         Sink node broadcasts the sync_ pulse per round in the specified sensor
         area.

4.3 Fuzzy C-Means (FCM) Clustering Approach

The Once the lifetime of a network is preserved, a cluster head node has been selected. Here, Fuzzy C-mean clustering approach is employed as an energy-efficient routing process in WSNs. The objective of the FCM approach is to elect the CH from the set of sensor nodes. In association with the above process, the nodes are aware of when to initiate the clustering process. Let assume, set of clusters is represented as $C = \{c_1, c_2, ...c_n\}$. The deployment of FCM is to efficiently utilize the energy of the sensor nodes. Thus, objective function of energy minimization is given as:

$$E_M = \sum_{i=1}^{c} \sum_{j=1}^{N} degree_{ij}^{M} \, distance_{ij}^{2} \tag{2}$$

Where,

$degree_{ij}$ is the j's degree of a node on cluster i.

$distance_{ij}$ is the distance between node j and the midpoint of cluster i.

Once the distance between nodes are minimized, then the energy consumption is also minimized. Since the CH election is done by rounds, each round performs data transmission operation. Initially, the nodes are assigned with the degree and thus, it is used for the formation of a clustering process. The proximity value of each node is estimated. If the value of a node is closer to the proximity value, then it is labelled as cluster C_1. Likewise, all sensor nodes in the application area are examined. Once the clusters are formed, the sink node selects the nearest cluster center C_i to become CH and then the information of Cluster head node broadcasted. The total number of clusters C is computed as:

$$C = \frac{\sqrt{N}}{\sqrt{2 * \pi}} \sqrt{\frac{E_{fu}}{E_M}} \frac{M}{distance_{Sink \, Node}^{2}} \tag{3}$$

In some cases, the non-cluster head at particular time T may attempt to transmit the data packets to sink nodes. Thus, it can deplete the energy. In order to eliminate this scenario, an optimal cluster head has been elected by the present cluster head at each round. At the initial round, the sink node selects the CH. Since HHCA is followed, then a new CH generates the TDMA for each clustering process. The CH is selected, then the data transmission process is scheduled by TDMA.

4.4 Reliable Cluster-Based Energy-Aware Routing (RCER) Protocol

It is found that some rounds are suffering from overloads of cluster members, which leads to heavy congestions as well as unwanted energy consumption. Therefore, the need of finding the shortest route path has reduced the effects of energy consumption rate. The below algorithm 2 explains the working of RCER protocols. The proposed protocols reduced the cluster heads overload, and also enhanced the lifetime of the network.

Algorithm 2: RCER (Reliable Cluster-based Energy-aware Routing protocol)

Input: $C = \{c_1, c_2, \ldots c_n\}$, Neighborhood Hop & Simulation measurements
Begin
Function Clusters
    Estimate the cluster midpoint by getting spatial information of nodes
for (n = 1; n< = S ; n++)
Do
        If $\Re_N$ >optimum $_{threshold}$
set list_of_cluster_centers []
End if
end for
for (n = 1; n< = list_of_cluste_centers;n++)
        Estimate the midpoint of N_n
        set the max(proximal value of N_n) as CH_n
end for
end procedure
Function Route detection
for each N_n
    Do
          call optimized_routing_protocol ()
  while (y! = destination)
          N_i sets next-hop by using highest FP_n
           FP_n Reply to y
               y = FP_n
  if y_n. sync_pulse is high
call route_restore ()
end if
end while
end for
end procedure
Function Nodes_status
        if (RE_c <μ*RE_{init})
             initiate re-routing ()
        if (Δt _ Sync_Pulse _expired) then
             initiate re-routing ()
end if
end if
end procedure
End

---

5 Experimental Results and Analysis

This section presents the experimental results of the proposed protocols and the significant observations from the results are also discussed.

    The Table 1 presents the simulation analysis of the IoT based WSNs environment.

Table 1. Simulation parameters

Parameters	Ranges
Channel	Wireless channel
Propagation	Two ray ground
Layer	Physical layer; Data link layer
Sensor area (M)	670 * 670
Packet length	5000 ifq
Routing protocol	RCER
No. of sensor nodes	28
Simulation time	15 s

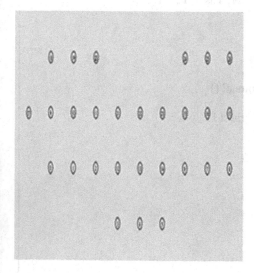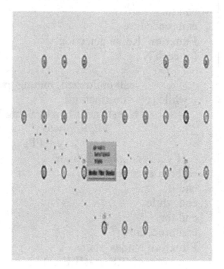

Fig. 2. (a) Deployment of sensor nodes; (b). Traffic analysis of the data packets in cbr.

The above Fig. 2(a) and Fig. 2(b) presents the deployment of the sensor nodes and the transmission data packets. In our study, the sensor nodes are deployed in an IoT environment. Therefore, a flat based topology is created. Then, initial transmission of data packets are observed in constant bit rate (cbr). The ability of the deployed sensor nodes are analyzed.

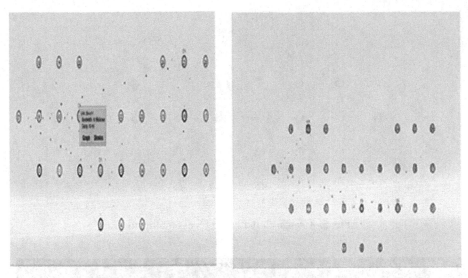

Fig. 3. (a). Creating the link between cluster and sensor nodes; (b). Formation of cluster head

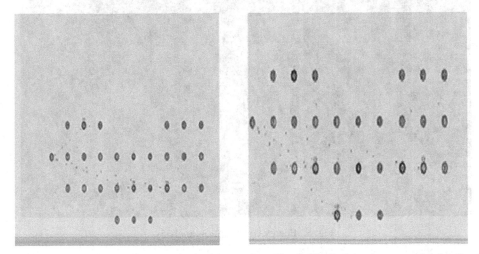

Fig. 4. (a). Data sending from cluster member to cluster head; (b). Find the shortest path selection

The above Fig. 3(a) presents the formation of clusters among the deployed sensor nodes. The proximity value of each sensor node is computed. The closer the proximity value, the nearest cluster center is elected. The Fig. 3(b) presents the formation of cluster heads. With the help of FCM approach, the cluster heads are elected under TDMA scheduling process.

The Fig. 4(a) presents the data transmission between cluster member and cluster head. Here, each CH is aware of residual energy of the cluster member node. By encountering the residual energy value, the data packets are collected, aggregated and transmitted to the sink node. The Fig. 4(b) presents the findings of optimal shortest path using RCER

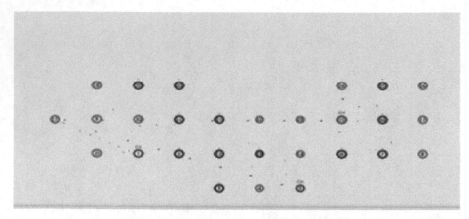

Fig. 5. Change cluster head process multipath

Fig. 6. Energy consumption analysis

protocol. In each round, the clustering approach is instantiated on the received request. However, the data collision prevails due to the oversampling (or) under sampling of the cluster heads and cluster members.

The above Fig. 5 presents the working of the second cluster heads via multipath approach. After completing the first round, the next round is initialized with new cluster head elected by current cluster head.

The Fig. 6 and Table 2 presents the analysis of energy consumption rate. The objective of the study is to minimize the energy consumption rate by proposing RCER protocols. It explored the better consumption rate than the existing AODV protocol. It is achieved by using an on-demand clustering approach which lacks in AODV protocol.

Table 2. Comparison of energy consumption

Nodes	AODV	RCER
0	100	100
5	98.8	99
10	94	97.5
15	91	94.8

Table 3. Comparison of end to end delay

Time (in seconds)	AODV	RCER
0	0	0
2	0.7	0.1
4	1.1	0.2
6	2.9	1.1
8	3.24	1.29

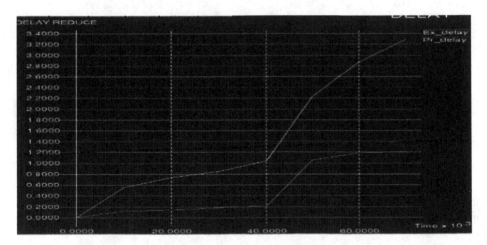

Fig. 7. End-to- End delay analysis

The Fig. 7 and Table 3 presents the analysis of end-to-end delay. Generally, delay occurs due to the network congestion (or) data congestion. In this system, both the congestion prevailing among the sensor area is eradicated by applying FCM approach.

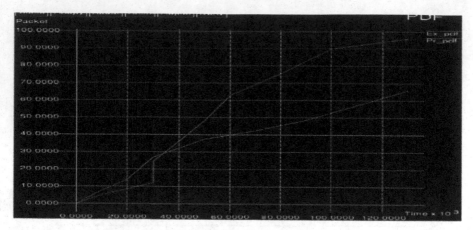

Fig. 8. Packet delivery ratio

Table 4. Comparison of packet delivery ratio

Time (in Seconds)	AODV (Packets kbps)	RCER (Packets kbps)
0.0	0	0
2.0	15	9
4.0	32	35
6.0	40	65
8.0	45	75
10.0	52	90
12.0	63	95

The above Fig. 8 and Table 4 represents the analysis of packet delivery ratio. The obtained results stated that the proposed routing protocols have significantly increased the PDR than the existing protocol. Due to the minimization utility factor of energy model, the higher data packets are communicated.

6   Conclusion

Rapid innovations of wireless technologies has impressed the researchers to delve into the study of wireless based IoT systems. In this study, an energy-efficient Hybrid Hierarchical Clustering Algorithm for Wireless Sensor Devices in IoT is designed. It is explored by two phases, namely, cluster head selection using FCM approach and shortest route path finding using Reliable Cluster-based Energy-aware Routing protocol. Our main novelty is, the clustering process is initiated on the received request from the sensor nodes. It eliminates the traffic analysis caused during clustering analysis. The proposed protocol is simulated using NS2 programming language, and the performance measures studied are packet delivery ratio, end-to-end delay, and energy utilization. From the results, the proposed HHCA approach outperforms better than the AODV protocol.

References

1. Thein, M.C.M., Thein, T.: An energy efficient cluster-head selection for wireless sensor networks. In: International Conference on Intelligent Systems, Modeling and Simulation, pp. 287–291 (2010)
2. Lee, J.S., Cheng, W.L.: Fuzzy-logic-based clustering approach for wireless sensor networks using energy prediction. IEEE Sens. J. **12**, 2891–2897 (2012)
3. Attea, B.A.A., Khalil, E.A.: A new evolutionary based routing protocol for clustered heterogeneous wireless sensor networks. Appl. Soft Comput. **12**, 1950–1957 (2012)
4. Kuila, P., Jana, P.K.: A novel differential evolution based clustering algorithm for wireless sensor networks. Appl. Soft Comput. **25**, 414–423 (2014)
5. Shokouhifar, M., Jalali, A.: A new evolutionary based application specific routing protocol for clustered wireless sensor networks, AEU-Int. J. Electron. Commun. **69**, 432–441 (2015)
6. Rohini, S., Lobiyal, D.K.: Proficiency analysis of AODV, DSR and TORA Ad- hoc routing protocols for energy holes problem in wireless sensor networks. Procedia Comput. Sci. **57**(2015), 1057–1066 (2015)
7. Mohemed, R.E., Saleh, A.I., Abdelrazzak, M., Samra, A.S.: Energy-efficient routing protocols for solving energy hole problem in wireless sensor networks. Comput. Netw. **114**(2017), 51–66 (2017)
8. Arumugam, G.S., Ponnuchamy, T.: EE-LEACH: development of energy-efficient LEACH protocol for data gathering in WSN. EURASIP J. Wireless Commun. Netw. **2015**(1), 1–9 (2015)
9. Mazaheri, M.R., Homayounfar, B., Mazinani, S.M.: QoS based and energy aware multipath hierarchical routing algorithms in WSNs. Wireless Sensor Netw. **4**, 31–39 (2012)
10. Sharma, S., Jena, S.K.: Cluster based multipath routing protocol for wireless sensor networks. ACM SIGCOMM Comput. Commun. Rev. **45**(2), 15–20 (2015)
11. Elhabyan, R.S., Yagoub, M.C.E.: PSO-HC: particle swarm optimization protocol for hierarchical clustering in wireless sensor networks 978-1-63190-043-3. In: 10th IEEE International Conference on Collaborative Computing: Networking, Applications and Worksharing (CollaborateCom 2014), pp. 417–424 (2014)
12. P., C.S.R., Banka, H., Jana, P.K.: PSO-based multiple-sink placement algorithm for protracting the lifetime of Wireless Sensor Networks. In: Satapathy, S.C., Raju, K.S., Mandal, J.K., Bhateja, V. (eds.) Proceedings of the Second International Conference on Computer and Communication Technologies. AISC, vol. 379, pp. 605–616. Springer, New Delhi (2016). https://doi.org/10.1007/978-81-322-2517-1_58
13. Vimalarani, C., Subramanian, R., Sivanandam, S.N.: An enhanced PSO-based clustering energy optimization algorithm for wireless sensor network. Sci. World J. **2016**, 1–11 (2016)
14. Hyadi, A., Afify, L., Shihada, B.: End-to-end delay analysis in wireless sensor networks with service vacation. In: IEEE Conference on Wireless Communications and Networking, Istanbul, Turkey (2014)
15. Abu-Baker, A.K.: Energy-efficient routing in cluster-based wireless sensor networks: optimization and analysis. Jordan J. Electr. Eng. **2**(2), 146–159 (2016)
16. Kumbhar A.D., Chavan, M.K.: An energy efficient ring routing protocol for wireless sensor network. In: International conference on I-SMAC (IoT in Social, Mobile, Analytics and Cloud) (I-SMAC 2017) (2017)

17. Baranidharan, B., Santhi, B.: GAECH: genetic algorithm based energy efficient clustering hierarchy in wireless sensor networks. Hindawi Publ. Corp. J. Sens. **15**, 20–35 (2015)
18. Dou, C., Chang, Y.-H., Ruan, J.-S.: On the Performance of weakly connected dominating set and loosely coupled dominating set for wireless sensor/mesh networks. Appl. Mech. Mater. **764–765**, 929–935 (2015)
19. Habib, M.A., Moh, S.: Game theory-based routing for wireless sensor networks: a comparative survey. Appl. Sci. **9**, 2896 (2019)
20. Behera, T.M., Mohapatra, S.K., Samal, U.C., Khan, M.S., Daneshmand, M., Gandomi, A.H.: Residual energy based cluster-head selection in WSNs for IoT application. IEEE Internet Things J. (2019)
21. Lin, D., Wang, Q.: A game theory based energy efficient clustering routing protocol for WSNs. J. Wireless Netw. **23**(4), 1101–1111 (2017)
22. Roy, N.R., Chandra, P.: EEDAC-WSN: energy efficient data aggregation in clustered WSN. In: International Conference on Automation, Computational and Technology Management (ICACTM) (2019)
23. Zhang, J., Xu, L., Ye, X.: An efficient connected dominating set algorithm in WSNs based on the induced tree of the crossed cube. Appl. Math. Comput. Sci. **25**(2), 295–309 (2015)
24. Bello, A.D., Lamba, O.S.: Energy efficient for data aggregation in wireless sensor networks. Int. J/ Eng. Res. Technol. (IJERT) **9**(1), 110–120 (2020)
25. Bhatlavande, A., Phatak, A.: Energy efficient approach for in-network aggregation in wireless sensor networks. Int. J. Curr. Eng. Technol. **5**(4), 2874–2879 (2015)

Fund Utilization Under Parliament Local Development Scheme: Machine Learning Base Approach

Arun Sharma$^{(\boxtimes)}$ and Deepa Paliwal

Indira Gandhi Delhi Technical University for Women, Delhi, India
arunsharma@igdtuw.ac.in

Abstract. In Many parts of the country the life of people in rural as well as in urban is quite miserable and below average. Government has initiated many reforms and schemes for the upliftment of the citizen of the country. One such scheme is Members of Parliament Local Area Development Scheme (MPLADS). Ministry of Statistic and Programme Implementation (MoSPI) is entrusted with the responsibility of implementation of MPLAD Scheme in the entire country. Under the scheme, each Member of Parliament of Lok Sabha and Rajya Sabha has the choice to suggest works, listed as per the guidelines of the scheme to the District Collector in their elected constituency or State. The motive behind this research is to analyze the selection of scheme in each sector by MP's during 16th Lok Sabha tenure i.e. year (2014 to 2019) and priorities the work using Machine Language. Different Machine Learning (ML) Models were implemented to find the best Classification Models which can be used to find the high and low priority Sector/Subsectors implemented in state. It was further analyzed that allocation of funds in sectors is not equally distributed among all states. This study try to improvise the Work Management System (WMS) by forecasting the undermine sectors and schemes in each state so that uniform distribution of fund can take place in the entire sector which may optimize the benefit of the MPLAD scheme. For this purpose data of 16th Lok Sabha present on the MPLADS portal in public domain is used.

Keywords: Data analytics · MPLADS · Decision Tree · Random Forest

1 Introduction

The MPLADS scheme started on 23rd December 1993, it provisions a mechanism for elected Lok Sabha and Rajya Sabha Members of Parliament to recommend work of development nature in their constituency/State. The work should be of durable community assets as per the Sectors and subsectors provided in the MPLADS Guidelines The annual allocation of funds under MPLADS scheme is shown in Table 1. It has been increased over the years.

Under the Scheme, the annual entitlement of Rs. 5 crore is released, in two equal installments of Rs 2.5 crore each, by Government of India. This fund is released to the

© Springer Nature Switzerland AG 2021
M. Singh et al. (Eds.): ICACDS 2021, CCIS 1440, pp. 15–25, 2021.
https://doi.org/10.1007/978-3-030-81462-5_2

Table 1. Annual allocation of Fund [3].

Year	1993–94	1994–95 to 1997–98	1998–99 to 2010–11	2011–12 onwards
Entitlement (Rs.Crore)	0.05	1.00	2.00	5.00

Nodal District account of Member of Parliament, who take care of its entire fund and work related information and ensures the completion of work. The release of MPLADS funds is made as per the provisions of Para 4.1 to 4.3 of the extant Guidelines [2].The year-wise fund released under the scheme is given below (Fig. 1):

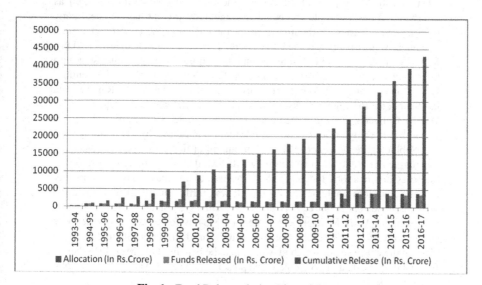

Fig. 1. Fund Release during Financial Year

The major sector and their corresponding subsectors present in the guidelines are (Table 2).

Table 2. Distribution of sector and subsectors [2]

S.No	Sector	Sub Sector
1	Drinking Water Facility	6
2	Education	5
3	Electricity Facility	3
4	Health and Family Welfare	9

(continued)

Table 2. (*continued*)

S.No	Sector	Sub Sector
5	Irrigation Facilities	5
6	Non-Conventional Energy Sources	3
7	Other Public Facilities	17
8	Railways, Road, Pathways and Bridges	20
9	Sanitation and Public Health	4
10	Sports	12
11	Works Relating to Animal Husbandry, Diary and Fisheries	5
12	Works Related to Agriculture	2
13	Works Relating to Cluster Dev for Handloom Weavers	1
14	Works Related to Urban Development	4

During the data analysis of 16th Lok Sabha, it was found that there is an uneven distribution of works and funds and there is a scope of improvising the scheme by using ML to identify and priorities the work based on high and low priority sectors and subsectors. There are various models used to identify the high and low priority sectors and sub subsectors. Different classification models like Logistic Regression, K-Nearest Neighbors (KNN), Decision Tree (DT), Random Forest, Artificial Neural Network (ANN) were used for this purpose.

2 Literature Review

ML has many advantage over the rule based that they focus on the individual transaction to give predefined result which is not sufficient to detach complicated transactional problem where as ML has a tendency to adapt in the new environment and produce efficient and reliable result from its past learning experience. Although there are many literature documenting benefit of ML and Big Data on citizen centric scheme, but literature using ML in detecting classification problem in Govt schemes are very limited. Most of the relevant literatures are based on function specification. This may be due to the lack of awareness or Knowledge.

Kartika Yadav and Mahesh Parmer [4] analysis the payment of wages to the worker under Mahatma Gandhi National Rural Employment Guarantee Act. Using Decision Tree, J48 and classification technique, they analyzed and found four types of delay in payment of wages these were Payment delay, System Delay, FTO Generation Delay etc.

Prabhakar et al. [5] focus on public opinion mining to find the best advertisement medium for Govt. Scheme and they applied Boosting, Bagging and Random Forest Technique to provide that the age group, education and qualification requirement are highly influence factor for the best advertising medium. They proposed a new classification algorithm to predict the best advertisement medium MEAN ERROR BASED ENSAMBLE.

Zhang and Trubey [6] monitored the one year transaction of U.S Financial Agency from money Launderer. They studied Machine Learning algorithm Naïve Bayes, logistic regression, DT, RF, support vector Machine (SVM) and ANN and predicted ANN as best a performance algorithm.

Wanli Xing and Dongping Du [7] analysis the dropout prediction in Massive open online course (MOOS) and propose the use a Deep Learning algorithm to build dropout model and suggest individual student dropout probabilities for intervention personalization.

Sabyasachi et al. [8] propose an e-Government Model named Intelligent Government System Advisor (IGoSA) to facilitate decision support system with scheme based analysis for citizen and Govt agencies as if now there is no system in place by which a Govt. department may quickly look for similar existing scheme before floating any new scheme.

Rajagopalan and Pandiya [9] suggested that a big data can result in increased efficiency and effectiveness in government by engagement of citizen in decision making. They propose open source big data analytic stack to provide the cost effective solution and aid in effective server implementation.

Mehta et al. [10] worked to identify dealers in Goods and service tax. In this paper they design a technique to detect a group of people who commit evasion in Goods and service tax with motive to reduce the tax liability They used cluster based anomaly detection based on cluster size where large clustered correspond to normal data and small cluster corresponds to anonyms.

Tamilarasi and Rani [11] suggested Crime rate against women using different ML algorithm. They used algorithm such as KNN, Naïve Bayes, Linear Regression, Classification and Regression Tree, SVM. From there study they analyzed that KNN has performed better than other algorithm. They also expressed the higher crime region and which crime type is happened in India.

Sawant et al. [12] have studied the analysis using ML algorithm to integrate the best and intelligent interactive interface to assist farmers in agricultural activities. They compare the accuracy of the algorithm and observe that the Random Forest is marginally better than KNN and DT.

Mayra et al. [13] have suggested a model by implementing algorithm and data mining. They capture logs of the attack to the data network of the organization and asses it with various algorithm of intrusion detection and suggested J48 and REPTree as best algorithm. They applied DT Model for calculating information gain via Entropy and minimize the error that arises from the values.

Anirudh et al. [17] proposed a method 'DRAP': DT and RF based classification Model to Predict Diabetes. Hybrid of both the algorithm is used to construct method, Seven features like glucose level, BP, Insulin, BMI, age, sex etc. of patient was taken up for classification and Prediction of Diabetes. The results of both the algorithms are marginally different.

Yuri Nieto et al. [19]. They study three algorithm for Strategic decision making in Higher Education Institution and predict graduation rates from the data of undergraduate Engineering students in South America. They studied algorithms like Decision Tree, Logistic Regression, and Random Forest. In this study they found Random Forest with best outcome.

3 Methodology

3.1 Data

The dataset used in study came from MPLADS portal i.e. mplads.gov.in. The data is structured in nature, we used data from 16th Lok Sabha "Completed Work Report" present on the public domain of the portal dated June 2020. It provides completed work details information. The Report is downloaded in CSV format. It contains many explanatory Variables. Variable selection process include variable transformation, missing or invalid values. In this study features are selected on the basis of high correlation values and finalized 7 main input variable. These are State, Sector, Scheme, Estimated amount, Sanctioned amount, Duration, count_of_work.

The output parameter is binary class (0, 1) which gives Low and High priority work with in state. We used same Data Set in all models to get fair comparison between different model performances.

3.2 Data Pre-processing

Data pre-processing is the method of converting data into useful format with optimized data quality. The Data Quality issue that need to be consider while preparing the data for prediction of analysis including removal of noise, outliers, missing values, inconsistent or duplicate data and the data that is biased shall be remove.

3.3 Methodology

Classification ML models likes Logistic Regression, K-Nearest Neighbors, Decision Tree, Random Forest, Artificial Neural Network were used for the study. The Analysis is performed using the Spyder, the Scientific Python Development Environment, is a free integrated development environment (IDE).

3.4 Basic Statistics

3.4.1 Frequency Distribution of Sector

From the following chart it shows that the sector with id 7 and 8 has maximum count i.e. most of the state has higher contribution of work in these sectors. It also shows that sector with id 6,11,12,13 and 14 has been least selected (Fig. 2).

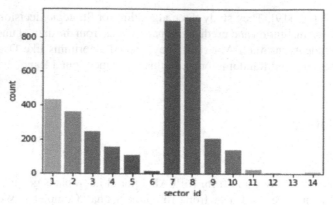

Fig. 2. Frequency distribution of sector

3.4.2 Frequency Distribution of Scheme

From the following chart it shows that the sector with sector id 40 has maximum count of works in the entire dataset.

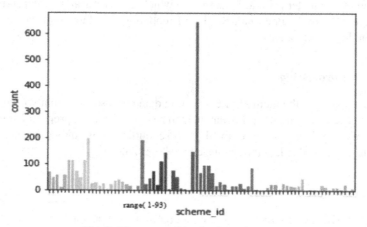

Fig. 3. Frequency Distribution of scheme

3.4.3 Histogram of the Dataset

Histogram of all seven independent variable is shown in Fig. 3

3.5 Association Rule-To Identify Relationship

Figure 4 provided the association between sector and the sanctioned amount in the entire states during the 16th Lok Sabha. Sector with id 8 has maximum value. It shows that maximum amount has been sanctioned in sector no 8 which has maximum no of subsectors or scheme present as per Table [2].

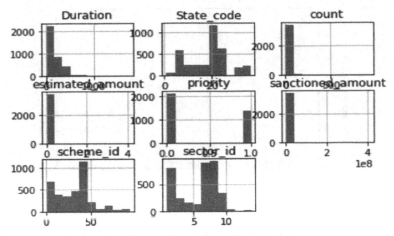

Fig. 4. Histogram of Dataset

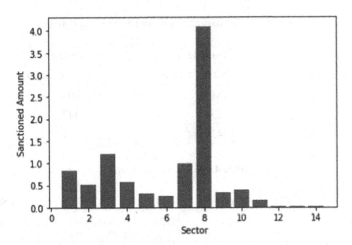

Fig. 5. Relationships between sector and sanctioned amount

Similarly when we visualize the association between scheme and Sanctioned Amount, the Fig. 5 shows that the scheme with Id 40 is the most implemented scheme in the entire tenure (Fig. 6).

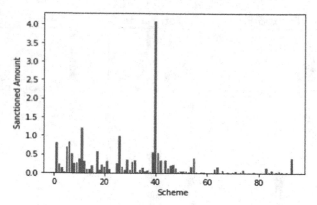

Fig. 6. Relationships between sector and sanctioned amount

4 Result

Table 3. Class distribution in training and testing dataset

Dataset		Training	Testing
Class	Total	No of Instances	No of Instances
0	2118	1514	604
1	1385	938	447
Total	3503	2452	1051

Table 4. Comparison report

Model	Class	Accuracy	Precision	Recall	F1	Support
Logistic Regression	0	0.43	0.83	0.01	0.02	604
	1		0.43	1	0.6	447
KNN	0	0.79	0.79	0.88	0.83	604
	1		0.8	0.68	0.74	447
Decision Tree	0	0.94	0.97	0.93	0.95	631
	1		0.90	0.95	0.92	420
Random Forest	0	0.89	0.89	0.95	0.91	628
	1		0.91	0.82	0.86	423
ANN	0	0.60	1	0.58	0.73	1049
	1		0	0.67	0.01	3

5 Discussion

The dataset consist of 3503 instances. Table 3 provided the information of distribution of class in training and testing dataset. In order to predict accuracy and classification summary, this research work is performed by tuning the parameter training and testing dataset For training data, all the ML algorithm were fitted with 70% of dataset while the remaining 30% will be used for testing.

Class with value 1 is interpreted as high.

Class with value 0 is interpreted with low.

Table 3 shows 2452 instances used for training the algorithm and 1051 instances were used testing purpose. In the testing dataset 604 instance were of high priority while 447 in lower priority.

Table 4 shows the comparative analysis of different machine learning algorithm used for this research. The dataset was trained to reflect 7 independent variable and one nominal variable as the dependent variable.

Five classification algorithms were used in the course for this research namely logistic regression, KNN, Decision Tree, Random Forest, ANN.Following performance metrics were considered for comparative analysis.

- Accuracy,
- Precision,
- Recall,
- F1-score,
- Support

It is observed that all the models were performed more that 60 percent of accuracy except for logistic regression. The Accuracy of Decision Tree is remarkably high while the accuracy of Random Forest is also good. But Generally Accuracy is not always considered enough information to make this decision. So we tried to find the precision, recall and F1 score of these ML models.

Now after comparing the Precision for high and low class, ANN shows the Maximum value 1 for low class and minimum 0 for high class so it cannot be considered as good Model while compared to all ML model. Decision tree again give the better Precision value, for low class it give .97 and for high class it gives .90 precision values.

Now comparing the values of Recall, it is observe that Decision Tree and Random Forest Provides good performance, Random Forest has .91 recall value for low class and .86 for high class while Decision Tree has .95 recall value for low class and .92 for high class. While considering both the classes it was concluded that decision tree Model is better that Random Forest.

Comparing F1 score which is the harmonic mean of the precision and recall we observed that the decision tree has the highest score for both the high and low class.

6 Conclusion and Future Scope

The research shows that data from https://mplads.gov.in is processed and analyzed with python which provides useful information and could be integrated to those applications having same set of features.

We observe that the best performance under Decision Tree and making Decision Tree an effective algorithm for finding the priority sector, schemes. It is found that Decision Tree perform well if we have limited feature and we have prior knowledge of important feature and data is non parametric and simple in nature. The Random Forest and Back propagation model also performed better as compared to Logistic Regression, KNN Model. It is worth noting that there are many factors which impact the performance of model. As it is supervised learning methodology so it is important that the data should assigned accurately. If preprocessing of data is not done then the biases or error or noise left in the data which will make the result erroneous. Model parameter should optimized otherwise it will lead to erroneous result.

For future scope we can further extend the work below the level of state like constituency, district and block so that the concrete information will be provided to the user. we will also try to extend the work with other model like Multi layer ANN (Deep Learning).we are quite optimistic that other model may exhibit better performance after thorough optimization.

Since this study is based on a single Lok Sabha Data, analysis data for more than one Lok Sabha and Rajya Sabha may optimize the performance of Machine learning models.

References

1. Tan, P.-N., Steinbach, M., Kumar, V.: Introduction to Data Mining, vol. 2, pp. 205–254. Pearson Education, United States (2006)
2. Member of Parliament Local Area Development Scheme Guidelines. https://www.mplads.gov.in/MPLADS/UploadedFiles/MPLADSGuidelines2016English_638.pdf
3. Member of Parliament Local Area Development Scheme Annual Report. https://www.mplads.gov.in/mplads/uploadedfiles/MPLADSAnnualReport2016-2017_284.pdf
4. Yadav, K., Parmer, M.: Mining better advertisement tool for government schemes using machine learning. Int. J. Comput. Trends Technol. (IJCTT), Int. J. Psychosoc. Rehabil. **23**(4) (2019)
5. Prabhakar, E., Suresh Kumar, V.S., Nandagopal, S., Dhivyaa, C.R.: Mining better advertisement tool for government schemes using machine learning. Int. J. Comput. Trends Technol. (IJCTT), Int. J. Psychosoc. Rehabil. **23**(4) (2019)
6. Zhang, Y., Trubey, P.: Machine learning and sampling scheme: an empirical study of money laundering detection. Computat. Econ. (2018)
7. Xing, W., Du, D.: Dropout prediction in MOOCs: using deep learning for personalized intervention. J. Educ. Comput. Res. 1–24 (2018)
8. Mohanty, S., MIshra, A.K., Panda, D.C.: IGoSA - a novel framework for analysis of and facilitating government schemes. In: IEEE 2nd International Conference on Recent Trends in Information Systems (ReTIS) (2015)
9. Rajagopalan, M.R., Pandiya, S.V.: Big data framework for national e-governance plan. In: Eleventh International Conference and Knowledge Engineering (2013)
10. Mehta, P., Mathews, J., Rao, S.V.K.V, Kumar, K.S., Suryamukhi, K., Babu, C.S.: Identifying malicious dealers in goods and services tax. In: The 4th IEEE International Conference on Big Data Analytics (2019)
11. Tamilarasi, P., Uma Rani, R.: Diagnosis of crime rate against women using kfold cross validation through machine learning algorithms. In: Proceedings of the Fourth International Conference on Computing Methodologies and Communication (2020)

12. Sawant, D., Jaiswal, A., Singh, J., Shah, P.: AgriBot, an intelligent interactive interface to assist farmers in agricultural activities. In: IEEE Bombay Section Signature Conference (IBSSC) (2019)
13. Macas, M., Lagla, L., Fuertes, W., Guerrero, G., Toulkeridis, T.: Data mining model in the discovery of trends and patterns of intruder attacks on the data network as a public-sector innovation. IEEE (2017)
14. Osisanwo, F.Y., Akinsola, J.E.T., Awodele, O., Hinmikaiye, J.O., Olakanmi, O., Akinjobi, J.: Supervised machine learning algorithm-classification and comparison. Int. J. Comput. Trends Technol. (IJCTT) **48**(3) (2017)
15. Kim, G.H., Trimi, S., Chung, J.-H.: Big-data applications in the government sector. Commun. ACM **57**(3) (2014)
16. Mohanty, S., Mangalapalli, V.K., Padhy, S.: GovSchemAna-a machine learning enabled android app for analysis of government schemes. Int. J. Comput. Sci. Mob. Appl. 5(10), 48–57 (2017)
17. Anirudh Hebber, P., Manoj Kumar, M.V., Sanjay H.A.: DRAP: decision tree and random forest based classification model to predict diabetes. In: 2019 1st International Conference on Advance in Information Technology. IEEE (2019)
18. Hwang, S., Yeo, H.G., Hong, J.S.: A new splitting criterion for better interrpretable trees. IEEE (2020)
19. Nieto, Y., Gacia-Diaz, V., Montenegro, C., Gonzalez, C.C., Crespo, R.G.: Usage of machine learning for strategic decision making at Higher Education Institutions. IEEE (2019)

Implementing Automatic Ontology Generation for the New Zealand Open Government Data: An Evaluative Approach

Paramjeet Kaur[✉] and Parma Nand

Auckland University of Technology, Auckland, New Zealand
{param.kaur,parma.nand}@aut.ac.nz

Abstract. Open Government Initiatives are increasingly gaining momentum and becoming important among developed and developing countries. Open government initiatives purpose is to provide transparency and data reuse. Governments around the globe are increasingly releasing data for public consumption, however, the drawback is that these data are disparate and in heterogeneous formats which makes it challenging to consume the data. This has given rise to a need for a framework that can transform these data into a form that is easily consumable and accessible to stakeholders as well as the general public. This paper introduces the design, implementation and usage of an approach driven by an ontology that captures the knowledge on a subset of data released by the government of New Zealand (NZ). This approach uses datasets from agriculture, and land use to generate ontologies where SPARQL queries are imposed to get the desired results. The implementation and evaluation outcomes demonstrated that the proposed approach is realistic to generate an Ontology Web Language (OWL) triple ontology from the given Comma Separated Value (CSV) datasets. In the future, this approach will further ease the semantic linking process of the multiple ontologies of different domains.

Keywords: Open government initiatives · Ontology Web Language · Linked open data · Semantic links · Ontology · SPARQL queries

1 Introduction

The information available in the open government initiatives are ineffective for sharing and reuse due to the huge volume and heterogeneity. By use of hyperlinks between web documents, the World Wide Web has connected the world. These hyperlinks are used to navigate free texts between hypertext markup language (HTML) pages. Data can thus be accessed via a single web link that integrates other connections in the original page [1]. Increased growth of knowledgeable social system has resulted in people are becoming more aware of their rights and want to involve and know about governance strategies. This has led to the need for transparency by governing bodies resulting in the release of data to the public to make the transparency rationale for policy decisions [2].

© Springer Nature Switzerland AG 2021
M. Singh et al. (Eds.): ICACDS 2021, CCIS 1440, pp. 26–36, 2021.
https://doi.org/10.1007/978-3-030-81462-5_3

The government can gain the trust of its citizens, by releasing data related to governance ministries. This gives the public a clearer picture of spending and policy decisions by increasing confidence among the public and government. The concept of open data was announced by United States (US) President Barrack Obama in 2009 [3]. According to him, it is the right of the individuals of a country to comprehend, what is happening within the Government and the way the government is investing their money. Many countries at that time were not in support of the concept, and Obama had faced several objections to this announcement. After this, however, the United Kingdom (UK) government introduced its Open Government Initiatives in 2012 [4]. Subsequently, many more developing and developed countries became proactive and embarked on their open data initiatives. Through numerous data portals, these open data initiatives publish raw data that are heterogeneous and complex to use, particularly by the general public.

During the past year's the increasing demand for ontologies in software system has gain momentum [5]. Ontologies are used in data sources for various purposes such as expanding upon resources for higher data retrieval, integrating knowledge from diverse sources and coupling intelligent systems automatically.

Despite several tools that have been presented to publish open data on the web, the recent tools and strategies are not compatible with all of the layers of data access. Thus, in this paper, we are proposing an approach for developing knowledge by representing and managing knowledge contained in government documents that are released as open data that can be represented in a single knowledge base. By utilizing suitable tools and the encoding technique, Resource Description Framework (RDF) and use of OWL to define complete ontologies to capture complex and disparate data.

The structure of the paper is as follows: section two provides the background information of approaches, tools and methods used for ontology design. Section three highlights the proposed methodology to create an ontology focusing on New Zealand Government Open data Initiatives. Section four demonstrate the evaluation and experiment results. Further, in section five the limitations and improvements are discussed for the proposed work and finally, section six includes the conclusion and future work.

2 Related Work

The first systematic study of open data ontology of public spending using triples was reported by Vafopoulos [6]. It was designed by using the data.gov.uk data portal. The input to the proposed architecture is given by Diavgeia which is an Extensible Markup Language (XML) based Application Programming Interface (API) and first Greek Government Open Data Portal. Output can be seen via SPARQL endpoints. However, ontology has a very basic class and relationship definitions. The data properties, objects, concept-restrictions and rules have not been created for the proposed ontology.

In another study, Theochairs, discussed an ontology [7] using protege 4.2 to link the public administration data. The built-in reasoner of protege is used to find out semantic errors in the designed ontology. This study is an attempt to present a part of the ontology concerning the characteristics of administrative acts. Therefore, the proposed ontology can be extended further by adding more concepts and their properties. It will contribute to forming a knowledge base for the management and development of open data. Furthermore, human evaluation is conducted in the form of posing questions to the ontology using SPARQL endpoints. However, human evaluation can be challenging because humans are disposed to make mistakes.

Another example is an attempt to enrich the Greek e-GIF ontology [8] where protege is used to designing the entities. Several entities are added in the existing ontology and comparison has done with Point of Single Contact (PSCs) of other European countries such as Cyprus, Malta, Spain and the Slovak Republic. However, more attempts are required for further enrichment of the entities. A refined version of ontology along with good comparison results can be used for semantic enrichment of higher elements of the web pages with Uniform Resource Identifiers (URI) properties. This is vital for the conversion of open government data into Linked Data.

Some further examples of linked open data are the exploratory study of Brazilian initiatives based on the principles of linked open data (LOD) is conducted by Ricardo et al. [9, 19] Brazilian portals are three stars which mean data sets are in XML, CSV and HTML.

Also, James et al. [10] have done discussion on the data.gov portal and the use of linked open data. They have focused on various sectors where linked data has utilized. In an analysis of linked open data, Zhao [11] retrieved graph-based ontology from the diverse data sources available publicly. The ontology alignment methods have applied to identify classes and properties of ontology from the data sets. Related classes and properties from different data sets are combined to find the missing "SameAs" links. This semi-automatically created cohesive approach solves the heterogeneity problem of ontology. Moreover, it finds out the missing and wrong properties in the data sets. However, only four datasets have been selected which is DBPedia, GeoNames, NYTimes and LinkedMDB. To extend the alignment process, more data sets are required and the Map-Reduce method can be used to deal with big data sets.

In an attempt to semantify open data, Khalifa [12] proposed a lightweight approach for re-using existing ontologies from Hoxha [13]. The main objective was to contribute to the knowledge of the semantic web and enabling data exchange and linking with other semantic sources over the web. However, the conversion process was not fully automated. Furthermore, there is no open data portal available in Saudi which gives rise to the problem of data extraction. If common vocabularies to access the data are developed, it will help to link open data initiatives world widely.

In an analysis of Open data for e-government Theocharis et al. [14] discussed the Greek open data initiatives opportunities and benefits. The authors have debated the challenges of opening the data for the public including the availability, accessibility, reuse, simplicity, global participation and redistribution. Moreover, the basic steps for converting data to open data have also highlighted by proposing an architecture for linked open government data.

Previous research typically only investigated the conversion of CSV to RDF, XML/HTML. There are an increasing need and scope for semantic enrichment of diverse data by opting for the OWL format [20]. More research is needed to understand the OWL conversion because RDF vocabularies provide the terms which are only helpful to create a basic description of the resources. However, the OWL provides a set of new terms targeting more detailed descriptions.

3 Proposed Methodology

In this section, we present the methodology to transform CSV data into OWL. The proposed methodology is different from the standard data transformation because it is a fully automated and hence less time-consuming process. It transforms the CSV datasets into OWL format so that an automatic ontology can be generated, and data can be extracted by imposing SPARQL queries. The core concepts of the approach are as follows:

- The CSV syntax and semantic follows the constraints and definition of dialect description RFC4180 document. Dialect description RFC4180 was used as it can recognize the CSV files format automatically. The DEFAULT method of the CSV Format Library is used to parse the CSV file [15]
- The CSV data is accompanied by Dublin Core metadata [16] annotations which offer interoperability for metadata vocabularies in OWL. The Dublin Core metadata enables information to be accurate and equally organized and enriched within various schemas.
- Protégé tool is used to visualize the generated ontologies.
- Apache Jena is used to converting the OWL files to RDF/Turtle format so that SPARQL queries can be imposed on the framework.

3.1 Process Flow and Architecture

Based on the proposed method, a prototype has designed. We have proposed an architecture for CSV to OWL and the corresponding converting process. The converting process takes a CSV file and meta vocabulary Dublin core as input and produces the converted OWL as output. The generated OWL file is stored in the local memory of the system. The resulting OWL file is then converted to Turtle format so that SPARQL Queries can be imposed. The whole process of generating ontology is divided into the two phases such as 1) CSV to OWL Conversion and visualization using Protégé, 2). SPARQL interface to Query the generated Ontology. The overall architecture and process flow along with important components are described in Fig. 1.

Here the first phase implements the generation of an automatic ontology from a given CSV data file. User can either upload the CSV file directly or can enter the available uniform resource locator (URL) of the CSV file. The current process can only take CSV data stream as input; however, this can easily be extended to other formats such as Portable document format (PDF), keyhole markup language (KML), HTML and JavaScript object notation (JSON).

Apache Commons CSV library is used to parse the CSV file where the CSV file follows the constraints and definition of dialect description RFC4180 [15] document. This defines the format of the CSV file such as header, end of the line and escaped character etc. and assist in handling the text-based fields of CSV. Based on the dialect description, the DEFAULT mode ensures that the non-Unicode characters of the CSV file are replaced by Unicode. The CSV Parser performs various functions to read and parse the row, cell values and quoted values of the CSV data.

The conversion of CSV data to OWL mainly requires additional metadata annotation which depicts the process of data interpretation. Dublin Core RDFS vocabulary was used for the description of generic metadata because it has been known as a tool that can be used by a non-expert to easily generate clear and descriptive records for information resources, while at the same time offering efficient search of the resources in the interconnected environment.

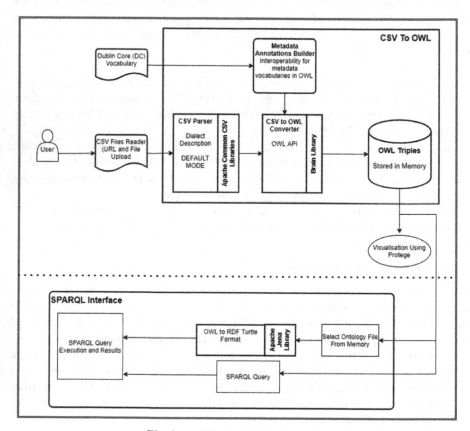

Fig. 1. Architecture and process flow

In the conversion process, the CSV datasets are transformed, and an Ontology is created from these datasets. Firstly, a name is assigned to the CSV file which is treated as the name of the ontology. After that, the column headers of the CSV file are assigned as data properties such as farm_type, area_ha, year, FID, region. These data properties have values that are considered as the individuals (instances) for the ontology. Each row under the column headers are treated as an individual or an instance and a unique identifier name is assigned to each record. These individuals hold the values for the data properties (such as farm_type "Forestry", area_ha 146236, the year 2002, FID "table-95343.100", the region "Hawke's Bay Region"). The ontology of the OWL format contains a set of axioms that include clear logical statements regarding three kinds of entities, namely, classes, individuals and data properties. Where class is a collection of concepts, things or objects. Individuals are the object or instances and data properties connect the individuals to the literals where the literals represent data value such as a particular integer or string.

Further, the axioms are created between the data properties and individuals so that the meaning of the class and its relationship can be explained, and it will further help to link multiple ontologies in future. After adding the corresponding axioms, the ontology is saved in OWL format at the local memory location. To visualize the generated ontology, Protégé 5.5.0 version was used where the OntoGraf feature is used to visualize the ontology.

Finally, the OWL-XML files are generated using the class and properties from the CSV files. These OWL-XML files contain information about the model of the ontology as well as the syntax of the individuals. To query in SPARQL, the generated OWL ontology are transformed into RDF format such as Turtle. Apache Jena library supported this conversion process.

3.2 SPARQL Interface to Query the Generated Ontology

SPARQL is a language that can be executed on RDF formatted Ontology triple stores which represents a subject (E.g. an individual), a predicate (a data or an object property) and an object (the resulting value of applying the predicate on the subject).

Once our ontology is attained. The final stage is to query the generated ontology with the help of the SPARQL interface. The SPARQL interface is designed so that the user can select the desired ontology file to query. To query in SPARQL we need to transform the generated OWL ontology to RDF format such as Turtle. Writing SPARQL queries that involve complex OWL expressions ranges from challenging to unpleasant because SPARQL query syntax is based on Turtle [17], which isn't intended for OWL. SPARQL queries against OWL data have to encode the RDF serialization of OWL expressions: these queries are typically verbose, difficult to write, and understand.

For the conversion process, we have utilized Apache Jena a Java library that can be used to convert the OWL files to RDF Turtle format and provides APIs to Query SPARQL from within a Java application. Figure 1 highlights the overall process flow of the SPARQL interface.

4   Case Study

The approach described above has been evaluated using the open data sets of agriculture, and land from the New Zealand Government website. Firstly, ontologies for agriculture and land datasets are created. For this purpose, the URL[1] of the datasets is entered into the system. The entered URL is parsed using the file parser and an OWL converter is used to convert the CSV datasets to OWL files. These OWL files are stored in the local memory of the system.

The OntoGraf tab of Protégé is used to visualize the ontology. Figure 2 highlights the link between the Owl: Thing and agriculture class. Accordingly, a snapshot and a

---

[1] https://data.mfe.govt.nz/services;key=554f7f05f1eb4f2584f356c01ea6073e/wfs?service=
WFS&version=2.0.0&request=GetFeature&typeNames=table-95343&outputFormat=csv.

schematic of the Agriculture and land ontologies are shown in Fig. 3. where, (A) refers to agriculture which is the subclass of owl: Thing having 5 data properties (area_ha, farm_type, FID, region, year) and 620 individuals (from Individual_1_159956422651 to Individual_602_159956422961). When the mouse pointer has hovered on the individual it highlights the data property assertions for that individual. And (B) refers to the land ontology where the red label shows the link between land class and the individual.

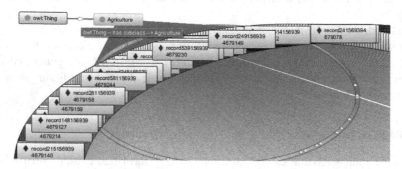

Fig. 2. The connection between owl: thing and agriculture class

To evaluate our approach, it is crucial to analyses the accuracy of the generated ontology. For this purpose, we have taken some sample SPARQL queries and fed those queries to the system. The results of those queries are recorded, and a manual evaluation is conducted for consistency.

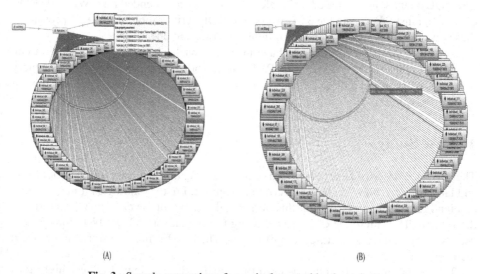

(A)   (B)

Fig. 3. Sample screenshots for agriculture and land ontologies.

The SPARQL queries are imposed on the datasets to extract useful information. The traditional data extraction methods are time-consuming as one needs to read through the datasets thoroughly but, an RDF query language makes the task easy as all records are stored as triples. One can easily find knowledgeable data by imposing the SPARQL queries. For testing purpose, we have imposed the following SPARQL queries on the ontology:

1) Find the Farm_types in Auckland for the year 2012. The corresponding SPARQL query is below:

```
PREFIX owl: http://www.w3.org/2002/07/owl#
PREFIX rdf: http://www.w3.org/1999/02/22-rdf-syntax ns#
SELECT *
WHERE {
?ind1 rdf:type owl:NamedIndividual.
?ind1rdf:type <http://www.ontogen.org/Agriculture/Agriculture>.
?ind1<http://www.ontogen.org/Agriculture/farm_type>    ?farm_type
.
?ind1<http://www.ontogen.org/Agriculture/year>2012 .
?ind1<http://www.ontogen.org/Agriculture/region>       "Auckland
Region".
}
```

2) Find the area_ha for urban type in the year 2008 for Bay_of_Plenty. The corresponding SPARQL query is below:

```
PREFIX owl: http://www.w3.org/2002/07/owl#
PREFIX rdf: http://www.w3.org/1999/02/22-rdf-syntax-ns#
SELECT  ? urban type
WHERE
{
   ?ind1 rdf:type owl:NamedIndividual .
   ?ind1 rdf:type <http://www.ontogen.org/Land/Land>.
   ?ind1 <http://www.ontogen.org/Land/type> ?type .
   ?ind1 <http://www.ontogen.org/Land/year> 2008 .
   ?ind1http://www.ontogen.org/Land/region "Bay_of_Plenty".}
```

The above SPARQL queries are simple yet effective as compare to traditional relational database queries. SPARQL queries can be executed to any database and results can be viewed as RDF by using any middleware. Whereas, relational database queries are specific to one particular database. SPARQL is a hypertext transport protocol (HTTP) protocol that provides access to any endpoints of SPARQL through a structural layer of transport. The results returned by RDF can be in different data formats and the entities of it can be identified as URIs [18]. Data forging with URIs enable data to be referenced directly throughout applications and resolves local search constraints. Therefore, supplementary APIs can be created and referenced. Later, the SPARQL queries will help to semantically link the multiple resources so that knowledge can be extracted without any constraints.

As per the first query 9 records were identified by the system, after analyzing the datasets manually, we found that the records identified by the system are correct. The second query identified 4 records that were found accurate during the manual analysis of the data. However, at this stage, we are querying only one ontology and using simple queries. In future, when multiple ontologies will be added and more complex SPARQL queries will be added, it can impact the overall accuracy of the results. The focus of this paper is to find the opportunity to generate an ontology for the open data sets of the NZ government. Once the ontology is generated to check the accuracy of the ontology and generated triples, some queries are imposed on the ontology.

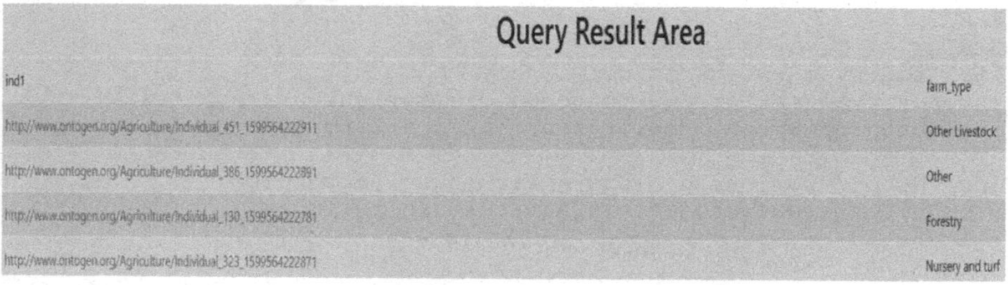

Fig. 4. Query results from agriculture datasets

Figure 4 shows the results captured for Agriculture datasets. The clickable links on the left will highlight the corresponding data properties for that individual. Figure 5 illustrates the details of the data properties for the first individual.

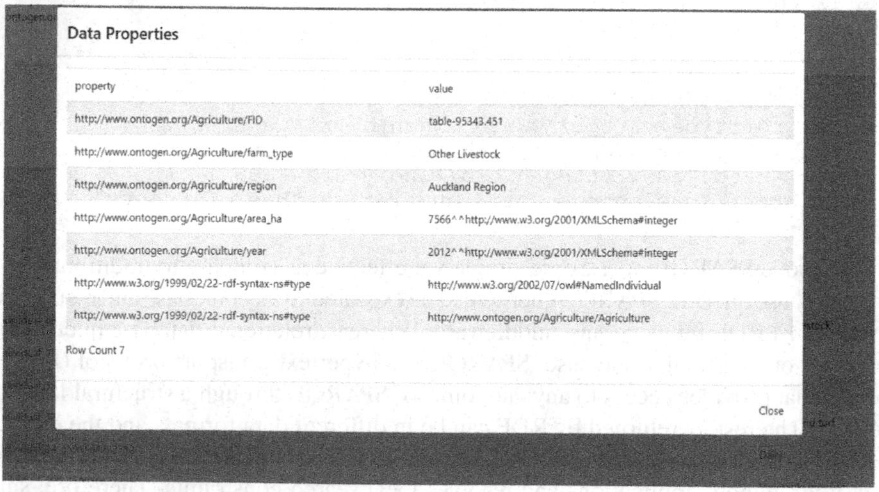

Fig. 5. The data properties of the first individual under ind1

5 Discussion

Data is one of the vital aspects of our day to day life. These days, everything on the web is governed by data. However, presently data is facing challenges of being machine unreadable and inaccessible. Several efforts have been undertaken to view and resolve the structured format issues of data. But, most of the times the semantic relationships between data are often not considered. This paper has proposed an approach to generate an Ontology by using the CSV open government datasets of the NZ government. Once the ontology is generated SPARQL queries are imposed to extract the useful information and to check the consistency of the system. The CSV to OWL conversion used here is different as compared to the normal data conversion as it is considering the RDF triples and we are converting the CSV datasets into OWL triples so that we can form an Ontology. As OWL vocabularies are the most updated ones and help to generate the semantically enriched data when multiple ontologies are linked.

To Query an ontology, it should be in Turtle form, OWL ontologies don't support the turtle format. To impose the Queries, the OWL ontologies are converted to RDF/turtle format. In future, this conversion will be used as a basis to generate multiple ontologies of different domain and we can semantically link different domain ontologies so that a big knowledgebase can be created.

6 Conclusion and Future Work

In this paper, we have introduced a new methodology for ontology generation by utilizing the datasets of the New Zealand Government. Moreover, the SPARQL interface is used to evaluate and query the ontology. The generated ontologies are visualized by using the Protégé tool. However, due to the big size of the generated ontology, protégé is unable to graphically represent it. The size of the ontology is also impacting the processing time of the system. Further, we have only considered two datasets that are agriculture and land sectors of the government. In future, we plan to semantically link the generated OWL files and extract the knowledge from the complete open data set using SPARQL queries. Furthermore, a data visualizing process will be implemented as a part of the experiment for better consumption of the results from the open data.

Data Availability. The data that support the findings of this study is publicly available at https://data.mfe.govt.nz/table/95343-agricultural-and-horticultural-land-use-200216/data/.

References

1. Bizer, C., Heath, T., Berners-Lee, T.: Linked data: the story so far. In: Semantic Services, Interoperability and Web Applications: Emerging Concepts, pp. 205–227. IGI Global (2011)
2. Ubaldi, B.: Open government data: towards empirical analysis of open government data initiatives. OECD Working Papers on Public Governance, no. 22, OECD Publishing, Paris (2013). https://doi.org/10.1787/5k46bj4f03s7-en
3. Orszag, P.R.: Open Government Directive (2009). https://obamawhitehouse.archives.gov/open/documents/open-government-directive (Accessed 27 Aug 2020)

4. Cabinet Office: Open Data Charter. 2010 to 2015 Conservative and Liberal Democrat Coalition Government (2013). https://www.gov.uk/government/publications/open-data-cha rter (Accessed 27Aug 27, 2020)

5. Corcho, O., Fernández-López, M., Gómez-Pérez, A.: Methodologies, tools and languages for building ontologies. Where is their meeting point?. Data Knowl. Eng. **46**(1), 41–64 (2003)

6. Vafopoulos, M.N., et al.: Public spending: Interconnecting and visualizing greek public expenditure following linked open data directives. (2012) SSRN 2064517

7. Theocharis, S., Tsihrintzis, G.A.: Ontology development to support the Open Public data-the Greek case. In: IISA 2014, The 5th International Conference on Information, Intelligence, Systems and Applications, pp. 385–390. IEEE, July 2014

8. Fragkou, P., Galiotou, E., Matsakas, M.: Enriching the e-GIF ontology for an improved application of linking data technologies to greek open government data. Procedia Soc. Behav. Sci. **147**, 167–174 (2014)

9. Matheus, R., Ribeiro, M.M., Vaz, J.C.: New perspectives for electronic government in Brazil: the adoption of open government data in national and subnational governments of Brazil. In: Proceedings of the 6th International Conference on Theory and Practice of Electronic Governance, pp. 22–29, Oct 2012

10. Hendler, J., Holm, J., Musialek, C., Thomas, G.: US government-linked open data: semantic. data. gov. IEEE Ann. Hist. Comput. **27**(3), 25–31 (2012)

11. Zhao, L., Ichise, R.: Ontology integration for linked data. J. Data Semant. **3**(4), 237–254 (2014)

12. Al-Khalifa, H.: A lightweight approach to semantify Saudi open government data. In: 2013 16th International Conference on Network-Based Information Systems (NBiS), pp. 594–596. IEEE Computer Society, Sept 2013

13. Hoxha, J., Brahaj, A., Vrandečić, D.: Open. data. al: increasing the utilization of government data in Albania. In: Proceedings of the 7th International Conference on Semantic Systems, pp. 237–240, Sept 2011

14. Theocharis, S.A., Tsihrintzis, G.A.: Open data for e-government the Greek case. In IISA 2013, pp. 1–6. IEEE, July 2013

15. Tennison, J., Kellogg, G., Herman, I.: Model for tabular data and metadata on the web (2015). https://www.w3.org/TR/2015/REC-tabular-data-model-20151217/#bib-RFC 4180/. Accessed Oct 29 2020

16. DCMI Usage Board: Dublin Core Metadata Initiative (2020). https://www.dublincore.org/ specifications/dublin-core/dcmi-terms/. Accessed 29 Oct 2020

17. Beckett, D., Berners-Lee, T.: Turtle-terse RDF triple language-W3C team submission, World Wide Web Consort. (WC) (2011). https://www.w3.org/TeamSubmission/2011/SUBM-turtle-20110328/. Accessed 29 Oct 2020

18. OntoText.: What is SPARQL? (2020). https://www.ontotext.com/knowledgehub/fundament als/what-is-sparql/. Accessed 9 Dec 2020

19. Alvite-Díez, M.L.: Linked open data portals: functionalities and user experience in semantic catalogues. Online Information Review (2021)

20. Almendros-Jiménez, J.M., Becerra-Terón, A.: Discovery and diagnosis of wrong SPARQL queries with ontology and constraint reasoning. Expert Syst. Appl. **165**, 113772 (2021)

Blockchain Based Framework to Maintain Chain of Custody (CoC) in a Forensic Investigation

Sarishma$^{(\boxtimes)}$, Abhishek Gupta, and Preeti Mishra

Department of Computer Science and Engineering, Graphic Era Deemed to be University, Dehradun, India

Abstract. Cybercrimes are exponentially rising in number and the forensic investigations are now being actively conducted to get to the root of problem. There are many challenges to conduct a smooth forensics investigation. It suffers from the problem of maintaining integrity, ownership, auditability and authenticity of digital evidence. In this work, we elaborately cover as to how blockchain can be used to tackle the challenges faced by forensic investigations. In particular, we propose a framework based on blockchain which can assist in maintaining Chain of Custody while preserving integrity, accountability and authenticity of the acquired digital evidence. We devise our own smart contracts for the execution of scripts under different circumstances as governed by the phases of forensic investigation. We conclude our work by discussing future research directions and open challenges for the same.

Keywords: Blockchain · Forensics · Decentralization · Chain of Custody · Digital evidence

1 Introduction

The emergence and advancement of technologies like machine learning, cloud computing, artificial intelligence, big data etc. along with the breakthrough in connectivity has led to an exponential increase in migration of workloads to technical platforms. This includes stakeholders from all areas of life including business, government, military and research based work.

Counter side of this adoption is the vulnerability to which users and their data is exposed which includes various cyber threats, advanced malware, root kits, deep web related issues etc. Once a cyberattack happens, forensics based analysis is done to know about the causes and culprits of the crime in a timely and effective investigation [1]. Acquiring, maintaining and handling digital evidence from such cases is a major issue in conducting a forensics based investigation. Maintaining Chain of Custody (CoC) in digital forensics is obligatory to maintain the integrity of the collected data during the investigation until it is submitted in the court of law. CoC in forensics investigation is defined as a trail of investigation

© Springer Nature Switzerland AG 2021
M. Singh et al. (Eds.): ICACDS 2021, CCIS 1440, pp. 37–46, 2021.
https://doi.org/10.1007/978-3-030-81462-5_4

in which evidence found at any stage is maintained along with the investigators who analyze the evidence. CoC is very important as it ensures the client and the court that a transparent and foolproof investigation has been done [2].

Blockchain is defined as a public persistent, transparent append-only ledger where each added data or transaction(s) is known as a block [3]. The first block of the chain is known as the parent block or the genesis block. All the blocks in Blockchain are time-stamped which helps in the sequential ordering of the blocks, and can only be identified by uniquely generated cryptographic hashes. Each block in the blockchain stores the cryptographic hash of the previous block and the chain is maintained sequentially. An unalterable transaction log is generated; any node which is connected to this network can review the data from anywhere, anytime [4].

Blockchain is a promising technology which has the potential to transform the cyber-security landscape for the current era. With the emergence of cloud, big data, deep web and dark web; the frequency as well as scope of cyber-crime has exponentially increased. Blockchain offers some key features including accountability, immutability, transparency and auditability which make it an attractive solution for many problem areas. Forensic investigation landscape is prone to attacks by multiple parties and blockchain can securely assist in solving problems faced by investigators to maintain their CoC.

In this work, we review the related work done in this field. We then propose a framework which uses permissioned blockchain as the underlying technology. Block structure and the working of components is explained in detail with the help of figures. The key contributions of this work are listed as follows:

- To propose a framework for maintaining integrity, authenticity, auditability and Chain of Custody for forensic based evidence
- To provide and discuss open research issues and challenges for digital forensic investigation under the context of proposed work.

The rest of the paper is organized as: Sect. 2 covers the related work in this area where blockchain is integrated with forensics. In particular we will focus on proposed models and frameworks. Section 3 covers the phases of a forensics process along with the intersection area with blockchain. It discusses the issues, challenges, benefits and limitations which come with incorporation of blockchain. A blockchain based framework is proposed to maintain chain of custody for digital evidence in Sect. 4 along with actors, system model and working. Lastly, we conclude our work along with open research directions.

2   Related Work

In this section, we outline the related work previously done in the area of blockchain and forensics. The key works are outlined as follows:

- Michael Köohn [5] generated a digital forensics process model (DFPM) from the Krause model which later coalesce with US DOJ, featured with collecting stage, authentication stage, examination stage, analyze stage, and final reporting stage. The major cast includes first responder, investigator, prosecutor, defense, the court.

- A.H Lone et al. [6] proposed a solution for the perpetuation of digital data for Chain of Custody (CoC) in forensics investigation where they had used Hyperledger Fabric which is a permissioned Blockchain, enabling the ability to switch between public blockchain and private blockchain via making consensus and membership service as plug and play feature. Hyperledger Composer is a development toolset which transforms the overall development into a plain sailing process. Hyperledger Caliper is a staging benchmark tool with pre-set use cases to enhance the final report.

- Y li et al. [7] presented the two decentralized methods for Public-key Infrastructure (PKI). Having Central Authority (CA) in PKI schema leads to some security issues like IP spoofing, DNS spoofing, single point of failure etc. Conventionally X.509 PKI CA-framework is being used. First is Blockchain-based PKI with CA which uses the distributed network for storing the data and for checking the integrity DPDP (Decentralized Provable Data Possession), DPoR (Decentralized Proof of Retrievability) and data storage privacy.

- Revocation of the certificate can be done through the use of a Distributed Hash Table (DHT) and Dynamic Bloom Filter data structures (DBF) [8]. The second is Blockchain-based PKI without CA, in which CA is removed and work is emanating between users and verifiers. The cryptographic technique like proof of commitment, zero-knowledge proof is used and for best performance Merkle tree is used.

- K Yamashita et al. [9] specified the security threats in Hyperldeger Fabric and smart contracts and then proposed a model written in Go language for detecting the vulnerabilities present in the Hyperledger Fabric. These technologies are vulnerable to attacks like DoA's, parity wallet, non-determinism arising from language instructions etc. Commending the vulnerability diagnostic tools like Oyente, ZEUS, securify for smart contracts, and Chainsecurity provides Chaincode Scanner for Hyperledger Fabric.

- J Behlet al. [10] proposed Hybster model, concocted with a well-trusted subsystem TrInc and Intel SGX (Intel Software Guard Extension) which results in the advanced version of PBFT (Practical Byzantine Fault Tolerance). Intel SGX is a sought-after Trusted Execution Environment (TEE), owner of this platform can use the feature enclave which is a secured folder that ensures integrity and concealment of data, with a memory of 128 MB, however, the size of the memory can be increased by certain protocols. Another feature enriched and blend of TrInc, a trusted subsystem formed called TrInx, to fulfill the demand of Hybster and optimize the performance. It also features parallelizable replication protocol which achieved 1 million operations per/second on the four-core machine.

- Z Bao et al. [11] divided the blockchain layer into six major parts: application layer, contact layer, incentive layer, consensus layer, data layer, and network layer, each layer conflated with Intel SGX and provided remarkable results.

3   The Proposed Framework

Forensic investigations take place under restricted or closed environment with known authorized participants from intelligence organizations. Key challenge is to prevent unnecessary interference by known/unknown participants. Core operating participants in our proposed framework are outlined as follows:

- Asset: The digital evidence when acquired is allotted one unique evidenceID i.e. eID which is calculated as a hash of the unique contents of memory and the digital signature of the participant who acquired the evidence from the scene.
- Participants: Supervisor, co-supervisor, analyst, prosecutor and investigators; these are the participants who are allowed to make changes to the blockchain where evidence in stored. Number of participants can change with no restriction; however every participant's role and permissions will be governed using specific smart contracts.
- Transactions: A very limited set of transactions will be allowed as minimum tampering is needed with the evidence. Creating a copy of evidence, migrating evidence from one place to another, preparing report for the evidence and presentation in a court of law – such transactions ensure that under no circumstances original evidence is affected.
- Owner: Supervisor of the case who is legally charged with carrying out the investigation will be assigned the role of owner.
- Genesis block: The starting block of the blockchain comprises of warrant details, case details, investigative team details and legal investigative authorities. Every event will be trailed as a block after the genesis block. However the permission to decode the blocks will only be provided to a select few participants to avoid unnecessary disclosure of information.

Hyperledger Fabric is a consortium (permissioned) blockchain framework, which comes with plug and play mechanism for achieving consensus. Blockchain is a very secure network in which data cannot be altered due to its feature of immutability. Consensus ensures security as well as the transparency of data stored in distributed ledger. To ensure the reliability of the system, a distributed PKI system is proposed. The distributed system helps in controlling the permissions of participants on evidence and also helps in achieving transparency during the investigation via decentralization.

In the investigation phase, first, all the IDs are issued by the higher authority to the investigating officers and the data is being stored in the hyperledger fabric blockchain network. The network is divided into two parts: private and public. The public network is for all the people in the investigation, and the private blockchain for higher root authority. To ensure further confidentiality of data, hybster a state machine replication protocol is used which comprises the TEE subsystem aimed at achieving rapid operation with data. Data structures like DHT and DBF are used to help easy revocation of data from the blockchain network.

3.1 Hyperledger Fabric and Blockchain Network

We have proposed to use a permissioned blockchain in which the authority will find it easier to share confidential data with higher authorities in a closed environment. Hyperledger Fabric is the core technology that is being used, which enables us to make use of permissioned blockchain. It is highly scalable and convenient to use, enabling us to make consensus a plug and play feature with the help of channels. It also helps in building confidential contracts between the higher authority and the investigating officers. All the chaincode written for Hyperledger Fabric will have to go through the web-based chaincode scanner tool at least once to detect the unwanted defects or bugs. These chaincodes are generally known as *smart contracts*. Smart contracts are a block of code that are stored in blockchain and get executed automatically when certain terms and conditions are satisfied. It is very useful as it terminates the unnecessary involvement of any outside party.

3.2 Consensus Mechanism

For high-security and fast operation in consensus mechanism, we have used Hybster [10] - a hybrid fault model which is a concoction of different technologies like TrInc TEE (Trusted Execution Environment) subsystem, Intel SGX TEE subsystem, PBFT consensus protocol, and others. In Hybster (hybrid protocol replication state machine model), we have the privilege to use consensus instances in parallel with the help of a consensus-oriented protocol mechanism. This protocol provides great efficiency in the management system of records and protection from cyber threats.

TrInx is an abstract cognate version of the TrInc [8] subsystem, which helps to tackle equivocation in the distributed system on a large scale. Unlike A2M (Append to Memory) Trusted Execution Environment subsystem which works on trusted log entries, whenever there is an outgoing message, it is completely working on the increment counter for every certification as same as the TrInc.

The Intel SGX is a TEE (Trusted Execution Environment) hardware technology integrated into the CPU, privileged with a secure container viz. Intel Enclave [11], which ensures confidentiality and integrity of the data. The introduction of Intel SGX based subsystem results in the reduction of complexity in the cryptography tools and makes it easy for protocol to use. The use of this subsystem had magnificently increased the overall efficiency of the system, it can be observed that it can operate over 1 million operations per second in multicore machines.

3.3 PKI System and Distributed Storage

Traditional PKI system which is centralized has an interception capability from threats like a single point of failure, IP and DNS spoofing. To tackle such problems we have used the Blockchain-based PKI with CA (Certification authority) [7] model, this will completely remove the third party interference during the

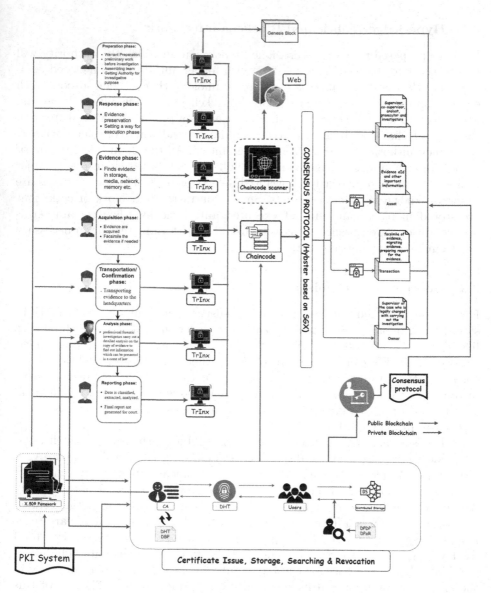

Fig. 1. Blockchain based framework for maintaining Chain of Custody

whole forensics investigation. In the blockchain network, CA is self-certified and registered first in the blockchain; which signs and issues certificates.

Registration, certificate issue, certificate storage, certificate search and certificate revocation are the main functions of the PKI system. The whole PKI system works on the traditional X.509 framework. A peer-to-peer off-chain network is used for the storage of these certificates (IPFS or SWARM). Unlike the IP address used in HTTPS protocol (which first finds the location at which data

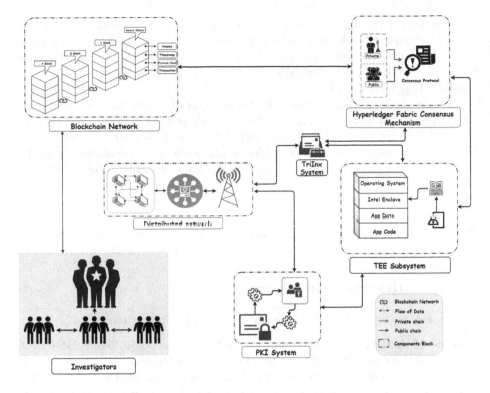

Fig. 2. Working of the components

is being stored and then retrieve it to the user) this distributed storage system uses the concept of content addressing, it does not matter at which location the data is being stored whosoever has the data can provide it to the user, this key feature of this distributed storage helps in the retrieval of data at a very high rate.

With the use of the TrInx subsystem, the security is augmented further and we can wield the other techniques regarding data security delineated in [8]. DPDP (Decentralized Provable Data Possession) or DPoR (Decentralized proof of Retrievialblity) protocols are used to scrutinize the integrity of stored data. With the help of data structures like DHT (Distributed Hash Table) or which have more efficient search and have dynamic operations is DBF (Dynamic Bloom Filter), we can tackle the immutability of blockchain and revocate the certificates.

3.4   Investigators Role

All the data uploaded on the blockchain network first have to go through the TrInx subsystem for verification to check the authenticity of the data, and then data will be checked by chaincode which will direct the flow according to the commandment. During the preparation phase, all the important data will be saved to the asset block after confirming the identity through smart contracts. After that in the response phase, evidence will be preserved. In the evidence phase, all the data will be sorted and collected in an asset block with the permission of higher authorities.

During the acquisition phase, the facsimile of the data will be done with the permission of higher authorities. The analysis process is carried out by professional investigators which are connected to two blockchain network, one is private for accessing the protected blocks like asset and transaction, and the other is public with all participants, among the professional investigator there is one CA for the PKI system whose role is to issue certificates for working officers during the forensic investigation. At last final report is submitted in the court of law. Whenever there is a flaw of data in the asset block or the transaction block the participants need access control from higher authorities before making any further changes.

4   Open Research Directions

In this section, some of the open research challenges are outlined.

- Some of the Hybrid Fault model protocols were proposed, like A2M-PBFT [10] based on the Trusted execution environment subsystem A2M (appened to memory) which kept the record of all outgoing messages in distinct log entries (for each log entry, A2M issue certificate for verification)which can be appended in increasing order only. Another hybrid model is MinBFT, which is based on the root feature of the counters, it increases the counter for each certification explicitly. Instead of using trusted logs like in the A2M-PBFT model, the MinBFT model uses a trusted incrementor counter. The limitation of these hybrid models is that they are largely dependent on working within a sequential processing manner which limits and interfaces with the performance in parallelization for the latest generation multi-core processor machines.
- The latest generation processor provides a Trusted Execution environment that protects the components of the software from the trustless operating system i.e. Intel SGX (enclaves) and ARM (trusted execution environments). To solve all these problems a new hybrid-fault model Hybster is proposed. Hybster is a hybrid state-machine replication protocol that can play consensus instances in parallelization with the mechanism of the consensus-oriented protocol. The uses of the Trusted Execution environment subsystem Intel SGX can greatly enhance the operating speed of over 1 million operations per second with use of only on four core processor. This is a hybrid state-machine protocol which solves all recent hybrid protocol problem.
- To tackle threats like a single point of failure, IP and DNS spoofing attacks, a decentralized blockchain-based PKI with CA and blockchain-based PKI without CA [13] methods are proposed. Distributed storage networks are used for effective retrieval of data and data structures like DHF(Data Hash

Table), BF(Bloom Filter), SBF(Simple Bloom Filter), and DBF(Dynamic Bloom filter are being) proposed for revocation of data from the immutability of Blockchain.
- Applying these efficient Hybster protocols will greatly help to increase data processing and management speed. With the use of Intel SGX, a Trusted Execution Environment subsystem will provides the enhanced security to confidential data. By using the Blockchain-based PKI system we can reduce the need for working peoples used in traditional central based PKI system. With the enriching property of blockchain, the data are safe from most cyber attacks.
- Further, the Intel SGX system is vulnerable to side-channel attacks and has only 128 MB of reserved memory however we can extend this number by taking some measures. However they are not efficient and thus an efficient solution is still required. If we talk about the chaincode (smart contracts in ethereum) there no such efficient software or tools to checks and detect security bugs in the chaincode, so it is a mandatory need for one.

5 Conclusion

Maintaining Chain of Custody while handling digital evidence is one of the prominent challenges faced by forensic investigators. In this work, we utilized the features provided by permissioned blockchain to address the challenges of integrity, authentication and accountability of handling digital evidence. A detailed framework along with working of components is provided. In the future, the work can be extended for finding the efficient algorithm for the revocation process from the blockchain.

References

1. Richter, J., Kuntze, N., Rudolph, C.: Security digital evidence. In: Fifth IEEE International Workshop on Systematic Approaches to Digital Forensic Engineering 2010, pp. 119–130 (2010)
2. Lone, A.H., Mir, R.N.: Forensic-chain: Ethereum blockchain based digital forensics chain of custody. Sci. Pract. Cyber Secur. J 1, 21–27 (2018)
3. Sharma, A., Tomar, R., Chilamkurti, N., Kim, B.-G.: Blockchain based smart contracts for internet of medical things in e-healthcare. Electronics 9(10), 1609 (2020)
4. Tomar, R.: Maintaining Trust in VANETs using blockchain. ACM SIGAda Ada Lett. 40(1), 91–96 (2020)
5. Köhn, M., Eloff, J.H.P., Olivier, M.S.: UML Modelling of digital forensic process models (DFPMs). In: ISSA 2008, pp. 1–13 (2008)
6. Lone, A.H., Mir, R.N.: Forensic-chain: blockchain based digital forensics chain of custody with PoC in Hyperledger Composer. Digit. Investig. 28, 44–55 (2019)
7. Li, Y., Yu, Y., Lou, C., Guizani, N., Wang, L.: Decentralized public key infrastructures atop blockchain. IEEE Netw. 34, 133–139 (2020)

8. Ali, M.S., Dolui, K., Antonelli, F.: IoT data privacy via blockchains and IPFS. In: Proceedings of the Seventh International Conference on the Internet of Things, pp. 1–7 (2017)
9. Yamashita, K., Nomura, Y., Zhou, E., Pi, B., Jun, S.: Potential risks of hyperledger fabric smart contracts. In: IEEE International Workshop on Blockchain Oriented Software Engineering (IWBOSE) 2019, pp. 1–10 (2019)
10. Behl, J., Distler, T., Kapitza, R.: Hybrids on steroids: SGX-based high performance BFT. In: Proceedings of the Twelfth European Conference on Computer Systems, pp. 222–237 (2017)
11. Bao, Z., Wang, Q., Shi, W., Wang, L., Lei, H., Chen, B.: When blockchain meets SGX: an overview, challenges, and open issues. IEEE Access (2020)
12. DigitalForensics 2019. IAICT, vol. 569. Springer, Cham (2019). https://doi.org/10.1007/978-3-030-28752-8_14
13. Levin, D., Douceur, J.R., Lorch, J.R., Moscibroda, T.: TrInc: small trusted hardware for large distributed systems. NSDI **9**, 1–14 (2009)

Parameters Extraction of the Double Diode Model for the Polycrystalline Silicon Solar Cells

T. Suganya[1(✉)], V. Rajendran[2], and P. Mangaiyarkarasi[3]

[1] Government Polytechnic College for Women, Coimbatore 441, India
[2] Government Polytechnic College for Women, Coimbatore 442, India
[3] Government College of Technology, Coimbatore 13, India

Abstract. In this paper, a dual diode model of solar powered photovoltaic is exploited to improve the efficiency of generation transforming structures of solar photovoltaic force. The double diode model is simple and straightforward to produce and offers greater precision which allows a more precise prediction of the presentation of photovoltaic structures. The survey depends on the reading results and it is for this reason that the MATLAB device is used. The reconstructions are completed by the fluctuation of the different limits of the model such as the oriented solar radiation, the temperature, the value of the parasitic screen, the ideality factor of the diode and the quantity of solar cells also associated which serve to accumulate the photovoltaic cluster. An obvious demonstration will be conducted to analyze the impacts of these details on the productivity curve and resistance/voltage performance of the PV cell for explicit models. Characterizing another frontier (α), a practically best solution is suppressed for the I-V curve of the photovoltaic model, which is better than the conventional outlier model due to its high computational efficiency in the constant supply of photovoltaic meadow.

Keywords: Parameters extraction · Short-circuit current · Open-circuit voltage · Output power · Solar cells

1 Introduction

Photovoltaic sun powered energy influences heat obstruction. While this component doesn't without help from anyone else convert heat into power, it doesn't dissipate. The yield force of a silicon sun powered cell is touchy to high temperatures [1–3]. Polycrystalline silicon is straightforward and reasonable. Rather than going through a lethargic and costly cycle to make a solitary precious stone; Meg-C sun based cells are a similar size and more costly than pc-si. Meg-C cells are 25% more proficient in standard research center and mechanical cells. In research facilities, 20% in PV-C cells is characterized as 14 to 18% and in stores 11 to 15% [4–6]. Every one of these qualities accessible available are the working states of a sunlight based cell, the ghastly conveyance of radians and the states of the standard temperature signal (SRC: light = 1000 W/m2, temperature = 25 °C, range from reference AM1.5).

© Springer Nature Switzerland AG 2021
M. Singh et al. (Eds.): ICACDS 2021, CCIS 1440, pp. 47–55, 2021.
https://doi.org/10.1007/978-3-030-81462-5_5

The I-V bend isn't just valuable for cell clusters and framework recreations, yet additionally as an examination instrument to comprehend the inside actual properties of the photovoltaic sun oriented cell [7]. Different models of comparable circuit charts have been created and proposed to portray the idea of the PV cell and the normally utilized single and double diode models [8]. In a solitary diode model, five example boundaries (called five limited boundaries) portray the overall properties of the photovoltaic cell: the current created by the light, the spillage current or converse focus, the quality factor of the diode, the arrangement obstruction and protection from shunt. The force and converse fixation that light makes can be called outside impacts, while others are inward impacts. The effortlessness of the five-boundary light model mirrors the idea of the photovoltaic framework [9]. Despite the fact that there are various reports in the writing where unequivocal portrayals for single diode models have been presented, clear answers for double diode models have just been looked for in two elite cases: first, since the philosophical part of a diode is twice as huge of a diode without equal opposition; No reasonable articulation was found for the cell current, yet the terminal voltage was unmistakably gotten without the requirement for the Lambert W work [10]. This article portrays the properties of the two PV module advances (I-V-P), for example H. Polycrystalline silicon (gem innovation: ASE-100 sort) and indistinct silicon (dainty film innovation: DS-40 sort) can be adjusted to all working conditions (any contrariness module temperature).

2   Literature Review

A few examination works are arising in the writing based on the double diode solar system. A combine of works are explored right now.

Ahmed et al. [11] gave gadget boundaries light energy to research the exhibition of a monocrystalline silicon sun powered module. The current-voltage properties (I-V) of the gadget were estimated with various lighting energies. The outcomes showed that the ideality factor (n), the arrangement obstruction (Rs) and the immersion current (Io) mostly rely upon the lighting energy, while the shunt opposition (Rsh) and the photocurrent (Iph) rely upon the illuminance. The expansion in light energy prompted a reduction in the arrangement obstruction, however expanded the immersion current and the ideality factor.

Lekouaghet et al. [12] introduced a selection of the boundaries of photovoltaic sunlight based cells/modules of different advances, which gives a precise portrayal of the current-voltage and force voltage bends. For this reason, the proposed streamlining calculation depends on a tumultuous generator, which is joined with the recently created Rao-1 advancement calculation to remove the photovoltaic boundaries.

Raya-Armenta et al. [13] analyzed and improved the precision of the generally utilized five-boundary single diode model. Two new actual conditions are acquainted with address shunt arrangement and protections, while different boundaries are addressed by set up actual terms. In the proposed model, a large portion of the boundaries identify with cell temperature, light and information sheet esteems, while a few boundaries should be changed. The model is contrasted with four known strategies for separating the boundaries of single and twofold diode models.

Parida et al. [14] fostered a differential development (DE) with dynamic control factors (DEDCF). DEDCF considers test IV informational indexes to gauge the boundaries of PV cell models and PV modules with one and two diodes. Control factors incorporate transformation and hybrid components, the two of which should be changed powerfully to improve arrangements.

3 Proposed PV System

The photovoltaic system is one of the renewable energy sources that generate electricity from the available renewable solar energy. During load power, the PV system requires a maximum power that exceeds the accessible power generation and requires limited power of the PV system in the load power below the required space. Traditional P&O values (Maximum Power Point Perturb and Observation) have been used here for the maximum performance of photovoltaic systems [15]. The following Eq. (1) defines the PV scheme current.

$$pv_i = i_g - i_o \left[\exp^{\left(\frac{qv_d}{K_B f T_C} \right)} - 1 \right] - \frac{v_d}{P_r} \tag{1}$$

Where the light produced current is named as i_g; the dull immersion current dependant on the cell temperature is connoted as i_o; the electron charge is addressed as $(1.6 \times 10^{-19} C)q$; Boltzmann's steady is implied as K_B, T is the cell supreme temperature; f is the cell glorifying factor; v_d is the diode voltage and P_r is the equal opposition.

3.1 Modeling of Proposed PV Double Diode Configurations

In general, the prediction strategy calculates the optimal situation in terms of voltage and current under the following irregular conditions. This view is used by manufacturers to misuse the PV line. Thus, it provides a simple strategy for determining the power generated by the methods for the photovoltaic regions specified in Eq. (2).

$$P^m = V^m . I^m \tag{2}$$

In this state, V^m the position of the most extreme voltage of the photovoltaic part, which appears in condition (3), is indicated indiscriminately,

$$V^m = V_n^m \left[1 + 0.0539 \ \ln \left(\frac{G}{G^{ref}} \right) \right] + \lambda_0 \Delta T \tag{3}$$

Below, we understand V_n^m the maximum voltage of the component under constant conditions (v), λ_0 explain the voltage ratio when setting the temperature (V/K), T means temperature, G mean solar radiation, G^{ref} show the orientation of solar radiation

(W/m^2) and I^m indicate the maximum value. The operating current of the photovoltaic components [16], referred to in Eq. (4),

$$I^m = I_{sc}\left\{1 - C_1\left[\exp\left(\frac{V^m}{C_2 V_{oc}}\right) - 1\right]\right\} + \Delta I \tag{4}$$

In this section, V_{oc} is represent the open circuit voltage of the component (V), I_{sc} is represent the minimum circuit current of the component (A), $T = T_C - T_0$ and C_1 and C_2 they represent the scale constraint [17], which is calculated in Eqs. (5) and (6),

$$C_1 = \left(1 - \frac{I_{mp}}{I_{sc}}\right)\exp\left(1 - \frac{V_{mp}}{V_{oc}}\right) \tag{5}$$

$$C_2 = \frac{(V_{mp}/V_{oc} - 1)}{\ln(1 - I_{mp}/I_{sc})} \tag{6}$$

This segment, I_{mp} and V_{mp} contains the maximum current and maximum voltage of the photovoltaic component. Depending on the characteristics of the PV component, the current (I_{mod}) is shown in Eq. (7).

$$I_{mod} = a_0\left[\frac{G}{G^{ref}}\right]\Delta T + \left[\frac{G}{G^{ref}} - 1\right]I_{oc} \tag{7}$$

This segment a_0 contains the current ratio to temperature (A K) [18]. Next, the equivalent model of a two-diode PV module circuit and its characteristics are shown in Fig. 1.

Furthermore, the maximum output power for an M N component in series and in parallel is found, which is defined in Eq. (8).

$$P^m_{mn} = FF\left(I_{sc}\frac{G}{G^{ref}}\right)\left(V_{oc}\frac{\ln(k_1 G)}{\ln(k_1 G^{ref})}\frac{T_0}{T}\right) \tag{8}$$

This k_1 is referred as constant, $k_1 = K/I_0$ (around 10^6 m²/W) and form factor (FF). The proportion of the square foundation of intend to the standard worth of the unpredictable trademark (current or voltage) is known as the structure factor [3, 19]. The norm of immediate upsides of current and voltage in full succession is appeared as the standard worth of the sporadic trademark, which is deductively alluded to in condition (8).

$$FF = \frac{P^m}{V_{oc} I_{sc}} \tag{9}$$

The calculation of the initial estimates, including solving three nonlinear equations, is performed by another MATLAB script.

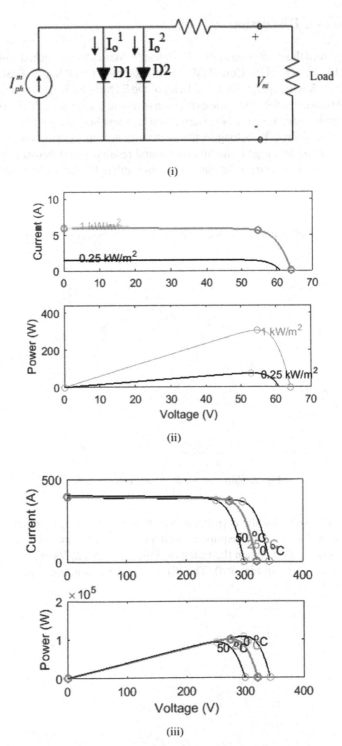

Fig. 1. The equivalent circuit of (i) PV module and the V-I and P-V characteristics in (ii) irradiance (iii) temperature variation conditions

4   Results and Discussion

This part improved the exhibition of the PID-P test framework is refreshed with Simulink 7.10.0 (R 2015A) and Intel (R) Core (DM), i5 processor with 5GB RAM. The objective of the proposed work is to separate the boundaries of the STM6–40/36 sun powered modules for the double diode model. The basic extraction measure comprises of three obligations. The initial step is to separate the boundary under the new boundary without feeling any estimation vulnerability. To belong to the parameter search range, then extracting the target parameters under target is the first step found results. The difference immediately from the third step is to extract the parameters immediately; the second step will be to include those search ranges. The parameter search ranges for both panels are evaluated. As described in the current composite material, the dual diode equivalent model of polysilicon solar cells at 24 °C or higher (50 °C) is not observed even under 5 h of light, making the cell more flammable. The simulation model of the proposed strategy is shown in Fig. 2.

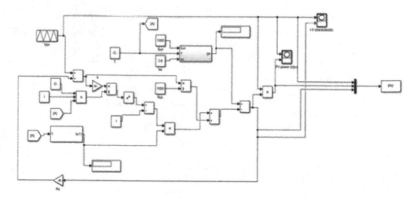

Fig. 2. Simulink model of PV cell structure

In Fig. 3 (i) shows that the correlation was tested. The design temperature here is between 25 °C and 45 °C and is equipped with a PV shunt. Figure 3(ii to iv) shows that the correlation is verified. When the radiation voltage is given, Vs is calculated on the basis of 3.5 A as shown in Fig. 3(ii). The PID-S panel has a nominal power of 360 W and a voltage of 120 V.

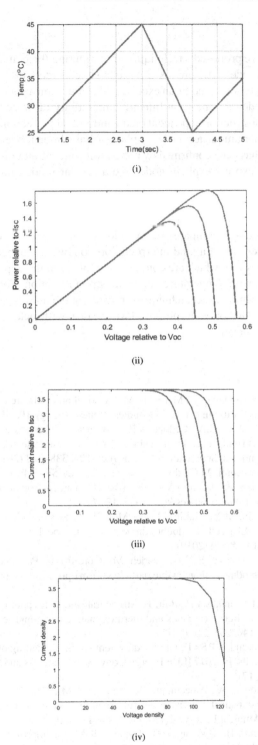

Fig. 3. Output of PV (i) Temperature (ii) Power Vs voltage (iii) Current Vs voltage and (iv) current Vs voltage in PV cell

5 Conclusion

In this article, we have proposed a technique for computing the boundaries of a twofold diode photovoltaic model. The proposed hypothesis considers the issue of estimation vulnerability. All the more correctly, an exceptional calculation decides the prompt worth of a boundary dependent on the plentifulness vulnerability of every boundary. Thevine computes the same obstruction dependent on the individual protections of the diode legs, bringing about another unique term I-V for photovoltaic diode models. The precision of the proposed model has been confirmed by recreation concentrates outside the STC. The blunder between an express explicit model and a regular model, found in exploratory examination and by reenactment, demonstrates the presentation of a particular model, which is satisfactory for recuperating I-V from photovoltaic cells. A significant consequence of this work is that albeit a reasonable two-diode model is presently accessible, the proposed model doesn't compare to conventional actual ideas and boundary esteems. Subsequently, another one of a kind interpretation condition is expected to refresh the introduced boundaries (α), which are completely evolved in this report. The end is that (α) it acts straightly with temperature and relies upon radiation. For future work, it is fascinating to foster a more exact technique for deciding the quest stretch for vulnerability boundaries (α) and to apply this proposed way to deal with tackle other improvement issues in the field of energy.

References

1. Dehghanzadeh, A., Farahani, G., Maboodi, M.: A novel approximate explicit double-diode model of solar cells for use in simulation studies. Renew. Energy **103**, 468–477 (2017)
2. Cappelletti, M.A., Casas, G.A., Cedola, A.P., y Blancá, E.P., Soucase, B.M.: Study of the reverse saturation current and series resistance of ppn perovskite solar cells using the single and double-diode models. Superlattices Microstruct. **123**, 338–348 (2018)
3. Bradaschia, F., Cavalcanti, M.C., do Nascimento, A.J., da Silva, E.A., de Souza Azevedo, G.M.: Parameter identification for PV modules based on an environment-dependent double-diode model. IEEE J. Photovoltaics **9**(5), 1388–1397 (2019)
4. Chandel, T.A., Yasin, M.Y., Mallick, M.A.: Modeling and simulation of photovoltaic cell using single diode solar cell and double diode solar cell model. Int. J. Innovative Technol. Explor. Eng. (IJITEE) **8**(10) (2019)
5. Ebrahimi, S.M., Salahshour, E., Malekzadeh, M., Gordillo, F.: Parameters identification of PV solar cells and modules using flexible particle swarm optimization algorithm. Energy **179**, 358–372 (2019)
6. Cuce, E., Cuce, P.M., Karakas, I.H., Bali, T.: An accurate model for photovoltaic (PV) modules to determine electrical characteristics and thermodynamic performance parameters. Energy Convers. Manage. **146**, 205–216 (2017)
7. Olayiwola, O.I., Barendse, P.S.: Dynamic equivalent circuit modelling of polycrystalline silicon photovoltaic cells. In: 2017 IEEE Energy Conversion Congress and Exposition (ECCE), pp. 2310–2317 (2017)
8. Louzazni, M., Khouya, A., Amechnoue, K., Mussetta, M., Crăciunescu, A.: Comparison and evaluation of statistical criteria in the solar cell and photovoltaic module parameters' extraction. Int. J. Ambient Energy **41**(13), 1482–1494 (2020)
9. Fadliondi, F., Isyanto, H., Budiyanto, B.: Bypass diodes for improving solar panel performance. Int. J. Electric. Comput. Eng. **8**(5), 2703 (2018)

10. Cavalcanti, M.C., Bradaschia, F., Junior, A.J.N., Azevedo, G.M., Barbosa, E.J.: Hybrid maximum power point tracking technique for PV modules based on a double-diode model. IEEE Trans. Ind. Electron. (2020)

11. Ahmed, D.R., et al.: The correlation of device parameters with illumination energy to explore the performance of a monocrystalline silicon solar module. Silicon 1–7 (2021)

12. Lekouaghet, B., Abdelkrim, B., Chabane, B.: Estimation of the photovoltaic cells/modules parameters using an improved Rao-based chaotic optimization technique. Energy Convers. Manage. **229**, 113722 (2021)

13. Raya-Armenta, J.M., Ortega, P.R., Bazmohammadi, N., Spataru, S.V., Vasquez, J.C., Guerrero, J.M.: An accurate physical model for PV modules with improved approximations of series-shunt resistances. IEEE J. Photovoltaics (2021)

14. Parida, S.M., Rout, P.K.: Differential evolution with dynamic control factors for parameter estimation of photovoltaic models. J. Computat. Electron. 1–14 (2021)

15. Ramzi, B.M.: Extraction of uncertain parameters of double-diode model of a photovoltaic panel using simulated annealing optimization. J. Phys. Chem. C **123**(48), 29096–29103 (2019)

16. Qais, M.H., Hasanien, H.M., Alghuwainem, S.: Identification of electrical parameters for three-diode photovoltaic model using analytical and sunflower optimization algorithm. Appl. Energy **250**, 109–117 (2020)

17. Hamid, N., Abounacer, R., Idali Oumhand, M., Feddaoui, M.B., Agliz, D.: Parameters identification of photovoltaic solar cells and module using the genetic algorithm with convex combination crossover. Int. J. Ambient Energy **40**(5), 517–524 (2019)

18. Muhammadsharif, F.F., et al.: Brent's algorithm based new computational approach for accurate determination of single-diode model parameters to simulate solar cells and modules. Solar Energy **193**, 782–798 (2019)

19. Wei, D., Wei, M., Cai, H., Zhang, X., Chen, L.: Parameters extraction method of PV model based on key points of IV curve. Energy Convers. Manage. **209**, 112656 (2020)

A Light SRGAN for Up-Scaling of Low Resolution and High Latency Images

Archan Ghosh[✉] [iD], Kalporoop Goswami[iD], Riju Chatterjee[iD],
and Paramita Sarkar[iD]

Calcutta Institute of Engineering and Management, Kolkata, West Bengal, India
https://ciem.ac.in/

Abstract. In the past few years Single Image Super-Resolution (SISR) has been one of the most researched topics in the field of AI. Super-Resolution Generative Adversarial Nets in short SRGAN paved the way to achieve Super-Resolution (SR) of images while hallucinating a lot of details. Deriving from the main components from SRGAN, i.e. Architecture, Loss and Adversarial nature, we have refined a model that works for very small images, and tries to make out as much information as possible in a short amount of time. The main things being focused are to create a fast Generator which also tries to keep a good SSIM score with the ground truth images, tries to recover as much of the information from relative pixels and also gets close enough to benchmark performance with as limited resources as possible. The core objective of having a simple, fast and light model, is not only to enlarge images but fill in as many missing details as it can from simple pixels, to fully defined and distinct features within that image that might have double or quadruple resolution than the Low-Resolution Images.

Keywords: Deep learning · Neural networks · Super-resolution · Image processing · Generative network · UpScaling · Perception · PSNR · SSIM · Adversarial nature

1 Introduction

Up-scaling or Super-Resolution (SR) of Images have been an intriguing problem in multiple research communities. In the past few years there have been some breakthrough results [12,20]. Generative Adversarial Nets (GANs) [6] have found several different applications after their introduction. Similarly SRGANs were substituted in place of SRCNNs [4] and SR-ResNet [22]. Creating High-Resolution (HR) Images from their corresponding Low-Resolution (LR) images can be tricky, but SRGAN [12] proved that GANs is very good at it. However GANs also has its own limitations. In the case of SR, an LR image might have some features that are not distinguishable, therefore the network can either

---

P. Sarkar—Co-Author and Supervisor for the Project.

M. Singh et al. (Eds.): ICACDS 2021, CCIS 1440, pp. 56–67, 2021.
https://doi.org/10.1007/978-3-030-81462-5_6

remove it or sometimes it can over-hallucinate a particular feature thereby completely disorienting the Generated image. Typically when we are dealing with SR problem, we cannot work with very low resolution images, let's say in the scale of 32px × 32px up to 128px × 128px, because there is very less information to work with and when a very high resolution image is downscaled to such a small size a lot of the features within the image gets condensed to such a factor that they are practically not present in the LR image and reconstructing such details can be a huge problem as we are trying to recreate features which are absent. Given below in Fig. 1 is an example from Set14 [21] of an LR image that was used for training and its original counterpart. In SR the general idea is to produce an image that not only has a good PSNR and SSIM value but at the same time can also be visually accurate. Previously measured MSE techniques can produce exceptional models with very low loss but when images are used they are visually not very accurate. The fundamental objective that we want to showcase is that we can in fact recreate visually accurate images from LR images that have good PSNR and SSIM scores. In this study, to produce a comparatively good result, we have used Residual Networks as suggested in the original SRGAN [12] and fine tuned them. For the model itself, we have used an adversarial approach accompanied with the perceptual loss based on VGG network instead of MSE which gives us better visually accurate images. Alternatively in our model, the ultra compressed images have very little information, and cannot compete with outcomes based on other SR solutions, rather it fills the void that we thought was necessary since this was an area that was relatively undermined.

(a) 32x32 Pixel LR for training

(b) 128x128 GT image for Verifying

(c) Original Image that was Compressed

Fig. 1. We took a sample from Set14 and reduced its size to compare with the working scale of our network. From the left 32 × 32 pixel image that is being used as LR, followed by 128 × 128 which is being used as Ground Truth and then farthest right is the original image with a resolution 2.5 times that of ground truth.

2   Related Work

Traditionally in image processing filter based approaches were used for both upscaling and downscaling of images like Bicubic, Linear and Lanczos. Filter based methods are generally very fast but the output given by them can sometimes have over-smoothed edges. Incase of LR images, using filter based methods does not yield very good results. On the contrary, deep learning based approaches to the SR problem using neural network techniques have had many honourable works like Dong et al. [4] who proposed the SR-CNN, Zhang et al. [22], that proposed the use of Residual Networks, unlike the prior which used simple CNN networks for an end to end approach of LR to HR. Other techniques that have been presented include Laplacian Pyramid Structures [11], deep back projection [7], densely connected networks [9] etc. Works like SRGAN [12] and ESRGAN [19] have achieved state-of-the-art PSNR results. SRGAN [12] was the first major breakthrough in GAN based Super-Resolution(SR) that had produced benchmark results. ESRGAN [19] was a follow-up on the SRGAN [12] architecture that used RRDB [20] networks instead of Batch-Normalization for the Generators. The perceptual quality that was set by [12] and [19] was incomparable and stratified the look for a SR solution based on GAN [6]. The techniques developed in [12] and used on benchmark datasets like [2] and [21] are just a few points off the scores set by SR-ResNet.

3   Method

The main objective for us, is to improve the overall visual quality of upscaled images along with the work of filling in as many missing details as it can and maintaining a stable PSNR and SSIM score. In this section we will focus on the architecture used, the adversarial component, loss functions, and feature extraction.

3.1   Network Architecture

Our architecture is derived from Ledig et al. [12], which has a generator that takes LR images and creates HR images from them. Since we are working on a much smaller scale we have further modified the architecture and used different versions of it to have a broad category of outcomes. We have also used different sizes of LR to HR mapping, ranging from 32px–128px upto 200px–800px. For base Architecture see Fig. 2.

As represented in Fig. 2, the Generator takes an LR image as Input, passes it to the basic convolution layer followed by a PReLU [8,12] which determines the basic features and channels. After these, we have used Residual blocks [10] that has an additive layer which adds the computed information from the previous block to the end of the current block, which helps in better information retention and minimizes the chance of Explosive Gradients [5]. Along with this, the information from the primary layers before residuals blocks is also added at the end. Finally we use UpSampling2d that comprehensively increases the resolution and finally a Convolution based output which reorganizes the final image into the 3 (RGB) channels and corrects resolution with respect to the HR image. Each of the residual blocks makes the connection deeper and increases performance. Tai et al. [16] nature of Residual Blocks [10] not only provides the needed deeper [9] and Dense connection but also helps in transmitting the required information in between the layers without the risk of collapsing information. To improve the performance [15] we also use different numbers of residual blocks and test them to obtain the best performance with respect to scaling pairs.

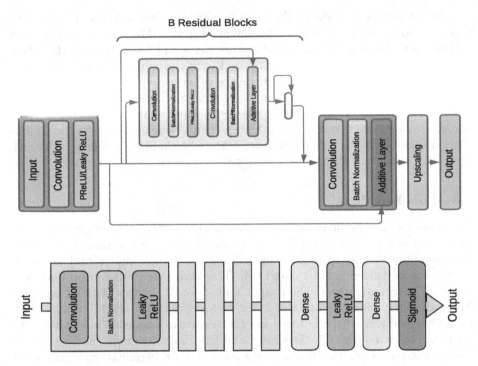

Fig. 2. The first image is the general architecture of the generator used, where B number of residuals blocks are used. The second Image represents the general architecture of the discriminator which compares the features extracted and tells whether the generated Image is closer to fake or real.

3.2 Adversarial Component

The general idea is derived from Goodfellow et al. [6], which has two networks that simultaneously compete with each other to produce better results. We have a Generator G and a discriminator D, the task of the Generator is to create fake images that can fool the Discriminator, and the task of the Discriminator is to detect whether the given image is fake or real. This action makes each of the networks a competitor to each other and ultimately a better outcome is obtained from this model.

The min-max problem from [6] which is modified and derived at [12] stands like this:

$$\min_{\theta_G} \max_{\theta_D} V(D, G) = \mathbb{E}_{I^{HR} \sim P_{train}(I^{HR})}[\log D_{\theta_D}(I^{HR})]$$
$$+ \mathbb{E}_{I^{LR} \sim P_G(I^{LR})}[\log(1 - D_{\theta_D}(G_{\theta_G}(I^{LR})))] \tag{1}$$

The Generator G works on the minimization problem as given in Eq. (1), to produce more realistic pictures with defined features. We use the convolutions and filters to increase the number of feature maps and for every layer the retention and feature optimization is maximized with the help of the architecture as shown in Fig. 2. The Discriminator D on the other hand works on the minimization problem as given in Eq. (1). In the architecture as shown in Fig. 2, we use strided convolutions to minimize the size of the input image and extract as many features as we can. The size of the feature maps increases from 64 to 512 like [12] and then connects to a dense layer followed by a Sigmoid function like [6], producing an output that either dictates the image to be fake or real.

3.3 Loss Functions

The basis for the loss function is to have a high density, feature rich image output as we are working with LR images having a size of less than 100 px. For the perceptual quality to be good enough, we use a combination of 3 main loss techniques as defined in [12,15] and [14]. The MSE based loss can produce perceptually inaccurate images and cannot reproduce HR images of high frequency. To produce HR images of high frequency, we are using VGG-19 [15] weight layers to analyse and extract the features and then create a completely new adversarial structure accompanying the Generator and Discriminator. Here the VGG based loss is transferred into the discriminator for a better determination, significantly increasing its performance. In Eq. 2, we have shown how we obtain Perceptual adversarial loss based on MSE and VGG content loss (Tables 1 and 2).

Table 1. Generator and discriminator loss over different dataset

Final loss over 32-128px Upscaling[a]	Div2k	COIL-100
Generator	0.343	0.382
Discriminator	0.372	0.374

[a]Some of the Numbers Provided might vary from actual experimental values

$$I^{SR}_{perceptual\ adversarial\ loss} = I^{SR}_{MSE} + I^{SR}_{VGG\ based\ content\ loss} \tag{2}$$

Table 2. PSNR and SSIM scores

	PSNR[a]	SSIM[a]
32px to 128px over COIL-100 (5000 Epochs)	17.0	0.252
32px to 128px over Div2k (35000 Epochs)	20.2	0.589
64px to 128 px over Div2k (35000 Epochs)	22.6	0.699

[a]Some of the Numbers Provided might vary from actual experimental values

3.4 Feature Extraction

Working with images of such scale was challenging, especially the scaling of features and also imitation of absent features from LR to HR. For this we have used VGG [15] with the weights of imagenet [3] and extracted the output from the 10th and 20th layer. VGG provides us with an array of feature creation and extensive feature recapture. To further improve the performance, we have used VGG based content loss while training, to improve the perceptual loss. Along with VGG based loss, another architecture was employed for feature extraction [17], but the overall performance of VGG was higher and provided better results over a variety of images.

4 Experiment

4.1 Data Used

The training was done on Div2K dataset [18] with a scaling factor of 4x as in [12,19]. The images are taken and resized using the PIL library to form the ground truth images for an LR-HR pair. Following this we apply augmentation to the image like flip and rotate. We are also using COIL-100 [13] dataset, as it primarily matches the target dimension for this experiment and also features single object images with a dynamic feature range. We are using a batch-size of 2 that was fed to the Generator and Discriminator. Evaluation was done on Set5 [2], and Set14 [21]. Both training and evaluation consisted of 3-coloured channel images in the order of RGB.

4.2 Training Details

The entire training was done using an NVIDIA GTX 1060 GPU. We ran multiple sessions, each having a different LR-HR pair ranging from 32px–128px to 200px–800px. As mentioned earlier, we are using a 4x upscaling factor. We used multiple number of epochs ranging from 5000 to 35000 epochs with a gap of 5000 in between each other. This was done to determine the training time and performance. Another factor that was changed over the course of the experiment was the number of Residual blocks that were used in the generator. The number of epochs and residual blocks were always kept in check, both to utilize the

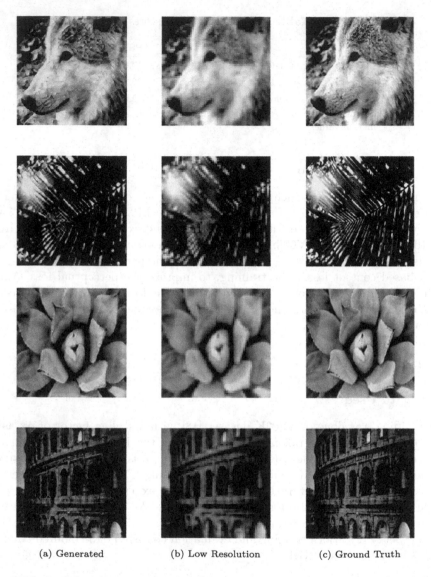

(a) Generated   (b) Low Resolution   (c) Ground Truth

Fig. 3. Each of these images were chosen to represent the different lighting conditions and color palettes and how each can be reproduced by the model.

memory and create a model that had the best results in the shortest possible training time.

Adam optimizer with $\beta^1 = 0.0002$ and $\beta^2 = 0.5$ has been used. LeakyReLU alpha was set to 0.2 and the momentum for Batch Normalization was set to 0.8. Each of the PReLU layers have their "alpha_initializers" set to "ZEROS". For the VGG model we use the resolution of the HR image as the shape of the

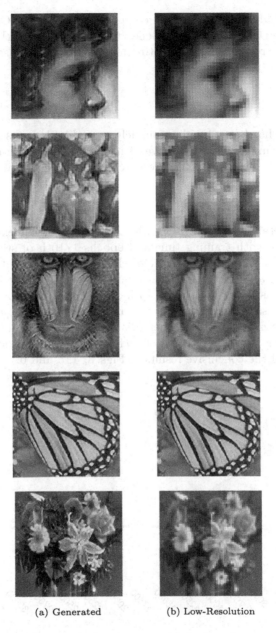

(a) Generated   (b) Low-Resolution

Fig. 4. Some examples from Set5 and Set14 Datasets. The first two images are on the scale of 32–128 and the last 3 are on the scale of 64–128

input tensor and the weight of the imagenet was imported into the model. The output was taken from the 10th or 20th layer depending on the activation being satisfied.

Finally, the adversarial model is compiled with Binary Crossentropy and MSE loss for the Discriminator and Generator respectively, with loss weights set to "1e–3" and "1" respectively. The entire implementation is based on Tensorflow 2.0 [1].

4.3   Analysis

The final model has been tested on publicly available datasets like Set5 [2] and Set14 [21]. The fundamental objective of having a near State-of-The-Art level performance with limited available resources was achieved by the model. Our network performs extremely well when it comes to recreation of missing or incomplete details from an LR image. The network has outperformed our expectations despite the resolution range of LR images on which it was trained and tested lacking many details even at sub-pixel level. The model has exceptional PSNR and SSIM scores, with training time and epochs 1/10th of standard SR models. The network also quantifies the color channels that it thinks suits the style of the image and adjusts the levels that might have been lost during compression. It can also assume the correct depth of field and focus range in a picture, making the necessary features stand out and out of focus objects blurred like the original. The model also can differentiate between hard-surface and organic structures making the SR recreation of such artifacts more prominent. We have provided some of the conclusive results in Figs. 3, 4, 5 and 6.

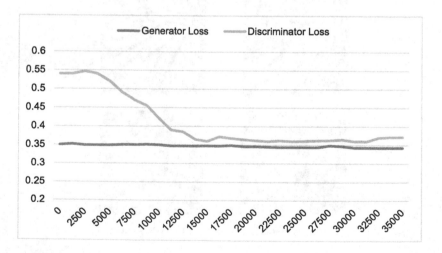

Fig. 5. Represents the average loss of generator and discriminator over the entire training process.

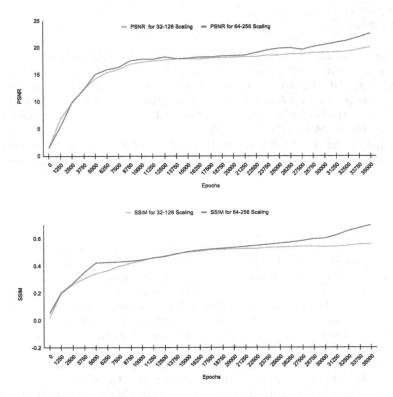

Fig. 6. Upper graph shows the average PSNR score over the two main resolution ranges and the lower graph shows the average SSIM score over the two main resolution ranges

5   Conclusion

The light SRGAN model that we presented in this paper accomplishes the goal of having near State-of-The-Art performance with restricted amount of resources and data. The model has achieved exceptional scores during its stress testing. The primary focus of recreation during up-scaling and having a proper perceptual quality was fulfilled. The smaller batch size also lets us controls the output of the model and how accurately it can reconstruct the features and also adjust the hue and depth of field accordingly. We have also designed the Generator in such a way that it can be used readily and produces results in a lightning fast manner while keeping a good perceptual quality over a range of images.

Acknowledgements. We would like to acknowledge Sergey Gladysh for making his batch randomization code available to us. Divakar Devarajan for the transcription of the base paper and Turbasu Chatterjee for the LaTeX typesetting. Google Colaboratory that was used for early trials. And finally Abhijit Mitra and Paramita Sarkar for guiding us during the culmination of this Paper.

References

1. Abadi, M., et al.: TensorFlow: large-scale machine learning on heterogeneous distributed systems. CoRR abs/1603.04467 (2016). http://arxiv.org/abs/1603.04467
2. Bevilacqua, M., Roumy, A., Guillemot, C., Alberi-Morel, M.L.: Low-complexity single-image super-resolution based on nonnegative neighbor embedding. In: Proceedings of the British Machine Vision Conference, pp. 135.1–135.10. BMVA Press (2012). https://doi.org/10.5244/C.26.135
3. Deng, J., Dong, W., Socher, R., Li, L., Li, K., Fei-Fei, L.: ImageNet: a large-scale hierarchical image database. In: 2009 IEEE Conference on Computer Vision and Pattern Recognition, pp. 248–255 (2009)
4. Dong, C., Loy, C.C., He, K., Tang, X.: Learning a deep convolutional network for image super-resolution. In: Fleet, D., Pajdla, T., Schiele, B., Tuytelaars, T. (eds.) ECCV 2014. LNCS, vol. 8692, pp. 184–199. Springer, Cham (2014). https://doi.org/10.1007/978-3-319-10593-2_13
5. Glorot, X., Bengio, Y.: Understanding the difficulty of training deep feedforward neural networks. J. Mach. Learn. Res. Proc. Track **9**, 249–256 (2010)
6. Goodfellow, I.J., et al.: Generative adversarial networks (2014)
7. Haris, M., Shakhnarovich, G., Ukita, N.: Deep back-projection networks for super-resolution. CoRR abs/1803.02735 (2018). http://arxiv.org/abs/1803.02735
8. He, K., Zhang, X., Ren, S., Sun, J.: Delving deep into rectifiers: surpassing human-level performance on ImageNet classification. CoRR abs/1502.01852 (2015). http://arxiv.org/abs/1502.01852
9. Huang, G., Liu, Z., Weinberger, K.Q.: Densely connected convolutional networks. CoRR abs/1608.06993 (2016). http://arxiv.org/abs/1608.06993
10. Ioffe, S., Szegedy, C.: Batch normalization: accelerating deep network training by reducing internal covariate shift. CoRR abs/1502.03167 (2015). http://arxiv.org/abs/1502.03167
11. Lai, W., Huang, J., Ahuja, N., Yang, M.: Deep Laplacian pyramid networks for fast and accurate super-resolution. CoRR abs/1704.03915 (2017). http://arxiv.org/abs/1704.03915
12. Ledig, C., et al.: Photo-realistic single image super-resolution using a generative adversarial network. CoRR abs/1609.04802 (2016). http://arxiv.org/abs/1609.04802
13. Nene, S.A., Nayar, S.K., Murase, H.: object image library (coil-100. Technical report (1996)
14. Shi, W., et al.: Real-time single image and video super-resolution using an efficient sub-pixel convolutional neural network. CoRR abs/1609.05158 (2016). http://arxiv.org/abs/1609.05158
15. Simonyan, K., Zisserman, A.: Very deep convolutional networks for large-scale image recognition (2015)
16. Tai, Y., Yang, J., Liu, X., Xu, C.: MemNet: a persistent memory network for image restoration. CoRR abs/1708.02209 (2017). http://arxiv.org/abs/1708.02209
17. Tan, M., Le, Q.V.: EfficientNet: rethinking model scaling for convolutional neural networks. CoRR abs/1905.11946 (2019). http://arxiv.org/abs/1905.11946
18. Timofte, R., et al.: NTIRE 2018 challenge on single image super-resolution: methods and results. In: 2018 IEEE/CVF Conference on Computer Vision and Pattern Recognition Workshops (CVPRW), pp. 965–96511 (2018)
19. Wang, X., et al.: ESRGAN: enhanced super-resolution generative adversarial networks. CoRR abs/1809.00219 (2018). http://arxiv.org/abs/1809.00219

20. Wang, Z., Simoncelli, E.P., Bovik, A.C.: Multiscale structural similarity for image quality assessment. In: The Thirty-Seventh Asilomar Conference on Signals, Systems Computers, 2003, vol. 2, pp. 1398–1402 (2003)
21. Zeyde, R., Elad, M., Protter, M.: On single image scale-up using sparse-representations. In: Boissonnat, J., et al. (eds.) Curves and Surfaces - 7th International Conference, 24–30 June 2010, Avignon, France, Revised Selected Papers. Lecture Notes in Computer Science, vol. 6920, pp. 711–730. Springer (2010). https://doi.org/10.1007/978-3-642-27413-8_47
22. Zhang, Y., Tian, Y., Kong, Y., Zhong, B., Fu, Y.: Residual dense network for image super-resolution. CoRR abs/1802.08797 (2018). http://arxiv.org/abs/1802.08797

Energy Efficient Clustering Routing Protocol and ACO Algorithm in WSN

Shalini Subramani[1], M. Selvi[1](✉), S. V. N. Santhosh Kumar[2], and A. Kannan[1]

[1] School of Computer Science and Engineering, VIT University, Vellore, India
{selvi.m,kannan.a}@vit.ac.in
[2] School of Information Technology and Engineering, VIT University, Vellore, India
santhoshkumar.svn@vit.ac.in

Abstract. In a Wireless Sensor Network (WSN), there are some energy constrained tiny sensor devices called nodes. These nodes can sense the environment,
capture the data and routing them optimally to the base station. It is used in
many applications including agriculture, medicine and transportation. However,
the existing routing algorithms for WSN are not energy efficient and hence it
reduces the network lifetime. Therefore, it is necessary to enhance the network
lifetime by applying clustering of nodes and the routing with cluster head nodes.
In the proposed model, the cluster heads are selected based on distance from other
member nodes, residual energy and low mobility. The cluster heads are rotated
periodically since they are made to accomplish the routing tasks. Consequently,
selecting the optimal clusters, cluster heads and routing through an energy efficient reliable routing is the challenge to be considered in the design of WSN.
In this paper, a new reliable as well as energy-efficient routing algorithm has
been proposed by performing clustering, cluster head selection, reliable routing
algorithm with failure identification based routing is considered in this work. For
achieving this, the nodes are clustered and then an intelligent routing process by
the application of Ant Colony Optimization (ACO) is proposed for effective data
delivery. The major findings are the improvement in communication reliability,
data delivery services and reduction of energy requirements.

Keywords: WSN · K-means clustering · LEACH · Energy efficiency · Routing
protocol and ACO algorithm

1 Introduction

A Wireless Sensor Network (WSN) is a large collection of low-powered, smart and
malfunctioning sensor devices connected to other neighbour nodes and the Base Station
(BS). The large numbers of distributed sensor nodes present in a WSN are used in a
specific environment for sensing and collecting the data. In most cases, these tiny sensor
nodes that include an antenna, radio transceiver, processor, memory and a battery [3].
The sensor unit tracks the atmosphere to gather data, processes it and transmitting them
through the communication network for sending them to the base station. However, the
capacity, bandwidth, memory and measurement are restricted in the sensor nodes and

© Springer Nature Switzerland AG 2021
M. Singh et al. (Eds.): ICACDS 2021, CCIS 1440, pp. 68–80, 2021.
https://doi.org/10.1007/978-3-030-81462-5_7

hence the network life time must be taken care of by applying suitable methods for energy conservation. Therefore, energy optimization and reliable secured routing are the issues to be tackled in the design of intelligent routing protocols for WSN [1, 39].

Many different day-to-day applications including weather management, agriculture data collection and healthcare data maintenance need the support of WSN [2]. The purpose of organizing an intelligent WSN is for obtaining intelligence about the conceptual events in the monitoring of the sensing area, translate the monitored data into an electric signal and transmitting it to the BS. The sensor nodes are using more amounts of resources for performing effective data collection and dissemination to the BS [4]. The data are transmitted by applying a cluster-based routing algorithm which works based on the application of shortest optimal path and multi-hop routing protocols. Clustering strategies divide the adjacent nodes into clusters and group them into groups. From the member nodes, one node with high credentials with respect to distance as well as energy is chosen as the cluster head (CH) and the rest are called as normal nodes [5].

When the data is in the nodes near the base station can send the data with single hop and the nodes having the higher distance will send the data through multi-hop communication with the support of CH nodes. Therefore, the energy requirements for the nearby nodes are different from the energy requirements for farthest nodes. A CH is chosen based on a certain parameter namely distance, energy and behavior. The CH is responsible for three operations which are the compilation, aggregation and transfer of data to BS by cluster members. CH is also a relay node for the transfer of data to BS by other CH. Member nodes must only interact for a short time with their respective cluster heads and absorb fewer resources. In the Clustering Protocols [40, 41], the optimum cluster heads are chosen and CH positions to the nodes are rotated periodically to match the node energy consumption [6]. The custom energy through sensor devices are significantly reduced by using clustering techniques. Moreover, the main motivation for designing a cluster data based routing model is to improve the network throughput, reduction of delay and energy consumption.

In this paper, a new energy efficient clustered and rule based routing methods are proposed to perform an intelligent Ant Colony Optimization(ACO) in WSN. The main motivations for this proposed work is to propose an intelligent routing algorithm which applies heuristic search approach for finding the optimal route through the behavior of ants. Moreover, the proposed work reduces the excessive energy depletion by the existing CH through the distribution of data delivery to all the nodes in every cluster heads optimally. Moreover, the cluster heads are rotated periodically based on energy and distance constraints. The key benefits of the proposed model are (i) It provides improved packet distribution and delivery at the cluster head nodes and the base station. (ii) The proposed routing algorithm increases the overall network performance by enhancing the number of rounds which is used to measure the network lifetime. (iii) The ACO based routing algorithm finds the optimal routes rather than shortest route and hence it achieves latency reduction. (iv) Finally, the clustering and ACO based intelligent routing algorithm proposed in this work to reduces the energy consumption in the WSN by finding the congestion less optimal paths. The remaining part of this manuscript is organized as follows: Sect. 2 highlight the detailed literature survey on routing protocols to provide the optimal solutions. The detailed proposed system architecture is provided in Sect. 3 which explains the functional modules of the proposed system. Section 4 gives the simulation results of proposed system with other existing systems. Finally, Sect. 5 provides the conclusion and future work of the proposed system.

2 Related Works

In WSNs, the consequence of the link quality is playing an important role on the reliability of data communication. The consumption of energy is one of the key constraints of WSN, so many energy perception routing studies have been performed in the past and many different routing protocols were developed. The clustering protocols for routing decrease the use of power in the network while making control simpler. The most critical optimizing topic in WSN is clustering and routing. The routing protocol provides the optimal solution in the network to provide the energy efficient routing [7] in WSN. In [8], the authors suggested that the link quality has a considerable impact on data transmission and they examined the links and observed that better quality of link can provide efficiency in network communication. At the same time, the bad quality of link will result in packet loss and a drop in massive packet throughput in the network. The first clustering approach used to perform cluster based routing in WSN was called "Low Energy Adaptive Clustering Hierarchy" (LEACH), which is the most common and well-known cluster-based routing protocol [9]. In the literature [34–38]. Many authors have proposed various works on energy efficient clustered based routing in WSN.

In [10], the authors have provided the solution for unequal cluster heads distribution to improve by the use of the LEACH protocol which is based on non-uniform clustering. The communication which is based on single-hop and multi-hop combinations help to improves the load of network and balance it to prevent the issue of network hot spots. A clustering strategy that uses local node data awareness to choose cluster heads was proposed in [11]. The method to create a cluster and to determine a cluster head and finally to disperse information was proposed in [12] by making the BS as energy-efficient route selector. A cluster-based routing strategy that implements a combination of the distance, double CH and dormancy of the nodes was proposed in [13]. In [14], a clustering algorithm which can works with two cluster was proposed to optimize the routing in the network. In their system, two nodes are chosen as the member nodes on every cluster based the nodes residual energy and their Euclidian distance from the Base Station. One cluster head is used to aggregate the data collected and the other is used for the transmission of data. For CH based data collection, the weighted likelihood is used as the decision parameter [15]. In [16], the authors recommended a scheme in where the nodes are divided into many clusters based on the K-means clustering approach and by employing fluid logical system, the corresponding cluster head is chosen to provide an optimal routing in WSN. In another work, author proposes an enhanced energy efficient, time-consuming and conscious routing system which shapes clusters in the initial stage and connects the number of Hamiltonian nodes using a greedy data communication strategy was proposed in [17]. In [18], the authors proposed a system which can able to prolong the life time of the network. This method uses different CH-selection thresholds. It takes into account the mean and variance of the distance of the nodes from the sink and as well as the available node capacity. A heterogeneous cluster route protocol focused on a probability model using the availability of the efficiency node and cluster head was proposed in [19].

In this model, if certain network nodes are powerless and the system changes as well as in the device. The relevance of sensor devices for energy-saving and effective protocols for the architecture of networking to optimize network life was discussed in [20]. In this approach, the CH uses the contact multi-hop approach for transferring the fused data to the BS. Many clustering protocols use the intra-cluster coordination framework Time Division Multiple Access where all nodes still have the power to collect relevant data [21]. Within the access point, the author in [22] recommended that a centralized classification routing protocol be executed. The routing algorithm rotates every definite quantity of sequences to upgrade the signal track and match the power consumption of the nodes in proximity to the BS. In [23], an ACO algorithm is used for improving the energy-efficient clustering routing protocol. A modern pheromone update framework is planned for rendering the energy and distance between devices. In [24], the authors suggested an approach to pick the subset of the device and control its sample rate and interval of transmission. In [25], the authors proposed an energy-efficient area routing protocol based on source routing, which decreases the energy usage of data transfer and balances the power consumption between the various nodes. All these works are focusing on minimizing the consumption of resources in WSN. However, the overall network performance needs further improvement. In this paper, they have proposed cluster based energy efficient intelligent routing protocol that uses ACO for routing by finding an optimum transmission path for the effective routing of collected data. The major differences between the proposed method and other existing clustered based routing in WSN includes heuristic route discovery, probability based route selection and efficient route maintenance techniques provided in the proposed work through the effective application of ant colony meta heuristics. Also, the proposed work uses a rule based decision making approach by firing the rules present in the knowledge base for performing route discovery, route selection and data routing.

3 Proposed Work

In this paper, we propose an energy-efficient and consistent routing algorithm by applying clustering and Ant Colony Optimization (ACO) algorithm to route the data efficiently and reliably and also to enhance the network life time. Moreover, the power consumption and delay with respect to data delivery in WSN is also reduced by the application of the proposed routing protocol. Figure 1 represents the architecture of the proposed routing and data delivery system. The nine major components of the proposed system are the sensor nodes, the data collection and consolidation module, the clustering module for node grouping, the intelligent ACO based data delivery module, the intelligent rule based decision manager, the optimal energy manager, the energy aware routing module, the rule based repository called knowledge base and the base station. The sensor nodes are deployed in the sensing area for data collection. Here, the sensed data are collected by the data collection module and it is sent to the routing module through clustering module. The decision manager takes care of every activity which is happening in the network through continuous monitoring and control. The clustering module is responsible for the formation of clusters from the nodes, identification of CH and its rotation. The energy module checks whether a given node has the necessary energy for taking part in the routing process. The ACO module finds the shortest path and optimal path for routing. The routing module sending the data through the clusters in the optimal routes to the sink. The rules are stored in the knowledge base that are fired for making decisions on data collection and routing. The proposed approaches have been explained in depth.

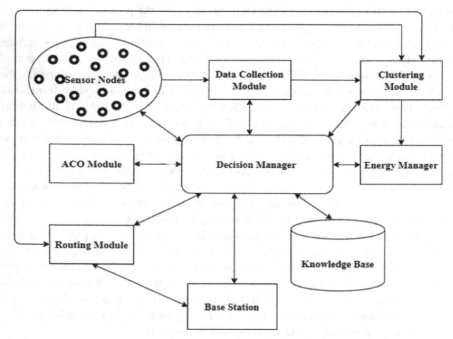

Fig. 1. System architecture

3.1 Clustering Formation in the Proposed Wireless Sensor Network

In this work, a new clustering and reliable routing algorithm has been proposed. Here the routing is carried out through CH nodes up to the sink node. The clusters are dynamically changed using a rotation policy leading to increase in the number of rounds for all the CH nodes. Cluster formation is applied to decisions on how the cluster system can be built at the initial phase. The cluster maintenance module is responsible for updating the cluster configuration for the node functionality or path violation, according to the corresponding network topology changes. The foremost important operation in the development of the groups as clusters and to perform cluster based routing is the selection of cluster heads [42].

Clustering is used in this work since it is very efficient and useful in achieving scalability, minimizing energy demand and extending the service life of a system with multiple mobile nodes and high mobility. During the management of clusters, the CHs frequently give its members hello messages to learn about the network changes [26]. The added traffic contributes to congestion and crashes with increased geographical density. Therefore, we have to take seriously the extra costs for sharing of clustering control packets, re-clustering and self-modification during the classification and maintenance stage, while evaluating or improving the clustering routing protocols. The selected CH transmits a cluster-join message to all the sensor nodes which further calculate the distance cost computed as

$$
\begin{aligned}
\text{Total Distance Cost} = \text{Cost of (Member node to CH distance} \\
+ \text{Distance from CH to BS)}
\end{aligned}
\tag{1}
$$

3.2 Route Discovery Using ACO

Route Discovery is the mechanism by which a minimum energy route is created between the source node and the destination. It uses two types of controls, called front and back. An ant from the path moves forwards but the same ant is then called backward from the target server into the source node. The route discovery is by forward ants and route reply is taken care of by backward ants.

A forward ant unit finds a neighbouring node with the neighbouring node matrix and measures the connections probability function values in the route. The ant is being forwarded to another more likely value-connection device and attaches its current device ID to the routing table. If a forward ant is reachable by the BS, it sends a reverse ant, which will follow the reverse route and deposits the pheromone on the route and it tests and changes the pheromone value in the route ties. Therefore, a new and optimal path is found for effective data dissemination [27]. The working of the algorithm starts with ant based route discovery by flooding of packets in the form of forward ants and then finding the optimal path based on the first ant that reaches the destination. The route maintenance is based on backward ants with pheromone values.

3.3 Routing Using Cluster and ACO-Based Routing Algorithm

In this work, a new intelligent routing algorithm with clustering and ACO is proposed. The first step is cluster formation based on distance metrics and energy availability. After clustering CH nodes are elected and ACO is applied through the CH nodes for finding the shortest and optimal path.

Cluster creation helps boost global network efficiency through data convergence, which balances the network load, small intermediate nodes and improves network life [28]. Only the CH nodes are used to transmit the data among them and finally to the BS. This helps to conserve the resources for cluster heads and it also reduces the network latency. In [29], many routing protocols were discussed which are based on clustering and cluster based routing.

In [30], the authors proposed the LEACH protocol which consists of two-phases namely the called cluster formation phase and the routing phase. In the setup phase, clusters are formed and whereas in the routing phase transmission of data takes place. In this proposed work, we select the CH according to the distance to reach the BS, not a

CH that is closest to it. Distance D between two nodes with locations (x1, y1) and (x2, y2) in this proposed approach is calculated according to the Eq. (2).

$$D(x, y) = \sqrt{(x_1 - x_2)^2 + (y_1 - y_2)^2} \qquad (2)$$

The distance is calculated from a node to every CH and from CH to BS and the minimum distance is chosen [31]. According to this protocol, we use the ACO algorithm for better efficiency of finding the minimum optimal path for increases the lifetime of wireless sensor networks [32].

3.4 Route Maintenance

Route maintenance level is accountable for managing the route developed and defined during the route discovery period. As data packets between the source and the destination nodes are passed along the identified path, the pheromone has increased the route between the sender and receiver such that an optimum path is maintained in the road discovery phase [33]. A receiving of the transmissions is sent to the source by the sink node. If the device of each node has not been received the recognition pattern within a span of time-off, a path error message would be sent to the previous node.

It is also needed and preserves the path of network topology changes over time and when it needs to be reformed between the nodes. If the network topology changes, the current source location from the destination is obtained and the track discovery phase is restarted. By reducing the energy use and latency of the WSN network, the track is continued by the clustering process.

The clustering protocol using intelligent ACO are proposed with following steps in the routing algorithm are as follows:

Input: Sensor nodes and rules.

Output: Clusters, Optimal routes and collection of routed data.

Step 1: Read the sensor nodes.

Step 2: Create the knowledge base by storing rules.

Step 3: Read the value of k to form k clusters.

Step 4: Measure Energy levels of nodes.

Step 5: Identify k centroid nodes.

Step 6: Form k clusters by applying distances and energy to admit the member nodes.

Step 7: Perform cluster head selection.

Step 8: Apply ACO and perform the route discovery through cluster heads.

Step 9: Route the packets.

Step 10: If more data are collected, perform cluster head rotation and go to step 7.

Step 11: Energy levels of nodes are less than threshold STOP.

In this algorithm, the forward ants are used for route discovery and the backward ants are used for route reply. The sink node gathers every routed data.

4 Results and Discussions

This proposed model was simulated with the NS2 simulator. The simulation parameter is shown in Table 1.

Table 1. Simulation parameters

Parameters	Values
Network area	$500 \times 500 \text{ m}^2$
Number of nodes	Up to 500 nodes
Basic routing protocol	LEACH
Initial energy	2J
Packets size	4096 bits
Mobility speed	10 m/s to 50 m/s
Mobility model	Random way

Figure 2 shows the packet delivery ratio analysis between the proposed ACO Based Routing algorithm named as ACOBR and with related algorithms including fuzzy logic based unequal clustering algorithm (FLBUC) [35], the most famous algorithms namely LEACH algorithm [9] and similar type of algorithm called HEED [34].

From Fig. 2, we perceived the proposed algorithm using ACO Based Routing (ACOBR) has increased the packet data delivery ratio by more than 2% and we compared with existing clustering algorithm of HEED [34], the related routing protocols LEACH [9] and parallel type of algorithm are called FLBUC [35].

It increases the packet This performance improvement with respect to increase in packet delivery ratio has been achieved in the proposed work by the use of ACO meta heuristics in the proposed routing technique namely ACOBR algorithm. The use of ACO metaheuristics has made the routing path to be optimal and without congestions. Therefore, the packet delivery ratio is increased in the proposed work by the optimal routing.

Figure 3 shows the delay analysis between the existing algorithms namely HEED, LEACH, FLBUC and the proposed algorithm namely the ACOBR algorithm by using 100, 200, 300, 400 and nodes in the network.

From Fig. 3, it is observed that the proposed ACO Based Routing algorithm (ACOBR) has decreased the delay by more than 10 milli seconds when it is compared with the existing cluster based routing algorithms namely HEED [34], LEACH [9] and FLBUC [35]. The reason of better communication delay is that the proposed system uses ACO meta heuristics algorithm which provides an optimal clustered based routing which can reduces the congestion in the network.

The energy consumption analysis of the existing systems and the proposed algorithm namely the ACOBR algorithm for five experiments namely E1, E2, E3, E4 and E5 is shown in Fig. 4.

From Fig. 4, it shows that the proposed ACO Based Routing algorithm (ACOBR) has decreased the energy consumption by more than 5 J in all the five experiments. The energy consumption has been decreased in the proposed work by using the ACO meta heuristics in which the packets are not made to wait at nodes by avoiding the congested routes along with rule based intelligent cluster formation and by providing cluster based routing with optimal routes.

76   S. Subramani et al.

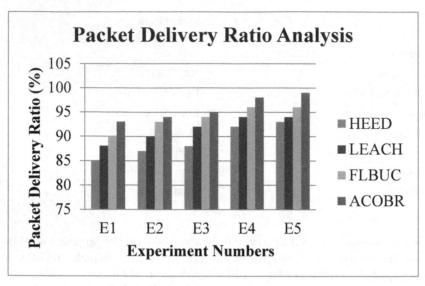

Fig. 2. Packet delivery ratio analysis

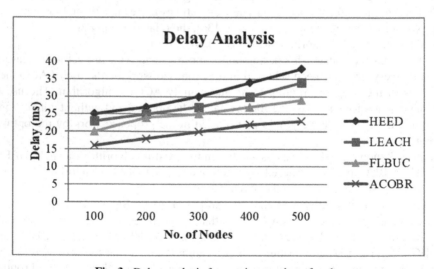

Fig. 3. Delay analysis for varying number of nodes

Figure 5 gives the analysis of network life time of the existing approaches and the proposed ACOBR algorithm for 100, 200, 300, 400 and 500 nodes.

From Fig. 5, it shows that, the proposed ACOBR has increased the network life time by more than 100 rounds in all the five experiments conducted with 100, 200, 300, 400 and 500 nodes when it is compared with existing approaches. The network life time has increased in the proposed work due to the finding of multiple paths and then the selection of optimal path by the ACO meta heuristics to perform energy aware and rule as well as cluster based routing with probabilistic reasoning.

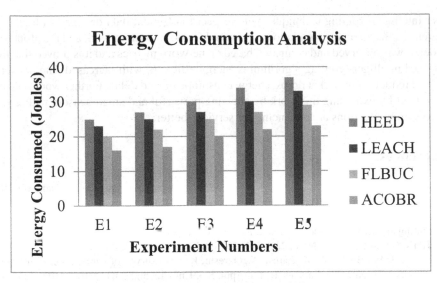

Fig. 4. Energy consumption analysis

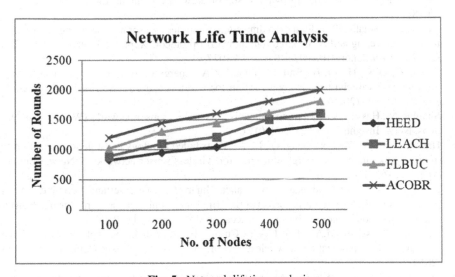

Fig. 5. Network lifetime analysis

5 Conclusion

In this paper, a new energy aware intelligent protocol for optimal data delivery has been proposed by applying ACO approach. For this purpose, a cluster based sensing field with clusters of sensor nodes were formed and for each cluster a leader was elected for taking the responsibility of reliable data delivery. A rule based approach has been introduced in this paper that uses pheromone values for making routing decisions that will lead to optimal routing with reduced energy. This proposed model has been compared with other intelligent routing techniques with respect to efficient data delivery with energy awareness. Based on the simulations performed for both existing work and the proposed model, it was observed and compared based on network metrics and it is proved that the proposed intelligent routing algorithm is energy efficient, with increased data delivery rate and reduction in packet drops, energy consumption and delay. Further works can be carried out by extending this work by applying powerful and smart sensors with agent based communications are enhancing the services better.

References

1. Kang, J., Kim, J., Kim, M., Sohn, M.: Machine learning-based energy-saving framework for environmental states-adaptive wireless sensor network. IEEE Access **8**, 69359–69367 (2020)
2. Al-Kiyumi, R.M., Foh, C.H., Vural, S., Chatzimisios, P., Tafazolli, R: Fuzzy logic-based routing algorithm for lifetime enhancement in heterogeneous wireless sensor networks. IEEE Trans. Green Commun. Netw. **2**(2), 517–532 (2018)
3. Anwar, R.W., Bakhtiari, M., Zainal, A., Qureshi, K.N.: A survey of wireless sensor network security and routing techniques. Res. J. Appl. Sci. Eng. Technol. **9**(11), 1016–1026 (2015)
4. Logambigai, R., Ganapathy, S., Kannan, A.: Energy-efficient grid-based routing algorithm using intelligent fuzzy rules for wireless sensor networks. Comput. Electr. Eng. **68**, 62–75 (2018)
5. Arjunan, S., Sujatha, P.: Lifetime maximization of wireless sensor network using fuzzy based unequal clustering and ACO based routing hybrid protocol. Appl. Intell. **48**(8), 2229–2246 (2017). https://doi.org/10.1007/s10489-017-1077-y
6. Wang, Z., Ding, H., Li, B., Bao, L., Yang, Z.: An energy efficient routing protocol based on improved artificial bee colony algorithm for wireless sensor networks. IEEE Access **8**, 133577–133596 (2020)
7. Shabbir, N., Hassan, S.R.: Routing protocols for wireless sensor networks (WSNs). Wireless Sens. Netw.-Insights Innov. 22–26 (2017)
8. Hao, X., Liu, W., Yao, N., Geng, D., Li, X.: Distributed topology construction algorithm to improve link quality and energy efficiency for wireless sensor networks. J. Netw. Comput. Appl. **72**, 162–170 (2016)
9. Heinzelman, W.R., Chandrakasan, A., Balakrishnan, H.: Energy-efficient communication protocol for wireless microsensor networks. In: Proceedings of the 33rd Annual Hawaii International Conference on System Sciences, pp. 1–10. IEEE, Jan 2000
10. Saad, E., Elhosseini, M.A., Haikal, A.Y.: Culture-based artificial bee colony with heritage mechanism for optimization of wireless sensors network. Appl. Soft Comput. **79**, 59–73 (2019)
11. Osamy, W., Salim, A., Khedr, A.M.: An information entropy based-clustering algorithm for heterogeneous wireless sensor networks. Wireless Netw. **26**(3), 1869–1886 (2018). https://doi.org/10.1007/s11276-018-1877-y
12. Marappan, P., Rodrigues, P.: An energy efficient routing protocol for correlated data using CL-LEACH in WSN. Wireless Netw. **22**(4), 1415–1423 (2015). https://doi.org/10.1007/s11276-015-1063-4
13. Hamamreh, R.A., Haji, M.M., Qutob, A.A.: An energy-efficient clustering routing protocol for WSN based on MRHC. 214–222 (2018)
14. Panag, T.S., Dhillon, J.S.: Dual head static clustering algorithm for wireless sensor networks. AEU-Int. J. Electron. Commun. **88**, 148–156 (2018)
15. Singh, R., Verma, A.K.: Energy efficient cross layer based adaptive threshold routing protocol for WSN. AEU-Int. J. Electron. Commun. **72**, 166–173 (2017)

16. Li, L., Li, D.: An energy-balanced routing protocol for a wireless sensor network. J. Sens. 1–12 (2018)
17. Yi, D., Yang, H.: HEER–a delay-aware and energy-efficient routing protocol for wireless sensor networks. Comput. Netw. **104**, 155–173 (2016)
18. Kia, G., Hassanzadeh, A.: A multi-threshold long life time protocol with consistent performance for wireless sensor networks. AEU-Int. J. Electron. Commun. **101**, 114–127 (2019)
19. Rawat, P., Chauhan, S.: Probability based cluster routing protocol for wireless sensor network. J. Ambient Intell. Humanized Comput. 1–13 (2020)
20. Elsmany, E.F.A., Omar, M.A., Wan, T.C., Altahir, A.A.: EESRA: energy efficient scalable routing algorithm for wireless sensor networks. IEEE Access **7**, 96974–96983 (2019)
21. Sarkar, A., Murugan, T.S.: Cluster head selection for energy efficient and delay- less routing in wireless sensor network. Wireless Netw. **25**(1), 303–320 (2019)
22. Yarinezhad, R., Hashemi, S.N.: Increasing the lifetime of sensor networks by a data dissemination model based on a new approximation algorithm. Ad Hoc Netw. **100**, 102084 (2020)
23. Li, X., Keegan, B., Mtenzi, F., Weise, T., Tan, M.: Energy-efficient load balancing ant based routing algorithm for wireless sensor networks. IEEE Access **7**, 113182–113196 (2019)
24. Kang, J., Kim, J., Sohn, M.M.: Supervised learning-based lifetime extension of wireless sensor network nodes. J. Internet Services Inf. Secur. **9**(4), 59–67 (2019)
25. Xu, C., Xiong, Z., Zhao, G., Yu, S.: An energy-efficient region source routing protocol for lifetime maximization in WSN. IEEE Access **7**, 135277–135289 (2019)
26. Wan, J., Yuan, D., Xu, X.: A review of cluster formation mechanism for clustering routing protocols. In: 2008 11th IEEE International Conference on Communication Technology, pp. 611–616. IEEE, Nov 2008
27. Srinivas, M.B.: Cluster based energy efficient routing protocol using ANT colony optimization and breadth first search. Procedia Comput. Sci. **89**, 124–133 (2016)
28. Bhola, J., Soni, S., Cheema, G.K.: Genetic algorithm based optimized leach protocol for energy efficient wireless sensor networks. J. Ambient. Intell. Humaniz. Comput. **11**(3), 1281–1288 (2019). https://doi.org/10.1007/s12652-019-01382-3
29. Samara, G., Blaou, K.M.: Wireless sensor networks hierarchical protocols. In: 2017 8th International Conference on Information Technology (ICIT), IEEE, pp. 998–1001, May 2017
30. Fanian, F., Rafsanjani, M.K.: Cluster-based routing protocols in wireless sensor networks: a survey based on methodology. J. Netw. Comput. Appl. **142**, 111–142 (2019)
31. Abu Salem, A.O., Shudifat, N.: Enhanced LEACH protocol for increasing a lifetime of WSNs. Pers. Ubiquit. Comput. **23**(5–6), 901–907 (2019). https://doi.org/10.1007/s00779-019-012 05-4
32. Mohajerani, A., Gharavian, D.: An ant colony optimization based routing algorithm for extending network lifetime in wireless sensor networks. Wireless Netw. **22**(8), 2637–2647 (2015). https://doi.org/10.1007/s11276-015-1061-6
33. Xiao, X., Huang, H.: A clustering routing algorithm based on improved ant colony optimization algorithms for underwater wireless sensor networks. Algorithms **13**(10), 250–255 (2020)
34. Younis, O., Fahmy, S.: HEED: a hybrid, energy-efficient, distributed clustering approach for ad hoc sensor networks. IEEE Trans. Mob. Comput. **3**(4), 366–379 (2004)
35. Logambigai, R., Kannan, A.: Fuzzy logic based unequal clustering for wireless sensor networks. Wireless Netw. **22**(3), 945–957 (2015). https://doi.org/10.1007/s11276-015-1013-1
36. Selvi, M, Thangaramya, K, Ganapathy, S., Kulothungan, K., Nehemiah, H.K., Kannan, A.: An energy aware trust based secure routing algorithm for effective communication in wireless sensor networks. Wireless Pers. Commun. **105**(4), 1475–1490 (2019)

37. Selvi, M., Velvizhy, M., Ganapathy, S., Nehemiah, H.K., Kannan, A.: A rule based delay constrained energy efficient routing technique for wireless sensor networks'. Clust. Comput. **22**, 10839–10848 (2019)
38. Munuswamy Selvi, S.V.N., Kumar, S., Ganapathy, S., Ayyanar, A., Nehemiah, H.K., Kannan, A.: An energy efficient clustered gravitational and fuzzy based routing algorithm in WSNs. Wireless. Pers. Commun. **116**(1), 61–90 (2021)
39. Nancy, P., Muthurajkumar, S., Ganapathy, S., Kumar, S.S., Selvi, M., Arputharaj, K.: Intrusion detection using dynamic feature selection and fuzzy temporal decision tree classification for wireless sensor networks. IET Commun. **14**(5), 888–895 (2020)
40. Thangaramya, K., Kulothungan, K., Indira Gandhi, S., Selvi, M., Santhosh Kumar, S.V.N., Arputharaj, K.: Intelligent fuzzy rule-based approach with outlier detection for secured routing in WSN. Soft. Comput. **24**(21), 16483–16497 (2020). https://doi.org/10.1007/s00500-020-04955-z
41. Thangaramya, K., Kulothungan, K., Logambigai, R., Selvi, M., Ganapathy, S., Kannan, A.: Energy aware cluster and neuro-fuzzy based routing algorithm for wireless sensor networks in IoT. Comput. Netw. **151**, 211–223 (2019)
42. Maheshwari, P., Sharma, A.K., Verma, K.: Energy efficient cluster based routing protocol for WSN using butterfly optimization algorithm and ant colony optimization. Ad Hoc Netw. **110**, 102317 (2021)

Efficient Social Distancing Detection Using Object Detection and Triangle Similarity

Vidya Zope, Nikhil Joshi[✉], Srivatsan Iyengar, Krish Mahadevan, and Meher Singh

Computer Engineering Department, Vivekanand Education Society's Institute of Technology, Chembur, Mumbai, India

{vidya.zope,2017.nikhil.joshi,2017.srivatsan.iyengar,
2017.mahadevan.krishvenkatteshwaran,2017.meher.singh}@ves.ac.in

Abstract. COVID-19 or the Coronavirus Disease has been wreaking havoc all around the globe. People have lost their lives and livelihoods because of this contiguous disease that multiplies at a very fast rate. Although countries have prepared vaccines and have started with the process of vaccination, it has been advised by the government to follow the norms of wearing face masks, following social distancing and hand sanitization for atleast a few months from now so that the vaccine can be effective against the virus. Following Social Distancing is one of the ways we prevent the mass spreading of the virus. The proposed system uses Object Detection and Triangle Similarity techniques to check if people are following Social Distancing or not in Images, Videos and Webcam feeds. If any individual is found not following the norms, an alarm will be sounded to alert the person and the police officials.

Keywords: COVID-19 · Object Detection · Triangle Similarity · YOLO · Social distancing · Euclidean distance

1 Introduction

COVID-19 or the SARS-CoV-2 virus has affected the entire globe since December 2019 when it began in Wuhan, China. It is still an ongoing pandemic which affects the respiratory system and causes problems while breathing and mild symptoms like cold and cough [1]. For some people, though, it can cause acute symptoms and has even led to deaths of many people. It has affected the entire globe with the US, India, Russia and Brazil among the most affected countries. Although people have started developing herd immunity against the virus and vaccination has also begun in some countries, the government has instructed the people to follow the rules of hand sanitization, face masks and social distancing atleast for the next few months for the vaccine to show its effect. Social Distancing is one of the methods by which the spread of the virus can be controlled. Social Distancing is a method of maintaining a minimum distance (in this case 2 m or 6 ft.) between 2 people to avoid close contacts so that the disease does not spread from one person to another. It is a very effective form of precaution. But in countries like India, where the population is massive, it is not practically possible to check if each and every

© Springer Nature Switzerland AG 2021
M. Singh et al. (Eds.): ICACDS 2021, CCIS 1440, pp. 81–89, 2021.
https://doi.org/10.1007/978-3-030-81462-5_8

individual is following the rules or not [2, 3]. Hence, we have designed this system which can be used in CCTV cameras to check if people are following the Social Distancing norms or not in Images, Videos and Webcam feeds. We have used the concepts of Object Detection (more specifically Person Detection) and Triangle Similarity to calculate the distance between every pair of individuals detected in the input feed [10]. If this distance is less than 6 ft. or 2 m then we infer that those individuals are not following the rules. If any person is not following the norms, an alarm will be rung to alert the person and the police or other government officials so that the officials can take the necessary action against that person [4, 5].

2 Problem Definition

The objective of this system is to ensure that people are following Social Distancing in all public places to avoid the spread of the Corona virus. An alert will be given if any person is found not following the rules in the input feed. The input to the system can be an image, video or webcam feed.

3 Scope of the Project

The proposed system has a wide scope since it is independent of the place where it is used. Typically, it can be used in CCTV cameras which are situated in public places like shopping complex, playgrounds, schools and colleges and so on.

4 Literature Survey

1. Mulia Pratama, Widodo Budi, Santoso Ahmad Dimyani, Achmad Praptijanto, Arifin Nur, Yanuandri Putrasari, "Performance of Inter-vehicular Distance Estimation: Pose from Orthography and Triangle Similarity" - This paper presents a comparison between 2 methods of measuring distances from a digital camera i.e. Pose from Orthographic Projection and Triangle Similarity. Both the methods incorporate a computer vision algorithm to proper functionality. Specifically, the paper describes the utilization of such methods for vehicular application like inter-vehicle distance measurement [6].
2. Narinder Singh Punn, Sanjay Kumar Sonbhadra, Sonali Agarwal, "Monitoring COVID-19 social distancing with person detection and tracking via fine-tuned YOLO v3 and Deepsort techniques" - This paper presents an automated framework to monitor social distancing using surveillance video. It uses YOLO v3 object detection for detecting pedestrians. It also compares the results with faster RCNN and SSD models through parameters like loss and FPS [7].
3. Rajesh Kannan Megalingam, Vignesh Shriram, Bommu Likhith, Gangireddy Rajesh, Sriharsha Ghanta, "Monocular Distance Estimation using Pinhole Camera Approximation to Avoid Vehicle Crash and Back-over Accidents" - This paper uses the Triangle Similarity approach to calculate the distance between an object and the camera using the focal length. It applies this technique to calculate the distance between the 2 vehicles and the vehicle and an obstacle to avoid accidents while parking and backing up [8].

4. Md Asgar Hossain; Md Mukit, "A real-time face to camera distance measurement algorithm using object classification" - This paper presents an estimation method based on feature detection, to calculate the distance from a camera to a human face, where detection of eyes, face and iris in an image sequence is described. It uses an architecture for face detection-based system on AdaBoost algorithm using Haar features and Canny and Hugo Transform for edge and circular iris estimation. Wrongly detected faces are removed by analyzing the disparity map. From the estimated face, Canny and Hugo transform is used to determine the iris, and to calculate the distance between the centroid of iris. Lastly, for distance estimation, Pythagoras theorem or Triangle Similarity can be used [9].

5. Mahdi Rezaei, Mohsen Azarmi, "DeepSOCIAL: Social Distancing Monitoring and Infection Risk Assessment in COVID-19 Pandemic" - In this paper, a DeepSocial network system is implemented which utilizes webcam as the source and detects whether people are following social distancing or not. This paper provides a great visualization using tools like heatmaps, moving trajectory etc. It also performs well in different situations like occlusion, lighting variations, shades, and partial visibility [2].

5 Proposed Solution

The proposed solution uses Object Detection and Triangle similarity for calculating the distance between 2 people. Object Detection is achieved by using pre-trained YOLO v3 weights which is trained on the COCO dataset [11, 12]. Through Object detection we will detect all the human bodies in the given input feed. The output of this phase is the diagonal coordinates of the bounding boxes formed around the human bodies. Using the diagonal coordinates, we can calculate the centroid of the bounding box. In the first part we will find the pixel wise distance between the 2 bounding boxes formed using Euclidean distance. This distance is measured in pixels, but we need the distance in feet or inches since all our distance measures are measured in feet or inches. To convert the distance in pixels to inches we have used the concept of PPI or Pixels per Inch. This metric gives the number of pixels covered in one inch. Hence, through this metric we can find the distance in inches or feet [13].

At this stage we have got the pairwise distance between the bounding boxes. But here we have covered only 2 coordinates. Hence at this stage we cannot calculate the distance accurately if the input image or video is from the front view where the people are behind each other and not side-on. This distance only gives the distance from one angle which is not sufficient. To incorporate the 3rd coordinate, 2 techniques can be used: Triangle Similarity or the Bird's Eye View. We have used the Triangle Similarity method which is an effective and efficient method to calculate the distance between an object and the camera. It can also be used on machines which do not have high-end hardware without any compromise on the performance. Then by using Euclidean distance again, we can find we actual distance between the 2 persons. Then we can compare this to the threshold distance which is 2 m or 6 ft. If the distance is less, the persons are not following social distancing and an alarm will be sounded (Fig. 1).

6 Proposed System Flow Diagram

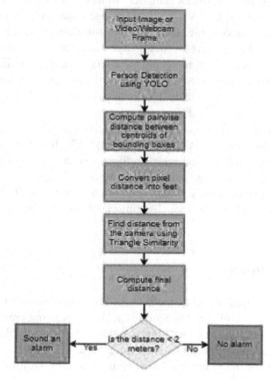

Fig. 1. Block diagram

7 Methodology Employed

The methodology used in the proposed system comprises of 4 phases. Each of the phases is explained in detail below:

7.1 Person Detection

The first part of the Social Distancing detector is to locate human bodies in the given input image/video. For this part, we have used the YOLO v3 pretrained weights which is trained on the COCO dataset and is one of the state-of-the-art models for Object Detection. The biggest advantage YOLO offers, is its speed which is close to 45 frames per second without much compromise in its accuracy. It uses a single fully Convolutional Neural Network (CNN) with only one forward propagation step to detect all the objects. It also uses Non max Suppression to suppress the low probability bounding boxes. Moreover, the YOLO v3 version consists of 53 convolutional layers for feature extraction and uses a new metric called "Objectness score" for each bounding box which uses Logistic

Regression and Binary Crossentropy loss while training. Hence it runs significantly fast and does not struggle with small objects as well. The output of this phase is the diagonal coordinates of the bounding boxes of all the detected human bodies. We can find the centroid coordinates from the diagonal coordinates by taking the mean of the diagonal coordinates. Once we know the centroids, we can find the pixel wise distance between each pair of the centroids using the Euclidean distance formula:

$$e = \sqrt{(x_1 - x_2)^2 + (y_1 - y_2)^2}$$

where (x_1, y_1) and (x_2, y_2) denote the centroid coordinates of the 2 bounding boxes.

7.2 Pixel to Inch Conversion

The distance calculated in the first phase is the pixel wise distance but we need the distance in inches or feet to apply the threshold condition. So, we need to convert the pixel distance to inch distance using the concept of PPI or Pixels per Inch. This is a metric which indicates the number of pixels per inch for the chosen screen or monitor. This is calculated as follows:

1. For any monitor or screen the total number of pixels or the pixel resolution is mentioned in the user manual or can be found out easily.
2. Using Pythagoras theorem, we can the number of pixels in the diagonal of the screen (Fig. 2).
3. Then, the PPI is found out using the formula given below:

$$PPI = \frac{diagonal\ length\ in\ pixels}{diagonal\ length\ in\ inches}$$

Fig. 2. Bird's eye view demonstration (Image source: landing.ai)

7.3 Bird's Eye View

Now, the next phase is to calculate the distance taking into consideration the third dimension i.e. distance from the camera as well. Bird's Eye View is one of the ways of getting the distance between 2 people in 3 dimensions rather than 2 dimensions. Through this method we can construct the Bird's Eye View or the top view of an image from its side-on angle. This can be done with the help of perspective transformation using the OpenCV library of Python. Once we get the top view, we can directly use Euclidean distance to get the final distance. This method is accurate but it is computationally expensive for low hardware machines. Also, it may not give real time results while taking the webcam feed as an input. Hence, instead of using this method we used the Triangle Similarity method which is described in the next section.

7.4 Triangle Similarity

This is another method which can be used to find the distance between an object and a camera. This method is slightly less accurate but efficient and computationally less expensive and hence it can be used on low hardware machines as well. This method involves 2 steps. The first step is to calculate the focal length of the camera which we are using. To do this, let us assume that we have placed an object of a known width 'W' in front of our camera at a distance of 'D'. Let the apparent width observed through the camera be 'P' pixels. From this data, we can calculate the perceived focal length using the formula:

$$F = \frac{P * D}{W}$$

Remember to take the accurate units while using the formula. The next step is to calculate the distance of the object from the camera using this perceived focal length. Now assume that we have placed the same object of width 'W' in front of the camera whose apparent width is 'P' pixels. Then using the focal length 'F' we can calculate the

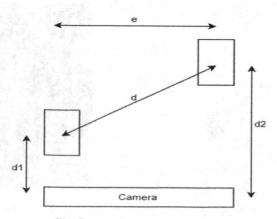

Fig. 3. Top view of the input feed

distance as (Fig. 3):

$$D' = \frac{W * F}{P}$$

7.5 Distance Calculation

Now we are ready to calculate the actual distance 'd' between the 2 people. We have already calculated the distance 'e' which is the pixel wise distance between the 2 people from the side-on view and converted it into inches. And we also have the distances d_1 and d_2 which is the distance calculated using Triangle Similarity. The final distance 'd' is given as:

$$d = \sqrt{e^2 + (d_2 - d_1)^2}$$

Once we have this distance, we can compare it with the threshold distance (2 m). If the distance is lesser, an alarm will be sounded to alert the people.

8 Results

We have tested our system on different types of inputs and the results are as follows (Figs. 4, 5, 6):

Fig. 4. Output 1: People are following social distancing (Image source: Google)

As we can see in the obtained results, the approach of Triangle Similarity is an efficient method and it correctly identifies most of the people who are not following social distancing and we get the correct results almost in most of the scenarios.

Fig. 5. Output 2: Few people are following social distancing and few are not (Image source: Google)

Fig. 6. Output 3: People are not following social distancing (Image source: Google)

9 Conclusion

In these tough times of COVID-19, Social Distancing is one of the effective ways of preventing the massive spread of the virus. This paper presents an efficient and effective way of detecting social distancing and checking whether people are following the norms or not. If any person is found violating the rules, an alarm is rung to alert the person and the police officials.

10 Future Work

One of the limitations of the proposed system is that camera calibration is essential on every system to find the system-specific parameters like PPI before using the system, which is an additional overhead. In the future, a generic system may be built to remove this overhead and the system can be used directly on every machine.

References

1. Zope, V., Joshi, N., Iyengar, S., Mahadevan, K.: COVID-19 care: checking whether people are following social distancing and wearing face masks or not using deep learning. In: International Conference on IoT Based Control Networks and Intelligent Systems, December 2020
2. Rezaei, M., Azarmi, M.: DeepSOCIAL: social distancing monitoring and infection risk assessment in COVID-19 Pandemic, MDPI article, October 2020
3. Yadav, S.: Deep learning based safe social distancing and face mask detection in public areas for COVID19 safety guidelines adherence. Int. J. Res. Appl. Sci. Eng. Technol. (IJRASET) 8(VII), July 2020
4. Yang, D., Yurtsever, E., Renganathan, V., Redmill, K.A., Ozguner, U.: A vision-based social distancing and critical density detection system for COVID-19. arXiv:2007.03578, July 2020
5. Küṛsat Çevik, K.: Computer vision based distance measurement system using stereo camera view. In: 2019 3rd International Symposium on Multidisciplinary Studies and Innovative Technologies (ISMSIT), October 2019
6. Keniya, R., Mehendale, N.: Real-time social distancing detector using SocialdistancingNet19 deep learning network. SSRN preprint, August 2020
7. Pratama, M., Budi, W., Dimyani, S.A., Praptijanto, A., Nur, A., Putrasari, Y.: performance of inter-vehicular distance estimation: pose from orthography and triangle similarity. In: 2019 International Conference on Sustainable Energy Engineering and Application (ICSEEA), October 2019
8. Megalingam, R.K., Shriram, V., Likhith, B., Rajesh, G., Ghanta, S.: Monocular distance estimation using pinhole camera approximation to avoid vehicle crash and back-over accidents. In: 2016 10th International Conference on Intelligent Systems and Control (ISCO), January 2016
9. Hossain, Md A., Mukit, Md.: A real-time face to camera distance measurement algorithm using object classification. In: 2015 International Conference on Computer and Information Engineering (ICCIE), November 2015
10. Satoh, K., Uchiyama, S., Yamamoto, H.: A head tracking method using bird's-eye view camera and gyroscope. In: Third IEEE and ACM International Symposium on Mixed and Augmented Reality, November 2004
11. Tyagi, V.: Understanding Digital Image Processing. CRC Press, Boca Raton (2018). https://doi.org/10.1201/9781315123905
12. More information on Triangle Similarity. https://www.pyimagesearch.com/2015/01/19/find-distance-camera-objectmarker-using-python-opencv/
13. Social Distancing Detection using Bird's Eye View. https://towardsdatascience.com/a-social-distancing-detector-using-atensorflow-objectdetection-model-python-and-opencv-4450a431238

Explaining a Black-Box Sentiment Analysis Model with Local Interpretable Model Diagnostics Explanation (LIME)

Kounteyo Roy Chowdhury[✉], Arpan Sil, and Sharvari Rahul Shukla

Symbiosis Statistical Institute, Symbiosis International (Deemed University), Pune, Maharashtra, India

Abstract. With the increase in usage of social media, sentiment analysis has emerged as an area of upcoming inter-disciplinary research. Twitter, a popular micro-blogging site has a rich source of text data in the form of tweets, which can be used to understand public perception on various domains like news, pandemics, motilities, diverse policies, legislations, personalities, and many more. This paper aims at understanding the sentiment of people using twitter data using Bi-directional Long Short-Term Memory (LSTM) networks with 72% accuracy and decrypt the black box deep learning model with Local Interpretable Model Diagnostics Explanation (LIME) to extract the important features and their interaction taken during prediction. In contemporary research work, LIME is used to decrypt the sequential neural network models mostly. This study takes a step forward in using LIME to decrypt the recurrent neural networks (LSTM to be specific) of a sentiment classification model.

Keywords: Sentiment analysis · Text-analysis · Natural language processing (NLP) · Bi-directional LSTM · Local Interpretable Model Diagnostics Explanation (LIME)

1 Introduction

Opinions are fundamental to practically all human exercises since they have a critical effect on individuals' conduct. Each time a choice is made; people search for others' opinions. In the past, individuals searched for opinions from their loved ones, while associations made surveys or coordinated center gatherings. All things considered, with the abrupt development of informal communities, for example, Twitter and Facebook, people and associations use information given by these ways to help their decision-making process. The field of sentiment analysis gain significance in this specific circumstance.

Sentiment analysis is ongoing zone in the field of information mining. There are various strategies for separating, handling, and looking for target information in texts. These parts including sentiments, opinions, and feelings, among others, are the focal point of sentiment analysis.

During this unprecedented time of the COVID-19 pandemic, people around the globe are going through challenging times with lot of uncertainties. Millions of people lost

M. Singh et al. (Eds.): ICACDS 2021, CCIS 1440, pp. 90–101, 2021.
https://doi.org/10.1007/978-3-030-81462-5_9

their jobs, while many others were obliged to accept subsidized wages to keep working. Also, the mode of working has been changed to online (or remote) and also for college and university students, it has been down to online classes.

In an attempt to understand the Sentiments prevailing among the people during unprecedented times, anonymized tweets are used for the analysis. Twitter is a microblogging site. When a person sends a message on Twitter, it is termed as a tweet.

Review was conducted in the domain of sentiment analysis using social media data. After that, we proceeded with giving an overview of the process of extracting tweets and pre-processing and sentiment analysis procedure using text mining and natural language processing.

2 Literature Survey

With regards to the pandemic, many studies explored the topics in tweets about public perception during the pandemic like COVID-19 situations [1]. A study from Nepal on sentiment analysis concluded that people of Nepal are mostly taking a positive and hopeful approach towards COVID-19, although there are instances of fear, disgust and anger [2]. Some studies similar study [2] was also conducted in New South Wales, Australia [3], where it was found that the sentiment of the people is significantly changing from positive to negative. This gives us a platform to discuss the techniques used in twitter sentiment analysis.

Most of the twitter sentiment analysis studies done till date either uses the supervised learning algorithm methods like ensemble classification or unsupervised learning using lexicon-based methods. A study [4] published in 2014 had shown how resembling classifiers formed using Random Forest, SVM, Multinomial Naive Bayes and Logistic Regression can improve accuracy of identifying sentiments from tweets. Another paper [5] on this domain discusses how pre-processing improves the accuracy and F1 score of classifiers used in twitter sentiment analysis. Few of the recent studies also use deep convolutional neural networks [6] for understanding the sentiment of the tweets. Numerous studies were also undertaken using the attention-based LSTM (Long Short-Term Memory) Networks [7] and combined LSTM-CNN models [8–10] to classify tweets for the purpose of sentiment analysis.

This paper focuses on first data sources, pre-processing and model building are explained in detail and then use of Bi-directional LSTM networks for twitter sentiment analysis. Further, we also explain the Deep learning model using the LIME [11].

3 Materials and Methods

3.1 Data Sources

Data was collected for a period of 2 years from twitter API. The training and testing data was obtained from the open source[1].

---

[1] https://github.com/lukasgarbas/nlp-text-emotion/tree/master/data.

3.2 Data Preparation

3.2.1 Data Cleaning

Almost of the types of natural language processing tasks like sentiment analysis, emotion detection etc. relies heavily on cleaning the data and this is the very first and most crucial stage. We touch upon the various techniques we used for data cleaning before moving on to discuss the model in detail. Various libraries are available in Python to achieve the data cleaning stage. In this study, we used the following libraries with user-defined functions for data cleaning:

User-defined functions:

1. Replacing the short forms by dictionary forms
2. Removing elongated words.
3. Replacing abbreviations (Like: bff→best friends forever)

The other important steps involved in data cleaning are

1. Removing references made to someone
2. Replacing the URL used in tweets
3. Removing special characters

An instance of the above operations (see Fig. 1):

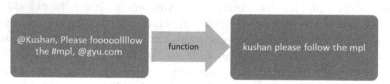

Fig. 1. Instance of the above operations

The above steps are of paramount importance to obtain homogeneity in the data. For example, 'Going' and 'going' will be stored as two separate vectors but they convey the same meaning.

3.2.2 Data Pre-processing

Tokenization: It is a standard process of breaking or splitting a collection of text into individual words called tokens (Fig. 2).

For example:

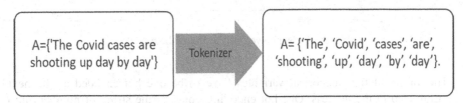

Fig. 2. Explaining tokenization and how it works

In a top-down approach, any problem is broken down into small pieces for ease of analysis. Tokenization helps to perform this objective. When a collection of texts is broken down into individual words and then converted into tokens, computation using it becomes more-manageable.

Among the various tokenizers, White Space Tokenizer is used in this study.

The pre-processed data is then passed through the emotion classifier

Sequencing: It is performed to convert the given text in an integer sequence. It transforms each text into an integer corresponding to the internal vocabulary prepared in the process of Tokenization.

Padding: The main objective behind Padding is to each input of the same length. We defined maximum number of words for our texts and input size to our model has to be fixed - padding with zeros to keep the same input length (longest input in our dataset is ~250 words).

Deployment of the Above Steps: The above approaches are compiled within a ML-Pipeline which takes input of raw text from the user and gives the output of word suggestions and emotion predictions. As shown below:

A pipeline mainly consists of 2 main steps:

i) A **featurization** pipeline that enables flexible definition and selection of features. A typical NLP task requires a great deal of feature engineering, a process that involves preparing the proper input data for training a model.
ii) A **training** pipeline that incorporates the output of the featurization pipeline for subsequent steps: vectorization and model training. The output of this featurization pipeline naturally serves as input to the next component, the training pipeline. The two pipelines are isomorphic in form, as the latter, too, contains a base design (class) from which more specific training pipelines can be derived to meet different training requirements (Fig. 3).

Target variable transformation: This is done to change the output labels to categorical values as computers works on numerical data and not texts. The following encoding is used:

encoding = {'joy': 0, 'fear': 1, 'anger': 2, 'sadness': 3, 'neutral': 4}

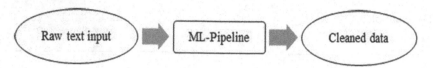

Fig. 3. Steps involved before fitting the model

The values of the categorical variables have to be one hot encoded as the next important step in the process. One-hot encoding converts the string of numbers into combinations of 0 and 1, the binary system which makes it possible for the algorithms to work on (Fig. 4).

3.3 Deep Learning Model Using Bidirectional LSTM

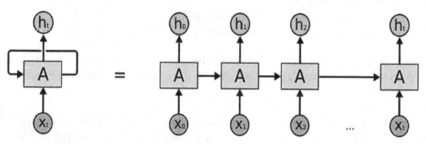

Fig. 4. Basic structure of Recurrent Neural Networks (RNNs) [12]

Due to the looping structure of Recurrent neural networks (RNNs), it is able to re-measure data to monitor the sequences in them, permitting data to endure. We utilize a LSTM rather than an ordinary RNN to eliminate the long-term dependencies (Fig. 5).

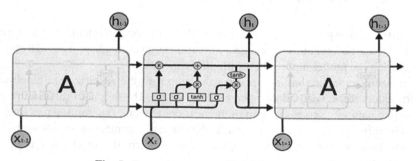

Fig. 5. Internal structure of LSTM-RNNs [12]

Conventional LSTMs has 3 gates for security and controlling the cell state: It takes a gander at h_{t-1} and x_t, and yields a number of 0 and 1 corresponding to every other number with in cell state denoted by C_{t-1}. '1' addresses "Totally Keep This" whereas

'0' addresses "Totally Dispose This." The "Input Gate Layer" chooses which values will be updated. The next layer chooses which values will be modified. Then, the tanh layer will create a vector of new competitor values, $\tilde{C}_t$, that will be added to the state.

Bidirectional LSTMs [13] are trained on two rather than one LSTMs on the sequence of inputs. The concept is to part the state neurons of a customary RNN in a section that is liable for the positive time bearing (forward states) and a section for the negative time heading (in reverse states). This can give extra setting to the network and result in quicker and considerably fuller learning on the issue.

LSTM with 1st hidden layer as word embeddings will eventually store the information and the sequence of the sentences are taken into consideration.

3.3.1 Choice of Embeddings for Input Layer

The word embeddings for each word are obtained utilizing a few strategies like: Word2Vec [14], Sentiment Specific Word Embedding (SSWE) [15], GloVe [16] just as FastText [17]. In the nonstop representation of terms, SSWE aims to encrypt sentiment data. Using each of the embeddings, a simple long-short-term memory (LSTM) model was trained to verify the performance each of these word embeddings for detecting emotions. An improved version of Recurrent Neural Networks (RNN) [18] in the form of LSTMs can catch hold of long-term dependencies in a sequence of inputs, and subsequently are useful for our project. Cross validation was utilized to decide the viability of various embeddings.

Our outcomes showed that FastText obtains the best mean F1 score, marginally above SSWE's F1 score. SSWE obtains high cosine similarity value while GloVe obtains a lower score for "depression", although the fact is that the 2 words convey similar sentiment. SSWE correctly offers a low score for the "sad" and "happy" pair, but FastText delivers a sensibly high score. In light of these perceptions, we pick FastText as our inserting for the Semantic LSTM layer. The specifications are given below:

- Vocabulary size: 11442
- Maximum Length: 500
- Embedding dimension: 300

3.3.2 Model Architecture

The model architecture is explained as follows:

- The first N words of each text, with proper padding, forms the input.
- Given embedding size and dimension of vocabulary, word embedding is created in the first level.
- LSTM/GRU layer receives word embeddings for each token in the tweet as inputs. This is done so that the output tokens store information of both the initial token and any previous tokens. Put differently, for original input, a new encoding is generated by the LSTM layer.
- The output layer contains equal number of neurons as the number of classes we want the model to classify into using a "SoftMax" activation function

- Bidirectional LSTMs are really just putting two independent LSTMs (RNNs) together. So, for every time step, this structure can make both backward and forward flow of information about the sequence possible.
- Unidirectional LSTMs can preserve information (use context) only from the past. While in case of bi-directional LSTMs, information can be preserved for both past to future as well as from future to past. Since, context from both past and future is used to classify the sentiment, the accuracy for bi-directional LSTMs will be more than uni-directional, and hence it is used in this model (Fig. 6).

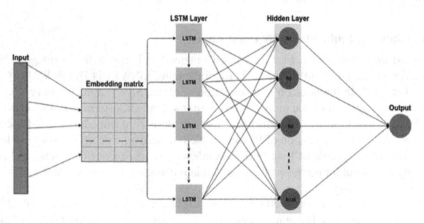

Fig. 6. Internal structure of the deep learning model used

3.3.3 Model Training

The model is trained by choosing a batch size of 128 with 35 number of epochs. From the model-accuracy graph shown below, (Fig. 7 and Fig. 8) it is observed that while training, the training accuracy starts from around 45% and the validation accuracy starts from around 47%. At the end of training, both the training and validation accuracies are

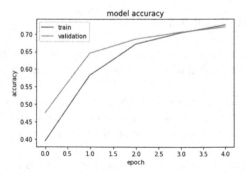

Fig. 7. Plots showing training and validation accuracy

very close to 72% which indicates that the data is not over-fitted by the model. It can be also seen that as we increase the number of epochs, the training accuracy starts getting ahead of the validation accuracy, which is an obvious sign of over-fitting of model.

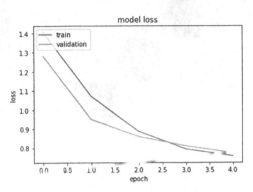

Fig. 8. Plots showing training and validation losses

The above figure (graph) shows how the train and validation losses are varying with the number of epochs. It is observed that the train and validation losses are high in the beginning, but get reduced to around 0.5 at around 40 epochs. The epochs are not increased further to prevent over-fitting.

4   Results

4.1   Model Performance

Further to model testing, we tried to assess the model performance. We calculate the accuracy using (1) [19] and F1 score using (2) [19] as the metrics to evaluate model performance.

$$\text{Accuracy} = \frac{\text{Number of right predictions}}{\text{Total number of predictions}}, \tag{1}$$

$$= \frac{\text{True Positives} + \text{True Negatives}}{\text{True Positives} + \text{True Negatives} + \text{False Postives} + \text{False Negatives}},$$

$$\text{F1 score} = 2 \times \frac{\text{Precision} \times \text{Recall}}{\text{Precision} + \text{Recall}} \tag{2}$$

$$\text{Precision} = \frac{\text{True Positives}}{\text{True Postives} + \text{False Positives}} \tag{3}$$

$$\text{Recall} = \frac{\text{True Positives}}{\text{True Postives} + \text{False Negatives}} \tag{4}$$

The overall accuracy is 72.06% and the F1 score also comes out to be 72.06% (Figs. 9, 10).

The non-normalized confusion matrix gives the actual number of correctly classified and miss-classified cases for the five emotions considered for prediction. The normalized confusion matrix gives the percentage for the same.

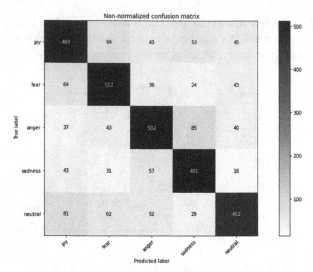

Fig. 9. Non-normalized confusion matrix

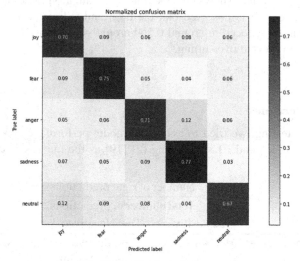

Fig. 10. Normalized confusion matrix

4.2 Local Interpretable Model Diagnostics Explanation (LIME)

In most of the machine learning models used for predictive analytics, the working of a model is often viewed as a black box which takes some input and gives certain output but how it is arriving at the output is not clearly comprehensible. To illustrate, in this project, exactly how the model is classifying the mood (or emotion) from a tweet is not understood from the complex model.

To solve this problem, a simpler model that mimics the working of the original model is built, which explains how the model is actually working and how it is making predictions. This process is referred to as LIME. It displays the predicted probabilities of a particular emotions on the basis of the words in a particular tweet as demonstrated below (Figs. 11, 12):

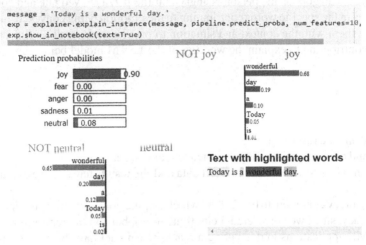

Fig. 11. Model explanation using LIME (Instance 1)

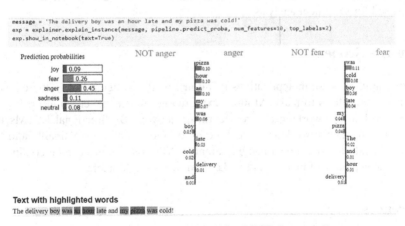

Fig. 12. Model explanation using LIME (Instance 2)

Above mentioned both examples helped us further to understand how LIME is helping us in understanding the working of the complex model. The LIME considers all the words in a sentence (message or tweet) and gives a probability score to each of the five emotions – joy, fear, anger, sadness and neutral, depending on the words, the context and sense it is trying to convey. The sentiment with the highest probability score is marked as the final sentiment of the tweet.

In the first instance, LIME captures the word 'wonderful' and gives it a 0.68 probability score of getting classified as a happy sentiment, which is true. In the second instance, 'anger' gets the highest probability score of 0.45 and fear gets a score of 0.26. This example illustrates the working of LIME. The word 'cold' is associated with 'fear' and hence, it is contributing to the 'fear' category, but given the overall context of the tweet (or message), the model understands and the LIME explains that this sentiment belongs to the 'anger' category as the person is angry that the pizza arrived late and cold. Thus, LIME takes in the specific words contributing to the emotion being conveyed, along with combining them with the context and situation to predict the overall sentiment. Hence, LIME is contributing to explain the working of the LSTM model being deployed here.

5 Conclusion

In this paper, sentiment analysis using bidirectional LSTMs (deep learning) on tweets is performed and the results have been explained using the concept of LIME. Initially, data cleaning and pre-processing was performed to make the data suitable for model creation. Then, this model is used to predict on test data and the results have been explained using LIME.

The accuracy comes out to be 72.06%, which is quite good (given any model will have 20% accuracy, since we considered 5 emotions here), but can be improved upon. Most of the LIME applications in decrypting a black-box model have been so far restricted to sequential Neural Network models. Our paper proposes a way in which LIME can be used to visualize the working of a bi-directional LSTM models, which are nowadays widely used for sentiment analysis.

6 Future Scope

With the rapid increase in applications using artificial intelligence (AI), there has also been increased efforts to make AI more explainable. In this paper, an effort has been made to explain the working of deep learning network Bi-directional LSTMs using LIME. This makes the working of such complex models more explainable and trust-worthy. Considering our model and explainable AI, more studies can be conducted to improve the accuracy of models and make them more explainable.

References

1. Boon-Itt, S., Skunkan, Y.: Public perception of the COVID-19 pandemic on Twitter: sentiment analysis and topic modeling study. JMIR Public Health Surveill. 6(4), e21978 (2020)
2. Pokharel, B.P.: Twitter sentiment analysis during covid-19 outbreak in nepal. Available at SSRN 3624719 (2020)
3. Zhou, J., Yang, S., Xiao, C., Chen, F.: Examination of community sentiment dynamics due to covid-19 pandemic: a case study from Australia. arXiv preprint arXiv:2006.12185 (2020)
4. Da Silva, N.F., Hruschka, E.R., Hruschka, E.R., Jr.: Tweet sentiment analysis with classifier ensembles. Decis. Support Syst. 66, 170–179 (2014)
5. Jianqiang, Z., Xiaolin, G.: Comparison research on text pre-processing methods on twitter sentiment analysis. IEEE Access 5, 2870–2879 (2017)

6. Jianqiang, Z., Xiaolin, G., Xuejun, Z.: Deep convolution neural networks for twitter sentiment analysis. IEEE Access **6**, 23253–23260 (2018)
7. Chen, Y., Yuan, J., You, Q., Luo, J.: Twitter sentiment analysis via bi-sense emoji embedding and attention-based LSTM. In: Proceedings of the 26th ACM international conference on Multimedia, pp. 117–125, October 2018
8. Sosa, P.M. Twitter sentiment analysis using combined lstm-cnn models (2017). Zugriff am, 10 (2019)
9. Chen, N., Wang, P.: Advanced combined lstm-cnn model for twitter sentiment analysis. In: 2018 5th IEEE International Conference on Cloud Computing and Intelligence Systems (CCIS), IEEE, pp. 684–687, November 2018
10. Minaee, S., Azimi, E., Abdolrashidi, A.: Deep-sentiment: sentiment analysis using ensemble of cnn and bi-lstm models. arXiv preprint arXiv:1904.04206 (2019)
11. Ribeiro, M.T., Singh, S., Guestrin, C.: "Why should i trust you?" Explaining the predictions of any classifier. In: Proceedings of the 22nd ACM SIGKDD international conference on knowledge discovery and data mining, pp. 1135–1144, August 2016
12. Olah, C.: Understanding LSTM Networks. https://colah.github.io/posts/2015-08-Understanding-LSTMs/
13. Schuster, M., Paliwal, K.K.: Bidirectional recurrent neural networks. IEEE Trans. Signal Process. **45**(11), 2673–2681 (1997)
14. Mikolov, T., Sutskever, I., Chen, K., Corrado, G.S., Dean, J.: Distributed representations of words and phrases and their compositionality. In: Advances in neural information processing systems, 3111–3119 (2013)
15. Tang, D., Wei, F., Yang, N., Zhou, M., Liu, T., Qin, B.: Learning sentiment-specific word embedding for twitter sentiment classification. In: Proceedings of the 52nd Annual Meeting of the Association for Computational Linguistics, pp. 1555–1565 (2014)
16. Pennington, J., Socher, R., Manning, C.D.: Glove: Global vectors for word representation. In EMNLP **14**, 1532–1543 (2014)
17. Joulin, A., Grave, E., Bojanowski, P., Mikolov, T.: Bag of tricks for efficient text classification. arXiv preprint arXiv:1607.01759 (2016)
18. Hochreiter, S., Schmidhuber, J.: Long short-term memory. Neural Comput. **9**, 1735–1780 (1997)
19. Hossin, M., Sulaiman, M.N.: A review on evaluation metrics for data classification evaluations. Int. J. Data Mining Knowl. Manage. Process **5.2** (2015), 1

Spelling Checking and Error Corrector System for Marathi Language Text Using Minimum Edit Distance Algorithm

Kavita. T. Patil[✉], R. P. Bhavsar, and B. V. Pawar

School of Computer Sciences, Kavayitri Bahinabai Chaudhari North Maharashtra University,
Jalgaon 425001, Maharashtra, India
{rpbhavsar,bvpawar}@nmu.ac.in

Abstract. The performance of any search engine, social media word processor depends deeply on the spelling checkers, grammar checkers etc. Spelling checker is the application used to correct the spelling mistakes done by users unintentionally. The minimum edit distance is one of the string-matching algorithms used in various applications like text mining, spell checking, bioinformatics and so on. In this paper, we proposed the minimum edit distance algorithm (MED) which correct the spelling mistakes in Marathi language text. It corrects the spelling errors by performing various operations like substitution, insertion, and deletion of characters. The algorithm detects non-word spelling errors and generates a suitable suggestion set for misspelled words by matching them with the corpus. While doing this, the misspelled word length is a key component to searching in the corpus so that, the searching complexity is minimum. In this paper, we have implemented a minimum edit distance algorithm and evaluated their performance with accuracy measure. The accuracy of the system is 85.5%. The performance of this algorithm is evaluated by suggestion generation accuracy for given misspelled words.

Keywords: Minimum edit distance algorithm · Error correction · Dynamic programming · Spelling checking

1 Introduction

Spelling checking is significance application of natural language processing. It was born in the early 60 [1]. However, various approaches have been proposed to make a perfect spelling checker like rule based, data driven, neural network etc. Some of this approach depends upon string similarity while others depend upon edit distance. In this paper, we proposed a data driven approach called the minimum edit distance algorithm (MED), also known as Levenshtein distance algorithm, which is used to calculate edit distance between two strings, to generate the suggestion for misspelled word [2].

The word error can be a non-word error or a real word error. The non-word error occurs when the word is not there in the dataset whereas the real word error refers to the error occurring due to a spelling mistake.

© Springer Nature Switzerland AG 2021
M. Singh et al. (Eds.): ICACDS 2021, CCIS 1440, pp. 102–111, 2021.
https://doi.org/10.1007/978-3-030-81462-5_10

1.1 Minimum Edit Distance Algorithm Preliminaries

In the minimum edit distance algorithm, input is taken as a string which returns the distance between the input string and the relevant string as computed by edit distance. In spelling checking application distance is fundamental part to be taken into consideration [3]. In this, a source string s with a set of characters, $x = [p_1, p_2, p_3, \ldots p_n]$ and target string $y = [q_1, q_2, q_3, \ldots q_m]$. The minimum cost of string transformation from x to y with three basic operations like insertion, deletion, and substitution. Each operation deliberates single cost.

In this paper, we deal with the outcome of edit distance between two strings x and y. If x is " "अमेरिका" " and y is " "अमेरिका" " and cost of operation is 0 as no operation is needed due to same string. If x is " "अमेरिकन" " and y is " "अमेरिका" ", then edit distance cost is 1 due to single substitution operation is required to transform the string x into y as shown in Fig. 1.

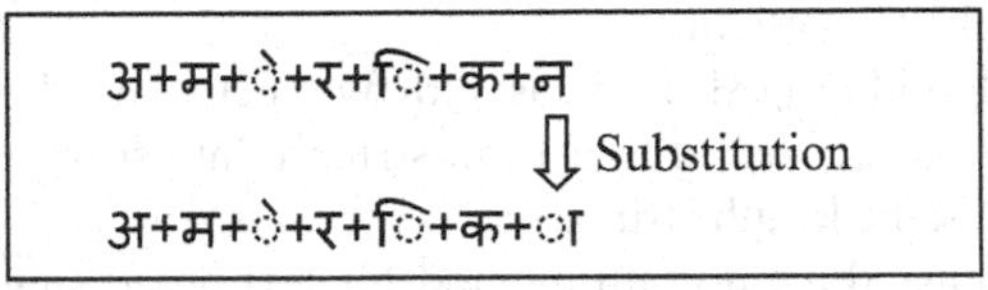

अ+म+ े+र+ि+क+न
⇩ Substitution
अ+म+ े+र+ि+क+ा

Fig. 1. Transformation of अमेरिकन to अमेरिका using Minimum Edit Distance Algorithm

Let n be the length of string x and m be the length of string y. For every alphabet of x and y the cost is determined by the checking the equality of two strings i.e. if x[i] == y[j] returns string with edit distance cost of 0 else if x[i] not equal to y[j] that is a different character mean x i.e. it measures the cost of every edit operation and returns the smallest edit distance. For different length string, sequence alignment method is applied, and string-matching procedure is repeated as above [4].

The paper is organized into six main sections. The Sect. 1 gives brief introduction about minimum edit distance algorithm. The literature survey of minimum edit distance algorithm is described in Sect. 2. Methodology and implementation for minimum edit distance is described in Sect. 3. The results and discussion are described in the Sect. 4. Section 5 discusses conclusion of the paper.

2 Related Work

In this field, several studies have been carried out for various Indian as well as international languages. Every method has its own significance and limitation. Here, we have discussed a few of them to express efforts done in this field [5].

Damerau, Levenshtein [6], considered only three basic operations of edit distance among a set of operations, namely insertion, deletion, and permutation. The distance equates two words by computing the amount of editing operations which help to change misspelled word into relevant word.

Oflazer [7] proposed a novel dictionary-based method termed as Error Tolerant Recognition. In this approach, erroneous words are corrected by importing words from the automata corpus. At each transition distance is computed and compared with the extreme threshold error, the obtained distance is termed as cut-off edit distance. All the transitions are set. Savary [8] proposed a modified method which omitted the use of cut-off edit distance.

Pollock and Zamora [9] proposed different ways to characterize a spelling error by manipulative the alpha-code. To correct the wrong word, the alpha-code is extracted and compared with the closest alpha-codes. The proposed approach handles permutation error excellently.

Ndiaye and Faltin [10] proposed a substitute approach to alpha-code. The defined method is responsible for handling errors in French language text. It is inspired by the alpha-code approach and revised the approach by merging with other methods such as phonetic reinterpretation, lest where the previous approach was unable to find the solutions.

Souque [11] and Mitton [12] proposed an approach to perform a critical analysis of the present approach for spelling checking but, sure that some type of erroneous word has not been handled by the system.

The Hamming algorithm [13], is measured by calculating the minimum number of substitution operations needed to transform string p into string q. This algorithm is applicable only to the same length string.

Shannon [14] discussed the n-gram method for text processing. It is applicable in many applications, for instance spelling correction, predicting words, etc. The most important benefit of the n-gram is that it is language independent. Retrieving of the n-gram relies on the number of characters n-grams illustrated in the method. The n-gram model is widely employed in text retrieval.

Neural net [15] used a hybrid method to surmount phonetic spelling errors and the typographic errors. Generally, phonetic spelling errors are tough for identification and correction. In this approach, the insertion and deletion errors are handled by the n-gram model and the substitution and transposition errors are handled by the Hamming distance algorithm.

3   Methodology and Implementation

The minimum edit distance algorithm uses a dynamic programming approach to find edit distance between two strings using weighted matrix [16].

3.1   Levenshtein Distance Algorithm

The Minimum Edit Distance Algorithm, commonly known as the Levenshtein distance algorithm, was developed by Vladimir Levenshtein in 1965, who is a scientist from Russia [17]. This algorithm assigns the cost of every operation as 1. It considers edit operations as insertion, deletion, and substitution [18]. The Levenshtein algorithm assists to describe the number of adjustments; insertions and deletions in a string str1 to be the same as a string str2 [19]. Finally, it calculates the minimum number of basic operations required to transform one string into another. The algorithm accomplishes this by employing a matrix of dimension $(a + 1) * (a + 1)$, where n and m are the lengths of the two previous strings [20]. The dist(a,b) is nothing but minimum MED value between two strings [21].

$X = x_1, x_2, ..., x_m$ and $Y = y_1, y_2, ..., y_n$ where m and m are length of matrix. In this, we calculate the edit distance between two string step by step [22]. It can be calculated by the following recurrence relation:

1. **Base conditions:**

$Dist(i,\phi) = i, Dist(\phi, j) = j,$

2. **Recurrence Relation:**

$$Dist(a, b) = \min \begin{cases} Dist(a-1, j) + cost\_ins(T[a-1]), & ....Insertion \\ Dist(a, b-1) + cost\_del(S[b-1]), & ....Deletion \\ Dist(a-1, b-1) + cost\_sub(S[b-1], T[a-1]) & ...Substitution \end{cases} \quad (1)$$

Where,

$$cost\_sub(S[b-1], T[a-1]) = \begin{cases} 0 \text{ if } (Dist(S[b-1] == Dist(T[a-1])) \\ 1 \text{ if } (Dist(S[b-1] \neq Dist(T[a-1])) \end{cases} \quad (2)$$

and,

$$cost\_ins(T[a-1]) = cost\_del(S[b-1]) = 1$$

$$Ptr(a, b) = \begin{cases} Ptr_{LEFT} ... \text{ Insertion} \\ Ptr_{DOWN} ... \text{ Deletion} \\ Ptr_{DIAGONAL} ... \text{ Substitution} \end{cases} \quad (3)$$

3. **Termination:**

Dist (a,b) is distance

3.2 Computing Methods

- $Dist(a,b) =$ best alignment score from dist11..dist1i to dist21.....dist2j
- $Dist[a-1,b] + cost\_ins(T[a-1])$
- $Dist(a-1, b-1) + cost\_sub(S[b-1],T[a-1]),$
  $cost\_sub(S[b-1],T[a-1])=0$ if $(Dist(S[b-1]==Dist(T[a-1]))$ else 1

- $Dist[0, b-1] + cost\_del(S[b-1])$
  Where, $cost\_ins(T[a-1])= cost\_del(S[b-1])=1$

3.3 Factors Affecting the Cost

- Dist(0, 0) = 0 # Initial cost of matrix
- Dist(a, 0) = Dist(a–1, 0) + 1 # Cost of insertion
- Dist(0, b) = Dist(0, b–1) + 1 # Cost of deletion

In dynamic programming the factors like cost of insertion and cost of deletion affect the value of the character in matrix.

Algorithm 1: Proposed Minimum Edit Distance Algorithm

Input: The Marathi word/Text

Output: Edit Distance of word with word present in the dictionary

Definition: MED (input S : array[1..m] of char, input T : array[1..n] of char)

Begin

1. Input the word/Documenti
2. If input is Document then tokenize the word
3. MED (T, S) return min_edit_dist
4. Create a dist_matrix [n+1,m+1]
5. dist [0,0]=0 #Initialization of Distance Matrix
6. For a=1 to n do   #for every Column
7.   dist[a,0]←dist[a-1,0]+ins-cost(target[a])
8. For b=1 to m do   #for every Row
9.   dist[a,0]←dist[0, b-1]+del-cost(source[b])
10. For a=1 to n do
11. For b=1 to m do
12.   dist[a,b]←Min [dist[a-1,b]+ins-cost(target[a-1]),
13.   dist[a-1,b-1]+sub-cost(source[b-1],target[a-1]),
14.   dist[0, b-1]+del-cost(source[b-1])]
15. Return dist[n, m]
16. Generate the suggestions

End

Example: The minimum edit distance between two string जवळपा and जवळपास can be calculated as shown in Fig. 2:

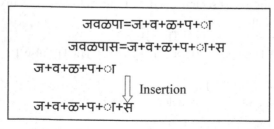

Fig. 2. Transformation of string जवळपा to जवळपास using minimum edit distance between

Table 1. Minimum edit distance using dynamic programming

	#	ज	व	ळ	पा	ल	स
#	0	1	2	3	4	5	6
ज	1	0	1	2	3	4	5
व	2	1	0	1	2	3	4
ळ	3	2	1	0	1	2	3
पा	4	3	2	1	0	1	2
ल	5	4	3	2	1	0	1

Table 1 illustrates the minimum edit distance between two strings जवळपा and जवळपास, calculated using dynamic programming. Table 2 describes number of suggestions and corresponding MED value produced by the proposed approach.

Table 2. List of suggestions generated by the proposed algorithm

Misspelled Word	Sr. No.	Suggestions	MED Values
जवळपा	1	जवळपास	1
	2	जपा	2
	3	जवळ	2
	4	जवळचे	2
	5	जवळीक	2

Figure 3 illustrates the proposed architecture of spell checking by a MED algorithm. In that, misspelled words are taken as input. The word is first checked in the dictionary or database as it is the wrong word, the word is highlighted and if we right click on the highlighted word we will get the proper suggestions with ranking in ascending order of their minimum edit distance. Among the generated suggestions, user will select the proper suggestion for the wrongly spelled word with the respective sentence.

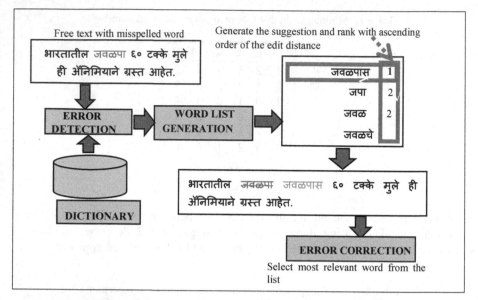

Fig. 3. Architecture of spell checking by minimum edit distance algorithm

4   Result and Discussion

We have collected and cleaned data about 9, 29, 663 unique words which is created from various resources like website http://www.tdil-dc.in [23] as well Hunspell Dictionary (54,300) [24]. More than 20,000 words have been created from book of "Marathi Lekhan Kosh" by Arun Phadke [25], extracted 8, 55, 363 words from Forum for Information Retrieval and Evaluation (FIRE) [26]. We have tested the results using 20% of data. As, Marathi is a morphologically reach language. The word may go under different transformations when used in a sentence. Among the eight PoS categories, the noun, pronoun, adjective, adverb, and verb undergo inflectional changes and that inflection is due to gender, number, person, and case (GNPC feature). For testing, we have used the data category, namely noun, verb, adverb, and adjectives, as these are four pillars of Marathi grammar, i.e. PoS, and tested our system for unknown words. Table 3. Describes the Category wise suggestion generation accuracy for wrongly spelled words. It is observed that, our system has a suggestion generation accuracy of about 85.56% and is calculated by Eq. 4. The suggestions are nothing, but the suffixes attached to the word. So, morphology is one of the key features which helps to generate the suggestion for wrongly spelled word.

Figure 4 illustrates the result analysis of the Minimum edit distance algorithm for manual and automated average MED value. It is observed that the suggestions generated by the manual system and automated system are almost similar. Only in a few cases, it gives different suggestions based on their MED. It means our system gives better performance. In Fig. 4 we have calculated the average minimum edit distance. X-axis indicates the number of words and Y-axis indicates the average minimum edit distance.

$$\text{Suggestion generation Accuracy}(\%) = \frac{\textit{No. of unknown words for which suggestions are Correctly generated}}{\textit{Total no. of words}} \times 100$$

(4)

$$\text{Suggestion generation Accuracy}(\%) = \frac{1,58,900}{1,85,000} \times 100 = 85.56\%$$

Table 3. Category wise suggestion generation accuracy for wrongly spelled words

Sr,No.	PoS Category	No.of Unique words	No. of word for which suggestions are correctly generated	Suggestion generation accuracy (%)
1	Noun	69,500	59,865	86.14
2	Adjective	47,900	41,855	87.38
3	Verb	42,500	36,240	85.27
4	Adverb	25,100	20,940	83.43

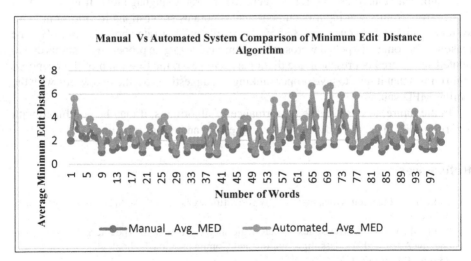

Fig. 4. Result analysis of minimum edit distance algorithm

4.1 Error Analysis of Proposed Algorithm

The proposed algorithm has generated errors of about 14.44%, which is due to mistyping or words that are not present in the dictionary. Here, we have considered the unknown about 1,85,000 words among which our algorithm is able to generate good suggestions for 1,58,900 words. For remaining words, the proper suggestions are not available, which results in an increase in error rate. Equation 5 calculates the root mean squared error. It is so because it gives same error value as predicted by the algorithm.

$$\text{Root mean squared error} \ = \ \sqrt{\frac{\sum_{i=1}^{d}(y_i-y_i')^2}{d}} \qquad (5)$$

Where,

y_i = expected output
y'_i = actual output.
d = total number of words.

5 Conclusion and Future Scope

In this paper, we have done extensive study of related work on the minimum edit distance algorithm. The study and implementation of a minimum edit distance algorithm has been carried out which is used to detect and correct the spelling errors in the Marathi language. We have collected and cleaned about 9,29,663 unique words corpus from various sources. We have tested our system manually as well as automatically.

The proposed algorithm is good for finding the similarity between two words. This algorithm can be used as an error corrector for Marathi language text. It handles non-word errors from given input document and generates suggestions for the same. Our system results in an accuracy of 85.56%. It results in error rate about the 14.44%. The reason is for some misspelled words our system unable to give proper suggestions due to related words are not present in the dictionary. However, the limitation of the proposed algorithm is that it fails to give proper ranking of suggestions for the two words with the similar MED value.

In the future, to improve the performance of algorithms, it may be combined with algorithms like n-gram model for better suggestion and error correction.

References

1. Kukich, K.: Techniques for automatically correcting words in text. ACM Comput. Surv. 24(4), 377–439 (1992)
2. Nejja, M., Yousfi, A.: The context in automatic spell correction. In: The International Conference on Advanced Wireless, Information, and Communication Technologies (AWICT), vol. 73, pp. 109–114 (2015)
3. Hall, P.A.V., Dowling, G.R.: Approximate string matching. Comput. Surveys 12, 381–402 (1980)
4. Putra, M.E.W., Suwardi, I.S.: Structural off-line handwriting character recognition using approximate subgraph matching and Levenshtein distance. Procedia Comput. Sci. 59, 340–349 (2015)
5. Damerau, F.J.: A technique for computer detection and correction of spelling errors. Communications of the Association for Computing Machinery (1964)
6. Levenshtein, V.: Binary codes capable of correcting deletions, insertions, and reversals. SOL Phys Dokl, 707–710 (1966)

7. Oflazer, K.: Error-tolerant finite-state recognition with applications to morphological analysis and spelling correction. Comput. Linguistics Archive **22**(1), 73–89 (1996)
8. Savary: Recensement et description des mots composés–méthodes et applications, version 1, 24 September 2011, 149–158 (2000)
9. Pollock, J.J., Zamora, A.: Automatic spelling correction in scientific and scholarly text. Commun. ACM **27**(4), 358–368 (1984)
10. Ndiaye, M., Faltin, A.V.: Correcteur Orthographique Adapté à Apprentissage du Français. Revue Bulag (29), 117–134 (2004)
11. SOUQUE: Approche critique des produits IdL:analyse comparative des correcteurs orthographiques de Word 2000 et OpenOffice 2. Master 1 Industries de la Langue, Université Stendhal-Grenoble 3 (2006)
12. Mitton, R.: Ordering the suggestions of a spellchecker without using context. Nat. Lang. Eng. **15**(2), 173–192 (2009)
13. Hamming, R.W.: Error detecting and error correcting codes. Bell Syst. Tech, J, **29**(?), 147–160 (1950)
14. Shannon, C.E.: Prediction and entropy of printed English. Bell Sys. Tec. J. **30**, 50–64 (1951)
15. Hodge, J., Austin, J.: A comparison of standard spell-checking algorithms and novel binary neural approach. IEEE Trans. Know. Dat. Eng. **15**(5), 1073–1081 (2003)
16. Desai, N., Narvekar, M.: Normalization of noisy text data. In: International Conference on Advanced Computing Technologies and Applications (ICACTA), vol. 45, 127-132 (2015)
17. Mary, R., Nishikant, A.S., Iyengar, N.C.S.: Use of edit distance algorithm to search a keyword in cloud environment. Int. J. Database Theor. Appl. **7**(6), 223–232 (2014)
18. Nejja, M., Yousfi, A.: The context in automatic spell correction. In: The International Conference on Advanced Wireless, Information, and Communication Technologies (AWICT), vol. 73, 109–114 (2015)
19. Awny, S., Amal, A.M.: IBRI-CASONTO: ontology-based semantic search engine. Egypt. Inf. J. **18**, 181–192 (2017)
20. Yao, Z.: Implementation of the autocomplete feature of the textbox based on Ajax and web service. J. Comput. **9**(8), 2197–2203 (2013)
21. Rubio, M.: A consensus algorithm for approximate string matching. In: Iberoamerican Conference on Electronics Engineering and Computer Science, vol. 7, 322-327 (2015)
22. Umar, R., Hendriana, Y., Budiyono, E.: Implementation of Levenshtein distance algorithm for E-Commerce of bravoisitees distro. Int. J. Comput. Trends Technol. (IJCTT). **27**(3), 131–136 (2015)
23. http://www.tdil.dc.in/index.php?option=com_download&task=fsearch&Itemid=547&lang=en. Accessed 05 Marc, 2019
24. https://code.google.com/archive/p/hunspell-marathidictionary/downloads.___Accessed 18 November 2020
25. Arun, P.: Marathi Lekhan Kosh, p. 2001. Keshav Bhikaji Dhavale Publishers, Mumbai (2001)
26. Forum for Information Retrieval (FIRE), Information Retrieval Society of India **12**, 2–4, Mumbai, Maharashtra, India. http://www.isical.ac.in/~fire/2010/index.html

A Study on Morphological Analyser for Indian Languages: A Literature Perspective

Jayashree Nair, L. S. Aiswarya$^{(\boxtimes)}$, and P. R. Sruthy

Department of Computer Science and Applications, Amrita Vishwa Vidyapeetham,
Amritapuri, India
jayashree@am.amrita.edu

Abstract. India is the home to a very large number of languages. The Indian languages are rich in literature and has been studied by native and foreign Linguists. Unlike English, Indian languages are Morphologically rich and follows free word-order. Even though there have been efforts towards building morphological analyser for Malayalam and Sanskrit, until now an efficient one is not available. In order to solve this problem we come up with the study on morphological analyzer in Indian languages. Morphological analyser is a linguistic tool that would generate the morphemes of a given word. These rules are based on Indian language linguistics. This paper gives a brief description of the approach used for morphological analyser. With the development of a Python Package that make use of Rule Based Approach for developing Morphological Analyzer. It mainly focusing on noun and this analyzer can be used for Information Retrieval, search engines, Machine Translation, speech recognizer, Text Processing etc.

Keywords: Morphological analyser · Morphemes · Rule based approach · Linguists · Indian languages

1 Introduction

India is a linguistically rich area which had different languages, caste, beliefs, art forms, cultures, religions etc. and it is the home for an uncountable number of different type of lingual families. There are 18 constitutional languages are there in this country, which are written in 10 different scripts. Different states of the country usually speak a different languages in India. English is known a universal language because it is widely spoken language in all over the world, because of that there is a large scope for translation between English and the Indian languages. There is a huge amount of different word forms are there in Indian language therefore it is known as Morphologically rich language consider an example i.e., in English the word 'Tree' has only one form that is 'Trees' similarly when we consider same word in Malayalam. In Fig. 1, there are different forms of Malayalam words for the English word 'Tree'.

© Springer Nature Switzerland AG 2021
M. Singh et al. (Eds.): ICACDS 2021, CCIS 1440, pp. 112–123, 2021.
https://doi.org/10.1007/978-3-030-81462-5_11

'മരം' it has different forms as follows:

മരത്തിന്റെ

മരത്തിൽനിന്ന്

മരത്തിലേക്ക്

മരങ്ങളുടെ

Fig. 1. Different forms of Malayalam word 'maram'

Simple lexical mapping will not help for retrieving and mapping of Morpho-syntactic information from English language. Morphology is a branch of linguistics unit that form a word in Natural Language i.e., it is the field of linguists that are concentrated on the study of formation of words. Every language has a set of words, combining this words along with their grammar with respect to their language to form meaningful sentences. To consider the arrangement of words we need Identification, analysis, and description of structure of a given morphemes of a language likewise to realize how words are worked from more modest part and other linguistic units, for example, root words, suffixes, affixes and so on. A single word in a language is the combination of one or more morphemes and this can be either a root word, suffix or prefix, for example happy, unhappy, happily. Morphological analyser is a linguistic tool that would generate the morphemes of a given word. For morphological analysis most of the NLP system make use of simple linguistic theories. The syntax of Morphological analyzer is:

$$Word = stem/root + suffix$$

The major use of this Morphological Analysers is in search engines, speech recognizer, spell and grammar checker and machine translation [2]. Malayalam is a member of Dravidian Language family because of a highly inflectional and Agglutinative character. This has posed a challenge for all kind of language processing. If we look at the Sanskrit literature, we see that many attempts have been made to render the learning of Sanskrit word formation easier [10]. A dictionary is an arrangement of different words with respect to their meanings, usage, origins, pronunciations etc. For data collection a root dictionary is there which contains root words and its suffix of Sanskrit and Malayalam. Rule based approach is used here, A rule-based approach applies human-made principles to store, sort and control information. In doing as such, it emulates human insight. To work, rule-based system require a bunch of realities or wellspring of information, and a bunch of rules for data manipulation [9]. This paper is the study to build an efficient Morphological Analyzer using rule based approach.

2 Literature Review

This section include different terms in Literature Review such as NLP(Natural language processing), Indian Languages, Malayalam, Sanskrit, Morphological Analyser, Morphology, Root Dictionary.

2.1 Natural Language Processing

Natural language processing (NLP) is both a contemporary computational technology and a way of investigating and evaluating claims about human language itself. Some choose the time period computational linguistics for you to seize this latter characteristic, however NLP is a term that links again into the records of Artificial Intelligence (AI).

Natural Language Processing is an interdisciplinary field where linguistic and computer science merge. To build computational models of natural language for its analysis and generation is the ultimate aim of NLP and is mainly focused on the study of language for communication [7]. It is the process of making the human language more easier and understandable to machines and performing different operations on it to retrieve useful information.

2.2 Indian Languages

India is the place having a large variety of languages, religions, cultures ets. Each languages have different word formation, grammatical features and script. India gain 4 th position in the world for having large number of languages. There are 18 constitutional languages are there in this country, which are written in 10 different scripts. Different states of the country usually speak a different languages in India. Hindi is known as the official language of India.

2.3 Malayalam

Malayalam is one of the 22 scheduled Indian language, over 34 million people spoke this language. It is the official language in Kerala Malayalam language comes under Dravidian family which have characteristic feature of agglutinative language which consisting of 15 vowels known as swarah akshara and 36 consonants which is vyenjana akshara also have symbols like chillu akshara. Malayalam language comprising of free consonant and vowel likewise has its own particular content, a syllabic letters in order, [4].

2.4 Sanskrit

Sanskrit is known as mother of all languages in India because it is the traditional means of communication and is used in ancient poetry, drama, religious and also in philosophical texts. This language is rich inflectional as well as derivational morphology [10]. The writing script used for Sanskrit is Brahmi script and this language is belong to Devanagari language family. There are 46 alphabets are there in this language which contains 16 vowels it is known as swaras.

2.5 Morphology

The study of morphemes is known Morphology, a morpheme means smallest unit of meaning in a language that form a word and we cannot divide further [8]. When we take a meaningful word that must have at least 1 morpheme, and also there are words with many morphemes too. Free and Bound are the two type of morphemes Fig. 2. Free morpheme has a meaning in the language ie independent word whereas bound is dependent word which are meaningless.

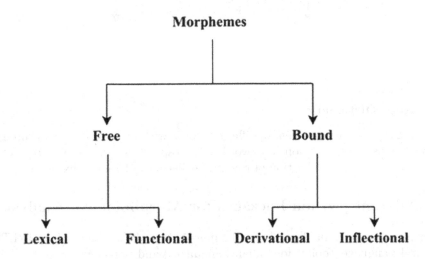

Fig. 2. Types of morpheme

Let us look it with an example Fig. 3, consider the word "Socialist" there are only 2 morphemes Social+ist and this cannot further divide. Here Social is free morpheme and ist is bound morpheme.

2.6 Morphological Analyser

Morphological analyser take a word as input and give an output as grammatical information which include root word and suffix. It is intended to dissect the constituents of the words and it will help for the division of words into stems. The Morphological analyser which return root/stem word alongside its syntactic data relying on its word category [9]. The general format for the morphological analyser is:

$$Word = stem/root + suffix$$

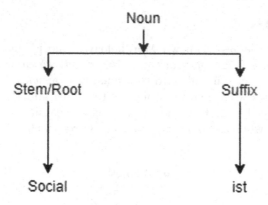

Fig. 3. Example of morpheme

2.7 Root Dictionary

A dictionary is an arrangement of different words with respect to their meanings, usage, origins, pronunciations etc. words are wrapped in a single resource. This root dictionary is the collection of romanized list of root and suffix words.

3 Different Python Packages for Morphological Analysis

Description of all available packages for morphological analyser such as INLTK (Natural Language Toolkit for Indian Languages) and polyglot.

3.1 INLTK (Natural Language Toolkit for Indian Languages)

iNLTK presents assist for numerous NLP applications in indic languages. The languages supported are Hindi denoted as (hi), Punjabi denoted as (pa), Sanskrit denoted as (sa), Gujarati denoted as (gu), Kannada denoted as (kn), Malayalam denoted as (ml), Nepali denoted as (ne), Odia denoted as (or), Marathi denoted as (mr), Bengali denoted as (bn), Tamil denoted as (ta), Urdu denoted as (ur), English denoted as (en). INLTK is similar to the nltk python bundle and that they offers same features for nlp along with tokenisation and vector embedding for enter text with an clean API interface. The Indian languages have some difficulties which come from sharing a lot of similarity in terms of script, phonology, language syntax, etc., and this library provides a general solution. Indic NLP Library provides functionalities like text normalisation, script normalization, tokenization, word segmentation, romanization, indicisation, script conversion, transliteration and translation.

3.2 Polyglot

Polyglot is used to perform different NLP operations and is an open-source python library. Different types of python libraries are available that can help us in performing NLP assignments. All libraries have certain novel highlights and which make them unique in relation to one another. This polyglot has an enormous variety of dedicated commands which makes it stand apart of the group also computation is depends on NumPy which is the reason it is known as quick and faster. It is similar to spacy and as compare to spacy this polyglot is better i.e., it can be used for languages that are not support by spacy. Polyglot provide different functionalities and can imported as and when required. Polyglot can recognize the language of the content passed to it using language function. In Morphological analysis it characterizes the consistencies behind word arrangement in human language. It can be used to identify the language in a specific text, followed by the tokenization in words and sentences. It is easy to use and can be used for a variety of NLP operations depending on our need.

Polyglot has proved to be a completely beneficial device for experimenting with new language features and for building different language-processing gear. Polyglot isn't only a preprocessor it supports the development of complicated language extensions that add new features to the java language, consisting of to its type device. This library is not a well-known library but it offers a wide range of analysis and splendid language coverage. And this library stands out from the crowd also because it requests the usage of a dedicated command in the command line through the pipeline mechanisms. It also works really fast and also it's very efficient, straightforward, and basically an excellent choice for projects involving a language Spacy doesn't support.

4   Comparative Study of Existing Morphological Analyzer for Indian Languages

A comparative study of different python packages had done which include Spacy, Inltk and Polyglot. In Spacy there is no morphological Analyser for malayalam, and for malayalam they have only limited functionalities like wordnet and lematization. In Inltk there is no morphological analyser instead of that they have stemmer and lemmatization and this stemmer removes only the root word and lemmatization combines words with similar words into one. In polyglot it have morphological analyzer for Malayalam but it doesn't provide an accurate result, for that we had done a test Fig. 4, For the test we had randomly select 100 words with suffix, which contain both nouns and verbs. When we import this file of 100 words in CSV into this polyglot. It will give output by splitting words into its root and suffix and many of them are wrong. we can clearly understand this from this chart. Here 100 words are there among that 37 of them are nouns and in that nouns 21 of them are correct and 16 are wrong, similarly 63 of them are verbs and 42 of them are correct and remaining 21 of them are wrong.In the comparative study we are trying to say that even though we had an analyser it doesn't give an accurate result i.e., for Malayalam until now there is not an efficient morphological analyser. This knowledge will help to develop an efficient morphological analyser for Malayalam.

Input Type	Total count	Correct Result	Error Result
Nouns	37	21	16
Verbs	63	42	21
Result	100	63	37

Fig. 4. comparative study

5   Related Works

Morphological analysis is the division of words into their part morphemes and the task of linguistic data to syntactic classifications. It contain identification of parts of the words, or more technically, constituents of the words and it will return its roots/stem of a word along with its grammatical information depending upon its word category [9]. There are different morphological analyzer developed for various languagaes using different approaches like hybrid approach, Finate state transducer, rule based approach, suffix stripping approach and so on.

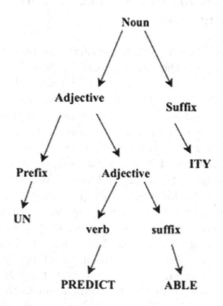

Fig. 5. Morphological structure of a word

Here the word 'Unpredictability' Fig. 5 is formed from an adjective and a suffix ie adjective is 'unpredictable' and suffix is 'ity' then this adjective is formed from a prefix and another adjective ie 'predictable' is the adjective and it is formed from a verb 'predict' and suffix 'able'.

5.1 Hybrid Approach

For Indian languages, hybrid approach which combines the advantages of the rule based approach and Data driven approach. This approach is simple because all the inflections are stored in the database and the inflections can be found directly from the database based on the given word. The morphological inflections of all the words of a Morphologically rich language is highly difficult to pre-record.

5.2 Finite State Transducer

The finite state transducer(FST) is proposed for morphological analyzer for Malayalam language and this FST approach which maps strings from one regular language into strings from another regular language and this process is reversible too [6]. FST is used to represent the lexicon computationally and it can be done by accepting the principle of two level morphology. That is, it which represents a word as the correspondence between lexical and surface level of morphology. An FST is represented as a two tape automation. By combining the lexicon, orthographic rules and spelling variations in the FST we can build a morphological analyzer [11]. The transducers is a kind of translating machine which read from one tape and write into the other. The FST method which acts as a two level morphology and this method is used for both analysis and generation.

5.3 Suffix Stripping

The suffix stripping method is used for the development of the morphological analyzer for Indian languages and it make use of a stem dictionary. This dictionary which contain all possible suffixes that nouns/verbs in the language can have morphotatic rule and morphophonemic rule [4]. This technique which identifies the suffix first and then the stem by way of making use of morphophonemic rules. The suffix stripping approach which is simpler to maintain and the searching process are relatively fast as the search is only done on suffixes. As we know, the Words are formed by adding suffixes to the root words. So this property can be appropriate for suffix stripping method. Once the suffix is identified, the root of the whole word can be acquired by removing the suffix from the root word and applying proper orthographic (sandhi) rules. There are set of dictionaries like root dictionary, suffix dictionary and also using morphotactics and sandhi rules, and they are used in suffix stripping algorithm [12].

5.4 Rule Based Approach

The rule based approach and this approach is generally used for building morphological analyser. This approach is very effective for the morphological analysis of Indian languages and it predicts correct grammatical features. This approach which works based on some set of guidelines and dictionary that includes roots and morphemes [2]. There are set of rules are there in this approach and these rules are directly or indirectly depends on each other.

Here we using rule based approach because this approach is mainly used for building the morphological analyser. In this approach, which include a set rules and dictionary that includes root and morphemes and there rules are both manually created or extracted from a large corpus that based on a few common features. The rules in this approach is both right away or indirectly depends on previous rule i.e., all of the rules which is depends on every different. This is used because, it is straightforward and this make it ideal and less time consuming. For the working of this rule based system requires a set of facts or source of data, and a set of rules for manipulating that data. And it also proves very effective and provides better accuracy than the existing ones.

In Fig. 6 paper [3] 'Morphological Analyzer for Malayalam Using Machine Learning' they have used rule based method for the analyser by using SVM tool and these rules are automatically learned from the data, for learning models and make predictions they used learning and classification algorithms. This gives Efficient output and also it correctly predict the grammatical features of words which are not available in the training set.

'FST Based Morphological Analyser for Hindi Language' is another paper [1] mentioned above, in this paper they utilizes Stuttgart Finite State Transducer (SFST) device for creating FST. A root word dictionary is made here. Rules are then added for creating inflectional and derivational words from these root words. The Morph Analyser created was utilized in a Part Of Speech (POS) Tagger dependent on Stanford POS Tagger. The framework was first prepared utilizing a physically labeled corpus and for labeling input sentences MAXENT (Maximum Entropy) approach of Stanford POS tagger was used.

The paper 'Morphological Analyser for Hindi using Rule Based Approach' [11] which is mentioned above uses the Rule Based approach. It uses lemmatize to extract the root word properly and a corpus which stores the exceptional words which does not match with the rule made. For the development of the corpus, commonly used words are used. Hindi morphological structure which consists of various word classes in which their derivational and inflectional forms are described. The rules are made to comprise almost all the phrase formations available after a deep evaluation and observe of the dictionary and different expertise assets to be had.

The paper 'A Graph Based Semi-Supervised Approach For analysis of derivational Nouns in Sanskrit' uses the Semi-Supervised graph based approach for morpho-syntactic lexicon induction. The Modified Adsorption(MAD) algorithm is used for the task. This approach is used for the analysis of derivational nouns in Sanskrit.

Sl.No	Paper	Method	Language	Contributions	Scope or limitation
1	Morphological Analyzer for Malayalam Using Machine Learning	Rule based method	Malayalam	SVMTool is applied to the Analyzer for predicts right syntactic highlights	Major rules of rule based approach is used because of that if one rule fails it will affect the entire rule that follows.
2	Morphological Analyser and Morphological Generator for Malayalam - Tamil Machine Translation	Suffix Joining and Suffix Stripping method is used.	Malayalam and Tamil	Bilingual word dictionary reference for Malayalam and Tamil comprise of the root/stem of the words with its linguistic class.	Malayalam tends to join two words, this is one of the significant issue.
3	Morphological analyzer and generator for Tulu language: a novel approach	Rule based method	Tulu	In view of the inflections and differences, all conceivable Morphotactic and Sandhi rules were composed.	This framework can be improved by adding more rules, eg: rules for complex morphology, rules for transitive structures.
4	FST Based Morphological Analyzer for Hindi Language	Stuttgart Finite State Transducer approach	Hindi	The Analyzer created was utilized the feature (POS) Tagger dependent on Stanford POS Tagger.	The Morph Analyzer can be enhanced by combining the paradigm approach with the FST approach.
5	A hybrid approach to Tamil morphological generation.	Hybrid approach	Tamil	This Morphological Generation combines the advantages of the rule based approach and Data driven approach.	This algorithm can be used to generate morphological generator for pronouns and adverbs.
6	A Graph Based Semi-Supervised Approach for Analysis of Derivational Nouns in Sanskrit	Graph based Semi-Supervised approach	Sanskrit	MAD(Modified Adsorption) algorithm used for the task.	There exists no analyser for Sanskrit and this prompts issues with tremendous extension getting ready of compositions in Sanskrit.
7	Morphology analysis for Malayalam Language using FST	Finite State Transducer(FST) approach	Malayalam	morphology model is compiled base using HFST toolkit. MI-morph uses its own POS tagging schema	The collected words are manually assure and clean up because of that this task is tedious, however is incredibly vital to the standard of analyser.
8	Morphological Analyser for Hindi -A Rule Based implementation	Rule Based Approach	Hindi	uses lemmatize to extract the root words properly and a corpus which stores the exceptional words which does not match with the rules made	It will integrate the word sense elucidation with this analyser therefore that the words having multiple senses can be analyzed accurately.

Fig. 6. A consolidated report charting all the important Morphological Analyzer Papers.

6   Research Gaps and Novelty

Morphological analyzer is a linguistic tool that would generate the morphemes of a given word. It is designed to analyse the constituents of the words and it will help for the segmentation of words into stems. The morphological analyzer which go back root/stem phrase in conjunction with its grammatical facts relying upon its phrase class [9]. The general format for the morphological analyzer is:

$$Word = stem/root + suffix$$

Example:of seetha => സീതയുടെ=സീത+ഉടെ

This example give clear view of what this morphological analyzer does in Malayalam language i.e., the Morphological analyzer will split the given input word into its own root and suffix. morphology is the study of word formation, how words are constructed up from smaller portions. i.e., to identification, analysis, and description of the shape of a given language's morphemes and other linguistic gadgets, which include root words, affixes, components of speech or implied context. Analysis of phrase or word structure (morphology) is split into fundamental fields as inflection and derivation. Consequently, the morphological shape of each phrase might also encompass factors such as prefix, suffix, infix, or even a separate root, and those factors can modify the meaning of the simple root or stern of the phrase. If the resultant phrase is only a paradigmatic application of its base shape, this modification of the phrase is known as inflection; however if the ensuing word is a different phrase or a compound, which is formed of two or more roots, it is known as derivation. At the same time as derivation is a word-creating method, inflection constitutes exclusive styles of any phrase or word [8]. For the development of the morphological analysis, rule based approach is mainly used, because it is a system that applies human made rule to store, sort and manipulate data. This approach is based on set of fixed rules and dictionary that contains root and morphemes and the rules which contained in this approach are made to incorporate and other knowledge resources available. The rules are made to comprise almost all the phrase formations available after a deep evaluation and observe of the dictionary and different expertise assets to be had.

6.1   Need of Morphological Analysis

Morphological Analysis is a decent organized technique and approach that assists with finding new connections or arrangements which may be disregarded by other less organized strategies [3]. It is a foundational and centered strategy, which permits coordinating existing data and producing new innovative thoughts for planning new items, advances and services.

7   Conclusion

This paper presents a short study on Morphological Analyzer for Indian Languages which is mainly concentrated to Malayalam and Sanskrit language. These are morphologically rich as compared with English. Morphological analyzers can be used for Information Retrieval, search engines, Machine Translation, speech recognizer, Text Processing etc. This paper drafts a brief presentation on the existing python packages that include MA. The comparison report on each of them is also depicted.

References

1. Kumar, D., Singh, M., Shukla, S.: FST based morphological analyzer for Hindi language. arXiv preprint arXiv:1207.5409 (2012)
2. Antony, P.J., Raj, H.B., Sahana, B.S., Alvares, D.S., Raj, A.: Morphological analyzer and generator for Tulu language: a novel approach. In: Proceedings of the International Conference on Advances in Computing, Communications and Informatics, pp. 828–834, August 2012
3. Abeera, V.P., et al.: Morphological analyzer for Malayalam using machine learning. In: Kannan, Rajkumar, Andres, Frederic (eds.) ICDEM 2010. LNCS, vol. 6411, pp. 252–254. Springer, Heidelberg (2012). https://doi.org/10.1007/978-3-642-27872-3_38
4. Jayan, J.P., Rajeev, R.R., Rajendran, S.: Morphological analyser and morphological generator for Malayalam-Tamil machine translation. Int. J. Comput. Appl. 13(8), 0975–8887 (2011)
5. Krishna, A., et al.: A graph based semi-supervised approach for analysis of derivational nouns in Sanskrit. In: Proceedings of TextGraphs-11: the Workshop on Graph-based Methods for Natural Language Processing, pp. 66–75, August 2017
6. Thottingal, S.: Finite State Transducer based Morphology analysis for Malayalam Language. In: Proceedings of the 2nd Workshop on Technologies for MT of Low Resource Languages, pp. 1–5 (2019)
7. Chaitanya, V., Sangal, R., Bharati, A.: Natural Language Processing: A Paninian Perspective. Prentice-Hall of India, Delhi (1996)
8. Saranya, S.K.: Morphological analyzer for Malayalam verbs. Unpublished M. Tech Thesis, Amrita School of Engineering, Coimbatore (2008)
9. Jayan, J.P., Rajeev, R.R., Rajendran, S.: Morphological analyser for Malayalam-a comparison of different approaches. IJCSIT 2(2), 155–160 (2009)
10. Kulkarni, A., Shukl, D.: Sanskrit morphological analyser: some issues. Indian Linguist. 70(1–4), 169–177 (2009)
11. Agarwal, A., Singh, S.P., Kumar, A., Darbari, H.: Morphological Analyser for Hindi-A rule based implementation. Int. J. Adv. Comput. Res. 4(1), 19 (2014)
12. Antony, P.J., Soman, K.P.: Computational morphology and natural language parsing for Indian languages: a literature survey. Int. J. Sci. Eng. Res. 3 (2012)

Cyber Safety Against Social Media Abusing

Yuvraj Anil Jadhav[(✉)], Sakshi Jitendra Jain, Bhushan Sanjay More,
Mayur Sunil Jadhav, and Bhushan Chaudhari

Department of Information Technology, SVKM's Institute of Technology, DBATU University
Lonere, Dhule, Maharashtra, India

Abstract. In few years of development Social Media has been one of the remarkable innovations, which have made communication a lot easier and accessible to all. But, it has some drawbacks associated with it as every other useful appliance has. Cyberbullying is one of its major drawbacks in which a group or individuals are embarrassed or harassed with messages from people. To deal with it, this research firstly surveys the state of social media networks. Then, a system is proposed where cyberbullying will be controlled also there is a knowledge intelligence to gather information about trollers and report them accordingly. The system uses JavaScript and deep learning technology to detect abusive texts and images. This paper proposes the simplest and new technology to spot social media abuse which will help in solving all the cyberbullying issues of all age groups.

Keywords: Social media · Cyberbullying · Abuse · Javascript · Deep learning

1 Introduction

Social media is a web based technology to facilitate social interaction between large groups of individuals through some sort of network. It works as a special vehicle to generate and facilitate rich data of multidimensional domains created by the people to the people. Social media is a data-driven innovation and use as a tool to manage knowledge flows within and across organizational boundaries in the process of innovation [1]. It allows them to communicate with people across the world through text; audio and video, to share their thoughts and opinions freely through articles, blogs, posts, etc. Along with them, these platforms have now become an extensive medium for building businesses and promoting oneself [2].

But still being such a fantastic and helpful technological innovation person uses those to do a kind of crime called cyberbullying. The term "cyberbullying" is defined as insulting or attacking others by the use of awful language. Cyberbullying gives a bully the ability to embarrass or hurt a victim before a whole community online [3]. Due to these the perspective of people is changing towards such social media platforms from a platform of communication and source of information to a platform with negative opinions, trolls, abuse, and cyberbullying.

In today's modern era of digital life communication are inevitable, rather than being a luxury now, it has become a need for day-to-day life. And thus online or cyber safety

© Springer Nature Switzerland AG 2021
M. Singh et al. (Eds.): ICACDS 2021, CCIS 1440, pp. 124–136, 2021.
https://doi.org/10.1007/978-3-030-81462-5_12

is becoming a lot more important [4]. According to statista.com, in 2020, as predicted a population of 3.6 billion were using social media websites worldwide, this number is estimated to increase to almost 4.41 billion in 2025. As of January 2020, 49% have reported that they have faced cyberbullying and social media penetration of one or another kind. Thus, we can analyze that this is going to exist and also increase in upcoming times. By seeing the number of internet users growing worldwide, the internets, governments, and organizations have expressed concerns about the safety of children and teenagers using the Internet [5].

Cyberbullying and its effects can trouble victims all day as it is not possible to remove it completely from the internet. The Cyberbullying research center reports that in 2016, approximately 34% of students had been a victim of it to some extent. It also reports that 36.7% of adolescent girls had experienced cyberbullying at some point in their lives, compared to 30.5% of adolescent boys [6]. Cyberbullying or abusing the victim on online platforms is performed in the ways of harassment, impersonation, website creation, video shaming, sub-tweeting, vague-booking, threatening, [7]. Some of the examples of abuses are:-

1. Humiliating/embarrassing content posted online about the victim.
2. Threatening a person to commit an act of violence
3. Child pornography or threats of child pornography
4. Cyberstalking
5. Outing means sharing private messages, pictures, or other information about the victim on the internet
6. Fraping is that the act of logging into some other person's social media handle and posting inappropriate content under their name
7. Fake identity.

The research shows that cyberbullying leaves some short-term or long-term effects on victim's mental, physical and emotional health such as they feel Overwhelmed, Powerless, Humiliated, Worthless, Vengeful, Isolated, Depressed, Physically Sick, Suicidal, etc. It is noted that 20% of victims choose suicide to get rid of it [8]. The government of India also provisions Section 507 IPC and Section 66 E of the IT Act to deal with cyberbullying. These analyses show a necessity to deal with this issue to create a safe and abusing-free social media environment.

1.1 Aims and Objectives of the Research

The medium of doing Cyberbullying is abusive texts, images, or videos. A system is developed using JavaScript and Deep learning technology to identify and collect information about Cyberbullying happening on social media websites. The system will also notify the intended victim of the abusive content and handle the whole situation appropriately. And on the victim's request, it will block that abusive content from floating or being permanent on the internet later on. The idea aims to create a trustworthy, safe, and abuse-free social interaction ecosystem.

2 Literature Survey

Cyberbullying is one of the fastest-growing crimes in India due to drastic growth in social media usage over a decade. Some artworks have tried to understand the overall review of social media based on its use or abuse; few of them were used to know the statistics of social media crimes. According, to backlinko.com the study of social media users shows that in 2020, there are 3.81 billion people active users which were 3.48 billion in 2019 showing a growth of 9.2% from year to year which is considerable. Asia was first in this survey of social media users with a base growth of 16.98% [9].

In 2014, Syed Zulkarnain Syed Idrus, and Nor Azizah Hitam while presenting a review on social media use or abuse have dropped a light on the analyses of use and abuse happening through social media particularly Facebook. Their study shows that a lot of college students are using social networking sites for a larger portion of their day for informal purposes such as making new friends, sharing, posting, surfing, etc. [10]. Their research found that a user who uses these sites more frequently are more positive and have more self-belief. But another study presented a negative relationship between self-esteem and social media activity. In terms of comments, views, likes, untagging, unknown threats, undesirable or photo-shopped pictures, comparison, abuse, etc.

Dianne L. Hoff and Sidney N. Mitchell [6] to understand the causes, effects, and possible remedies from the point of view of students had undertaken a survey at school. While mentioning the causes of cyberbullying in teenagers' authors said that online bullying is often deliberate, also it is more uncomforting due to the hidden identity of the abuser. The abuser can easily hide their identity with the help of a strong identity-hiding algorithm. As teenagers are emotionally week in comparison with other age peoples, the attacks can be psychologically vicious. For example, students might take some personal pictures which later can be used to dramatically alter and post on social media sites once relationships sour, leaving the target uncomfortable before millions to see. Cyber-bullies also create chat boards, or share, or online news boards that invite others to contribute hateful and malicious remarks.

It was recorded that cyberbullying is amazingly ubiquitous in individuals' lives and, increasingly in companies [11], affecting 56.1% of the students in this study, with quite a large difference in the victimization of females. Examples were sorted consistent with four specific relationship tensions that emerged during this study, categories including break-ups cases of almost 41%, the envy of 20%, intolerance issues of 16%, or ganging up against one of nearly 14%. Among the population studied in this research, sexual point of reference was targeted most often, but some students also mentioned instances where their disability, religion, caste, society, or gender was attacked.

Mostly teenagers have been the target of cyberbullying all over the world. As per the result of a study by the Pew Research Center recorded on September 27, 2018, shows the type of cyberbullying that mostly happens with teenagers on social media. The most common type of harassment youth encountered is abusive or offensive name-calling [12]. Almost 42% say they have been called offensive names, 32% said that someone has spread false rumors about them on the internet. A vast majority of teens 90% think that online harassment is a problem that affects teenagers majorly.

In 2019, Bertie Vidgen, Alex Harris, Dong Nguyen, Scott Hale, Rebekah Tromble, and Helen Margetts have studied the challenges and frontiers in detecting abusive content.

Authors have presented detailed research about the issues faced in designing an algorithm for detecting logically correct abusive content. According, to Waseem one of the chief variations between sub-tasks, is whether the bullying is intended towards a particular person or towards a group of people. Abuses can be categorized into three types based on the above differences. The first type is Individual-directed abuse that is the abuse directed against only one person. It involves directly abusing an individual by mentioning identity or tagging the victim. The second type is Identity-directed abuses that are directed against an identity, such as a religion, community, or affiliations. The third type is Concept-directed abuse it is aimed against a concept, a belief, a system, country, or ideology [13]. Some examples of the above types are given below:-

1. @Username is such an A*shole
2. [name] is a f**uck!**g who*e
3. According to me [color/social group/caste] peoples should be thrown out of our country

The other most used medium for cyberbullying is through images or videos our study also focuses on the detection of these types of abusive content. Image-based abuse happens when an inappropriate, personal, nude, or sexual image of a victim is shared without their consent or permission to insult the victim before millions of audiences on social media [14]. This includes real, photoshopped, and drawn pictures and videos. The effects of image-based abuse are also similar to text-based abuse [15]. Women and children are twice likely to be victims of image-based abuse [16].

According to staista.com statistics of "Cyber Stalking and Bullying Cases Reported in India 2018 by leading state" published by Sandhay Keelery. Maharashtra state had 400 cases of cyberstalking and bullying registered, which were highest in comparison with the other states of the country. A total of 739 cases were registered in the whole country. The National Crime Records Bureau data have recorded there has been a 25% increase in the number of cybercrime cases from 2017–2018. Where cases of cyberstalking or bullying of ladies and youngsters have drastically raised by 36% from 542 in 2017 to 739 in 2018. On the other hand, the rate of cyber blackmailing or threatening was fallen by 15% from 311 in 2017 to 223 in 2018 [18].

Child sexual abuse images are one of the increasing cyberbullying issues. In 2014, to study the causes, effects of child sexual abuse images on victim's author Jennifer Martin has undergone a survey which proposes experiences and views of people working in this field. According to them, online abuse has more effects on the child than physical contact sexual abuse because it is hard to delete them from the internet it is everlasting and also to stop further sharing of it [19].

In 2016, M. Medvedeva, & Agbozo, Ebenezer & Nalivayko, D have presented a model using machine learning techniques in building an efficient algorithm that would be able to filter out abusive Twitter posts (tweets). The initial step involved gathering an already existing dataset, training it by applying various data mining techniques, and finally, testing the built model. The tools and software made of use were the R programming language, RStudio IDE (Integrated Development Environment), and several R packages. To make it automatic, it is then stored as a script and executed in a cloud computing infrastructure via Microsoft Azure's machine learning module.

In 2018, another methodology to detect abusive content from text, images, audio, or videos was proposed by Antigoni-Maria Fountaz, Despoina Chatzakouz, Nicolas Kourtellis, Jeremy Blackburny, Athina Vakaliz, Ilias Leontiadis. They have applied this methodology to detect abuse on Twitter. It is a unified architecture in a seamless, transparent fashion to detect different types of abusive behavior e.g. hate speech, sexism vs. racism, bullying, sarcasm, etc. The deep learning architecture works on text classification networks that consider the raw text as input and used Recurrent Neural Networks (RNN) [20].

3 Limitations in Existing Systems

This section presents all the limitations observed in the existing work that needed to be kept in mind before creating a fully-fledged solution.

1. The classification of abuses to be detected is not carried out properly till now. In terms of effects, causes, intent, etc.
2. An effective way to identify and detect only logically valid abusive contents is not discovered till now.
3. Solution on detecting text-based abusive contents even after abuser using hiding techniques of replacing letters with symbols are not presented.
4. Detection of abuses is only available in the English language, no method for Hinglish or content in any other language.
5. There is no facility or portal to easily file a complaint about abuses or cyberbullying happening on social media.
6. Provision to stop or block abusers on primary stages from posting and sharing abusive comments is not available.
7. Lack of effective solution to permanently delete the abusive content about the victim from the internet.

4 Solutions to Limitations

Till now there are no artworks which studied or identified all type of abusive media and proposed a solution to detect them all. The rate of image-based abuse is also increasing day by day. So we can't ignore it and employ only text-based abuse detecting solutions. The current research is proposing a fully feasible solution for detecting all forms of abuse.

Before solving any problem the most important task is to know all the aspects of the issue and classify the subtasks properly. The major drawback is no complete classification of abuses is available which can help in identifying subtasks or subcategories of abuses. It is very complicated because abusive content is so diverse and the most confusing is to determine the intended victim. Let us know the solution to this by identifying the subtasks or subcategories of abusive content. The thing to be the focus is the expression behind the abuse which can include different kinds of emotions and consent with which it is said e.g. hatefulness, offensive, insults, taunts, derogation, lies, misrepresentations, stereotypes, finger-pointing, and threatening comments.

As stated above before characterizing abuses we should also consider the purpose of the abuser because the reason for posting such content mainly depends upon the speakers' feeling or purpose (e.g. 'anger', which suggests a specific orientation of disappointment). Then the effect of abuse, every person gets differently affected by a certain comment, for example, the one content or comment which is found derogatory or disrespectful or objectionable by one person, or in one situation, might not have the same meaning to others or in other situations. Recognizing abusive texts is not also an effortless task so to make it easier to differentiate we have identified four linguistics difficulties that increase the challenge of detecting abusive content.

1. Spelling alteration: - Spelling alteration is mostly used in abusive content to avoid detection. They are ubiquitous; it includes elongation of words, use of alternatives, nearly-identical content, use or symbol, etc.
2. Emotions: - Emotions behind the content are a very important aspect as discussed above. Generally, it is one of the most confusing types of abuse detection.
3. Multiple meanings: - One spelling with multiple meanings exists also called polysemous. It is challenging as users can express hate more cunningly.
4. Language switch: - Abuser uses different languages to avoid detection from automatic systems of detecting abuse in standard social media language English (e.g. Hinglish). Also the syntax, grammar, and lexicons of language changes over time.

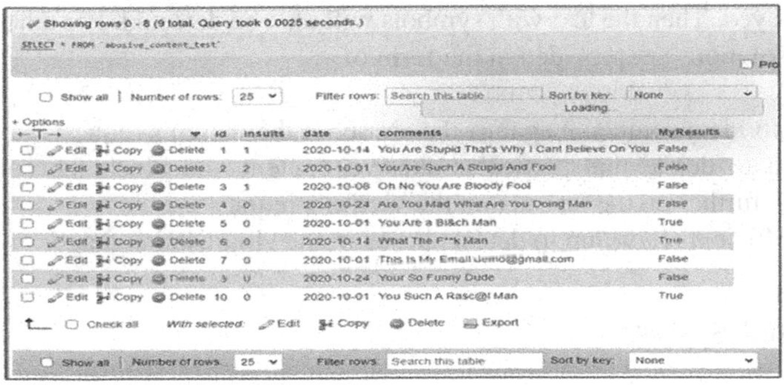

Showing rows 0 - 8 (9 total, Query took 0.0025 seconds.)

SELECT * FROM `abusive_content_test`

id	insults	date	comments	MyResults
1	1	2020-10-14	You Are Stupid That's Why I Cant Believe On You	False
2	2	2020-10-01	You Are Such A Stupid And Fool	False
3	1	2020-10-08	Oh No You Are Bloody Fool	False
4	0	2020-10-24	Are You Mad What Are You Doing Man	False
5	0	2020-10-01	You Are a Bi&ch Man	True
6	0	2020-10-14	What The F**k Man	True
7	0	2020-10-01	This Is My Email demo@gmail.com	False
8	0	2020-10-24	Your So Funny Dude	False
10	0	2020-10-01	You Such A Rasc@l Man	True

Fig. 1. Database result 1

Another most difficult drawback is to detect only logically valid abusive contents as shown in Fig. 1. Because only comparing the text with a dictionary of abusive words will not be sufficient to detect appropriate abusive content. It may happen that the algorithm will give false results because it will include the abusive content which contains abusive words directly but is considered non-abuse logically as preset in Fig. 1. In column 'insult' the value is true but it is false in the column 'my result'. The former column contains the result obtained through the algorithm after comparing it with a JavaScript-based array of words. And the latter one results from the self-analysis. To solve this issue we will use Normalization and Tokenization methods in cleaning data. We need to perform some transformation and Data cleaning. Then a new big data set will be created and trained with the logically abusive contents and updated into the algorithm. After undergoing these transformation procedures and comparing those with a new dataset a much-better algorithm with increased accuracy will be gained.

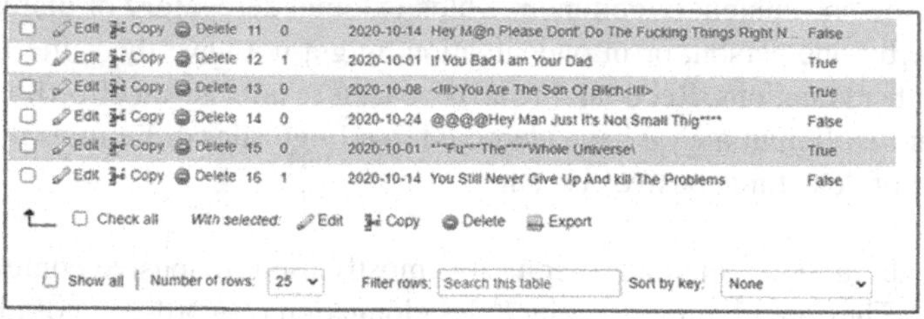

☐	Edit	Copy	Delete	11	0	2020-10-14	Hey M@n Please Dont Do The Fucking Things Right N...	False
☐	Edit	Copy	Delete	12	1	2020-10-01	If You Bad I am Your Dad	True
☐	Edit	Copy	Delete	13	0	2020-10-08	<lll>You Are The Son Of Bitch<lll>	True
☐	Edit	Copy	Delete	14	0	2020-10-24	@@@@Hey Man Just It's Not Small Thig****	False
☐	Edit	Copy	Delete	15	0	2020-10-01	***Fu***The****Whole Universe\	True
☐	Edit	Copy	Delete	16	1	2020-10-14	You Still Never Give Up And kill The Problems	False

Check all With selected: Edit Copy Delete Export

Show all | Number of rows: 25 Filter rows: Search this table Sort by key: None

Fig. 2. Database result 2

Figure 2 depicts the next issue which is most of the abusers use symbols or spelling modification intentionally to pass undetected through an automatic abuse detecting system. This is the easiest way to stay unidentified, we have also identified and explained it in linguistic difficulties as spelling variations (e.g. @username is a fu$k&ng idiot). These will also be detected in a new algorithm with the use of a dataset containing more possible abusive words as similar to the above-shown results. Despite applying this solution if all such contents will not be detected, then a method with a system for detecting such texts which are matching 50% to the abusive texts mentioned in the dictionary will be employed. Then the text with symbols will also be detected. For other motioned limitations, solutions are provided in the form of:-

1. A portal to file a complaint against the abuser at the victim's request.
2. A system to delete that particular abusive content and block the content from spreading further on the internet on the victim's request.
3. A more efficient algorithm to detect abuses in the Hinglish language.

5 Proposed System

After undergoing all the surveys and studying all the existing methodologies we came to understand that the issue needs a system with cyber safety against online abuse which can detect both textual and visual content. The system is integrated with the existing social media platforms (e.g. Facebook, Instagram, Twitter, etc.) as they are already so much popular and are effective in decreasing cyberbullying. The idea is to provide a system to obtain complete cyber safety against social media abusing. Following are the modules provided by the system:-

1. An environment of the existing social media sites for the user to post content.
2. Arrangement to detect abusive words from the post.
3. System to detect abusive images from the post.
4. Identifying bots and attackers.
5. Notification sending system to the victim or target.

6. Automatic complaint filing algorithm on the victim's request.
7. Abusive post deletion or blocking authority to the admin on the victim's request.
8. Abusive comments detection system.
9. Analyzing system for historical and predictive analysis.

5.1 System Architecture

Figure 3 is the UML diagram of the whole system architecture. Once the users enter the web portal, if they are already a user of that social media site then they can login directly, and if not then they need to sign in into the system. The system checks into the DB for authentication and after verification, the user is allowed to access the site. User's credentials, metadata, and info from that session are stored in the DB.

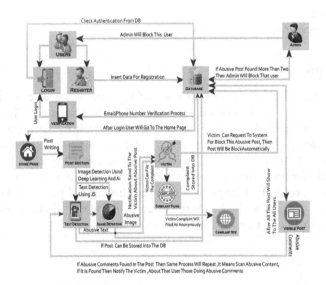

Fig. 3. DFD of system architecture

After that on the homepage, the user is provided with the tab of the post section. Once the user posts any content in the form of text or images the system is initiated to work. Based on the type of data contained in the post the algorithm is selected. If the posted content is textual then the text-based detection algorithm is selected and starts to work otherwise if it's visual then the image-based detection algorithm gets selected to work. In case, a post contains both types of data that are textual and visual then first the post is first directed towards the text detection algorithm and then towards the other one. Then one of these three cases is followed:-

1. If the post contains abusive text or images then the notifications are sent to the targeted persons and the post with abusive content and user info from that session are stored in the database in the table abusive content along with the number of abusive contents.

2. If the post contains both abusive images and abusive text then only one notification is sent to victims and corresponding data is stored in the database in the above manner.
3. If the post doesn't contain any kind of abusive content then only the post and details are stored in the post table.

Later, on notification to the victim, if the victims want to file a complaint against this cyberbullying act then they can request the system to file the complaint automatically. A victim can also request the system to block this abusive post and terminate it from floating on the internet. If abusive posts are found more than twice from a particular user then the system will automatically block that user for a few days. All these actions are taken once they are validated by the admin. Admin has the full authority to monitor the whole system. After all this process all the viable posts present in the post table are allowed to post on social media and can be seen by anyone on the internet. The same procedure is followed for every comment, story, feeds, etc. posted on the internet and the same cycle goes on.

5.2 Algorithms Used

We have seen the whole system architecture of the idea along with its working. To detect abusive content we have used some algorithms and solutions. The initial step involved is gathering an already existing data-set, training it by applying various data mining techniques, and finally, testing the built model. As a source to detect bad words, by comparison, a dictionary of bad words which was compiled by Google in 2011 was utilized. This dictionary is a text file that contains a list of abusive words in English which was a result of a project by Google to censor searches made by users. This dictionary of bad words is compiled as the JavaScript array of bad words in the system. These all words stored in the dictionary are also called word clouds.

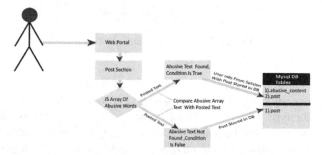

Fig. 4. UML of abusive text detection

Figure 4 shows the working of the text detection algorithm along with the proposed technology being used. As the user enters the web portal of a social media site the user is directed towards the post section or any alternative system available for posting users' content. The instance the user uploads any kind of text-based content the algorithm starts to work. It compares the posted content with the JavaScript Array of abusive words available in the system. If any kind of abusive words is found in the content the condition becomes true and a true value is assigned to the variable 'abuse'. If no such content is found the condition becomes false and a value false is assigned to the variable 'abuse'. The number of times such words are encountered is also updated in the variable 'index'. The system keeps on checking in the loop until all the abusive text is recorded. Along with this to make the algorithm more efficient we have employed the above-mentioned two solutions. The first one is for detecting only logical errors and avoiding false alerts and the second one to detect the intentionally hided abuses with the use of symbols to pass undetected from the automatic abuse checking systems.

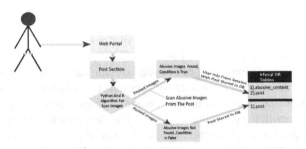

Fig. 5. UML of abusive image detection

The second most used medium for abusing is images. As we have seen earlier that recognizing and detecting images includes a lot of conditions and constraints. Figure 5 demonstrates the working of an image-based detection algorithm. Once the user enters the web portal and is directed towards the post section, if the user posts any kind of image-based content, the image-based algorithm begins to work. All the images are scanned one by one for abusive content. If an abusive image is found then the condition becomes true and variable 'abuse' is updated with a true value. If not, the variable 'abuse' is updated with a false value. All this data is also saved into the above database in the two tables post and abusive_content. Post containing any kind of abusive content out of two forms is stored in both tables post and abusive_content. While the post with no abusive content is only stored in the post table.

To identify abusive images we are deploying a deep learning and AI-based solution. We have used APIs of Keras and Tensorflow to detect abusive images from the posts through image mining, image data generator techniques. The whole system is designed with the help of Python and R languages. Anaconda platform is also used to compile the complete solution by using Jupyter notebooks. To solve the image categorization issue, we are using a CNN-based method that extracts image characteristics and also combines them with an abusive text-based detection algorithm to spot texts, tags, and comments on that particular image. Image metadata are also recorded i.e. time, place, IP, social media id, etc. All these metadata are stored in the database for future analysis.

A high-level database is designed with the help of Django(a web framework written in a python programming language) and MySQL database which is an open-source relational database tool. This design helps in structuring data properly in a customized manner. In the future, these databases can also be used to obtain an analysis of cyberbullying happening through social media. For front-end development, HTML/CS/JS-based framework is used.

6 Results

In this section, we have presented the results of the proposed idea with the help of obtained sample outputs. The whole system architecture is implemented on a newly designed social media website for demonstrating the complete idea. This website is taken as an example for any other social media website and can be replaced by anyone for example Facebook, Instagram, Twitter, etc.

To test the effectiveness of the proposed system the results of the currently used Social media app are compared with the results of above mentioned Social media app on which the system is implemented. For analysis purposes, the data of one month is recorded in both cases. Different types of cyberbullying cases are taken as parameters. Figure 6 presents the results in the form of a graph. It is evident from the graph that the proposed system has decreased the percentage of cases considerably. Also, the total percentage of cyberbullying cases that occurred in the first case has decreased from 51% to 21% in the second case. A drastic increase is noticed in the number of complaints filed officially of cyberbullying in the latter case due to the automatic complaint filing interface.

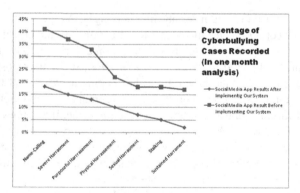

Fig. 6. Comparative analysis of cyberbullying cases recorded with and without implementing our system

The whole idea decreases the rate of abuse by detecting them with the help of both algorithms and then after detection treating them appropriately. It also tries to minimize the abuse in the first place by warnings to the user. Our experiment with a lot of different data sets and situations proved that the architecture works efficiently.

7 Conclusion

The goal of the current research is to investigate the automatic detection of cyberbullying-related posts on social media. This idea sought to model an algorithm for automatic detection of abuse in social media by proposing a simple and accurate methodology. It is trying to utilize the easiest way to detect Cyber Bullying crimes. With this attempt, it is working on solving major issues of depression, isolation, and suicidal thoughts which have seen increasing day by day. The system is remarkable for creating a safe social communication medium and preserving social media laws, in addition to designing a trustworthy and abuse-free social media environment.

References

1. Medvedeva, M., Agbozo, E., Nalivayko, D.: Automatic detection of abuse on social media. In: 16th International Multidisciplinary Scientific Geoconference (SGEM 2016) - Albena, Bulgaria, 30 June 2016
2. Social Media, Email, Blog, & Chat: Study.com. https://study.com/academy/lesson/internet-communication-twitter-email-im blogging-rss-newsgroups.html, 6 June 2018
3. Van Hee, C., Jacobs, G., Emmery, C., Desmet, B., Lefever, E., Verhoeven, B.: Automatic detection of cyberbullying in social media text. PLoS ONE **13**(10), e0203794, 8 October 2018
4. Internet safety. https://en.wikipedia.org/wiki/Internet_safety, 6 January 2021
5. Hoff, D., Mitchell, S.: Cyberbullying: causes, effects, and remedies. J. Educ. Admin. **47**, 652–665 (2009)
6. Clement, J.: Number of social network users worldwide from 2017 to 2025. Statista.com. https://www.statista.com/statistics/278414/number-of-worldwide-social-network-users/, 15 July 2020
7. What Is Cyberbullying? An Overview for Students, Parents, and Teachers: Maryvillie Uiversity (2018). https://online.maryville.edu/blog/what-is-cyberbullying-an-overview-for-students-parents-and-teachers
8. He, K., Zhang, X., Ren, S., Sun, J.: Deep Residual Learning for Image Recognition. CVPR paper, 10 December 2015
9. Dean, B.: Social Network Usage & Growth Statistics: How Many People Use Social Media in 2020?. backlinko.com. https://backlinko.com/social-media-users, 12 August 2020
10. SyedIdrus, S.Z., Hitam, N.A.: Social media use or abuse: a review. J. Hum. Dev. Commun., 48–58 (2014)
11. Bhimania, H., Mentionb, A.L., Barlatierc, P.J.: Social media and innovation: a systematic literature review and future research directions. J. Elesvier Technol. Forecast. Soc. Change, 11 October 2018
12. A Majority of Teens Have Experienced Some Form of Cyberbullying: Pew Research Center Internet & Technology. https://www.pewresearch.org/internet/2018/09/27/amajority-of-teens-have-experienced-some-form-of-cyberbullying/, 27 September 2020
13. Vidgen, B., Harris, A., Nguyen, D., Hale, S., Tromble, R., Margetts, H.: Challenges and frontiers in abusive content detection. In: Proceedings of the Third Workshop on Abusive Language Online, 1 August 2019
14. Innovation of Communication and Information Technologies: Packtpub. https://hub.packtpub.com/innovation-communication-and-information-technologies/, 5 June 2013
15. Image-Based Abuse: 1800 Respect National Sexual Assault Domestic Family Violence Counselling Service (2020). https://www.1800respect.org.au/violence-and-abuse/image-based-abuse
16. Sanderson, J., Weathers, M.: Snapchat and child sexual abuse in sport: Protecting child athletes in the social media age. Elsevier, Sport Manage. Rev., 81–94.6 (2020)

17. Number of cyber stalking and bullying incidents against women and children across India in 2018, by leading state: Statista.com. https://www.statista.com/statistics/1097724/india-cyber-stalking-bullying-cases-against-women-children-by-leading-state/#statisticContainer, 16 October 2020

18. In one year alone, cyberbullying of Indian women and teenagers rose by 36 Percent: Scroll.in. https://amp.scroll.in/article/956085/in-one-year-alone-cyberbullying-of-indian-women-and-teenagers-rose-by-36, 16 March 2020

19. Martin, J.: It's just an image, right?: Practitioners' understanding of child sexual abuse images online and effects on victims. Child Youth Serv. (2014)

20. Fountaz, A.M., Chatzakouz, D., Kourtellis, N., Blackburny, J., Vakaliz, A., Leontiadis, I.: A Unified Deep Learning Architecture for Abuse Detection, 21 Feburary 2018

Predictive Rood Pattern Search for Efficient Video Compression

Hussain Ahmed Choudhury(⊠)

CVR College of Engineering, Hyderabad 501510, India

Abstract. The modern era of digital communication heavily relied on video for information, communication, and education. Videos are becoming an integral part of the digital world. The tremendous use of videos leads to the production of a high amount of data sometimes referred to as Big Data. So to process this much data with limited resources is impossible if videos are not compressed properly and efficiently. The heart of video compression is the process that predicts the movement of an object from one frame to another by using different Block Matching Algorithms and is called motion estimation. Many Block Matching Algorithms were reported to carry out the motion estimation to date but none of the research work considered the direction of movement of Macro Blocks i.e. objects while applying any algorithm. The proposed work titled Predictive Rood Pattern Search (PRPS), here will first identify the direction of movement of the object and apply the best suitable algorithms out of the existing but best among the reported works which consider the correlation between the current macroblock and the neighboring macroblock. The obtained results show the efficiency of the new algorithms over the other algorithms while providing the option of selecting the proper algorithm based on direction.

Keywords: Block matching algorithm · Mean absolute difference · DI · PSNR

1 Introduction

With the expansion of telecommunication technology & IoT in the present scenario, the transformation happened from voice over phone & voice over the internet to video on demand, video conferencing and Video has emerged as a powerful and most available medium of learning and education. When the growth of the use of video increased exponentially, then transmission, processing, and storing of such a huge amount of data became difficult with a limited amount of resources. Here comes the importance of motion estimation which is the heart of video compression and the most critical concept to deal with. The complexity of space, time, and quality of Motion Estimation decide the complexity and easiness of the whole video coding cum video compression process. Over the years' research is going on how to make ME faster and efficient. Due to huge growth in the use of video, the area and field of research is still a hot topic and tremendous research is carried out to find better and better techniques to do ME. The process of ME is a way to estimate the motion of an object from one frame to another which involves the

© Springer Nature Switzerland AG 2021
M. Singh et al. (Eds.): ICACDS 2021, CCIS 1440, pp. 137–150, 2021.
https://doi.org/10.1007/978-3-030-81462-5_13

comparison of Macro Blocks (MBs) of two consecutive frames commonly refereed as a current frame and a reference frame. The comparison, if done on a pixel basis, leads to a great amount of computational complexity. So the concept of block-based comparison for estimating the motion of an object comes to a picture which is called Block-Based Motion Estimation (BBME). Due to the importance and key role played by ME, it is used in most of the popular video compression techniques such a MPEG-2, MPEG-4 & standards like H.264.

In BM, each frame is first divided into blocks with NxN dimension which are not overlapped. For every MB of the current frame, a search window is pre-decided in the reference frame with dimension $(2p + 1)x(2p + 1)$ where the p is window size or the maximum allowed displacement. The MB in the current frame is referred to as current MB and that too in the reference frame is called reference MB. The reference frame may be of two types i.e. either past frame or future frame. The first frame for there is no past or future frame encoded directly which is referred to as an Intra-coded frame or I-frame. The frames which have either past frame or future frame as reference are called Predicted or P-Frame and are interceded. The frames which are encoded with the help of both past and future frame are designated as bidirectional frame or B-Frame.

Based on the extensive study of motion vector probability (MVP), all these algorithms are trying to make the search patterns approximately around a circle and the selected search pattern heavily affects the search results. Moreover, all the algorithms are designed and focused on how the number of search points i.e. computational complexity will be reduced without much degradation in the quality of the reconstructed image. But after a thorough study of the existing works and implementing a few of the BMAs, it is observed that no algorithm addresses the issues like type of videos, slow-motion type, average motion type videos, fast motion type videos, and videos where the movement of objects are high or different objects move in both/different directions which also play a major role in the results obtained by BMAs. Moreover, reported works failed to answer if all available algorithms are suitable for the movement of an object in all possible directions within a video or not. Existing algorithms didn't consider the direction of movement of MB while applying the BMAs. In this paper, a dynamic BMA is projected which selects the appropriate neighbors to predict the MV of current MB based on MB direction and is titled Predictive Rood Pattern Search (PRPS).

This paper is organized into 8 sections which are as follows: Sect. 2 elaborates the literature survey, Sect. 3 explains the motivation behind the proposed work. The process of Block matching and the methodologies used in it will be briefed in Sect. 4. Different block matching criteria and parameters; Sect. 5 discusses classification videos; Sect. 4 narrates the proposed algorithms. Simulation results and analysis are tabulated and an explanation is given in Sect. 7. The comparative results with existing techniques are shown in Sect. 8 which is followed by the conclusion.

2 Literature Survey

All of the existing BMAs use only one-directional prediction to reduce the computational cost. The research on video compression is a very old area but the increasing use of videos never let the area go out of scope. Accordingly, Jain & Jain (1981) first put forward the concept of Full Search BMA [1] which paves the way for all other developed algorithms, and still today research is going on. The variable size block matching algorithm was introduced by Chan et al. [3] which considered different sized macroblocks (MB) for matching and to find the best match. The BMAs usually follow some fixed initial search patterns and gradually the search window and search patterns change based on different criteria which are mostly the location of minimum Block Distortion Measure (BDM). Gradually different BMAs are developed, namely Three Step Search (TSS) [2], New Three-Step Search (NTSS) [4], Four-Step Search (4SS) [5], Diamond Search (DS) ([6, 7]), New Cross Diamond Search (NCDS) [8], Cross Diamond Search (CDS) [9], Reduced Diamond Search (RDS) [10], Three-Step Logarithmic Search (TSLS) [11], Cross Three-step Logarithmic Search (CTSLS) [12], Simple and Efficient Search (SES) [13] and few other during the last decade. Some of the algorithms like Adaptive Rood Pattern Search (ARPS) ([13]–[15]), Correlation-based Search Pattern (CSP) used inter-block correlation for estimating MV. The Correlation Based Rood Pattern Search (CBRPS) [34] was found to be one of the best BMAs which not only reduces the search point i.e. computational complexity but also maintained the quality of the compensated image. It has been observed over the years that DS became very popular amongst the researchers and this leads to the development of various new BMAs which used the DS algorithm directly or used the concepts like RDS, Modified Diamond Search (MDS) [16], Cross Diamond Hexagonal Search (CDHS) [17], Orthogonal Diamond Search (ODS) [18], Hexagonal Diamond Search ([19, 20]), Kite Search Pattern (KSP) [21], Enhanced Cross Diamond Search (ECDS) [22], Hexagonal Diamond Search (HDS) [24], Orthogonal-Diamond Search (ODS) ([25, 26]), New Horizontal Diamond Search (NHDS) [27], etc. In addition to all such pattern-based BMAs, biological inspired algorithms like Genetic algorithm (GA) ([28]–[31]), Particle swarm optimization (PSO) ([29]–[31]), Artificial Bee Colony optimization ([32–34]) are also implemented for Block matching.

3 Motivation

The technological advancements, modern tools, and techniques made it possible to easily identify and categorize videos based on different criteria. The criteria may be based on the speed of MB, the number of moving MB, and even the direction of movement of MB. But neither the existing pattern-based BMA nor Nature-inspired BMA considered the quality of videos. Even though it is proved from experiments that few are good for fast videos, few are good for slow videos, few are good for large motion and few are for small motion i.e. videos with smaller values of mv between two consecutive frames but there is no option in any of the existing BMAs to check the quality and category of videos before implementing any BMA. Moreover, the movement of MBs in all frames and all videos is not in the same direction. So, there is a need to have the provision of a selection of neighbouring MB which will be used as a predictor block and identify the direction of movement of MBs. So based on direction of movement and neighbouring MB, algorithms are selected so that the best suitable algorithm will get selected based on the type of videos.

4 Block-Based Motion Estimation

The process of block matching is a searching technique that is achieved through comparison between the reference frame and candidate frame. The comparison can be done at the pixel level or macroblock level. As pixel-level comparison may result in the wrong result due to spatial redundancy and high computations are involved, the block-level comparison is preferred. In block-level comparison, each frame is divided into equal-sized Macro Blocks and then comparisons are done to find out the best matching MB in a reference frame. With the help of a block matching algorithm, the difference between coordinate locations of MB in the reference frame to that of candidate frame is found which is known as Motion Vector (MV) which in turn helps in the estimation of motion of MB. This process is referred to as Block-Based Motion Estimation (BBME or BME). Based on the reference frames used, the motion estimation process can be classified into two categories -first type known as backward motion estimation as it uses past frame or older frames for reference and the 2nd type is considered as forwarding motion estimation which uses the future frame as a reference in motion estimation. The technique of ME is shown in Fig. 1.

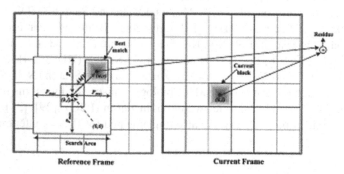

Fig. 1. ME process

4.1 Block Matching Parameters

The matching of macroblocks involves the comparison of MBs in two consecutive frames and the difference i.e. MV is used to reconstruct the original image. In doing so, the efficiency of algorithms varies depending on what parameters are used in finding MV and the quality of the reconstructed image. As per the available literature, most of the research works used Mean Absolute Difference (MAD) for calculating MV, and Peak Signal to Noise ratio (PSNR) is used for checking the quality of the reconstructed image. The same parameters are also used in this paper.

4.2 Complexity Reduction Step

As the search starts for minimal complication resulted from the number of searching MB dictated by BMA, the early detection of the role played by MB towards the complexity of BMA becomes crucial. The role of MB is decided by the position and value of the minimal Block distortion measure (BDM) value which is taken as the MAD value in the current scenario. If the position of minimal BDM is located at the center of the search window, the MB is regarded as "inactive" and has no role in complications of BMAs. Whereas MB for which the lowest BDM is found to be at any other position except center, they contribute to the overall complexity of BMA and are projected as "Active" MB. So while searching is carried out for the best match, inactive MBs are ignored and only active MB and their locations are searched. This step helps in further reduction of intricacy.

5 Classification of Videos Based on Speed

The videos have both moving parts and non-moving parts. The difference in the content of a frame with its neighboring one is due to the movement of an object. Most portion of a video remains constant and only a small portion moves for which spatial redundancy occurs in a video as videos are collections of images/frames. Here comes confusion that how to identify videos if it is slow video, or moderate video, or fast-moving video because few BMAs are only suitable for slow video, few BMAs are designed for moderate video and the rest are for fast-moving videos. Also, there are questions on how to decide the nearest neighbors out of all neighboring MBs of current MB for predicting the motion of current MB.

To determine the category of videos depending on the speed of moving object in a video and the amount of displacement it undergoes from one frame to another, two parameters can be considered. The first one will show how much displacement an MB undergoes between two consecutive frames called displacement indicator (DI) and the other one called Stationary (St) will denote the ratio of static MBs to that of total MBs in a frame. The calculation expression is given by Eq. (1) through Eq. (3).

$$DI_i = \frac{\sum_{i=1}^{n}\{abs(MV_x) + abs(MV_y)\}}{n} \tag{1}$$

$$St = \frac{No.\ of\ static\ MB}{Total\ no.\ of\ MB\ per\ frame(n)} \tag{2}$$

$$DI_{avg} = \sum_{i=1}^{i=n}\frac{DI_i}{n} \tag{3}$$

Based on Table 1 and criteria, the frames are first classified, and then frame by frame the algorithms will check the speed of an object and static MBs.

6 Proposed Methodology

The methodology can be understood through the following steps: The first step explains the selection of neighbouring MB which will be considered for the frame of reference and the 2nd step discusses the steps of the algorithm.

Table 1. Frame types based on speed

Sl no.	DI values	St (in %)	Frame category
1.	$0 \leq DI \leq 2$	$0 < St \leq 30$	SH
2.		$30 < St \leq 70$	SM
3.		$70 < St \leq 100$	SL
4.	$2 < DI \leq 4$	$0 < St \leq 30$	MH
5.		$30 < St \leq 70$	MM
6.		$70 < St \leq 100$	ML
7.	$4 < DI \leq 6$	$0 < St \leq 30$	FH
8.		$30 < St \leq 70$	FM
9.		$70 < St \leq 100$	FL
10.	$6 < DI \leq p$	$0 < St \leq 30$	VFH
11.		$30 < St \leq 70$	VFM
12.		$70 < St \leq 100$	VFL

6.1 Neighbourhood Selection Step

The videos are nothing but a pool of frames moving at very high speed. Due to the speed at which the frames appear say 30–33 frames per second, the amount of information or content change from one frame to another is quite a negligee i.e. most of the contents remain the same. This repeated information is called Temporal Redundancy. The other type of repeated or redundant information that exists is due to spatial locations. The neighboring pixel values are also observed to be very similar and so as the MB values which are called spatially redundant information. In the video compression process, the temporal redundancies are eradicated as the contents or change in neighboring pixel/MB can be well predicted using the content/change in contents/information in neighboring MB due to the existence of coherent information.

While considering the neighboring MB, it is found that each MB is surrounded by 8 MB in 8 directions. So the selection of appropriate MB for prediction of the movement of its neighborhood MB is quite difficult if we are not sure about the direction of movement of blocks e.g. while a body (MB) is moving in the horizontal right direction, the use of MV of its immediate left MB will be the best choice to predict its movement. So based on the different Neighbourhood/adjacent MB, 8 different MV prediction types/schemes are defined as specified below and the selection of each type will solely depend on the direction of MB movement (Table 2).

Table 2. 8 Neighbouring MBs of current MB.

ULB	UB	URB
LB	CB	RB
LLB	LB	LRB

Table 3. Selection of neighbouring MBs

MV_x values	MV_y values	Predictor block	Direction of movement
$0 < MVx_0 \leq +p$	$-1 \leq MVy_i \leq 1$	LB	Towards East
	$-p \leq MVy_i < -1$	LLB	North East Direction
	$1 < MVyi \leq 1$	ULB	southeast Direction
$0 > MVx_i \geq -p$	$-1 \leq MV_{yi} \leq 1$	RB	West Direction
	$-p \leq MV_{yi} < -1$	LRB	northwest Direction
	$1 < MV_{yi} \leq p$	URB	South West Direction
$-1 < MVx_0 \leq +1$	$0 < MVy_0 \leq +p$	UB	South Direction
	$0 > MVy_0 \geq -p$	LB	North Direction

A. For the direction of movement:

i)   +v direction

For $\forall$ MB from i= 0: N-1 where N=no. of MB in a frame, check If $0<MVx_0\leq+p$

(a) If $-1\leq MVyi\leq1$ then reference MB=Left MB(LB)----------------------+ve horizontal.
(b) If $-p<MVy_i<-1$ then reference MB=Left Lower MB (LLB) ------------NE direction.
(c) $1<MVyi\leq1$ then reference MB=Upper Left MB (ULB) ----------------SE Direction.

ii)   -ve direction:

Check If $0> MVx_i\geq-p$
(a)$-1\leq MV_{yi}\leq1$, then reference MB=right MB(RB) -----------------------'-'ve horizontal.
(b)$-p\leq MV_{yi}<-1$, then reference MB=Left right MB (LRB)------------------NW direction.
(b)$1<MV_{yi}\leq p$, then reference MB=upper right MB (URB)------------------SW direction.

iii) For Vertical movement

if $\forall$ MB   check If $-1<MVx_0\leq+1$,
a)   If $0<MVy_0\leq+p$, reference block=Upper block (UB)---------vertically downwards.
b)   If $0>MVy_0\geq-p$, reference block=Lower block (LB)--------------vertically upward.

6.2 Suggested Methodology

The objects in videos have movements in all possible directions. The possibility also is there that the number of MBs of the same object is moving in different directions in different frames. So the proposed algorithm determines the direction of movement of the current MB and then selects the appropriate MB as a predictor block for determining the MV of the current frame based on the following rules.

Therefore, the PRPS will start the implementation of the sequence of BMA and will try to explore the results obtained by the combination of the three most efficient algorithms namely ARPS, CBRPS, and CSP. During the experiment, all possible combinations of three BMAs are implemented. For our convenience, the first letter of each BMA mentioned above to represent the sequence i.e. ACbC is used for the above sequence ("A" for ARPS, "Cb" for CBRPS, and "C" for CSP) are taken. Since the motion estimation is done by switching between 3 existing BMAs depending upon the frame criteria for which it also can be regarded as Dynamic and since MV of neighboring MB is used to predict the MV of the current MB, it is called Predictive Rood Pattern Search (PRPS). The algorithm has the following steps:

Algorithm:

Input: Test video sequences.
Output: MV, PSNR

Steps of Algorithm:

1. Apply CBRPS $\forall$ MB of 1st frame and Store all MV $\forall$ MB and calculate the DI and St for all MBs.
2. For frame 2:29
   Check the MV of all MB of 1st frame and the above-mentioned rules in Table 1 and Table 3 are applied for MB$_i$ to select appropriate reference block and identify frame type.
3. For each frame 2:29, carry out the following steps
   a) if $0 \leq DI \leq 2$ check the conditions (i) through (iii)
      i. If reference blocks are LB, LLB, or ULB: Choose the algorithm "ARPS".
      ii. If the reference block is RB, LRB, or URB: Choose the algorithm "CBRPS".
      iii. If the reference block is UB or LB: Choose the algorithm "CSP"
   b) If the value of $2<DI<=4$, check the conditions (i) through (iii) again.
   c) If the value of $4<DI<=6$, check the conditions (i) through (iii) again.
   d) If the value of $6<DI<=p$, apply "CBRPS".
4. If the BDM is at the center, stop the search else shift the center to the new minimum and follow SDSP till the center will have minimum BDM i.e. cost.

7 Results and Discussion

The experiments were carried out using MATLAB R2013a on Windows 10 Machine to gauze the efficiency and fallouts of the proposed BMA. The videos selected for our experiments include Stefan, News, Coastguard, Salesman, Clairo, etc. The first 30 frames

Table 4. BMAs based on the direction of movement

Sl. No.	Direction of movement	Algorithm type
1	Towards East	**PRPS which is a combination**
2	North East Direction	**of CBRPS, ARPS, and CSP**
3	southeast Direction	
4	West Direction	
5	northwest Direction	
6	South West Direction	
7	South Direction	
8	North Direction	

are our input to the BMA. The output of BMA will be MV. The frames are split into MB of size 16×16 where N = 16 and the maximum pixels in all directions which defines the boundary within which the MB will be searched for its best-matched ones are taken to be ± 7. So the considerable window size will be $15 \times 15 = 225$.

The comparative results with other existing algorithms show that PRPS performs better than all other pattern-based BMAs. Table 5 results highlight the dynamic behavior of suggested BMA which not only calculates Displacement Indicator(DI) and St(%) for a video sequence "Coastguard.qcif" but also selects the best and suitable BMA among the three efficient algorithms. The Computational complexity obtained by PRPS for different video sequences is listed in Table 6 whereas corresponding PSNR values for the reconstructed image are shown in Table 7.

Table 5. Proposed PRPS results for the video "Coastguard.qcif"

Parameters	Frames									
	F1	F2	F3	F4	F5	F6	F7	F8	F9	F10
DI	1.311	1.007	1.243	1.128	1.207	1.112	1.101	1.217	1.416	1.58
St (%)	32.071	27.273	39.899	27.525	27.525	40.657	38.636	45.707	46.212	35.606
Frame type	SH	SH	SH	SH	SH	SM	SH	SM	SM	SH
BMA to be used	ARPS	ARPS	ARPS	ARPS	ARPS	CBRPS	ARPS	CBRPS	CBRPS	CBRPS
Comp	5.917	7.24	6.44	8.34	6.394	7.46	9.73	1.394	17.088	17.058
PSNR	35.316	34.581	34.61	35.16	34.25	33.95	34.211	33.431	33.217	33.891

The results obtained by implementing the proposed algorithm on different video sequences are compared with the results of other reported algorithms like SES, CBRPS, TSS, DS, NTSS, etc. in terms of Average Search Point (ASP) taken by search algorithms i.e. known as computational complexity and the corresponding PSNR values to judge the quality of the reconstructed image is shown in Fig. 2 and Fig. 3 for "Cricket Ball.avi".

Table 6. Computational complexities obtained by proposed PRPS on different videos

Videos	Frames									
	F1	F2	F3	F4	F5	F6	F7	F8	F9	F10
Cricket Ball	6.061	6.479	6.415	6.541	6.347	6.577	6.395	6.319	6.293	6.942
Stefan	11.042	12.706	12.711	12.713	12.698	12.702	12.704	12.711	12.698	12.705
Claire	7.586	16.359	16.745	6.740	7.346	16.682	6.823	7.409	16.735	6.694
Cricket Batting	8.942	16.756	8.870	8.888	8.951	16.786	9.128	16.663	9.064	16.696
Bus	10.520	12.563	12.573	12.558	12.545	12.518	12.500	12.518	12.543	12.548
Garden	10.162	12.689	12.698	12.703	12.698	12.699	12.692	12.694	12.700	12.692
Take Off	11.349	12.808	12.806	12.804	12.802	12.807	12.805	12.804	12.808	12.805
Coastguard	6.917	12.394	12.404	12.394	12.394	12.396	17.073	12.394	17.088	17.058
Carphone	8.790	12.396	16.116	12.414	16.902	8.962	12.394	16.321	12.407	12.394

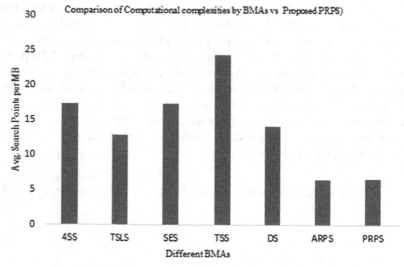

Fig. 2. ASP for comparison by proposed PRPS vs other BMAs

The results obtained by above mentioned BMAs and proposed BMAs are shown in Table 8 and Table 9 which also justifies the superiority of the proposed BMA.

Table 7. PSNR values obtained by the proposed PRPS on different videos

Videos	Frames									
	F1	F2	F3	F4	F5	F6	F7	F8	F9	F10
Cricket ball	36.346	35.668	35.808	35.489	36.282	35.045	36.349	35.275	37.263	32.268
Stefan	25.378	22.701	21.984	21.346	20.608	19.986	19.658	19.561	19.411	19.504
Claire	41.405	36.963	39.605	42.696	38.834	42.556	40.176	39.514	39.174	40.046
Cricket batting	39.766	41.289	41.651	41.540	41.485	40.952	39.366	39.962	40.010	38.317
Bus	21.364	19.769	20.481	20.329	18.728	20.081	20.244	20.270	20.722	20.789
Garden	23.712	22.083	21.830	21.809	21.291	21.587	21.918	21.948	21.543	21.701
TakeOff	45.448	41.326	45.847	44.253	42.011	45.579	44.771	40.823	39.723	44.548
Coastguard	34.206	35.581	33.600	36.167	35.052	32.595	34.821	32.531	34.117	32.891
Carphone	31.019	32.809	31.774	32.379	35.190	31.702	33.355	29.782	31.844	32.049

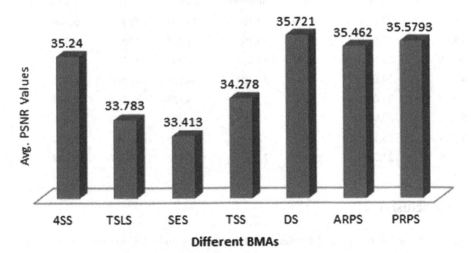

Fig. 3. Avg. PSNR by proposed PRPS vs All algorithm for Cricket Ball.avi

Table 8. ASP comparisons between the existing BMA with proposed PRPS

Algorithm	Videos					
	Stefan	Claire	Salesman	News	Akiyo	Coastguard
FS	202.23	204.25	218.6	204.28	204.12	206.178
TSS	23.34	23.22	22.5	22.75	22.859	24.453

(continued)

Table 8. (*continued*)

Algorithm	Videos					
	Stefan	Claire	Salesman	News	Akiyo	Coastguard
NTSS	20.89	16.02	19.6	23.05	17.467	17.475
4SS	17.82	19.66	25.76	18.82	26.761	17.413
DS	16.5	12.5	13.6	12.67	12.073	14.256
SES	16.34	17.49	16.64	16.97	16.587	17.314
TSLS	12.86	12.69	11.87	12.05	12.998	12.783
PRPS	23.73	20.23	21.48	22.45	23.889	23.398

Table 9. PSNR comparisons between the existing BMA with Proposed versions of PRPS

Algorithm	Videos					
	Stefan	Claire	Salesman	News	Akiyo	Coastguard
FS	24.772	38.2799	33.392	34.362	42.813	28.146
TSS	23.624	38.0247	33.202	34.235	42.565	27.747
NTSS	23.666	37.779	33.055	34.087	41.744	27.556
4SS	24.0412	38.0594	33.284	34.208	42.564	26.411
DS	24.067	38.1039	33.332	34.228	42.564	27.034
SES	23.95	37.565	31.729	33.977	41.889	25.431
TSLS	23.515	38.7155	32.691	34.098	41.98	25.022
PRPS	23.278	32.67	29.012	26.033	31.103	30.482

8   Conclusion

The BMAs are designed and implemented for making the ME process more fast and efficient. This ultimately makes the video compression efficient. This paper discusses a very novel idea of identifying the type of videos based on speed and then selecting the best neighboring MB which will be used for predicting the direction of movement of the current MB. The suggested technique takes the advantage of three best BMAs namely ARPS, CBRPS, and CPS for which it becomes dynamic as BMAs are getting selected based on the properties of the frame. The proposed PRPS has been implemented on various experimental video sequences and results are compared with various existing BMAs. The results show many folds improvements in the case of Computational cost and time over Full search and a lot of improvements over many reported BMAs. At the same time, the proposed BMA is found to have almost similar PSNR values for the reconstructed image which is the most critical aspect while designing a BMA for accurately performing ME.

Acknowledgement. The author would like to thank Dr. Nidul Sinha sir and Dr. Monjul Saikia sir for their guidance and help to carry out the research work. The author would also like to thank CVR College of Engineering and NIT Silchar for providing infrastructure to carry out the work. The author would also extend his gratitude to Ms. Nasima Aktar Lakar for her continuous motivation and feedback during the research work.

References

1. Jain, J., Jain, A.: Displacement measurement and its application in interframe image coding. IEEE Trans. Commun. **29**(12), 1799–1808 (1981)
2. Chan, M.H., Yu, Y.B., Constantinides, A.G.: Variable size block matching motion compensation with application to video coding. IEE Proceedings I (Commun., Speech, Vision). **137**(4), 205–212 (1990)
3. Koga, T., Iinuma, K., Hirano, A., Iijima, Y., Ishiguro, T.: Motion-compensated interframe coding for video conferencing. In: Proc. NTC81, pp. C9.6.1–9.6.5, November 1981
4. Kim, J.N., Choi, T.S.: A fast three-step search algorithm with minimum checking points using unimodal error surface assumption. IEEE Trans. Consum. Electron. **44**(3), 638–648 (1998)
5. Choudhury, H.A., Saikia, M.: Comparative study of block matching algorithm for motion estimation. In: International Conference on Advanced Computer Science and Information Technology, Bhubaneswar, India. vol. 17 (2013)
6. Chau, L.P., Jing, X.: Efficient three-step search algorithm for block motion estimation in video coding. In: 2003 IEEE International Conference on Acoustics, Speech, and Signal Processing, 2003. Proceedings. (ICASSP 2003), vol. 3, pp. III-421. IEEE
7. Ghanbari, M.: The cross-search algorithm for motion estimation (image coding). IEEE Trans. Commun. **38**(7), 950–953 (1990)
8. Li, R., Zeng, B., Liou, M.L.: A new three-step search algorithm for block motion estimation. IEEE Trans. Circuits Syst. Video Technol. **4**(4), 438–442 (1994)
9. Po, L.M., Ma, W.C.: A novel four-step search algorithm for fast block motion estimation. IEEE Trans. Circuits Syst. Video Technol. **6**(3), 313–317 (1996)
10. Zhu, S., Ma, K.-K.: A new diamond search algorithm for fast block-matching motion estimation. In: 1997 Proc. Int. Conf. Information, Communications and Signal Processing (ICICS), vol. 1, pp. 292–296, 9–12 September 1997.
11. Jia, H., Zhang, L.: A new cross diamond search algorithm for block motion estimation. In: 2004 IEEE International Conference on Acoustics, Speech, and Signal Processing 2004 May 17, vol. 3, pp. iii-357. IEEE
12. Cheung, C.H., Po, L.M.: A novel cross-diamond search algorithm for fast block motion estimation. IEEE Trans. Circuits Syst. Video Technol. **12**(12), 1168–1177 (2002)
13. Tham, J.Y., Ranganath, S., Ranganath, M., Kassim, A.A.: A novel unrestricted center-biased diamond search algorithm for block motion estimation. In: IEEE transactions on Circuits and Systems for Video Technology, vol. 8, no. 4, pp. 369–377 (1998)
14. Cheung, C.H., Po, L.M.: A novel small-cross-diamond search algorithm for fast video coding and videoconferencing application. In: Proceedings. International Conference on Image Processing, vol. 1, pp. I-I. IEEE (2002)
15. Long, L., Zhe, F.Z.: A square-diamond search algorithm for block motion estimation. Chin. J. Comput. **7** (2002)
16. Choi, W.I.L., Jeon, B., Jeong, J.: Fast motion estimation with the modified diamond search for variable motion block sizes. In: Proceedings 2003 International Conference on Image Processing (Cat. No. 03CH37429), vol. 2, pp. II-371. IEEE (2003)

17. Hashad, A., Sadek, R., Mandour, S.: A novel reduced diamond search algorithm with early termination for fast motion estimation. Int. J. Video Image Process. Netw. Secur. (IJVIPNS) **10**(04) (2010)

18. Choudhury, H.A., Saikia, M.: Reduced three steps logarithmic search for motion estimation. In: International Conference on Information Communication and Embedded Systems (ICICES2014), pp. 1–5. IEEE, 2014 February

19. Barjatya, A.: Block matching algorithms for motion estimation. IEEE Trans. Evol. Comput. **8**(3), 225–239 (2004)

20. Zhu, C., Lin, X., Chau, L.P.: Hexagon-based search pattern for fast block motion estimation. IEEE Trans. Circuits Syst. Video Technol. **12**(5), 349–355 (2002)

21. Cheung, C.H., Po, L.-M.: Novel cross-diamond-hexagonal search algorithms for fast block motion estimation. IEEE Trans. Multimedia **7**(1), 16–22 (2005)

22. Choudhury, H.A., Saikia, M.: Block matching algorithms for motion estimation: a performance-based study. In: Bora, P.K., Prasanna, S.R.M., Sarma, K.K., Saikia, N. (eds.) Advances in Communication and Computing. LNEE, vol. 347, pp. 149–160. Springer, New Delhi (2015). https://doi.org/10.1007/978-81-322-2464-8_12

23. Nie, Y., Ma, K.K.: Adaptive rood pattern search for fast block-matching motion estimation. IEEE Trans. Image Process. **11**(12), 1442–1449 (2002)

24. Hsieh, C.H., Lu, P.C., Shyn, J.S., Lu, E.H.: Motion estimation algorithm using inter-block correlation. Electron. Lett. **26**(5), 276–277 (1990)

25. Rsai, J.C., Hsieh, C.H., Weng, S.K., Lai, M.F.: Block-matching motion estimation using a correlation search algorithm. Signal Process. Image Commun. **13**(2), 119–33, 1 August 1998

26. Choudhury, H.A., Sinha, D., Saikia, M.: Correlation based rood pattern search (CBRPS) for motion estimation in video processing. J. Intell. Fuzzy Syst., (Preprint), 1–11

27. Gorpuni, P.: Development of fast motion estimation algorithms for video compression. Diss (2009)

28. Chow, K.H.K., Liou, M.L.: Genetic motion search algorithm for video compression. IEEE Trans. Circ. Syst. Video Technol. **3**(6), 440–445 (1993)

29. Kennedy, J.: Particle swarm optimization. In: Encyclopedia of Machine Learning 2011, pp. 760–766. Springer, Boston, MA

30. Du, G.Y., Huang, T.S., Song, L.X., Zhao, B.J.: A novel fast motion estimation method based on particle swarm optimization. In: Machine Learning and Cybernetics, 2005. Proceedings of 2005 International Conference on 2005 August 18, vol. 8, pp. 5038–5042). IEEE

31. Yuan, X., Shen, X.: Block matching algorithm based on particle swarm optimization for motion estimation. In: The 2008 International Conference on Embedded Software and Systems (ICESS2008), 29 July 2008, pp. 191–195. IEEE

32. Cuevas, E., Zaldívar, D., Pérez-Cisneros, M., Sossa, H., Osuna, V.: Block matching algorithm for motion estimation based on Artificial Bee Colony (ABC). Appl. Soft Comput. **13**(6), 3047–3059 (2013)

33. Choudhury, H.A., Sinha Nidul, S.N.: Application of nature-inspired algorithms (NIA) for optimization of video compression. J. Intell. Fuzzy Syst. **38**(3), 3419-3443 (2020)

34. Tyagi, V.: Understanding Digital Image Processing. CRC Press, Boca Raton (2018). https://doi.org/10.1201/9781315123905

35. Choudhury, H.A., Sinha Nidul, S.N.: Correlation Based Rood Pattern Search (CBRPS) for motion estimation in video processing. J. Intell. Fuzzy Syst. **36**(6), 5989-5999 (2019)

An Effective Approach for Classifying Acute Lymphoblastic Leukemia Using Hybrid Hierarchical Classifiers

Sharath Sunil[✉], P. Sonu, S. Sarath, R. Rahul Nath, and Vivek Viswan

Department of Computer Science and Applications, Amrita Vishwa Vidyapeetham, Amrita School of Engineering, Vallikavu, Amritapuri, India

{sharathsunil,ssarath,rahulnathr,
vivekviswan}@am.students.amrita.edu, sonu@am.amrita.edu

Abstract. Acute Lymphoblastic Leukemia is an anomaly that affects White Blood Cells. This type of cancer occurs when there is an error in blood cell DNA. Children are prone to this type of cancer. After the initial stages, it spreads to other organs like the liver and spleen. The problem with this type of cancer is that, unlike other forms of cancer, it doesn't cause any tumors hence it is very hard to detect. Manual testing methods were used before the automation, but it was time-consuming and very much prone to errors. To solve that problem, automated testing methods were introduced. Different systems [1, 3, 4] were introduced in the past, but most of them have variable accuracies. These Automated Systems [1, 3, 4] used image processing and unsupervised machine learning techniques to classify the images into cancerous and healthy. The proposed system uses Hybrid Hierarchical Classifiers to classify the cancer cells, which will be an improvement over the previous systems and solves the problem of variable accuracies.

Keywords: Image processing · Hybrid hierarchical classifiers · Acute lymphoblastic leukemia · Machine learning · Medical image analysis

1 Introduction

Acute Lymphoblastic Leukemia is a type of cancer that mainly targets white blood cells. This type of cancer mainly occurs due to some errors(s) in the blood cell DNA. After its initial stages, ALL affects the liver, spleen, and other organs. Unlike other forms of cancer ALL is comparatively very hard to detect, as it doesn't cause any tumors. The main risk factors of ALL are exposure to chemical solvents, radiation(s), electromagnetic fields, etc. Some promising treatments include chemotherapy and radiation therapy and stem cell therapies. The blood cell samples have to be tested manually to detect the disease, which is both time-consuming and a painstaking process. There are both the WHO and the FAB forms of classification for the ALL cancer cells. According to Fig. 1. Leukemia can be classified into different subtypes, which include Acute Lymphoblastic Leukemia, Acute Myeloid Leukemia, Chronic Myeloid Leukemia, and Chronic Lymphocytic Leukemia. Acute leukemia tends to spread faster than chronic leukemia, hence

© Springer Nature Switzerland AG 2021
M. Singh et al. (Eds.): ICACDS 2021, CCIS 1440, pp. 151–161, 2021.
https://doi.org/10.1007/978-3-030-81462-5_14

more lethal. In the proposed system, FAB or the French American British form of classification will be used. To tackle the problems of speed and accuracy in manual testing, automated systems were introduced. These automated systems use image processing, features extraction, and Machine Learning techniques to segment/Classify Cancer Cells. They [1, 3, 4] used various algorithms like Naïve Bayes, K- Nearest Neighbor, SVM, etc. to classify the cancer cells. The proposed system uses Hybrid Hierarchical classifiers to classify the cells. The aforementioned algorithm/technique will be an improvement over existing machine learning algorithms in terms of accuracy. The system consists of three modules namely – Segmentation, Feature Extraction, and Classifier module. To address the problem of subtype classification, we hierarchically arrange the system, so that it will classify the cells at each stage. The images are obtained from a high-powered microscope. Since we are dealing with medical data, we have to clean the data. In this case, we have to process the images, so that we can remove the noises and imperfections. So image processing is a very crucial stage. The datasets are to be used are ALL-IDB 1 and ADB-IDB 2.

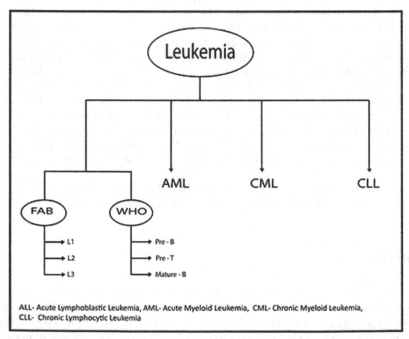

ALL- Acute lymphoblastic leukemia, AML- Acute Myeloid Leukemia, CML- Chronic Myeloid Leukemia, CLL- Chronic Lymphocytic Leukemia

Fig. 1. Different types of Leukemia and their *subtypes*, according to FAB (French American British) and WHO (World Health Organization) other forms of Leukemia include CLL (Chronic Lymphocytic Leukemia), CML (Chronic Myeloid Leukemia), and AML (Acute Myeloid Leukemia).

ALL, unlike other forms of cancer, is very hard to detect because it doesn't cause any tumors. This cancer mainly affects the WBC s and then spreads to organs like the liver and spleen; if it is left untreated then it is very fatal. So the proposed automated system will make use of image processing and unsupervised machine learning techniques to segment the cancer cells. To classify the subtypes we use binary classifiers at each step. In that way, we can detect cancer at a very early stage and will be able to provide the needed treatments. Children are more prone to this type of cancer, so we can provide them with prior treatment.

2 Similar Systems

A similar system was proposed by Jyoti Rawat, Annapurna Singh. S. Bhadauria, Jitendra Virmani, J. S. Devgun [1], which classifies the blood cells into different forms of cancer cells using the Hybrid Hierarchical Classifiers. The same image dataset ALL-IDB was used for their purpose. Their system consisted of three modules Segmentation, Feature Extraction, and Classification Module. The blood cell images were processed before extracting the necessary features, as the dataset images contain irregularities and noises. After the image processing, the essential feature is extracted, in the feature extraction module. The final and the most important phase of the system is the classification module where the cells undergo three classifications to segment the cells into Healthy cells, L1, L2, and L3 forms of cells. Several algorithms like SVM, KNN, etc. were used and they calculated and compared the accuracies of the algorithms used.

The system proposed by Sriram Selvaraj and Bommannaraja Kanakaraj [4], used the Naïve Bayers algorithm to classify the lymphocytes. They took into account the following features for feature extraction, namely Shape, area, compactness, form factor, and perimeter. Besides these morphological features, they also noted the features such as entropy, energy, variance, and correlation. Naïve Bayesian algorithm was used to classify the cells into normal and abnormal lymphocytes. The specified algorithm was used because of its simplicity. Performance of the system is evaluated by calculating the performance metrics such as Accuracy, Sensitivity, and specificity from the confusion matrix.

3 Datasets

The dataset used in this system is the ALL-IDB1 and ALL-IDB 2. The datasets are private datasets that belong to Fabio Scotti [2], prior permissions were obtained for experimental purposes. They are images obtained from a powerful microscope and are focused on the area of interest i.e. the WBC. The WBCs are stained with Leiban stain so that they are visible and stand apart from the background. The dataset contains the lymphocyte elements labeled by expert oncologists. The ALL-IDB 1 dataset can be used for testing the segmentation capabilities of classifier system(s) and also for image processing. The ALL-IDB 2 dataset has been designed for testing the performances of the classification systems. The images of the ALL-IDB2 dataset are labeled "ImXYZ 0.jpg" if the central cell is a probable blast, and "ImXYZ 1.jpg" in the other cases.

4 Morphological Analysis of the Cells

The Acute Lymphoblastic cancer cells are classified according to the FAB and WHO standards for classification. In the proposed system, FAB (French American and British) classification will be used. According to FAB standards the cells are classified into L1, L2, and L3 cancer cells. Normally lymphocytes have e regular shape and a compact nucleus, with smooth and continuous edges. But in the case of lymphoblast, they exhibit irregularities in their shape. A detailed description of cancer has been given below:

- L1 - Blasts are small and homogeneous. The nuclei are circular and regular with little clefting and the cytoplasm lack vacuoles.
- L2 - blasts are large and heterogeneous. The cytoplasm size is variable and may contain vacuoles. Sometimes large nuclei may be present.
- L3 - blasts abundant in size and homogeneous. Oval/Circular shaped nucleus may be present, with a prominent nucleus. Vacuoles are prominent.

5 The Proposed System

The proposed system consists of three modules- namely segmentation, feature extraction, and classification module. The system uses image processing and machine learning algorithms to classify the cancer cells into L1, L2, and L3 cells. Different algorithms [1, 3, 4] were used in the past, Hybrid Hierarchical classifiers will be used in this system which will be an improvement over existing algorithms. Before classification, the image dataset must be cleaned. So to clean the dataset, we must process the image dataset. Image processing is the most crucial step in the proposed system, since the images may contain noise(s) that needs to be removed. Moreover, we need to obtain the binary image to extract the needed features. After the segmentation, comes the feature extraction phase, where we extract the features needed for classifying the cancer cells. They include features like the area, perimeter, texture details, chromatic and morphological features, etc. The final module is the classifier module where the cancer cells are clustered/classified into L1, L2, and L3 forms of cancer cells. Hybrid hierarchical classifiers will be used to cluster the cancer cells. First, they are classified into Healthy cells and cancer cells, then they are classified into L1 and other forms of cancer cells, and finally into l2 and L3 cells.

6 The Modules

- Segmentation module- this is the first module of the system. The noises and irregularities in the images must be removed before the classification phase. In this phase, several image processing techniques are used to clean the image dataset. First of all the image is converted into greyscale, then it undergoes 2D-order statistic filtering and histogram equalization. After that, the pre-processed image (image obtained after Histogram equalization), should be converted into a binary image. The complement of the image is obtained, and then it undergoes Otsu's thresholding to finally obtain a binary image. To obtain the nucleus, Fig. 2 we have to apply morphological opening on the binary image.

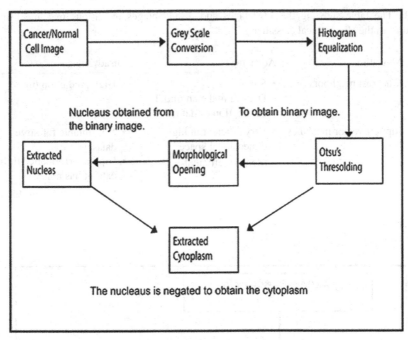

Fig. 2. The above diagram shows the flowchart of image preprocess and nucleus extraction. The Cancer cell undergoes grayscale conversion, histogram equalization to equalize the channels and pixel values, otsu's thresholding to convert the previous image into a binary image, and finally, the nucleus is extracted by applying morphological opening.

- Feature Extraction - From the binary image obtained from the above segmentation phase, we have to obtain the necessary features needed for classification. The essential features include the area, perimeter, chromatic, shape, and other morphological features. The irregularities in the overall shape of the cell and nucleus help to detect an anomaly.
- The classification module- is the final module of the system. This module consists of unsupervised machine learning algorithms to classify cancer cells. There are three classifiers which at first classify the cells into Healthy and cancerous cells, and then further classifies them into L1 and other forms of cancer cells, and finally into L2 and L3. The module makes use of hybrid hierarchical classifiers to classify the cells. Hierarchical classifiers are used since they have much better accuracy as compared to previous algorithms. Previous algorithms [1, 3, 4] used in the past showed variable accuracies and each of them has its pros and cons. The working of the classification module is shown in Fig. 3.

Table 1. The table showing the advantages and disadvantages of various machine learning algorithms in the Hierarchical classifiers.

S.no	Algorithm	Advantages	Disadvantages
1.	K nearest neighbors	• Simple • Doesn't make an initial assumption of data	• High prediction time
2.	Support vector machines	• Very efficient in high dimensional space • Memory efficient	• Not suitable for large datasets • Dip in performance if the dataset has noise
3.	Naïve Bayes	• Effective in multiclass prediction problems • Fast and easy to implement	• Vanishing value problem • No regression

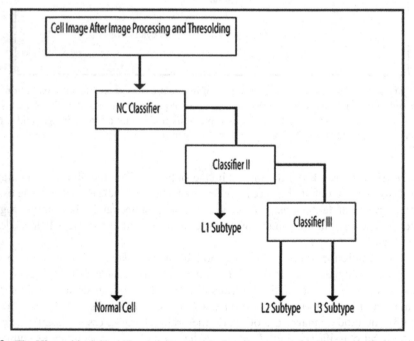

Fig. 3. The Hierarchical Classification System, At first the system classifies the cells into normal and cancerous, then into L1 and other subtypes, and finally L2 and L3.

7 Hybrid Hierarchical Classifiers

Hybrid Hierarchical Classification is an unsupervised machine learning classification algorithm that begins with the hierarchical classification of data. The classifiers map the data into subtypes and categories. A hybrid Hierarchical Classifier is usually used to solve multi-class problems, with the help of binary classifiers arranged as in the form of a tree

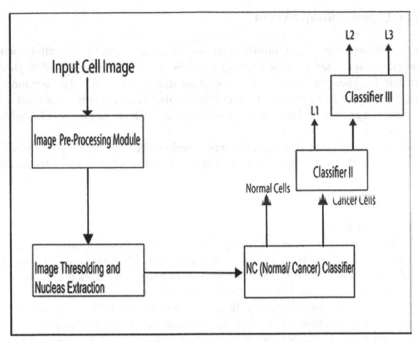

Fig. 4. The overall working of the Classification system, the image undergoes pre-processing and thresholding to obtaining the binary image finally undergoes hierarchical classification to check whether the cell is normal or is of type L1, L2, or L3.

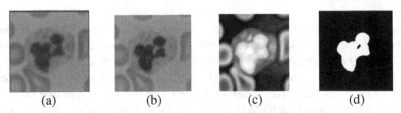

Fig. 5. (a, b, c, and d) Cropped image from ALL-IDB2 dataset(RGB) the images are obtained from a powerful microscope, Grayscale image, the inverted image after undergoing grey scale transformation, the binary image obtained after global thresholding.

in the space of different distinct classes/categories. The classification first begins at the low levels and gradually combines to give one final output. Hierarchical classifiers have a wide variety of applications, including Computer Vision. When compared to traditional classifiers; they offer much more accuracy and complexity. Features extracted from the system belong to certain distinct classes. The classification module consists of two modules namely – Classification module one, which segregates the healthy cells from the cancerous one, and The Classification module two which segregates the cancerous cells into L1, L2 and, L3 cells. For the development of classifiers for normal/cancer cells, five classifiers are arranged hierarchically at three levels.

8 The Classification System

- KNN- K Nearest Neighbor is one of the simplest Machine Learning algorithms, which is based on supervised machine learning. In KNN, we select the number of neighbors, which is K. Then we calculate the Euclidean distance between the K-number of neighbors. In the K- neighbors, we count the number of data points in each category. Then we assign the new data points to the category, in which the number of neighbors is maximum.
- SVM – Support Vector Machines, is a supervised machine learning algorithm, which is used for both regression and classification. The SVM classifier is a frontier that segregates the data into classes via a hyper lane. A Kernel is also used in the case where we have to classify non-linear data. Hyperplanes are partitions that help in the classification of data points, the points which fall into either side of the hyperplane are allotted to distinct classes.
- Naïve Bayes – they are a collection of algorithms, which is based on the Bayes theorem. It is a family of algorithms rather than a single algorithm. The Bayes theorem is used for calculation problems relating to conditional probability. It assumes that the presence of certain in a class is unrelated to the presence of any other feature. The Gaussian Naïve Bayes is an efficient and simple extension of the Naïve Bayes algorithm. The aforementioned algorithm only takes into account the mean and standard deviation of the training data. The other extension of Naïve Bayes is the Multinomial variant, which is used classification of discrete values/data.

9 Result Analysis

Multiple algorithms were implemented for comparing the accuracies and finding out the best among them. Out of the tested algorithms, KNN Fig. 5 and Naïve Bayes Multinomial got the highest accuracy of 60.8%. On testing the data with a confusion matrix, we got the following results; for KNN, SVM (Linear Kernel), and SVM (Gaussian Kernel) Seven True positives, zero true negatives, eleven false positives, and five false negatives. The accuracy values for SVM (Linear) and SVM (Gaussian) are 52.17 and 52.17 respectively. For Naïve Bayes (Gaussian), we got five true positives, two true negatives, eleven false positives, and five false negatives. For Naïve Bayes (Multinomial), we got five true positives, two true negatives, seven false positives, and nine false negatives. The accuracies for Naïve Bayes multinomial and Gaussian were 60.8 and 43.4 respectively. The accuracy measurement is the ratio of the sum of true positives and true negatives and the true negatives, true positives, false negatives, and false positives. Mathematically it can be expressed as:

$$Accuracy \% = (TN + TP/TN + TP + FP + FN) \times 100 \qquad (1)$$

Table 2. Remarks for the TP, FP, FN, TN regarding the cancer cell classifier.

S. no	Results	Remarks
1.	True positives (TP)	The system has identified the healthy cells correctly
2.	True negatives (TN)	The system has identified the cancerous cells correctly
3.	False positives (FP)	The system has incorrectly identified the normal cells
4.	False negatives (FN)	The system has incorrectly identified the cancerous cells

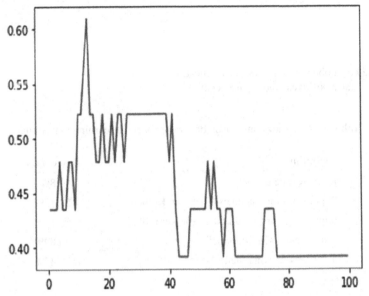

Fig. 6. Graph showing maximum accuracy obtained When k nearest neighbors are 13.

```
[[4 3]
 [8 8]]
              precision    recall  f1-score   support

        0.0       0.33      0.57      0.42         7
        1.0       0.73      0.50      0.59        16

   accuracy                          0.52        23
  macro avg       0.53      0.54      0.51        23
weighted avg      0.61      0.52      0.54        23
```

Fig. 7. The result obtained after SVM classification (Linear) Showing 4 True positives, 3 true negatives, 8 false positives, and 8 false negatives.

```
[[ 7  0]
 [11  5]]
              precision    recall  f1-score    support

        0.0      0.39       1.00      0.56          7
        1.0      1.00       0.31      0.48         16

   accuracy                           0.52         23
  macro avg      0.69       0.66      0.52         23
weighted avg     0.81       0.52      0.50         23
```

Fig. 8. The result obtained after SVM classification (Gaussian) Showing 7 True positives, 0 true negatives, 11 false positives, and 5 false negatives

Table 3. Accuracy comparison of various machine learning algorithms

S.no	Algorithm	Accuracy %
1	K Nearest neighbor	60.8%
2	Support vector machines (Gaussian Kernel)	52.17%
3	Support vector machines (Linear Kernel)	52.17%
4	Naïve bayers (Multinomial)	60.89%
5	Naïve bayers (Gaussian)	43.47%

10 Conclusions

Before automate testing, manual testing methods were used to segment the cancer cells. This was a painstaking and time-consuming process, so to tackle that problem automated systems were introduced. These systems used machine learning algorithms to classify the data. The proposed system will be an improvement over other existing algorithms, as it uses hybrid hierarchical classifiers to classify the cancer cells. The system will predict the cancer cells with much better accuracy and also help the medical practitioners and researchers to provide the patients, especially children, with prior treatment. The system also solves the problem of variable accuracies. This system is one of the many machine learning models to predict cancer cells. Since we're dealing with medical data, precision and high accuracy must be at our disposal. So, in the proposed system, deep learning techniques will be used to enhance performance and improve accuracy. For that purpose, Convolutional Neural Networks will be used, which will also include the subtype classification.

References

1. Rawat, J., Singh, A., Bhadauria, H.S., Virmani, J., Devgun, J.S.: Classification of acute lymphoblastic leukemia using hybrid hierarchical classifiers. Multimedia Tools Appl. Int. J. ISSN 1380-7501 (30 August 2016). Clerk Maxwell, J.: A Treatise on Electricity and Magnetism, 3rd ed., vol. 2, pp.68–73. Oxford, Clarendon (1892)
2. Donida Labati, R., Piuri, V., Scotti, F.: ALL-IDB: the acute lymphoblastic leukemia image database for image processing. In: Proceeding of the 2011 IEEE International Conference on Image Processing (ICIP 2011), Brussels, Belgium, September 11–14, pp. 2045–2048 (2011). ISBN: 978-1-4577-1302-6
3. Al-Jaboriy, S.S., Sjarif, N.N.A., Chuprat, S.: Segmentation and Detection of acute leukemia using image processing and machine learning techniques: a review, September 2019. In: K. Elissa, Title of Paper if Known (unpublished)
4. Selvaraj, S., Kanakaraj, B.: Näive Bayesian classifier for acute lymphocytic leukemia detection. In: Department of Biomedical Engineering, P.S.N.A College of Engineering and Technology, Dindigul, Tamilnadu, India 2K. P.R. Institute of Engineering and Technology, Coimbatore, India.
5. Bhuvana, D., Dr. Bhagavathi Sivakumar P.: Brain tumor detection and classification in MRI images using probabilistic neural networks. In: Proceedings of the Second International Conference on Emerging Research in Computing, Information, Communication, and Applications (ERICA-14) (2014)
6. Poorna, S.S., et al.: Computer vision aided study for melanoma detection: a deep learning versus conventional supervised learning approach. In: Advanced Computing and Intelligent Engineering, Advances in Intelligent Systems and Computing (2020)
7. Bhagya, T., Anand, K., Kanchana, D.S., Remya Ajai, A.S.: Analysis of image segmentation algorithms for the effective detection of leukemic cells. In: 2019 3rd International Conference on Trends in Electronics and Informatics (ICOEI) (2019)
8. Saiprasath, G.B., Babu, N., Arun Priyan, J., Vinaya kumar, R., Sowmya, V., Dr. Soman, K.P.: Performance comparison of machine learning algorithms for malaria detection using microscopic images. IJRAR19RP014 Int. J. Res. Anal. Rev. (IJRAR) (2019)
9. Bradley, J., Erickson, MD, Ph.D., Panagiotis Korfiatis, Ph.D., Zeynettin Akkus, Ph.D., Timothy L. Kline, Ph.D.: Machine Learning for Medical Imaging
10. Bahadure, N.B., Ray, A.K., Thethi, H.P.: Image analysis for mri based brain tumor detection and feature extraction using biologically inspired BWT and SVM. Int. J. Biomed. Imaging **2017**, 12 (2017)
11. Anilkumar, K.K. Manoj, V.J., Sagi, T.M.: A survey on image segmentation of blood and bone marrow smear images with emphasis to automated detection of leukemia. In: Biocybernetics and Biomedical Engineering, Elsevier, October–December 2020
12. https://homes.di.unimi.it/scotti/all/. Accessed 31 Jan 2021
13. http://www.nlm.nih.gov/medlineplus/leukemia.html. Accessed 31 Jan 2021
14. https://cbica.github.io/CaPTk/tr_Feature. Accessed 31 Jan 2021
15. http://www.ausrevista.com/26-2.1.html. Accessed 31 Jan 2021

Abnormal Blood Vessels Segmentation for Proliferative Diabetic Retinopathy Screening Using Convolutional Neural Network

Vasavi Agarwal, Ridhi Sipani, and P. Saranya(✉)

Department of Computer Science and Engineering, SRM Institute of Science and Technology,
Kattankulathur, Tamilnadu 603203, India
saranyap@srmist.edu.in

Abstract. Diabetic Retinopathy is an ailment that influences the eyes. It is brought about by uncontrolled diabetes. When Diabetic Retinopathy progresses to severe type, it is known as Proliferative Diabetic Retinopathy. It involves the growth of new, abnormal blood vessels. These blood vessels interrupt the normal fluid flow of the eye, which builds up pressure in the eye. This might damage the optic nerve, which may lead to a disease called glaucoma. The objective of the proposed model is to perform blood vessel segmentation and assist in detecting whether a patient has Proliferative Diabetic Retinopathy or not. At first, Image Pre-processing is performed on the retinal fundus images which will include Image Filtering, then Feature Extraction to extract blood vessels from the fundus images and finally various Convolution Neural Networks techniques are used to detect blood vessel segmentation thereby informing whether a person has Proliferative DR or not and achieved the maximum specificity of 97.55%.

Keywords: Blood vessels · Proliferative Diabetic Retinopathy · Fundus images · Dense net

1 Introduction

There are two sorts of Diabetic Retinopathy - Proliferative Diabetic Retinopathy (PDR) and Non-Proliferative Diabetic Retinopathy (NPDR). During Non-Proliferative DR, the walls of the vessels in the retina debilitate. Proliferative DR happens when Diabetic Retinopathy has progressed to serious kind. It primes to the progression of new abnormal blood vessels in the retina [1]. If this is not detected and prevented in earlier stages, it may lead to glaucoma. Diabetic Retinopathy as the name suggests is a diabetes complication that affects eyes. It has become one of the important cases of blindness and impairment worldwide. Automated diagnosis has great potential to detect DR. This will help to reduce the number of blindness caused due to DR [2]. The model emphases on helping in the detection of severe diabetic retinopathy. It is also called Proliferative DR. This occurs when the retina starts growing new blood vessels [3]. This process is called neovascularization.

© Springer Nature Switzerland AG 2021
M. Singh et al. (Eds.): ICACDS 2021, CCIS 1440, pp. 162–170, 2021.
https://doi.org/10.1007/978-3-030-81462-5_15

The objective of the proposed work is to take a step further towards automating the detection of severe DR, thereby making the process less exhausting and time consuming [4]. This model concentrates on the use of Image Processing techniques and CNN to detect whether a patient has Proliferative DR using retinal fundus images. The model is trained using DRIVE dataset and test the model using HRF datasets. Supervised methods like convoluted neural networks (CNN), support vector machines (SVM) and various other algorithms can be used to perform blood vessel segmentation [5]. A convolutional neural network helps in analysing data images. CNN is one of the most accurate and most widely used techniques to classify images and has been used for various such processes [6]. Various CNN layers of neural networks such as Conv2D layer, MaxPooling2D layer, ReLU layer has been used in the proposed model. It is very significant to detect which layer should be used and when as it drastically affects the accuracy of the model [7].

The expected outcome of this project is to be able to assist in the detection of whether the patient has severe DR or not, by producing blood vessel segmentation images when it is fed with retinal fundus images. This will help with doctors and patients to take prevention measures and take cautions to prevent further spread of the disease.

2 Related Works

Asra Momeni et al. [5] produced a model using Contrast Limited Adaptive Histogram Equalization, Convolution Neural Network and Efficient Net-B 5 architecture for the classification step. The model is trained on a mixture of two datasets and evaluated on the Messidor dataset. The advantage of this model is that when compared with state of art models, this model produced better AUC.

Shuang Yu et al. [3] help us with concepts of Optic Disc Detection, Pre-processing, Feature extraction includes extraction of vessel morphological and texture-based features, Vessel Segmentation and SVM. This model has a very fast speed of computation but it does not focus on Neovascularization Elsewhere (NVE) detection.

L. Ngo and J.-H. Han [7] resorted to a multi-level CNN model. They used max-resizing techniques. The model reduced the resolution of the input image. This increased the training course generalisation. The advantage of this model is that even though it is a simple model, it can be compared to complicated models.

Yun Jiang et al. [8] used a Dense Convolutional Neural Network structure named D-Net. This model outperformed some really efficient methods, such as N4-fields, U-Net and DRIU on the basis of accuracy, sensitivity, specificity and AUC ROC. The performance of the model is unstable because the network structure used is unbalanced and the feature extraction capability is relatively weak.

Mona Leeza and Humera Farooq [9] made use of extraction of visual features of each image. The advantage of this system is that the model uses a simple classifier which does not focuses on characterisation of lesions. However, the sensitivity and specificity obtained by the proposed model allows the model to be integrated by a CAD system for future enhancements of the model.

Rubina Sarki et al. [10] use CNN (Softmax) to produce results. This model provides complete information about Diabetic Eye Disease (DED) detection methods. The model can be enhanced in the future by developing stronger DL models and performing training on minimal data. The enhancements that can be brought about in this model is developing strong deep learning models, automating choosing the optimum values for deep learning architectures, training on minimal data and integrating deep learning, cloud computing and telehealth.

X. Zeng, H. Chen, Y. Luo and W. Ye [2] study CNN model focusses on Deep Learning method. This model beats the existing monocular model dependent on Inception V3. The proposed model will deal with issues while preparing or testing with other datasets in which matched fundus pictures are inaccessible.

Sheikh, Sarah and Qidwai, Uvais [6] use four different pre-trained models like DenseNet, VGGNet, ResNet and InceptionV3. The smartphone-based screening tools helps in generating plan for the patients thereby skipping the entire requirement of a health care provider to just classify the disease.

Wahid F.F., Raju G. [12] provide a comparison of two models. The first model has Deep CNN- based DR detection whereas the second model has CNN model for DR detection. The only drawback of this model is that it requires a large number of training samples in CNN.

S. Qummar et al. [1] uses a combination of five deep CNN model and performs optimization using SGD and Adam. Future enhancements of this model would involve using specific models for specific stages to increase the accuracy at earlier stages.

M. Mateen, J. Wen, M. Hassan, N. Nasrullah, S. Sun and S. Hayat [4] uses techniques like median filtering, homomorphism, AHE, average filtering and contrast adjustment. This technique proves the approach of deep learning to be more effective than traditional approach.

In the above models, although some of the models could be optimised in future by trial-and-error method to get better accuracy which is not a very efficient method. The performance of the model can be unstable when DCNN is used because the structure is modest and the feature extraction capability is relatively weak. Some of these models can be enhanced by developing stronger deep learning models and training on minimal data.

3   Proposed Model

The model starts from image pre-processing to remove noise and to enhance the contrast illumination etc. Then the image filtering is done to upgrade the specific highlights of the picture. Followed by the pre-processing and image filtering, feature extraction is done to extract the abnormal blood vessels from the image. Then the image segmentation is done using UNet and DenseNet architecture of the CNN is used for the further classification. The workflow of the proposed model is shown in Fig. 1.

3.1   Image Pre-processing

Image Pre-processing allows transformation of data before it is fed into the model. Normalization is performed to make sure all the features are in the scale. After the base size is set for all images, noise is removed. HE was used to remove noise and increase the quality of image. AHE proved to be a better technique to increase the quality of image

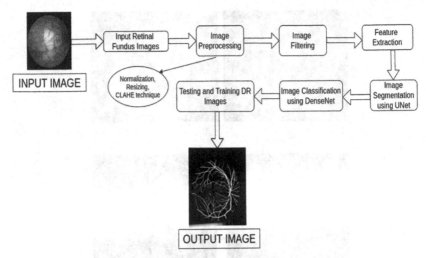

Fig. 1. Work-flow diagram of the model

[13, 14]. Finally, CLAHE proved to be the most efficient technique to eliminate noise and increase the quality of images. The size of the images obtained after pre-processing is 565 × 565 pixels. Figure 2 (a) indicates a retinal fundus grayscale image which is obtained from the DRIVE dataset. Figure 2 (b) indicates the image after HE has been applied to it. Figure 2 (c), shows the Intensity vs Frequency histogram for an original grayscale image and in Fig. 2. (d) It can be noticed the changes in the Intensity vs Frequency histogram after applying Adaptive Histogram Equalization. Figure 3 (a) and Fig. 3 (b) shows the difference in retinal fundus before and after applying CLAHE.

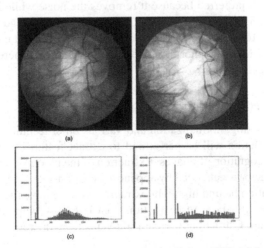

Fig. 2. (a) Original grayscale image, (b) Image after applying AHE, (c) Intensity vs Frequency histogram for original grayscale image, (d) Image after applying adaptive histogram equalization

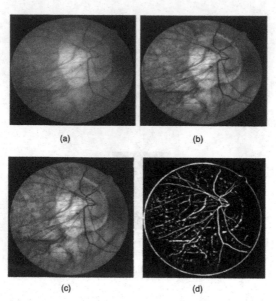

Fig. 3. (a) Original grayscale image, (b) Image after applying CLAHE, (c) Image before applying filters, (d) Image after applying filters

3.1.1 Filtering

Filtering is performed to enhance the images which have been fed to model by different datasets. It is done in order to remove or enhance certain features. Application of different filters such as Gaussian/Mean Filtering is used to enhance the image features. Gaussian Filter is used to blur the image or reduce noise furthermore [15, 16]. Median Filter also reduces noise but it is preferred because it removes the noise while keeping the edges sharp. Through Fig. 3 (c) and Fig. 3 (d), it can be noted why Image Filtering is a very important technique for blood vessel segmentation. In Fig. 3 (d) it can be seen that the use of Gaussian and Median Filter has helped us in enhancing thereby helping in the detection of neovascularization.

3.2 Feature Extraction

After performing resizing, normalization and various other techniques on the sampled images, feature extraction will be performed to extract features that will be used for classification and recognition of images. Extraction of the blood vessels from the retinal fundus. A feature vector is generated by appending every pixel value one after the other. The primary motivation behind highlight extraction is to remove the veins from retinal pictures. The image after Feature Extraction, that is Fig. 4, helps in recognizing the significance of Feature Extraction in separating veins from the retinal fundus pictures.

3.3 Convolution Neural Network

Convolution Neural Networks play a very important role in image classification and recognition considering that they provide a very high accuracy. CNN helps in extraction

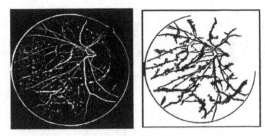

Fig. 4. Image before feature extraction vs after feature extraction

of vessels from the retinal fundus pictures correctly. A combination of two CNN architectures to extract all the features and produce maximum accuracy. DenseNet and UNet are the two models that we use in this model.

3.3.1 Image Segmentation

Image segmentation helps us in detecting the presence of Automatic Diabetic Retinopathy more accurately. Segmentation partitions the images into regions. UNet is one of the most successful CNN models for pixel-based image segmentation due to its architecture. UNet helps in producing accurate results even with limited data. In UNet architecture, a group of pixels are reduced to a single pixel [17]. These pooling layers help in increasing the accuracy of the output. The main feature of using UNet architecture is input size is equal to output size. The output size is therefore preserved. These features of UNet architecture also serve as the major reasons for using UNet for Image Segmentation.

3.3.2 Classification

In this paper, we use DenseNet architecture for classification. In DenseNet, each layer obtains additional input from all the already existing layers and passes on its feature maps to all the subsequent layers thereby making DenseNet the most suitable architecture for classification. In DenseNet, the identity function is directly added. It is one of the most efficient CNN architectures to calculate accuracy correctly. The Fig. 5 below shows a deep DenseNet architecture with three dense blocks. The layers between two contiguous dense blocks are alluded to as feature maps and transition layers sizes are changed with convolution and pooling layers.

4   Results and Discussions

In this model, DRIVE database images are used for training the model and HRF database fundus images are used for testing the model. This helped in achieving optimum accuracy. The categorical accuracy obtained after training the model is 97.32%, as in Fig. 6 (a). This helps in checking whether the index of the maximum true value is the same as the index of the maximum predicted value.

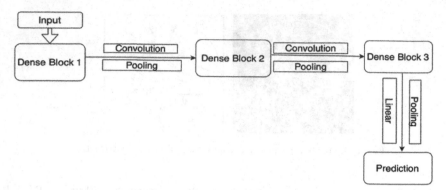

Fig. 5. Dense-Net architecture

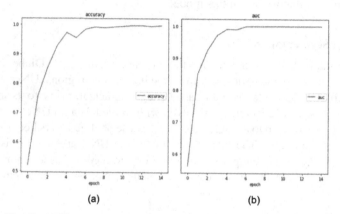

(a)   (b)

Fig. 6. (a) Shows the categorical accuracy, (b) shows the AUC graph

The validation loss was improved from 0.6023 to 0.5910. This indicates that no over-fitting takes place since regularization of the model is done during training. Table 1 shows the Accuracy and Specificity obtained after testing. The testing accuracy and specificity achieved by the proposed model is 83.30% and 97.55% respectively.

Table 1. Accuracy and specificity for HRF dataset

Dataset	Accuracy	Specificity
HRF	83.30%	97.55%

Table 2 shows the performance evaluation with existing models. It is evident that the proposed model has achieved the better sensitivity than the existing models. Hence the results achieved by the existing model is satisfactory.

Table 2. Performance evaluation with existing models

	Specificity
S. Yu et al. (2020) [3]	96.30%
R. Sarki et al. (2020) [10]	94.50%
D.A. Dharmawan et al. (2019) [11]	97.0%
Our Model (Dense – UNet)	97.55%

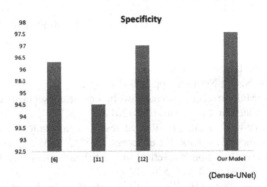

Fig. 7. Visual comparison with existing models

Figure 7 depicts the visual comparison of the results with the existing models. The work contributes towards developing an automated model to detect the proliferative diabetic retinopathy which can help the clinical specialists to predict the disease in a short time and the false positive rate.

5 Conclusion

In this paper, a very useful technique called CLAHE is presented to amplify retinal fundus images during image pre-processing. This technique helps in reducing any noise which might deviate our model from producing correct results. Feature extraction is performed to extract blood vessels from the retinal fundus images. Applications like OpenCV and Keras are used for this purpose. CNN models like DenseNet and UNet are used to extract blood vessel segmentation images from retinal fundus images. This helps in producing a higher accuracy while assisting in the detection of whether a patient has Proliferative DR or not. The model can be enhanced in future by using classification techniques like binary classification which would directly detect whether a person has Proliferative DR or not. The future work of this research will be to increase the accuracy obtained from the proposed model. This can be done by training various models with different images and then ensemble the outcome in order to increase the accuracy obtained.

References

1. Qummar, S., et al.: A deep learning ensemble approach for diabetic retinopathy detection. IEEE Access 7, 150530–150539 (2019)
2. Zeng, X., Chen, H., Luo, Y., Ye, W.: Automated diabetic retinopathy detection based on binocular Siamese-like convolutional neural network. IEEE Access 7, 30744–30753 (2019)
3. Yu, S., Xiao, D., Kanagasingam, Y.: Machine learning based automatic neovascularization detection on optic disc region. IEEE J. Biomed. Health Inform. 22(3), 886–894 (2018)
4. Mateen, M., Wen, J., Hassan, M., Nasrullah, N., Sun, S., Hayat, S.: Automatic detection of diabetic retinopathy: a review on datasets, methods and evaluation metrics. IEEE Access 8, 48784–48811 (2020)
5. Momeni Pour, A., Seyedarabi, H., Abbasi Jahromi, S.H., Javadzadeh, A.: Automatic detection and monitoring of diabetic retinopathy using efficient convolutional neural networks and contrast limited adaptive histogram equalization. IEEE Access 8, 136668–136673 (2020)
6. Sheikh, S., Qidwai, U.: Smartphone-Based Diabetic Retinopathy Severity Classification Using Convolution Neural Networks, pp. 469–481 (2021)
7. Ngo, L., Han, J.H.: Multi-level deep neural network for efficient segmentation of blood vessels in fundus images. Electron. Lett. 1096–1098 (2017)
8. Jiang, Y., Tan, N., Peng, T., Zhang, H.: Retinal vessels segmentation based on dilated multi-scale convolutional neural network. IEEE Access 7, 76342–76352 (2019)
9. Leeza, M., Farooq, H.: Detection of severity level of diabetic retinopathy using bag of features model. IET Comput. Vision 13, 523–530 (2019)
10. Sarki, R., Ahmed, K., Wang, H., Zhang, Y.: Automatic detection of diabetic eye disease through deep learning using fundus images: a survey. IEEE Access 8, 151133–151149 (2020)
11. Dharmawan, D.A., Li, D., Ng, B.P., Rahardja, S.: A new hybrid algorithm for retinal vessels segmentation on fundus images. IEEE Access 7, 41885–41896 (2019)
12. Wahid, F., Raju, G.: Diabetic retinopathy detection using convolutional neural network—a study. In: Jat, D.S., Shukla, S., Unal, A., Mishra, D.K. (eds.) Data Science and Security. LNNS, vol. 132, pp. 127–133. Springer, Singapore (2021). https://doi.org/10.1007/978-981-15-5309-7_13
13. Maison, M., Luthfi, A.: Retinal blood vessel segmentation using Gaussian filter. J. Phys. Conf. Ser. 1376, 012023 (2019)
14. Venkatachalam, K., Devipriya, A., Maniraj, J., Sivaram, M., Ambikapathy, A., Iraj, S.A.: A novel method of motor imagery classification using EEG signal. Artif. Intell. Med. 103 (2020)
15. Uma Maheswari, K.M., Pranesh, A., Govindarajan, S.: Network anomaly detector using machine learning. Int. J. Eng. Technol. 7, 178–179 (2018)
16. Samikannu, R., Ravi, R., Murugan, S., Diarra, B.: An efficient image analysis framework for the classification of glioma brain images using CNN approach. Comput. Mater. Continua 63, 1133–1142 (2020)
17. Pandian, A., Venkata Subramanian, J., Ramalingam, V.V., Uma Maheswari, K.M.: Authorship analysis for regional languages using machine learning approach. ARPN J. Eng. Appl. Sci. 11, 11244–11249 (2016)

Predictive Programmatic Classification Model to Improve Ad-Campaign Click Through Rate

Nisheel Saseendran[(✉)] and C. Sneha

Research and Development, MiQ Digital India, Bangalore 560001, KA, India
{nisheel.s,sneha}@miqdigital.com
http://www.wearemiq.com

Abstract. Programmatic advertising is a component of digital advertising where software is used to buy digital ad space in real time connecting advertiser to a specific consumer. Gaining more clicks and improving Click Through Rate (CTR) is considered as a goal for many of the programmatic advertising campaigns. The average CTR across the internet hovers around 0.2% but generally fluctuates somewhere between 0.1%–0.3% making clicks a rare event. Improving the CTR helps to manage the campaign budget efficiently. So, it's very important to get high CTR values for the campaign to be effective. In this paper we present a classification model that will predict whether the user will click an impression or not. Most of the state-of-the-art works in this area are focused on learning the feature interactions and capturing the user interest evolving process. As majority of the features in ad-techspace are categorical type and with high cardinality, it's very important to encode the features to best relate it with the class label and reduce the dimensionality. Less attention is being paid to the encoding part by most of the state-of-the-art works. The proposed method uses a custom feature encoding which is underpinned on class label distribution. A hybrid approach comprising class weight, stratified cross validation and probability thresholding is used for dealing with class imbalance. f1-score is set as the model evaluation metric and proposed encoding scheme showed a 68.55% test f1-score whereas the classical one hot encoding showed just 60.92% f1-score.

Keywords: Programmatic advertising · Real time bidding · CTR · Artificial intelligence · Machine learning · Digital marketing

1 Introduction

The digital advertising [1] landscape spans across diverse devices like TV, desktop, mobile, DOOH etc. and various mediums like email, SEO, social advertising [2], and programmatic advertising. Recently programmatic advertising [3] with RTB has gained more attention and became the central part of the ad tech space [4]. Programmatic advertising [5] is when software is used to connect brands with

© Springer Nature Switzerland AG 2021
M. Singh et al. (Eds.): ICACDS 2021, CCIS 1440, pp. 171–180, 2021.
https://doi.org/10.1007/978-3-030-81462-5_16

end users via real time bidding of ad slots [6]. Real time bidding (RTB) [7] is a server-to-server buying process that allows inventory (ad space on websites) to be bought and sold on a per-impression basis [8]. It happens instantaneous through an auction that determines who gets to buy a specific impression [9].

One of the most common goal of advertisers is to drive new users to their website through their advertising investments [10]. Impact of such ad campaigns is measured by the metric - Click Through Rate - CTR [11] which indicates that the user saw the ad and engaged with it [12]. Predicting the user will click or not will helps a lot to improve the CTR which has direct impact on the revenue and plays a key role in the ad-techspace.

Recently CTR improvement has gained more attention and many research works have been done on this area. Deep Interest Evolution Network (DIEN) [13] model was developed to capture users interest evolving process relative to the target item. This network also has the attention mechanism and auxiliary loss function added to strengthen the interest evolving process. A feature generation convolution neural network (FGCNN) [14] is developed with two major components. The first component is feature generation component that leverages CNN to generate the local patterns in the data. The second component is the deep classifier which adopts the IPNN structure to learn the feature interactions from the augmented feature space. A feature interaction graph neural network (Fi-GNN) [15] leveraging the representative power of the graphs was developed to improve CTR. In this model nodes represent the features and the edges between the nodes represent the interaction between the features. Fi-GNN is not only good in terms of learning sophisticated feature interactions but also provides an explanation for CTR.

A Field-Leveraged Embedding Network (FLEN) [16] for recommender systems was developed. This work discusses a field wise bi-interaction pooling method which captures both inter-field and intra-filed conjunction. This work also contributed a novel drop-out mechanism to address the problem of independent latent features co-adaption. A dynamic neural network (DNN) [17] was developed to deal with the two key challenges namely the dynamic nature of RTB and the user information scarcity. An interaction layer is put forward in the network to fuse the explicit user responses. This work also discusses and address some of the limitations faced in the big data ecosystem. Hybrid models consisting multiple components have also been developed to improve CTR. A fuzzy deep neural network [18] with three components was also developed to improve CTR. First component fuzzy Gaussian Bernoulli restricted Boltzmann machine is applied to the raw dataset followed by the second component fuzzy restricted Boltzmann machine to construct the fuzzy deep belief network. Finally, the third component fuzzy logistic regression model is used for the click prediction.

A Multiplex Target-Behavior Relation enhanced Network (MTBRN) [19] has been developed to improve the CTR. This work is based on leveraging multiplex relations between user behavior and the target item to improve the CTR. Multiplex relationship is composed of semantics which helps in grasping user's interest. To model the multiplex relations various graphs like item-item similarity graph and knowledge graphs have been used. For the feature encoding, Bi-LSTM is applied at each path in the feature extraction layer. A path activation and fusion network are devised to aggregate and learn the representation of all paths in the prediction model. A neural network based model called AutoFeature [20] has been proposed to improve the CTR. This network automatically finds essential feature interactions and select a nearly optimal structure to model the feature interactions. This work also contributed a search algorithm that recursively refine the search space by partitioning into sub search spaces. This model has an advantage that it uses fewer flops parameters, so it can incorporate the higher order feature interactions more efficiently. These works mainly exploit attention mechanism that uses the product embedding to encode the user interactions.

The state-of-the-art works mentioned in the literature for improving CTR lacks in paying much attention to the encoding of categorical features and the dimension of categorical features. Based on all observations on the limitations and gap in the works, a custom categorical feature encoding scheme has been proposed in this paper.

The rest of this paper is organized as follows. Section 2 describes the data preparation done and Sect. 3 describes the proposed encoding scheme. The results and validation are detailed and presented in Sect. 4. Section 5 concludes this paper with observations and limitations of the work which needs to be considered in further research on this topic.

2 Data Preparation

Each time your ad appears on a web page it's counted as one impression. Impressions are the number of times your content is displayed, no matter if it was clicked or not. The details of an impressions are recorded on impression feed data aka standard feed. For building the model, an impression feed data is used that has 22 columns which includes data about user identifier, creative and domain info, geo location info, operating system, browser, device info etc. A sample impression feed data is given on Table 1 and the description of each field is given on Table 2.

Table 1. Sample impression feed data

Gender	Dt	Width	Height	Fold_Position	Geo_Country	Geo_Region
M	2020-09-09 12:15:45	320	50	0	US	CA
F	2020-09-09 14:12:33	320	50	1	US	CA
F	2020-09-09 19:25:29	320	50	1	US	CA
M	2020-09-09 02:33:58	728	90	0	US	TX
F	2020-09-09 12:05:12	300	50	0	CA	ON
F	2020-09-09 08:19:54	320	50	1	US	CA
F	2020-09-09 01:40:22	300	250	0	CA	ON
M	2020-09-09 03:45:29	320	50	1	US	CA
F	2020-09-09 02:23:01	728	90	1	US	TX
F	2020-09-09 17:12:00	300	250	0	US	CA

Table 2. Description of fields of impression feed data

Feature	Description
Age	Age of the user to whom the ad was served
Gender	Gender of user to whom the ad was served
User_Id	Cookie id of the user to whom the ad was served
Dt	Time stamp at which the ad was shown
Width	Width of the creative
Height	Height of the creative
Fold_Position	Position of the creative on the page
Country	Country where the ad was shown
Region	Region where the ad was shown
City	City where the ad was shown
DMA	A parameters of geo location where the ad was shown
Postal_Code	postal code of location where the ad was shown
Device Type	Device type of the user to whom the ad was served
OS	Operating system of the user to whom the ad was served
Browser	Browser of the user to whom the ad was served
Seller	The ID of seller related to the ad served
Publisher_Id	The ID of publisher related to the ad served
Language	The language of the creative message
Click	Indicates the impression was clicked or not

A basic analysis of this dataset showed that the gender column had more than 97% missing values. Geo region field had 2% missing data. For handling the missing values for gender, missing itself is considered as another category. We have male, female and unknown categories. The missing values under geo region were substituted with the most frequent category of that feature while building the training dataset.

Most of the features are categorical in nature, so it's important to check the number of distinct values that each categorical column takes. Features like geo-city and postal code has high cardinality more than fifty thousand and these features are correlated with geo DMA, geo country and geo region which have relatively low cardinality. So, the features Geo city and postal code were dropped while building our dataset. These columns were dropped because of its high cardinality and also these columns are at too granular level.

The International Advertising Bureau (IAB) has set certain standard sizes for the creative of an impression. There are hardly 20 different sizes for the creative. So, we merged height and width as height x width and treated it as a categorical column. From the Dt column, we extracted the hour of the day as a feature and treated it as a numeric column. So, hour of the day and age are the only two numeric columns and all other features are categorical.

3   Proposed Encoding Scheme

Impression feed data set consists of both numeric and categorical features. Numeric features can be used as it is while building the model but the categorical ones. Categorical features must be encoded before we feed that into the model. Since majority of the features are categorical in nature, a proper encoding scheme is needed for it. If the encoding scheme doesn't contain enough information about the class label, the prediction may not be accurate and the CTR will drop.

In this paper a probability distribution-based encoding scheme is proposed for the categorical features. Proposed encoding scheme not only helps to improve the CTR values but also helps to rank the features based on its importance.

Predicting whether the user will click an ad or not is a binary classification problem. To better relate a category to the class label, category to target label relation needs to be captured. To encode a category, we check what percentage of its occurrence in the dataset belongs to click vs what percentage belongs to non-click. Say out of 100 records of a category 72 belongs to non-click and 28 belongs to click, then the category can be represented as a 2-dimensional value based on its probability distribution as [0.78, 0.28]. Figure 1 shows the probability distribution-based encoding for the category Windows 7 of the operating system feature.

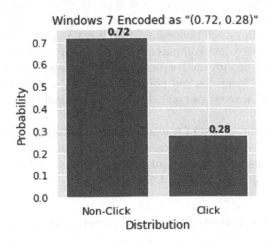

Fig. 1. Encoding of windows 7 category

Once the encoding is done, the next step is to identify the importance of each feature in predicting a class label. The proposed encoding scheme can be used to rank the features based on its importance. A categorical feature can be represented as the mean probability distribution of all its categories. To rank the features, the absolute difference between click and non-click percentage values of mean probability distribution can be considered. Higher value of this metric indicates that the feature is better in separating out click and non-click instances. Table 3 shows the top 10 features based on the absolute value difference between first two dimensions of the mean probability distribution.

Table 3. Top 10 features

Feature	Mean_Distribution	Distribution_Difference	#Distinct_Values
Device_Type	[0.81, 0.19]	0.63	6
Fold_Position	[0.75, 0.25]	0.49	3
Site_Domain	[0.74, 0.26]	0.49	9817
Publisher_Id	[0.73, 0.27]	0.46	6319
Seller_Member_Id	[0.72, 0.28]	0.44	718
Creative_Size	[0.71, 0.29]	0.42	20
Geo_Dma	[0.69, 0.31]	0.39	593
Gender	[0.69, 0.31]	0.38	3
Browser	[0.66, 0.34]	0.32	32
Language	[0.65, 0.35]	0.29	40

To account the rareness or frequency of a category the proposed scheme adds one more dimension, the overall percentage of the category to the 2-dimensional probability distribution-based encoding. Say if a particular category is present in 10% of the records when compared with all the available categories of that feature then we can denote the frequency of this category as 0.10. So, the final encoding becomes a 3-dimensional point as [0.78, 0.28, 0.10] for the windows 7 example. So, in this way we can encode each categorical value as 3-dimensional value.

We named first dimension of the encoding as feature_pos represents the distribution for click, feature_neg represents the distribution for non-click and feature_count represents the frequency or rareness of that feature. Table 4 summarizes the 3-dimensional encoding of OS feature.

Table 4. Encoded dimensions of operating system feature

Dimension#	Dimension name	Description
1	OS_Pos	Distribution of click class for the category
2	OS_Neg	Distribution of non-click class for the category
3	OS_Count	Frequency of the category

4   Results and Validation

In this paper, the model performance for the proposed encoding scheme is compared with the classical one hot encoding scheme. As click is a rare event and the dataset is highly imbalanced, f1-score is set as the performance evaluation metric. A random forest model is trained with the proposed encoding scheme and another random forest model is trained with classical one-hot encoding scheme. The f1-scores of these models are compared for validating the performance of proposed scheme. To handle the imbalance on the dataset a hybrid approach consisting of probability thresholding moving, stratified k fold cross validation and class weight scheme is used.

Figure 2 shows the test confusion matrix for one-hot encoding method and Fig. 3 shows the proposed encoding method confusion matrix From the results it's clear that the proposed encoding method performed well in terms of low false positives and low false negatives. The one-hot encoded version showed only 60.92% f1-score whereas the proposed encoding showed an f1-score of 68.55. The proposed scheme outperformed the one-hot encoding scheme by 7.63%.

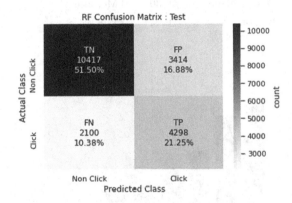

Fig. 2. One-hot encoding - model confusion matrix

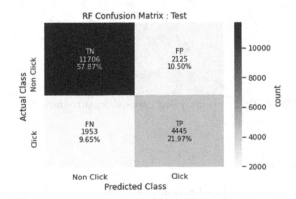

Fig. 3. Proposed encoding - model confusion matrix

Python 3 and sklearn libraries were used to build the random forest model. The important features identified by the model were extracted from the RF model. Figure 4 shows the important features picked up by the model for predicting the class label. 8 out of 10 top features picked up by the model are from the features which were ranked under top 10 by the proposed encoding scheme. A weight of evidence analysis (WOE) was also done on the raw categorical features and the results were compared with the custom encoding feature importance. WOE results with top 5 features and its information value (IV) is given on Table 5. From the results its observed that the proposed encoding method feature importance is well aligned with the WOE method results.

Table 5. Weight of evidence analysis

Feature	IV value	Predictive power
Site_Domain	0.95	Very strong
Publisher_Id	0.86	Very strong
Operating_System	0.52	Strong
Device_Type	0.30	Medium
Region	0.04	Weak

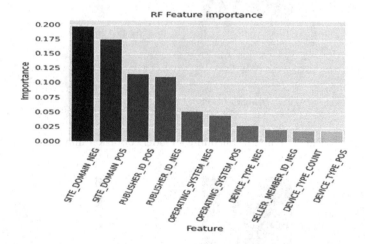

Fig. 4. Proposed encoding - model feature importance

5   Conclusion

In this paper a predictive classification model that helps to improve the CTR is presented. CTR measures the number of clicks advertisers receive on their ads per number of impressions. A high CTR means that a high percentage of people who see your ad click it. High CTR leads to high quality scores which helps to improve or maintain ad position for lower costs.

The proposed feature encoding scheme outperformed when compared with the classical one-hot encoding scheme by 7.63% f1-score that helps in managing the campaign budget efficiently.

There are some limitations to the proposed method that it cannot deal with features having very high cardinality and some features were dropped because it had very high cardinality. In continued research, we will also have to consider fraud/bot clicks which biases our data.

References

1. Raza, S.H., Zaman, U.: Effect of cultural distinctiveness and perception of digital advertising appeals on online purchase intention of clothing brands: moderation of gender egalitarianism. Information **12**, 72 (2021)
2. Winter, S., Maslowska, E., Vos, A.L.: The effects of trait-based personalization in social media advertising. Comput. Hum. Behav. **114**, 106525 (2021)
3. Samuel, A., White, G.R., Thomas, R., Jones, P.: Programmatic advertising: an exegesis of consumer concerns. Comput. Hum. Behav. **116**, 106657 (2021)
4. Golrezaei, N., Javanmard, A., Mirrokni, V.S.: Dynamic incentive-aware learning: robust pricing in contextual auctions. Oper. Res. **69**, 297–314 (2021)
5. Ding, Y., Luo, D., Xiang, H., Liu, W., Wang, Y.: Design and implementation of blockchain-based digital advertising media promotion system. Peer-to-Peer Network. Appl. **14**(2), 482–496 (2020). https://doi.org/10.1007/s12083-020-00984-5
6. Kingsley, S., Wang, C., Mikhalenko, A., Sinha, P., Kulkarni, C.: Auditing Digital Platforms for Discrimination in Economic Opportunity Advertising, CoRR, abs/2008.09656 (2020)
7. Mehta, S., Dawande, M., Janakiraman, G., Mookerjee, V.S.: Sustaining a good impression: mechanisms for selling partitioned impressions at Ad exchanges. Inf. Syst. Res. **31**, 126–147 (2020)
8. Gitomer, A., Oleinikov, P.V., Baum, L.M., Fowler, E.F., Shai, S.: Geographic impressions in Facebook political ads. Appl. Netw. Sci. **6**, 18 (2021)
9. Liu, M., Li, J., Zhengning, H., Liu, J., Nie, X.: A dynamic bidding strategy based on model-free reinforcement learning in display advertising. IEEE Access **8**, 213587–213601 (2020)
10. Liu, S., Yong, Yu.: Bid-aware active learning in real-time bidding for display advertising. IEEE Access **8**, 26561–26572 (2020)
11. Kim, K., Kwon, E., Park, J.: Deep user segment interest network modeling for click-through rate prediction of online advertising. IEEE Access **9**, 9812–9821 (2021)
12. Li, D., et al.: Attentive capsule network for click-through rate and conversion rate prediction in online advertising. Knowl. Based Syst. **211**, 106522 (2021)
13. Zhou, G., et al.: Deep Interest Evolution Network for Click-Through Rate Prediction, CoRR, abs/1809.03672 (2018)
14. Liu, B., Tang, R., Chen, Y., Yu, J., Guo, H., Zhang, Y.: Feature Generation by Convolutional Neural Network for Click-Through Rate Prediction, CoRR, abs/1904.04447 (2019)

15. Li, Z., Cui, Z., Wu, S., Zhang, X., Wang, L.: Fi-GNN: Modeling Feature Inter-actions via Graph Neural Networks for CTR Prediction, CoRR, abs/1910.05552 (2019)
16. Chen, W., Zhan, L., Ci, Y., Lin, C.: FLEN: Leveraging Field for Scalable CTR Prediction, CoRR, abs/1911.04690 (2019)
17. Qu, X., Li, L., Liu, X., Chen, R., Ge, Y., Choi, S.-H.: A dynamic neural network model for click-through rate prediction in real-time bidding. In: IEEE International Conference on Big Data, 978-1-7281-0858-2 (2019)
18. Jiang, Z., Gao, S., Li, M.: An improved advertising CTR prediction approach based on the fuzzy deep neural network. PLoS ONE **13**(5), e0190831 (2018). https://doi.org/10.1371/journal.pone.0190831. Accessed 24 March 2021
19. Feng, Y., et al.: MTBRN: Multiplex Target-Behavior Relation Enhanced Network for Click-Through Rate Prediction, pp. 2421–2428. ACM (2020)
20. Khawar, F., Hang, X., Tang, R., Liu, B., Li, Z., He, X.: AutoFeature: Searching for Feature Interactions and Their Architectures for Click-through Rate Prediction, pp. 625–634. ACM (2020)

Live Stream Processing Techniques to Assist Unmanned, Regulated Railway Crossings

Jacob John[✉], Mariam Varkey, and M. Selvi

School of Computer Science and Engineering, Vellore Institute of Technology, Tiruvalam Road, Vellore, Tamil Nadu 632014, India
{jacob.john2016,mariamsunil.varkey2016}@vitalum.ac.in,
selvi.m@vit.ac.in

Abstract. The Indian railways play a vital role in the lifestyle of a commuter. Serving close to 23 million passengers a day, it is considered quintessential in an Indian setting. However, due to track placements in rural and urban areas, they tend to threaten animals and human beings that traverse it. This paper alleviates the danger by suggesting a preventative measure that involves signaling the train engineer prior to an accident. Furthermore, it presents a novel method consisting of a microcontroller structure to detect and identify objects as they intersect the railway. Additionally, six state-of-the-art algorithms were evaluated, trained, and tested on the MS COCO dataset using robust metrics. Subsequently, this paper weighs inference times, processing frame rates, object recognition details, and localization accuracy to suggest an appropriate algorithm. This paper also integrates the object detection algorithm to form an end-to-end system to supplement the Indian railways.

Keywords: Object detection · Automatic warning system · Unmanned railway level crossings · Internet of Things · Deep learning · Computer vision

1 Introduction

The railway transportation system is an essential form of transportation. Based on the Indian Railway Catering and Tourism Corporation (IRCTC) [1], there are over 31800 operational level crossings in India, where about 18300 of them are manned while about 13500 of them are unmanned level crossings. Furthermore, the railway tracks alone spread across 64000 km in distance. Hence, they serve as a significant contributor to accidents. Over 60% of accidents take place on unmanned railway crossings, while 5% occur on manned level crossings. 63% of such individual accidents lead to death while 33% lead to fatal injuries making level crossings an extremely vulnerable locality. According to Indian Express [2], level crossing accidents in India increased by 20% in 2019. This paper focuses on accidents that occur in level crossings and proposes an end-to-end system for the same.

Object detection has been deemed one of the most challenging branches in the realm of computer vision. With advancements in technology, object detection algorithms have

© Springer Nature Switzerland AG 2021
M. Singh et al. (Eds.): ICACDS 2021, CCIS 1440, pp. 181–192, 2021.
https://doi.org/10.1007/978-3-030-81462-5_17

gained paramount importance in the upcoming years. Fields such as security, military, transportation, medicine, manufacturing, and home automation have adopted object detection algorithms to enhance and boost the performance of pre-existing systems. Progress in the past decade in Graphics Processing Units (GPUs) computing power, Central Processing Units (CPUs) computing power, computational algorithms, high-performance computing, and the development of deep neural networks have all contributed to the development of these algorithms. Needless to say, the availability of larger data sets such as COCO, PASCAL VOC, DeepFashion2, and PanNuke has also significantly impacted the research and development in this field. Moreover, the introduction of benchmarks for detecting objects in images such as MS COCO, KITI, Caltech, Open Images V4, and Changedetection.net have streamlined and inspired research innovations.

2 Related Work

While unmanned level crossings are unsafe, human errors have caused accidents at manned level crossings as well. Accidents at railway crossings contribute to almost half of all train accidents and more than half of all train fatalities [3].

Authors of [4–6] automated the railway gate using sensors and various other components to detect an incoming train. [4] and [6] additionally provided an audio-visual warning to the commuters about the approaching train. In [6], the data can be collected to analyze traffic patterns and model an algorithm to help prevent accidents. The methodology in [7] focuses on grade crossings and trespassing in near-miss accidents. The authors presented an algorithm to detect trespassing using surveillance video footage, computer vision, and AI.

An earlier system had automated railway crossings based on the timings of the arriving trains. In [8–11], similar systems have been proposed that combine automating the gate with an obstacle detection system for detecting obstacles such as vehicles, people, or animals at the level crossing. Time, manpower, and possibility of errors are reduced with these fully automated systems. In [12], the authors review existing systems and suggest a novel one that can alert a person even if he/she is using headphones or is indulged in any application on their phone. This system must detect an incoming train and measure its distance by its horn, even in noisy environments.

A multi-sensor detection system is proposed in [13]. Optical passive sensors (video cameras) and optical active sensors (Light Detection and Ranging or LIDAR) are used. The system fuses active and passive optical sensors along with a railway track database to provide good obstacle detection efficiency. The authors of [14] present a technique to detect static and moving objects at railway crossings that work well in outdoor environments. The method involves three steps–initially, three non-parametric background models with different learning rates are created to detect static and moving objects. To decrease the number of false positives, an object tracking strategy is introduced. The last step involves a feedback mechanism to update the background models selectively when static objects are removed.

The authors of [15] suggest a model using RFID readers and RFID tags to prevent subway train accidents. In this system, the train is equipped with an RFID reader and a control circuit with a microcontroller. RFID active tags are placed at several hazardous points in the route. RFID tags for commuters can be placed in their phones, wristbands, or watches. Three algorithms introduced in this paper determine what and how much adjustment needs to be made by the control circuit to reduce the train speed. A unique approach is taken in [16]. An Anti-Collision System (ACS) for the train that detects "obstacles in front of it" is implemented. If set in automatic mode, the emergency brake is pressed upon seeing an obstacle. In manual mode, the train driver is warned about cars, animals, personnel, and other such hindrances.

3 System Architecture

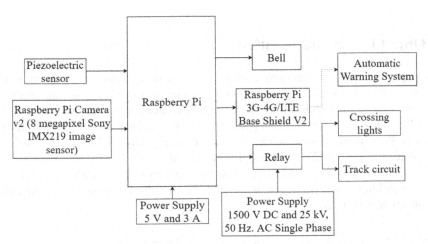

Fig. 1. A Raspberry Pi based architecture displaying how the modules are interconnected. Power supply details are based on the Indian railway guidelines published online [17].

As displayed in Fig. 1, the Raspberry Pi runs Raspbian (a native Linux-based OS) and contains GPIO (General Purpose Input/Output) pins that allow users to control electronic components for physical computing and Internet of Things (IoT) applications. The Pi will be onboarded with a pre-trained neural network that will be used to analyze the live stream from the camera. The neural network is the principal software used for object detection. The Raspberry Pi Camera v2 is a custom-built, high-quality 8-megapixel Sony IMX219 image sensor fabricated as an add-on board with a fixed-focus lens for Raspberry Pi. This lens is capable of supporting video resolutions up to 1080p at 30 fps.

The piezoelectric sensors utilize the piezoelectric effect to measure changes in pressure, acceleration, temperature, strain, or force by converting them to an electrical charge. The piezoelectric sensors can wake up a Pi from a low power state if it detects an object crossing the track. This arrangement enables the Pi to save power when operating in a battery mode. Additionally, Raspberry Pi 3G-4G/LTE Base Shield V2 is a cellular add-on board for Raspberry Pi. It can operate at a data rate up to 100 Mbps and 50 Mbps, depending on the installed cellular module. The LTE Base Shield provides the system with the capability to warn the train's Automatic Warning System. This module also allows the board to send a dying gasp signal in case of a power failure, enabling the railway board to take necessary precautions proactively.

The relay is an electromagnetic system that conveys messages electrically from one circuit to another via a series of contacts (back or front contacts). This component enables the Pi to communicate with the bell and crossing lights and powers them on in case of an object crossing the railway track. Furthermore, the Automatic Warning System provides a train driver with an audible alert and visual reminder that they are approaching a distant signal. As per the railways' guidelines, its service was subsequently expanded to include warnings for a reduction in allowable speed and temporary or emergency speed limitation. Finally, the track circuit is an electrical system that can identify whether a train is present on the rail tracks. It is used to warn commuters and control the necessary signals.

4 Object Detection Algorithms

This paper segregates pre-existing neural network architectures into three broad categories – one-stage object detectors, two-stage object detectors, and state-of-the-art detectors. One-stage detectors can achieve high inference speed, while two-stage detectors, albeit slower, benefit from improved object recognition and higher localization accuracy. Fully convolutional architectures are typically adopted by several novel one-stage detectors [18–21, 27]. Outputs of all three categories of networks are bounding boxes, spatially positioned and centered around an anchor point with their classification probabilities; some academic works prefer to use the confidence level, i.e., class probability $\times$ IOU [22].

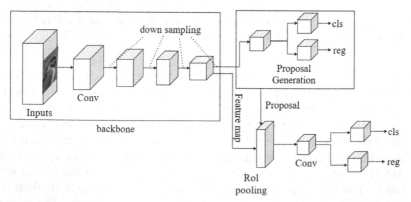

Fig. 2. The typical pipeline of a two-stage detector. *Reg* is the output from the box-regression layer, while *cls* is the output from the box-classification layer. *Conv* represents convolution layers.

Stages in two-stage object detectors are divided as per their Region of Interest (RoI) pooling layers. Additionally, the pipeline is more intricate and complex than single-stage detectors [23–25]. As displayed in Fig. 2, the first stage is to use a Regional Proposal

Network (RPN) to filter out regions with high probabilities. Finally, an unmanned, regulated railway crossing would require a time-efficient algorithm for real-time prediction from a continuous input stream. This paper can therefore hypothesize that in such a scenario, a single shot detector would be ideal. However, if the need is to operate reliably, a two-shot detector can be proposed instead.

4.1 Single-Shot Detectors for High Inference Speeds

YOLO. You Only Look Once (YOLO) [18] is built based on the Faster R-CNN architecture. YOLO's principal selling point is its capability to detect full-images and videos in real-time. However, one shortcoming YOLO presents is its ability to predict fewer than 100 bounding boxes per image. The pipeline can extract class probabilities and bounding boxes directly from input images.

YOLOv2. This algorithm [19] aims to build upon the pre-existing YOLO design by focusing on speed and precision improvements. YOLOv2's new backbone architecture, Darknet-19, employs five max-pooling layers and nineteen convolution layers. Useful priors for training set bounding boxes are automatically generated using K-means clustering. Furthermore, to improve YOLO's robustness in terms of image sizes, a new image dimension size is chosen randomly during every 10th batch of training. Finally, adjacent features from higher resolution feature maps are concatenated and stacked onto low-resolution feature maps into distinct channels, similar to that in ResNet [37].

YOLOv3. This algorithm [20] utilizes a novel backbone inspired by ResNet called Darknet-53. YOLOv3 possesses the capacity to process overlapping labels due to the use of independent logistic classifiers for multi-label classification. Medium and large size object detection is where this algorithm performs comparatively poorly. However, YOLOv3's new multi-label classifier allows it to detect smaller objects well. Therefore, YOLOv3 would be adequately fitted for camera angles that are further away from the object of interest.

4.2 Two-Stage Detectors for High Localization and Object Recognition Accuracy

Faster R-CNN. Faster Regional Proposal Network (RPN) based Convolution Neural Network [25] is an algorithm based on Fast R-CNN [26]. Faster R-CNN works with a boarder range of image scales and enables almost cost-free region proposals via a novel RPN that administers complete convolution features of images with the detection network. Multi-scale anchors are employed as references to detect objects of different sizes. Additionally, anchor-based connections are translation-invariant, eliminating the need for multiple scales of input images.

Mask R-CNN. This algorithm by He et al. [24] proposes minor overhead to Faster R-CNN by introducing a branch for predicting an object's mask. Despite trading off time, mask R-CNN improves accuracy by utilizing ResNet-FPN with Faster R-CNN. RoI features are extracted from various feature pyramid levels as per their scale using a backbone feature pyramid network. In the feature pyramid network (FPN), there are lateral connections among bottom-up pathways and top-down pathways. Features from both pathways are merged using element-wise addition. Small objects are detected using high-resolution feature maps, whereas lower-resolution feature maps are used for larger objects.

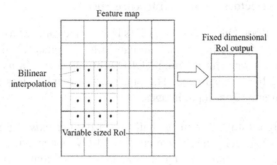

Fig. 3. The RoIAlign produces a fixed dimensional RoI output via sampling four regular spaced bins using bilinear interpolation.

The Mask R-CNN also uses a RoIAlign (Region of Interest Align) [27] operation rather than the conventional RoI pooling operation to improve accuracy. In RoI Pooling, misalignments ensue among the RoI bins and the extracted features, which deteriorate the predicted pixel-accurate masks. In RoIAlign, as displayed in Fig. 3, input features' exact values are calculated using bilinear interpolation sampled regularly in each RoI bin at four spaced locations.

4.3   State-of-the-art Detectors

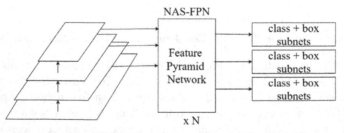

Fig. 4. The architecture of the NAS-FPN algorithm allows it to iterate N times over the search space of FPNs to capture the best hyperparameters.

NAS FPN. This state-of-the-art algorithm is proposed by Ghiasi et al. [21], authors from the Google Brain. An FPN [23] is presented with a novel architecture for object detection. Features are fused across a variety of different scales using both top-down and bottom-up connections. As illustrated in Fig. 4, NAS-FPN aims to build feature pyramid representations by iterating N times over a search space of FPNs. A reinforcement learning methodology called the Neural Architecture Search framework [28] is adopted to find hyperparameters for the backbone RetinaNet [29].

5 Results and Discussion

5.1 Evaluation Metrics

The PASCAL VOC dataset comprises about 11000 images spread across 20 different object classes [30]. Out of these 20 classes, three generalized categories–animals, people, and vehicles-play a significant role in this experiment's use case of detecting people crossing the railroad [31]. VOC2007 [32] uses novel criteria for evaluation that penalizes an algorithm for duplicate detection of instances, missing object instances, and false-positive detections. Two crucial metrics are utilized. Precision evaluates the number of correct predictions made by the algorithm out of the examples that belong to the positive class. Recall calculates the correct class predictions made by the algorithm out of all examples in the positive class in the dataset. Finally, the indicator function is defined as $[\phi(\beta_p, \beta_{gt}) \geq T]$ with a threshold T set as 0.5 in the VOC metric.

The intersection over union (IoU) or $\phi(\beta_p, \beta_{gt})$ is defined as the ratio of the area between the intersection of the bounding box of the predicted class (β_p) and the ground-truth class (β_{gt}) over the union of the bounding box of the predicted class (β_p) and the ground-truth class (β_{gt}). The precision-recall curves pose a challenge in measuring AUC as they are often a zig-zag pattern. Analytically taking the integral of this curve leads to inaccurate measurements. Two approaches can be presented to eliminate the zig-zag pattern before determining the AUC–an 11-point interpolation and an all-point interpolation [33]. In an 11-point interpolation estimate, a set of 11 equally spaced levels are expressed.

$$AP_{11} = \frac{1}{11} \sum_{R \in \{0.0, 0.1, 0.2..., 0.9, 1.0\}} Precision_{interp}(R) \tag{1}$$

$$Precision_{interp}(R) = \max_{\tilde{R}:\tilde{R} \geq R} Precision_{interp}(\tilde{R}) \tag{2}$$

In all-point interpolation, we interpolate through all the points rather than interpolating through 11 equally spaced points:

$$AP_{all} = \sum_{n} (R_{n+1} - R_n) \cdot Precision_{interp}(R_{n+1}) \tag{3}$$

$$Precision_{interp}(R_{n+1}) = \max_{\tilde{R}:\tilde{R} \geq R_{n+1}} Precision_{interp}(\tilde{R}) \tag{4}$$

The AP is determined for objects of three different sizes as a means to evaluate it across scales. Small size is considered as an area less than 322 pixels and represented as AP_S. The medium size has a pixel area between 322 and 962 pixels and is expressed as AP_M. Large size has greater than 962 pixels and is interpreted as AP_L. Furthermore, different IoUs are used to estimate the AP using the precision and recall metrics as defined in the preceding sections. AP_{50} and AP_{75} are used in this paper, indicating an IoU threshold of 50% and 75%, respectively. Finally, the mean Average Precision or mAP is the average AP over each class of the dataset.

This paper trains and tests the algorithms using the MS COCO (Microsoft Common Objects in Context) dataset. MS COCO contains objects in their natural environments [34], which is ideal for the use case of railway crossing object detection.

5.2 Existing Results and Discussion

YOLO makes significantly fewer false positives than Faster R-CNN, however, at the expense of accuracy. YOLO's main component of prediction error was localization error. The PASCAL VOC dataset resulted in a 63.4% mAP at 45 fps. Fast R-CNN had a higher mAP of 70.0% at 0.5 fps, while Faster R-CNN produced an mAP of 73.2% at 7 fps. YOLOv2 boasts a better mAP than Fast R-CNN and Faster R-CNN, which have an mAP of 70.0% at 0.5 fps and 76.4% at 5 fps, respectively. Furthermore, SSD500 has an mAP of 76.8% at 19 fps. This mAP and fps rate exhibit YOLOv2's effective processing rate and extraordinary detecting precision compared to some two-stage detectors.

Using MS COCO metrics, RetinaNet attains an mAP of 61.1%, YOLOv3 achieves an mAP of 57.9%, and DSSD514 [38] had an mAP of 53.3%, at IOU 0.5 or AP50. YOLOv3 had an AP of 33%, while DSSD513 delivered 33.2%. Additionally, DSSD513 is approximately three times slower than YOLOv3. However, this paper noted that RetinaNet performed significantly better than both algorithms, with an AP of 40.8%. NAS-FPN achieved an mAP of 44.9% on the MS COCO test-dev dataset, while the original FPN had an mAP of 42.1% while adopting ResNet-50 as the backbone. Results show that NAS-FPN improved the original FPN's mAP by 2.9%, with a feature dimension of 256. On the COCO test-dev dataset, this exemplary configuration achieves an mAP of 48.3%.

Fast R-CNN obtained an mAP of 66.9%, while Faster R-CNN performed an mAP of 69.9% on the PASCAL VOC 2007 test set. This algorithm also had an improved processing rate of 5 fps compared to the 0.5 fps of Fast R-CNN, using the same VGG-16 backbone [35]. Finally, Mask R-CNN with a backbone of ResNeXt-101-FPN scored an AP of 37.1 on the MS COCO dataset. It also outperforms the complex FCIS++++ OHEM [36], an algorithm based on the 2016 COCO segmentation winner challenge.

5.3 Experiment Results

NAS-FPN performs significantly better than even the two-stage object detectors as according to Table 1. NAS-FPN with the AmoebaNet backbone (stacked with 7 FPNs, DropBlock, and with a feature dimension of 384) had the highest AP of 47.9 out of all the tested algorithms. This high inference accuracy comes at the tradeoff of a reduced processing FPS. Mask R-CNN has the best AP for larger objects due to its lower-resolution feature maps. NAS-FPN is marginally worse than the Mask R-CNN in identifying objects with more robust semantics. YOLOv2 achieves the highest frame processing rate of 65.9 fps. For the use case of railway crossing object detection, a picamera operating at

30fps would be sufficient. Based on the proposed experiment using the Raspberry Pi 4, the picamera yields an fps of 15 on average. However, by dedicating a separate I/O thread, the processing rate can be extended by 250% or 37.5 fps. However, lower frame rates help preserve battery life in cases where power is not available. Hence, NAS-FPN's tradeoff would be negligible in this experiment's scenario.

Table 1. The detection results based on MS COCO metrics and the trainval35k dataset with RetinaNet as a reference

Model	Backbone/method	AP	AP_{50}	AP_{75}	AP_S	AP_M	AP_L	FPS
RetinaNet500	ResNet-101	34.4	53.1	36.8	14.7	38.5	49.1	25.1
RetinaNet800	ResNet-101-FPN	39.1	59.1	42.3	21.0	42.7	50.2	24.3
YOLO	-	15.4	32.7	15.6	4.0	16.3	27.2	46.3
YOLOv2 544	DarkNet-19	20.5	43.5	20.4	5.2	21.8	36.3	**65.9**
YOLOv3 608	DarkNet-53	34.2	56.4	35.2	19.1	34.9	42.1	38.3
NAS-FPN	RetinaNet	44.9	65.2	47.1	23.9	42.9	49.5	25.4
NAS-FPN	AmoebaNet	**47.9**	**69.5**	**49.3**	**25.6**	**45.9**	51.8	24.6
Faster R-CNN	VGG-16	22.3	42.1	32.4	12.5	34.1	47.6	8.1
Mask R-CNN	ResNeXt101-FPN	36.8	59.3	39.7	17.2	40.5	**53.8**	5.0

Fig. 5. Comparison of object detection using (a) Mask R-CNN (*top left*), (b) NAS-FPN (*top right*), (c) YOLOv3 (*bottom right*), and (d) YOLOv2 (*bottom left*). The image displays people crossing a railway line via the zebra crossing.

Figure 5 draws a comparison across the well-performing algorithms. The image displays a group of pedestrians being escorted across a standard railway crossing. This image serves as a valuable measure of how well each algorithm performs. It must be noted that this paper uses a confidence threshold of 0.55. YOLOv2 performed the worst as it missed identifying several individuals on track, especially those immediately in front of the train. However, it had the lowest inference time out of the four. Three out of the four algorithms misidentified the train as a bus. However, these false positives can be ignored in this experiment's use case as this paper focuses more on human beings and animals. YOLOv3 identified more commuters than YOLOv2, but not all users in the foreground of the train. NAS-FPN took the second least amount of time for inference. Furthermore, it was the only algorithm that accurately recognized the train while also distinguishing commuters in the train through the window. However, it still missed out on one individual in front of the train, which Mask R-CNN detected.

6 Conclusion and Future Work

This paper has discussed six different state-of-the-art object detection algorithms and their applicability in a level crossing system. This paper assessed and analyzed three single-stage detectors–YOLO, YOLOv2, and YOLOv3, two two-stage detectors–Faster R-CNN and Mask R-CNN, and one state-of-the-art detector–NAS-FPN. This paper evaluated these algorithms based on the metrics provided by the MS COCO dataset. Apart from the localization accuracy and object detection quality, this paper also looked at the inference time and processing fps. Such an evaluation enabled this research to understand that NAS-FPN would be a good fit for the proposed system.

This paper integrated the pre-trained NAS-FPN algorithm with the proposed system architecture to further understand its performance. Despite yielding reliable metrics during testing, it operates at a lower FPS than algorithms such as YOLOv3. However, this paper concluded that an FPS of 25 would be sufficient in the initially proposed use case. Furthermore, NAS-FPN is not as effective as Mask R-CNN at detecting larger objects due to its substandard low-resolution feature extraction. Needless to say, the proposed system would also bode well with existing automatic warning systems due to its wireless capabilities. Furthermore, the step-to-wakeup system allows the microcontroller to be in a low power state until activated, allowing for longevity on battery-operated power. Therefore, this setup can be incorporated with pre-existing level crossings and be utilized as a means to save lives.

Some challenges during moving object detection include video defocus and motion blur that result in false positives. As an extension of this research, this paper proposes leveraging spatio-temporal contexts in order to mitigate this issue. Furthermore, group behaviors of moving objects, such as those shown in Fig. 5, can also be utilized to empower detection results further. Intra-image contexts can be employed by linking frames and then conjecturing movements of objects in neighboring frames. When objects move, their overlap between adjacent frames would be weaker, and this behavior can be exploited.

References

1. Jain, S., Kumar, A.: Level crossing scenario of Indian railways. International Railway Safety Council (2017). http://international-railway-safety-council.com/wp-content/uploads/2017/09/jain-kumar-level-crossings-scenario-of-indian-railways.pdf

2. Indian Express, Level crossing accidents up 20% in 2019 (2020). https://indianexpress.com/article/india/level-crossing-accidents-up-20-in-2019-ncrb-6579574/
3. Singhal, V., et al.: Artificial intelligence enabled road vehicle-train collision risk assessment framework for unmanned railway level crossings. IEEE Access **8**, 113790–113806 (2020)
4. Ramkumar, M.S.: Unmanned automated railway level crossing system using zigbee. Int. J. Electron. Eng. Res. (IJEER) **9**, 1361–1371 (2017)
5. Reddy, E.A., Kavati, I., Rao, K.S., Kumar, G.K.: A secure railway crossing system using IoT. In: 2nd International Conference of Electronics, Communication and Aerospace Technology (ICECA), Coimbatore, India. IEEE, pp. 196–199 (2017)
6. Sharad, S., Sivakumar, P.B., Ananthanarayanan, V.: An automated system to mitigate loss of life at unmanned level crossings. Procedia. Comput. Sci. **92**, 404–409 (2016)
7. Zhang, Z., Trivedi, C., Liu, X.: Automated detection of grade-crossing-trespassing near misses based on computer vision analysis of surveillance video data. Saf. Sci. **110**, 276–285 (2018)
8. Athavale, C., Athavale, M.: Obstacle detection & gate automation at railway crossings. Int. J. Multidiscip Res Sci. Eng. Technol. (IJMRSET) **2**, 422–428 (2019)
9. Shetty, R., Patel, P., Sampat, A., Shukla, S., Singh, A.K., Deshmukh, P.: Automated railway crossing and obstacle detection. In: Somaiya, K.J. (ed.) 2nd International Conference on Advances in Science & Technology (ICAST), Mumbai, India. Institute of Engineering and Information Technology (2019)
10. John, J., Varkey, M.S., Selvi, M.: Security attacks in S-WBANs on IoT based healthcare applications. Int. J. Innovat. Technol. Explor. Eng. (IJITEE) **9**, 2088–2097 (2019)
11. Thangaramya, K., Kulothungan, K., Logambigai, R., Selvi, M., Ganapathy, S., Kannan, A.: Energy aware cluster and neuro-fuzzy based routing algorithm for wireless sensor networks in IoT. Comput. Network **151**, 211–223 (2019)
12. Gakkhar, S., Panchal, B.: A review on accident prevention methods at railway line crossings. Int. Res. J. Eng. Technol. (IRJET) **5**, 1102–1107 (2018)
13. Mockel, S., Scherer, F., Schuster, P.F.: Multi-sensor obstacle detection on railway tracks. In: IEEE IV2003 Intelligent Vehicles Symposium. Proceedings (Cat. No. 03TH8683), Columbus, Ohio. IEEE, pp. 42–46 (2003)
14. Cai, N., Chen, H., Li, Y., Peng, Y.: Intrusion detection and tracking at railway crossing. In: Proceedings of the 2019 International Conference on Artificial Intelligence and Advanced Manufacturing, Dublin, Ireland. ACM, pp. 1–6 (2019)
15. Sahba, F., Sahba, R.: Prevention of metro rail accidents and incidents in stations using RFID technology. In: 2018 World Automation Congress (WAC), Stevenson, Washington, USA. IEEE, pp. 1–5 (2018)
16. Majumder, N.S., Hossain, M.S., Abdullah, D.M.: Collision Object Detection and Prevention of Train Accident Dynamically by Using Ultrasound and Embedded System (2017)
17. International Railways Civil Engineering Portal. Regulations for Power Line Crossings of Railway Tracks (1987). https://ircep.gov.in/WLRMS/Regulation.pdf
18. Redmon, J., Divvala, S., Girshick, R., Farhadi, A.: You only look once: unified, real-time object detection. In: Proceedings of the IEEE Conference on Computer Vision and Pattern Recognition, Las Vegas, Nevada. IEEE, pp. 779–788 (2016)
19. Redmon, J., Farhadi, A.: YOLO9000: better, faster, stronger. In: Proceedings of the IEEE Conference on Computer Vision and Pattern Recognition, Honolulu, Hawaii. IEEE, pp. 7263–7271 (2017)
20. Redmon, J., Farhadi, A.: Yolov3: an incremental improvement. arXiv preprint arXiv:1804.02767 (2018)
21. Ghiasi, G., Lin, T.Y., Le, Q.V.: NAS-FPN: learning scalable feature pyramid architecture for object detection. In: Proceedings of the IEEE/CVF Conference on Computer Vision and Pattern Recognition, Long Beach, California, USA, pp. 7036–7045 (2019)

22. Lu, X., Li, Q., Li, B., Yan, J.: MimicDet: bridging the gap between one-stage and two-stage object detection. In: Vedaldi, A., Bischof, H., Brox, T., Frahm, J.-M. (eds.) ECCV 2020. LNCS, vol. 12359, pp. 541–557. Springer, Cham (2020). https://doi.org/10.1007/978-3-030-58568-6_32

23. Lin, T.Y., Dollár, P., Girshick, R., He, K., Hariharan, B., Belongie, S.: Feature pyramid networks for object detection. In: Proceedings of the IEEE Conference on Computer Vision and Pattern Recognition, Honolulu, Hawaii. IEEE, pp. 2117–2125 (2017)

24. He, K., Gkioxari, G., Dollár, P., Girshick, R.: Mask R-CNN. In: Proceedings of the IEEE International Conference on Computer Vision, Venice, Italy. IEEE, pp. 2961–2969 (2017)

25. Ren, S., He, K., Girshick, R., Sun, J.: Faster R-CNN: Towards real-time object detection with region proposal networks. arXiv preprint arXiv:1506.01497 (2015)

26. Girshick, R.: Fast R-CNN. In: Proceedings of the IEEE International Conference on Computer Vision, Santiago, Chile. IEEE, pp. 1440–1448 (2015)

27. Chen, Y., Han, C.,Wang, N., Zhang, Z.: Revisiting Feature Alignment for One-stage Object Detection. arXiv preprint arXiv:1908.01570 (2019)

28. Zoph, B., Le, Q.V.: Neural architecture search with reinforcement learning. arXiv preprint arXiv:1611.01578 (2016)

29. Lin, T.Y., Goyal, P., Girshick, R., He, K., Dollár, P: Focal loss for dense object detection. In: Proceedings of the IEEE International Conference on Computer Vision, Venice, Italy. IEEE, pp. 2980–2988 (2017)

30. Everingham, M., Van Gool, L., Williams, C.K., Winn, J., Zisserman, A.: The PASCAL visual object classes (VOC) challenge. Int. J. Comput. Vis. (IJCV) **88**, 303–338 (2010)

31. Jiao, L., et al.: A survey of deep learning-based object detection. IEEE Access **7**, 128837–128868 (2019)

32. Everingham, M., Van Gool, L., Williams, C.K., Winn, J., Zisserman, A.:The PASCAL Visual Object Classes Challenge 2007 (VOC2007) Results (2007)

33. Padilla, R., Netto, S.L., da Silva, E.A.: A survey on performance metrics for object-detection algorithms. In: 2020 International Conference on Systems, Signals and Image Processing (IWSSIP), Niteroi, Brazil. IEEE, pp. 237–242 (2020)

34. Lin, T.Y., Maire, M., Belongie, S., Hays, J., Perona, P., Ramanan, D., Dollár, P., Zitnick, C.L.: Microsoft COCO: common objects in context. In: European Conference on Computer Vision, Zurich, Switzerland. Springer, pp. 740–755 (2014)

35. Simonyan, K., Zisserman, A.: Very deep convolutional networks for large-scale image recognition. arXiv preprint arXiv:1409.1556 (2014)

36. Shrivastava, A., Gupta, A., Girshick, R.: Training region-based object detectors with online hard example mining. In: Proceedings of the IEEE Conference on Computer Vision and Pattern Recognition, Las Vegas, Nevada. IEEE, pp. 761–769 (2016)

37. He, K., Zhang, X., Ren, S., Sun, J.: Deep residual learning for image recognition. In: Proceedings of the IEEE Conference on Computer Vision and Pattern Recognition, Las Vegas, Nevada. IEEE, pp. 770–778 (2016)

38. Fu, C.Y., Liu, W., Ranga, A., Tyagi, A., Berg, A. C.: DSSD: Deconvolutional single shot detector. arXiv preprint arXiv:1701.06659 (2017)

Most Significant Bit-Plane Based Local Ternary Pattern for Biomedical Image Retrieval

Nilima Mohite[(⊠)], Manisha Patil, Anil Gonde, and Laxman Waghmare

SGGS Institute of Engineering and Technology, Nanded, Maharashtra, India

Abstract. In this paper, Most Significant Bit-plane using Local Ternary Pattern is proposed for retrieval of biomedical images. Biomedical images have been evolved in hospitals and medical institutions for disease diagnosis of patients and to detect and record the patient's history. Algorithm uses transformation scheme to calculate transformed value of each neighboring pixel present in a local bit plane. The proposed method is generated by calculating two binary pattern from a ternary pattern using difference of local biplane transformed values with the intensity value of center pixel. The retrieval efficiency of proposed method is tested on two standard bio-medical databases OASIS-MRI and MESSIDOR database by using Average Retrieval Precision (ARP). We get 66.17% result on OASIS database and 56.27% on MESSIDOR dataset. The MSBPLTP is evaluated using ARP and compared with existing image retrieval methods.

Keywords: Average Retrieval Precision · Local Ternary Pattern (LTP) · Local bit-plane transformation · MSBPLTP

1 Introduction

Patients with chronic diseases like Diabetes, Cancer, and Arthritis etc. are in-creasing day by day. To treat patients with these types of diseases experts have to make keen observation and maintain the record of each and every patient as a history for further reference. The Biomedical images have been evolved in hospitals and medical institutions for disease diagnosis of patients and to detect and record the patient's history. The various formats in which Medical images exist are, Computer Tomography, X-ray, Fundus imaging, Ultrasound, Magnetic Resonance Imaging, etc. Many hospitals at present have enormous data of patients available at their records and to handle such a Meta data by human eyes (such images need an expert eye) is not feasible. To address this problem, evolution of field like Digital Image processing (DIP) in medical applications is significant and using this DIP the system come into existence called as Content Based Image Retrieval (CBIR), which is extensively used in biomedical field. CBIR follows three different stages: The first stage is used to process the image using different pre-processing techniques of image. Then carry out feature extraction and feature vector formation. The third stage makes use of some distance measure to carry out similarity measurement to find minimum distance between database images and the query image.

© Springer Nature Switzerland AG 2021
M. Singh et al. (Eds.): ICACDS 2021, CCIS 1440, pp. 193–203, 2021.
https://doi.org/10.1007/978-3-030-81462-5_18

1.1 Literature Survey

W. M. Smeulders et al. [1] mentioned the extensive literature survey about CBIR. A review of CBIR system in medical application for the benefits of medical institutes is carried out by different authors. Texture classification can be done with Local binary pattern is proved by Ojala et al. [2], further they extended this method for rotation invariant and gray-scale texture classification patterns [3]. Another application of LBP comes into existence such as analysis of pulmonary emphysema presented by L. Sorensen et al. [4], then face recognition using LBP [5], quantitative analysis of facial paralysis [6] etc. B. Zhang et al. [7] introduced local derivative pattern (LDP) vs local binary pattern for face recognition application. S. Liao et al. [8] and Z. Guo et al. [9] used LBP for texture classification. Using a local diagonal component of images S. R. Dubey et al. [10] introduced a new biomedical image retrieval feature descriptor.

S. Murala et al. [11] presented directional binary wavelet pattern, the local mesh pattern (LMeP) [12], Peak valley edge pattern (PVEP) [13] and Local mesh peak valley edge pattern (LMePVEP) [14] for indexing and retrieval of biomedical images. The local ternary pattern (LTP) is proposed by X. Tan and B. Triggs [15] to overcome the difficulties of illumination changes by enhancing texture feature for face recognition application. S. Murala, et al. [16] further has extended LTP for biomedical images retrieval by introducing local ternary co-occurrence pattern and effectiveness of system has been tested on medical images dataset such as OASIS-MRI and ELCAP-CT etc. Jinsa Kuruvilla and Gunavathi [17] proposed an effective method for content based retrieval of Computer Tomographic images of Lungs. Using dual-tree complex wavelet transform, Baby and Candy [18] proposed a retrieval method for retinal images using a MESSIDOR database. Dubey et al. [19] presented a Local Bit-plane Decoded Pattern (LBDP) for retrieval of biomedical images.

1.2 Main Contribution

The motivation behind MSBPLTP algorithm for retrieval of biomedical images is LBDP. The presented method differs from the existing LBDP method as it makes use of LTP [15] to overcome the drawbacks of lighting variations, hard thresholding and noise present in the LBP [2]. Also, in proposed method we used four MSB planes instead of complete eight planes so memory required to store features is less and length of feature vector also get reduced. In proposed method we consider the local neighbors of a reference pixel and convert intensity values of each neighbor into 8 bit binary number and then decompose 8-bits into the 8-planes. Later encode the relationship between transformed values and neighbors, further generate MSBPLTP and extract features, and then efficiency of algorithm is confirmed by observing results of experiments done on different biomedical image dataset.

Paper is organized as: Sect. 2 contains short introduction of existing local patterns. Proposed methodology, feature vector calculation and similarity measurement criteria is explained in Sect. 3. Experimental result and discussion is presented in Sect. 4. Section 5 concludes the paper.

2 Local Patterns

2.1 Local Binary Patterns (LBP)

The pattern is proposed by Ojala et al. [2] for classification of texture and to encode the correlation between the reference pixel and surrounding neighboring pixels. LBP can be represented by following equations:

$$LBP_{N,R} = \sum_{N=1}^{N} 2^{(N-1)} \times f(X_N, X_C) \quad f(a,b) = \begin{cases} 1, & a = b \\ 0, & else \end{cases} \quad (1)$$

Where X_N and X_C are the intensity values of Nth neighboring pixel and center pixel C having radius R. Number of neighbors distributed along a circular radius ($R = 1$). After analyzing complete pattern the whole image is represented in terms of histogram to generate a feature vector.

2.2 Local Ternary Patterns (LTP)

Tan et al. [15] converted the two value code to three value code and name is given as local ternary pattern. They certify that local binary pattern is more reactive to the lighting variations and noise, to address this problem local ternary pattern came into existence. Three valued function for LTP as shown in Eq. 2

$$f(n, X_C, t) = \begin{cases} +1, & n \geq X_C + t \\ 0, & |n - X_C| < t \\ -1, & n \leq X_C - t \end{cases} \quad (2)$$

Where n is the $(X_N - X_C)$. LTP is calculated by considering threshold value t, if n is greater than or equal to $(X_C + t)$ then pattern value is $+1$, else if it is less than or equal to $(X_C - t)$ then pattern value is -1, otherwise pattern value is 0 if n is in between $[(X_C + t), (X_C - t)]$.

3 Methodology

3.1 Neighboring Pixel Decomposition

Local features efficiently describe an image as compared to global features so the design approach for image retrieval uses local feature, extracted by considering 3×3 local region of an image. In proposed method, we first decomposed intensity values in to binary values for further processing. Let I is a grayscale image with dimension $m \times n$, (x, y) is the pixel of image I and its intensity value is $I(x, y)$. intensity values of N neighbors (local neighbors of center pixel $c(x, y)$) of image I are decomposed into number of bits having bit-depth of K-bits. Equation 3 shows decomposition of gray scale image into K number of P planes (Fig. 2).

$$P_K^{x,y} = B_K(I(x, y)) \quad (3)$$

K	8	7	6	5	4	3	2	1
B	0	0	1	0	1	1	1	1

Fig. 1. B represent the binary value of $I(x, y)$ ($I (1, 1) = 47$) and K is the bits location.

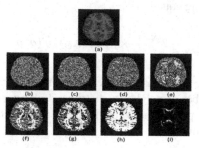

Fig. 2. Decomposition of local bit-planes images (a) Input image from OASIS database. (b) − (i) Images of 8 bit plane from LSB to MSB.

Where, $B_K(I(x, y))$ is the eight bit binary value of pixels with intensity $I(x, y)$ and K (position of bits) ranges from 1(LSB) to 8(MSB) as shown in Fig. 1. For eight bits we have mapped eight bit planes i.e. from MSB plane to LSB plane, and stored binary bit in corresponding K^{th} planes. Figure 3 shows $P = 8$, $N = 8$ and $R = 1$ bit planes which consists of the binary values of each neighbor. Each bit-plane P_K ($K \in [1,8]$) is separated from the local neighbors of the pixel $C(x,y)$ using decomposition step. Using the decomposition process, the binary values for each bit-plane are calculated and which is applied on N neighbors except center pixel $C(x,y)$.

3.2   Local Bit-Map Transformation

After bit plane decomposition step, the transformation step uses the binary map to generate different planes. Transformed values (*i.e.* $T_K{}^{x,y}$) for local bit plane is calculated by using transformation Eq. 4.

$$T_K^{x,y} = \sum_{N=1}^{8} P_K^{x,y} \times w \quad w = 2^{N-1} \tag{4}$$

Where N is local neighbors in bit-planes and K is the number of biplanes and w multiplied with each neighbor (N) of a reference pixel in bit-planes as normalization called as weighting factor. The range of $T_K^{x,y}$ is between 0 to 2^{N-1}.

3.3   MSBPLTP

In this section, a binary pattern is generated called most significant bit-plane based local ternary pattern (MSBPLTP) by traversing the relationship between intensity value of reference pixel $C(x, y)$ with the transformed values ($T_K^{x,y}$) for each bit-plane K. LTP

[15] is used to calculate the patterns for proposed method by exploring the relationship between N neighbors of image I at radius R and is defined as follows:

$$f(z, C(x, y), t) = \begin{cases} +1, & T_K^{x,y} \geq C(x, y) + t \\ 0, & |T_K^{x,y} - C(x, y)| < t, \\ -1, & T_K^{x,y} \leq C(x, y) - t \end{cases} \quad z = \left(T_K^{x,y} - C(x, y)\right) \quad (5)$$

Where t is user-specified threshold value, z is computed using the transformed value for K bit-planes Eq. 7, where $T_{K,N}^{x,y}$ the transformed value of N neighbors in bit-planes and $C(x, y)$ is the reference value for N neighbors in image I. From survey it is observed that local ternary pattern is used for solving the problems in local binary pattern, such as noise coherency in flat areas, effect of illumination changes and problem of hard thresholding etc. LTP is a three value function hence feature vector length is increased as compared to the LBP. To solve the problem of dimension we decomposed original three values LTP function into two value function such as lower LTP (L_ltp) and upper LTP (U_ltp) as shown in following Eq. 6 and Eq. 7.

$$L\_MSBPLTP_{K,N}^{x,y} = f(\beta) \times w^{N-1} \quad (6)$$

$$U\_MSBPLTP_{K,N}^{x,y} = f(\beta) \times w^{N-1} \quad (7)$$

Where N varies from 1 to 8, so w is the weighting factor ranges in between [1 to 128].$L\_MSBPLTP_{K,N}^{x,y}$ and $U\_MSBPLTP_{K,N}^{x,y}$ are the lower ternary pattern and upper ternary pattern of N neighbors for K^{th} plane respectively and it is represented by concatenating all $L\_MSBPLTP_{K,N}^{x,y}$ and $U\_MSBPLTP_{K,N}^{x,y}$ as follows:

$$MSBPLTP_{K,N}^{x,y} = \begin{cases} L\_MSBPLTP_{1,N}^{x,y}, U\_MSBPLTP_{1,N}^{x,y}, L\_MSBPLTP_{2,N}^{x,y}, U\_MSBPLTP_{2,N}^{x,y} \ldots \\ \ldots\ldots\ldots\ldots\ldots\ldots\ldots U\_MSBPLTP_{K,N}^{x,y} \end{cases} \quad (8)$$

Where $MSBPLTP_{k,N}^{x,y}$ is the most significant bit plane based local ternary pattern We have considered only 5th to 8th MSB planes, because it contains significant information as compared to the LSB planes hence called as most significant bit plane based pattern. An example of bit plane based local ternary has been illustrated in Fig. 3.

3.4 Feature Vector Calculation

In previous sections, we observe that we get eight decoded ternary patterns. After analyzing complete local pattern image is represented in the form of histogram as follow:

$$H_{MSBPLTP}(\psi) = \sum_{x=1}^{m} \sum_{y=1}^{n} f\left(MSBPLTP_{s,N}^{x,y}(x, y), \psi\right) \quad \psi \in \left[0, 2^{N-1}\right] \quad (9)$$

$$f(x, y) = \begin{cases} 1, & x=y \\ 0, & else \end{cases} \quad (10)$$

Where m and n represents size of image, (x,y) are the pixel co-ordinates of transformed image and it's variations depend on the size of image, N is the number of

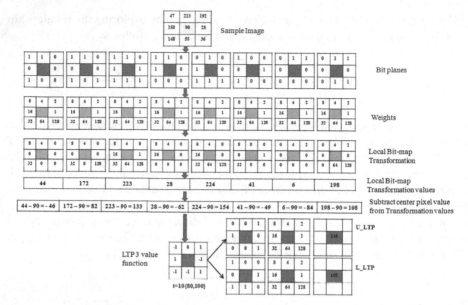

Fig. 3. Example to obtain the bit plane based ternary pattern.

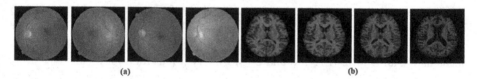

Fig. 4. Sample images from (a) MESSIDOR database (b) OASIS database.

neighboring pixel ($N \in 8$) evenly distributed along a circle with radius of R, s is number of selected bit planes varies from five to eight. In proposed method we calculate features using LTP and represented them using a histogram as shown in Fig. 5. Ternary pattern get converted into two binary patterns, and these two binary patterns apply on four bit-planes (*P5, P6, P7, P8*) hence length of feature vector become ($2 \times 4 \times 256 = 2048$).

3.5 Similarity Measurement criteria

In evaluation, query image used, is the image from database (covering all the images one by one) and retrieved top (t) matching images. Similarity measurement is done by matching distance between feature vector of each image in the database and query image feature vector. Distance can be measured using Canberra distance measure, d1 distance measure, chi-square distance measure, Euclidean distance measure etc., we evaluate the proposed methods performance using d1 distance measure [20] and is given by:

$$d1(q, db) = \left| \sum_{\beta=1}^{L} \frac{(Z_q(\beta) - Z_{dbi}(\beta))}{1 + Z_q(\beta) + Z_{dbi}(\beta)} \right| \tag{11}$$

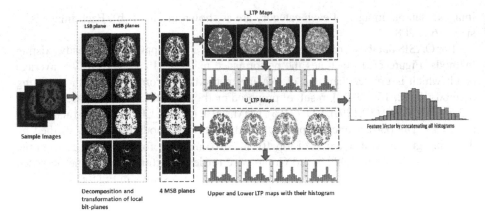

Fig. 5. Framework for feature extraction.

Input image and other images in the database are represented as q and db respectively. Query image feature vector (q) and database images is represented as $Z_q = \{Z_{q1}, Z_{q2}, \ldots, Z_{qL}\}$ and $Z_{db} = \{Z_{dbi1}, Z_{dbi2}, \ldots, Z_{dbiL}\}$; where $i = 1, 2, \ldots, db$ respectively. We calculated the ARP by mean measuring of the average precision over each group of the database by considering each image as the query image. The precision P (Img_q) is given as follows:

$$\text{Precision}: \ P\left(\text{Img}_q\right) = \frac{\#relevant\ images\ retrieved}{Total\#images retrieved} \tag{12}$$

Average retrieval precision is given as follows:

$$\text{ARP}: \ ARP = \frac{1}{db}\sum_{l=1}^{db} P\left(\text{Img}_q\right); \quad (for\ n \leq 10) \tag{13}$$

Where, n is the number of images considered for query matching.

4 Results and Observation

We evaluated result of the presented system on two biomedical image databases using Average retrieval precision (ARP). The two experiments are performed over MESSIDOR retinal fundus image database [23], and Open Access Series of Imaging Studies (OASIS) [22]. We compared results of the MSBPLTP with the results of the existing algorithms.

4.1 Experiment 1

For this experiment, we have used the OASIS magnetic resonance imaging (MRI) database that is available publicly for study, research and analysis purpose. This database includes a series of MRI images and a collection of 421 subjects in the age group of 18 to 96 years and all the images are categorized into four groups (124, 102, 89, and 106 images), sample images of each category are as shown in Fig. 4 (b). Each image is of size 176 × 208.

For OASIS database, results of proposed method are compared with already existing methods. Figure 6(a) illustrates the performance using ARP, we get 66.17% average result which is greater than that of existing methods. As compared to other existing methods [2, 12, 15, 19, 22], result of proposed method MSBPLTP is increased by LBP (23.54%), DBWP (19.12%), LTP (21%), LDMaMEP (8.3%), LBDP (7.17%) and LMEP (15.77%). The performance of proposed and existing methods have been verified by depicting graphs as a function of top 10 (n = 10) matches for query image from the database as shown in Table 1 and Group wise retrieval performance of the proposed method and other existing methods on OASIS database is shown in Fig. 6(b).

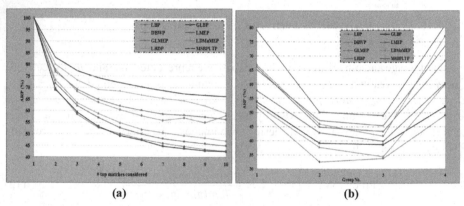

(a)   (b)

Fig. 6. Graph of (a) ARP as a function of # top matches considered (b) Group wise average retrieval precision for proposed and existing methods for OASIS database

Table 1. Group wise ARP comparison on OASIS database using proposed and existing algorithms

Method	Group1	Group2	Group3	Group4	Average
LBP	51.77	32.54	33.82	49.06	42.63
GLBP	53.79	39.21	38.73	52.16	46.55
DBWP	52.74	37.74	34.38	60.00	47.05
LMEP	57.82	42.84	39.89	60.37	50.4
GLMEP	66.12	44.90	43.38	68.30	56.34
LDMaMEP	66.77	45.98	41.72	76.99	57.87
LBDP	65.18	47.26	45.43	73.52	59
3DLCDWP [24]	78.38	48.72	43.25	73.96	62.66
3D-LCDP [24]	74.51	49.21	45.95	73.30	62.04
MSBPLTP	79.11	50.16	48.98	80.84	66.17

4.2 Experiment 2

The MESSIDOR database has been proposed to facilitate study on computer aided diagnosis of diabetic retinopathy. The database includes total 1200 retinal fundus color images. Dataset contains three different sizes of images 2240×1488, 2304×1536 and 1440×960. For retrieval purpose, these 1200 images are categorized into 4 groups (547, 155, 247 & 252 images) as shown in Fig. 4 (a). Table 2 shows the comparative retrieval result of proposed method (MSBPLTP) and the other existing method for this dataset in terms of ARP. The retrieval performance of the proposed method and existing method has been verified from Table 2 and get 69.89%, 36.38%, 43.07% and 56.27% result for group 1, 2, 3 and 4 respectively in terms of Average retrieval precision.

Table 2. Group wise ARP comparison on OASIS database using proposed and existing algorithms

Method	Group1	Group2	Group3	Group4	Average
LBP	65.78	36.9	44.85	43.88	53.15
LDMaMEP	65.79	37.42	41.62	46.67	53.13
CWT_GCD	64	35	40	56	53.70
LBDP	63.07	35.48	39.43	45.31	50.94
MSBPLTP	69.89	36.38	43.07	58.51	56.27

5 Conclusion

We have proposed a feature descriptor method MSBPLTP for retrieval and indexing of biomedical images. The presented method is different than other feature descriptor methods in terms of; it makes use of LTP to overcome the drawbacks of lighting variations, hard thresholding and noise present in the LBP. Also, in proposed method we used four MSB planes instead of complete eight planes so memory required to store features is less and length of feature vector also gets reduced. The local ternary pattern is used for extraction of features from the local bit-plane decoded pattern. Two experiments were performed on two biomedical image databases to observe the retrieval efficiency of MSBPLTP in terms of ARP. The OASIS (MRI) and MESSIDOR (Retinal) database gave effective result as compared to existing methods.

References

1. Smeulders, A.W.M., Worring, M., Santini, S., Gupta, A., Jain, R.: Content-based image retrieval at the end of the early years. IEEE Trans. Pattern Anal. Mach. Intell. **22**(12), 1349–1380 (2000)
2. Ojala, T., Pietikainen, M., Harwood, D.: A comparative study of texture measures with classification based on feature distributions. Pattern Recogn. **29**(1), 51–59 (1996)

3. Ojala, T., Pietikainen, M., Maenpaa, T.: Multiresolution gray-scale and rotation invariant texture classification with local binary patterns. IEEE Trans. Pattern Anal. Mach. Intell. **24**(7), 971–987 (2002)
4. Sorensen, L., Shaker, S.B., de Bruijne, M.: Quantitative analysis of pulmonary emphysema using local binary patterns. IEEE Trans. Med. Imaging **29**(2), 559–569 (2010)
5. Ahonen, T., Hadid, A., Pietikainen, M.: Face description with local binary patterns: application to face recognition. IEEE Trans. Pattern Anal. Mach. Intell. **28**(12), 2037–2041 (2006)
6. He, S., Soraghan, J.J., O'Reilly, B.F., Xing, D.: Quantitative analysis of facial paralysis using local binary patterns in biomedical videos. IEEE Trans. Biomed. Eng. **56**(7), 1864–1870 (2009)
7. Zhang, B., Gao, Y., Zhao, S., Liu, J.: Local derivative pattern versus local binary pattern: face recognition with higher-order local pattern descriptor. IEEE Trans. Image Process. **19**(2), 533–544 (2010)
8. Liao, S., Law, M.W.K., Chung, A.C.S.: Dominant local binary patterns for texture classification. IEEE Trans. Image Process. **18**(5), 1107–1118 (2009)
9. Guo, Z., Zhang, L., Zhang, D.: A completed modeling of local binary pattern operator for texture classification. IEEE Trans. Image Process. **19**(6), 1657–1663 (2010)
10. Dubey, S.R., Singh, S.K., Singh, R.K.: Local diagonal extrema pattern: a new and efficient feature descriptor for CT image retrieval. IEEE Signal Process. Lett. **22**(9), 1215–1219 (2015)
11. Murala, S., Maheshwari, R.P., Balasubramanian, R.: Directional binary wavelet patterns for biomedical image indexing and retrieval. J. Med. Syst. **36**(5), 2865–2879 (2012)
12. Murala, S., Wu, Q.M.J.: Local mesh patterns versus local binary patterns: biomedical image indexing and retrieval. IEEE J. Biomed. Health Informat. **18**(3), 929–938 (2014)
13. Murala, S., Wu, Q.M.J.: Peak valley edge patterns: a new descriptor for biomedical image indexing and retrieval. In: IEEE Conference on Computer Vision and Pattern Recognition, pp. 444–449 (2013)
14. Murala, S., Wu, Q.M.J.: MRI and CT image indexing and retrieval using local mesh peak valley edge patterns. Signal Process. Image Commun. **29**(3), 400–409 (2014)
15. Tan, X., Triggs, B.: Enhanced local texture feature sets for face recognition under difficult lighting conditions. IEEE Trans. Image Process. **19**(6), 1635–1650 (2010)
16. Murala, S., Wu, Q.M.J.: Local ternary co-occurrence patterns: a new feature descriptor for MRI and CT image retrieval. Neurocomputing **119**, 399–412 (2013)
17. Kuruvilla, J., Gunavathi, K.: Lung cancer classification using neural networks for CT images. Comp. Methods Prog. Biomed. **113**, 202–209 (2014)
18. Baby, C.G., Chandy, D.A.: Content based retinal image retrieval using dual tree complex wavelet transform. In: IEEE International Conference on SIPR, pp. 4673–4862 (2013)
19. Dubey, S.R., Singh, S.K., Singh, R.K.: Local bit-plane decoded pattern: a novel feature descriptor for biomedical image retrieval. IEEE J. Biomed. Health Inf. **20**(4), 1139–1147 (2016)
20. Murala, S., Maheshwari, R.P., Balasubramanian, R.: Local maximum edge binary patterns: a new descriptor for image retrieval and object tracking. Signal Process. **92**(6), 1467–1479 (2012)
21. Vipparthi, S.K., Murala, S., Gonde, A.B., Jonathan, Q.M.: Local directional mask maximum edge pattern for image retrieval and face recognition. IET Comput. Vision **10**(3), 182–192 (2016)
22. Marcus, D.S., Wang, T.H., Parker, J., Csernansky, J., Morris, J.C., Buckner, R.L.: Open access series of imaging studies (OASIS): crosssectional MRI data in young, middle aged, nondemented, and demented older adults. J. Cogn. Neurosci. **19**(9), 1498–1507 (2007)

23. MESSIDOR database is available at http://messidor.crihan.fr
24. Mohite, N., Waghmare, L., Gonde, A., Vipparthi, S.: 3D local circular difference patterns for biomedical image retrieval. Int. J. Multimedia Inf. Retrieval **8**(2), 115–125 (2019). https://doi.org/10.1007/s13735-019-00170-1

Facial Monitoring Using Gradient Based Approach

Arush Jain, Mani Sachdeva, and Paramita De[(⊠)]

G. L. Bajaj Institute of Technology and Management, Greater Noida, India

Abstract. Security is a major concern for every organization and various security methodologies are applied to keep the work area safe and secure. The proposed algorithm computes the local gradients from an input facial image and detects the 68 landmark points to identify any unknown person. The images of all recognized and authorized persons are stored in the database after performing some encryption of the images to make database data secure. During the monitoring process, the faces encountered in the camera frame in real time need to be matched with the database images. For matching, the retrieved database images need to be decrypted and encrypted again to make the format same as used by matching algorithm. After the matching process, a notification is sent to the concerned authority for some unknown face which is also highlighted by a rectangular box. The performance and robustness of the algorithm are evaluated by testing images with various postures, background, lighting condition that is shown in experimental results.

Keywords: Face recognition · Video surveillance · Histogram of gradients

1 Introduction

Now-a-days, computer vision has become versatile and used in various domains like real time monitoring, face recognition, face comparison, biometric scanning, and many more. In computer vision literature, face detection, face monitoring in the surveillance system is one of the heavily studied areas of research.

In an existing surveillance system, every camera is being monitored by a monitoring personnel, commonly known as a security monitoring person (SMP). So, it is necessary for the SMP to know each person in the organization to identify any unknown person who comes into the screen and take necessary action. This is not an easy task, as human have a high probability of mistaking in recognizing a person's face. Thus, to decrease the chances of mistakes, the task is divided into teams which work on different shifts. This has reduced the chances of mistakes, but it has also decreased the reliability. Human can easily manipulate the data and thus more number of people in security leads to less privacy of the organization. So, with these flaws the organization has either need to compromise on lack of privacy or lack of accuracy in security system.

© Springer Nature Switzerland AG 2021
M. Singh et al. (Eds.): ICACDS 2021, CCIS 1440, pp. 204–213, 2021.
https://doi.org/10.1007/978-3-030-81462-5_19

In smart surveillance systems, computer vision plays a very important role that includes the methods like acquiring, processing and analyzing images to produce appropriate information. Automatically detecting and tracking people are important in security, surveillance and the health/independent living domains. It can also be applied to identify any individuals at a particular place, to control the access to secure environments and intruder detection. Recently, as the surveillance system continuously improving and expanding, there is a big challenge for us to store, analyse, and retrieve the huge volume of monitoring data. Now, in the smart monitoring system, intelligent video analysis technologies are used to monitor and generate a signal in case of any abnormal behavior or event.

In this paper, the human faces are captured and identified by the system in real time by matching with the faces stored in the database using the histogram based gradient features. During the monitoring, upon detecting any unknown face a notification is sent to the concerned authority. The captured images are encrypted before storing into the database, to make the data secure in the database. Database encrypted image is decrypted for matching with the faces that were coming in real time. If any unknown face is captured by the camera then that face is framed in a box and in parallel to that, a notification regarding this is sent to the concerned person's phone. In this way, the security of the organization can be increased considering the database security too. An instant notification will help in getting things monitored more easily and with less manpower. If the activities of the framed face are found malicious then proper action can be taken at the correct time. The proposed work is capable to identify the face in the dim light, exposure to light, when the person was attending a call on a mobile phone, when the person was wearing the headphones, in different apparel and in other situations also. The proposed work focuses only on the face of a person rather than the whole body of the individuals. The proposed method can be very useful to track an unidentified person even if there were multiple cameras like in big societies as all the cameras are connected to the same database and for any unauthorized individual a separate database is maintained to monitor their activities which can help to strengthen the security system.

The rest of the paper is organized as follows. The literature review is presented in Sect. 2. Section 3 describes the proposed approach. The experimental results are given in Sect. 4. Finally, we conclude with discussions in Sect. 5.

2 Related Work

In recent times almost every organization uses surveillance system for security and a lot of research work has been done on face recognition. A method of facial recognition on video surveillance is proposed in [1], where the facial recognition was done with the help of a facial tracker ensemble for robust spatial temporal recognition. Pagano et al. [2] proposed different method of facial monitoring. It was proposed that ARTMAP neural classifiers and Dynamic Particle Swarm Optimization (DPSO) can be used for video surveillance. It is more accurate and efficient than Dynamic Niching Particle Swarm Optimization (DNSPO) approach. A cost effective (cross-correlation matching) CCM-CNN architecture is proposed in [3]. It is accurate and specialized for single reference still to video face recognition. A set of algorithms was presented in [4] to detect motion in real time with the help of a robot. The method was tested in different outdoor environments. A method for unconstrained environment facial recognition for a single face and real time video surveillance is proposed in [5]. The proposed system is automatic and robust to illumination and poses variation in facial images. Qinghan Xiao et al. [6] proposed a method to prevent unwanted person to use a system. The proposed system will lock the system when the authorized person will move out of the camera frame. Thus help in securing the information present in the system. Arandjelovic et al. [7] proposed an approach to remove particular image distortion systematically and the method has reached high recall and precision rate. A hybrid facial recognition algorithm is presented in [8], which is fully automatic and multi-modal. The model included spherical facial recognition in 3D faces and an automatic pose correction algorithm. Zhang et al. [10] proposed a system which recognizes faces in real time video streaming. The system used FPCA (Fuzzy Principal Component Analysis) to recognize the faces in different scenarios and was giving accurate results in both good and bad quality video images. Awais et al. [11] proposed a surveillance system which works on real time video streaming. The system used HOG and feed forward neural network for face matching which take more time. Also, the method worked for constant environment only and it can handle a limited light variations and shadow distortion. In [12], FaceNet and Multi-Task Cascade Convolutional Neural Network (MTCNN) algorithms has been used to detect multiple faces in video images. The system is hardware dependent and require Jetson TX2 to implement the system and use multiple cameras to accomplish the task. They maintain the geo location of all the known and unknown faces with timestamp into the log directory database for future reference which need a huge storage for the log database and also the locations are not actual geographical location. In [13] a facial detection system has been proposed where the head and shoulder of the person are detected first and it combine the RGB and YCbCr approaches to figure out the presence of skin in the image. An extended MTCNN approach has been proposed in [14], where 68 facial landmarks are extracted and the method is compared with MTCNN and Dlib. They conclude that MTCNN extracts 5 facial landmarks which are not accurate while working with real time applications and Dlib which extracts the 68 facial landmarks is comparatively slow. The extended MTCNN on the other side was capable to detect the 68 facial landmarks in less time. A system to recognize facial emotions has been presented in [15] by the help of the facial landmarks and multiclass SVM with an RGB kernel. The accuracy of the proposed method was observed to be about 70.65%. A analysis on Eigenface and Fisherface approach has been carried out in [16], where it has been stated that Eigenface work on PCA to decrease the dimensionality of the face data and it is limited to number of faces. The maximum accuracy observed in Eigenface system is about 90%. Several approaches like Fisherface, Local Binary

Pattern, Eigenface can be applied to the face recognition task and a performance comparison is performed for all these approaches that is given in Table 1.

Table 1. Performance comparison of different face recognition approaches

Features	Eigenface	Fisherface	LBPH
Time complexity	Less	More	Less
Accuracy	Less	More than Eigenface	High
Computation	Less	More	Less

3   Proposed Method

In facial monitoring, the person's face is captured by the camera, analyzed, and compared based on the person's facial details to distinguish between authorized and unauthorized person. The proposed system can help to build the security system more secure for any organizations like, schools, colleges, universities, etc. where many are people gathered at a time and this can detect and inform the security person about any intruder in real time. The flow diagram of the proposed security system is shown in Fig. 1 and the various important steps of the proposed work are explained in detail in the following sections.

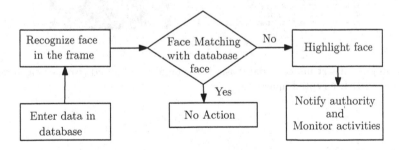

Fig. 1. Steps of the proposed surveillance system

3.1   Enter Data in Database

In the proposed system, first the faces of the authorized persons are taken and stored in the database. For this the database administrator needs to take a photo of each person and enter the name of the person in the database. In the database the name and corresponding photo of each authorized persons are saved. Since, the image is in three planes that are red, green, blue, due to which the size of the image is large. The image size is reduced by converting it into gray scale. The image is not stored as the normal image format in the database, whereas it is being encrypted in the format of base 64 and then stored in the database. Figure 2 shows the result of different steps of conversion before storing the image in the database. The main objective of this is to decrease the size of the image and hence optimize the space complexity of the database as we know that it required a large storage cost to store huge volumes of surveillance data and also to make the database data secure.

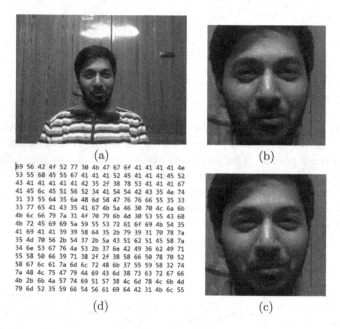

Fig. 2. Steps to insert image in database. (a) Image encountered (b) Face captured (c) Gray image (d) Encrypted image.

3.2   Face Recognition

To detect the face the standard Haar Cascade classifier is used, which is basically an object detection method where using the negative and positive images a cascade function is trained to detect objects in images. The images that have the images which the user want to identify by the classifier is known as positive images, whereas the negative images means where the target image is not present in the image. Facial recognition is done with the help of different points on the face. These points are known as 68 facial landmark points [9]. With the help of the 68 landmark points the face is being recognized and matched with other faces. The 17 landmark points are used to obtain the jawline of the face, 5 points are for each eyebrow i.e., total 10 landmarks are for right and left eyebrow. For the nose there are 9 landmark points. To recognize both the eyes 12 landmark points are being considered, 6 landmarks for each eye. Mouth recognition is done with the help of 13 facial points. For lip recognition 7 facial landmarks are being used. Hence, the facial landmark points are left and right eyebrows, left and right eyes, nose, mouth, lips and the jaw lines. The facial landmark points are identified by using the Dlib face landmark detector. Figure 3(a) and (b), all the 68 landmarks and the identified landmarks are shown respectively.

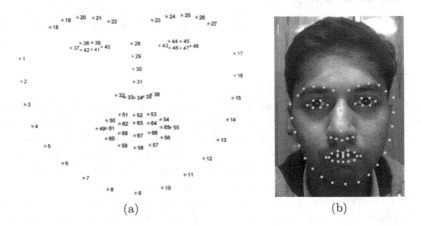

(a)   (b)

Fig. 3. Highlighting the facial landmarks. (a) All 68 landmarks (b) Highlighted landmarks on the captured image.

3.3   Face Matching

Before the face matching process starts, during the data retrieved from the database two lists are being created; one containing all the names in the sequence that are fed into the database and the second list initially contains all the encrypted images that were fed into the database in the correct sequence as that of the names. The images are first decrypted from base 64 encoding to the normal image file, and then the normal image file is again encrypted for face matching process as this module uses separate encryption technique. After that, the following operations are performed on the elements of the second list. When the camera has been turned on, first in each frame each face location has been calculated. Every input encoded face has been appended to a list. For matching the camera encountered face and face in the database, an algorithm based on the Histogram of Gradients (HOG) feature is used. For the captured face a vector of HOG feature is extracted. The gradients for each pixel of the image, i.e., the magnitude (g) and directions (θ) are calculated by applying Eq. 1 and Eq. 2.

$$g = \sqrt{G_x^2 + G_y^2} \tag{1}$$

$$\theta = arctan\frac{G_y}{G_x} \tag{2}$$

where, G_x and G_y are the horizontal and vertical components of the change in the pixel intensity, respectively. If the encountered face does matches with any face of the database, then that face is highlighted and monitored by putting a rectangular box over that face on the camera screen.

3.4 Face Monitoring

In the face monitoring process, each face that comes in camera frame is being checked and matched with the faces from the database. After matching the captured image with the database images, the face which matches the most with the input image is given as the output. No further action is required if the face is matched with any database image, but if the input image does not match with any images of the database, then that face is being captured and highlighted within the frame. The unknown face is outlined with a red rectangular box. Thus, it helps the person in the surveillance room to easily identify the unknown face. The person in surveillance room can easily monitor the activities of the unknown person because that face will remain outlined with a rectangular box till it is in the camera frame. Parallel to this another important activity of sending an alert message to the concerned authority is also performed. Through this process the SPOC (Single Point of Contact) received a notification of capturing an unknown face by the system camera and can take necessary actions at the correct time in case of any major issue. Thus, with the proposed system in real time security can be accomplished and give an effective solution to the problem related to traditional surveillance system.

4   Experimental Results

The proposed work is implemented in the system with Intel Core i3 processor having RAM of 4 GB with a 64 bit operating system and Python is used as a programming language. Performance of the algorithm has been tested on the images from the Flickr-Faces-HQ dataset and also by taking images in real time.

In the face matching algorithm, when the input image matches with any of the faces in the database, then the detected face is being highlighted by a frame across it. The name associated with that face is also fetched from the database to display with the detected known person. The purpose of this is to help the surveillance authority to monitor the face as well as the added names give more advantages in monitoring a person without remembering every persons face with their names. During the matching process, if the encountered face does not match with any of the faces stored in the database, then that face is being subtitled as *Unknown*. The subtitled text is written within the red frame created across that face. The result of this is shown in Fig. 4(a), where initially the face is not stored in the database and the facial matching algorithm fails to find a match and subtitled the face as *Unknown*. After adding the data of the same person in the database, the facial matching algorithm successfully finds the match and fetches the corresponding data of the person from the database. The result of successful matching is shown in Fig. 4(b), where the known face is being subtitled by the corresponding name. The facial matching algorithm can also detect the face successfully in other directions like facing towards the left and right, which is shown in Fig. 4(c) and (d) respectively.

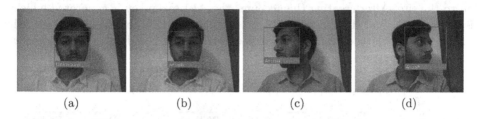

(a)   (b)   (c)   (d)

Fig. 4. Results of face matching algorithm. (a) Unknown face encountered (b) Known face encountered (c) Known face facing towards left (d) Known face facing towards right.

The algorithm can identify the face even if the person is not looking into the camera. The proposed face detection algorithm can perfectly identify the face in different posture as well as different light condition. It also works well if the person is doing some work like talking on the phone or listening music with a headphone. All the result of different scenarios is shown in Fig. 5.

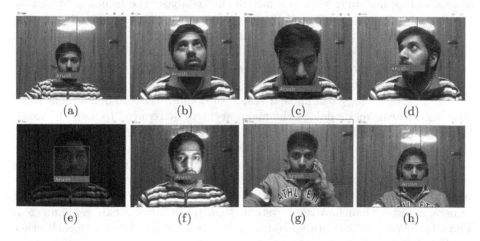

(a)   (b)   (c)   (d)

(e)   (f)   (g)   (h)

Fig. 5. Results of face matching algorithm in different condition. (a) Different posture with different background (b) Up side face (c) Down side face (d) Tilt upside face (e) In dim light (f) In exposure to more light (g) Taking phone call (h) With headphone.

The system is efficient and robust in finding out authorized and unauthorized person in a frame. In Table 2 the results of detection of the face at different angles are shown. The performance of the recognition algorithm is measured by finding the number of correctly identified faces among the total number of faces. This system is able to recognize the faces, irrespective of the hairs and features that lie outside the face boundary. This system focused only on the features that lie within the face boundary. To send the notification to the authorized person, the SMS broker library is used that can send the same message to multiple devices. To implement this the devices must be logged in with the same account as logging on to the site. After creating the account, the API token key is generated which is required to send the message. When an unknown face has been encountered in the camera the alert message has been sent to all the devices in which the account was logged in.

Table 2. Performance of the proposed recognition algorithm.

Face position	Front view	Left view				Right view				Down face	Up face
Angle	0	30°	45°	60°	75°	30°	45°	60°	75°	30°	30°
Recognition rate	100%	100%	98%	71%	0%	100%	97%	74%	0%	96%	97%

5 Conclusion

In the proposed work the data from the surveillance camera are taken and the algorithm for face recognition is executed on it. Each face that comes in the frame of the camera is being matched with the faces of database that are stored earlier. In case of face matching with any face in the database, the face is subtitled by the corresponding name. But, if the face does not match with any face in the database, then a notification will be sent to the concerned authority, and that face will also be marked with a rectangular box to monitor the person's activity in each frame. The proposed algorithm can perfectly identify the face in different posture and light condition.

The new technology will make it easier for organizations to make their workplace safe. In this methodology the manpower will be reduced to a great extent. When human being involvement will be very less then chances of mistakes will also be reduced. Thus, it can be concluded that it is a much efficient method than the traditional method which is being used currently. In this system, if the speed of processing of input images by the camera is slower than the hardware system, proper result cannot be obtained. Also, if the processing speed of the camera is faster than hardware system, then the hardware can get damaged. Thus, hardware and camera should be properly configured to meet each other specification.

References

1. De-la-Torre, M., Granger, E., Radtke, P.V., Sabourin, R., Gorodnichy, D.O.: Partially supervised learning from facial trajectories for face recognition in video surveillance. Inf. Fusion **24**, 3–53 (2015)

2. Pagano, C., Granger, E., Sabourin, R., Gorodnichy, D.O.: Detector ensembles for face recognition in video surveillance. In: International Joint Conference on Neural Networks (IJCNN-2012), pp. 1–8. IEEE (2012)
3. Parchami, M., Bashbaghi, S., Granger, E.: CNNs with cross-correlation matching for face recognition in video surveillance using a single training sample per person. In: 14th IEEE International Conference on Advanced Video and Signal Based Surveillance (AVSS-2017), pp. 1–6 (2017)
4. Jung, B., Sukhatme, G.S.: Real-time motion tracking from a mobile robot. Int. J. Soc. Robot. **2**(1), 63–78 (2010)
5. Haghighat, M., Abdel-Mottaleb, M., Alhalabi, W.: Fully automatic face normalization and single sample face recognition in unconstrained environments. Expert Syst. Appl. **47**, 23–34 (2016)
6. Xiao, Q., Yang, X.D.: A facial presence monitoring system for information security. In: IEEE Workshop on Computational Intelligence in Biometrics: Theory, Algorithms, and Applications, pp. 69–76 (2009)
7. Arandjelovic, O., Zisserman, A.: Automatic face recognition for film character retrieval in feature-length films. In: IEEE Computer Society Conference on Computer Vision and Pattern Recognition (CVPR 2005), vol. 1, pp. 860–867 (2005)
8. Mian, A., Bennamoun, M., Owens, R.: An efficient multimodal 2D–3D hybrid approach to automatic face recognition. IEEE Trans. Pattern Anal. Mach. Intell. **29**(11), 1927–1943 (2007)
9. Qiu, Y., Wan, Y.: Facial expression recognition based on landmarks. In: IEEE 4th Advanced Information Technology, Electronic and Automation Control Conference (IAEAC), pp. 1356–1360 (2019)
10. Zhang, D., An, P., Zhang, H.: Application of robust face recognition in video surveillance systems. Optoelectron. Lett. **14**(2), 152–155 (2018). https://doi.org/10.1007/s11801-018-7199-6
11. Awais, M., et al.: Real-time surveillance through face recognition using HOG and feedforward neural networks. IEEE Access **7**, 121236–121244 (2019)
12. Jose, E., Greeshma, M., Haridas, M.T., Supriya, M.H.: Face recognition based surveillance system using FaceNet and MTCNN on Jetson TX2. In: 2019 5th International Conference on Advanced Computing and Communication Systems (ICACCS). IEEE, pp. 608–613, March 2019
13. Hosni Mahmoud, H.A., Mengash, H.A.: A novel technique for automated concealed face detection in surveillance videos. Pers. Ubiquit. Comput. **25**(1), 129–140 (2020). https://doi.org/10.1007/s00779-020-01419-x
14. Kim, H., Kim, H., Hwang, E.: Real-time facial feature extraction scheme using cascaded networks. In: 2019 IEEE International Conference on Big Data and Smart Computing (BigComp), pp. 1–7. IEEE (2019)
15. Nguyen, B.T., Trinh, M.H., Phan, T.V., Nguyen, H.D.: An efficient real-time emotion detection using camera and facial landmarks. In: 2017 Seventh International Conference on Information Science and Technology (ICIST), pp. 251–255. IEEE (2017)
16. Singh, G., Goel, A.K.: Face detection and recognition system using digital image processing. In: 2020 2nd International Conference on Innovative Mechanisms for Industry Applications (ICIMIA), pp. 348–352. IEEE (2020)

Overlapped Circular Convolution Based Feature Extraction Algorithm for Classification of High Dimensional Datasets

Rupali Tajanpure[(✉)] and Akkalakshmi Muddana

GITAM University, Hyderabad, Telangana, India
tajanpure.rupali@kbtcoe.org, amuddana@gitam.edu

Abstract. Analysis of massive data collected from various sources is the demand of today's world for different purposes like health monitoring, business, sentiment analysis etc . It is a need for real-time data processing that the processing algorithm would have less computational time and space. The feature reduction plays a vital role in this regard for high dimensional data. This paper aims to introduce the Overlapped Circular Convolution (OCC) method as a better feature reduction method. The results obtained show that the proposed method gives more than a 30% reduction in the datasets under consideration. The proposed system's computation time is less as it works on reducing features and uses Fast Fourier Transform (FFT) algorithm to find convolution. The storage space is also less for high dimensional datasets due to the reduction in input features.

Keywords: Overlapped Circular Convolution (OCC) method · Feature extraction · High dimensional datasets · FFT

1 Introduction

Nowadays, IoT-based health information systems are in demand because of their high importance application area and simplicity in recording many parameters using different sensors. The basic techniques Internet of Things (IoT) and Artificial Intelligence (AI) prove helpful in such a healthcare system. The IoT devices here records several parameters related to a patient's health and then process them by AI techniques to assist doctors to monitor a patient's health continuously [1].

The biomedical data generated using sensors have to be recorded daily or monthly as per the severity and type of disease. B. Tarle et al. presented a survey of different data mining algorithms for medical data classification [2]. The author presented a comparative study of different techniques like support vector machine, Neural Network, Decision tree etc. Generated large data needs to be processed in real-time applications to conclude. Data analytics has become a boon for today's world of real-time data processing.

Extensive data slows down the processing. Also, the performance of the classifier may get decreased. While processing the data, one must keep in mind that massive data does not lead to good analysis as many irrelevant features may be present in the data. One needs to select the features carefully in data, which directly affects the accuracy of

© Springer Nature Switzerland AG 2021
M. Singh et al. (Eds.): ICACDS 2021, CCIS 1440, pp. 214–223, 2021.
https://doi.org/10.1007/978-3-030-81462-5_20

data classification. Here the presence of irrelevant or redundant parameters may lead to misclassification of data. Feature reduction has become necessary for this reason [3].

Large amounts of data generated every day, constituting a volume, are challenging tasks to analyze. Techniques such as feature selection are advisable while working with large datasets. Parallel processing proves helpful to work with extensive dataset processing to alleviate this problem [4].

The problem in widely used minimum-Redundancy-Maximum-Relevance (mRMR) algorithm of feature reduction algorithm is overcome by Jorge Gonzalez-Dominguez et al. [5] with proposal of fast-mRMR-MPI. Here author uses MPI and OpenMP to achieve speed while feature selection.

Feature Subset Selection Problem selects a relevant subset of features from the initial set to classify future instances. Felix Garcia Lopez et al. proposed a metaheuristic algorithm to solve the problem with feature subset selection. Here the author proposed two methods for combining solutions in the Scatter Search metaheuristic. A comparison is presented between the Genetic Algorithm and the proposed method [6].

Many dimensionality reduction algorithms are present in the literature. For feature selection, filter, wrapper, and embedded or hybrid techniques are more common. Popular feature reduction algorithms include PCA, discrete wavelet transform, attribute subset selection, histogram, sampling, and clustering. Here for high dimensional datasets, it is crucial to reduce the dimensions without losing data analytics accuracy. Any feature reduction algorithm's desirable property includes retaining the accuracy of data despite a reduction in features.

This paper proposed a feature reduction algorithm that uses a convolution technique to extract the original set of features into a reduced set of features. The use of the FFT algorithm ensures less computation time. The combination of feature convolution and FFT algorithms confirms less computation time and space [7].

The paper is organized as follows. Literature survey in the dimensionality reduction domain is depicted in Sect. 2. Section 3 gives details of the proposed methodology. Section 4 elaborates on the results obtained, and Sect. 5 presents the concluding remark.

2 Literature Survey

From the partition viewpoint in a rough set, Fachao Li et al. proposed a partition differentiation entropy algorithm [8]. The proposed method finds features based on the measurement of the importance of the attributes.

Jianwen Xie et al. [9] proposed feature selection algorithm using association rule mining. In this work author found closely related attributes using association rule mining. Many real and artificial data sets are tested with the proposed strategy. This algorithm presents reduction in number of features at the cost of time complexity.

Massive data processing requires long processing time, large data storage, and some issues with high dimensional data analysis. As a consequence, it became essential to reduce datasets with minimum loss of information. Renato Cordeiro de Amorim et al. introduced the Web-Scale Minkowski weighted k-means (WSMW k-means) algorithm for data reduction. The proposed feature reduction method is a clustering algorithm that works on the weight of features [10].

Osman Gokalp et al. proposed an Iterated Greedy (IG) wrapper algorithm for sentiment analysis [11]. The author also developed a selection procedure IG algorithm. The approach uses pre-calculated filter scores. The proposed algorithm is tested on datasets related to public sentiment and some product reviews.

Lokeswari Venkataramana et al. identified two challenges parallelizing computational methods in biological data: it is multi-dimensional, and one gets reduced execution time at the cost of a reduction in the accuracy. The author proposed two approaches like vertical and horizontal partitioning. The proposed algorithm selects optimal features for classifying cancer subtypes. Parallel Random Forest and the parallel M-FS algorithms are compared with existing parallel feature selection algorithms in the paper [12].

Heng Liu et al. proposed a semi-parallel optimizing method that provides a similar feature set with a greedy feature selection approach [13]. The author focused on the computational complexity of existing algorithms and hence suggested a greedy selection-based algorithm.

Zhen Liu et al. presented a new feature extraction method that focuses on extracting irrelevant features [14]. The extracted feature-based samples are used directly on cross-platform applications. Also, the proposed method requires less computing.

Feature selection is an essential preprocessing technique for improving the learning algorithm's performance on a large-scale dataset. Majid Soheili et al. proposed Distributed quadratic programming-based feature selection (DQPFS) algorithm for large scale data [15].

Feature selection in high dimensional data with fewer samples is always a crucial data preprocessing task. Chih-Fong Tsai et al. proved that ensemble feature selection performs better in terms of classification accuracy compared to single feature selection [16].

Yonghong Peng et al. proposed a combination of filter and wrapper methods and a sequential search procedure. This proposed method differentiates individual and a subset of features using the Receiver operating characteristics (ROC) curve. This method gives better accuracy than sequential forward floating search (SFS). Also, it solves the problem of overfitting [17].

A multistage classifier system approach, proposed by Rupali Tajanpure, was based on feature decimation as per their level of processing need. The first level features are processed first, and as per the output of the first classifier, it decides to go for processing the next level of attributes or stop there itself [18].

Yi Wanga et al. proposed an artificial immune system-based feature selection algorithm [19]. The proposed algorithm forms a cluster of relevant features in the sample space. Due to the local clustering, inter-class distance is maximized, and intra-class distance is minimized for a sample's features.

K. Keerthi Vasan et al. [20] focused on efficiency of PCA in intrusion detection system. Author experimented based on reduction ratio, impact of noisy data on PCA and ideal number of principal components. Experiments on benchmark dataset shows that first 10 principal components are important for classification.

For high dimensional datasets, feature selection and classification become challenging. Since all features do not contribute to classification, some features even degrade the performance. A feature subset containing essential features increases the classification accuracy and overall performance. Bin Cao et al. [21] constructed a model that simultaneously works on the classification error and feature redundancy.

Machine learning and data mining face the issue of feature selection. Demand is that a feature selection algorithm must be with minor error and high performance. Ivan Smetannikov [22] proposed the MeLiF algorithm using ensembles of ranking filters.

Georgios Kontonatsios et al. [23] worked on the citation screening process with a neural network-based feature extraction algorithm. The proposed feature extraction method leads to average workload savings of 56% across 23 medical datasets.

Jinghua Liu proposed a feature selection method based on the specific ability of the features. The maximum nearest neighbour concept is used to distinguish between the features. Benchmarked datasets are tested using proposed algorithms and compared with some feature selection algorithms like Cart, LSVM, KNN etc. [24].

Researchers for different applications propose several dimensionality reduction algorithms. These algorithms work with some advantages and limitations like overfitting, higher time complexity etc. Also, relevant feature selection improves the performance of the classification algorithm. Therefore, concentrating on classifier and feature reduction performance, we proposed a feature reduction method based on circular convolution. This proposed system's main aim is to develop feature reduction strategy, which will prove helpful over high dimensional datasets in computation time and space.

3 Proposed System Design

We proposed an overlapped circular convolution feature reduction strategy based on convolution. This strategy uses the overlapping of results of circular convolution to obtain a twofold reduction in features. The work started with the aim of feature reduction. While studying the property of Discrete Fourier Transform, the circular convolution property found interesting. With this circular convolution approach, the output contains samples equal to samples in one input sequence. So, we started with experimentation and proposed the feature reduction FrbyCC (Feature reduction using circular convolution), which uses the circular convolution technique. In FrbyCC, we divided the input tuple into two parts so that each part will contain values nearly equal to 2^n. Then Discrete Fourier Transform (DFT) is used to find out the circular convolution of the two parts of the input tuple. Since this algorithm uses DFT, the number of input values should be in the range of 2^n. Hence Table 1 takes the values 4, 8, 16 etc. The application of the FrbyCC algorithm shows similar accuracy despite feature reduction. The result is compared with the Principal Component Analysis (PCA) algorithm.

In the FrbyCC system, calculating the input tuple's circular convolution, Discrete Fourier transform (DFT) is used. But in the proposed method, instead of DFT, Fast Fourier transform (FFT) is used to reduce the proposed algorithm's computational time. FFT algorithms have less computational complexity, as shown in Table 1, [7].

Table 1. Comparison of computations needed for DFT versus FFT algorithm

Number of values (N)	No. of multiplications in DFT (N^2)	No. of complex multiplications in FFT ($N/2 \log_2 N$)	Speed improvement factor
4	16	4	4.0
8	64	12	5.3
16	256	32	8.0
32	1024	80	12.8

And so on.

The architecture of the proposed system based on convolution is as follows:

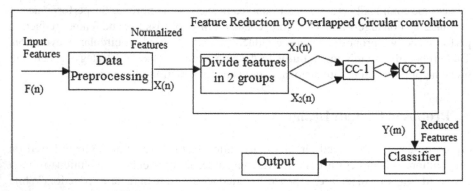

Fig. 1. Proposed feature reduction system

The proposed system is built by keeping a focus on feature reduction. This work aims to check an effect on accuracy of the classifier system when convolution is applied twice on the input data set to get the feature reduction to 30–50%. As shown in Fig. 1, the input dataset undergoes the data preprocessing. In data preprocessing, the attributes with missing values are replaced with the average of the attribute columns. Also, the tuples are normalized so that they have the numerical value lying between 0–1.

After data preprocessing, the input tuple is broken into two small length sequences with the nearest 2^n points in each. This leads to the two sequences to be convolved circularly. The resultant sequence contains the extracted attributes which is a circular convolution of input. The result contains number of attributes equal to samples in one input sequence. This gives the reduction in features. To achieve more reduction, this output is again broken into two parts. Then these short length sequences are convolved to get the final result. Hence the system is named as overlapped circular convolution (OCC). The procedure of feature reduction by OCC is summarized as follows,

Algorithm: Feature reduction by overlapped Circular Convolution
Input: F(n) -set of input features.
Output: Y(n)- reduced set of features.

Begin

1. Preprocess the input dataset. The categorical values, if present, are first converted in numerical form. The average of the attribute column replaces missing values. Then normalization is carried out to bring all data on one platform of range (0-1). The formula applied for normalization is

$$X' = (X - Xmin) / (Xmax - Xmin) \qquad (1)$$

Where,

X' is calculated normalized value
X is value considered to be normalized
Xmin is the minimum value of dimension
Xmax is the maximum value of dimension

2. Break the input dataset tuple F(n) in two sets $x_1(n)$ and $x_2(n)$ containing nearly equal elements
3. Each block is zero-padded to have a number of values equal to the selected upper nearest value of 2^n
4. Compute Circular convolution using FFT algorithm [25]
   a. Find the DFT of the first input sequence $x_1(n)$ to get $X_1(K)$
   b. Find the DFT of second input sequence $x_2(n)$ to get $X_2(K)$
   c. Multiply the two DFTs $X_1(K)$ and $X_2(K)$ using point by point multiplication to get $Y(K)$
   d. Find Inverse Discrete Fourier Transform (IDFT) of $Y(K)$ to get $Y(m)$.
5. Steps 2-4 are repeated on the output of step 4 to get the final output y(n)
6. The output y(n) of step 5 is given as input to the classifier to judge the proposed feature reduction algorithm's performance.

End

4 Experimental Setup and Evaluation

Five datasets with hundreds of features are used to evaluate the proposed algorithm from the UCI repository [26]. Here the purpose of selecting a large attribute dataset is to show the reduction in features dominantly. The basic information of the datasets is as in Table 2.

Table 2. Datasets under consideration.

Dataset	No. of tuples	No. of features
Parkinson's disease	756	754
Arrhythmia	452	280
Internet Ads	3279	782
QSAR	1687	1025
SCADI	70	206

In the proposed method, we divided each dataset tuple into two set. The circular convolution of two sets is calculated. The resultant set is again divided and circularly convolved to get the final output, and then tested for accuracy with different classifiers. The percent of reduction achieved with the OCC method is given in Table 3.

Table 3. % feature reduction with OCC.

Name of dataset	Tuples	Attributes	OCC stage1 extracted attributes	OCC stage2 extracted attributes	% Reduction in attributes	% reduction in attributes after stage 2
Parkinson's disease	756	754	512	256	32.09	66.05
Arrhrythmia	452	280	128	64	54.28	77.14
Internet Ads	3279	782	512	256	34.52	67.26
QSAR	1687	1025	512	256	50.04	75.02
SCADI	70	206	64	32	68.93	84.47

The performance of OCC output is tested by using different classifiers. Support vector machine (SVM), Decision Tree (DT) and K- Nearest Neighbour classifier is used for the same. The output of classifiers is compared with the existing PCA algorithm. Also, it is compared with a single application of circular convolution, i.e., the FrbyCC algorithm as a base of the OCC algorithm. SVM classifier results after feature reduction by the OCC method for different datasets are as in Table 4.

Table 4. Comparison of SVM Classifier accuracy obtained with FrbyCC, OCC and PCA feature reduction algorithms.

Name of dataset	% Accuracy of SVM classifier		
	CC	OCC	PCA
Parkinson's disease	74.60	78.04	74.6
Arrhrythmia	54.20	58.19	54.2
Internet Ads	90.88	90.48	91
QSAR	88.10	88.38	88.14
SCADI	77.14	77.14	77.14

Table 5. Comparison of KNN Classifier accuracy obtained with FrbyCC, OCC and PCA feature reduction algorithms.

Name of dataset	% Accuracy of KNN classifier		
	CC	OCC	PCA
Parkinson's disease	82.9	81.22	74.34
Arrhrythmia	60.18	54.87	54.2
Internet Ads	97.62	55.87	46.35
QSAR	90.10	87.08	88.2
SCADI	87.14	81.43	77,14

From the above, it is clear that SVM with OCC feature reduction retains or gives better accuracy for different datasets. So SVM classifier can work well with this transformed attribute set after the application of Overlapped Circular Convolution.

Similarly, the OCC output is evaluated with KNN and DT classifiers. Table 5 and Table 6 shows the results of accuracy for the KNN and DT classifier, respectively.

From Table 5, it is clear that KNN also works well for all the datasets under consideration since after convolution, the dataset obtained is categorized/ clustered well.

From Table 6, it is clear that the DT classifier cannot decide on the convolved dataset. The output of OCC is extracted with convolution of features hence the DT classifier is unable to split well on the attributes. The Table 6 shows that on SCADI dataset the accuracy decreases drastically since the features get more interlinked while extraction.

Table 6. Comparison of DT Classifier accuracy obtained with FrbyCC, OCC and PCA feature reduction algorithms.

Name of dataset	% Accuracy of DT classifier		
	CC	OCC	PCA
Parkinson's disease	79.76	79.23	72.2
Arrhrythmia	62.39	55.75	61.28
Internet Ads	100	90.24	90.11
QSAR	85.54	81.33	85.36
SCADI	98.57	70	95.71

It is observed from Table 3 that feature reduction is more if the number of attributes in a dataset is near to any 2^n value. Otherwise, due to zero padding, the reduction in attributes is limited.

5 Conclusion

Feature reduction is an important aspect for high dimensional data analysis, and feature reduction without loss of accuracy is an essential and desirable parameter for every feature reduction algorithm. In this paper, we proposed a feature reduction algorithm based on circular convolution. The algorithm is applied twice to get more reduction in features. This algorithm works very well, resulting in more than 30% feature reduction. The proposed algorithm gives inline accuracy when compared with the PCA and single application of the same algorithm. The FFT algorithm and twice reduction in features make the proposed system to have less classifier computation time. Less storage space is achieved with feature reduction too. So, the proposed method proves to be better for high dimensional data analysis.

References

1. Fouad, H., Hassanein, A.S., Soliman, A.M., Al-Feel, H.: Analyzing patient health information based on IoT sensor with AI for improving patient assistance in the future direction. Measurement **159**, 107757, ISSN 0263–2241 (2020). https://doi.org/10.1016/j.measurement. 2020.107757. http://www.sciencedirect.com/science/article/pii/S0263224120302955
2. Tarle, B., Tajanpure, R., Jena, S.: Medical data classification using different optimization techniques: a survey. Int. J. Res. Eng. Tech. (2016)
3. Han, J., Kamber, M.: Data Mining: Concepts and Techniques, 3rd edn. Morgan Kaufmann Publishers (2006)
4. Eiras-Franco, C., Bolón-Canedo, V., Ramos, S., González-Domínguez, J., Alonso-Betanzos, A., Touriño, J.: Multithreaded and spark parallelization of feature selection filters. J. Comput. Sci. **17**, 609–619 (2016)
5. Gonzalez-Dominguez, J., Bolon-Canedo, V., Freire, B., Tourino, J.: Parallel feature selection for distributed-memory clusters. Inf. Sci. **496**, 399–409, ISSN 0020-0255 (2019). https://doi. org/10.1016/j.ins.2019.01.050
6. Garcia Lopez, F., Garcia Torres, M., Melian Batista, B., Moreno Perez, J.A., Marcos Moreno-Vega, J.: Solving feature subset selection problem by a parallel scatter search. Eur. J. Oper. Res. **169**, 477–489 (2006)
7. Proakis, J.G., Manolakis, D.K.: Digital Signal Processing: Principles, Algorithms, and Applications, 3rd edn. Pearson Publications (1996)
8. Li, F., Zhang, Z., Jina, C.: Feature selection with partition differentiation entropy for large-scale datasets. Inf. Sci. **329**, 690–700 (2016)
9. Xie, J., Wu, J., Qian, Q.: Feature selection algorithm based on association rules mining method. In: 2009 8th IEEE/ACIS International Conference on Computer and Information Science, Shanghai, China, pp. 357–362 (2009). https://doi.org/10.1109/ICIS.2009.103
10. Cordeiro de Amorim, R.: Unsupervised feature selection for large data sets. Pattern Recogn. Lett. **128**, 183–189 (2019)
11. Gokalp, O., Tasci, E., Ugur, A.: A novel wrapper feature selection algorithm based on iterated greedy metaheuristic for sentiment classification. Expert Syst. Appl. **146**, 113176
12. Venkataramana, L., Gracia Jacob, S., Ramadoss, R.: A parallel multilevel feature selection algorithm for improved cancer classification. J. Parallel Distrib. Comput. **138**, 78–98 (2020)
13. Liu, H., Ditzler, G.: A semi-parallel framework for greedy information-theoretic feature selection. Inf. Sci. **492**, 13–28 (2019)

14. Liu, Z., Japkowicz, N., Wang, R., Cai, Y., Tang D., Cai, X.: A statistical pattern-based feature extraction method on system call traces for anomaly detection information and Software. Technology **126**, 106348 (2020)
15. Soheili, M., Eftekhari-Moghadam, A.M.: DQPFS: distributed quadratic programming-based feature selection for big data. J. Parallel Distrib. Comput. **138**, 1–14 (2020)
16. Tsai, C.F., Sung, Y.T.: Ensemble feature selection in high dimension, low sample size datasets: parallel and serial combination approaches. Knowl. Based Syst. **203**, 106097 (2020)
17. Peng, Y., Wu, Z., Jiang, J.: A novel feature selection approach for biomedical data classification. J. Biomed. Inform. **43**, 15–23 (2010)
18. Tajanpure, R.R., Jena, S.: Diagnosis of disease using feature decimation with multiple classifier system, In: Dash, S., Das, S., Panigrahi, B. (eds.) International Conference on Intelligent Computing and Applications. Advances in Intelligent Systems and Computing, vol. 632, Springer, Singapore (2018)
19. Wanga, Y., Li, T.: Local feature selection based on the artificial immune system for classification. Appl. Soft Comput. J. **87**, 105989 (2020)
20. Keerthi Vasan, K., Surendiran, B.: Dimensionality reduction using principal component analysis for network intrusion detection. Perspect. Sci. **8**, 510–512 (2016)
21. Cao, B., et al.: Multiobjective feature selection for microarray data via distributed parallel algorithms. Future Gener. Comp. Syst. **100**, 952–981 (2019)
22. Smetannikov, I., Isaev, I., Filchenkov, A.: New approaches to parallelization in filters aggregation based feature selection algorithms. Procedia Comput. Sci. **101**, 45–52 (2016)
23. Kontonatsios, G., Spencer, S., Matthew, P., Korkontzelos, I.: Using a neural network-based feature extraction method to facilitate citation screening for systematic reviews. Expert Syst. Appl. **X**(6), 100030 (2020)
24. Liu, J., Lin, Y., Lin, M., Wu, S., Zhang, J.: Feature selection based on quality of information. Neurocomputing **225**, 11–22 (2017)
25. Oppenheim, A.V., Schafer, R.W.: Digital Signal Processing, 1st edn. Pearson (1975)
26. Dua, D., Graff, C.: UCI Machine Learning Repository. University of California, School of Information and Computer Science, Irvine, CA (2019). http://archive.ics.uci.edu/ml

Binary Decision Tree Based Packet Queuing Schema for Next Generation Firewall

Manthan Patel[✉] and P. P. Amritha[✉]

TIFAC-CORE in Cyber Security, Amrita School of Engineering, Coimbatore,
Amrita Vishwa Vidyapeetham, Coimbatore, India
cb.en.p2cys19011@cb.students.amrita.edu, pp.amritha@cb.amrita.edu

Abstract. Technology is drastically increasing and internet is most important part of this change. Security is very critical parameter in the internet environment. Network security is first concern in every industrial world. To protect data and network availability traditional firewall is not capable. To secure network environment Next Generation Firewall (NGFW) is the powerful device. NGFW is having various feature which will protect network from modern attacks. Firewall efficiency and accuracy are major factor which will affect the network performance. In this paper we proposed Binary decision tree based packet queuing model for Next Generation Firewall. This model is prioritizing the UDP packer over TCP. In VoIP (Voice over IP) service UDP ports are most frequently used. In this model we proposed two buffers, First one is High priority (HP) buffer and second one is Low priority (LP) buffer. When network packet will come to the NGFW first it will labeled according to predefined rules. Then that packet will store in HP or LP after passing through training dataset. This model will check duplicate packets in buffer which will help firewall process at initial stage for removing duplicate packets.

Keywords: Next Generation Firewall · Packet queuing schema · Binary decision tree algorithm · Packet prioritization

1 Introduction

Network security is the primary element for every industry's defenses system, the foundation on which other protection services such as endpoint, server, mobile and encryption are layered. The expanding complexity of networks and the need to make them increasingly open due to the growing emphasis on and attractiveness of the Internet as a mode for business transactions, imply that networks are exposed to increasingly more attacks, both from without and from within. One of the protective mechanisms under serious consideration is the firewall which protects a network by guarding the points of entry to it [5]. Traditional and stateful firewalls will filter network packets based on the predefined rules which is not sufficient to protect the network against DDoS, MITM (Man in the middle)

© Springer Nature Switzerland AG 2021
M. Singh et al. (Eds.): ICACDS 2021, CCIS 1440, pp. 224–233, 2021.
https://doi.org/10.1007/978-3-030-81462-5_21

and many more powerful attacks. Next generation firewall will use deep packet inspection and other different features to protect network environment from modern and powerful attacks. Stateful firewalls are providing stateful inspection on the outgoing and incoming network packets but with NGFW is providing extra feature like application integrated intrusion prevention, cloud-delivered threat intelligence and awareness and control [14]. On the network traffic traditional firewall will give stateful protection which will allows or blocks network flow depends on the port detail, protocol and state but in NGFW every packet will go through the deep packet inspection which will check all the parameters of network packet.Role mining helps in creating a Role-Based Access Control (RBAC) system from a discretionary access control (DAC) system [16].

1.1   Key Points of Next Generation Firewall

1. Application Awareness: NGWF will give all the visibility over the application security. It will allow and block traffic according to customer requirements. This feature will give control on each applications.
2. Identity Awareness: With use of this feature, administrator can check every activity of the users. Which will show how much bandwidth is used by users.
3. Centralized Management, Administration, Logging and Reporting: NGFW is used centralized management to maintain every firewall which is connected into the network. This feature will help administrator to do health check of every devices through remote service.
4. State-full Inspection: NGFW will do traffic inspection from layer 1 to layer 7 which will give more control on the network traffic flows. Traditional firewall is used to control traffic flow from the layer 4 to layer 7.
5. Integrated IPS: Intrusion Prevention System is main advantage of the next generation firewall. IPS can prevent network from the modern attacks. IPS is detecting and preventing attacks. It will send the event logs to the administrator. Intrusion detection and Privation system (IDS/IPS) is important layer of security in every IT environment due to an urge towards cyber safety in the day-to-day world [18].
6. Sandboxing: Next Generation Firewall is providing isolated environment for the research and development testing. In this environment admin can practice different rules and observed the outcome of it.

1.2   Binary Decision Tree Algorithms

Binary decision trees based non-parametric data statistical models offer a solution to difficult decision issues where many groups and many characteristics linked in a complex way are available. Based on the defined rules binary decision tree will classifies a pair of cases into the different classes [1]. Decision tree's nodes are having attribute names. These attributes are used for labelling the possible values on edges and various classes are represented by leaf nodes. Every parent node have two child nodes with their respective values.

1.3 Packet Queuing Schema

Next Generation Firewall is most important packet filtering device which will redirect the data flow over a network. NGFWs is having more input/output nodes which is transmitting and receiving packets in network environment. NGFW's memory is finite, So When packets will arrive more then allocated memory to the firewall in such situation NGFW will act according to the algorithm which is set primarily in it [7]. This situation will occur when firewall's memory is lesser then the arrival of network packet rate. At that time firewall will ignore the newly arrived packet or it will drop the older packets which is stored in the buffer. As part of the resource allocation mechanisms, Firewalls must implement some queuing discipline that governs how packets are buffered or dropped when required.

1. First-In, First-Out Queuing (FIFO): In this packet queuing schema, Firewall will give priority to the older packets first which is arrived earlier and it will drop the newly arrived packets. This algorithm is working on first come fist serve manner (Fig. 1).

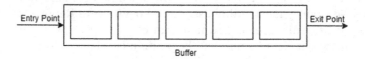

Fig. 1. FIFO method

2. Priority Queuing (PQ): This queuing schema will give priority to the network packet according to predefined rules. If arrived packet rate will get higher then the firewall's memory, this packet queuing schema will give priority to the important network packet according to predefined rules set in it (Fig. 2).

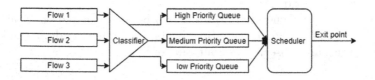

Fig. 2. Priority queue method

3. Weighted Fair Queuing (WFQ): According to assigned weights this packet scheduling schema is giving fair output bandwidth sharing. Weighted fair queue is a variant of fair queue equipped with a weighted bandwidth allocation. The bandwidth allocation among packet queues involves not only the traffic discrimination but the weightings assigned to packet queues (Fig. 3).

2   Proposed Model

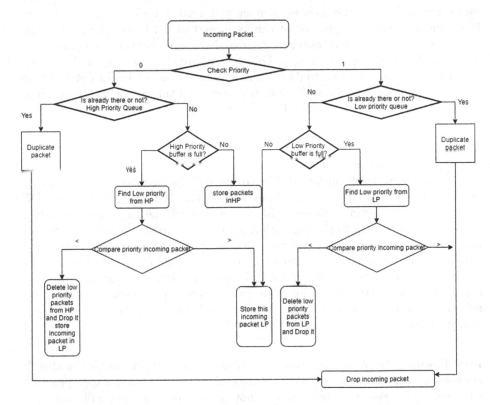

Fig. 3. Binary decision tree based packet queuing schema

Network firewall is very important security device in industrial environment. Our purpose is to make that device more secure, faster and accurate so that it can prevent whole network environment and industrial data. We proposed Genetic Binary Decision Tree based Packet queuing schema to prioritize the network packets. This model is prioritising the UDP packet over TCP. In this proposed model, We mostly focused on VoIP applications like Microsoft teams, Skype, Whatsapp calling, etc. This applications are using UDP packets to create communication. For an example Microsoft Team application is using UDP ports 6579-6581. TCP packets has the feature of Acknowledgement so that when sender is sending any information to the receiver, Receiver will respond with the acknowledgement. Here sender will get to know that receiver got that packet (message) or not. There is predefined timeout already set on the sender side, If in case sender is not receiving any acknowledgment from the receiver side at that moment sender will assume that packet is not received properly and sender will resend the same packet to the receiver. This feature is not available in UDP communication. So that if any packet will drop in between UPD communication. It will take time to reconnect that communication back. In case of buffer over flow or in very high traffic flow we can drop TCP packet because if sender won't get ACK from the receiver, Sender will send it back but we can't drop UDP packet. Our main purpose of this proposed model is to amplify the time efficiency and accuracy in queuing schema. Now a days every companies are accepting IoT (Internet of Things), This devices are also working on UDP port. In VoIP call we can't afford to drop a single network packet because it will take time to reconnect the communication. Because of all this reason we proposed to prioritize the UDP ports over TCP.

High Priority Buffer. In proposed model, we choose binary decision tree algorithm to prioritize UDP packet over TCP. When packet will arrive to the firewall first it will get labeled with set of predefined rules. Here we are giving label 0,1 to the UDP packets and label 2, 3 to the TCP packets. If it is UDP packet or If the packet is labeled with 0 it will check the High priority buffer. If space is available in to the high priority buffer it will simply put that packet into the buffer. But if there is no space available into the High priority buffer it will check all the packets in the buffer if any packets are there with low priority compare to arrived packet algorithm will drop that stored packet and put this arrived packet into the high priority buffer. If there is no packet which is low in priority compare to arrived packet, program will put that packet into the low priority buffer.

Low Priority Buffer. Same process will work for Low priority buffer also. If network packet is low prioritized on bases of predefined rules. First, program will check low priority buffer is full or not. If it is full program will check the other network packet priority which is already in queue of buffer. If program will find any low priority network packet into the queue compare to newly arrived network packet. It will drop that stored packet from the queue and store this newly arrived network packet into the buffer. If program is not able to find any low priority network packet into the queue of buffer at that time it will drop newly arrived packet.

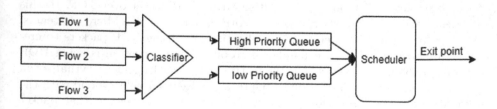

Fig. 4. UDP and TCP buffers

TCP and UDP Buffer Queue. As shown in the Fig. 4 there will be two buffer, first one is High priority UDP buffer and another one is low priority TCP buffer. Proposed model will put network packet into this buffer as per the binary decision tree based classification.

3 Experiments and Results

We have implemented this proposed binary decision tree based algorithm. For this practical we tried two different approaches. In first approach we used one dataset and we split it into the training & testing datasets. In second approach we captured live packets for testing dataset.

3.1 Experimental Phase 1

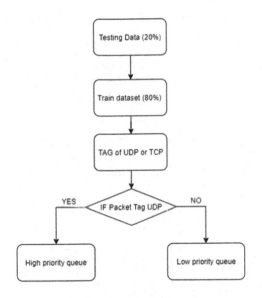

Fig. 5. Experimental phase 1

In the first phase of implementation, we took one dataset and we segregate that dataset into the two parts. We took 80% of that dataset as a training dataset and 20% of that dataset as a testing dataset. When we start predicting binary decision tree algorithm, it is giving 99.98% accuracy and for one predication it is taking 0.0001 s which is more efficient according with the time parameter. As we can see in the Fig. 5, From the 715460 packets program predicted 291413 packets as a TCP and 424047 packets as UDP packets. Confusion matrix also showing us 99.98% of accuracy to predict and put network packet into the correct buffer.

```
[[424047        0]
 [      0 291413]]
                precision      recall  f1-score      support

            0       1.00        1.00      1.00        424047
            1       1.00        1.00      1.00        291413

    accuracy                              1.00        715460
   macro avg        1.00        1.00      1.00        715460
weighted avg        1.00        1.00      1.00        715460
```

Fig. 6. Result of phase 1

3.2   Experimental Phase 2

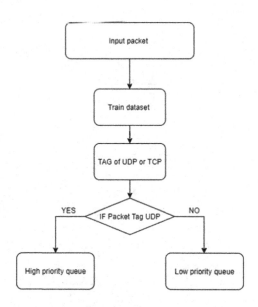

Fig. 7. Experimental phase 2

In second phase we captured the network packet through wireshark while calling in Microsoft teams. We stored that dataset into the excel file. We used that dataset as a testing dataset. We applied this testing dataset to the trained model of binary decision tree. In this phase 2 we got prediction accuracy 99.99% and time efficiency also we got as a 0.0001 s. as shown in the Fig. 6 we can say that the confusion matrix also having zero wrong prediction and accuracy also we got as expected (Fig. 7).

```
[[937    0]
 [  0 503]]
              precision     recall  f1-score    support

           0       1.00       1.00      1.00        937
           1       1.00       1.00      1.00        503

    accuracy                            1.00       1440
   macro avg       1.00       1.00      1.00       1440
weighted avg       1.00       1.00      1.00       1440
```

Fig. 8. Result of phase 2

3.3 Time Efficiency

Time is one of the major factor in the experiments. Next Generation Firewall is use to protect network environment so that this device should be faster as much as it possible. When we started apply testing dataset to the training dataset. As shown in Fig. 8, for one prediction it is taking average 0.1 ms which is equal to 0.0001 s. When this code will implement in hardware level it will give 10 to 20 time faster speed. So with this model we can utilized the bandwidth of network transmission line. This binary decision tree based proposed model will give more efficiency in terms of time and speed (Fig. 9).

```
time==> 0.31830000079935417 milliseconds
time==> 0.25630000163801014 milliseconds
time==> 0.19109999993816018 milliseconds
time==> 0.23949999922479037 milliseconds
time==> 0.3015000002051238 milliseconds
time==> 0.3128000007563969 milliseconds
time==> 0.29180000092310365 milliseconds
time==> 0.26389999948150944 milliseconds
time==> 0.149200001033023 milliseconds
time==> 0.23689999943599105 milliseconds
time==> 0.18310000086785294 milliseconds
time==> 0.09759999920788687 milliseconds
time==> 0.08120000165945385 milliseconds
time==> 0.08010000055946875 milliseconds
time==> 0.07860000005166512 milliseconds
time==> 0.0777000004745787 milliseconds
time==> 0.07669999831705354 milliseconds
time==> 0.07689999983995222 milliseconds
time==> 0.07619999996677507 milliseconds
time==> 0.0751000068577938 milliseconds
```

Fig. 9. Time efficiency

4   Conclusion and Future Work

We proposed Binary decision tree algorithm for network packet queuing model in next generation firewall to improve the efficiency and accuracy of packet queuing model. We mainly focused on prioritization of the network packet which will give important to UDP packet over TCP. In implementation we got 99.98% of accuracy to classify the packet into the UDP and TCP in very less amount of time. In this model we are removing duplicate packets at initial state of the NGFW which is giving extra security layer in firewall. With use of this model we can make Next Generation Firewall more effective and accurate. In future work, we can make this model more robust and faster.

References

1. Kathuria, M., Gambhir, S.: Genetic binary decision tree based packet handling schema for WBAN system. In: 2014 Recent Advances in Engineering and Computational Sciences (RAECS). IEEE (2014)
2. Indira, B., Valarmathi, K., Devaraj, D.: An approach to enhance packet classification performance of software-defined network using deep learning. Soft. Comput. **23**(18), 8609–8619 (2019). https://doi.org/10.1007/s00500-019-03975-8
3. Zapechnikov, S., Miloslavskaya, N., Tolstoy, A.: Modeling of next-generation firewalls as queueing services. In: Proceedings of the 8th International Conference on Security of Information and Networks (2015)
4. Chakraborty, P., Rahman, M.Z., Rahman, S.: Building new generation firewall including artificial intelligence. Int. J. Comput. Appl. **975**, 8887
5. Neupane, K., Haddad, R., Chen, L.: Next generation firewall for network security: a survey. In: SoutheastCon 2018. IEEE (2018)
6. Liu, E.: Firewall control system based on a next generation network service and method thereof. U.S. Patent No. 7,987,503, 26 July 2011
7. Belz, D., et al.: Pipelined packet switching and queuing architecture. U.S. Patent No. 6,980,552, 27 December 2005
8. Wyatt, R.M.: Port packet queuing. U.S. Patent No. 7,236,489, 26 June 2007
9. Tsigas, P., Zhang, Y.: A simple, fast and scalable non-blocking concurrent FIFO queue for shared memory multiprocessor systems. In: Proceedings of the Thirteenth Annual ACM Symposium on Parallel Algorithms and Architectures (2001)
10. Devassy, R., et al.: Reliable transmission of short packets through queues and noisy channels under latency and peak-age violation guarantees. IEEE J. Sel. Areas Commun. **37**(4), 721–734 (2019)
11. Ni, S.-h.: Algorithm for time based queuing in network traffic engineering. U.S. Patent No. 7,042,843, 9 May 2006
12. Ma, Y., et al.: Queuing model and delay analysis on network coding. In: IEEE International Symposium on Communications and Information Technology 2005. ISCIT 2005, vol. 1. IEEE (2005)
13. Bouchard, G.A., et al.: Packet queuing, scheduling and ordering. U.S. Patent No. 7,895,431, 22 February 2011
14. Soewito, B., Andhika, C.E.: Next generation firewall for improving security in company and IoT network. In: 2019 International Seminar on Intelligent Technology and Its Applications (ISITIA). IEEE (2019)

15. Freet, D., Agrawal, R.: Network security and next-generation firewalls. In: Proceedings of International Conference on Technology Management (ICTM 2016) (2016)
16. Vinayakumar, R., Soman, K.P., Poornachandran, P.: Evaluating effectiveness of shallow and deep networks to intrusion detection system. In: 2017 International Conference on Advances in Computing, Communications and Informatics (ICACCI), pp. 1282–1289. IEEE (2017)
17. Dath, A., Praveen, K.: Role mining in distributed firewall using matrix factorization methods. In: 2020 4th International Conference on Trends in Electronics and Informatics (ICOEI), no. 48184, pp. 625–629. IEEE (2020)
18. Vigneswaran, R.K., Vinayakumar, R., Soman, K.P., Poornachandran, P.: Evaluating shallow and deep neural networks for network intrusion detection systems in cyber security. In: 2018 9th International Conference on Computing, Communication and Networking Technologies (ICCCNT), pp. 1–6. IEEE (2018)

Automatic Tabla Stroke Source Separation Using Machine Learning

Shambhavi Shete[1]($\boxtimes$) and Saurabh Deshmukh[2]($\boxtimes$)

[1] Career Point University, Kota, Rajasthan, India
[2] Maharashtra Institute of Technology, Aurangabad, Maharashtra, India

Abstract. Percussion instrument Tabla plays a vital role in North Indian Classical Music. It is one of the most significant accompanying rhythm instruments used in classical and semi-classical forms. For musicians and instrument players, it is important to be precise about the exact Tabla stroke being played in the music. In the computer-music interface, automatic recognition of such Tabla stroke is useful for automatic score generation, music transcription, and music synthesis. In this paper, we have addressed the less explored problem of automatic Tabla stroke source separation. The system detects the onsets of audio that contains continuous Tabla strokes and segments them to represent a single Tabla stroke. Mel frequency cepstral coefficient (MFCC), Timbral, attack, decay, sustain, release, and fundamental frequency features are extracted from each audio segment. Finally, multi-support vector machine (MSVM) and feed-forward back propagation neural network (FFBPNN) machine learning algorithms are compared to detect the Tabla stroke source. From the results, it is concluded that the sound source separation of the left drum is inferior to the right drum and both the drums simultaneously due to damping applied to its production. On the other hand, the Timbral audio descriptors are very useful to separate the right drum strokes which are harmonic. As a result, the highest drum stroke source separation accuracy is 91.15% for right drum strokes.

Keywords: Music information retrieval · North Indian Classical Music · Tabla · Timbre · Machine learning

1 Introduction

Sound is a physical phenomenon that is generated by hitting one entity over the other. The entity could be metal, wood, or skin. Every day we hear different types of sounds. We are surrounded by sounds like a car horn, railway engine sound, airplane sound, sound of a river, trees, birds, animals, etc. The sound is majorly classified as pleasant and unpleasant based on the perceptual beliefs and intensity of the sound. The pleasant sounds are organized frequencies called music. Music contains a blend of different audio frequencies that gives a pleasant experience to the human ear. With the advancements of technology, computer music analysis and synthesis have become simple.

In computer music analysis, different technologies evolved that can analyze a piece of music with different perspectives. A lot of musical information could be retrieved and

© Springer Nature Switzerland AG 2021
M. Singh et al. (Eds.): ICACDS 2021, CCIS 1440, pp. 234–243, 2021.
https://doi.org/10.1007/978-3-030-81462-5_22

preserved for further synthesis of the sound. In this paper, we are presenting a method that automatically identifies and separates the source of Tabla stroke. Tabla is a musical instrument that has two drums. The sound source origin out of two drums is identified and separated to label the Tabla stroke type. Musical instrument identification requires analysis and classification of sound based on frequencies and energies involved in the sound [1].

There are majorly three types of musical note textures. Monophonic music consists of a single melody line. This type of music has no harmony or counterpoint. It may consist of rhythm instruments but that also containing only one specific pitch throughout. Indian Classical music may be termed as monophonic when a singer practicing along with the only Tabla. But the involvement of the Tanpura instrument always brings harmony to the singing. An example of monophonic music could be a person whistling a song or simply a tune [2].

Homophonic music involves other entities such as other singers or a group of singers singing in different pitches and volumes giving a feel of harmony in the music. This type of music can also involve other musical instruments. Here the main singer singing a melody line is accompanied by other instruments such that together there is a harmony in the music. It is much relevant to North Indian classical music in that it involves a continuous play of Tanpura and accompanying instruments such as a harmonium or a violin giving the feel of Homophonic music containing harmony [3].

The musical instrument Tabla also exhibits properties of resonating percussion instruments producing homophonic texture of the sound. It is important to note that, in contrast with western rhythm instruments, the sound produced from the left drum, right drum, and both the drums together generates homophonic texture of sound. For the polyphonic type of music, it is comparatively simple to bifurcate the frequencies of different musical instruments. On the other hand, in North Indian Classical Music (NICM), homophonic sound version, the frequencies blend such that they are difficult to separate. Hence a major research gap is found in the source separation of North Indian Classical Music's Tabla instrument sound.

2 Literature Survey

Percussion instrument plays an important role in Indian Music concerts. A Tabla is useful as an accompanying musical instrument for various vocal recitals such as Raga, Tappa, Ghazal, Thumri, Bhajan, etc. A rhythmic structure of the strokes played periodically on Tabla is called Tala. Depending upon the strokes being played, different versions of Tala are generated. The Musical Instrument Tabla constitutes of two pieces of instruments, namely Tabla (right drum) and Dagga (left drum) but the collective name for both the drums is Tabla [4]. The instrument is played by two hands and strokes are generated by hitting the surface of each drum with fingers. Dagga is made up of steel in the shape of a pumpkin. Animal skin is tightly applied on the opening side of the drum. A black colored paste of rice flour and applied in the middle of the drum. The right drum Tabla is a spherically shaped drum made up of wood. Typically, this wood is teak wood. The size of this drum is smaller than the left drum. Both the drums are laced. The Tabla is tuned using wooden dowels attached to the laces.

The Tabla stroke types based on lifting of the hand are open hits and closed hits. When the fingers of the Tabla players are kept stuck on the skin of the drum producing closed hit strokes. The sustain time of these strokes is abruptly close to zero. On the other hand, when the fingers that hit the skin of the drum are lifted after hit producing a sound that has various sustaining harmonics in it. Many times, when alternately two open hits are played a sustained sound of previous strokes mixes with the hit sound of the next stroke. The strokes produced out of the left drum exhibit lower frequencies as compared to the right drum frequencies. The left drum generates bass frequencies. Table 1 shows nine basic Tabla strokes, their sources, and method to produce the strokes.

Table1. Basic Tabla strokes

Sr. No	Stroke name	Stroke source	Stroke type	Method of production
1	Na	Tabla (Right Drum)	Open	Played by using forefinger on 'Rim'
2	Ta	Tabla (Right Drum)	Open	Played by using forefinger on 'Between Rim and Ink'
3	Te	Tabla (Right Drum)	Closed	Played by using palm on 'Ink'
4	Tun	Tabla (Right Drum)	Open	Played by using forefinger on 'Ink'
5	Ga	Dagga (Left Drum)	Closed	Played by using middle finger at the center
6	Ka	Dagga (Left Drum)	Closed	Played by using Palm at the center
7	Dha	Both (Tabla + Dagga)	Open	Na + Ga
8	Dhin	Both (Tabla + Dagga)	Open	Ta + Ga
9	Tin	Both (Tabla + Dagga)	Open	Ta + Ka

The Tabla strokes sources are classified into three parts. Strokes produced from the left drum, right drum, and simultaneously produced from both the drums. Some special strokes are generated by alternately hitting both these drums which are not considered here. The Tabla strokes could be produced by various combinations and patterns of playing the drums. A total of nine basic Tabla strokes are considered here namely, 'Na', 'Ta', 'Te', 'Tun' from the right drum, 'Ga', 'Ka' from the left drum, and 'Dha', 'Dhin', 'Tin' from both the drums simultaneously. Open strokes have a sharp attack, a clear sense of pitch, and long sustain. 'Na', 'Ta', 'Tun' played on Tabla are examples of open strokes. Closed sound is dumped and has sharp attack and decay. 'Ka' played on Dagga and 'Te' played on Tabla are examples of closed sound. Tabla player changes the pitch of the stroke by controlling the tension of the surface of the Dagga by using the base of the palm. 'Dha' (Na+Ga), 'Dhin' (Ta+Ga), and 'Tin' (Ta+Ka) are the combined strokes that give resonating sound. Tabla produces harmonic overtones for different Tabla strokes [5].

The information retrieved out of a musical sound is helpful for various applications. For example, the musical information extracted could be categorized into sounds that originated from different musical instruments. This is one form of musical instrument identification by identifying the source of the sound. These musical instruments sounds can then be identified, separated, or even removed [6]. For individual musical instruments, it is comparatively easy to identify the source instrument of the sound since they exhibit different Timbral attributes. On the other hand, Tabla itself consists of two different drums, and in contrast with other rhythm instruments, and music theories, NICM considers simultaneous production of strokes from both the drums as a single stroke. Thus, it is challenging to separate the Tabla stroke source.

The singing voice has a different attribute for frequency range and Timbre. The extracted musical information is useful to separate singing voices from other background accompanying musical instruments [7]. Two different singers also differ in singing stylization, voice production, vocal cavity, and psychoacoustic attributes of their voices. This difference helps in identifying a singer [8]. The gender of a singer is identified based on the fundamental frequency, frequency range, and timbrel attributes of the voice [9]. Music signals could be labeled quickly through the identification of the musical note frequencies [10].

To classify the Tabla strokes the audio is segmented based on onsets detected. Features are extracted from the segmented audio files of isolated strokes. Therefore, it becomes essential to correctly identify the onsets [11]. The segmentation and detection of strokes are based on onsets. For the detection of strokes from the stroke locations, the onsets are generated. Beat tracking is done based on the analysis of musical signal onsets. The amplitude thresholding method is used to detect onsets that are used to segment and detect the signal. The Timbral features of audio files are extracted for the classification of Tabla strokes and to identify the source of Tabla stroke from the left drum, right drum, or both the drums simultaneously.

An audio descriptor is a basic component of an audio file that is used to detect and classify the strokes in the prescribed format. Audio descriptor plays a crucial role in the genre recognition process [12]. To identify the source of a musical instrument like drum sound detection [13], drum source separation [14], automatic drum transcription [15], detection of Tala, and rhythm analysis of Tabla signal [16]. Audio descriptor plays important role in it. From the pool of hundreds of audio descriptors available, it is critical to select the appropriate type of audio descriptor. Filtering the required audio descriptors without a classifier is one way to reduce the number of audio descriptors. On the other hand, a wrapper approach called 'Hybrid Selection Algorithm' is a promising solution that considers the output of a classifier in the selection of the audio descriptor [17]. The common audio descriptors extracted from an audio file are Mel Frequency Cepstral Coefficient (MFCC), Timbrel features like, Roll-off, roughness, brightness, irregularity. Timbrel audio descriptors play a vital role to identify the Tabla strokes [18]. The audio features have a diversified range of values therefore the feature values are normalized between zeros and one.

Machine learning is a strong tool useful in classifying the audio source of Tabla stroke. In this research, we have used two popular machine learning algorithms namely Multi-Support Vector Machine (MSVM) and Feed Forward Back Propagation Neural

Network (FFBPNN). Support Vector Machine is a supervised machine learning algorithm where the support vectors are coordinates of the individual observation. Tabla stroke source separation is to be done into three classes hence MSVM is used here. SVM is a proven technology useful for discrimination of Sitar and Tabla strokes [19]. The one versus rest all classes method used in MSVM classifies the source of Tabla stroke into one source versus rest all sources. The FFBPNN learns the audio features of each source and classifies the test sample into one of the three sources. The Tabla stroke source separation is useful in many applications such as Tabla stroke and Tala detection [20], to recognize different percussion patterns [21], to label the Tabla signals [22] and to make the rhythm analysis of Tabla stroke excerpts [23].

3 Proposed System

The automatic Tabla stroke source separation system is shown in Fig. 1. The audio database was recorded in a studio environment with a sampling frequency of 44100 Hz with 16-bit pulse code modulation.wav representation. The Tabla player was asked to play different Tala in loops. Then all the Tabla strokes were separated and manually labeled according to their stroke types. This way a training dataset was created for nine basic Tabla strokes. The Tala considered here were selected such that they contain all nine basic Tabla strokes. The Tabla strokes sources considered here are from the left drum, right drum, and both the drums played simultaneously. The duration of the Tabla strokes varies from one to three second.

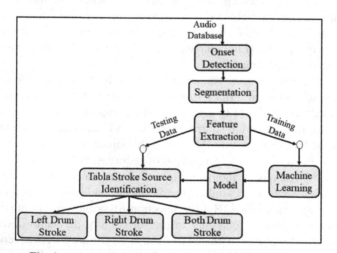

Fig. 1. Automatic Tabla stroke Source Separation System

Onsets are detected from each audio excerpt based on their highest value of amplitude. The continuous Tabla audio is segmented based on these onsets detected. Audio features are extracted from each of these segmented audio excerpts. The audio descriptors are the characteristic audio features that uniquely describe audio. All the Tabla strokes differ from each other concerning their attack, decay, sustain and release time along with energy, fundamental frequency, and Timbral attributes. Mel Frequency Cepstral Coefficient (MFCC) (20 coefficients), Zero Cross Rate, Roll Off, Roughness, Brightness, Irregularity (Timbral 5) and attack, decay, sustain, release (ADSR 4), and Fundamental Frequency (F0) is the 30 audio features considered here. Two machine learning algorithms namely MSVM and FFBPNN are trained using these audio features. The MSVM uses the logic of one versus rest all classes for classification. A three-layer FFBPNN is used to classify the Tabla stroke source into three classes.

The training dataset consists of nine basic Tabla strokes where, two strokes were considered from the left drum, four from the right, and three from both the drums. The system was trained using 105 Tabla excerpts originated from three sources (35 samples per source), tested, and cross-validated for 45 Tabla strokes (15 samples per source). The accuracy of the Tabla stroke source separation system was calculated from the ratio of correctly identified Tabla source with total testing samples.

4   Experiments and Results

In North Indian Classical Music Tabla instrument the stroke differs by a slight change in the location of the finger hitting the Tabla surface. For example, the right drum stroke 'Na' is produced by hitting on the outer surface of the drum while, stroke 'Tun' is produced by hitting the same finger on the black ink. Therefore, it is challenging to separate the source of each Tabla stroke. The strokes produced by the left drum are damped. The strokes produced from the right drum are harmonic. The ADSR values of the strokes, the fundamental frequency along with Timbral audio feature constitutes the feature matrix of size 30 by 1. Machine learning algorithms namely MSVM and FFBPNN were trained and tested.

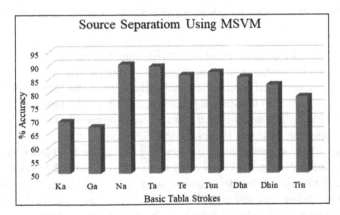

Fig. 2. Basic Tabla stroke Source Separation Accuracy using MSVM

The source separation accuracies for nine basic Tabla strokes using MSVM are as shown in Fig. 2. It is observed that the source separation accuracy for the left drum is low as compared to the right drum and both the drum strokes. Also, due to higher values of brightness and sustain, the right drum strokes source separation accuracy such as 'Na',

'Ta', 'Te', 'Tun' is higher than the combined stroke 'Dha', being closed stroke. However, the open stroke from both the drums, 'Dhin', exhibits higher values of harmonicity than 'Dha'. Thus, the stroke separation accuracy for 'Dha' is like the right drum stroke source separation accuracy.

Another machine learning algorithm FFBPNN was trained and tested for the classification of Tabla stroke sources. The results are shown in Fig. 3. The neural network used consists of thirty neurons in input layers represents thirty attributes of the Tabla stroke sound. At the output layer, nine neurons represent nine basic Tabla strokes. The Levenberg-Marquardt algorithm was used to train the network. The results show that the Tabla stroke source separation accuracy for the left drum is lower than the right drum. However, in contrast with MSVM the left drum stroke source separation accuracy is higher.

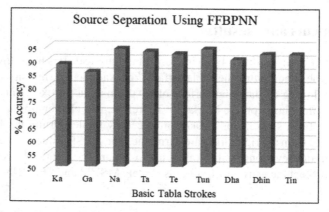

Fig. 3. Basic Tabla stroke Source Separation Accuracy using FFBPNN

Figure 4 shows the comparison of accuracies obtained for Tabla stroke source separation. Tabla stroke source separation accuracy is obtained lowest for the left drum strokes. The possible reason behind the misclassification of the Tabla stroke source for the left drum is because all the strokes played on this drum are damped. Thus, the previous stroke of the right drum or both the drums are prominently identified in the onset detection skipping the onset location of the left Tabla drum.

Right drum Tabla strokes produce a resonating sound that has higher ADSR values. The Timbral audio descriptor plays an important role to identify the stroke source. It is observed that the highest Tabla stroke source separation could be achieved for the right drum due to the brightness and inharmonicity of the sound. The Tabla stroke source separation of the strokes produced from both the drums simultaneously is challenging since it involves a combined sound produced from both the drums.

It is observed that the left Tabla drum strokes exhibit less duration, energy, and lower values of ADSR. Therefore, the highest source separation accuracy obtained for the left drum using FFBPNN is 86.75%. In the Tabla performance, the Tabla is played using continuous stroke originated from any of these sources. The performer may use any combination and repetition of the strokes that originated from these three sources. Due to higher values of release and sustain for the right drum, the right drum stroke release may be still ringing when the left drum stroke reaches its highest values of attacks. Thus, making it difficult to recognize the left drum strokes. This challenge is solved by carefully designing the onset detection and segmentation routine by using appropriate bandpass filters.

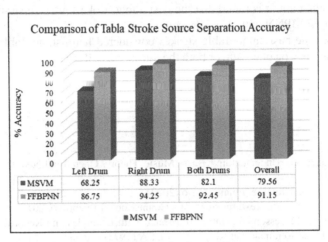

Fig. 4. Comparison of Tabla stroke Source Separation Accuracy

The use of MFCC along with Timbrel audio descriptors makes the system sensitive to small changes in the sound produced by Tabla strokes. The stroke produced by the right drum usually has harmonics in it. The audio descriptors used here along with fundamental frequency help the system to correctly classify Tabla stroke sources giving 94.25% accuracy for the right drum. The stroke originated from both the drums simultaneously contains a stroke from the left and right drum. Recall that the left drum strokes exhibit a damping nature while the right drum produces relatively sustainable harmonic sound. Therefore, the Tabla stroke source separation accuracy for both the drums is 92.45%. Due to the one versus all classification strategy used in MSVM, the Tabla stroke source separation accuracy of MSVM is inferior to FFBPNN. Overall, Tabla stroke source separation accuracy for all the sources using FFBPNN is 91.15%. The Tabla stroke source separation accuracies for strokes originated from the right and both the Tabla drums are higher than strokes originated from the left drum.

5 Conclusion

Automatic Tabla stroke source separation is a challenging research area with primary application in transcription. By identifying the Tabla stroke source, it is comparatively easy to transcript the series of Tabla solo performances. In this paper, a method for the separation of Tabla stroke from three sources is presented. Total 9 Tabla strokes, two from the left drum, four from the right drum, and three from both the drums being played simultaneously are considered here. Other Tabla strokes that are played alternately from the left and right drum are neglected. Out of 150 audio excerpts 105 are used for training and 45 are used for testing.

The audio descriptors play an important role in the identification and separation of the source of Tabla strokes. For this research, we have used MFCC, Timbral audio descriptors, fundamental frequency, and ADSR audio features. A total of 30 audio features and two machine learning classifiers are used to evaluate the system performance. After experimentation, it is found that the best Tabla stroke source separation could be achieved using FFBPNN with 91.15% accuracy.

Apart from the basic nine Tabla strokes considered here, some Tabla strokes are played continuously on one drum or both the drums but still are treated as a single stroke in NICM. Identification of such complex Tabla stroke and separation of their sources is left as future scope for this research.

References

1. Schmidt-Jones, C.: Understanding Basic Music Theory. Open Textbooks for Hong Kong, February 2015
2. Sharlene: Musical Texture. Epianostudio (2007)
3. Schmidt-Jones, C.: The Textures of Music. Connexions, 15 February 2013
4. Gottlieb, R.S.: 42 Lessons for Tabla: Illustrated with Examples and Recordings of Master Performer Ustad Keramatullah Khan, NYC, USA (1973)
5. Patranabis, A., Banerjee, K., Midya, V.: Harmonic and timber analysis of Tabla strokes (2015)
6. Cano, E., Schuller, G., Dittmar, C.: Pitch-informed solo and accompaniment separation towards its use in music education applications. EURASIP J. Adv. Signal Process. **2014**, 23 (2014). https://doi.org/10.1186/1687-6180-2014-23
7. Mavaddati, S.: A novel singing voice separation method based on a learnable decomposition technique. Circuits Syst. Signal Process. **39**(7), 3652–3681 (2020). https://doi.org/10.1007/s00034-019-01338-0
8. Deshmukh, S., Bhirud, S.: North Indian classical music's singer identification by timbre recognition using MIR toolbox. Int. J. Comput. Appl. **91**(4), 1–5 (2014)
9. Deshmukh, S., Bhirud, S.: A novel method to identify audio descriptors, useful in gender identification from North Indian classical music vocal. Int. J. Comput. Sci. Inf. Technol. **5**(2), 1139–1143 (2014)
10. Bello, J.P., Plumbley, M.: Fast labeling of notes in music signals. In: 5th International Conference on Music Information Retrieval, Barcelona, Barcelona (2004)
11. Miguel, A., Bertrand, D., Gael, R.: Tempo and beat estimation of musical signals. In: ISMIR (2004)
12. Rosner, A., Kostek, B.: Automatic music genre classification based on musical instrument track separation. J. Intell. Inf. Syst. **50**(2), 363–384 (2017). https://doi.org/10.1007/s10844-017-0464-5
13. Scarpiniti, M., Scardapane, S., Comminiello, D., Parisi, R., Uncini, A.: Separation of drum and bass from monaural tracks. In: Esposito, A., Faundez-Zanuy, M., Morabito, F.C., Pasero, E. (eds.) WIRN 2017 2017. SIST, vol. 102, pp. 141–151. Springer, Cham (2019). https://doi.org/10.1007/978-3-319-95098-3_13
14. FitzGerald, D.: Automatic drum transcription and source separation. ARROW@TU Dublin, Dublin (2004)
15. Vogl, R., Dorfer, M., Knees, P.: Recurrent neural networks for drum transcription. In: ISMIR Conference, New York, USA (2016)

16. Bhaduri, S., Saha, S., Mazumdar, C.: Matra and tempo detection for INDIC Tala-s. In: Kumar Kundu, M., Mohapatra, D.P., Konar, A., Chakraborty, A. (eds.) Advanced Computing, Networking and Informatics- Volume 1. SIST, vol. 27, pp. 213–220. Springer, Cham (2014). https://doi.org/10.1007/978-3-319-07353-8_25
17. Deshmukh, S.H.: A hybrid selection method of audio descriptors for singer identification in North Indian classical music. In: IEEE Explorer, Himeji, Japan (2012)
18. Shete, S., Deshmukh, S.: Analysis and comparison of timbral audio descriptors with traditional audio descriptors used in automatic Tabla Bol identification of North Indian classical music. In: Bhalla, S., Kwan, P., Bedekar, M., Phalnikar, R., Sirsikar, S. (eds.) Proceeding of International Conference on Computational Science and Applications. AIS, pp. 295–307. Springer, Singapore (2020). https://doi.org/10.1007/978-981-15-0790-8_29
19. Dhanvini, G., Vinutha, T.P., Rao, P.: Discrimination of Sitar and Tabla strokes in instrumental concerts using spectral features. In: Proceedings of Frontiers Research on Speech and Music (FRSM), Orissa, India (2017)
20. Shete, S., Deshmukh, S.: North Indian classical music Tabla Tala (rhythm) prediction system using machine learning. In: Biswas, A., Wennekes, E., Hong, T.-P., Wieczorkowska, A. (eds.) Advances in Speech and Music Technology. AISC, vol. 1320, pp. 187–197. Springer, Singapore (2021). https://doi.org/10.1007/978-981-33-6881-1_16
21. Gupta, S.: Discovery of percussion patterns from Tabla solo recordings, 20 September 2015
22. Sarkar, R., Mondal, A., Singh, A., Saha, S.: Automatic identification of Tala from Tabla signal. In: Gavrilova, M.L., Tan, C.J.K., Chaki, N., Saeed, K. (eds.) Transactions on Computational Science XXXI. LNCS, vol. 10730, pp. 20–30. Springer, Heidelberg (2018). https://doi.org/10.1007/978-3-662-56499-8_2
23. Srinivasamurthy, A., Holzapfel, K., Ganguli, K., Serra, X.: Aspects of tempo and rhythmic elaboration in Hindustani music: a corpus study. Front. Digit. Humanit. 4, 1–16 (2017)

Classification of Immunity Booster Medicinal Plants Using CNN: A Deep Learning Approach

Md. Musa$^{(\boxtimes)}$, Md. Shohel Arman$^{(\boxtimes)}$, Md. Ekram Hossain$^{(\boxtimes)}$,
Ashraful Hossen Thusar$^{(\boxtimes)}$, Nahid Kawsar Nisat$^{(\boxtimes)}$, and Arni Islam$^{(\boxtimes)}$

Daffodil International University, Dhaka 1207, Bangladesh
{musa35-1870,arman.swe,ekram35-1936,ashraful35-1908,
nahid35-1889,arni15-8789}@diu.edu.bd

Abstract. Environment has blessed us with various kinds of plants. Some of them uses as resources of medicines as it is called medicinal plant. In Bangladesh medicinal plants are also known as Ayurveda, Homeopathy and Unani. Experts says medicinal plants can be very useful in the fight with recent pandemic which is Covid-19. As we know health of a body depends on its immune system, so it is important to keep immunity stronger. Strong immune system can be influential to any infectious virus, bacteria and pathogens. On the other hand, inactive one can get easily infected with virus and other illness. There are certain medicinal plants which reinforce our immunity. Therefore, classification of these medical plants is very important. For this classification we have collected leaf images for six different classes which's local names are Darchini, Tulshi, Tejpata, Sojne, Neem, Pathorkuchi. In this article we introduced a famous algorithm for classification named CNN (Convolutional neural network). We used CNN (Convolutional neural network) to recognize the plant from leaf images and got 95.58% accuracy. In future infectious virus can appear which can be more threatening than others, our research will help people to know about immune system and medicinal plants which reinforce our immunity, so that they can fight with diseases and viruses.

Keywords: Immunity system · Medicinal plant · Plant classification · Convolutional neural network

1 Introduction

Covid-19 has shown us the importance of our immune system. Immune is our most centric benefactor and its prime duty is to keep us healthy and strong. A strong immune system depends on which foods we are taking and if we take foods which are fresh and contain vitamins and minerals, our immune system will be capable of to do its battle against virus, harmful bacteria, toxins and other pathogens. The immune is not a single entity, it's a system. Balance and harmony are main requirements for function it well. Nature has blessed mankind with enormous medicinal herbs which provide timely and suitable remedies to various health disorders. In the time of health crisis like novel Corona virus, the medicinal herbs are enabling the people to boost their immunity

© Springer Nature Switzerland AG 2021
M. Singh et al. (Eds.): ICACDS 2021, CCIS 1440, pp. 244–254, 2021.
https://doi.org/10.1007/978-3-030-81462-5_23

[1]. Understanding the correlation among immune system, medicinal herbs and Covid-19 in the present time is very necessary. Adequate nutrition is the backbone for the development, maintenance and optional functioning of immune cells in this outbreak of Covid-19.

Based on the principle of plant classification, categorization of plants occurs with plant parts such as roots, flowers, fruits and leaves. However, in various incidents characteristics like roots, flowers and fruits don't show much difference in computationally and mathematically. However, for species identification leaves are reach resources and some of them boost our immunity and also used in medicinal treatment. World Health Organization (WHO) stated that on the purpose of remedies medicinal plants are used by 65% to 80% of the world population and there are around 17810 of different medicinal species in the world [2].

Images are collected from different nurseries and botanical garden and the number of images are 920. This image has six different classes. The local and scientific names of these plants are:

Table 1. Dataset information

No.	Scientific name	Local name	No. of samples
1	Cinnamomumverum	Darchini	150
2	Ocimumtenuiflorum	Tulsi	150
3	Cinnamomumtamala	Tejpatta	150
4	Moringaoleifera	Sojne	150
5	Azadirachtaindica	Neem	150
6	Bryophyllum pinnatum	Pathorkuchi	150
		Total	900

Around the world high percentage of people use medicinal plants leaves for their primary healthcare. Leaves has no side effects and enhancement use of the medicinal plants can reduce medicine side effects. Plants are not only used in medical science but it also uses as resources of cosmetics and other products also. Large numbers of industries are heavily dependent on medicinal plants as plants are their main resources of manufacturing. In this time these plants are in the threat of extinction because of deforestation, urbanization and lack of awareness in medicinal plants. According to the Food and Agriculture organization in 2002 the number of medicinal plants were 50,000 and in 2016 according to the Royal Botanic gardens the number is decreasing in high rate and the estimate number is now 17,810 [3]. So, it is high time to save these plants for our beneficial concern.

2 Literature Review

A lot of work is being done with medicinal plant images. In medical science plants has huge impact and it is considered as great asset. In 2016 D Venkatraman and mangayarkarasi N proposed an automated system based on vision approach which helps a common man on identifying the medicinal plants. Usually, plants are classified on leaves features like – shape, color and texture. Their classification was based on leaves texture. They used GLCM method for classifying the leaves and to find the dissimilarity between the leaves [4].

In year 2017 Adams Begue and VemitheKowlessur collected leaves of 24 medicinal plants species and exterted each leaf on their width, shape, color, length, perimeter, area and vertices number. They used Random Forest algorithm for their classification and gets 90.1% accuracy. Their anticipation for this automated recognition was encourage researchers to develop more techniques on species identification and help common people to know more about medicinal plants [5].

Plants are not only used in medical science it also used in several types of cosmetics and other products as well. In 2016, Prabhakar Poudel and Shyamdew Kumar proposed an automation system for leaf detection. Leaf extraction was based on diverse feature, categorization and pattern identification. Support Vector Machine (SVM) was applied for the classification and for extracting the features they utilized Scale Invariant Feature Transformation (SIFT) [6].

In 2019 C. Amuthalingeswaran had built a Deep Neural Networks model for the identification of medicinal plants. They used 8000 images for four classes and got accuracy of 85%. They collected their images from the open field land areas [7]. Nazish Tunio, For extraction of the region of the interest (ROI) and an algorithm which is image based was proposed by Abdul Latif Memon. It was also used for identify the plant species and to recognize the disease of certain plants as well. Support Vector Machine (SVM) classifier was applied for better outcomes and they got their expected result of 93.5% accuracy with their leaves dataset which had four different classes [8].

A. Aakif proposed an algorithm for identification of plants with three ways i) Preprocessing ii) Feature extraction iii) Classification in 2015. Several leaf feature was extracted and these features became input vector for their build model Artificial Neural Network (ANN). They trained their model with 817 leaf images from 14 different classes and got accuracy of 96% and for checking the effectiveness of the model they trained it with another dataset and also got accuracy of 96% [9].

In 2017 MM Ghazi used Deep Convolutional Network for identification the plants species. They applied popular deep learning architectures like GoogLeNet, AlexNet, and VGGNet and trained it with Life CLEF 2015 dataset. Their combined system had overall accuracy of 80% on validation set [10].

On the other hand, in 2013, R. Janani and A Gopal Artificial Neural Network (ANU) classifier to identify the medicinal plants from leaf images Artificial Neural Network (ANN) was used for less computational complexity and to gain high efficiency. The models give 94.4% of accuracy after testing on 63 leaf images and which was impressive [12].

C. Ananth and Azha. Proposed an automation system of plants identification because manual classification has huge chance of human error. In this research they use MATLAB. They extract leaf images and used models like segmentation, thresholding and applied to neural network [13].

In recent times (2019) digital image Processing for identification of medicinal plants was proposed by P. Chitra and S. JanesPushparani. Speeded Up Robust Feature Transform (SURF) and Scale Invariant Feature Transform (SIFT) both was used for leaf extraction and for distinguishing the structure. They used Support Vector Machine (SVM) classifier to get correct plant identification [14].

RAJANI S and VEENA M.N proposed an automatic identification and classification of medicinal plants. Create awareness and encourage people to know more about medicinal plant is their main goal of this automatic identification. Like many others they only used plant leaves images for the classification but they also used plant flowers, fruits and seeds images for their classification [15].

3 Proposed Methodology

For this study we use convolutional neural network (CNN) for plant detection. Our working procedure is in Fig. 1.

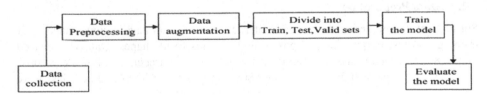

Fig. 1. Work flow

3.1 Dataset Collection

Various sorts of medicinal plants leaves are gathered from several nurseries and botanical garden from Dhaka city. Our dataset contains 900 images with 6 unique classes. For testing we utilize 200 images and other 700 images for training and validation. Now every one of them have 150 images for each. We have named the medicinal plants as their native names for our study. Dataset classes are (Fig. 2):

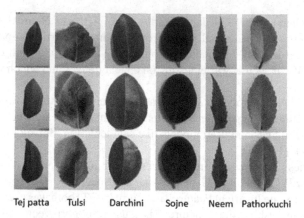

Tej patta   Tulsi   Darchini   Sojne   Neem   Pathorkuchi

Fig. 2. Dataset sample

3.2 Data Preprocessing

3.2.1 Data Augmentation

In limited dataset overfitting is a typical issue. We may face overfitting problem as well as our dataset is also limited. we implement data augmentation to eliminate overfitting. Augmentation is actually an artificial process which extends the dataset. In our dataset Augmentation Technique modified version of images is absence. Augmentation was done four way and they are height shift range, rotate 90° left, flip horizontally and shear by certain amount.

3.2.2 Data Preparation

Subsequent to completing the increase interaction these data have various shapes. For training and test purpose our proposed model needs fixed shapes. Data was resized into 128 × 128 pixels and we used RGB for getting decent accuracy. For the purpose of testing, we separated 20% images and 80% images for training and per class were individual.

3.3 Convolutional Neural Network

The classification of Artificial Neural Network is an unimaginable mainstream approach to solving pattern recognition problem. Convolutional Neural Network (CNN) is named fully connected neural network and it was one of the vital elements adding to these tests. Out of many advantages of CNN there is a principal advantage which is without human guidance important features are automatically detected. CNN model has two segments starting fragment is for include extraction and the ensuing portion works for classification. The central layer will endeavor to recognize edges and shape a demonstrate to distinguish the edge. At that point or maybe layers may endeavor to solidify them in less complex ways. Include channel to the picture and attempt to extricate the picture edges are the primary layer's work. The second layers are called the pooling layers and just like the first layer, it too includes channels to the image. This layer finds highlights more profoundly from the figure and the layer has ReLU usefulness that makes a difference to associate the taking after neuron. At that point, the flatten layer changes over 3D picture data to 2D data for classification [16].

3.3.1 Convolution Layer

This is called CNN's heart structure. The convolution layer computes the dab item among width, the tallness of the input layer, and the channels. It also contains filters. If the dimension of image is $128 \times 128 \times x3$ and number of filters is 24 then dimension of output is $128 \times 128 \times 24$ (Fig. 3).

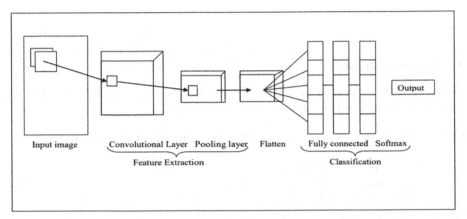

Fig. 3. Architecture of CNN

3.3.2 Activation Function Layer

Activation could be a sort of function and gives the related yield that's bolstered from any input. Functions like Direct and Nonlinear are out. Relu may be a nonlinear enactment work that's utilized in our consideration. ReLU ordinarily changes over negative numbers by setting them to zero.

$$F(x) = max(0, x)$$

3.3.3 Pool Layer

After the activation layer comes the Pool layer. A few sorts of pool layers are there. Max pooling is for the most part utilized pooling layer additionally utilized for our errand. The essential work of max-pooling is lessening the measurement of the pictures. In the event, that pool estimate 2×2 and walk 2 utilizing the max-pooling layer, at that point the yield measurement is $64 \times 64 \times 24$. Figure 4 appears the visualization of a max-pooling operation.

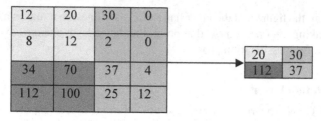

Fig. 4. 2 × 2 Max polling

3.3.4 Fully Connected Layer

This is the final layer of CNN. It composes a complete combination of all former layer and it takes input from the previous layers then makes into id vector.

3.4 Proposed Model

We have created an architecture to recognize the medicinal plants and it is a 12-layered network. The developed layers are –

Convolution Layer: In our model we apply four convolution layer and they are:

i.   Convolution (32 – 3 × 3) filter.
ii. Convolution (64 – 3 × 3) filter.
iii. Convolution (128 – 3 × 3) filter.
iv. Convolution (256 – 3 × 3) filter.

Pooling Layer: Pooling layer come after convolutional layer. We use four max pooling layers with a pool size of 2 × 2.

Dropout Layer: In neural network there are dropping out some units and it is called dropout. Dropout is use for avoid overfitting. Our proposed model has two dropout layers with layer rate of 0.25 and 0.5.

Flatten Layer: The 2D vector includes a pool highlight outline and this vector makes a continuous 1D vector where the method is called flatting. Our show contains a flatten layer.

Dense Layer: The dense layer utilized as a classifier.

3.5 Training the Model

We utilized an optimizer known as Adam [17] optimizer to plan our show with a little learning rate. It is likely the foremost utilized optimizer that's quick and dependable to utilize. For 15 epochs of 35 batch, the size we first ought to compile our model at that point begin our model utilizing the fit () strategy. The issue is that multi-class classification is for classification cross-entropy loss (Fig. 5).

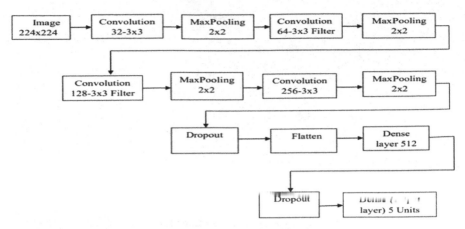

Fig. 5. Proposed model architecture

4 Performance Evaluation

Model assessment is exceptionally imperative in machine learning. This appears how proficient the scoring of the dataset has been by a prepared model. We accomplished our result by completing preparation and approval. This is often called preparing exactness when we apply the show to the preparing data and it is called validity accuracy when we apply the show to the test data with distinctive classes.

Figure 6 appears that the training accuracy of our model is 95.58% and we kept up our approval exactness from 92% to 94.5%.

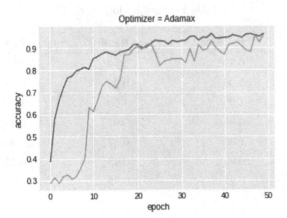

Fig. 6. Training vs validation accuracy

In Fig. 7 we can see the training and validation loss of our model. 0.08 is our validation loss.

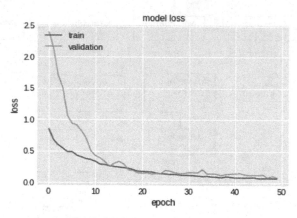

Fig. 7. Training vs validation loss

5 Result and Discussion

The classification report visualizer appears the accuracy, precision, F1-score, and bolster scores for the model. To assist less complex interpretation and issue revelation, the report arranges scientific scores with a shading coded heatmap. All heatmaps are within the reach (0.0, 1.0) to empower a straightforward comparison of classification models over different arrange reports. After completing the training strategy our model has been tested with 200 data. Table 2 appears the classification report which is computed from our model. We assessed precision, recall, and f1-score.

Table 2. Classification report

Class	Precision	Recall	F1-score
Darchini	0.87	1.00	0.93
Tulsi	1.00	0.93	0.96
Tej patta	1.00	1.00	1.00
Neem	0.94	0.97	0.95
Pathorkuchi	1.00	0.87	0.93
Sojne	0.90	0.93	0.92
Average	0.95	0.95	0.95

Confusion matrix of our model is shown in Table 3.

Table 3. Confusion matrix

	Darchini	Tulsi	Tej patta	Neem	Pathorkuchi	Sojne
Darchini	27	0	0	0	0	0
Tulsi	0	27	0	2	0	0
Tej patta	0	0	31	0	0	0
Neem	0	0	0	30	0	1
Pathorkuchi	2	0	0	0	27	2
Sojne	2	0	0	0	0	28

6 Conclusion

In this study, our presented model appears great precision for helpful plant leaf arrangements. We advertised a model by our claim. This method worked well within the field pictures that we attempted. This work makes a difference Botanists get its plants related to their long-term treatment. It can at that point exceptionally well be utilized for reestablishing takes off and bunches in pertinent ranges where leaf arrangement is required.

7 Future Work

For ages, various sorts of therapeutic plants have been found all over the world. Each one of them has its characteristics that remedy an assortment of afflictions. Along these lines, within the future, we have to make this dataset greater with diverse medicinal plant leaf pictures and proposed a (Convolutional Neural Organize) CNN with more upgrades. We make superior treatment solutions utilizing modern propels and movement in computer vision. Within the future individuals will utilize more medicinal herbs clears out as contradicted to medication for the treatment.

References

1. http://www.nytimes.com/2010/07/06/health/06real.html?pagew
2. Arun, C.H., Sam Emmanuel, W.R., Christopher Duraira, D.: Texture feature extraction for identification of medicinal plants and comparison of different classifiers. Int. J. Comput. Appl. **62**(12) (2013). 0975 - 8887
3. O'Donnell, K., Sharrock, S.: Botanic gardens complement agricultural gene bank in collecting and conserving plant genetic diversity. Biopreserv Biobank. **16**(5), 384–390 (2018). https://doi.org/10.1089/bio.2018.0028
4. Venkataraman, D., Mangayarkarasi, N.: Computer vision based feature extraction of leaves for identification of medicinal values of plants. In: IEEE International Conference on Computational Intelligence and Computing Research. IEEE (2016). 978-1-5090-0612-0/16/$31.00 ©2016

5. Begue, A., Kowlessur, V., Mahomoodally, F., Singh, U., Pudaruth, S.: Automatic recognition of medicinal plants using machine learning techniques. Int. J. Adv. Comput. Sci. Appl. **8**(4) (2017)
6. Poudel, P., Kumar, S.: Robust recognition and classification of herbal leaves. **05**(04) (2016). ICESMART-2016
7. Amuthalingeswaran, C., Sivakumar, M., Renuga, P., Alexpandi, S., Elamathi, J., Hari, S.S.: Identification of medicinal plant's and their usage by using deep learning. In: 2019 3rd International Conference on Trends in Electronics and Informatics (ICOEI), Tirunelveli, India, pp. 886–890 (2019). https://doi.org/10.1109/ICOEI.2019.8862765
8. Tunio, N., Memon, A., Khuhawar, F., Abro, G.M.: Detection of infected leaves and botanical diseases using curvelet transform (2019)
9. Aakif, A., Khan, M.: Automatic classification of plants based on their leaves. Biosyst. Eng. **139**, 66–75 (2015). https://doi.org/10.1016/j.biosystemseng.2015.08.003
10. Ghazi, M.M., Yanikoglu, B., Aptoula, E.: Plant identification using deep neural networks via optimization of transfer learning parameters. Neurocomputing **235**(C), 228–235 (2017)
11. Jain, G., Mittal, D.: Prototype designing of computer-aided classification system for leaf images of medicinal plants. J. Biomed. Eng. Med. Imaging **4**(2), 115 (2017). https://doi.org/10.14738/jbemi.42.3053
12. Janani, R., Gopal, A.: Identification of selected medicinal plants leaves using image features and ANN. In: 2013 International Conference on Advanced Electronic System (ICAES) (2013). 978-1-4799-1441-8
13. Ananthi, C., Periasamy, A., Muruganand, S.: Pattern recognition of medicinal leaves using image processing techniques. J. Nano Sci. Nano Technol. **2**(2) (2014). Spring Edition, ISSN 2279 – 0381
14. Chitra, P.L., James Pushparani, S.: SURF points versus SIFT points in identification of medicinal plants. Int. J. Eng. Adv. Technol. (IJEAT) **9**(2) (2019). ISSN : 2249 -8958
15. Rajani, S., Veena, M.N.: Study on identification and classification of medical plants. Int. J. Adv. Sci. Eng. Technol. **6**(2), Spl. Issue-2 (2018). ISSN(p): 2321 –8991, ISSN(e): 2321 –9009
16. Bengio, Y., Lecun, Y.: Convolutional networks for images, speech, and time-series (1997)
17. Kingma, D.P., Ba, J.: Adam: a method for stochastic optimization. CoRR, abs/1412.6980 (2014)

Machine Learning Model Interpretability in NLP and Computer Vision Applications

Navoneel Chakrabarty[✉]

International Institute of Information Technology (IIIT) Bangalore, Bengaluru, India
navoneelchakrab.dml16@iiitb.net, nc2012@cse.jgec.ac.in

Abstract. In light of the recent advancements in Artificial Intelligence (AI), the application of Machine Learning in the domains of Natural Language Processing and Computer Vision is increasing by leaps and bounds. Deployment of Machine Learning Models in applications (apps) have been rampant with the aim of achieving automation, mainly involving textual and image data. Textual data indulges the subject of Text Analytics (also known as NLP) into action and image data indulges the subject of Computer Vision into play. But, the performance of Machine Learning in the domains of Text Analytics or Vision needs to be judged before deployment. Now, performance analysis of ML Models are done with the help of performance metrics, most importantly AUC Score in Classification Problems, but justification by means of numerical scores only, can't establish the relevance of the Model Performance with Domain Knowledge. In this paper, 1 standard NLP use-case and 4 Computer Vision use-cases are considered for ML Model Interpretability Enhancement that can throw light on the relevance with the concerned Domain Knowledge, the use-case deals with.

Keywords: Model Interpretability · NLP · Computer Vision · Wordcloud · Feature Map Visualization

1 Introduction

In Natural Language Processing, there are several stages of Text Analytics, starting from Lexical Processing, Syntactic Processing to finally Semantic Processing. In this work, only Lexical Processing is involved in a Classification Problem. After Lexical Processing, a Machine Learning Model is trained with the word-features-engineered data (tf-idf) from raw text data. Now, the ML Model can make predictions, relevant to the tags or the labels concerned, but in order to elucidate the type and amount of contribution of the driver variables (word features) in making the prediction, Model Interpretability is very essential. In order to achieve this, Model Wordclouds are constructed, taking into account the feature coefficients and importance scores generated by the trained ML Models. Now, in Computer Vision, among the 4 use-cases considered, two are from the domains of Medical Imaging and the other 2 hail from the domain of Hand

© Springer Nature Switzerland AG 2021
M. Singh et al. (Eds.): ICACDS 2021, CCIS 1440, pp. 255–267, 2021.
https://doi.org/10.1007/978-3-030-81462-5_24

Anatomy. They are also Classification Problems, but the classification output (though correct) is justified by Feature Map Visualizations in all the intermediate layers of the Convolutional Neural Network. Such procedures of ML Model Interpretability Enhancement can establish direct relevance of use-cases with the respective domain knowledge.

2   Literature Review and Proposed Work

2.1   Natural Language Processing (NLP)

In the domain of Text Analytics, one use-case, **Toxic Comment Classification** is selected for Machine Learning Model Interpretability Enhancement.

In this use-case, given a user-comment, the Machine Learning Model will tag the comment with tags, **toxic**, **severe_toxic**, **obscene**, **threat**, **insult** and **identity_hate**. So, it is a multi-class and multi-label problem. This means, for an instance, there are multiple classes and among the classes, more than one of them can be true. For the purpose, the published approach by Chakrabarty [1] is adopted in which the Wikipedia Comment Dataset prepared by Jigsaw [2] (snapshot of the dataset is shown in Fig. 1) has been used.

id	comment_text	toxic	severe_toxic	obscene	threat	insult	identity_hate
0044cf18cc2655b3	What page shoudld there be for important characters that DON'T reoccur?	0	0	0	0	0	0
00472b8e2d38d1ea	Void, Black Doom, Mephiles, etc	1	0	1	0	1	1

Fig. 1. Snapshot of the Wikipedia Comment Dataset

The methodology proposed by Chakrabarty [1] is shown in Fig. 2.

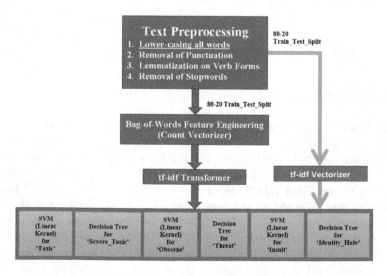

Fig. 2. Flow-chart of the comment toxicity classification model by Chakrabarty [1] denoted by only blue blocks and blue arrows and the modifications done on the model in this paper are indicated by green arrows and underlined text (Color figure online)

From the paper by Chakrabarty [1] the following are the observations:

1. All the words are not lower-cased. This can prove fatal, from the interpretation perspective as the same word may be treated as different features due to case-sensitivity.
2. The ROC Curves for each of the 6 sub-models are plotted by taking the predicted target values (0/1) instead of the predicted probabilities which is not the norm, as the Receiver Operator Characteristic Curve (TPR by FPR Curve) is plotted for different thresholds applied on predicted probabilities, not labels.

The work-flow is modified and re-proposed in this paper (shown in Fig. 2 by green arrows and underlined text). This proposed methodology is termed as **SVM_Decision_Tree_EnsembleV1** and the changes are as follows.

(i) All the words are lower-cased.
(ii) The work-flow is shortened by using tf-idf Vectorizer instead of using Count Vectorizer followed by tf-idf Transformer, though yeilding the same resultant feature-engineered data.
(iii) For label, **severe_toxic**, for a comment, if the Sub-model 1 for **Toxic** classifies it as Non-Toxic, then it can't be **severe_toxic** at all. So, the Test Accuracy, Precision, Recall and F1-Score are given in pairs of values for the label **severe_toxic** in which the 1st value is without Sub-model 1 check and the 2nd value is after Sub-model 1 check. Also Chakrabarty [1] has reported the results for Precision, Recall and F1-Score as weighted averages. On the contrary, in this paper, the results for the aforementioned metrics are reported as unweighted with (1) as positive label and (0) as negative label. All the results of the 6 sub-models are tabulated and shown in Table 1 with the corresponding Confusion Matrix (after Sub-model 1 check) and ROC Curves are shown in Fig. 3.

Table 1. Performance analysis for **SVM_Decision_Tree_EnsembleV1** with unweighted Precision, Recall and F1-Score

Sub-model (Label)	Training accuracy	Test accuracy	Precision	Recall	F1-score
1 (Toxic)	0.9873	0.9603	0.8629	0.6875	0.7653
2 (Severe_Toxic)	0.9997	0.9877, 0.9885	0.3188, 0.3502	0.2795, 0.2795	0.2975, 0.3109
3 (Obscene)	0.9944	0.9797	0.8597	0.7146	0.7805
4 (Threat)	0.9999	0.9964	0.3658	0.3297	0.3468
5 (Insult)	0.991	0.9712	0.7535	0.5726	0.6507
6 (Identity_Hate)	0.9999	0.9897	0.3785	0.2924	0.3299

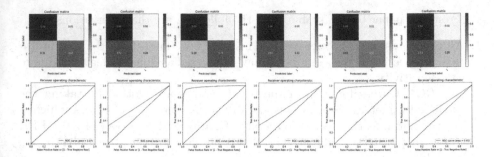

Fig. 3. Confusion matrix (1st row) and ROC curves (2nd row) for the 6 sub-models of SVM_Decision_Tree_EnsembleV1

So, the Macro F1 (unweighted) for the model is **0.5229** and Macro AUC is **0.81**. Now, in **SVM_Decision_Tree_EnsembleV1**, the AUROC or AUC for 2nd sub-model (for severe_toxic), 4th sub-model (for threat) and 6th sub-model (for identity_hate) are 0.65, 0.66 and 0.65 respectively. This indicates that these 3 sub-models are biased towards majority class (0). To reduce the biased nature of the sub-models or to enhance the discriminative power of these sub-models, class weights in the Gini-impurity of the tree models are modified as shown in Table 2.

Table 2. Class-weights in the Gini-impurity for the decision tree classifier based sub-models

Sub-models	Label 0 weight	Label 1 weight
2nd Sub-model (for severe_toxic)	0.2	0.8
4th Sub-model (for threat)	0.1	0.9
6th Sub-model (for identity_hate)	0.15	0.85

The resulting model (with modified sub-models) is termed as **SVM_Decision_Tree_EnsembleV2**. All the results of the models are tabulated in Table 3 and corresponding Confusion Matrix and ROC Curves are shown in Fig. 4.

Table 3. Performance Analysis for **SVM_Decision_Tree_EnsembleV2** with unweighted Precision, Recall and F1-Score

Sub-model (Label)	Training accuracy	Test accuracy	Precision	Recall	F1-score
1 (Toxic)	0.9873	0.9603	0.8629	0.6875	0.7653
2 (Severe_Toxic)	0.9996	0.9866, 0.9872	0.3271, 0.3446	0.4108, 0.4108	0.3642, 0.3748
3 (Obscene)	0.9944	0.9797	0.8597	0.7146	0.7805
4 (Threat)	0.9999	0.9956	0.3064	0.4176	0.3535
5 (Insult)	0.991	0.9712	0.7535	0.5726	0.6507
6 (Identity_Hate)	0.9998	0.9881	0.3482	0.4224	0.3817

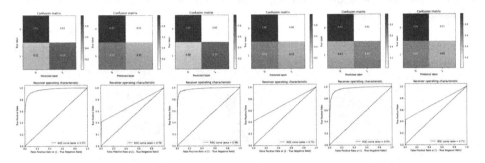

Fig. 4. Confusion matrix (1st row) and ROC curves (2nd row) for the 6 sub-models of SVM_Decision_Tree_EnsembleV2

So, the Macro F1 (unweighted) for the model has increased from 0.5229 to **0.5521** and so is the Macro AUC from 0.81 to **0.84**.

(iv) **Model Interpretability:** Now, for each of the 6 sub-models i.e., for each of the labels, Toxic, Severe_Toxic, Obscene, Threat, Insult and Identity Hate, interpretability enhancement is done by highlighting the words (here features) responsible for the comment to belong to each of the labels and tags in the form of Model Wordclouds. The model wordclouds are based on the coefficient/importance value for each and every word feature generated by the 6 sub-models of SVM and Decision Tree Variant in the ensemble, **SVM_Decision_Tree_EnsembleV2** and are shown in Fig. 5.

Fig. 5. Model Wordclouds for all the 6 sub-models (labels)

2.2 Computer Vision

In the domain of Computer Vision, 4 use-cases are considered for Model Interpretability Enhancement:

- Gender Identification from Hand Dorsal Images
- Age Classification from Hand Dorsal Images
- Diabetic Retinopathy Classification from fundus retinal images
- Glaucoma Screening or Diagnosis from fundus retinal images

2.2.1 Gender Identification from Hand Dorsal Images

In this problem, given an RGB image of the Dorsal side of hand, the Model should identify the gender of the person, whose image of the dorsal side of hand is fed into. So, there are 2 labels, Male and Female and hence, is a Binary Classification Problem.

For this use-case, the state-of-the-art model by Chakrabarty [3] is reviewed for consideration, in which the 11k Hands Dataset (image samples are shown in Fig. 6) [4] is used.

Fig. 6. Sample images from the 11k hands dataset

The flow-diagram for the proposed model by Chakrabarty [3] is shown in Fig. 7.

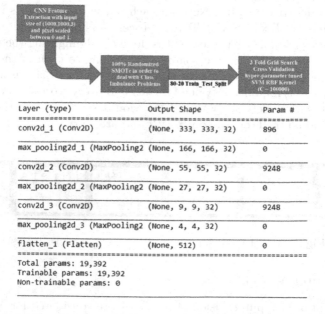

Fig. 7. Horizontal flow-diagram and CNN architecture by Chakrabarty [3] for gender identification from hand dorsal images

So, its a Hybrid Model with pre-initialized weights found in the link [5] for CNN Feature Extraction. After Feature Extraction, train_test_split is done in 80–20 ratio. Then, SVM (RBF Kernel) model is trained (using the best obtained value of C parameter from 3-Fold Grid-Search Cross Validation) and when evaluated on the Test Set, it yielded a 100% Test Accuracy with Precision, Recall, F1-Score and AUC each equal to 1.0, resulting in a full-proof model. But, though numerically, the model seems full-proof, but no method or visualization is followed or even prescribed by the author for checking the authenticity of the model in terms of feature extraction. In this paper, a Visualization Technique known as Intermediate Activation Visualization or Visualization of Feature Maps is performed. This gives an idea of how, the feature maps in each and every layer of the CNN looks, as the CNN extracts generic features to specific features (specific to the problem statement) from the early layers to the later layers of the network. So, as per the CNN Architecture followed, there are 6 layers, so the feature maps extracted in every channel of each of the 6-layer activations are shown in Fig. 8.

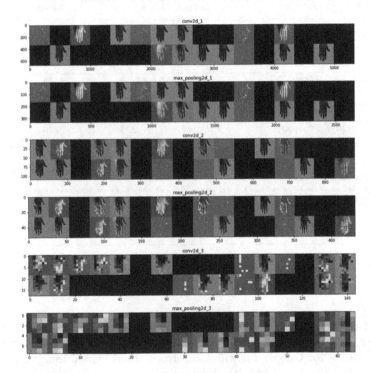

Fig. 8. Feature map visualization in every channel of each of the 6 layers of the CNN for gender identification

2.2.2 Age Classification from Hand Dorsal Images
In this use-case, given an RGB Image of the Dorsal side of hand, the Model classifies the age-group of the person, whose dorsal side of hand image is fed as input to the model. The defined age-groups are, **less than 22 as Label 0, 22**

to 26 (both inclusive) as **Label 1** and **greater than 26 as Label 2**. Hence, this is a multi-class classification problem with 3 classes. The state-of-the-art model by Chakrabarty et al. [6] that uses the 11k Hands Dataset (snapshot in Fig. 6) [2], is reviewed for interpretability enhancement. The flow-diagram for the proposed model by Chakrabarty et al. [6] is shown in Fig. 9.

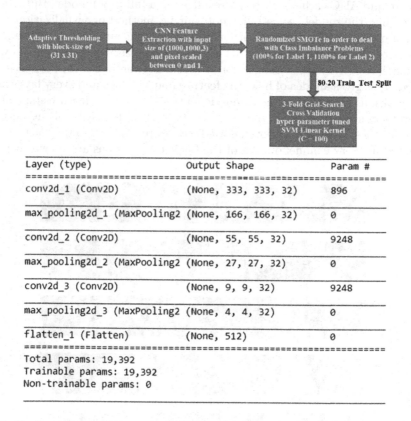

Layer (type)	Output Shape	Param #
conv2d_1 (Conv2D)	(None, 333, 333, 32)	896
max_pooling2d_1 (MaxPooling2	(None, 166, 166, 32)	0
conv2d_2 (Conv2D)	(None, 55, 55, 32)	9248
max_pooling2d_2 (MaxPooling2	(None, 27, 27, 32)	0
conv2d_3 (Conv2D)	(None, 9, 9, 32)	9248
max_pooling2d_3 (MaxPooling2	(None, 4, 4, 32)	0
flatten_1 (Flatten)	(None, 512)	0

Total params: 19,392
Trainable params: 19,392
Non-trainable params: 0

Fig. 9. Horizontal flow-diagram and CNN architecture by Chakrabarty et al. [6] for age classification from hand dorsal images

So, the methodology involves Adaptive Thresholding as a major pre-processing step before training a hybrid-model similar to the model by Chakrabarty [3] for Gender Identification from Hand Dorsal Images. It has the same CNN Architecture but with different pre-initialized weights, that can be found in the link [7] and the SVM Linear Kernel (using the best obtained value of C parameter from 3-Fold Grid-Search Cross Validation) is used for model training. This also clocked a 100% Test Accuracy with Precision, Recall, F1-Score, AUC each equal to 1.0 resulting in a full-proof model, but the authors didn't propose any Visualization Technique for Feature Extraction Verification.

So, the Visualization of the Intermediate Activations is done in this paper, to validate the Feature Extraction and is shown in Fig. 10.

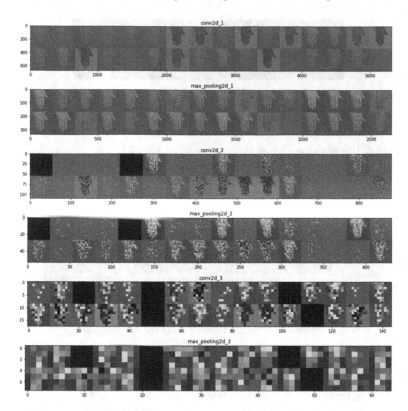

Fig. 10. Feature map visualization in every channel of each of the 6 layers of the CNN for age classification

Both Gender Identification and Age Classification use cases demand skin features of the dorsal side of the hand and hand texture features including shape of the hand (including fingers), hand nails etc. to be extracted and the same is confirmed by the respective Feature Map Visualizations in Fig. 8 and Fig. 10.

The models developed in the subsequent use-cases have the same issue of no validation approach for Feature Extraction. So, Feature Map Visualizations are done for each one of them by elucidating the problem statement and the dataset.

2.2.3 Diabetic Retinopathy Classification

In this application of Medical Imaging, on providing an HRF Retina Image to the Model, the Model classifies whether the patient is suffering from Diabetic Retinopathy or not. So, this is a Binary Classification Problem. The state-of-the-art model by Chakrabarty et al. [8] is reviewed for Intermediate Activation Visualization that forms the basis of interpretability enhancement for Computer Vision Applications. The model uses the HRF Database (sample images are shown in Fig. 11) prepared by the University of Erlangen-Nuremberg [9].

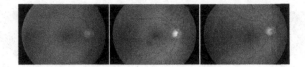

Fig. 11. Sample images from HRF database

The flow-diagram for the model is shown in Fig. 12.

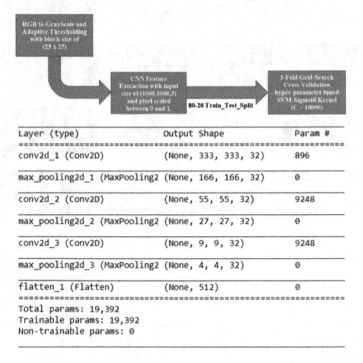

Layer (type)	Output Shape	Param #
conv2d_1 (Conv2D)	(None, 333, 333, 32)	896
max_pooling2d_1 (MaxPooling2	(None, 166, 166, 32)	0
conv2d_2 (Conv2D)	(None, 55, 55, 32)	9248
max_pooling2d_2 (MaxPooling2	(None, 27, 27, 32)	0
conv2d_3 (Conv2D)	(None, 9, 9, 32)	9248
max_pooling2d_3 (MaxPooling2	(None, 4, 4, 32)	0
flatten_1 (Flatten)	(None, 512)	0

Total params: 19,392
Trainable params: 19,392
Non-trainable params: 0

Fig. 12. Horizontal flow-diagram and CNN architecture by Chakrabarty et al. [8] for diabetic retinopathy classification

This methodology also involves Adaptive Thresholding followed by Hybrid Modelling in CNN and SVM with Sigmoid kernel (using the best obtained value of C parameter from 3-Fold Grid-Search Cross Validation). As far as the CNN Architecture is concerned, its the same as in 2.2.1 and 2.2.2 but with different pre-initialized weights that can be found in the link [10]. This model also achieved a 100% Test Accuracy with Precision, Recall, F1-Score, AUC each equal to 1.0 resulting in a full-proof model. The feature map visualizations for the same are shown in Fig. 13.

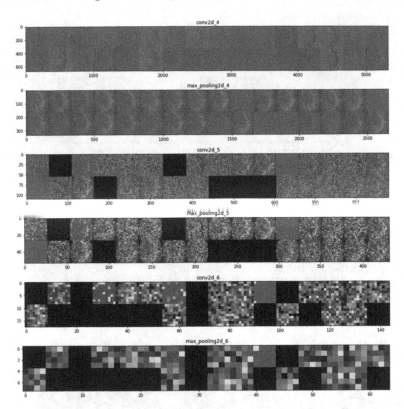

Fig. 13. Feature map visualization in every channel of each of the 6 layers of the CNN for diabetic retinopathy classification

The vital portions of the retinal fundus image responsible for Diabetic Retinopathy are extracted and highlighted by the Intermediate Layer Activations in Fig. 13.

2.2.4 Glaucoma Screening or Diagnosis

This is also a use-case of Applied Medical Imaging in which, given an HRF Retina Image as input to the Model, the Model classifies whether the person is suffering from Glaucoma or not. So, this is also a Binary Classification Problem. For the purpose, the state-of-the-art model by Chakrabarty et al. [11] is considered, which also uses the HRF Dataset [9] (sample images are shown in Fig. 11) and the corresponding work-flow diagram for the model is given in Fig. 14. In this model, Adaptive Thresholding is followed by SVM-RBF Kernel for Model Training (using the best obtained value of C parameter from 3-Fold Grid-Search Cross Validation) and the CNN weights can be found in the link [12]. This model also reached a 100% Test Accuracy with Precision, Recall, F1-Score, AUC each equal to 1.0 resulting in a full-proof model. The Intermediate Layer Activation Visualizations are shown in Fig. 15.

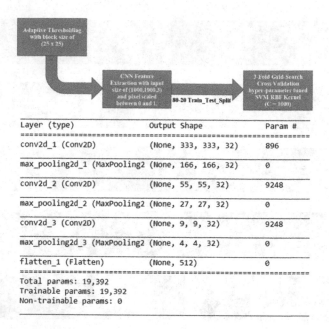

Fig. 14. Horizontal flow-diagram and CNN architecture by Chakrabarty et al. [11] for Glaucoma screening or diagnosis

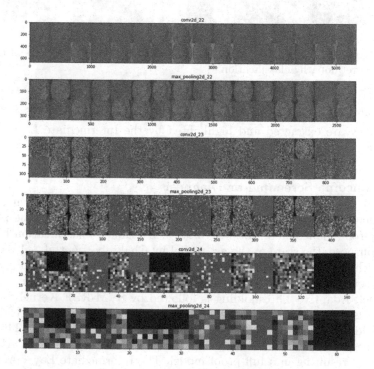

Fig. 15. Feature map visualization in every channel of each of the 6 layers of the CNN for Glaucoma screening

The crucial portions of the retinal fundus image responsible for Glaucoma Diagnosis are extracted and highlighted by the Feature Map Visualizations in Fig. 15.

3   Conclusion

This paper proposed Machine Learning Model Interpretation Visualization Methods like feature coefficient/importance based Model Wordclouds in NLP Applications and Intermediate Layer Activations for CNN Feature Extraction in Computer Vision Applications. These are highly essential for relating ML Model Performance with the concerned Domain Knowledge relevance in order to get the acceptance for final deployment.

References

1. Chakrabarty, N.: A machine learning approach to comment toxicity classification. In: Das, A.K., Nayak, J., Naik, B., Pati, S.K., Pelusi, D. (eds.) Computational Intelligence in Pattern Recognition. AISC, vol. 999, pp. 183–193. Springer, Singapore (2020). https://doi.org/10.1007/978-981-13-9042-5_16
2. https://www.kaggle.com/c/jigsaw-toxic-comment-classification-challenge/data
3. Chakrabarty, N.: A novel strategy for gender identification from hand dorsal images using computer vision. In: 2019 3rd International Conference on Computing Methodologies and Communication (ICCMC) 2019, 27 March 2019, pp. 108–113. IEEE (2019)
4. 11k Hands Dataset at https://sites.google.com/view/11khands/home
5. gender.h5 at https://github.com/navoneel1092283/Gender_Recognition_CNN_weights
6. Chakrabarty, N., Chatterjee, S.: A novel approach to age classification from hand dorsal images using computer vision. In: 2019 3rd International Conference on Computing Methodologies and Communication (ICCMC), 27 March 2019, pp. 198–202. IEEE (2019)
7. conv_age.h5 at https://github.com/navoneel1092283/Age_Classification_CNN_weights
8. Chakrabarty, N., Chatterjee, S.: An offbeat technique for diabetic retinopathy detection using computer vision. In: 2019 10th International Conference on Computing, Communication and Networking Technologies (ICCCNT), 6 July 2019, pp. 1–5. IEEE (2019)
9. HRF Dataset at https://www5.cs.fau.de/research/data/fundus-images/
10. new_cdr.h5 at https://github.com/navoneel1092283/Diabetic_Retinopathy_CNN_weights
11. Chakrabarty, N., Chatterjee, S.: A novel approach to glaucoma screening using computer vision. In: 2019 International Conference on Smart Systems and Inventive Technology (ICSSIT), 27 November 2019, pp. 881–884. IEEE (2019)
12. conv3.h5 at https://github.com/navoneel1092283/Glaucoma_Screening_CNN_weights

Optimal Sizing and Siting of Multiple Dispersed Generation System Using Metaheuristic Algorithm

Lokesh Kumar Yadav[1,2]([✉]), Mitresh Kumar Verma[2], and Puneet Joshi[1]

[1] Electrical Engineering Department, Rajkiya Engineering College,
Ambedkar Nagar 224122, India
lokeshky@recabn.ac.in
[2] Electrical Engineering Department, IIT (BHU) Varanasi, Varanasi 221005, India

Abstract. Optimally sized and sited DGs improve efficiency & the voltage scenario of any radial distribution system (RDS). Optimal DG allocation and sizing problem may be solved by selecting a suitable meta heuristic optimization technique. In this article a hybrid particle swarm optimization-gray wolf optimization (PSO-GWO) is used for optimal placement and sizing of DG units. A multi-objective function is framed utilizing weighted sum of voltage stability index, real power loss and voltage deviation whose weights are optimized using Genetic Algorithm. The proposed methodology is tested on IEEE 33 bus RDS for Type-I DGs. The results obtained demonstrate that PSO-GWO has a clear advantage over other similar techniques.

Keywords: Voltage Stability Index (VSI) · PSO · GWO · DGS · Active power loss (P_{Loss})

1 Introduction

Power system structure is generally classified in to three divisions namely Generation, Transmission and Distribution [1]. Power is generated at centralized generation end and received at substation through inter connected transmission infrastructure. The end user receives power from substation through RDS. Unidirectional power flow and passiveness is the characteristics of RDS due to high R/X ratio [2]. The high R/X ratio introduces P_{Loss} and voltage drop at each node. Due to this, under critical loading, system may reach voltage stability limit.

In modern days, the problem of voltage instability and P_{Loss} in RDS may be mitigated by injecting DGs of appropriate capacity at optimal location. The allocation of DGs with suitable size can be calculated by taking either single objective or multi-objective fitness function (MOFF) comprising of various objectives *viz.*, real power loss, emissions, voltage stability index, voltage deviation, reliability, etc. Henceforth, in this article a multi-objective problem for optimal siting and sizing of DGs in a RDS has been formulated considering real power loss, voltage stability index and voltage deviation.

© Springer Nature Switzerland AG 2021
M. Singh et al. (Eds.): ICACDS 2021, CCIS 1440, pp. 268–278, 2021.
https://doi.org/10.1007/978-3-030-81462-5_25

The penalty coefficients are determined using GA. A MOFF comprising of Real Power Loss, voltage deviation (VD) and VSI is used to find optimal size and site of multiple type I DGs in IEEE 33-bus RDS.

In this script PSO-GWO algorithm is used for optimal assignment and rating of type I DGs. The rest part of the paper constitutes of following sections. Section 2 is devoted to state-of-the-art findings and explaining & enlisting of the constraints related to it. Section 3 discusses the details of problem formulation and Sect. 4 briefly explains the adopted methodology. Result and Discussion is covered in Sect. 5 and the last Sect. 6 concludes the article.

2 Literature Work

There are several metaheuristics and hybrid optimization techniques found in literature to solve the said problem. In [3] Taboo Search algorithm is used to solve optimal site and size of multiple DGs and results shows its superiority in respect of processing time, which will becomes more critical parameter for complex system. Optimal planning of multiple DGs in 13-bus RDS for the objective of improving VD, minimizing P_{Loss} and total harmonic distortion is done in [4] with the help of PSO. In addition, the results are compared with GA. In [5] Ant colony optimizer (ACO) determines the optimal site and size of DGs for minimum investment cost of DG and operation cost of system. Higher quality solutions obtained from multi-objective function as compared to single objective for optimal siting and sizing of DG with the help of Simulated Annealing is shown in [6]. [7] proposed a novel optimization algorithm (i.e., water cycle algorithm (WCA)) for determining the optimal configuration of DGs over multi-objective function. WCA give more flexibility in terms of power factor (pf) of DGs which shows its advantage over DGs with fixed pf. Apart from this many other heuristic techniques like hybrid of Artificial Bee Colony and Cuckoo search (ABC–CS) algorithm [8], hybrid of ABC and (ACO) [9], Krill Herd Algorithm (KHA) [10] and Modified-PSO [11] have been used for DG configuration in RDS.

3 Problem Formulation

This section formulates the objective of optimal DG sizing and siting problem along with all constraints.

3.1 Active Power Loss [12]

$$P_{LOSS} = \sum_{a=1}^{n} \sum_{b=1}^{n} (\alpha_{ab}(P_a P_b + Q_a Q_b) + \beta_{ab}(Q_a P_b - P_a Q_b)) \quad (1)$$

where,

$$\alpha_{ab} = \frac{r_{ab}}{V_a V_b} \cos(\delta_a - \delta_b) \quad (2)$$

$$\beta_{ab} = \frac{r_{ab}}{V_a V_b} \sin(\delta_a - \delta_b) \quad (3)$$

$$Z_{ab} = r_{ab} + jx_{ab} \tag{4}$$

where,

r_{ab}, Z_{ab} and x_{ab}: line resistance, impedance and reactance respectively.
V_a, V_b: voltages; δ_a, δ_b: voltage angles.
Q_a, P_a: reactive and active power injections respectively.
Q_b, P_b: reactive and active power at receiving ends respectively.

The first objective for optimal assignment and rating problem of DG is minimization of P_{Losss} as shown in the Eq. (5).

$$OF_1 = \text{minmization}(P_{LOSS}) \tag{5}$$

3.2 Voltage Deviation

Voltage deviation [12] from the reference voltage can be expressed as

$$\Delta VD = \sum_{i=1}^{n_b} (v_i - v_b)^2 \tag{6}$$

where,

v_i: bus voltage; v_b: base/reference voltage

The second objective is minimization of VD at each bus given in Eq. (7)

$$OF_2 = \text{minimization}(\Delta VD) \tag{7}$$

3.3 Voltage Stability Index

Stability index (SI) [12] is used to find the level of stability of RDS. Since SI is based on the feasible solution of voltage quadratic equation of RDS, we may select it as suitable VSI to calculate level of stability of any RDS.

Fig. 1. Diagram for the line connecting n_1 and n_2 buses.

Figure 1 may be considered as the electrical equivalent of n-bus radial distribution system.

The voltage stability is expressed as

$$|V_{n_1}|^4 - 4\{P_{n_2}x_{n_1n_2} - Q_{n_2}r_{n_1n_2}\}^2 - 4\{P_{n_2}r_{n_1n_2} + Q_{n_2}x_{n_1n_2}\}|V_{n_1}|^2 \geq 0 \qquad (8)$$

or

$$SI(n2) = |V_{n1}|^4 - 4.0\{P_{n2}x_{n_1n_2} - Q_{n2})r_{n_1n_2}\}^2 - 4.0\{P_{n2}r_{n_1n_2} + Q_{n2}x_{n_1n_2}\}V_{n1}^2 \qquad (9)$$

where,

$SI(n_2)$: stability index of bus/node n_2 (where $n_2 = 2, 3, 4\ldots\ldots n_b$) and the condition for stability is $SI(n_2) \geq 0$.

The most severe node/bus is the one that has the lowest value of SI. Therefore, for stable operation, SI should be maximum. The last objective taken in this paper is maximization of SI as given in the Eq. (10).

$$OF_3 = \text{minimization}(^1/_{SI}) \qquad (10)$$

3.4 Objective Function

All objectives illustrated earlier in Eq. (5), (7) and (10) can be combined together to form MOFF given as Eq. (11).

$$Ob\_Fun = \text{minmization}(\lambda_1 \times OF_1 + \lambda_2 \times OF_2 + \lambda_3 \times OF_3) \qquad (11)$$

Where λ_1 is the penalty coefficient of value 0.17; λ_2 is the penalty coefficient of value 0.7 and λ_3 is the penalty coefficient of value 0.13.

Initially the random values are generated between 0.1 and 1. To avoid the ineffectiveness of any penalty coefficient the lower limit of all lambdas is selected 0.1 instead of 0. Optimized sets of values are selected through GA. Here in this paper the set, corresponding to which MOFF is found minimum is selected.

3.5 System Constraints

System Equation/Power Balance Equation

$$P_{Ga} - P_{Da} = \sum_{b=1}^{N} V_a V_a [G_{ab}\cos(\delta_a - \delta_b) + B_{ab}\sin(\delta_a - \delta_b)] \qquad (12)$$

$$Q_{Ga} - Q_{Db} = \sum_{b=1}^{N} V_a V_a [G_{ab}\sin(\delta_a - \delta_b) + B_{ab}\cos(\delta_a - \delta_b)] \qquad (13)$$

Voltage Constraints

$$V_{\min} \leq V_a \leq V_{\max} \tag{14}$$

Thermal Limit/Current Constraints

$$I_a \leq I_a^{Rated} \tag{15}$$

DG Real Power Limit

$$P_{i\min}^{DG} \leq P_i^{DG} \leq P_{i\max}^{DG} \tag{16}$$

DG Reactive Power Limit

$$Q_{i\min}^{DG} \leq Q_i^{DG} \leq Q_{i\max}^{DG} \tag{17}$$

4 Optimization Techniques

In this article amalgamation of PSO and GWO is used to detect optimal site and rating of multiple Type I DGs.

4.1 Particle Swarm Optimization (PSO)

PSO algorithm is inspired by flocks [13]. It gives optimal solution based on population. The population of particles searches for better solution through their own experience and also by that of the others.

PSO can be mathematically represented as

$$POS_m = (pos_{m,1}, pos_{m,2}, pos_{m,3}, pos_{m,4}, \ldots\ldots pos_{m,n}) \tag{18}$$

$$Velo_m = (Velo_{m,1}, Velo_{m,2}, Velo_{m,3}, Velo_{m,4}, \ldots\ldots Velo_{m,n}) \tag{19}$$

The modified position (POS) and velocity (Velo) of each particle is given as:

$$POS_{xd}^{k+1} = POS_{xd}^k + Velo_{xd}^{k+1} \tag{20}$$

$$Velo_{xd}^{k+1} = \omega_x Velo_{xd}^k + c_1 rand \times \left(pbest_{xd} - x_{xd}^k\right) + c_2 rand \times \left(gbest_{xd} - POS_{xd}^k\right) \tag{21}$$

where,

$$x = 1, 2, ..., n$$
$$d = 1, 2, ..., m \tag{22}$$

$$\omega_k = \omega_{max} - \frac{\omega_{max} - \omega_{min}}{k_{max}}.k \tag{23}$$

The values of c_1, c_2, ω_{max} and ω_{min} has been optimized through hit and trial. k_{max} is maximum iterations.

1.3 Grey Wolf Optimization (GWO)

GWO technique [14] is inspired from Grey wolves (Canis lupus) that belong to Canidae family. They have a very disciplined social dominance hierarchy as shown in the Fig. 2.

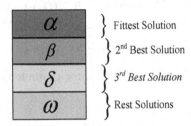

Fig. 2. Social order of grey wolf

The various phases of grey wolves hunting are shown in Fig. 3.

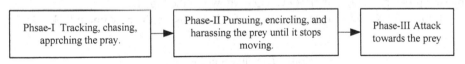

Fig. 3. Phases of grey wolves hunting

Mathematical Model. Mathematical representation of encircling behaviour of grey wolves to pray is given below.

$$D = \left| E \cdot POS_{prey}(t) - POS(t) \right| \tag{24}$$

$$POS(t + 1) = POS_{prey}(t) - B \cdot D \tag{25}$$

$$B = 2b \cdot r_1 - b \tag{26}$$

$$E = 2 \cdot r_2 \qquad (27)$$

where,

D: location of prey; t: current iteration; B & E: coefficient vectors; $0 < b < 2$ in decreasing order.; $0 \le r_1 \le 1, 0 \le r_2 \le 1$.

The mathematical equation for finding position of prey by α, β and δ wolves is given here.

$$\left.\begin{aligned} D_\alpha &= |E_1 \cdot POS_\alpha - POS| \\ D_\beta &= |E_2 \cdot POS_\beta - POS| \\ D_\delta &= |E_3 \cdot POS_\delta - POS| \end{aligned}\right\} \qquad (28)$$

$$\left.\begin{aligned} POS_1 &= POS_\alpha - B_1 \cdot (D_\alpha) \\ POS_2 &= POS_\beta - B_2 \cdot (D_\beta) \\ POS_3 &= POS_\delta - B_3 \cdot (D_\delta) \end{aligned}\right\} \qquad (29)$$

According to Eq. (28) and (29) (i.e. the best three positions) remaining wolves update their position.

4.3 Hybrid PSO-GWO

The hybrid PSO-GWO algorithm is formed in order to utilize the better exploitation capability of PSO and good exploration capability of GWO [15]. To have the values of P, Q, V & δ at various stages, backward/forward load flow has been applied. The step by step problem solution by PSO-GWO is shown in the Fig. 4.

Table 1 shows the voltages at each bus for both case with and without DG.

Table 1. Voltages at each bus

Bus no.	Without DG	With DG	Bus no.	Without DG	With DG
1	1.0000	1.0000	18	**0.9131**	0.9856
2	0.9970	0.9989	19	0.9965	0.9983
3	0.9829	0.9946	20	0.9929	0.9948
4	0.9755	0.9943	22	0.9922	0.9941
5	0.9681	0.9945	23	0.9916	0.9934
6	0.9497	0.9926	24	0.9794	0.9910
7	0.9462	0.9907	25	0.9727	0.9844
8	0.9413	0.9916	26	0.9694	**0.9811**

(*continued*)

Table 1. (*continued*)

Bus no.	Without DG	With DG	Bus no.	Without DG	With DG
9	0.9351	0.9936	27	0.9477	0.9930
10	0.9292	0.9962	28	0.9452	0.9923
11	0.9284	0.9969	29	0.9337	0.9878
12	0.9269	0.9984	30	0.9255	0.9848
13	0.9208	0.9927	31	0.9220	0.9845
14	0.9185	0.9906	32	0.9178	0.9863
15	0.9171	0.9893	33	0.9169	0.9855
16	0.9157	0.9881			
17	0.9137	0.9862			

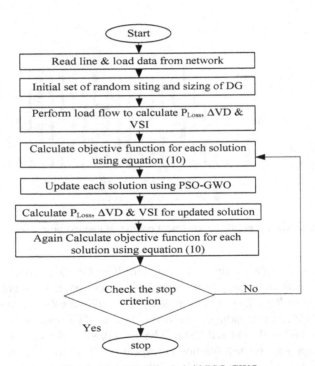

Fig. 4. Flow chart for hybrid PSO-GWO

5 Results and Discussion

The effectiveness and validation of PSO-GWO technique is tested on IEEE 33 bus RDS shown in Fig. 4 to determine the optimal siting and sizing of Type-I DGs in order to minimize P_{Loss}, VD and maximize VSI. The population size and iterations are 50 and 100 respectively to find best results. The best results are shown in the Table 2.

Table 2. Simulation results of different optimization techniques of IEEE 33 bus RDS

GA [16]		PSO [16]		GA-PSO [16]		PSO-GWO	
ODGL	ODGS	ODGL	ODGS	ODGL	ODGS	ODGL	ODGS
11	1.5	8	1.1768	11	0.925	31	0.9343
29	0.4228	13	0.9816	16	0.863	26	0.8663
30	1.0714	32	0.8297	32	1.2	12	1.2252
	GA [16]		PSO [16]		GA-PSO [16]		PSO-GWO
P_{Loss} (KW)	106.30		105.30		103.40		**92.7686**
VD (pu)	0.0407		0.0335		0.0124		**0.0034**
VSI (pu)	0.9490		0.9256		0.9508		**0.9560**

ODGL-Optimal DG location, ODGS-Optimal DG size.

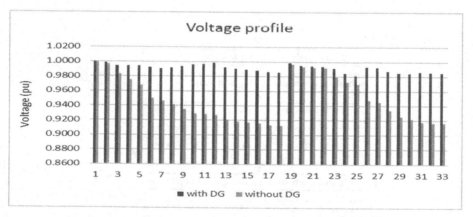

Fig. 5. Voltage profile improvement corresponding to optimal solution

The simulation results obtained by the hybrid PSO-GWO approach are compared with other optimization techniques in Table 2. The multi-objective problem discussed in this paper for multiple DG placement, there are multiple solutions possible for the FF of value 0.2156. corresponding to this optimal value of FF optimal P_{Loss} is shown in the Table 2, VD will be 0.0034 and VSI will be 0.9560. Since the weight corresponding to the VD is higher so the best solution in respect of VD minimization is cited in the Table 2. It can clearly be observed in Fig. 5 that the voltage of each bus has improved significantly. The minimum value of voltages with and without DG are 0.9811 pu and 0.9131 pu at buses 26 and 18 respectively shown in Table 1.

6   Conclusion

Type I DGs have been optimally placed with suitable size by using PSO-GWO algorithm with minimum active power losses, voltage deviation and maximum VSI in standard

IEEE 33-bus RDS. Weights are optimized through GA optimization tool box in MAT-LAB environment. The amalgamation of GWO with PSO provides better convergence performance to PSO and also avoid results to stuck in local minima. The result and discussion section guaranties better results provided by PSO-GWO than conventional PSO.

Acknowledgement. I am highly thankful to TEQIP-III, Rajkiya Engineering College, Ambedkarnagar for funding this project.

References

1. Hanson, A., Grigsby, L.: Power system analysis. In: Systems, Controls, Embedded Systems, Energy, and Machines (2017)
2. Aman, M.M., Jasmon, G.B., Mokhlis, H., Bakar, A.H.A.: Optimal placement and sizing of a DG based on a new power stability index and line losses. Int. J. Electr. Power Energy Syst. **43**, 1296–1304 (2012)
3. Maciel, R.S., Padilha-Feltrin, A.: Distributed generation impact evaluation using a multi-objective tabu search. In: 2009 15th International Conference on Intelligent System Applications to Power Systems. ISAP 2009 (2009)
4. Alinejad-Beromi, Y., Sedighizadeh, M., Sadighi, M.: A particle swarm optimization for sitting and sizing of distributed generation in distribution network to improve voltage profile and reduce THD and losses. In: Proceedings of the Universities Power Engineering Conference (2008)
5. Falaghi, H., Haghifam, M.R.: ACO based algorithm for distributed generation sources allocation and sizing in distribution systems. In: 2007 IEEE Lausanne POWERTECH, Proceedings, pp. 555–560 (2007)
6. Aly, A.I., Hegazy, Y.G., Alsharkawy, M.A.: A simulated annealing algorithm for multi-objective distributed generation planning. In: IEEE PES General Meeting PES, pp. 1–7 (2010)
7. El-Ela, A.A.A., El-Sehiemy, R.A., Abbas, A.S.: Optimal placement and sizing of distributed generation and capacitor banks in distribution systems using water cycle algorithm. IEEE Syst. J. **12**, 3629–3636 (2018)
8. Bala, R., Ghosh, S.: Optimal position and rating of DG in distribution networks by ABC–CS from load flow solutions illustrated by fuzzy-PSO. Neural Comput. Appl. **31**(2), 489–507 (2017). https://doi.org/10.1007/s00521-017-3084-7
9. Kefayat, M., Lashkar Ara, A., Nabavi Niaki, S.A.: A hybrid of ant colony optimization and artificial bee colony algorithm for probabilistic optimal placement and sizing of distributed energy resources. Energy Convers. Manag. **92**, 49–161 (2015)
10. Sultana, S., Roy, P.K.: Krill herd algorithm for optimal location of distributed generator in radial distribution system. Appl. Soft Comput. J. **40**, 391–404 (2016)
11. Jain, N., Singh, S.N., Srivastava, S.C.: PSO based placement of multiple wind DGs and capacitors utilizing probabilistic load flow model. Swarm Evol. Comput. **19**, 15–24 (2014)
12. Moradi, M.H., Abedini, M.: A combination of genetic algorithm and particle swarm optimization for optimal DG location and sizing in distribution systems. Int. J. Electr. Power Energy Syst. **34**, 66–74 (2012)
13. Rao, S.S.: Engineering Optimization: Theory and Practice, 4th edn. Wiley & Sons Inc. (2009)
14. Mirjalili, S., Mirjalili, S.M., Lewis, A.: Grey wolf optimizer. Adv. Eng. Softw. **69**, 46–61 (2014)

15. Şenel, F.A., Gökçe, F., Yüksel, A.S., Yiğit, T.: A novel hybrid PSO–GWO algorithm for optimization problems. Eng. Comput. **35**(4), 1359–1373 (2018). https://doi.org/10.1007/s00 366-018-0668-5
16. Sultana, S., Roy, P.K.: Multi-objective quasi-oppositional teaching learning based optimization for optimal location of distributed generator in radial distribution systems. Int. J. Electr. Power Energy Syst. **63**, 534–545 (2014)

Design of a Fused Triple Convolutional Neural Network for Malware Detection: A Visual Classification Approach

Santosh K. Smmarwar$^{(\boxtimes)}$, Govind P. Gupta, and Sanjay Kumar

Department of Information Technology, National Institute of Technology, Raipur, India
{sksmmarwar.phd2019.it,gpgupta.it,skumar.it}@nitrr.ac.in

Abstract. Detection of malware signatures from executable files requires effective signal processing and sandboxing operations, wherein the executable file is scanned for any malignant behavior. The recent malware detection techniques are based on static approaches that use machine and deep learning for analyzing malware signatures from byte and assembly-level program data. The byte-patterns are based on outliers, and the program-data is classified as a malware. These methods are not capable of detecting new variants of malware with long patterns of codes and huge dataset to classify the benign or malicious files. The issue with pattern analysis of large byte code dataset needs effective classification performance. To overcome these drawbacks, this paper has proposed a novel fused-triple convolutional neural network (fCNN) based framework for malware detection. This framework improves the accuracy of malware classification by converting the byte and assembly information into image data. This framework obtained more than 98% accuracy on the Microsoft Malware Dataset.

Keywords: Malware detection · CNN · VGGNet · Byte · Assembly · Image · Classification

1 Introduction

The exponential growth in the rise of new malware poses a critical threat to the internet, the user's integrity, privacy, confidentiality and valuable resources of computer systems. The malwares can cause a major loss to the organizations, societies and nations. Malware is a software code having malicious intentions to the computer systems, mobile environment and cloud service platform *etc.* various types of malware exists based on their functionalities like viruses, worms, Trojans, adware, spyware, Ransomware and many more *etc.* [1]. A large number of internet viruses are malwares which inject themselves onto the victim's device or machine and try to run malicious programs on the host. These programs would either try to interrupt the normal working of the host machine or steal valuable information from the machine which can be later used for blackmailing or invading the user's privacy [2]. In addition, these malwares can analyse activities over the keyboard to steal passwords and other sensitive information. In current scenarios of

© Springer Nature Switzerland AG 2021
M. Singh et al. (Eds.): ICACDS 2021, CCIS 1440, pp. 279–289, 2021.
https://doi.org/10.1007/978-3-030-81462-5_26

technological development traditional machine learning and deep learning based methods have achieved better performance in detection and classification of malware based on signature patterns to detect known malware. However, these techniques are expensive in terms of time, resources and feature engineering and require large dataset for better detection and classification of malware and unable to detect new variants of malware [3].

To detect and classify these malwares, various researchers have proposed different approaches in the literature. A survey of such approaches is presented in the next section. From this survey, it is observed that most of these approaches depend on short-term pattern analysis, which limits their performance and applicability to real-time datasets. Moreover, these techniques require large training data sets, which again limit the system applicability for applications where limited historical data is available. To overcome these drawbacks, this research work proposes a novel fused triple CNN approach, which initially converts the byte-level and assembly-level information of the programs into images. Next, these images are processed individually and in a combined form via a triple CNN VGGNet classifier. The statistical analysis indicates that the classifier has superior performance when compared with other single-program approaches and has the flexibility to be applied for cross-platform malwares. This text also indicates some of the future research that can be taken up to achieve cross-platform analysis for the malwares using a modified version of the CNN architecture. In this paper the main contribution of our work as given below.

- In this work, a novel fused triple convolutional neural network (fCNN) design is proposed for detection of malware.
- Proposed design is able to improve the accuracy of malware classification by converting the byte and assembly information into image data and then by using a VGGNet inspired CNN design to obtain more than 98% accuracy on Microsoft malware dataset.
- These images are processed individually and in a combined form via a triple CNN VGGNet classifier.
- The statistical analysis indicates that the classifier has superior performance when compared with other single-program approaches and has the flexibility to be applied for cross-platform malwares.

The remaining part of the paper consists of the following sections as follows Sect. 2 presents the related work, Sect. 3 indicates the design of proposed triple CNN architecture, Sect. 4 shows the result evaluation, Sect. 5 describes the brief about the dataset and finally Sect. 6 presents the conclusion and future scope of our work.

2   Related Work

The process of converting byte and assembly file data into imagery for malware classification has proven to be an effective approach. This can be observed from [4], wherein space filling curve mapping (SCFM) and Markov dot plot (MDP) have been used to convert the byte-file data into image data to identify decompression bomb attacks. The research work in [5] converts the opcodes and application programming interface (API) calls into word-to-vector features and obtains the derived names from these features. The approach is found to be approximately 1.3% better in terms of accuracy performance when compared to a long-short-term-memory (LSTM) based approach. The work in [6] evaluates structural and behavioural features from the malware files to classify poly-morphic malwares. The work in [7] showcases that LSTM-based convolutional neural networks with transfer learning are most suited for this purpose. In [8], authors have suggested that Random Forests (RF) and k-Nearest Neighbours (kNN) classifiers are best suited for the purpose of malware classification. A specific file-access control vul-nerability detector using machine learning is proposed in [9]. In [10] the byte files are converted into images, and an eXtreme Gradient Boosting (XGBoost) classifier is used to classify these images into malware categories. The work in [11] uses soft-relevance value (s-value) to find out the relevance of one-type of malware signatures with non-malware signatures. Another approach for detection of unknown malwares is described in [12]. This approach uses graph-embedding wherein the byte files are converted into eigenspace graphs using power iteration method. Malwares can be injected in systems using portable document files (PDFs). The work in [13] scans these PDF files for malware patterns by converting the PDF file byte data into directed graphs. A generic approach for detection of malware on mobile devices is mentioned in [14]. It uses a skip-gram model for detecting different kinds of malwares on Android devices. Another byte-to-image conversion approach that utilizes Markov image creation is mentioned in [15]. Here, the input byte-level data is converted into image format using bytes transfer probability matrices. An accuracy of 97% is achieved on Microsoft dataset, which can be improved by using better CNN architecture design that involves grouping of byte files with similar execution structure. Such a work can be observed in [16], which claims to improve the accuracy of classification to 99% by simply using linear classifiers like SVM, RF and kNN. The work in [15] and [16] can be combined to develop a highly accurate malware classification system. The work in [15] can also be extended using the work in [17], wherein random forests are used for classification of malware image data for improving the accuracy of classification to nearly 99%. Moreover, a multi-view learning method can be adopted for making the malware detection system applicable to any kind of oper-ating system. This can be referred from the work in [18]. All these views are combined to form a master-feature vector and are given to a CNN classification engine. Due to the master feature vector, the final accuracy of the proposed system is nearly 99.6%, which is almost 1% higher than using only byte and operation code feature sets. This inspired the underlying research to develop a triple CNN architecture for obtaining better malware classification performance. The selection of a CNN engine for such a system design is also important. So, the work in [19] was referred. The use of image-based malware detection can also be confirmed from the work done in [20–22] and [23] which represent the byte-level and assembly level data into different visual formats and then applies machine learning methods to identify malware signatures. The work in [24] suggests the use of registry analysis and byte-level analysis as previously mentioned and combines it with machine learning approaches to obtain better classification performance even for cloud malwares.

To evaluate the underlying research a large-scale dataset is needed; therefore, this text uses the Microsoft Malware Classification dataset as described in [25]. This dataset consists of different kinds of files for each kind of OS operation. These operations are linked to malware activities, which assists in linking the files to malware activities. This dataset is also used in [26] and [27] for identification of malwares and prove that CNN-based approaches along with visual data representation are best suited for malware detection. The concept can be further extended to web-based malware detection and mobile-based malware detection.

Table 1. Microsoft malware dataset sample composition

Class ID	Malware family name	Number of samples	Type of attack
1	Ramnit	1487	Worms
2	Lollipop	2321	Adware
3	Kelihos_ver3	2880	Backdoor
4	Vundo	407	Trojan
5	Simda	42	Backdoor
6	Tracur	705	Trojan download malware
7	Kelihos_ver1	361	Backdoor
8	Obfuscation.ACY	1163	Any type of obfuscated malware
9	Gatak	983	Backdoor

3 Proposed Malware Detection Framework Using Fused-Triple CNN

In the proposed fused-Triple CNN based malware detection framework, both the byte file data and the assembly file data need to be converted into image data. Both set of image datasets are then merged to obtain a 3rd fused image. This 3rd image dataset is generated by appending the bytes from the assembly file to the byte file. Once these images are generated, then an individual CNN architecture is adopted to classify the images into one of 9 malware categories. These categories are listed in Table 1. The description of each of these steps can be observed in the following sub-sections, while the overall architecture diagram for the entire system can be observed from Fig. 1, wherein all the blocks and their connections are shown. The final fusion process evaluates the best of three selection processes to obtain the final malware class.

3.1 Conversion of Byte Files into Image Data

The byte files contain data in the form of operation codes, and operands. Each data is of one byte, and thus is directly added to an image array. Due to the variation in program size, each byte file has a different length. To normalize this length, the images are converted into 224 × 224 size using bicubic interpolation method as shown in Fig. 2.

3.2 Conversion of Assembly Files into Images

Assembly files contain the following attributes,

- Headers indicating addresses, read/write permissions, and other metadata
- Text attribute consisting of the text data needed by the assembly file
- RData consisting of the read data or operation codes in the assembly file
- IData consisting of import data, which indicates the different external programs imported by this assembly file
- Data consisting of memory and addressing data

From these data fields, the RData and IData fields are selected for image creation. This is due to the fact that other data which is added to the assembly file is verified by the OS and cannot contain any malicious code. While the data which will be continuously read and imported by the program might contain malicious code bytes. All these bytes are combined together to form the final image file using the same process as discussed in Sect. 3.1. The final image is converted into 224 × 224 sizes for normalization as depicted in Fig. 3.

3.3 Generation of the Fused Image

The fused image is created by adding the bytes from both byte file and assembly file one-after-the-other. All these bytes when combined create a new image file, which consists of both assembly and byte file patterns as shown in Fig. 4. The resulting image is converted into 448 × 448 sizes for normalization and further processing.

3.4 CNN Design for Byte File, Assembly File, and Fused File Classification

The VGGNet classifier is selected for classification of these images. Architecture of the VGGNet is illustrated in Fig. 5. The input layer is 224 × 224 × N in size, wherein 224 × 224 is the image size, while N is the number of images in the training set. A sample CNN architecture with 224 × 224 × N configuration can be observed from Fig. 5, wherein the input image is given to the CNN, and the CNN uses a rectilinear unit (ReLU) to convert this data into 112 × 112 form using Max pooling. This process is repeated 6 times, wherein the final features of size 7 × 7 × 512 are obtained. The final fully connected layer is capable of classifying the data into 1 of 'M' classes; in this case the value of M is kept as 9, due to the need of 9 malware classes at the output. Two such CNNs are deployed in order to classify the assembly file and the byte files individually.

For the fused file an additional convolutional + ReLU layer along with a Max pooling layer is attached in order to obtain the final class. This is done because the size of input imagery for byte and assembly files is twice the size of the fused file. These networks are trained, and the final class is obtained via fusion of the outputs of these networks.

In order to perform class fusion, the resulting outputs are merged in such that a manner, that if a minimum 2 out of 3 networks produce the same output, then the network agrees on the final result. If none of the outputs are the same, then the output is marked to be a combination of these malwares, and thus assists in discovery of new malwares from the system. Description of the dataset and result evaluation of the proposed framework is discussed in the next section.

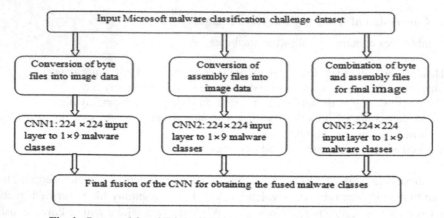

Fig. 1. Proposed fused Triple CNN based malware detection Framework

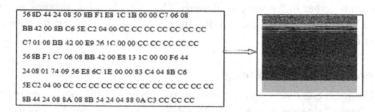

Fig. 2. Hexadecimal view of Byte files to image conversion

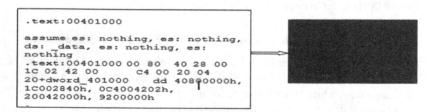

Fig. 3. Views of assembly file to image conversion

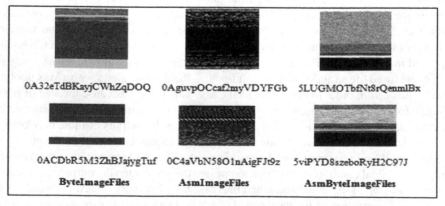

Fig. 4. Generation of the fused image files from byte files and asm files

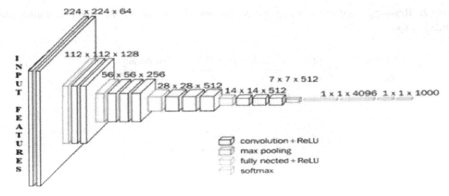

Fig. 5. Design of VGG16 architecture

4 Dataset Description

The proposed model used the Microsoft malware classification dataset BIG 2015 challenge from Kaggle provided by Microsoft. The dataset contains the 10,349 malware sample files of nine types of malware families shown in Table 1, each sample having the twenty-character class ID hash value of binary files and assembly files. The experiment is performed on windows 10 home, G7 core i9 10th Gen-(16 GB/1 TB SD/8 GB graphics/NVIDIA RTX 2070/300 Hz). The proposed model used python programming language with keras and Tensor flow packages.

5 Result Analysis

To evaluate the system, the entire Microsoft malware classification dataset is used. The dataset is divided into the following parts for training, testing and validation for incremental evaluation.

- 50:25:25
- 60:20:20
- 70:15:15
- 80:10:10

The classification results from each of these evaluations are tabulated using Table 2, wherein it can be observed that the network produces 99.9% classification accuracy even for 60:20:20 training, testing and validation split. Due to a combination of triple CNN, the entire dataset can be classified with highest possible accuracy with limited training set information. The selection of training set requires careful segregation of records such that the final data split is capable of classifying testing and validation sets with high accuracy.

These results when compared with other implementations also showcase superior performance in terms of accuracy of classification. This performance can be observed from Table 3, wherein five state-of-art techniques are used for comparison. From these

Table 2. Resulting accuracy of the proposed framework for each of the training, testing and validation splits

Train	Test	Validation	Accuracy (%)
50	25	25	95.40
60	20	20	99.90
70	15	15	99.91
80	10	10	99.98

Table 3. Accuracy comparison of proposed method with existing methods

Method	Accuracy (%)
TCN [5]	97.52
SCFM & MDP [4]	99.36
XGBoost [10]	99.77
HYDRA [3]	99.75
Soft relevance [11]	99.8
Proposed (fCNN)	99.98

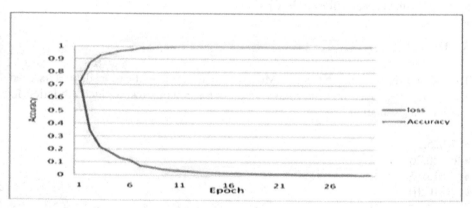

Fig. 6. Training loss and training accuracy comparison of our proposed model

results shown below it can be observed that the proposed framework is highly accurate and can be used for real-time malware classification on Microsoft malware classification dataset. Figure 6 showing the training loss, training accuracy and Fig. 7 showing the validation loss and validation accuracy of our proposed method.

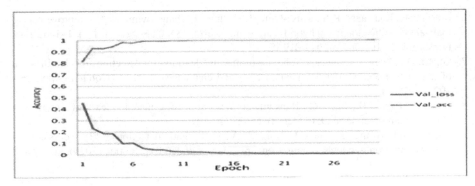

Fig. 7. Validation loss and validation accuracy comparison of our proposed model

6 Conclusion and Future Scope

This paper proposed a novel fused-triple convolutional neural network (fCNN)-based malware detection framework where first byte and assembly information are converted into image data and then VGGNet inspired fused-triple CNN is employed for classification of malwares. Performance evaluation of the proposed framework is evaluated using Microsoft malware dataset. Performance Comparisons of the proposed framework are done using five existing state-of-art techniques. Result analysis confirms the superiority of the proposed framework over the existing state-of-art framework. Proposed framework has achieved 99.98% accuracy. In future, the proposed framework is planned to be evaluated on other datasets like Android malware datasets, iOS malware datasets and network malware datasets. Converting byte data to images requires proper image sizing, which might create accuracy limitations while designing for larger datasets. To remove this limitation, the work can also be extended by addition of gated recurrent unit (GRU) and long-short-term-memory (LSTM) models, which will allow the system to skip byte to image conversion, and obtain the final classes with increased accuracy.

References

1. Namanya, A.P., Cullen, A., Awan, I.U., Disso, J.P.: The world of malware: an overview. In: 2018 IEEE 6th International Conference on Future Internet of Things and Cloud (FiCloud), pp. 420–427. IEEE, August 2018
2. Aslan, Ö.A., Samet, R.: A comprehensive review on malware detection approaches. IEEE Access **8**, 6249–6271 (2020)
3. Gibert, D., Mateu, C., Planes, J.: HYDRA: a multimodal deep learning framework for malware classification. Comput. Secur. **95**, 101873 (2020)
4. Ren, Z., Chen, G., Lu, W.: Malware visualization methods based on deep convolution neural networks. Multimedia Tools Appl. **79**(15–16), 10975–10993 (2019). https://doi.org/10.1007/s11042-019-08310-9
5. Sun, J., Luo, X., Gao, H., Wang, W., Gao, Y., Yang, X.: Categorizing malware via a Word2Vec-based temporal convolutional network scheme. J. Cloud Comput. **9**(1), 1–14 (2020). https://doi.org/10.1186/s13677-020-00200-y

6. Masabo, E., Kaawaase, K.S., Sansa-Otim, J., Ngubiri, J., Hanyurwimfura, D.: Improvement of malware classification using hybrid feature engineering. SN Comput. Sci. 1(1), 1–14 (2019). https://doi.org/10.1007/s42979-019-0017-9
7. Gibert, D., Mateu, C., Planes, J.: The rise of machine learning for detection and classification of malware: Research developments, trends and challenges. J. Netw. Comput. Appl. 153, 102526 (2020)
8. Bae, S.I., Lee, G.B., Im, E.G.: Ransomware detection using machine learning algorithms. Concurr. Computat. Pract. Exp. 32(18), e5422 (2020)
9. Lu, J., Gu, F., Wang, Y., Chen, J., Peng, Z., Wen, S.: Static detection of file access control vulnerabilities on windows system. Concurr. Comput. Pract. Exp., e6004 (2020). https://doi.org/10.1002/cpe.6004
10. Ahmadi, M., Ulyanov, D., Semenov, S., Trofimov, M., Giacinto, G.: Novel feature extraction, selection and fusion for effective malware family classification. In: Proceedings of the Sixth ACM Conference on Data and Application Security and Privacy, pp. 183–194, March 2016
11. Zhang, Y., Liu, Z., Jiang, Y.: The classification and detection of malware using soft relevance evaluation. IEEE Trans. Reliab., 1–12 (2020). https://doi.org/10.1109/TR.2020.3020954
12. Hashemi, H., Azmoodeh, A., Hamzeh, A., Hashemi, S.: Graph embedding as a new approach for unknown malware detection. J. Comput. Virol. Hacking Tech. 13(3), 153–166 (2016). https://doi.org/10.1007/s11416-016-0278-y
13. Singh, P., Tapaswi, S., Gupta, S.: Malware detection in PDF and office documents: a survey. Inf. Secur. J. Glob. Perspect. 29(3), 134–153 (2020)
14. Egitmen, A., Bulut, I., Aygun, R., Gunduz, A.B., Seyrekbasan, O., Yavuz, A.G.: Combat mobile evasive malware via skip-gram-based malware detection. Secur. Commun. Netw. 2020, article ID 6726147, 10 p. (2020). https://doi.org/10.1155/2020/6726147
15. Yuan, B., Wang, J., Liu, D., Guo, W., Wu, P., Bao, X.: Byte-level malware classification based on Markov images and deep learning. Comput. Secur. 92, 101740 (2020)
16. Sahay, S.K., Sharma, A.: Grouping the executables to detect malware with high accuracy. arXiv preprint arXiv:1606.06908 (2016)
17. Roseline, S.A., Geetha, S., Kadry, S., Nam, Y.: Intelligent vision-based malware detection and classification using deep random forest paradigm. IEEE Access 8, 206303–206324 (2020)
18. Darabian, H., et al.: A multiview learning method for malware threat hunting: windows, IoT and android as case studies. World Wide Web 23(2), 1241–1260 (2020). https://doi.org/10.1007/s11280-019-00755-0
19. Khan, R.U., Zhang, X., Kumar, R.: Analysis of ResNet and GoogleNet models for malware detection. J. Comput. Virol. Hacking Tech. 15(1), 29–37 (2018). https://doi.org/10.1007/s11416-018-0324-z
20. Zhang, Z., Cheng, Y., Gao, Y., Nepal, S., Liu, D., Zou, Y.: Detecting hardware-assisted virtualization with inconspicuous features. IEEE Trans. Inf. Forensics Secur. 16, 16–27 (2020)
21. Bai, J., Shi, Q., Mu, S.: A malware and variant detection method using function call graph isomorphism. Secur. Commun. Netw. 2019, article ID 1043794, 12 p. (2019). https://doi.org/10.1155/2019/1043794
22. Gao, X., Hu, C., Shan, C., Liu, B., Niu, Z., Xie, H.: Malware classification for the cloud via semi-supervised transfer learning. J. Inf. Secur. Appl. 55, 102661 (2020)
23. Narouei, M., Ahmadi, M., Giacinto, G., Takabi, H., Sami, A.: DLLMiner: structural mining for malware detection. Secur. Commun. Netw. 8(18), 3311–3322 (2015)
24. Tien, C.W., Huang, T.Y., Tien, C.W., Huang, T.C., Kuo, S.Y.: KubAnomaly: anomaly detection for the Docker orchestration platform with neural network approaches. Eng. Rep. 1(5), e12080 (2019)
25. Ronen, R., Radu, M., Feuerstein, C., Yom-Tov, E., Ahmadi, M.: Microsoft malware classification challenge. arXiv preprint arXiv:1802.10135 (2018)

26. Sharma, S., Krishna, C.R., Sahay, S.K.: Detection of advanced malware by machine learning techniques. In: Ray, K., Sharma, T., Rawat, S., Saini, R., Bandyopadhyay, A. (eds.) Soft Computing: Theories and Applications. AISC, vol 742, pp. 333–342. Springer, Singapore (2019). https://doi.org/10.1007/978-981-13-0589-4_31
27. Ding, H., Sun, W., Chen, Y., Zhao, B., Gui, H. Malware detection and classification based on parallel sequence comparison. In: 2018 5th International Conference on Systems and Informatics (ICSAI), pp. 670–675. IEEE, November 2018

Mobile Agent Security Using Lagrange Interpolation with Multilayer Perception Neural Network

Pradeep Kumar[1]([✉]), Niraj Singhal[2], Mohammad Asim[3], Ajay Kumar[1], and Mahboob Alam[1]

[1] Department of Computer Science and Engineering, JSS Academy of Technical Education, Noida, Uttar Pradesh, India
pradeep8984@gmail.com, {ajaykrverma,mahboobalam}@jssaten.ac.in
[2] Shobhit Institute of Engineering & Technology (Deemed to be University), Meerut, India
niraj@shobhituniversity.ac.in
[3] Department of CSE, MGM's College of Engineering and Technology, Noida, Uttar Pradesh, India
asim@coet.in

Abstract. Mobile agents are an emerging computing area which replaces client server computing model. Mobile agents are small piece of code and data works automatically on the behalf of owner. Processed different type of activity during the life cycle of mobile agent and execute code on another host computer. Mobile agents execute their assigned function in malicious heterogeneous environment. Mobile agent is using in many applications like E-commerce, parallel computing, and Network management etc. To provide protection for mobile agents is one of the prime issues in broadens of the mobile agent computing. Today, the greatest challenge to the mobile agent technology is security. The numerous advantages accompanying its usage have been fettered by the security concerns/threats. In this article, propose a new framework for securing the secret key which is based on Shamir's Secret Share and Error Back propagation ANN. Shamir's SS has been employed for share generation and BP-ANN is used for secret retrieval. Some keystone features of this scheme are: the shareholders need not disclose their identities at any stage (anonymous), no imposition of threshold limit, liberty to use public communication channel, etc. Furthermore, the ongoing research and developments in ANN technology broadens the scope for improvements in the scheme.

Keywords: Mobile agents (MA) · Lagrange interpolation · Error Back propagation ANN

1 Introduction

The Mobile agent is software possessing the ability to carry data and migrate from one host to another independently and execute the required functions. It introduces numerous benefits to distributed systems like load reduction, overcoming network latency,

M. Singh et al. (Eds.): ICACDS 2021, CCIS 1440, pp. 290–302, 2021.
https://doi.org/10.1007/978-3-030-81462-5_27

executing asynchronously, dynamically, autonomously, and many more. Mobile Agent technology is a relatively new research field; it experienced a boom in research in the early nineties. As promising as it seems for the future distributed networks, it currently suffers from major setbacks. The mobile agent technology is susceptible to various active attacks as well as passive attacks such as eavesdropping, unauthorized access, etc. limiting its usage worldwide. The majority of attacks can be performed when the mobile agent traverses the communication channel. Secondly, we can never be sure that the platform which constitutes the environment in which the agent exits and executes is safe or venomous for it Because of the common attacks on the platforms of mobile agents and agents; there is a necessity for countermeasures to fulfill the numerous security requirements/objectives. A proper trust model is of utmost importance to build a robust security system. This paper aims to mitigate a number of the security threats of the mobile agent by providing an approach that is feasible along with being effective. This will help in realizing the goal for mobile agent technology to become the central technology for e-commerce and m-commerce applications rather than being seen just as a programming paradigm. As Mobile agents switch from host to host, security is a major concern for mobile agents. Security of mobile agent depends on the following parameter Confidentiality, Data integrity and Availability.

2 Related Work

The key types of security risks are leakage of information, denial of service and corruption of information. These classes of threats can be explored in more depth in many ways as they relate to the agent system. Mobile agents actually provide a greater potential for harassment and misuse, greatly widening the scope of risks.

Rakesh Kumar et al. [1] presented work aims at safe transmission of images where a random encryption algorithm is used to encrypt various stego image shares, which is generated when a hidden image and the cover image are inserted together and uses a secret sharing technique to generate shares. At the receiving end, an artificial neural network is used to decrypt the encrypted shares and thus remove the need for key exchange before the transmission of data, which is a prerequisite for most of the overall encryption algorithm. The artificial neural network is used to provide high data protection and to generate less decrypted images for distortion. Tsung-Chih Hsiao et al. [2] proposed Key management framework to strengthen data access protection issues. We can easily create a hierarchical access control with the technology of mobile agents. Features of simple calculation and compromise resistance are given by the Lagrange interpolation method. To effectively safe data access control, we implement sensitive files across the device with mobile agents. There are some kinds of attacks that are discussed. Our proposed method is shown to be safe against external attacks, reverse attacks, cooperative attacks, and attacks by equation. Lein Harn et al. [3] proposed the First, the CRT-based MTSS. The security of our proposed scheme is the same as the unconditionally protected Asmuth-Bloom SS. Shareholders are grouped into various security subsets in our proposed scheme and each subset has different thresholds. When there are enough shares available, the secret can be retrieved. In an MTSS, in order to recover the password, any share in the higher-level subset can be used as a share in the lower-level subset. One distinctive feature of our proposed MTSS is that only one private share is held by each shareholder. Hong Zhong *et al.* [4] investigate the topic of secret sharing anonymously. Since previous systems have drawbacks, such as their construction being inefficient and established boundaries, we are presenting a new anonymous scheme based on the BP Artificial Neural Network. It is simple to construct the system and the reconstruction is efficient. In particular, it is an ideal threshold scheme (t, n) and has no threshold limit for the t parameter. In addition, our scheme does not need a safe channel and has verifiable property as well.

Om Prakash Verma et al. [5]. The proposed algorithm shares between shareholders several secrets, where shareholders are often divided/classified into different levels. Therefore, multiple as well as multi-level secret sharing is included. Secrets can be retrieved at intra- or inter-level levels, where higher-level shareholders can lead to lower levels of their shares. The one-way hash function is used instead of the hard number-theoretical issues to decrease the complexity. The new scheme is opposed by dishonest dealers and shareholders. *J.K. Mandal et al.* [6] A secret key was developed in this paper using the Hopfield network, which is a perception network with a single layer. Since the Hopfield network has a function to form an input and output cycle, it is possible to select output from any intermediate iteration as the secret key and this logic will only be known to the sender and receiver. In this proposed HNBNKG technique, the probability of attack is very low for this reason.

Mayank Gupta *et al.* [7] proposed a model based Shamir's secret share and neural network. The encryption and decryption based on the XOR operation. Secret Shares are created using t−1 degree polynomial in the first phase, and shares are encoded using by applying neural cryptography in the second phase. The proposed model is using for the protection of shares on the public channel. The Shamir scheme is used to provide shares, but it is also essential to safeguard each share, which can only be achieved via a secure key mechanism. Secret keys are only exchanged between two participants when conversation is conducted correctly. Two participants can use their weights as the secret key when synchronization is achieved. Analysis of results show that the system is highly secure, does not disclose hidden data and regeneration is lossless.

Supriya K. Narad *et al.* [8] proposed Group authentication based on the Neural Network is intended for group-oriented applications using the Shamir Secret Sharing Scheme. It offers many-to-many forms of authentication where it is possible to safely execute group activities. The Back-propagation Neural Network is used for precision in experimental outcomes. M. Muhil *et al.* [9] proposed a secret sharing algorithm to study and protect the Multi-Cloud. This goal is accomplished using Shamir's algorithm of secret sharing. This hidden sharing scheme has a strong framework that offers an outstanding evidence and application forum. Discussed the drawbacks of single cloud and advantage of multi cloud computing. Cloud storage is one of the latest innovations that are addressed everywhere and by offering their specifications, it also meets the business and its client, but user and enterprise data should be protected and that the administrator must manage and ensure. By keeping in mind its ability to minimize breaches and other security problems, migration to multi cloud is promoted.

Lifeng Yuan *et al.* [10] proposed two enhanced dealer-free thresholds for changing secret sharing networks. Both systems can avoid collusion attacks conducted by actors who carry both historical and current hidden shadows by using two-variable one-way features. We also show that our schemes can safely modify the threshold in case the security policy and adversary structure shift. Michail Fragkakis *et al.* [11] proposed a comparison framework for MAS confidence and security models was developed in this paper, which was used to compare four systems. The findings highlighted the absence of standardization of confidence and security models and the absence of a confidence-granting authority that could be the focus of future study. Standardization, if adopted by the industry, could lead to safe and true interoperable systems and would enable the widespread use of mobile agent systems. Adri Jovin John Joseph *et al.* [12] proposed Trust Score based new metric to determine the honest platform and the path planning algorithm, which assists the Agent in Trust Score-based decision-making. By adding the Trust Co-efficient of Variance, the Trust Scoring model is improved. For server network, the Trust Scoring system creates multiple entries, but this system unifies the multiple entries using the Trust Co-efficient of Variance of the entries that are unified as a new metric. The order of ranking is performed for the server platforms on the basis of the Trustability Coefficient of Variation. On the basis of experimental findings, the proposed Trust Rating system has been found to be better compared to other current decision support systems of that kind. Shyamalendu Kandar *et al.* [13] proposed a verifiable secret sharing mechanism. Each shareholders is allocated a share in this scheme that eliminates the hazard of restoration of the hidden plot by combining the number of shareholders from the rest of the threshold. It offers immunity to various security threats as well. The agents are believed to be trusted by the shareholders in the presented technique.

3 Problem Statement

Security of mobile agents and platform significant issue for the utilizations of mobile agent's paradigm based application in any organization. At the hour of performing task mobile agents using the software and hardware of none trusted platform. So the mobile agents and platform are always in the threat of attack by malicious agents or platform respectively. So the security of migrating agents and platform in major issue in mobile agent based applications. A new model is proposed here for the protection of mobile agents and platform based on Lagrange interpolation and back propagation neural network.

4 Preliminaries

In this section, we present some crucial foundations of secret sharing scheme.

4.1 Shamir Secret Share

Choose arbitrary positive integer $d_0, d_1, d_2, \ldots \ldots d_{k-1} \in$ galosis field (p) $F(x) = (d_0 x^0 + d_1 x^1 + d_2 x^2 + \ldots \ldots \ldots \ldots + n_{t-1} x^{t-1}) \% p$ if we are putting $x = 0$ generate $d_0 =$ private key and p is a huge prime number and $d_1, d_2 \ldots$, and d_{k-1} are real number chosen from Z. Then each node have unique identity, on the basis of identity generate partial

secret key = f (idi). At the receiver end choose randomly t authenticate partially share
to regenerate key using lagrange polynomial

$$F(x) = \sum\nolimits_{i=1}^{t} Y_i \prod\nolimits_{1 \le j \le t, j \ne i} \frac{X - X_j}{X_i - X_j} \tag{1}$$

Since f (0) = d0 = S, secret share
The shared secret calculated by the following formula

$$S = \sum\nolimits_{i=1}^{t} q_i Y_i \tag{2}$$

$$q_i = \prod\nolimits_{1 \le j \le t, j \ne i} \frac{X_j}{X_j - X_i} \tag{3}$$

Secret key for execution is created by collecting t authentic share (minimum no of
threshold) by using formula F(0) = d_0%p.

4.2 Backpropgation Neural Network

Back-Propagation neural organization learning mechanism is basically made out of two
phases: feed forward and backword propagation. Figure 1 shows the organization for a
three layer Backpropgation (BP).

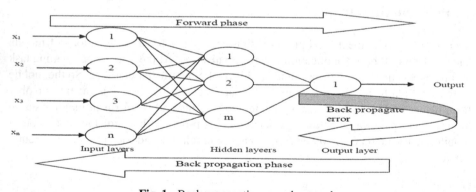

Fig. 1. Back propagation neural network

We use input vector vector{$x_1, x_2, \cdots, x_t$} with cardinality 't', Hidden layer vector{h_1,
$h_2, \cdots, h_m$} and one output. Where m = (2/3)t. Weight of the node input layer 'k' and
the hidden layer node 'j' presents by w_{jk}^h. Weight of connecting j and the output layer
node 'i' represent by w_{ij}^o. Where $1 \le k \le t$, $1 \le j \le (2/3)t$ and i = 1.

4.3 Principle Phases of Back Spread Neural Organization

Back propagation neural Network (BPN) comprising of various fundamental stages
examining as follow.

a. **Initialization of Network:** The neural network initializes the number of neurons in the input layer, the number of neurons in the hidden layer and the output layer in the first step of back propagation. Use a random function between 0 and 1 to initialize the weight of every layer and base value.

b. **Forward Stage:** Apply the randomly selected weight input in forward and compute output with the sigmoid activation function. At the input layer, there are n 'input neurons' and m 'hidden layer neurons'.

c. **Activation Function:** The activation feature is also known as the switch feature. Covert linear regression into non-linearity regression through the use of activation function in back propagation neural networks. The aim of the activation function is to make the neural network learn and perform complex tasks in real time.

Sigmoid: Figure 2 displays the sigmoid activation mechanism, known as the logistic activation function. The sigmoid activation function is used in the neural network Back Propagation.

$$\sigma(x) = \frac{1}{1 + e^{-x}} \quad \text{where } 0 < \sigma(x) < 1 \tag{4}$$

Differentiation of Sigmoid Function

$$\frac{d}{dx}(\sigma(x)) = \sigma(x) * (1 - \sigma(x)) \tag{5}$$

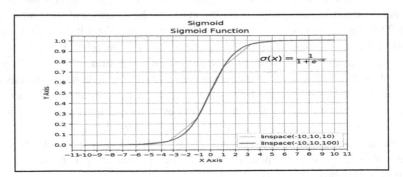

Fig. 2. Sigmoid activation function

d. **Error function:** After the forward step, measure the error difference between the real output and the desired output.

$$\textbf{Error} = \begin{cases} 0, & \textit{if output} = \textit{Desired output} \\ 1/2(\textit{output} - \textit{Desired output})^2 \end{cases} \tag{6}$$

Error is always positive because the square of any value is always positive.

e. **Back propagation (Minimization of error):** The primary aim is to mitigate the problem. The desired output is constant, so by updating the weight using gradient descent, we need to update the output. Gradient descent measures the error gradient with respect to the corresponding weight.

$$\text{New weight } W_x = \text{old weight} - \alpha \left(\frac{\partial error}{\partial wx} \right) \tag{7}$$

Where $W_x =$ old weight $\alpha =$ Training rate of algorithm.
$\left(\frac{\partial error}{\partial wx} \right) =$ Derivative of error function with respect to weight.

4.4   Proposed Model

A security model for secure key management based on the Lagrange interpolation and back propagation neural network proposed here shown in Fig. 3. Basically proposed model divided in to two phase in the first phase a random secret key 'S' if generated. Further secret 'S' is portioned in to 'n' share based on the Algorithm 1 using $t-1$ degree polynomial where 't' is threshold value. Algorithm 1 returns the n shares (xi, f(xi)) for $0 < i \leq n$. where xi is the identity of mobile agents. In second phase choose 't' share out on 'n' by using permutation method n_{p_t}. Considering all 't' permutation value as a input of back propagation neural network and trained the neural network using Algorithm 3. After training of neural network save this program and broadcast to all platforms. Mobile agents are migrating automatically and want to execute assigned task of different platform according to requirement. For execution of mobile agents want secret key. To generate secret key choose 't' share out of 'n' apply Lagrange interpolation and generate secret key using algorithm. But main task how can we identify authenticity of key. This will be achieved by applying t share on artificial back propagation neural network if the error generate by BPN is greater than the predefined threshold value. There will be cheating occurred during the secret key generation.

The algorithm makes a reliable environment in such a way that a malicious agent cannot uncover the agent process by reading straight the code of agent. The developed technique focuses on mobile agent security and evolved along the traditional lines of key generation security techniques of mobile agent framework.

A simple example below demonstrates how the proposed scheme works in different chosen threshold values for n number of distributed shares for a secret key generated using proposed approach.

Algorithm 1 Secret sharing using polynomial
1. Begin:
2. Create n share for specific secret key 'S'
3. B= [ ]
4. $f(x)=S+b_1x^1+b_2x^2+ b_3x^3 \ldots \ldots \ldots b_{t-1}x^{t-1}$
5. Start i=1 to n
6. Select random value xi to evaluate f(xi)
7. B= (xi, f(xi))
8. end

Algorithm 2 Secret sharing using polynomial
1. Begin:
2. Select random t share out of n secret share generate in algorithm 1.
3. $F(x) = \sum_{i=1}^{t} Y_i \prod_{1 \le j \le t, j \ne i} \frac{X - X_j}{X_i - X_j}$
4. $S = \sum_{i=1}^{t} q_i Y_i$
5. /* $q_i = \prod_{1 \le j \le t, j \ne i} \frac{X_j}{X_j - X_i}$ *\
4. end

Algorithm 3 Back propagation Learning Algorithm

1. Input: Apply 't' input xi, $1 \le i \le t$, select out of n permutation of secret share with predefined learning rate, desired output with activation function
2. Output: Back propagation Neural networks with updated weight $w_{jk}^h\ w_{ij}^o$.
    Where $1 \le k \le t$, $1 \le j \le (2/3)t$ and $i=1$.
3. Begin:
4. Initialize all weight randomly w_{jk}^h , w_{ij}^o
5. for maximum iteration
6. for all input sample 1to n
7. Forward phase
8. for $j = 1$, to (2/3)t do
9. $h_j = f(\sum_{j=1}^{t} x_k * w_{jk}^h)$
10. $O_i = f(\sum_{j=1}^{t} h_j * w_{1j}^o)$
11. if error$= \frac{1}{2}(t_1 - O_1)^2 > .001$
12. Back-Propagation phase:
13. $\delta w_{ij}^o = -(t_1 - O_1) * h_j$
14. $\delta w_{jk}^h = -h_j(1 - h_j) * x_k[(t_1 - O_1) * w_{ij}^o]$
15. $w_{1j} = w_{1j} - \alpha \delta w_{1j}$ /*α= learning rate*/
16. $w_{jk}^h = w_{jk}^h - \alpha \delta w_{jk}^h$
17. *else*
18. End of neural Network Training
19. end

The algorithm produces a reliable environment in such a way that the agent mechanism cannot be discovered by a malicious agent by explicitly reading the agent code. The developed methodology focuses on mobile agent protection and has grown along the conventional lines of the mobile agent framework's key generation security techniques. A basic example below illustrates how the proposed method functions in various ways.

Case 1: Test Case
Let us consider a small prime number p = 19 and secret share S = 16 host computer create 8 mobile agents.

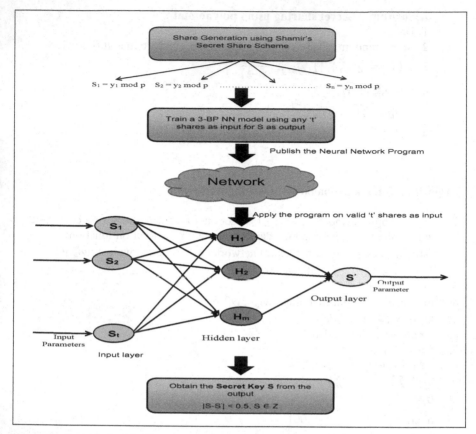

Fig. 3. Framework for Mobile agent Security using Lagrange interpolation with Back rogation Neural Network

$maj_{1 \leq j \leq t-1}$, each mobile agents have unique id 1, 2, 3, 4, 5, 6, 7, 8 respectively. On the basis of mobile agents ids Host create 8 partial key based on Lagrange interpolation polynomial

$$g(x) = \left(\beta_0 x^0 + \beta_1 x^1 + \beta_2 x^2 + \ldots\ldots\ldots + \beta_{t-1} x^{t-1} \right) \bmod p$$

Using Algorithm 1, Mobile host choose threshold value t = 4. So polynomial of degree 3. $g(x) = (14 + 3x + 7x^2 + 9x^3) \bmod 23$ (Table 1).

Table 1. Shares generated by Algorithm 2

Secret	K1	K2	K3	K4	K5	K6	K7	K8
14	10	5	7	1	18	20	15	11

Select any 't' value from given share. Let us consider here $t_1 = 3$, number of way select 3 different pair out of 7 values is equal to $7_{c_3} = 35$. Out of 35 combination select randomly any three combination are [(**10, 5, 7**), (**10, 1, 20**), (**5, 20, 11**)]. As a back propagation neural network input with 3 input neurons, 2 hidden layer neurons and 1 output neuron, pick any three combinations. During preparation, the desired performance value S = 14 steps in is 1500, set the accuracy of 0.002, and choose another parameter according to the back propagation algorithm. Save the final weight of this network program after neural network preparation. And sent to every mobile agent (Table 2).

Table 2. Simulated output

S. No	Apply input	Desired output in program	Output by simulation
1	(10, 5, 7)	14	[0.14390287]
2	(10, 1, 20)	14	[0.13483672]
3	(5, 20, 11)	14	[0.14118277]

There is no limit to the threshold value under this method at training of neural network. If we choose $t_2 = 3$, only 3 parameters in the simulation can produce the correct output. One secret share generated by Lagrange interpolation another calculated by back propagation neural network by using Algorithm 2. If both output are different then cheating will done by any mobile agent.

Case 2: Choose any value from a given share. Let us consider $t_2 = 4$ here, the way name selects 4 separate pairs out of 8 values is equal to. $7_{C_4} = 35$. Out of 35 combination select randomly any three combination are [(**10, 5, 7, 18**), (**10, 18, 15, 11**), (5, 18, 20, 15)]. As a back propagation neural network input with 4 input neurons, 2 hidden layer neurons, and 1 output neuron, pick any four combinations. The appropriate output value 'S = 14' is 1000 steps during preparation, set the accuracy to 0.002, and select other parameters according to the algorithm of back propagation. This network software is saved with final weight after neural network training and sent to all receiver systems, then the device applies selected three inputs to this program and takes output as shown in Table 3.

Table 3. Simulated output

S. No	Apply input	Desired output in program	Output by simulation
1	[(**10, 5, 7, 18**)]	14	[14.311832]
2	[(**10, 18, 15, 11**)]	14	[14.253221]
3	[(**5, 18, 20, 15**)]	14	[14.401256]

One secret share generated by Lagrange interpolation another calculated by back propagation neural network by using Algorithm 2. If both output are different then cheating will done by any mobile agent.

Case 3: Suppose some malicious mobile agent tries to use unauthenticated shares to access the secret key. Select three sets [10, 8, 9, 18], [8, 9, 15, 11] and [18, 10, 15, 6] at random and add the simulated output to the artificial neural network as shown in Table 4.

Table 4. Simulated output for unauthenticated share

S. No	Apply input	Desired output in program	Output by simulation
1	[10, 8, 9, 18]	14	13.436002 (wrong output)
2	[8, 9, 15, 11]	14	15.790711 (wrong output)
3	[18, 10, 15, 6]	14	17.969599 (wrong output)

If unauthorized mobile agent wants access the key of execution can not reveal the secret key. S. No 1, 2 and 3, using three different set of cardinality of four, unauthorized mobile agents and apply on trained neural network. But actual secret not generate by back rogation neural network. One secret share generated by Lagrange interpolation another calculated by back propagation neural network by using Algorithm 2. If both output are different then cheating will done by any forged mobile agent.

5   Implementation and Results

The proposed framework is based on Lagrange interpolation and neural network of back propagation implemented in python. Comparing the proposed scheme with the previous scheme is successful, as shown in Table 5 below.

It takes much less computational time to produce the secret using Lagrange interpolation on the basis of comparison at the initialization point. After training, a neural specified threshold value save the program with a fixed user and sent to the different number of platforms on which mobile agents want to perform tasks.

Choose the authenticated mobile agent share package for the session key generation on the executing network. Compared to conventional mechanisms shown in Table 5, it takes only O(1) time for secret key generation by using artificial neural networks program has better computational time.

There is no constraint on the threshold value during the training of artificial neural networks. This is another benefit of the proposed scheme because, as used in Shamir secret share and back propagation algorithm. The graph of artificial neural network training with 1000 training iterations with error deduction is shown in Fig. 4.

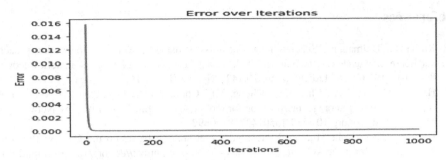

Fig. 4. Reducing error with iteration

Table 5. Comparison of proposed scheme with other scheme

Technique	Share generation complexity	Recreation time complexity	Threshold value 't'
Combinatorial technique	Big oh(n)	Big oh(n)	t = 1 or t = 1
Anonymous secret sharing schemes	Big oh(n)	Big oh(n)	t = 2 or t = 1
Providing anonymity in unconditionally secure secret sharing scheme	Big oh(n)	Big oh(n)	t = 2 or t = 1 and other cases
Proposed scheme	Big oh(n)	O(1)	Using two threshold one for Shamir secret share another have no limitation

6   Conclusion and Future Scope

The protection of agents in hostile environments during relocation is a major problem. Design of a fusion system for Lagrange interpolation and back Propagation Neural network in this scheme to provide authentication. It is proposed that dynamic threshold values and anonymous actions of the proposed approach based on the back propagation neural network generate greater mobile agent protection during the secret sharing and reconstruction of key agents. The proposed mobile agent migration approach improves the security of the execution key. In an open world, it focuses on enhancing stable mobile agent migration. The threshold parameter does not have a constraint. The threshold parameter does not have a constraint. In the future, we will try to design the Artificial Neural Network quick Back propagation (BP) to increase the proposed algorithm's efficiency and also design some new secret scheme instead of Shamir secret.

References

1. Kumar, R., Dhiman, M.: Secured image transmission using a novel neural network approach and secret image sharing technique. Int. J. Signal Process. Image Process. Pattern Recogn. **8**(1), 161–192 (2015). https://doi.org/10.14257/ijsip.2015.8.1.16.
2. Hsiao, T.C., Wu, Z.Y., Chen, T.L., Chung, Y.F., Chen, T.S.: A hierarchical access control scheme based on Lagrange interpolation for mobile agents. Int. J. Distrib. Sens. Netw. **14**(7) (2018). https://doi.org/10.1177/1550147718790892.
3. Harn, L., Fuyou, M.: Multilevel threshold secret sharing based on the Chinese Remainder Theorem. Inf. Process. Lett. **114**(9), 504–509 (2014). https://doi.org/10.1016/j.ipl.2014.04.006
4. Zhong, H., Wei, X., Shi, R.: A novel anonymous secret sharing scheme based on BP Artificial Neural Network. In: Proceedings of International Conference on Natural Computation, ICNC, pp. 366–370 (2012). https://doi.org/10.1109/ICNC.2012.6234550.
5. Verma, O.P., Jain, N., Pal, S.K.: A hybrid-based verifiable secret sharing scheme using Chinese Remainder Theorem. Arab. J. Sci. Eng. **45**(4), 2395–2406 (2019). https://doi.org/10.1007/s13369-019-03992-7
6. Satapathy, S.C., Biswal, B.N., Udgata, S.K., Mandal, J.K. (eds.): Proceedings of the 3rd International Conference on Frontiers of Intelligent Computing: Theory and Applications (FICTA) 2014. AISC, vol. 328. Springer, Cham (2015). https://doi.org/10.1007/978-3-319-12012-6
7. Gupta, M., Gupta, M., Deshmukh, M.: Single secret image sharing scheme using neural cryptography. Multimedia Tools Appl. **79**(17–18), 12183–12204 (2020). https://doi.org/10.1007/s11042-019-08454-8
8. Narad, M.S.K.: Group authentication using back-propagation neural network **6**(10), 272–278 (2017). https://doi.org/10.17148/IJARCCE.2017.61048
9. Muhil, M., Krishna, U.H., Kumar, R.K., Anita, E.A.M.: Securing multi-cloud using secret sharing algorithm. Procedia Comput. Sci. **50**, 421–426 (2015). https://doi.org/10.1016/j.procs.2015.04.011
10. Yuan, L., Li, M., Guo, C., Choo, K.K.R., Ren, Y.: Novel threshold changeable secret sharing schemes based on polynomial interpolation. PLoS ONE **11**(10), 1–19 (2016). https://doi.org/10.1371/journal.pone.0165512
11. Fragkakis, M., Alexandris, N., Georgiakodis, F.: A survey on the trust and security models of mobile agent platforms. J. Discrete Math. Sci. Cryptogr. **15**(1), 31–47 (2012). https://doi.org/10.1080/09720529.2012.10698362
12. John Joseph, A.J., Mariappan, M.: A novel trust-scoring system using trustability co-efficient of variation for identification of secure agent platforms. PLoS ONE **13**(8), 1–19 (2018). https://doi.org/10.1371/journal.pone.0201600
13. Kandar, S., Dhara, B.C.: A verifiable secret sharing scheme with combiner verification and cheater identification. J. Inf. Secur. Appl. **51**, 102430 (2020). https://doi.org/10.1016/j.jisa.2019.102430

Performance Analysis of Channel Coding Techniques for 5G Networks

Mrinmayi Patil[1]([⊠]), Sanjay Pawar[1], and Zia Saquib[2]

[1] Usha Mittal Institute of Technology, Mumbai, India
sanjay.pawar@umit.sndt.ac.in
[2] Reliance Jio Infocomm Limited, Mumbai, India

Abstract. Turbo Codes and Tail Biting Convolutional Codes are used for channel coding in LTE, but these codes are not pertinent in 5G. For Channel coding, advanced error correction techniques are required for the 5G. Polar codes are adopted for control information and LDPC codes are required for Data information for the 5G. Various coding techniques such as LDPC codes, Quasi Cyclic (QC)-LDPC codes, Non-Binary (NB)-LDPC codes, Spatially Coupled (SC)-LDPC codes, Irregular Repeat Accumulate (IRA) codes, Polar codes, Distributed CRC Aided (DCA) codes, and Fountain codes are illustrated. For Decoding Min-Sum Algorithm and Belief Propagation are used for LDPC codes. Min-Sum, OMS, NMS, BP methods are applied in QC-LDPC codes. SC, SCL, FSCL, CASCL, BP algorithms implemented for Polar codes. Simulation of QC-LDPC codes, Polar codes, and LDPC codes are generated by using MATLAB Software. The performance of all channel coding techniques with consideration to BER and BLER is mentioned.

Keywords: 5G · LDPC codes · Polar codes

1 Introduction

In the field of wireless communication, the demand for bandwidth is increasing from time to time. From 1G to 4G networks as per custom needs, the benefits were unusual. 5G technology has practically no limitations on wireless systems. 5th generation networks provide resilience, excessive speed, moderate latency, and peak throughput. 5G allows diverse applications for the organization since fully as alternative practice cases. Scientists, experts are taking their efforts, so that, this technology is constructive for the field. A lot of people have established 5G experiments; in many parts of the world plans for 5G formation have been created.

The main 5G standards groups are International Telecommunication Union (ITU), 3rd Generation Partnership Project (3GPP), and Internet Engineering Task Force (IETF). For IMT-2020, ITU is striving towards implementing stable universal principles, sufficient range, and proper measures, and the core network to set up outstanding 5G distributions at the local and global standards [1]. The initial stage of 3GPP 5G specifications in Release-15 was introduced in 2019 and the second stage will be wrapped up in 2020.

M. Singh et al. (Eds.): ICACDS 2021, CCIS 1440, pp. 303–316, 2021.
https://doi.org/10.1007/978-3-030-81462-5_28

5G NR (New Radio) can have two types of frequencies; smaller frequencies are (FR1) and greater frequencies (FR2). FR1 is below 6 GHz and FR2 is above 24GHz [2]. In IETF, mean requirements as virtualization capacities developing IP protocols to protect system virtualization. It establishes Service Function Chaining (SFC) that is joined with virtualized information of the 5G architecture is the handling gateway, base station, and packet data gateway in a particular direction. Other modern technology under progress by IETF includes protocols for a shared system, path computing, and slice routing and routing-detailed testing to reach the restraints of 5G Technology [3]. The 5G demands a round trip latency of about 1 ms and data count demands to rise by approximately 1000 points out of 4th generation to 5th generation, and the amount of data transmitted must be multiplied by a factor of ten thousand. The peak rate of 5G ranges from 100 Mbps to 1 Gigabit per second. Energy production must be increased by a factor of 100. The 5G system is expected to have a 10 to 100 time's greater number of connected devices. In terms of expansion, the devices must be placed in a systematic manner, have an appropriate level of protection, and be able to move between various types of radio access networks [4].

The following collection of 5G criteria is increasing industry cooperation, according to the majority of demands [5].

- Connections - 1–10 Gbps to edge times within the field
- Latency - 1 ms finish to finish round trip delay
- Bandwidth - 1000× per unit space
- Devices - 10–100× connected
- Throughput - 99.999%
- Notion - 100% coverage
- 90% reduction in network intensity usage.

There are three main applications of 5G such mMTC (massive Machine Type Communications), eMBB (enhanced Mobile Broadband), and URLLC (Ultra-Reliable Low Latency Communications). Deployment can be achieved by applying Standalone (SA) and Non-Standalone (NSA). 5G NR can be set up on small, mid, or large bands with no regulations. Bands are specified into two frequency spectra, First frequency spectrum is 450 MHz to 6000 MHz, and Bands are estimated among to 255, normally introduced since Sub-6 GHz. The second frequency spectrum is 24250 MHz to 52600 MHz, Bands estimated among 257 to 511, usually related as mm-Wave. Key factors of 5G NR such as Waveform and Multiple access, Physical Channel Bandwidth, Numerologies, Frame Structure, Modulation, and Channel Coding [6].

Channel coding is crucial for the communication field. It is applied for improving the transmission faults generated by turbulence, interference, and low signal stability. This is carried out by applying a channel encoder in the transmitter to deal with each then it is termed as a message block containing K message items and modify it toward a later encoded block containing N > K encoded samples that are transferred. At the receiver side, supplementary (N − K) encoded samples required to channel decoder with information so assigns it to reveal and regulate communication faults in the seminal K information bits. To correct the communication failures, cellular information strategies requiring channel codes. The turbo code was elected as the fundamental channel code in the 3G and 4G cellular specifications as a result of powerful error correction capability and resilience [7].

Various 5G channel coding techniques, as well as encoding, decoding methods, and simulations of LDPC, QC-LDPC, and Polar codes, are discussed in this article.

2 Literature Review

Turbo Codes and Tail Biting Convolutional Codes (TBCC) are accepted in LTE, but in 5G these codes cannot fulfill the conditions as per the requirement. In 5G advanced error control system is employed for channel coding, Turbo Codes are taken over by LDPC codes and TBCC is changed by Polar Codes. When compared to LDPC, turbo codes have a higher error floor BER. Polar codes are the first high-performance codes with low encoding and decoding complexity, having been introduced in 2009. It will provide a great deal of flexibility while remaining dependable in any code range above the error floor [8].

LDPC codes have various advantages such as significant coding gain, excessive throughput, and inadequate power dissipation. Min Sum Algorithm is used for decoding. It is a Class of soft-decision method. By using the traditional approach that is dependent on HDL code is time spending and more complex. In this, Simulink is used for designing these codes. The proposed technique requires less time, low power, and it reduces design complexity [9].

In [10] contrast with Convolutional, Turbo, LDPC, and Polar codes, In Polar codes, SCL decoding using little list range is a potential alternative for small information lengths less than or equal to 128 bits owing to the fault behavior and comparatively small complexity. Non-binary LDPC codes can give excellent behavior at an amount like extended complexity still among stronger spectral efficiency. In Polar Codes, SCL needed insignificant area and nominal power utilization contrasted to LDPC codes.

In [11] LDPC codes a hybrid decoding method are explained. A suggested algorithm involves the NMS interpretation and successive approximation to the test-node established on its strength. A suggested algorithm uses a multiplication component to load the channel improvement and a set of continuous performs for test-node modernizing. With the iterative inception to decoding simulations, this method can attain enhanced performance as near to belief propagation decoding. It gives increased complexity compared to others.

In Quasi-cyclic (QC-LDPC) codes at the transmitter part, code samples to an encoder are transferred to a rate matcher. At the receiver part, a de-rate matcher is used to associate code samples of various transportation undertakes as well as delivers to a decoder. The result of the QC-LDPC systematic bits is required. In this code, the first full-base matrix is shortened. Contrasted to old techniques suggested method increases the throughput of QC-LDPC codes [12].

The bit-flip technique was employed for the SC decoder to increase the BLER operation for polar codes. BP Algorithm is exhibited. Simulation results indicate that BLER of conventional BP decoder compared to suggested bit-flip decoder can produce a substantial SNR improvement that related to a CA-SCL decoder along with small list size. The latency of the given BP bit-flip decoder is greater compared to the traditional BP decoder in the intermediate and large SNR fields [13].

3   Channel Coding Techniques

Turbo codes adopted for 3rd generation and 4th generation are not at all fulfill the performance specifications of eMBB for every block length and code rate, where an implementation complexity is extremely large as larger data rates and an error floor are also present in the BER. TBCC is no more useful for the eMBB and it includes a weak behavior in huge block ranges and small code rates. The demands of eMBB, 5G requires obtaining substantial improvements in spectral performance, signaling capacity, bandwidth, and coverage correlated to 4G. To obtain significant spectral energy, channel coding plays notable appearances in the physical layer [14].

Turbo codes are unable to support 5G use cases and are generally noted for BLER 10^{-3} to 10^{-4}, which is not significant for URLLC. Coding techniques in 5G will correspond with BLER performance, latency, implementation complexity, and flexibility [15].

Turbo codes are serial in structure and LDPC Codes are parallel in a structure that allows LDPC to reinforce small latency applications than Turbo codes. Turbo Codes has greater error floor BER performance related to the BER performance of LDPC Codes. A Parity check matrix in LDPC codes is expanded to fewer rates than turbo codes, obtaining greater coding gains for poor rate applications showing significant accuracy. In the eMBB scenario with code block lengths exceeding 10,000 and the code rate of 8/9 are usual [8].

Turbo Codes and TBCC codes are suitable for LTE. Turbo codes use reasonable encoding complexity and large decoding complexity, where LDPC codes have a large encoding complexity and less decoding complexity. Turbo Codes is not supporting high peak rates and low latency and does not give better at small packet transmissions so LDPC codes are used.

The significant advancements of 5G LDPC codes related to turbo codes followed in 4G such as it possesses further area throughput capability and considerably larger peak throughput, it gets cut down decoding complexity and increased decoding latency owing to greater rate. It has increased behavior for all code lengths and code rates, with error floors about or below BLER 10^{-5} As a result of these upgrades, LDPC codes are now suitable for extremely high throughputs and URLLC required for 5G. Polar codes usually outrun the TBCCs for capacities of importance for control information, specifically at small code rates and the intricacy of polar codes is noticeably bigger than TBCCs [16]. In 5G, advanced channel coding techniques are used for large data block support with low complexity.

3.1   LDPC Codes

In LDPC Codes the number of 1's in the parity check matrix is small. Regular LDPC code has the same number of rows and columns whereas in irregular LDPC Codes the number of rows and columns is not the same. Irregular LDPC codes perform better than regular LDPC codes. It uses matrix depiction and graphical depiction. In the matrix depiction parity check matrix is used whereas in a graphical depiction Tanner graph is used. In this code, two base matrix is used. This code is represented by a parity check matrix H. The H for LDPC codes can be described by the shift size Z, shift coefficients, and base graph (BG).

The encoding can be suggested as follows:

$$C = uG \tag{1}$$

Whereas C is a codeword, u input block, and G is a generator matrix. For this code, the parity check matrix is an important parameter compared generator matrix it is even be achieved from a given parity check matrix [17].

3.1.1 Belief Propagation (BP)

These techniques are called as Sum-Product methods and Message Passing methods and iterative methods. The concept behind their mention is in every defeat of the techniques information is moved from variable nodes to check nodes and vice versa. The information moved beliefs or probabilities run around the edges in this form. The information sent from a variable node v to a check node c is the probability that v will take a particular amount given the variable node's marked amount and the number of the means sent to v in the preceding finish from check nodes circumstance that of c. Another part of the information moved the probability that v takes a certain amount of total information transferred to c in the previous finish from variable nodes other than v is called from c to v [18].

3.1.2 Min Sum Approximation (MSA)

The sum-product method is adapted to lessen the implementation intricacy like decoder. The Min-Sum algorithm adopts the same steps as the tanh rule SPA. The above method is simpler to achieve as it looks at a purge of the tanh calculation. Variable node equation is given by:

$$V_i = LLR_n + \sum_{j \neq i} C_j \tag{2}$$

And the Check node equation is expressed as follows:

$$C_k = \prod_{l \neq k} sign(Vl) \times \min_{l \neq k} |Vl| \tag{3}$$

In given equations l and k are degrees of check node [9].

3.2 QC-LDPC Codes

QC-LDPC codes are described by applying the theory of protograph codes. A circularly shifted identity matrix is a permutation matrix of these codes. This matrix was chosen as a permutation matrix since each permutation is unique characterized along with a separate sum. It reduces the requirement of memory for implementation and it uses simple steps as encoding and decoding. As per the 5G condition, a protograph is formally noted a base graph, two classes of base graphs are used and they are established by code rate or a range of message items [19]. Belief Propagation, Min-Sum, Normalized Min-Sum, Optimized Min-Sum methods used for decoding.

3.2.1 Belief Propagation (BP)

The BP method is generally applied in LDPC decoding also it is usually resolved in the log domain. It gives the finest decoding behavior in use. It consists of two phases of variable-to-check message passing and check to variable message passing. The main use of the revised method is that the computing complexity and computation delay are adjusted among two decoding stages [20].

3.3 Irregular Repeat Accumulate (IRA) Codes

In the encoding procedure of IRA codes, the exterior code is a variety of repetition codes of changing place and the bits at the output of the exterior code are interleaved, also thus they are separated into disjoint sets. The parity of each series of bits is figured out, and bits are collected [21].

3.4 Non-binary LDPC Codes

NBLDPC Codes can change binary LDPC Codes but these methods experience the issue of decoding complexity. In a non-binary LDPC Codes for transmission (AWGN) channel with (BPSK) is used. In the given theory, m BPSK patterns are transferred for each c_k where is a codeword symbol. The quantized received signal is present at the receiver side. In SPA (Sum-Product Algorithm) every variable node receives d_v probability vectors from its related check nodes, where d_v is the degree of the variable node [22].

3.5 Spatially Coupled LDPC Codes

SC-LDPC codes are generally recognized for their efficiency, achieving decoding behaviour via threshold concentration. These codes do not require any re-shape when the channel qualities vary. It is described by using protograph matrix B. Protograph can be regarded as a prototype of larger graphs, where S replicates of the protograph are casually associated by side permutations. Each non-zero entrance of the comparable base matrix B shows the number of related sides to this node [23].

3.6 Polar Codes

Polar codes are invented by Erdal Arikan. Polar Codes are the type of linear block codes that depends on the theory of channel polarization. Three Channels are used in Polar codes such as Binary Symmetric Channel (BSC), Binary Erasure Channel (BEC), and Additive White Gaussian Noise (AWGN). Polar Codes achieves Shannon limit capacity. It is used for eMBB in 5G Communication for Control channel and Physical Broadcast Channel (PBCH). A polar encoder consists of three successive entities such as message block conditioning, the polar encoder kernel, and encoded block conditioning. The encoder is basically the kernel's expression of the polarisation transform.

$$F2 = \begin{bmatrix} 1 & 0 \\ 1 & 1 \end{bmatrix}$$

A kronecker sum of this kernel among itself achieves the transform as a larger input set leading to polar codes has dimensions such as powers of 2. The code of length N, and $n = \log2 (N)$ [17]. A Generator Matrix is indicated as [28]:

$$G_N = B_N F_2^{\otimes n} \qquad (4)$$

3.6.1 Successive Cancellation (SC)

In SC Decoding a conflict generated by the old bits had eliminated in each subsequent step. Amongst the sections covered with the information bit, this method requires choosing a section with a greater expectation. In the process of improper conviction of an individual bit, the full scheme breaks down. This method is not useful for large codeword lengths [24].

3.6.2 Belief Propagation (BP)

The BP algorithm depends on soft results. It includes several advantages correlated to SC Algorithm. This gives a greater data rate as it uses a parallel structure and reduces delay.

3.6.3 Successive Cancellation List (SCL)

This process divides into two directions when a bit is determined. First, the path determines a bit like a '0', and a diverse as a '1'. Thus, at every bit evaluation, the total codeword paths increasing, until a list length (L) is achieved. In SCL with $L = 1$ is considered an SC Decoding. A path metric (PM) contains information about the path of the likelihood of the altered codeword. The list is updated as per the bit sequence released by their PM information. SCL decoding provides a stronger error-correction performance compared to SC Decoding [25].

3.6.4 Fast Successive Cancellation List (FSCL)

A fast decoding procedure for Rate-1 nodes and apply it to produce Fast-SCL. It is a reinforced technique to lessen a decoding latency for Rate-1 nodes, cut down to min $(L - 1, Nv)$ time steps with zero fault-correction behavior deterioration. When separating the directions up the Rate-1 node, the line split that appears not to contest as the type of the LLR will ever be disposed of after the $L - 1$-th step.

3.6.5 CRC-Aided Successive Cancellation List (CA-SCL)

CRC (Cyclic Redundancy Check) gives a stopping principle for the iterative decoding process. In this decoding method with the amount of list L represented by CA-SCL. The decoder gains the initial path crossing the CRC disclosure as the evaluation sequence. The behavior of polar codes can be better enhanced by this CRC-aided SCL decoding [26].

3.7 Distributed CRC-Aided (DCA) Polar Codes

In 5G this code consists of a distributed CRC as Downlink Control Information (DCI), where the input bit interleaver is permitted as PBCH payloads and PDCCH DCI. In the encoding of DCA Polar codes CRC encoder, Interleaver, sub-channel allocation, and polar code encoder blocks are employed. The interleaver input size is restricted to K ≤ 164 [27].

3.8 Fountain Codes

These codes are a form of code that includes variable code-rate. An absolute-range coded arrangement can be created with these codes, the assist of a fix-range input arrangement in a particular that suggests the restoration of input arrangement from a sub categorize of size the same or meagrely higher in addition to the input sequence With adequate encoding and decoding methods, these codes can be used for large amounts of data. The difficulty of encoding and decoding is high [24].

4 Performance Analysis

In this part, the analysis of LDPC Codes, QC-LDPC Codes, and Polar Codes are explored. In the initial results, the BER performance of LDPC Codes is investigated. In 5G URLLC channel code cannot express an error floor over a BLER of 10^{-5}. In modern theory, eMBB code sizes vary from one hundred to eight thousand and code rate differs from 1/5 to 8/9.

Polar codes are regarded as effective contenders for URLLC and mMTC at block lengths less than 1024, coding rate from 1/2 to 2/3 at targeted BLER of 10^{-4} to 10^{-5} [10]. For eMBB Communication in 5G system LDPC, QC-LDPC and Polar codes use BER and BLER at 10^{-5} [14].

Table 1. Parameters used for simulation

Parameters	LDPC	QC-LDPC	Polar
Channel	AWGN	AWGN	AWGN
Code rate	1/3,1/2, 2/3, 5/6	½	½
Code length	1024	8448	128, 256, 512, 1024, 2048
List size	NA	NA	1, 2, 4, 8, 16, 32
Modulation decoding algorithms	Min-Sum	BP, MS, OMS, NMS	SC, SCL, CA-SCL, FSCL

For LDPC Codes, AWGN (Additive White Gaussian Noise) Channel is required. Figure 1 shows BER (Bit Error Rate) performance of LDPC codes with two base graphs by using 20 iterations at R = 1/3, 1/2, 2/3, 5/6 and it uses BER at 10^{-5}. Min Sum Algorithm is used for decoding. In LDPC Codes R = 1/2 is mostly used [29].

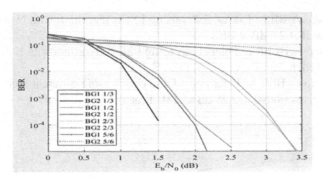

Fig. 1. BER of LDPC codes using different code rates

For QC-LDPC Codes different algorithms are used for decoding they are Belief Propagation (BP), Min-Sum (MS), Normalized Min-Sum (NMS) and Optimized MIN-Sum (OMS). Figure 2 shows BER performance of QC-LDPC Codes with different algorithms. In this Belief propagation gives better results compared to other. We get good performance at 2.5 dB and BER at 10^{-5}.

For Polar Codes Belief Propagation (BP), Successive Cancellation List (SCL), Successive Cancellation (SC), Fast Successive Cancellation List (FSCL), CRC- Aided Successive Cancellation List (CA-SCL) algorithms used for decoding. Figure 3 represents the SCL algorithm with different list sizes in that L = 32 is best among all. L = 32 performs well at 2.5 dB and BER at 10^{-5}.

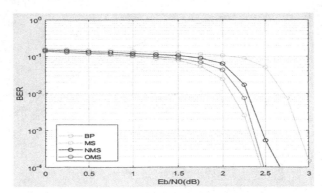

Fig. 2. BER of QC-LDPC codes using different algorithm

Figure 4 represents BLER of Polar codes using BP algorithm with different values of N such as N = 128, 256, 512, 1024, 2048. In that N = 2048 is best among all at 3.5 dB and BLER at 10^{-5}.

Figure 5 shows BLER of Polar codes using CA-SCL algorithm using different values of N with List size L = 32. In that N = 2048 at L = 32 performs well at 2 dB and BLER at 10^{-5}.

Figure 6 shows BER of Polar codes using FSCL algorithm with N = 1024 at L = 32 performs well at 2.5 dB at BER 10^{-5}.

Figure 7 shows BLER of Polar codes using FSCL algorithm with N = 1024 at L = 32 performs well at 2.5 dB and BLER at 10^{-5}.

Figure 8 shows BER of Polar and LDPC codes using different algorithm. In that LDPC Codes gives Eb/N0 = 2.25 dB and BER at 10^{-4}.

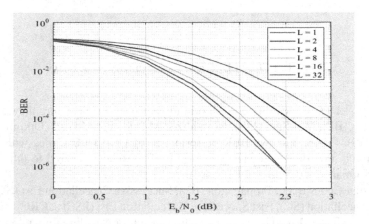

Fig. 3. BER of Polar codes using SCL algorithm with different list size

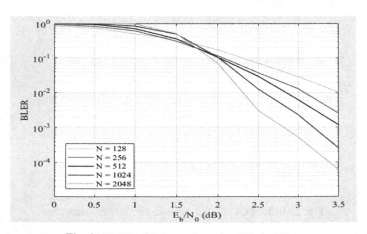

Fig. 4. BLER of Polar codes using BP algorithm

For Polar codes SC gives Eb/N0 = 2.5 dB and BER at 10^{-3}, SCL gives Eb/N0 = 2.5 dB and BER at 10^{-5} and FSCL gives Eb/N0 = 2.5 dB and BER at 10^{-5}. SCL, FSCL algorithms list size 32 is used for operation. In that Polar Codes gives better performance compared to LDPC codes.

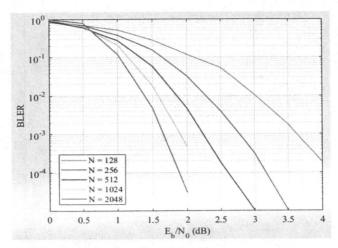

Fig. 5. BLER of Polar codes using CA-SCL algorithm with L = 32

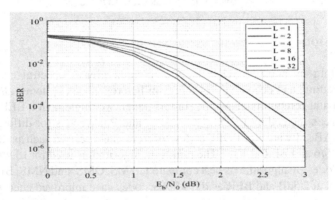

Fig. 6. BER of Polar codes using FSCL algorithm for N = 1024

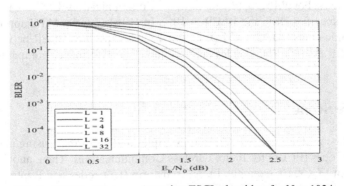

Fig. 7. BLER of Polar codes using FSCL algorithm for N = 1024

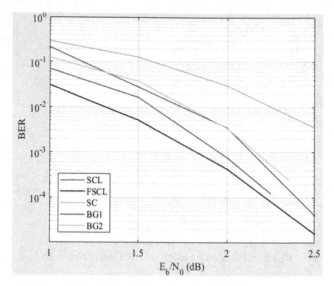

Fig. 8. BER of Polar codes and LDPC code

5 Conclusion

Error-correcting methods of 5G techniques contain greater spectral energy, multi-terminal encoding, and decoding procedures in heterogeneous systems, also cut down transmission and computing capability. In this paper, we implement a BER and BLER relation between the coding LDPC, QC-LDPC, and Polar codes for different plots by using MATLAB software. For LDPC codes different rates are used such as Rate R = 1/3, 1/2, 2/3, and 5/6 and BER at 10^{-5}. The min-Sum algorithm is used for LDPC Codes. In QC-LDPC Codes BP algorithm is best compared to MS, OMS, and NMS. BP Algorithm performs well at 2.5dB and BER at 10^{-5}. Polar codes have improved energy-efficiency for great block sizes compared to alternative codes. For Polar codes, SC, BP, SCL, FSCL, and CA-SCL decoding algorithms are used for simulation and it gives BLER at 10^{-5}. L = 32 is the best choice for Polar codes at 2.5 dB and it performs well at limited and wide block sizes. For eMBB applications in 5G Communications LDPC, QC-LDPC and Polar codes specifies BER and BLER at 10^{-5}. In URLLC and mMTC applications for 5G Communications LDPC, QC-LDPC, Polar Codes specifies BLER between 10^{-4} to 10^{-5} and BER at 10^{-4}. Polar Codes delivers superior performance compared to LDPC codes.

6 Future Scope

Polar codes beat LDPC codes in explicit examples and performs a rising class of Error Correcting Codes (ECCs) for future wireless Communications. Then, another predominant class of ECCs, for example, Luby Transform (LT) codes and Raptor codes, was excessively generally contemplated. In practical applications, rateless codes are exceptionally helpful for plans where the Channel State Information (CSI) is nonexistent at the transmitter terminal. Channel coding for 5G is an effective research area to deal with many great challenges in developing times.

References

1. https://www.itu.int/en/ITU-R/study-groups/rsg5/rwp5d/imt-2020/Pages/default.aspx. Accessed 15 Mar 2020
2. http://www.3gpp.org/news-events/3gpp-news/1614-sa_5g. Accessed 20 Mar 2020
3. https://www.sdxcentral.com/5g/definitions/5g-standards/. Accessed 25 Mar 2020
4. Andrews, J.G., et al.: What will 5G be? IEEE J. Sel. Areas Commun. **32**(6), 1065–1082 (2014)
5. https://www.electronics-notes.com/articles/connectivity/5g-mobile-wireless-cellular/requir ements.phn. Accessed 26 Mar 2020
6. https://cdn-www.mediatek.com/page/MediaTek-5G-NR-Summary.pdf. Accessed 27 Mar 2020
7. Maunder, R.: A Vision for 5G Channel Coding, AccelerComm White Paper (2016)
8. https://cdn-www.mediatek.com/page/MediaTek-5G-NR-White-Paper-PDF5GNRWP.pdf. Accessed 28 Mar 2020
9. Dhanorkar, P., Kalbande, M.: Design of LDPC decoder using message passing algorithm. In: International Conference on Communication and Signal Processing (ICCSP), Chennai, pp. 1923–1926 (2017)
10. Abdulwahab, W.K., Kadhim, A.A.: Comparative study of channel coding schemes for 5G. In: International Conference on Advanced Science and Engineering (ICOASE), Duhok, pp. 239–243 (2018)
11. Sun, K., Jiang, M.: A hybrid decoding algorithm for low-rate LDPC codes in 5G. In: 10th International Conference on Wireless Communications and Signal Processing (WCSP), Hangzhou, pp. 1–5 (2018)
12. Wu, H., Wang, H.: A High Throughput Implementation of QC-LDPC Codes for 5G NR. IEEE Access **7**, 185373–185384 (2019)
13. Yu, Y., Pan, Z., Liu, N., You, X.: Belief propagation bit-flip decoder for polar codes. IEEE Access **7**, 10937–10946 (2019)
14. Gamage, H., Rajatheva, N., Latva-aho, M.: Channel coding for enhanced mobile broadband communication in 5G systems. In: European Conference on Networks and Communications (EuCNC), Oulu, pp. 1–6 (2017)
15. Sybis, M., Wesolowski, K., Jayasinghe, K., Venkatasubramanian, V., Vukadinovic, V.: Channel Coding for ultra-reliable low-latency communication in 5G systems. In: IEEE 84th Vehicular Technology Conference (VTC-Fall), Montreal, QC, pp. 1–5 (2016)
16. Hui, D., Sandberg, S., Blankenship, Y., Andersson, M., Grosjean, L.: Channel coding in 5G new radio: a tutorial overview and performance comparison with 4G LTE. IEEE Veh. Technol. Mag. **13**(4), 60–69 (2018)
17. Tahir, B., Schwarz, S., Rupp, M.: BER comparison between Convolutional, Turbo, LDPC, and Polar codes. In: 24th International Conference on Telecommunications (ICT), Limassol, pp. 1–7 (2017)
18. https://www.ics.uci.edu/~welling/teaching/ICS279/LPCD.pdf. Accessed 28 Mar 2020
19. Bae, J.H., Abotabl, A., Lin, H.-P., Song, K.-B., Lee, J.: An overview of channel coding for 5G NR cellular communications. APSIPA Trans. Sig. Inf. Process. **8**, e17 (2019)
20. Z. Wang, Z. Cui: Low-complexity high-speed decoder design for quasi-cyclic LDPC codes. IEEE Trans. Very Large Scale Integr. (VLSI) Syst. **15**(1), 104–114 (2007)
21. Pfister, H.D., Sason, I., Urbanke, R.: Capacity-achieving ensembles for the binary erasure channel with bounded complexity. IEEE Trans. Inf. Theor. **51**(7), 2352–2379 (2005)

22. Stark, M., Bauch, G., Lewandowsky, J., Saha, S.: Decoding of non-binary LDPC codes using the information bottleneck method, pp. 1–6 (2019). https://doi.org/10.1109/ICC.2019.876 1712
23. Cammerer, S., Wang, X., Ma, Y., Brink, S.: Spatially coupled LDPC codes and the multiple access channel. In: Annual Conference on Information Sciences and Systems (CISS), Baltimore, MD, USA, pp. 1–6 (2019)
24. Arora, K., Singh, J., Randhawa, Y.S.: A survey on channel coding techniques for 5G wireless networks. Telecommun. Syst. **73**(4), 637–663 (2019). https://doi.org/10.1007/s11235-019-00630-3
25. Ercan, F., Condo, C., Hashemi, S.A., Gross, W.: On error-correction performance and implementation of polar code list decoders for 5G (2017)
26. Niu, K., Chen, K.: CRC-aided decoding of polar codes. IEEE Commun. Lett. **16**(10), 1668–1671 (2012)
27. Pillet, C., Bioglio, V., Condo, C.: On list decoding of 5G-NR polar codes (2019)
28. Chen, P., Xu, M., Bai, B., Wang, J.: Design and performance of polar codes for 5G communication under high mobility scenarios. In: IEEE 85th Vehicular Technology Conference (VTC Spring), Sydney, NSW (2017)
29. Patil, M.V., Pawar, S., Saquib, Z.: Coding techniques for 5G networks: a review. In: 2020 3rd International Conference on Communication System, Computing and IT Applications (CSCITA), Mumbai, India, pp. 208–213 (2020)

An Ensemble Learning Approach for Software Defect Prediction in Developing Quality Software Product

Yakub Kayode Saheed[1], Olumide Longe[1], Usman Ahmad Baba[1], Sandip Rakshit[1], and Narasimha Rao Vajjhala[2(✉)]

[1] American University of Nigeria, Yola, Nigeria
{yakubu.saheed,olumide.longe,usman.ahmad,
sandip.rakshit}@aun.edu.ng
[2] University of New York Tirana, Tirana, Albania
narasimharao@unyt.edu.al

Abstract. Software Defect Prediction (SDP) is a major research field in the software development life cycle. The accurate SDP would assist software developers and engineers in developing a reliable software product. Several machine learning techniques for SDP have been reported in the literature. Most of these studies suffered in terms of prediction accuracy and other performance metrics. Many of these studies focus only on accuracy and this is not enough in measuring the performance of SDP. In this research, we propose a seven-ensemble machine learning model for SDP. The Cat boost, Light Gradient Boosting Machine (LGBM), Extreme Gradient Boosting (XgBoost), boosted cat boost, bagged logistic regression, boosted LGBM, and boosted XgBoost were used for the experimental analysis. We also used the separate individual base model of logistic regression for the analysis on six datasets. This paper extends the performance metrics from only the accuracy, the Area Under Curve (AUC), precision, recall, F-measure, and Matthew Correlation Coefficient (MCC) were used as performance metrics. The results obtained showed that the proposed ensemble Cat boost model gave an outstanding performance for all the three defects datasets as a result of being able to decrease overfitting and reduce the training time.

Keywords: Software defect prediction · Software product · Cat boost · Light gradient boosting machine · NASA repository · Extreme gradient boosting · Area under curve · F-measure

1 Introduction

Software Defect Prediction (SDP) is a major part of the Software Quality Assurance (SQA) processes that intends to automatically predict the defect and fault prone-modules utilizing past software data from earlier deployment [1], for instance, bug file reports [2] and source code logs [3], before the beginning of software testing. An effective SDP could assist software tester to find bugs and ease the distribution of SQA assets

© Springer Nature Switzerland AG 2021
M. Singh et al. (Eds.): ICACDS 2021, CCIS 1440, pp. 317–326, 2021.
https://doi.org/10.1007/978-3-030-81462-5_29

economically and optimally. Therefore, it is an important research problem in software development [4–7]. A predictive model is employed popularly to predict the defective units in one of the three groups: classification based on binary [1, 8–10], the defects number [11, 12] and defect prediction severity [13].

In software engineering discipline, SDP in the early phases is important for software quality and reliability [14, 15]. The purpose of SDP is to assist in software products before they are released by predicting defect in the software, as spotting bugs after software are released is very time consuming and exhausting. Additionally, SDP methods have been shown to enhance the quality of software, as they assist developers spot the possible defective units [16, 17]. SDP is regarded as an important issue; therefore, many machine learning classification algorithms have been utilized to determine and predict units that are defective [18]. The goal is to expand the feasibility of software testing, the SDP can be used to differentiate defective part in subsequent and current forms of the software product. Thus, SDP methods are useful in assigning resources and the efforts for investigating, examining and testing most defective modules [19]. The modelling approach is based on mainly regression or classification methods [20]. Thirdly, the dependent variables are generated by the predictive model usually categorical and binary classification as defect prone or not defect prone. Lastly, a model is developed to measure the performance of a predictive model. Machine learning performance metrics such as confusion matrix are mostly utilized for categorical predictive models and metrics such as predictive errors are mostly utilized for continuous features predictions.

Software practitioners and researchers have utilized many machine learning and statistical approaches to spot faulty modules and lessen maintenance and software development costs. Several supervised machine learning SDP approaches have been introduced and presented in the past two decades like Support Vector Machine (SVM) [21, 22], K-Nearest Neighbor (KNN) [23, 24], Neural Network (NN) [25, 26], Naïve Bayes (NB) [27, 28], Random Forest (RF) [29], Decision Tree (DT) [30, 31], Association Rule Mining (ARM) [32], Artificial Immune System (AIS) [33], Logistics Regression (LR) [34], Genetic Programming (GP), and Fuzzy Logic [5]. A good number of studies in the SDP used the ensemble approaches like voting [35], bagging method [36], random tree [37], stacking method random forest [38] and boosting method [39]. The neural networks have also been used widely for SDP [40] and the deep neural network [41].

The SDP is seen as challenging and problematic in a machine learning perspective as a result of the unbalanced nature of the classes in the SDP datasets, where the non-defects are higher than the faulty ones. In this scenario, the traditional algorithms are inclined to categorize properly the majority class and, disregard the minority class. Thus, the defect or faulty modules are a significant class. So, this results in a poor classification of the classifier. Many of the previous studies also used accuracy alone as the performance metric. To address this problem, the authors proposed an ensemble learning models for spotting software flaws in three dissimilar datasets and the work extends the performance metrics as compared to the previous studies in this paper.

This paper is structured as follows. Section 2 presents the review of recent literature. We presented the methodology in Sect. 3. The results and discussion are given in Sect. 4 and we gave the conclusion together with recommendations for future work in Sect. 5.

2 Review of Literature

Challagulla, et al. [42] proposed a comparison among the different ML methods for SDP based on four (4) datasets from the NASA dataset. The results revealed that no classification algorithm gave the greatest result for all the datasets. However, it also showed that the instance learning based model and one rule outperformed other in accuracy terms. Wahono, et al. [20] presented a particle swarm optimization approach as a feature selection technique and the bagging approach for classification, the study employed nine datasets from NASA and used eleven classification algorithms, the comparison of results was made and it revealed that the classification algorithms were improved except the SVM that does show any significant improvement. Gray, et al. [22] utilized SVM as the classification algorithm and performed data cleansing on the NASA SDP datasets. The accuracy was used as performance metrics and analysis was done on eleven datasets. The accuracy range between 64% and 82% and the average accuracy gave 70% for the eleven datasets. The authors disregard MCC and ROC citing the only accuracy as the performance metric.

Rong et al. [43] proposed a novel hybrid method of the support vector machine and CBA which add the power of the two classifiers; the optimization capacity of the Bat algorithm incorporated with the centroid approach and the non-linear capacity in the later. The analysis was done on some popular datasets and results comparison was done in terms of accuracy with other reported methods and the results obtained performed better. The study neglects to consider another performance metric. Wahono, et al. [20] used PSO algorithms and genetic metaheuristics algorithms as the feature selection method. They used the bagging method to handle the imbalance issue using ten different classification algorithms. They discovered that no significant difference between the two algorithms used for the feature selection method, the incorporation of particle swarm optimization and genetic algorithm with the bagging method gave better results. Magal.R and Jacob [44] gave a comparison between the feature selection methods using correlation-based feature selection (CFS), information gain and gain ratio and discovered that the CFS gave higher accuracy and used five classification algorithms with results indicating higher accuracy for RF among the tested classifier. The CFS feature selection technique and RF gave 98.3% accuracy. The study paid no attention to other performance metrics.

Ibrahim, et al. [38] used the Bat algorithm for the feature selection method and RF for the classification SDP. The study also utilized a good number of feature selection strategies to see how effective they are in SDP. However, the only metric used to analyze the performance of the model was accuracy. Aquil and Wan Ishak [45] proposed both supervised and non-supervised machine learning approaches for SDP. Aljamaan and Alazba [46] proposed seven ensembles models in SDP, they classified the ensemble models as boosting and boosting ensembles. The work used eleven NASA defects datasets. The finding showed good results for all the ensemble models. However, the work was evaluated on accuracy and AUC as the performance metrics. These two metrics cannot be used alone to justify the SDP. As can be seen from these studies, it was noted that many of the studies used only accuracy as the performance metrics. Accuracy alone cannot guarantee the performance of SDP. This paper proposes to fill up this gap.

3 Methodology

In this section, we presented the methodology used in this paper. In the first segment of this research, we perform data normalization on the three public datasets obtained from the Promise repository [47]. We presented and utilized the individual base classifier Logistic Regression model and seven ensemble models which are Cat boost, light gradient boosting machine, extreme gradient boosting, bagged logistic regression, boosted light gradient boosting machine, and Boosted extreme gradient boosting machine. We used 70 percent of the dataset for training and 30% for testing the model. The feature selection was also performed before the classification with ensemble model and individual base classifier.

The three datasets used to conduct and verification of our experiments were obtained from the popular Promise repository [47]. These datasets were gotten from real Software Projects from NASA developed in C+ + /C for the Storage Management (kc1 dataset), Scientific Data Preprocessing (jm1 dataset), and Storage Management (kc2 dataset). The cat boost algorithm is an ensemble algorithm that has superior generalization and prediction ability [48]. Though Cat boost is developed to handle categorical features [49], however, it can still handle continuous or numerical attributes. Cat boost possesses the ability to decrease overfitting and reduce the training time [46]. Cat Boost model is a special feature that is added to the gradient-boosting decision tree algorithm [50]. The cat boost algorithm was recently used for SDP in the study [46]. However, the performance metrics used in the paper were accuracy and AUC. The Logistic Regression (LR) is a method in statistical machine learning used to evaluate a dataset that consists of more than one independent feature which is used to determine the class label [51]. The LR intends to investigate the model with good fitting which shows the association between the independent features and the class label. The LR can be expressed further in the Eq. 1.

$$\text{Logit } (p) = a_0 + a_1 * X_1 + a_2 * X_2 + a_3 * X_3 + a_4 * X_4 + \ldots \ldots + a_k * X_k \tag{1}$$

where, the feature p is the probability of the presence of characteristic of interest.

4 Results and Discussion

The experimental analysis is performed to compare the results of the individual base classifiers and the ensemble classification algorithms for SDP. The analysis was done on three datasets from NASA for SDP. The 10-fold cross-validation technique was used as the hold-out method where each of the datasets will be divided into 10 subsets with nine out of the ten subsets used for training each of the model and one for testing model obtained by the classifier. Many of the previous studies used accuracy as the only metric for evaluating the SDP model. However, accuracy alone cannot provide adequate information on the performance of SDP. We, therefore, extend the performance metrics by introducing the Area under Curve, recall, precision, F-measure, and MCC for SDP. Table 1 presents the results from applying the individual base classifier and the ensemble

Table 1. KC1 dataset

Metrics	Cat boost	LR	LGBM	XgBoost	Boosted Catboost	Bagged LR	Boosted LGBM	Boosted XgBoost
Accuracy	85.71	84.69	84.75	84.69	85.64	85.1	83.88	84.55
AUC	80.15	77.75	73.97	75.24	74.63	77.95	70.33	0
Recall	22	14.55	29.43	31.6	32.08	16.28	29.86	0
Precision	61.35	53.58	52.14	51.11	56.09	55.92	45.86	0
F-measure	31.88	22.18	37.05	38.82	40.22	24.51	35.78	0
MCC	30.38	21.52	31.02	31.93	34.8	23.95	28.15	0

models on the kc1 NASA dataset. Among the individual base classifier, the cat boost classifier gave the best results in terms of accuracy, AUC, and precision.

The boosted cat boost gave the best results for recall and f- measure. The boosted Xgboost model performs poorly in terms of AUC, recall, precision, and f-measure. As can be seen in Fig. 1, our proposed Cat boost algorithm outperformed all other proposed ensemble models. However, other proposed ensemble models also gave competitive performance than individual base model logistic regression.

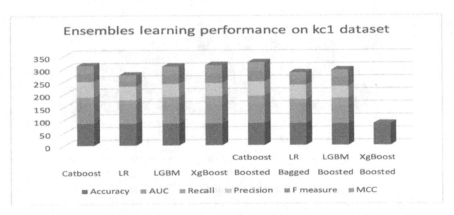

Fig. 1. Performance evaluation of kc1 dataset

Table 2 presents the results from applying the individual base classifier and the ensemble models on the jm1 NASA dataset. Among the individual base classifier, the cat boost classifier gave the best results in terms of accuracy, and AUC. Boosted LGBM gave the best in terms of recall. Bagged LR gave the best results in terms of precision. The Boosted LGBM gave the best results in terms of MCC. The XgBoost model gave the worst performance in terms of AUC, recall, precision, F-measure, and MCC. Boosted LGBM gave the best results in terms of F-measure.

Table 2. JM1 dataset

Metrics	Catboost	LR	LGBM	XgBoost	Boosted Catboost	Bagged LR	Boosted LGBM	Boosted XgBoost
Accuracy	81.14	80.68	81.31	80.9	81.1	80.9	80.39	80.65
AUC	73.24	51.87	72.71	70.89	70.52	59.64	68.31	0
Recall	14.66	5.77	18.79	20.56	24.01	6.58	27.4	0
Precision	54.88	51.6	55.08	51.74	52.7	55.16	48.92	0
F measure	23.09	10.3	27.95	29.38	32.97	11.68	35.04	0
MCC	16.26	11.96	24	23.63	26.21	13.84	26.05	0

As can be seen in Fig. 2, the proposed Cat boost ensemble model performed excellently than all other proposed ensemble models. The proposed ensemble models LGBM, Xgboost, boosted cat boost, bagged LR, and boosted LGBM performed excellently than the individual base model LR. Though, the individual model LR gave better results than the boosted Xgboost.

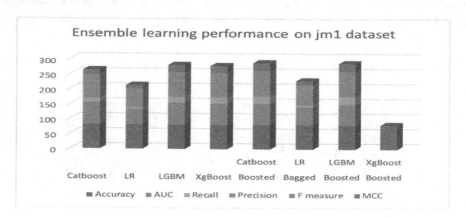

Fig. 2. Performance evaluation of JM1 dataset

4.1 Comparison with the State-of-the Art Techniques

In this section, we compared the results of our proposed ensemble models with the state-of-the-art results in Table 3. Our proposed ensembles model performances were excellent, competitive and very promising for SDP.

Table 3. Comparison with other state-of-the art methods

Authors	Algorithms	Datasets	Accuracy	AUC	Recall	Precision	F-measure	MCC
Aljamaan and Alazba [46]	DT	kc1	69.1	69	x	x	x	x
	RF	kc1	72.5	79	x	x	x	x
	GB	kc1	69	76	x	x	x	x
Alsawalqah et al. [1]	NB	Jm1	80.42	x	x	x	x	x
	J48	Jm1	79.73	x	x	x	x	x
	MLP	Jm1	81.00	x	x	x	x	x
Aljamaan and Alazba [46]	DT	pc1	88.59	89	x	x	x	x
	RF	pc1	93.79	99	x	x	x	x
	GB	pc1	92.24	98	x	x	x	x
Alsawalqah et al. [1]	NB	pc3	48.26	x	x	x	x	x
	J48	pc3	88.14	x	x	x	x	x
	MLP	pc3	88.20	x	x	x	x	x
Aquil and Wan Ishak [45]	ET	pc2	97.86	x	x	x	x	x
	RF	pc2	97.46	x	x	x	x	x
	ET	kc2	78.78	x	x	x	x	x
	RF	kc2	81.27	x	x	x	x	x
Our Ensemble	Cat boost	kc1	85.71	80.15	22	61.35	31.88	30.38
Our Individual Base model	LR	jm1	80.68	51.87	5.77	51.6	10.3	11.96
Our Ensemble	LGBM	pc1	93.3	82.43	0.3	52.33	36.46	35.74
Our Ensemble	XgBoost	pc2	99.54	79.57	x	x	x	x
Our Ensemble	Boosted Cat Boost	pc3	90.50	82.02	21.52	56.92	30.57	30.77
Our Ensemble	Bagged LR	kc2	84.39	76.8	45.18	67.32	53	44.35

5   Conclusion and Future Work

This paper investigated the performance of different ensemble models for SDP. The SDP is an active area of research in software development and the automation would greatly assist software engineers in developing a timely and efficient software product. Six ensemble machine learning model were presented in this paper and the analysis was done on the three datasets. The findings showed that the proposed ensemble method's performance was significantly better and competitive than the individual base classifier logistic regression model for all the six defect datasets. The results obtained were also benchmarked with the state-of-the-art and our results are superior in terms of the accuracy, AUC, recall, precision, and MCC. The performance metrics was also extended in this paper as compared to previous studies. The findings of the proposed ensemble models were very promising and very significant than the state-of-the-art techniques for SDP. The future work will be to extend the number of datasets used in this paper and a deep learning model can be used to evaluate the model. Thus, the results can be compared with this study.

References

1. Alsawalqah, H., et al.: Software defect prediction using heterogeneous ensemble classification based on segmented patterns. Appl. Sci. **10**(5), 1745 (2020)
2. Bhattacharya, P., et al.: Graph-based analysis and prediction for software evolution, pp. 419–429
3. Moser, R., Pedrycz, W., Succi, G.: A comparative analysis of the efficiency of change metrics and static code attributes for defect prediction, pp. 181–190
4. Abaei, G., Selamat, A.: A survey on software fault detection based on different prediction approaches. Vietnam J. Comput. Sci. **1**(2), 79–95 (2014)
5. Wang, S., Liu, T., Tan, L.: Automatically learning semantic features for defect prediction. In: Proceedings of the 38th International Conference on Software Engineering, Austin, Texas, pp. 297–308 (2016)
6. Hall, T., et al.: A systematic literature review on fault prediction performance in software engineering. IEEE Trans. Software Eng. **38**(6), 1276–1304 (2012)
7. Menzies, T., et al.: Defect prediction from static code features: current results, limitations, new approaches. Automated Softw. Eng. **17**(4), 375–407 (2010)
8. Li, Z., Reformat, M.: A practical method for the software fault-prediction, pp. 659–666
9. Vandecruys, O., et al.: Mining software repositories for comprehensible software fault prediction models. J. Syst. Softw. **81**(5), 823–839 (2008)
10. Mendes-Moreira, J., et al.: Ensemble approaches for regression: a survey. ACM Comput. Surv. **45**(1), (2012). Article 10
11. Rathore, S.S., Kuamr, S.: Comparative analysis of neural network and genetic programming for number of software faults prediction, pp. 328–332
12. Rathore, S.S., Kumar, S.: Linear and non-linear heterogeneous ensemble methods to predict the number of faults in software systems. Knowl. Based Syst. **119**, 232–256, (2017)
13. Shatnawi, R., Li, W.: The effectiveness of software metrics in identifying error-prone classes in post-release software evolution process. J. Syst. Softw. **81**(11), 1868–1882 (2008)
14. Bowes, D., Hall, T., Petrić, J.: Software defect prediction: do different classifiers find the same defects? Software Qual. J. **26**(2), 525–552 (2017). https://doi.org/10.1007/s11219-016-9353-3
15. Rawat, M., Dubey, S.: Software defect prediction models for quality improvement: a literature study. Int. J. Comput. Sci. Issues **9**, 288–296 (2012)
16. Singh, P.D., Chug, A.: Software defect prediction analysis using machine learning algorithms, pp. 775–781
17. Ge, J., Liu, J., Liu, W.: Comparative study on defect prediction algorithms of supervised learning software based on imbalanced classification data sets, pp. 399–406
18. Song, Q., Guo, Y., Shepperd, M.: A Comprehensive investigation of the role of imbalanced learning for software defect prediction. IEEE Trans. Softw. Eng. **45**(12), 1253–1269 (2019)
19. Chang, R., Mu, X., Zhang, L.: Software defect prediction using non-negative matrix factorization. JSW **6**, 2114–2120 (2011)
20. Wahono, R., Suryana, N., Ahmad, S.: Metaheuristic optimization based feature selection for software defect prediction. J. Softw. **9**, 1324–1333 (2014)
21. Elish, K.O., Elish, M.O.: Predicting defect-prone software modules using support vector machines. J. Syst. Softw. **81**(5), 649–660 (2008)

22. Gray, D., et al.: Using the support vector machine as a classification method for software defect prediction with static code metrics, pp. 223–234
23. Gong, L., et al.: Empirical evaluation of the impact of class overlap on software defect prediction, pp. 698–709
24. Mabayoje, M., et al.: Parameter tuning in KNN for software defect prediction: an empirical analysis. Jurnal Teknologi dan Sistem Komputer **7**, 121–126 (2019)
25. Tong-Seng, Q., Mie Mie Thet, T.: Application of neural networks for software quality prediction using object-oriented metrics, pp. 116–125
26. Thwin, M.M.T., Quah, T.-S.: Application of neural networks for software quality prediction using object-oriented metrics. J. Syst. Softw. **76**(2), 147–156 (2005)
27. Zhang, H., Zhang, X.: Comments on "data mining static code attributes to learn defect predictors." IEEE Trans. Softw. Eng. **33**(9), 635–637 (2007)
28. Mori, T., Uchihira, N.: Balancing the trade-off between accuracy and interpretability in software defect prediction. Empir. Softw. Eng. **24**(2), 779–825 (2018). https://doi.org/10.1007/s10664-018-9638-1
29. Ramler, R., et al.: Key questions in building defect prediction models in practice, pp. 14–27
30. Gayatri, N., Savarimuthu, N., Reddy, A.: Feature selection using decision tree induction in class level metrics dataset for software defect predictions, Lecture Notes in Engineering and Computer Science, vol. 1 (2010)
31. Pelayo, L., Dick, S.: Applying novel resampling strategies to software defect prediction, pp. 69–72
32. Czibula, G., Marian, Z., Czibula, I.G.: Software defect prediction using relational association rule mining. Inf. Sci. **264**, 260–278 (2014)
33. Catal, C., Diri, B.: Software fault prediction with object-oriented metrics based artificial immune recognition system, pp. 300–314
34. Aida, E., Nima Karimpour, D.: CBM-Of-TRaCE: an ontology-driven framework for the improvement of business service traceability, consistency management and reusability. Int. J. Soft Comput. Softw. Eng. [JSCSE], pp. 69–78
35. Moustafa, S., et al.: Software bug prediction using weighted majority voting techniques. Alexandria Eng. J. **57**(4), 2763–2774 (2018)
36. Mousavi, R., Eftekhari, M., Rahdari, F.: Omni-ensemble learning (OEL): utilizing over-bagging, static and dynamic ensemble selection approaches for software defect prediction. Int. J. Artif. Intell. Tools **27**(06), 1850024 (2018)
37. Tanwar, H., Kakkar, M.: A review of software defect prediction models. In: Proceedings of ICDMAI 2018, vol. 1, pp. 89–97 (2019)
38. Ibrahim, D.R., Ghnemat, R., Hudaib, A.: Software defect prediction using feature selection and random forest algorithm, pp. 252–257
39. Cai, X., et al.: An under-sampled software defect prediction method based on hybrid multi-objective cuckoo search. Concurr. Comput. Pract. Exp. **32**(5), e5478 (2020)
40. Jayanthi, R., Florence, L.: Software defect prediction techniques using metrics based on neural network classifier. Clust. Comput. **22**(1), 77–88 (2018). https://doi.org/10.1007/s10586-018-1730-1
41. Manjula, C., Florence, L.: Deep neural network based hybrid approach for software defect prediction using software metrics. Clust. Comput. **22**(4), 9847–9863 (2018). https://doi.org/10.1007/s10586-018-1696-z
42. Challagulla, V.U.B., et al.: Empirical assessment of machine learning based software defect prediction techniques. Int. J. Artif. Intell. Tools **17**(02), 389–400 (2008)
43. Rong, X., Li, F., Cui, Z.: A model for software defect prediction using support vector machine based on CBA. Int. J. Intell. Syst. Technol. Appl. **15**(1), 19–34 (2016)
44. Magal. K.R., Jacob, S.: Improved random forest algorithm for software defect prediction through data mining techniques. Int. J. Comput. Appl. **117**, 18–22 (2015)

45. Aquil, M.A.I., Wan Ishak, W.H.: Predicting software defects using machine learning techniques. Int. J. Adv. Trends Comput. Sci. Eng. **9**, 6609 (2020)
46. Aljamaan, H., Alazba, A.: Software defect prediction using tree-based ensembles. In: Proceedings of the 16th ACM International Conference on Predictive Models and Data Analytics in Software Engineering, pp. 1–10. Association for Computing Machinery (2020)
47. Shepperd, M., et al.: Data quality: some comments on the NASA software defect datasets. IEEE Trans. Softw. Eng. **39**(9), 1208–1215 (2013)
48. Deng, K., et al.: A remaining useful life prediction method with long-short term feature processing for aircraft engines. Appl. Soft Comput. **93**, 106344 (2020)
49. Dorogush, A., Ershov, V., Gulin, A.: CatBoost: gradient boosting with categorical features support (2018)
50. Kavitha, G., Elango, N.M.: An approach to feature selection in intrusion detection systems using machine learning algorithms. Int. J. e-Collaboration (IJeC) **16**(4), 48–58 (2020)
51. Peng, C.-Y.J., Lee, K.L., Ingersoll, G.M.: An introduction to logistic regression analysis and reporting. J. Educ. Res. **96**(1), 3–14 (2002)

A Study on Energy-Aware Virtual Machine Consolidation Policies in Cloud Data Centers Using Cloudsim Toolkit

Dipak Dabhi$^{(\boxtimes)}$ and Devendra Thakor

C. G. Patel Institute of Technology, Uka Tarsadia University, Bardoli,, India

Abstract. In cloud computing, many data centers are used to serve the user's service request and lead to high energy consumption. These vast data centers generate high carbon emissions in the environment and high operational costs for the cloud service provider. The method of reducing energy consumption by the data center is called Virtual Machine Consolidation (VMC). In VMC, one host's migration to another saves energy by putting some hosts to sleep mode. There are four VMC methods: overload detection of the host, underload detection host, VM selection, and VM placement. VMC improves the utilization of energy consumption in data centers with maintain Quality of Service (QoS). Since VM consolidation is an important area for research, many researchers are working on it. In this article, we implement and execute the different twenty VMC policies using PlanetLab workload traces in the CloudSim toolkit and perform a comparative analysis of simulation results.

Keywords: Cloud computing · VM consolidation · QoS · VM selection · Overload host · VM placement · Underload host

1 Introduction

Cloud computing follows pay only for what you used model. it also facilitates on-demand access of cloud resources to the users through data centers scattered over the globe. Cloud data centers consume substantial power sources to serve user's requests, which generate a large amount of carbon dioxide (CO_2) in the ecosystem, which is similar to the global aviation industry and harms our environment. Also, the energy usage by the data center will increase exponentially over time. Energy-efficient techniques need cloud data centers to resolve this problem [1, 2]. Initially, the workload assigns to the available Physical Machine (PM) through the Virtual Machine (VM). Allocation of VMs may change by increasing/decreasing the demand of users at the time. As the PM is limited to incorporate each user request, the mapping of VMs dynamically is required to serving the desired resource request [3]. This mapping between PM and VM per the requests done by cloud providers using VM live migration techniques [2, 5].

© Springer Nature Switzerland AG 2021
M. Singh et al. (Eds.): ICACDS 2021, CCIS 1440, pp. 327–337, 2021.
https://doi.org/10.1007/978-3-030-81462-5_30

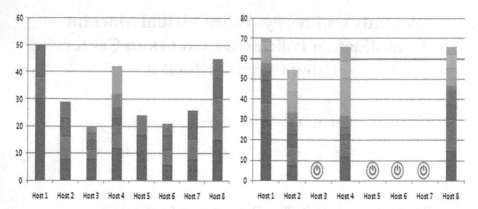

Fig. 1. VMC Before consolidation (left) and after consolidation (right) [2].

Figure 1 shows the number of hosts that are active before and after the VM consolidation process. Figure 1 shows the eight PM will require to serve the user request, but out of eight hosts, some of the host utilization significantly less, and the remaining host's utilization is medium. But irrespective of utilization level, all server consumes high energy, so we can sleep some of the less utilized hosts and migrate the VM's from underutilized hosts to other hosts using VM migration [8]. Figure 1, it will show how the lower utilized host consumes energy before consolidation and we put four hosts in sleep mode and transfer their loads to other hosts. In cloud computing, all the data center resources are heterogeneous, so that data center resources may have a different physical configuration. In that case, VM mapping from one host to another host is very difficult for service providers [2, 9]. In VMC, the resource allocation policies have various strategies to adaptively allocate the requested VM to complete with minimal active PM and save energy [10, 12].

2 Literature Work

This section of the paper focuses on various VM consolidation policies for resource allocation and management. There are two main goals of VM consolidation policies that need to achieve: avoidance of host overload and green computing. To handle the host overload, the resource requested by the VM should be lower than the available resource of the host. Otherwise, it will violate the service agreement between the user and service provider which also known as service level agreement (SLA). SLA is important for measure the Quality of Service (QoS) and availability of service to the end-user [11, 25, 26]. For green computing, cloud service providers serve the user's request with the minimum number of active hosts without SLA violation using switch the load from underutilized hosts to other hosts and put underutilized hosts in sleep mode so the energy consumption by that host can save energy [3].

2.1   VM Consolidation Methods

Host Overload Detection Algorithm	Static Threshold (Th)	Median Absolute Deviation (MAD)	Interquartile Range (IQR)	Local Regression (LR)	Robust Local Regression (LRR)
Host Underload Detection Algorithm	Host with Minimum Utilization				
VM Selection Algorithm	Minimum Migration Time (MMT)	Random Choice (RC)	Maximum Correlation (MC)	Minimum Utilization (MU)	
VM Placement Algorithm	First Fit (FF)	First Fit Decreasing (FFD)	Best Fit Decreasing (BFD)	Worst Fit Decreasing (WFD)	Power Aware Best Fit Decreasing (PABFD)

Fig. 2. VM consolidation algorithms [2]

Host Overload Detection Algorithm: Figure 2 shows the classification of VMC policy. Multiple VMs may run on a single PM as per user request, and because of the dynamic nature of user request, the utilization level of hosts changes dynamically. If the host's utilization level reached 100%, it will affect the performance and violate the SLA. Identifying the overloaded host is the first step of the VM consolidation process [16, 17]. After identifying the overloaded host, the system needs to find VMs responsible for host overload and select them for migration [20].

Host Underload Detection Algorithm: In the cloud data center, many of the hosts may significantly less utilize or not utilized; still, the host is not in sleep mode, so energy consumes by that host is also there. This algorithm uses to identify an underloaded host and after migration of load put an underutilized host in sleep mode to save energy [7, 28].

VM Selection Algorithm: If the overloaded host identifies the system needs to select the VMs responsible for host overload and put them into the migration list, we can avoid performance degradation. If the underload host identifies, select all the underload host VMs and put them into the migration list [29].

VM Placement Algorithm: The placement algorithm is responsible for selecting the destination host for VM migration from one server to another [4, 27]. In our experiment, we consider the power consumption aware Best Fit Decreasing (BFD) policy as our placement algorithm. experiment, we consider the power consumption aware Best Fit Decreasing (BFD) policy as our placement algorithm.

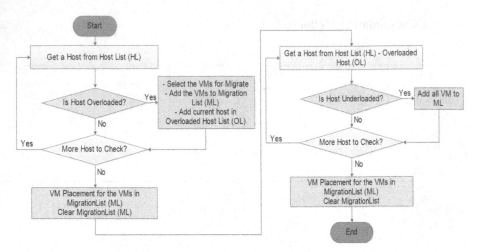

Fig. 3. VM consolidation process flowchart

2.2 VM Consolidation Process Flowchart

Figure 3 shows the flowchart of the VMC process. The process of VMC needs to follow four steps. First, need to identify the overloaded host if any available host as per current load. After selecting the VMs that need to switch the overloaded host to another host, the migration process occurs for selected VMs. In that process system again checks after the migration whether the host is still overloaded; if yes, then the same process is repeated. In underload detection, all VMs of selected host and find the destination host for migration. This process will continue to minimize energy consumption by maintaining SLA between cloud service providers and users [23, 24]. In that process system, again check after the migration whether the host is still overloaded; if yes, then the same process is repeated. In underload detection, all VMs of selected host and find the destination host for migration [18].

3 Experimental Setup and Dataset

To evaluate the different VM consolidation policies, we perform series of experiments. This section provides detail of the data center's experimental setup, dataset, and different evaluation parameters used to measure policy performance. For the simulation and modeling of virtualized data centers, we are using the CloudSim toolkit [8].

3.1 Experimental Setup

In our experiment, we used a data center that having 800 physical heterogeneous hosts. There are two types of host used; one is HP ProLiant ML110 G4 which contain 400 hosts and other HP ProLiant ML110 G5 also contain other 400 hosts. The detailed characteristics like host processor, cores of the host, memory, network bandwidth, and memory are shown in Table 1. The VM plays a crucial role in the migration process. VM

Configuration Table 2 shows VM configuration for this experiment based on varying speed and RAM. The equal number of VM based on types used for the experiment concerning the total number of VMs. The power consumption by the server is measured in Kilowatt per hour and the configuration of the selected servers is shown in Table 3.

Table 1. Configuration of hosts

Server	HP ProLiant ML110 G4	HP ProLiant ML110 G5
CPU model	Intel Xeon 3040	Intel Xeon 3075
Processor (MIPS)	1860	2660
Number of cores	2	?
Memory (GB)	4	4
Network bandwidth (GB/s)	1	1

Table 2. Configuration of VM Types

VM Type	Core	CPU (MIPS)	Memory (MB)
High-CPU Medium	1	2500	870
Extra Large	1	2000	1740
Small	1	1000	1740
Micro	1	500	613

Table 3. Servers at different load level Power consumption

Server	Sleep	0%	10%	20%	30%	40%	50%	60%	70%	80%	90%	100%
HP G4	10	86	89.4	92.6	96	99.5	102	106	108	112	114	117
HP G5	10	93.7	97	101	105	110	116	121	125	129	133	135

3.2 Datasets

The real dataset traces used in this series of the experiment collected from the CoMon project, which has data of PlanetLab's infrastructure monitoring program [4]. The PlanetLab dataset is having ten different days of workload traces. We consider the 3rd March 2011 dataset, which having 800 hosts with 1052 VMs [5]. The 3rd March 2011 dataset contains the VM utilization level of each VM at every 5-min interval, so each VM contains 288 records for 24 h.

4 Evaluation Parameters

In this experiment, we are using some of the well-known evaluation parameters used by most researchers to evaluate the VMC policy's performance.

Energy Consumption (EC): The energy consumption parameter is essential when we have our research gap to minimize energy consumption. We calculate the total MIPS of each host also allowed MIPS of each VMs, and time are used as input parameters to calculate the total energy consumption [1, 20, 23].

Number of VM Migrations (VMM): Using the VM overload detection and underload detection, we can find the overloaded or underloaded host, then VM selection policies identify the VMs for migration. If we migrate more VM, then it violates the SLA, and performance will degrade. The migration process also violates the service level agreement [21].

SLA Violation Metrics: SLA violation metrics are essential to evaluate the QoS to the system. The CPU needs of a VM frequently change over time, and SLA violations will increase if the host is overutilized. Two different metrics will be used to calculate the SLAV value. When the host utilization reaches 100%, the system cannot perform its CPU demands, leading to violating the SLA which know as Service Level Agreement Violation Time per Active Host (SLATAH). To measure the SLA violation due to the host's overutilization, we use the SLATAH metrics formula in Eq. 1 [1, 13]. Also, when we perform the migration, it will take time, and during this time, duration performance will be degraded which know as Performance Degradation due to Migration (PDM) [16, 25] and the formula for calculating PDM is in Eq. 2. Both metrics are equally important and independent from each other so overall SLA Violation calculated through Eq. 3. [6, 7].

$$SLATAH = \frac{1}{N} \sum_{i=1}^{N} \frac{Ts_i}{Ta_i} \qquad (1)$$

$$PDM = \frac{1}{M} \sum_{j=1}^{M} \frac{Cd_j}{Cr_j} \qquad (2)$$

$$SLAV = SLATAH*PDM \qquad (3)$$

The Eq. 1, Eq. 2 and Eq. 3 are used to calculate different SLA Violation metrics [1, 2, 7, 12]. Where N and M represent the number of VMs and PMs in a data center, respectively; Ts_i is the period when the PM utilization of CPU reaches 100%; Ta_i is the total active time of the host. Cd_j is the estimated performance degradation of VM_j caused by VM migrations. Cr_j is the total CPU capacity requested by VM_j [5, 14].

5 Experimental Results and Analysis

An experiment carried out using PlanetLab workload traces. In this paper, a total of twenty experiments were performed, first, two experiments were conducted for non-power aware policy and Dynamic Voltage and Frequency Scaling algorithm without VM consolidation. Next, we conduct a twenty VM consolidation-based experiment. We combining five PM overload detection algorithms (Thr, IQR, LR, LRR, and MAD) with four VM selection algorithms (MU, MMT, MC, and RS) and Single VM Placement (PABFD). Table 4 show the results of simulated experimental with mention evaluation parameters like EC, VMM, SLAV, PDM, and SLATAH.

Table 4. Experimental results

Method	EC	VMM	SLAV (%)	PDM (%)	SLATAH (%)
NPA	2410.8	0	0	0	0
DVFS	803.9	0	0	0	0
IQR_MC_1.5	177.1	23035	0.00701	0.1	6.89
IQR_MMT_1.5	188.86	26476	0.00315	0.06	4.96
IQR_MU1.5	204.22	29901	0.0048	0.06	7.41
IQR_RS_1.5	179.72	23930	0.00707	0.1	6.97
LR_MC_1.2	150.33	23004	0.00677	0.1	6.97
LR_MMT_1.2	163.15	27632	0.00463	0.08	5.84
LR_MU_1.2	174.24	29555	0.00592	0.07	8.18
LR_RS_1.2	150.48	23355	0.00715	0.1	7.1
LRR_MC_1.2	150.33	23004	0.00677	0.1	6.97
LRR_MMT_1.2	163.15	27632	0.00463	0.08	5.84
LRR_MU_1.2	174.24	29555	0.00592	0.07	8.18
LRR_RS_1.2	149.64	22888	0.00712	0.1	7.19
MAD_MC_2.5	176.13	23691	0.00739	0.1	7.06
MAD_MMT_2.5	184.88	26292	0.00331	0.07	5.03
MAD_MU_2.5	200.4	30051	0.0051	0.07	7.57
MAD_RS_2.5	176.74	24297	0.00744	0.1	7.12
Thr_MC_0.8	183.61	24235	0.00697	0.1	6.84
Thr_MMT_0.8	191.73	26634	0.00324	0.07	4.95
Thr_MU_0.8	206.73	30188	0.00481	0.06	7.42
Thr_RS_0.8	184.64	24296	0.00699	0.1	6.89

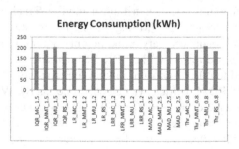

Fig. 4. EC

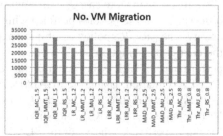

Fig. 5. VMM

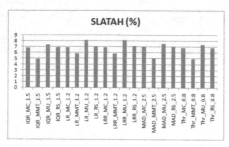

Fig. 6. SLATAH

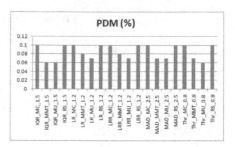

Fig. 7. PDM

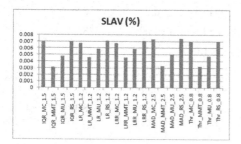

Fig. 8. SLAV

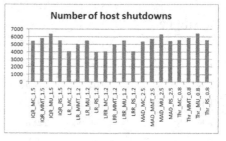

Fig. 9. Number of Host Shutdown

Energy Consumption (EC): These metrics are used to compare different policies energy consumption for the same dataset. We evaluate a total of twenty VM Consolidation policies combinations and the results shown in Table 4 and Fig. 4 shows that LRR-RS outperforms compare to other approaches. THR-MU consumed the maximum energy for the selected dataset.

Further analysis finds that VM consolidation policies that have minimum utilization (MU) algorithm consume higher power than other policies. The MU algorithm performs many undesired VM migrations and because a higher VMM affects the performance of CPU, resulting in more energy consumption, which is shown in Table 4 and Fig. 5 It also observes that those policies perform a higher number of VM migration consume more energy. The LRR-RS policy combination avoids unnecessary or inefficient migration so the energy consumption and VM migration are lowest in the LRR-RS method so the performance of LRR-RS is better compared to other approaches in our experiments.

SLA Violation: Using the SLATAH and PDM metrics, we can calculate the SLAV metrics. These metrics indicate the QoS service of the cloud provider. The Number of VM migration affects SLA violation. Table 4 and Fig. 6, 7, 8 show the SLA metrics for our experiment. The value of SLAV will reflect the QoS, like a higher slav value represent lower QoS.

The SLATAH indicates the overutilization of the PM for a given time. MU algorithm having a high percentage of SLATAH compare to another algorithm because, in MU the higher level of utilizing VM will be first migrated for any host, so the possibility of other host overloaded is high, so more migration is required.

The PDM evaluation parameter identifies the performance degradation because the system performs migration of VMs, and at that time, the host will not execute routine tasks. From the experiment result, we can identify that those policy combinations have RS algorithm having higher PDM value. RS algorithm will select randomly selects the VM for migration, and the purposeless VMM causes the increased wastage of computing resources. The number of the host shutdown for our experiment shown in Fig. 9 VMM and energy consumption are closely related to each other, and VM migration leads to degrading VM performance and causes high energy consumption.

The above simulation results indicate the better VM consolidation method can save a large amount of energy using VM consolidation, but the system also needs to care about SLA Violation. So, the balance between EC and SLAV needs to be maintained.

6 Conclusion

Virtual Machine Consolidation is a method that helps to improve energy consumption in the data center. In this article, we have simulated different twenty-plus VM consolidation policies on a real-time dataset (PlanetLab) in the CloudSim simulation toolkit. The comparative analysis of simulated results of implemented VM consolidation policies is done. The result shows that the LRR_RS policy outperformed in respect of energy consumption and the IQR_MMT policy outperformed in respect of SLAV. Our observation from the comparative analysis is that there is not a single VM consolidation policy that maintains the balance between EC and SLAV metrics. So, our conclusion from the result is that there is a necessity to develop efficient VM consolidation policies that maintain the balance between EC and SLAV.

References

1. Beloglazov, A.: Energy-efficient management of virtual machines in data centers for cloud computing (Doctoral dissertation) (2013)
2. Patel, N., Patel, H.: Energy efficient strategy for placement of virtual machines selected from underloaded servers in compute cloud. J. King Saud Univ. Comput. Inf. Sci. **32**(6), 700–708 (2020). https://doi.org/10.1016/j.jksuci.2017.11.003
3. Shetty, S.: Growth potential aware virtual machine consolidation framework (2020)
4. Mandal, R., Mondal, M.K., Banerjee, S., Biswas, U.: An approach toward design and development of an energy-aware VM selection policy with improved SLA violation in the domain of green cloud computing. J. Supercomput. **76**(9), 7374–7393 (2020). https://doi.org/10.1007/s11227-020-03165-6
5. Singh, J., Sidhu, J.: Comparative analysis of VM consolidation algorithms for cloud computing. Proc. Comput. Sci. **167**, 1390–1399 (2020)
6. Tarafdar, A., Debnath, M., Khatua, S., Das, R.K.: Energy and quality of service-aware virtual machine consolidation in a cloud data center. J. Supercomput. **76**(11), 9095–9126 (2020). https://doi.org/10.1007/s11227-020-03203-3
7. Saadi, Y., El Kafhali, S.: Energy-efficient strategy for virtual machine consolidation in cloud environment. Soft. Comput. **24**(19), 14845–14859 (2020). https://doi.org/10.1007/s00500-020-04839-2

8. Zhang, C., Wang, Y., Lv, Y., Wu, H., Guo, H.: An energy and SLA-aware resource management strategy in cloud data centers. Sci. Programm. **2019**, 3204346:1–3204346:16 (2019). https://doi.org/10.1155/2019/3204346

9. Abdelsamea, A., El-Moursy, A.A., Hemayed, E.E., Eldeeb, H.: Virtual machine consolidation enhancement using hybrid regression algorithms. Egyptian Inform. J. **18**(3), 161–170 (2017)

10. Yavari, M., Ghaffarpour Rahbar, A., Fathi, M.H.: Temperature and energy-aware consolidation algorithms in cloud computing. J. Cloud Comput. **8**(1), 1–16 (2019). https://doi.org/10.1186/s13677-019-0136-9

11. Hamdi, N., Chainbi, W.: A survey on energy aware VM consolidation strategies. Sustain. Comput. Inform. Syst. **23**, 80–87 (2019)

12. Zhou, Q., et al.: Energy efficient algorithms based on VM consolidation for cloud computing: comparisons and evaluations. In: 2020 20th IEEE/ACM International Symposium on Cluster, Cloud and Internet Computing (CCGRID), pp. 489–498. IEEE (2020)

13. Li, Z., Yu, X., Yu, L., Guo, S., Chang, V.: Energy-efficient and quality-aware VM consolidation method. Futur. Gener. Comput. Syst. **102**, 789–809 (2020)

14. John, N.P.: A review on dynamic consolidation of virtual machines for effective energy management and resource utilization in data centres of cloud computing. In: 2020 Fourth International Conference on Computing Methodologies and Communication (ICCMC), pp. 614–619. IEEE (2020)

15. Singh, B.P., Kumar, S.A., Gao, X.Z., Kohli, M., Katiyar, S.: A study on energy consumption of DVFS and Simple VM consolidation policies in cloud computing data centers using CloudSim Toolkit. Wireless Personal Commun. 1–13 (2020)

16. Khattar, N., Singh, J., Sidhu, J.: An energy efficient and adaptive threshold VM consolidation framework for cloud environment. Wireless Pers. Commun. **113**(1), 349–367 (2020)

17. Wang, J.V., Ganganath, N., Cheng, C.T., Chi, K.T.: Bio-inspired heuristics for vm consolidation in cloud data centers. IEEE Syst. J. **14**(1), 152–163 (2019)

18. Singh, P., Gupta, P., Jyoti, K.: Energy aware VM consolidation using dynamic threshold in cloud computing. In: 2019 International Conference on Intelligent Computing and Control Systems (ICCS), pp. 1098–1102. IEEE (2019)

19. Moges, F.F., Abebe, S.L.: Energy-aware VM placement algorithms for the OpenStack Neat consolidation framework. J. Cloud Comput. **8**(1), 1–14 (2019). https://doi.org/10.1186/s13677-019-0126-y

20. Li, H., Li, T., Shuhua, Z.: Energy-performance optimization for the dynamic consolidation of virtual machines in cloud computing. Int. J. Serv. Oper. Inf. **9**(1), 62–82 (2018)

21. Arockia, R.A., Arun, S.: Virtual machine consolidation framework for energy and performance efficient cloud data centers. In: 2019 IEEE International Conference on System, Computation, Automation and Networking (ICSCAN), pp. 1–7. IEEE (2019)

22. Zolfaghari, R., Rahmani, A.M.: Virtual machine consolidation in cloud computing systems: challenges and future trends. Wireless Pers. Commun. **115**(3), 2289–2326 (2020). https://doi.org/10.1007/s11277-020-07682-8

23. Zhou, Z., Hu, Z., Li, K.: Virtual machine placement algorithm for both energy-awareness and SLA violation reduction in cloud data centers. Sci. Programm. **2016**, 5612039:1–5612039:11 (2016). https://doi.org/10.1155/2016/5612039

24. Khosravi, A., Andrew, L.L., Buyya, R.: Dynamic vm placement method for minimizing energy and carbon cost in geographically distributed cloud data centers. IEEE Trans. Sustain. Comput. **2**(2), 183–196 (2017)

25. Mahadevamangalam, S.: Energy-aware adaptation in Cloud datacenters (2018)

26. Mosa, A.: Virtual machine consolidation in cloud data centres using a parameter based placement strategy. The University of Manchester (United Kingdom) (2019)

27. Sharma, O.: Energy efficient virtual machine consolidation for cloud environment (2019)

28. Zhao, D.M., Zhou, J.T., Li, K.: An energy-aware algorithm for virtual machine placement in cloud computing. IEEE Access **7**, 55659–55668 (2019)
29. Masdari, M., Gharehpasha, S., Ghobaei-Arani, M., Ghasemi, V.: Bio-inspired virtual machine placement schemes in cloud computing environment: taxonomy, review, and future research directions. Clust. Comput. **23**(4), 2533–2563 (2019). https://doi.org/10.1007/s10586-019-030 26-9

Predicting Insomnia Using Multilayer Stacked Ensemble Model

Md. Sabab Zulfiker[1]($\boxtimes$) (iD), Nasrin Kabir[2], Al Amin Biswas[1],
and Partha Chakraborty[3] (iD)

[1] Department of Computer Science and Engineering, Daffodil International University,
Dhaka, Bangladesh
[2] Department of Computer Science and Engineering, Jahangirnagar University,
Dhaka, Bangladesh
[3] Department of Computer Science and Engineering, Comilla University, Cumilla, Bangladesh
partha.chak@cou.ac.bd

Abstract. Different forms of sleep disorders have become major health problems among people around the world, and insomnia is one of them. It is a physical condition in which a person faces difficulties to fall asleep at night. It leads to various mental disorders, like anxiety and depression. One of the vital causes of substance abuse is insomnia. This study has proposed a machine learning approach to predict insomnia using different socio-demographic factors of the participants. A multilayer stacking model has been employed in this study to predict the appearance of insomnia in a person. For feature reduction, Principal Component Analysis (PCA) has been used. Our proposed ensemble model has attained an accuracy of 88.60%. The effectiveness of our proposed model has been compared to that of other state-of-the-art ensemble classifiers, like AdaBoost, Gradient Boost, Bagging, and Weighted Voting classifier. The proposed model stated in this study has surpassed the performance of the other ensemble classifiers in terms of different efficacy metrics, like sensitivity, precision, specificity, area under the curve (AUC), accuracy, and F1-score.

Keywords: Insomnia · Prediction · Machine learning · Ensemble classifier · SMOTE · PCA

1 Introduction

Sleep is an incredibly essential fact for functioning the human body properly. As a result, any kind of sleep disorder can result in several detrimental health consequences. Globally up to 40% of the general population is affected by sleep disorders while insomnia alone affects approximately 10% [1]. The most ubiquitous type of sleep disturbance is insomnia that makes it difficult to fall or stay asleep throughout the night, and the insomniac person wakes up earlier than desired in the morning [2]. Insomnia is considered chronic when it happens three times a week for at least three months or more [3]. Insomnia may lead to several issues such as daytime fatigue, irritability, anxiety, lack of concentration, mood disturbance, etc. It is also responsible for the development of chronic diseases

© Springer Nature Switzerland AG 2021
M. Singh et al. (Eds.): ICACDS 2021, CCIS 1440, pp. 338–350, 2021.
https://doi.org/10.1007/978-3-030-81462-5_31

like diabetes, heart disease, depression, obesity, paranoia, hallucinations, etc. It has been observed that the incidence of insomnia in the elderly population is even higher. About 20%–40% of the elderly population have been reported to have sleep disorders, where 25% of them suffer from insomnia [4]. As insomnia is considered a major public health concern, before getting medical treatment, the most crucial task is to identify whether an individual is clinically insomniac or not.

Nowadays, in the realm of computer and technology, machine learning is a trending topic. Though machine learning is commonly utilized in medical sciences, bioinformatics, biological sciences, and psychometric analysis, its use in identifying sleep disorders is relatively new.

Heretofore, a few research studies have been conducted using machine learning techniques to identify and predict insomnia. This work tries to fill this research gap. This study aims to determine whether a person is insomniac or not, and it has proposed a noble machine learning approach to identify insomnia. For assessing the presence of insomnia, experts use different types of insomnia screening scales. But an individual may feel some difficulties answering all the questions of these scales due to the complexity of these questions. But in this study, we have predicted the appearance of insomnia based on the socio-demographic information of the individuals. And an individual can easily provide this required socio-demographic information without facing any difficulties.

The key contributions of this research are:

I.   Generating a dataset for predicting insomnia that includes various socio-demographic information of a person.
II. Proposing a noble multilayer stacking classifier model to identify insomnia.
III. Analyzing the performance of different ensemble classifiers to determine the best machine learning technique to identify insomnia.

The remainder of the paper is arranged in the following structure: Sect. 2 discusses the related works. The methodology is thoroughly explained in Sect. 3. Section 4 presents the findings and results of the study, and Sect. 5 concludes the paper by outlining some plausible scopes for future study.

2   Related Works

Enormous studies have been performed to find out the association between insomnia and other diseases, but limited researches have been conducted using machine learning approaches in this aspect of research problems. In order to find out the research gap among the existing works, this section has presented a short review of the literature regarding similar research articles.

Angelova et al. [5] introduced a new machine learning approach for automatically detecting acute insomnia using nocturnal actigraphy time series data. They extracted statistical and dynamical features from the multi-night actigraphy data using traditional statistical operators, entropy measure, and Poincaré plot. Later, using two supervised algorithms, namely Random Forest (RF) and Support Vector Machine (SVM), those features were combined to differentiate people with acute insomnia from normal sleep controls. They found that RF outperformed SVM with an accuracy of 84.00%, where SVM achieved 73.00% accuracy.

A short-term insomnia detection system was suggested by Kuo and Chen [6], based on a single-channel sleep Electrooculography (EOG) with a Refined Composite Multiscale Entropy (RCMSE) analysis. For evaluating the performance of the proposed system, Polysomnography (PSG) data was taken from 16 insomniac patients and 16 healthy individuals and was fed to the Support Vector Machine (SVM) classifier, which bought out the average accuracy of 89.31%, specificity of 82.00%, sensitivity of 96.63%, F1-score of 90.04%, and kappa coefficient of 0.79.

A content analysis was presented by Jamison-Powel et al. [7] using a mixed methodology where a collection of 18901 tweets were analyzed to discover how people discuss sleep disorders and insomnia on the microblogging services. They revealed that tweets or messages having "insomnia" words contained considerably more negative health information, which implied that people were divulging their sleep disorders.

A comparative study of 15 machine learning algorithms, considering 14 main factors for predicting insomnia, was conducted by Ahuja et al. [8]. Among them, SVM outperformed all other algorithms with accuracy of 91.60%, kappa of 0.83 and F-measure of 92.13%. They further applied SVM on an additional dataset consisting of 100 patients and achieved 92.00% accuracy. According to this study, primarily insomnia depended on the following factors: sleep disorder, vision problem, and mobility problem.

Seth et al. [9] presented a noble probabilistic approach focused on anxiety, depression, stress, and social adjustment of individuals to predict the probability of insomnia. They used the neural network model and returned the probability of having insomnia in an individual, depending on their survey results, and the existing positive predictive values of the Athens Insomnia Scale (AIS) and Insomnia Severity Index (ISI).

An intelligent model utilizing seven distinct machine learning approaches to predict chronic insomnia was proposed by Islam et al. [10], and they found that the logistic regression model, with an accuracy of 98.00%, outperformed all other classifiers.

Mencar et al. [11] used questionnaire data and demographic information for predicting the severity of the Obstructive Sleep Apnea Syndrome (OSAS). PCA was applied to the original dataset of 313 patients to select features, and to assess the prediction ability of OSAS severity, seven classification models and five regression models were trained. They showed that SVM and RF models were best suited for classification, and in predicting apnea-hypopnea index, SVM and linear regression performed better.

Phan et al. [12] conducted their study in order to predict sleep disorders in asthma patients. They contrasted the efficacy of different standard machine learning models such as K-Nearest Neighbor (KNN), Random Forest (RF), and Support Vector Machine (SVM) with various deep learning models. Finally, they found that the Convolutional Neural Network (CNN) model had outperformed all other models in predicting asthmatic patients' sleep disorders.

Sathyanarayana et al. [13] compared the performance of traditional logistic regression with different deep learning models to predict the sleep quality of the teenagers. Wearable actigraphy sensors were used to capture the physical activity data of the individuals while they were awake. Based on this data, they predicted the quality of the sleep of the individuals. According to this study, CNN model showed the best performance in terms of various performance metrics.

Based on the above discussions, it can be inferred that only a few studies have been performed to date to assess the prevalence of insomnia using various socio-demographic factors. This study attempts to overcome this limitation by exploring the influence of different socio-demographic factors on insomnia.

3 Methodology

The methodology section has been divided into three subsections, namely: data description, stacking classifiers for classification and implementation procedure. The subsections are described below.

3.1 Data Description

An online survey has been conducted to gather data from citizens of various ages and occupations in Bangladesh. The dataset contains data of 453 participants. The survey was performed over the period from September 2020 to December 2020. The generated dataset contains different socio-demographic information of the respondents.

Here, 16.56% of participants are insomniac, and 83.44% of participants are not insomniac. Mean age of the participants is 23.09 years. 71.74% of participants of the gathered dataset are male. 28.04% and 0.22% of participants are female and transgender, respectively. The majority of the participants are from urban areas. The obtained dataset consists of twenty-four independent variables and one dependent variable. Table 1 describes the feature variables of the dataset including their types and potential values.

Table 1. Feature Variables for Predicting Insomnia.

Feature name	Feature type	Feature explanation	Potential values
AGE	Independent	Respondent's age	4–80 (in years)
GENDER	Independent	Respondent's gender	Male, Female, Other
PROFESS	Independent	Respondent's profession	Student, Housewife, Service holder (Government), Service holder (Private), Unemployed, Other
MARITALSTS	Independent	Respondent's marital status	Married, Unmarried, Widow/ Widower, Divorced
SMOKE	Independent	Whether the respondent has smoking habit or not	Yes, No

(continued)

Table 1. (*continued*)

Feature name	Feature type	Feature explanation	Potential values
DRINKALC	Independent	Whether the respondent consumes alcohol/ wine or not	Yes, No
DRINKTC	Independent	Whether the respondent drinks tea/ coffee more than two times a day	Yes, No
NAP	Independent	Whether the respondent takes nap in the noon/ afternoon	Everyday, Sometimes
PHYEX	Independent	Whether or not the respondent takes physical exercise/ perform yoga/ perform meditation	Yes, No
OUTGAME	Independent	Whether or not the respondent play any outdoor game	Yes, No
INTERNETBROWSING	Independent	Amount of time (in hours) the respondent spends by browsing social network or performing video streaming in a day	0–24
VIDEOGAME	Independent	Amount of time (in hours) the respondent spends in a day by playing video games	0–24
SCREENTIME	Independent	Amount of time (in hours) the respondent spends in front of the Computer, Smartphone in a day (for study/ job/ recreation/ other purposes)	0–24
ILLNESS	Independent	Whether or not the respondent is affected with a life-threatening disease	Yes, No
FAMILLNESS	Independent	Whether or not any family member of the respondent is affected with a life-threatening disease	Yes, No

(*continued*)

Table 1. (*continued*)

Feature name	Feature type	Feature explanation	Potential values
PRESMED	Independent	Whether or not the respondent is taking any medications that have been recommended/ prescribed by a doctor	Yes, No
EATDISORDER	Independent	Whether or not the respondent has any eating disorders like loss of appetite/ overeating	Yes, No
FINSTRESS	Independent	Whether or not the respondent has any financial stress	Yes, No
DEBT	Independent	Whether or not the respondent has any debt	Yes, No
ANXIETY	Independent	Whether or not the respondent feels anxiety for something recently	Yes, No
POSSAT	Independent	Whether or not the respondent is satisfied with his academic achievements/ current status/ current position	Yes, No
DEPRESSION	Independent	Whether or not the respondent recently feels depressed	Yes, No
WORKPRESS	Independent	Recent study/ work pressure of the respondent	No Pressure, Mild, Moderate, Severe
LOST	Independent	Whether or not the respondent has lost any close person recently	Yes, No
INSOMNIA	Dependent	It represents whether the respondent is insomniac or not	0 (Not Insomniac), 1 (Insomniac)

3.2 Stacking Classifiers for Classification

This ensemble method is also known as Stacked Generalization. In the case of the hard voting approach, the final prediction is derived by aggregating the predictions of base classifiers. But in stacking, a new classifier is trained for performing this aggregation. The final classifier is known as meta-learner or blender.

Here, the training dataset is split into two subsets. The base learners of the first layer are trained by using the first subset. Then the base learners are used to make predictions for the second subset. A new training set is created by taking the predictions of the base learners as predictor variables keeping the target values [14]. Later, the new training set is used to train the blender or the meta-learner. Figure 1 shows the basic structure of a single layer stacking model with three base learners.

We can extend a single layer stacking model to a multilayer stacking model. In a two layer stacking model, firstly we fit the base learners of the first layer. Then in the second layer, instead of fitting a single meta-learner with predictions of the first layer, $n-$ meta-learners are fitted. The predictions of meta-learners of the second layer are then used to fit the final meta-learner. Figure 2 shows a two layer stacking model with three base learners in the first layer and two meta-learners in the second layer.

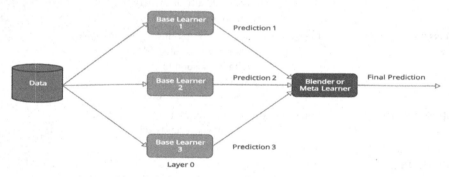

Fig. 1. A stacking model with a single layer.

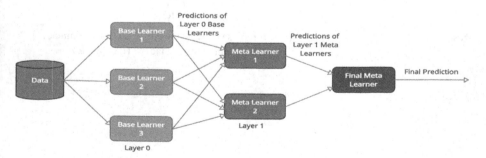

Fig. 2. A multilayer stacking model

3.3 Implementation Procedure

This study has been implemented using the Scikit-Learn library of python. The whole implementation procedure is demonstrated in Fig. 3. The next subsections explain the steps of the implementation procedure for predicting insomnia.

Data Splitting. We have collected the data of 453 individuals. Data splitting step has divided the dataset into training and testing datasets. 75% data of the dataset has been kept in the training dataset for training the classifier. The testing dataset has been constructed by taking the remaining 25% data of the dataset.

Data Encoding. In this step, the categorical data of the training and testing datasets have been converted into their numerical counterpart, as most of the classifiers show superior performance while playing with numerical data.

Data Scaling. The range of the values of different features of the dataset is different. So, we have performed data scaling to scale the value of different features in the same range.

This study has employed min-max scaling to scale the values of the dataset in a range from 0 to 1. The min-max scaling can be defined by the following equation:

$$v' = \frac{v - v_{min}}{v_{max} - v_{min}} \tag{1}$$

Here, v is the value before scaling. v_{min}, and v_{max} are the minimum and maximum values of that feature, respectively, and v' is the scaled value.

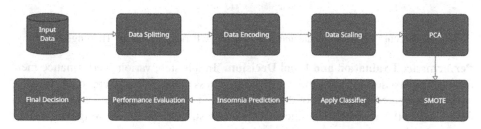

Fig. 3. Implementation Procedure for Predicting Insomnia.

Feature Reduction Using PCA. After scaling the dataset, we have performed PCA to reduce the dataset's dimensionality. PCA produces new features that are linear variations of the initial ones. In this study we have selected four principal components using PCA.

Applying SMOTE. Synthetic Minority Oversampling Technique (SMOTE) has been used to eliminate the class imbalance problem. The percentages of insomniac and not insomniac participants in the training dataset are 16.52% and 83.48%, respectively. As the training dataset is highly imbalanced, SMOTE has been performed on the training dataset for tackling the class imbalance problem. The percentages of insomniac and not insomniac participants in the training dataset, before and after performing SMOTE is shown in Fig. 4.

Applying Classifier for Predicting Insomnia. In this study, we have built a two layer stacking model classifier using Logistic Regression (LR) and Multilayer Perceptron (MLP). Table 2 shows the specification of the base classifiers used in this study for building the two layer stacking model.

As the base learners in the first layer, an LR1 and an MLP classifier have been used. And in the second layer, an LR2 and an MLP have been used as meta-learners. Finally, an LR2 has been used as the final blender. Figure 5 shows the graphical representation of the multilayer stacking model used in this study.

In this step, we have trained the stacked classifier with the training dataset. Then we have predicted the existence of insomnia in the participants for the testing dataset using the proposed multilayer stacked model.

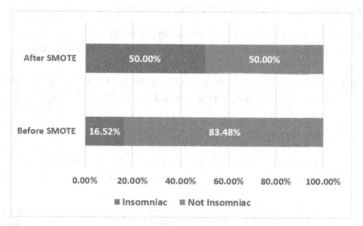

Fig. 4. Percentages of Insomniac and Not Insomniac Participants in the Training Dataset.

Performance Evaluation and Final Decision. In this step, various performance metrics have been assessed to determine the effectiveness of the proposed approach. Finally, we have compared the efficacy of the proposed multilayer stacking model with the performance of other ensemble classifiers like AdaBoost, Gradient Boost, Weighted Voting,

Table 2. Specification of the Base Classifiers.

Classifiers	Specifications
Logistic Regression 1 (LR1)	C: 1.0, class_weight: None, solver: 'lbfgs', intercept_scaling: 1, fit_intercept: True, dual: False, l1_ratio: None, max_iter: 2, multi_class: 'auto', n_jobs: None, random_state: None, penalty: 'l2', tol: 0.0001, warm_start: False, verbose: 0
Logistic Regression 2 (LR2)	C: 1.0, class_weight: None, solver: 'saga', intercept_scaling: 1, fit_intercept: True, dual: False, l1_ratio: 1, max_iter: 1, multi_class: 'auto', n_jobs: None, random_state: 0, penalty: 'l2', tol: 0.0001, warm_start: False, verbose: 0

(continued)

Table 2. (*continued*)

Classifiers	Specifications
Multilayer Perceptron (MLP)	activation: 'relu', batch_size: 'auto', alpha: 0.0001, beta_1: 0.9, beta_2: 0.999, epsilon: 1e-08, hidden_layer_sizes: (100,), early_stopping: False, learning_rate_init: 0.001, learning_rate: 'constant', max_fun: 15000, max_iter: 8, n_iter_no_change: 10, momentum: 0.9, power_t: 0.5, nesterovs_momentum: True, random_state: 0, solver: 'adam', shuffle: True, validation_fraction: 0.1, tol: 0.0001, warm_start: False, verbose: False

and Bagging classifier. Based on the performance metrics of these classifiers, the best model for predicting insomnia have been selected.

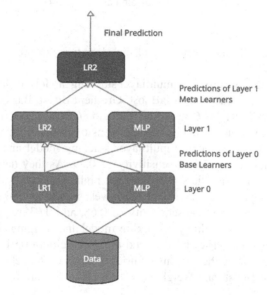

Fig. 5. Multilayer stacking model used in this study.

4   Results

For evaluating the proposed model's efficiency, we have used the data of 114 participants as testing data. Among the participants of the testing data, 16.67% suffer from insomnia, and the rest 83.33% of participants do not have insomnia. We have also compared the efficacy of our proposed model with other ensemble classifiers. The confusion matrices of the classifiers for predicting the existence of insomnia in the participants is delineated

Table 3. Confusion matrices of the classifiers to predict Insomnia.

Classifiers name	Predicted → actual ↓	Not insomniac	Insomniac
Multilayer stacking model	Not Insomniac	86	9
	Insomniac	4	15
AdaBoost	Not Insomniac	61	34
	Insomniac	6	13
Gradient boost	Not Insomniac	75	20
	Insomniac	7	12
Bagging	Not Insomniac	86	9
	Insomniac	12	7
Weighted voting (With LR1, LR2, MLP)	Not Insomniac	84	11
	Insomniac	4	15

in Table 3. Various performance metrics of the classifiers for predicting insomnia are shown in Table 4.

Table 4 exhibits that our proposed multilayer stacking model has shown the best accuracy of 88.60%. The accuracies of AdaBoost, Gradient Boost, Bagging, and Weighted Voting classifiers are 64.91%, 76.32%, 81.58%, and 86.84%, respectively.

For evaluating the efficacy of the models, sensitivity and specificity often play a significant role. Here, our proposed multilayer stacking model and Weighted Voting classifier have achieved the highest sensitivity of 0.79. As they have shown the best sensitivity than the other classifiers, they have the highest ability to identify the insomniac persons. Bagging classifier has shown the lowest sensitivity of 0.37. AdaBoost and Gradient Boost have achieved the sensitivity of 0.68, and 0.63, respectively. In terms of specificity, the proposed multilayer stacking model and Bagging classifier have outperformed other classifiers. They have gained the specificity of 0.91. So, they have the highest ability to recognize the persons without insomnia. The achieved specificity of AdaBoost, Gradient Boost, and Weighted Voting classifiers are 0.64, 0.79, and 0.88, respectively.

Table 4. Performance of the Classifiers for Predicting Insomnia.

Classifiers	Accuracy (%)	Sensitivity	Specificity	Precision	F1- score	AUC
Multilayer stacking model	88.60%	0.79	0.91	0.63	0.70	0.87
AdaBoost	64.91%	0.68	0.64	0.28	0.39	0.73

(continued)

Table 4. (*continued*)

Classifiers	Accuracy (%)	Sensitivity	Specificity	Precision	F1- score	AUC
Gradient boost	76.32%	0.63	0.79	0.38	0.47	0.76
Bagging	81.58%	0.37	0.91	0.44	0.40	0.63
Weighted voting (With LR1, LR2, MLP)	86.84%	0.79	0.88	0.58	0.67	0.86

The Receiver Operator Characteristic (ROC) curve represents the trade-off between the sensitivity and the specificity of a classifier. The classifier that gives the ROC curve closer to the top left corner of the graph is considered as a better classifier. Figure 6 represents the ROC curves of the classifiers. It depicts that the ROC curve of our proposed multilayer stacking model is closer to the upper left corner of the graph than the other classifiers.

Our proposed model has outperformed other ensemble classifiers in terms of area under the curve (AUC), F1-score, and precision also.

Based on the above discussions, it is apparent that the suggested multilayer stacking model has beaten other ensemble classifiers in terms of all the performance metrics.

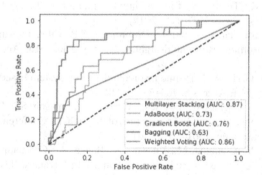

Fig. 6. ROC Curves of the Classifiers for Predicting Insomnia.

5 Conclusion

This study has employed different socio-demographic information of the participants to predict insomnia. For conducting this research, initially, we have generated a dataset that contains various socio-demographic information of 453 individuals. The dataset was then split into training and test datasets. These datasets were then subjected to various data preparation techniques such as data encoding, data scaling, and feature reduction. After performing data preprocessing, SMOTE was applied to the training data to handle the class imbalance problem. A multilayer stacking model has been used in this study which has efficiently identified the presence of insomnia in a person. Our proposed ensemble classifier has shown superior performance than other ensemble classifiers like AdaBoost, Gradient Boost, Bagging, and Weighted Voting classifiers on the basis of various performance metrics. The accuracy of our proposed model is 88.60%.

Various biological markers are also responsible for forming insomnia in a person. In this study, we have only considered the socio-demographic factors of the participants to predict insomnia. In the latter study, we can include different biological factors of a participant also. This study hasn't shown the relationship between different socio-demographic factors causing insomnia and hasn't extracted the most influential factors that form insomnia. In future, we can handle these issues using different feature selection methods.

References

1. Mulaffer, L., Shahin, M., Glos, M., Penzel, T., Ahmed, B.: Comparing two insomnia detection models of clinical diagnosis techniques. In: 2017 39th Annual International Conference of the IEEE Engineering in Medicine and Biology Society (EMBC), pp. 3749–3752 (2017)
2. Levenson, J.C., Kay, D.B., Buysse, D.J.: The pathophysiology of insomnia. Chest **147**(4), 1179–1192 (2015)
3. Erten-Uyumaz, B., et al.: Design and evaluation of a negotiation-based sleep scheduler app for insomnia treatment. In: Proceedings of the 13th EAI International Conference on Pervasive Computing Technologies for Healthcare, pp. 225–233 (2019)
4. Singh, A., Zhao, K.: Treatment of insomnia with traditional Chinese herbal medicine. Int. Rev. Neurobiol. Elsevier **135**, 97–115 (2017)
5. Angelova, M., Karmakar, C., Zhu, Y., Drummond, S.P., Ellis, J.: Automated method for detecting acute insomnia using multi-night actigraphy data. IEEE Access **8**, 74413–74422 (2020)
6. Kuo, C.-E., Chen, G.-T.: A short-time insomnia detection system based on sleep EOG with RCMSE analysis. IEEE Access **8**, 763–773 (2020)
7. Jamison-Powell, S., Linehan, C., Daley, L., Garbett, A., Lawson, S.: "I can't get no sleep": discussing# insomnia on twitter. In Proceedings of the SIGCHI Conference on Human Factors in Computing Systems, pp. 1501–1510 (2012)
8. Ahuja, R., Vivek, V., Chandna, M., Virmani, S., Banga, A.: Comparative study of various machine learning algorithms for prediction of insomnia. In Advanced classification techniques for healthcare analysis. IGI Global, pp. 234–257 (2019)
9. Seth, A., Babu, B.S., Iyenger, S.: Machine learning model for predicting insomnia levels in indian college students. In: 2019 4th International Conference on Computational Systems and Information Technology for Sustainable Solution (CSITSS), vol. 4. IEEE, pp. 1–6 (2019)
10. Islam, M.M., Masum, A.K.M, Abujar, S., Hossain, S.A.: Prediction of chronic insomnia using machine learning techniques. In: 2020 11th International Conference on Computing, Communication and Networking Technologies (ICCCNT), IEEE, pp. 1–7 (2020)
11. Mencar, C., et al.: Application of machine learning to predict obstructive sleep apnea syndrome severity. Health Informatics J. **26**(1), 298–317 (2020)
12. Phan, D.-V., Yang, N.-P., Kuo, C.-Y., Chan, C.-L.: Deep learning approaches for sleep disorder prediction in an asthma cohort. J. Asthma, 1–9 (2020)
13. Sathyanarayana, A., et al.: Sleep quality prediction from wearable data using deep learning. JMIR mHealth and uHealth **4**(4), e125 (2016)
14. Géron, A.: Hands-on machine learning with Scikit-Learn, Keras, and TensorFlow: Concepts, Tools, and Techniques to Build Intelligent Systems. O'Reilly Media, Newton (2019)

A Novel Encryption Scheme Based on Fully Homomorphic Encryption and RR-AES Along with Privacy Preservation for Vehicular Networks

Righa Tandon[(✉)] and P. K. Gupta

Department of Computer Science and Engineering,
Jaypee University of Information Technology, Solan 173234, HP, India

Abstract. With the advancements in vehicular networks the privacy protection of vehicles forms a crucial part. In this paper, we have proposed a privacy-preserving scheme that amalgamates Fully Homomorphic Encryption (FHE) with Reduced Round Advanced Encryption standard (RR-AES) to ensure that the private information of the vehicle is preserved and protected in a vehicular network. This proposed scheme is capable of handling several attacks such as denial of service (DoS), replay, non-repudiation, authentication and man-in-the-middle attack. The results show that the computational overhead is minimized by 5.76%, and 25% by the proposed scheme when compared with the existing schemes. Further, the operational cost of encryption and decryption has also been reduced with the proposed scheme.

Keywords: Privacy preservation · Homomorphic encryption · Vehicular network · RR-AES

1 Introduction

With the recent advancements in technology, the field of vehicular networks has been growing at a great pace. The main focus of vehicular network is on the vehicles and their communication. So, the security and privacy of vehicles that forms a network has become an important aspect. In [2], a secure framework has been proposed for preserving the privacy of the vehicle. This provides a mechanism for assigning and changing the pseudonyms of vehicles so as to preserve the privacy. In [26], a mix-group privacy scheme has been proposed for handling location privacy issue of vehicles. This scheme can prevent any adversary attack on vehicles. Results show that this scheme is secure and efficient when compared with other schemes. In [14], for considering privacy issues, a Pseudonymous authentication scheme is proposed. Using this scheme, privacy of vehicles is protected from the intruders. The general overview of a vehicular network is shown in Fig. 1.

In this paper, we have proposed a scheme in order to strengthen the privacy of vehicles along with the security. In the vehicular network, for preserving the

M. Singh et al. (Eds.): ICACDS 2021, CCIS 1440, pp. 351–360, 2021.
https://doi.org/10.1007/978-3-030-81462-5_32

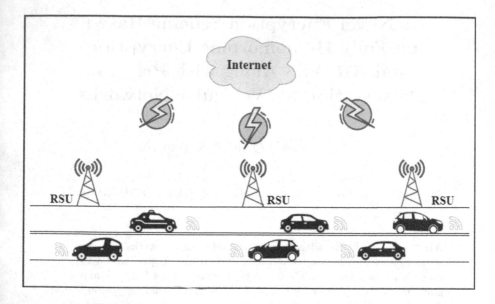

Fig. 1. Vehicular network

privacy of the vehicles, we have deployed pseudonyms. An encryption scheme has also been proposed for securing the information of the vehicles that are a part of the vehicular network. This proposed scheme is based on FHE and RR-AES. This research work contributes the following:

- Proposes a system model for vehicles in the network.
- Provides a privacy preservation scheme based on use of pseudonyms.
- Provides an encryption scheme based on FHE and RR-AES

Further, Sect. 2 includes the existing work related to the proposed scheme. Section 3 discusses the system model for the proposed vehicular network. Section 4 describes the overall methodology of the proposed work. In Sect. 5 results are represented and a detailed comparison of obtained results have been provided with other existing schemes. Finally, Sect. 6 concludes this work.

2   Related Work

This section categorizes the work that has been categorized into following subsections.

2.1   Privacy Preservation Schemes

In [9], the conditional privacy preservation scheme is proposed for vehicular ad-hoc network (VANET). This scheme has a better performance as it avoids the use of bilinear pairing. Further, it reduces the communication and computational costs. In [19], pseudonymous authentication scheme is proposed for preserving the privacy of vehicles. This scheme prevents any intrusion attack on the vehicles. Also, this scheme helps in minimizing the certificate and revocation cost. In [12], a protocol is proposed for preserving the privacy during vehicular communication. This protocol makes use of group and identity-based signature schemes. In [8], a security framework is proposed for vehicular communication. In this framework, a group signature technique is used for privacy preservation. This framework helps in filtering out the messages coming from an affected node. In [5], have designed a mechanism that overcomes the security and privacy issues in VANETs. The designed mechanism helps in improving the privacy and also reduces the security overhead. In [16], a dynamic protocol is proposed for privacy preservation that uses a dynamic topology. Results show that this protocol is efficient, secure, and privacy-preserving. In [3], an algorithm is proposed for privacy preservation of vehicles in a network. It assigns pseudonyms to vehicles using a cryptography technique that helps in maintaining their privacy on the network. In [13], privacy and security of message dissemination problem has been addressed and solution to this problem has been given. In [7], an emergent intelligent privacy preservation technique has been proposed for a vehicular network. This helps in preserving the privacy of vehicles in the network. This technique also handles various attacks that tend to modify and access the information of the vehicles. In [6], cooperative localization privacy preservation of the vehicles have been presented. The performance has been analysed on the basis of privacy preservation strength.

2.2   Security and Authentication Based Schemes

In [21], a secure framework is proposed for vehicular communication. This security framework is based on LIAU and LSMB schemes. These schemes helps in resisting various integrity attacks during vehicular communication. In [11], an authentication framework is proposed that uses public key cryptography. This helps in achieving higher security during vehicular communication. Framework implements the pseudonyms that helps in ensuring the non-repudiation of vehicles. In [25], a LIAU scheme is used for secure communication in vehicular network. This scheme results in handling several attacks that can affect the network. It also reduces the time required for communication and its execution. In [20], a certificate-less authentication scheme is proposed for vehicular network. This helps in preventing DoS attack when large number of vehicles are authenticated. Performance analysis represent that this scheme is efficient when compared with the existing schemes. In [15], full homomorphic encryption scheme is used for enhancing security and reducing overhead. This scheme also ensures privacy by making use of pseudonyms for different vehicles. In [1], have proposed a fully homomorphic encryption (FHE) system that is based on advanced encryption standard. It is capable of solving the large cipher text problem that causes an increase in the amount of noise when FHE is used. This reduces the overhead of the system resulting in a better performance. [24] have addressed the problem of revocation of the vehicles from the network by the use of tamper proof devices. A signature verification algorithms has also been proposed that makes the process of authentication and verification of vehicles more efficient. [23] have addressed the security issues in vehicular networks. A secure and efficient message authentication protocol has been proposed that overcomes the security issues and makes the network more efficient.

3   System Model

In this section, a system model has been proposed for a vehicular network. The main entities of this model are vehicles, roadside units (RSU), and registration authority which is shown in Fig. 2.

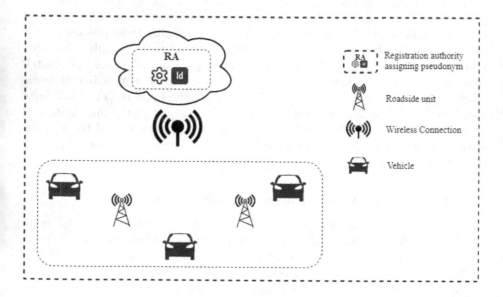

Fig. 2. System model

- *Registration authority (RA):* RA is the apex entity of the vehicular network. The RA is responsible for registering and assigning pseudonyms to the vehicles in the network. Thus, RA plays an integral part in privacy preservation of the vehicles.
- *Roadside unit (RSU):* RSU acts as an inter-mediator between RA and vehicles. RSU verifies the pseudonyms of the registered vehicles that are present in the network.
- *Vehicles:* Vehicles form the most integral part of the vehicular network. Only registered vehicles can be the part of the vehicular network.

In the proposed system model, the vehicles that are a part of the vehicular network are registered and assigned with a pseudonym by the RA. Further the vehicles are authenticated by the RSU to ensure that no malicious vehicle is allowed to enter the network. Further, a hybrid encryption scheme using FHE with RR-AES has also been proposed to encrypt the information of the vehicles. This prevents any intruder attacks on the vehicle's information.

The various notations used in this paper are provided in Table 1

Table 1. Primary notations used

Notation	Illustration
V_n	n^{th} Vehicle
RA	Registration Authority
RSU	Roadside unit
P_k	Key generated by RR-AES
R_{id}	Real identity of the vehicle
P_{id}	Assigned pseudonym
V_{inf}	Vehicle information
$Enc(V_{inf})$	Encrypted information
fun	Function fed to FHE
$Enc(V_{inf})'$	Translated information
$f(V_{inf})$	Output of FHE

4 Methodology

In this section, detailed process of vehicle registration and pseudonym assignment has been discussed and description about the proposed encryption scheme has been provided. Detailed methodology is presented in the following subsections:

4.1 Registration and Assigning Pseudonyms to the Vehicles

If a vehicle wants to enter in vehicular network, firstly, it registers itself. The registration process is carried out by the RA. RA registers the particular vehicle onto the network by using it's real identity. Secondly, It authenticates and assigns a pseudonym to the vehicle. This ensures the privacy of the vehicle. The overall process is shown in Algorithm 1.

Algorithm 1. Registration and Authentication Algorithm

---
1. Registration of V_n by RA:
   1.1. V_n registers using R_{id}.
   1.2. RA authenticates V_n
   1.3. Assign P_{id} to V_n.
   1.4. Forward the P_{id} to RSU.
2. Verification of V_n by RSU:
   2.1. Compare and verify the P_{id}.
   2.2. IF $RSU(P_{id})! = V_n(P_{id})$ THEN
   2.3. V_n is removed from the network.
   2.4. ELSE V_n is verified.

4.2 FHE with RR-AES

The vehicle information is most vulnerable to intruders at the time of encryption or decryption. A vehicle has to undergo a decryption/encryption cycle whenever it needs to update certain information. Therefore, to reduce these additional operations, we encrypt the data only once and any further changes that has to be performed are carried out using Fully homomorphic encryption scheme. This allows to update the vehicle's information inside the vehicular network without having the need to decrypt it. The use of FHE further helps in preventing any intruder attack on the private information of the vehicle and thus enhances the privacy preservation. Reduced round-Advanced encryption standard is used for generating key (128-bit) in the encryption process that uses 7 rounds for encryption process. Here, the same key is used for both encryption and decryption of the vehicle information. A function is generated and is fed to the FHE evaluator along with the vehicles' encrypted information. As a result, translated

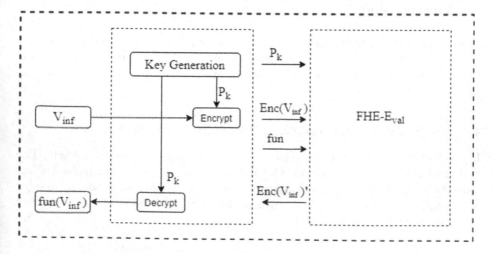

Fig. 3. Encryption using FHE with RR-AES

information is sent back to the vehicle by evaluator. The vehicle's information is then decrypted using the same key. The overall working is shown in Fig. 3. The process of encryption using FHE and RR-AES is shown in Algorithm 2.

Algorithm 2. Encryption Algorithm

Input: V_{inf}
1. Registered V_n enters the network.
2. P_k is generated using RR-AES.
3. V_{inf} is encrypted using P_k:
$$V_{inf} \rightarrow Enc(V_{inf}).$$
4. fun is generated.
5. $FHE - E_{val}$ is fed with fun and $Enc(V_{inf})$.
6. $FHE - E_{val}$ returns translated $Enc(V_{inf})$:
$$Enc(V_{inf}) \rightarrow Enc(V_{inf})'.$$
7. $Enc(V_{inf})'$ is decrypted using P_k.
Output: $fun(V_{inf})$

---

5 Results

In this section, the performance of the proposed scheme is evaluated and compared with other existing schemes. Figure 4 represents the comparison on the basis of time required for authentication of vehicles in the vehicular network. This can be seen that the proposed scheme has reduced the computational overhead by 5.76%, and 25% when compared with RA [10], and AC [17] respectively.

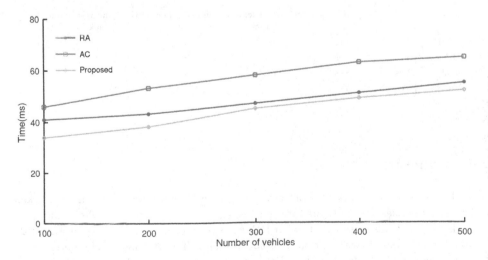

Fig. 4. Computational overhead for vehicle authentication

Figure 5 shows the time taken for encryption/decryption. The proposed scheme has the least operation overhead when compared with S-FHE [18], M-FHE [22], and Z-FHE [4].

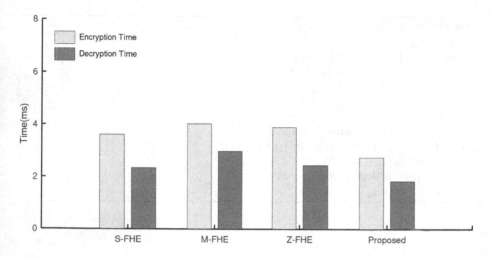

Fig. 5. Computational cost for encryption/decryption

Furthermore, the proposed scheme is capable of resisting various attacks and the comparison with the other schemes is shown in Table 2.

Table 2. Comparison of security attacks handled by various schemes

Scheme	DoS attack	Replay attack	Non-repudiation attack	Authentication attack	Man-in-the-middle attack
S-FHE	Yes	Yes	No	No	No
M-FHE	Yes	Yes	No	No	No
Z-FHE	Yes	Yes	Yes	No	Yes
Proposed	Yes	Yes	Yes	Yes	Yes

6   Conclusion

In this paper, we proposed a privacy preserving scheme for vehicular networks. The scheme uses pseudonyms for securing the identity of the vehicle in the network. We have also used FHE with RR-AES to encrypt the vehicle information that further increases the privacy of the vehicles. The proposed scheme is resistant to various attacks like DoS, Replay, Non-repudiation, Authentication, and Man-in-the-middle attack. Results show that the proposed scheme reduces the computational overhead by 5.76% and 25% and has a lower operation overhead when compared with existing schemes. In the future, we will try to improve the performance of the FHE to further enhance the efficiency of the proposed scheme.

References

1. Alkady, Y., Farouk, F., Rizk, R.: Fully homomorphic encryption with AES in cloud computing security. In: Hassanien, A.E., Tolba, M.F., Shaalan, K., Azar, A.T. (eds.) AISI 2018. AISC, vol. 845, pp. 370–382. Springer, Cham (2019). https://doi.org/10.1007/978-3-319-99010-1_34
2. Artail, H., Abbani, N.: A pseudonym management system to achieve anonymity in vehicular ad hoc networks. IEEE Comput. Archit. Lett. **13**(01), 106–119 (2016)
3. Bouchelaghem, S., Omar, M.: Secure and efficient pseudonymization for privacy-preserving vehicular communications in smart cities. Comput. Electr. Eng. **82**, 106557 (2020)
4. Brakerski, Z., Vaikuntanathan, V.: Efficient fully homomorphic encryption from (standard) LWE. SIAM J. Comput. **43**(2), 831–871 (2014)
5. Calandriello, G., Papadimitratos, P., Hubaux, J.P., Lioy, A.: Efficient and robust pseudonymous authentication in VANET. In: Proceedings of the 4th ACM International Workshop on Vehicular Ad Hoc Networks, pp. 19–28 (2007)
6. Chandra Shit, R., Sharma, S., Watters, P., et al.: Privacy-preserving cooperative localization in vehicular edge computing infrastructure. Concurr. Comput. Pract. Exp., e5827 (2020). https://doi.org/10.1002/cpe.5827
7. Chavhan, S., Gupta, D., Chandana, B., Khanna, A., Rodrigues, J.J.: Agent pseudonymous authentication-based conditional privacy preservation: an emergent intelligence technique. IEEE Syst. J. **14**(4), 5233–5244 (2020)
8. Guo, J., Baugh, J.P., Wang, S.: A group signature based secure and privacy-preserving vehicular communication framework. In: 2007 Mobile Networking for Vehicular Environments, pp. 103–108. IEEE (2007)
9. He, D., Zeadally, S., Xu, B., Huang, X.: An efficient identity-based conditional privacy-preserving authentication scheme for vehicular Ad Hoc networks. IEEE Trans. Inf. Forensics Secur. **10**(12), 2681–2691 (2015)
10. Jiang, W., Li, F., Lin, D., Bertino, E.: No one can track you: randomized authentication in vehicular Ad-Hoc networks. In: 2017 IEEE International Conference on Pervasive Computing and Communications (PerCom), pp. 197–206. IEEE (2017)
11. Li, J., Lu, H., Guizani, M.: ACPN: a novel authentication framework with conditional privacy-preservation and non-repudiation for VANETs. IEEE Trans. Parallel Distrib. Syst. **26**(4), 938–948 (2014)
12. Lin, X., Sun, X., Ho, P.H., Shen, X.: GSIS: a secure and privacy-preserving protocol for vehicular communications. IEEE Trans. Veh. Technol. **56**(6), 3442–3456 (2007)
13. Manivannan, D., Moni, S.S., Zeadally, S.: Secure authentication and privacy-preserving techniques in vehicular Ad-Hoc networks (VANETs). Veh. Commun. **25**, 100247 (2020)
14. Park, Y., Sur, C., Rhee, K.H.: Pseudonymous authentication for secure V2I services in cloud-based vehicular networks. J. Ambient. Intell. Hum. Comput. **7**(5), 661–671 (2016)
15. Prema, N.: Efficient secure aggregation in VANETs using fully homomorphic encryption (FHE). Mob. Netw. Appl. **24**(2), 434–442 (2019)

16. Shah, S.A., Gongliang, C., Jianhua, L., Glani, Y.: A dynamic privacy preserving authentication protocol in VANET using social network. In: Lee, R. (ed.) SNPD 2019. SCI, vol. 850, pp. 53–65. Springer, Cham (2020). https://doi.org/10.1007/978-3-030-26428-4_4

17. Singh, A., Fhom, H.C.S.: Restricted usage of anonymous credentials in vehicular Ad Hoc networks for misbehavior detection. Int. J. Inf. Secur. 16(2), 195–211 (2016). https://doi.org/10.1007/s10207-016-0328-y

18. Smart, N.P., Vercauteren, F.: Fully homomorphic encryption with relatively small key and ciphertext sizes. In: Nguyen, P.Q., Pointcheval, D. (eds.) PKC 2010. LNCS, vol. 6056, pp. 420–443. Springer, Heidelberg (2010). https://doi.org/10.1007/978-3-642-13013-7_25

19. Sun, Y., Lu, R., Lin, X., Shen, X., Su, J.: An efficient pseudonymous authentication scheme with strong privacy preservation for vehicular communications. IEEE Trans. Veh. Technol. 59(7), 3589–3603 (2010)

20. Tan, H., Gui, Z., Chung, I.: A secure and efficient certificateless authentication scheme with unsupervised anomaly detection in VANETs. IEEE Access 6, 74260–74276 (2018)

21. Tandon, R., Gupta, P.K.: SV2VCS: a secure vehicle-to-vehicle communication scheme based on lightweight authentication and concurrent data collection trees. J. Ambient Intell. Human. Comput. 12, 9791–9807 (2021). https://doi.org/10.1007/s12652-020-02721-5

22. van Dijk, M., Gentry, C., Halevi, S., Vaikuntanathan, V.: Fully homomorphic encryption over the integers. In: Gilbert, H. (ed.) EUROCRYPT 2010. LNCS, vol. 6110, pp. 24–43. Springer, Heidelberg (2010). https://doi.org/10.1007/978-3-642-13190-5_2

23. Wang, P., Liu, Y.: SEMA: Secure and efficient message authentication protocol for VANETs. IEEE Syst. J. 15(1), 846–855 (2021)

24. Wang, Y., Zhong, H., Xu, Y., Cui, J., Wu, G.: Enhanced security identity-based privacy-preserving authentication scheme supporting revocation for VANETs. IEEE Syst. J. 14(4), 5373–5383 (2020)

25. Xu, H., Zeng, M., Hu, W., Wang, J.: Authentication-based vehicle-to-vehicle secure communication for VANETs. Mob. Inf. Syst. 2019, 9 (2019)

26. Yu, R., Kang, J., Huang, X., Xie, S., Zhang, Y., Gjessing, S.: MixGroup: accumulative pseudonym exchanging for location privacy enhancement in vehicular social networks. IEEE Trans. Dependable Secure Comput. 13(1), 93–105 (2015)

Key-Based Decoding for Coded Modulation Schemes in the Presence of ISI

Vanaja Shivakumar$^{(\boxtimes)}$

Department of Electronics and Communication Engineering, CMR Institute of Technology,
Bengaluru 560037, Karnataka, India
vanaja.s@cmrit.ac.in

Abstract. Evolutions in the field of digital communications and information science has resulted in the global level communications at our finger tips, and, ample applications started using digital technology over the past decades. In support of this, various technological developments have come up with a goal of satisfying the user requirements globally. However, developing a decoding strategy for a digital transmission scheme has become one of the thrust areas for the researchers in the current scenario. With this motivation, this paper presents an innovative decoding strategy for a coded modulation scheme, based on a Key generation on the receiving side, and, the resulting methodology is called Key-Based Decoding Algorithm (KBDA). The motivating factor in the development of KBDA is to improve the performance of a digital transmission scheme in the presence of noise and interference. To aid the Key-code generation functionality on the receiving side, a Key is sent from the transmitter end on timely basis, and the specific time-interval is called state-set interval which is a variable parameter in the KBDA. Innovative processing steps are introduced into the KBDA to aid the data estimation functionality during demodulation and decoding so that the error involved in the detection process is minimized. The KBDA has the flexibility to decide on the state-set interval within the acceptable range of an application. In the present work, the KBDA has been implemented on the Four-state Sixteen-QAM coded modulation scheme through simulation, and improved decoding performances have been recorded by plotting the characteristic graphs.

Keywords: KBDA · Key-based · Key-code · State-transitions · Estimation

1 Introduction

Coded modulation technique, also called Trellis Coded Modulation (TCM) technique is a methodology to encode and modulate the digital data bits before transmission, as a combined approach. The TCM scheme is an invention by G. Ungerboeck [6] for a high gain and bandwidth efficient reliable data transmission over bandlimited ISI channels. Higher coding gain has been achieved over the uncoded modulation schemes without increasing the transmission power. These characteristics of the TCM schemes have attracted many researches to enhance its application in various fields of advanced

© Springer Nature Switzerland AG 2021
M. Singh et al. (Eds.): ICACDS 2021, CCIS 1440, pp. 361–372, 2021.
https://doi.org/10.1007/978-3-030-81462-5_33

technology requirements. Some of the advanced applications are: Satellite communications, Telemedicine, Biomedical, Optical Communications, Surveillance, Under Water Communication, MIMO, and digital data networks [3–5, 9, 12, 13].

Trellis encoder and modulator is designed for: 1. The required signal constellation size of the transmitter; 2. The required number of encoder states; 3. The required pattern of the trellis structure which decides the structure of the decoder. The trellis structure to be considered in the decoder design also decides the encoder states. The trellis encoder is internally connected to the convolutional encoding; accordingly, its logical internal structure is decided and the internal memory size or memory length is decided. The transmission rate and the signal constellation size will decide: 1. The bits-set processed by the encoder at a time; 2. The number of uncoded bits, and, 3. The encoded bits. Matching to this, the bits-set per symbol transmission is fixed. On this basis the serial to parallel converter of the encoder is designed.

The digital data under transmission arrives serially, and this has to be converted to parallel bits before encoding. The Trellis encoder of the scheme is designed for a specific rate of the requirement, which in turn represent the rate of the convolutional encoder. For 'm' bits of transmission per symbol interval, redundancy bits are added to the data during encoding, as a result, the encoder output contains a greater number of bits. Finally, the output of the Trellis encoder is mapped by the mapper into a signal point selected from the signal constellation of size M, which is greater than the size that has been considered for uncoded M-QAM modulation [6].

Developing a decoding strategy for a coded modulation scheme has become one of the thrust areas for researchers in the field of communication engineering. Though many optimal and suboptimal decoding strategies are existing for TCM schemes [6], there is a need for further developments with improved efficiency without increasing the power and demanding for additional bandwidth. Each application demands for a better performance under different constraints.

Our approach in this paper is that we are presenting an innovative Key-Based Decoding Algorithm (KBDA) for coded modulation schemes. The communication channel under consideration is the bandlimited channel wherein the effect of interference due to other symbols on the sequence of symbols transmitted is accounted. The KBDA has been implemented on the Four-state Sixteen-QAM coded modulation scheme through simulation, and the decoding performance has been recorded through characteristic graphs.

2 Literature Survey

The research in [1], a joint process has been considered in order to achieve robust and better performance for medical image transmission. Source coding and coded modulation schemes are considered together. A new hierarchical pattern of modulation has been implemented. Turbo Trellis Coded Modulation in under consideration. The objective is to perform compression of the digital data input from the source lossless and improve the efficiency of transmission suitably. Simulation has been done to send image data. Rayleigh fading model is simulated and the results are analyzed.

In paper [2], the authors worked on Turbo Trellis-Coded Modulation schemes and the performance has been analyzed over AWGN Channel. The information flow during decoding has been tracked and EXIT chart has been generated for the analysis.

The paper [3] is a survey on coding systems from the point of view of 4G and 5G communication systems are considered. It mainly emphasized on LDPC codes and Polar codes.

The paper [4] is about a codec. The objective in this paper is to design and develop an image codec for optimal performance. A trellis coded quantizer (TCQ) has been considered, and image compression framework with a deep learning approach. High quality performance has been demonstrated at low rate of transmission.

A new variant of the TCM has been presented in [5], for transmission of color images. The approach followed is based on combining the Ungerboeck's mapping concept in trellis coded modulation schemes with Gray coding. Simulation results are obtained on Rayleigh fading channel. Better performance has been demonstrated over Ungerboeck's TCM scheme.

In the research carried out in [7], a section of the Viterbi Algorithm (VA) for improved performance has been addressed. The Add-Compare-Select (ACS) of VA is developed using modified pipelined architecture; Optimization is done for the resource required, delay involved and power consumption.

Neural network approaches are used to enable computing machines to learn, recognize and make predictions to solve problems. In [8], a neural network-based decoder for trellis coded modulation (TCM) scheme has been presented. Decoding is performed with Radial Basis Function Networks and Multi-Layer Perceptron. Using neural network approach decoder, an adaptive approach has been introduced in the Viterbi algorithm which is known for TCM decoding. Using neural network, the adaptability has been brought into function effectively and the channel imperfections are made learn effectively. The algorithm has been tested for its performance through simulation over an imperfect channel, by transmitting trellis encoded 16-QAM signals and the decoding results are analyzed for the improved functionality.

An efficient methodology for reversible discrete cosine transforms (RDCT) based embedded image coder has been proposed in the paper [9]. The motivating factor is to design for a lossless Region of Interest coding. very high compression ratio is the main objective. Improved performance has been demonstrated over some of the state-of-art still image compression methods.

The authors in [10] presented about an efficient decoding structure for 4-dimensional 8-PSK Trellis-Coded Modulation scheme. A concept on sharing of a substructure of the decoder to develop a transition-metric architecture which is lower in complexity level has been considered, then, a hybrid structure has been developed to execute Viterbi algorithm which will consider T-Algorithm for branch metric as well as path metric computations. Reduction in computational complexity by about 50% has been demonstrated.

A new structure for trellis coded modulation technique has been presented in [11]. The new approach implemented is the consideration of rotation of the signal constellation locally in the 2dimensional plane. The PSK/QAM constellation has been used for transmission and the mapping technique is done using UGM. Improved performance of the scheme over Rayleigh fading channel has been documented through simulation results.

Considering the water as the communication media, various factors of interest which decides the underwater communication channel performance are: High path loss, Multipath ISI, Non-negligible doppler effect. For effective design of the underwater communication system, available channel models are: Basis Expansion Model, Polynomial Amplitude Variation Model and Radial Basis Function Model. AS in [12], impulse response of various channel models for under water communication is determined through simulation and the performances are analyzed using TCM-OFDM.

In [13], a research related to the antenna structure as applicable to MIMO is presented. The authors followed newer approach to create communication channels in MIMO applications, considering the cases: 1. independent path; 2. simultaneous paths. Also, with innovation, a symmetric antenna structure having four-port is presented. The innovative antenna structure designed fulfill the characteristic requirements such as: very low correlation coefficient and higher gain.

The paper [14] talks about intensity modulation. A novel approach has been implemented for trellis coded modulation schemes as applicable to intensity modulation and direct detection system. The concept of probabilistic shaping (PS) has been implemented. Performance improvement of 1.8dB has been demonstrated through simulation results. Simulation was done for PS-TCM-32QAM.

A new methodology has been implemented in paper [15] for soft-input soft-output decoding algorithm as suitable for turbo coding/decoding system. Specific case of interest is for lower complexity with high radix. Higher throughput has been demonstrated through simulation results over MLM algorithm.

The current paper discusses on an innovative decoding algorithm. We are presenting a Key-Based Decoding Algorithm (KBDA) for coded modulation schemes. The communication channel under consideration is the bandlimited channel wherein the effect of Intersymbol Influence (ISI) on the sequence of symbols transmitted is accounted. Referring to the Soft Output Viterbi Algorithm (SOVA) as a reference decoding algorithm, and reduced state sequence estimation methodology, innovative processing steps are introduced in KBDA based on a Key-code generated on the receiving side.

3 Encoder Structural Analysis

The TCM is a bandwidth efficient and integrated coded-modulation scheme. Its output characteristics are dependent on convolutional coding. Referring to the coded modulator structure in the Fig. 1, 'm' data bits enter the serial-to-parallel converter during the transmission interval t_n and 'R' is the size of the digital data which is encoded with convolutional coding by the encoder unit. The encoder rate is the ratio of the input bits(uncoded) to the output bits(encoded), which is R/K in the Fig. 1. The convolutional encoder output K-bits, and the data input bits which are uncoded and taken into consideration for mapping, (m-R)-bits, are representing a signal point in the signal constellation of the transmission system, and hence $M = 2^{K+m-R}$. The convolutional code along with the uncoded data bits form the trellis code. The coded K-bits will select one particular section of the signal set under consideration for transmission, and, from the selected segment, a signal point is selected by the uncoded (m-R)-bits. The above procedure is called signal mapping by set partitioning. The encoder adds redundant bits during the code generation, resulting in enhancement in the size of the signal constellation. Since the convolutional encoder has memory it has a definite number of states for its operation [6].

The trellis code generated at any discrete time interval 't_n' depends on two factors, namely, the encoder state at the time t_n and the input data bit-pattern fed at the time t_n. Next interval state transition of the convolutional encoder depends on two factors: 1. Its current state; 2. Its current input. Since the convolutional encoder adds redundant bits to the data and generates a sequence of codes which are inter-dependent, the mapped symbols in the sequence of transmission have inter-dependency. This redundant information has been carried by the trellis structure of the encoder, and, it will be used in the development of the decoding algorithms. If the communication channel is bandlimited and has ISI, then the combined code-ISI trellis structure is considered by the decoding algorithms [6]. Hence, as the number of symbols causing ISI increases, the decoder complexity becomes more and more complicated. This is one of the reasons that has attracted researchers strongly to work in this area.

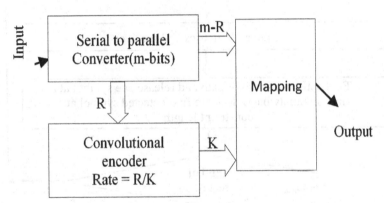

Fig. 1. Coded Modulation Encoder Structure with the following representations: m-Number of serial input bits entering the Trellis encoder; R- number of input bits out of m-bits entering the convolutional encoder; K-Convolutional encoder output bits; (m-R) – Number of input data bits entering the mapper directly; K + (m-R) – Number of bits in the Trellis code; M- Signal constellation size of the scheme, and is equal to $2^{(K+(m-R))}$; n- Number of redundant bits added, and is equal to (K-R)

4   Key-Based Decoding Algorithm (KBDA)

Decoding methodologies are application dependent and many constraints need to be considered for the development. In our approach we have assumed that the communication channel under consideration for transmission is bandlimited, and, affected by ISI additionally there is AWGN effect. It is known that received sequence estimation

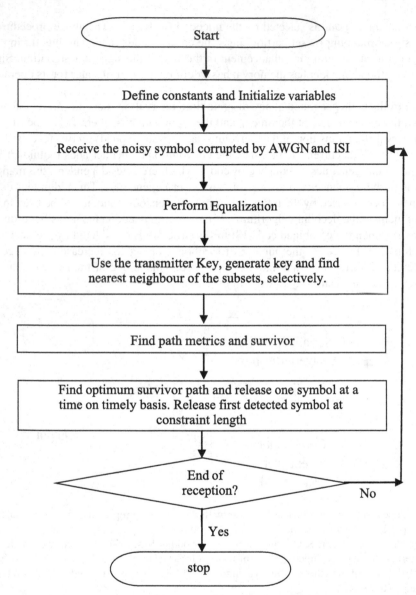

Fig. 2. Flow chart of KBDA

to the Maximum Likelihood extent is the optimal decoding algorithm for coded modulation schemes, and, its implication is unreliable in real time applications because of its complexity [6]. Accordingly, suboptimal decoding algorithms are being developed depending upon the applications and constraints under consideration.

We have developed the decoding algorithm, KBDA, for coded modulation schemes with an innovative approach to decode the received symbol sequences which are corrupted. An improved efficiency of KBDA in the sense of error event probability has been demonstrated. The process execution of KBDA is given in the flow chart of the Fig. 2.

Innovative processing steps are introduced by generating a Key-code on the receiving side for better estimation of the distorted symbol-sequences received, accordingly, the KBDA is executed and state-transitions are evaluated.

To aid the Key-code generation functionality on the receiving side, a Key is sent from the transmitter end on timely basis, and the interval has been identified as state-set interval. Improved error event probability of the system has been observed from the simulation results.

5 Result Analysis

The innovative KBDA has been simulated and implemented on Four-state Sixteen-QAM TCM-scheme transmission. Bandlimited ISI channel has been simulated with 1 symbol interval as well as 2 symbols interval of ISI, and characteristic graphs are plotted. The results are compared with the characteristics obtained from the reduced state sequence estimation of TCM transmission on bandlimited ISI channels [6].

The Fig. 3 is the graph of probability of error event Vs Signal to Noise ratio (SNR) in dB, obtained from the reduced state sequence estimation technique without considering KBDA. The results are obtained for the bandlimited ISI channels whose characteristic coefficients are: h_0, h_1 and h_2. Total number of symbols transmitted is 10^5.

The left most curve in the Fig. 3 is the error event probability characteristic for zero ISI effect. The second curve from the left is for the channel coefficients $h_0 = 0.7$, $h_1 = 0.2$ and $h_2 = 0.1$, and the right most curve is for the channel $h_0 = 0.6$, $h_1 = 0.3$ and h_2

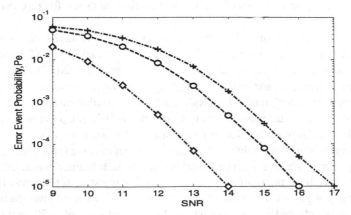

Fig. 3. The graph of probability of error event Vs SNR in dB, obtained from the reduced state sequence estimation methodology under the cases: 1. Zero ISI (Left most curve); 2. Coefficients: $h_0 = 0.7$, $h_1 = 0.2$, $h_2 = 0.1$ (Second curve from the Left); 3. $h_0 = 0.6$, $h_1 = 0.3$, $h_2 = 0.1$ (Right most curve).

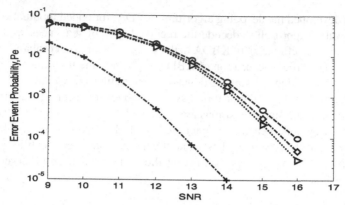

Fig. 4. Is the graph of probability of error event Vs SNR in dB, obtained for different ISI length: 1 (right most curve), 2 (second curve from the right) and 3 (Third curve from the right). Left most curve is for zero ISI.

$= 0.1$. It has been observed that as the amount of ISI decreases, the characteristic curves move towards the characteristic obtained for zero ISI (left most curve).

The Fig. 4 is the graph of probability of error event Vs SNR in dB, obtained for different ISI length, such as: 1, 2 and 3 symbol intervals.

The ISI channel characteristic coefficients are: case 1: h_0–1, h_1–0, h_2–0, h_3–0 (left most curve); case2: h_0–0.6, h_1–0.4, h_2–0, h_3–0 (right most curve); case3: h_0–0.6, h_1–0.3, h_2–0.1, h_3–0 (Second curve from the right); case4: h_0–0.6, h_1–0.2, h_2–0.15 h_3–0.05 (Second Curve from the left), without considering KBDA and only reduced state sequence estimation has been implemented. Total number of symbols transmitted is 10^5.

In Fig. 4, the coded modulation system simulated is for four-state sixteen-QAM scheme. It has been observed that as the amount of ISI effect increases, the characteristic curves move away (towards right) from the characteristic obtained for zero ISI condition (left most curve).

The Fig. 5 is the graph of probability of error event Vs SNR in dB, obtained with the innovative decoding algorithm KBDA. The Four-state sixteen-QAM TCM scheme has been implemented and the results are obtained for the bandlimited ISI channels whose characteristic coefficients are: h_0–0.6, h_1–0.3, h_2–0.1. The transmitter key-transmission interval, called 'state-set' interval considered for the simulation are: 25, 50 and 100 symbol intervals. Total number of symbols transmitted is 10^5, and, previously transmitted two symbols interference effect has been considered for simulation.

As seen in the Fig. 5 the right most characteristic curve is the error event probability characteristic for the state-set interval 0, and is taken as the reference characteristic curve. The second curve from the right is obtained by setting the state-set interval 100, which provides a coding gain of 0.1 dB at an error rate of 10^{-4}. The third characteristic curve from the right is obtained for the case where the state-set interval is 50, and it provides a coding gain of 0.3 dB at the error rate 10^{-4}. The fourth curve from the right is for state-set interval 25, provides a coding gain of 0.5 dB at the error rate 10^{-4}. It has been observed that as state-set interval improves (decreases), the characteristic curves move away (towards left) from the characteristic obtained for zero state-set interval (right most) which is the result obtained for the case without KBDA. As we move from the right most curves towards left, improved error event performance has been observed. The left most curve is for better improved performance with KBDA and a coding gain of 0.5 dB at an error rate of 10^{-4} has been observed.

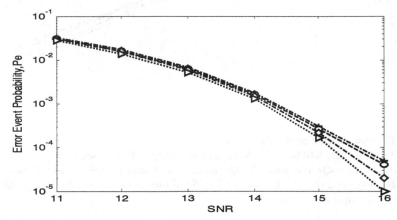

Fig. 5. The graph of probability of error event Vs Signal to Noise Ratio (SNR) in dB, obtained with the innovative decoding algorithm KBDA, for four-state sixteen-QAM TCM scheme; Bandlimited ISI channel coefficients are: h_0–0.6, h_1–0.3, h_2–0.1, and, "state-set' symbol intervals are: 25(for left most curve), 50 (for the second curve from the left), 100(for the second curve from the right) and 0 (for the right most curve).

The Fig. 6 shows the characteristic curves obtained for the ISI channel, with the coefficients: h_0–0.6 and h_1–0.4. The innovative KBDA has been implemented and the TCM scheme under consideration is a four-state sixteen-QAM scheme. Referring to the curves in the Fig. 6, the right most curve is the error event probability characteristic for the state-set interval 0. The second curve from the right is obtained by setting the state-set interval 100, and it provides the coding gain of 0.15 dB. The third curve from the right is obtained for the case where state-set interval is 50, and it provides a coding gain of 0.35 dB at the error rate 10^{-4}. The left most curve is for state-set interval 25, and improved performance with a coding gain of 0.6 dB has been observed at the error rate 10^{-4}.

It has been observed that as state-set interval improves (decreases), the characteristic curves move away (towards left) from the characteristic obtained for zero state-set interval. Observing the characteristic curves from right to left, improved error event performance has been documented. The left most curve is for better improved performance, and a coding gain of 0.6 dB at an error rate of 10^{-4} has been observed.

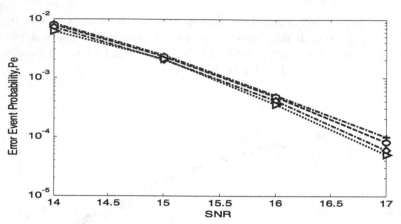

Fig. 6. The characteristic curves obtained for the ISI channel, with the coefficients: h_0–0.6 and h_1–0.4. The innovative KBDA has been implemented and the TCM scheme under consideration is a four-state sixteen-QAM scheme. The ISI channel coefficients: h_0–0.6 and h_1–0.4. and, "state-set' symbol intervals are: 25(for left most curve), 50 (for second from the left), 100(second from right) and 0 (for right most curve).

6   Limitations of the Work

The proposed KBDA is an innovative approach for coded modulation schemes. It has been tested through computer simulation on Four-state Sixteen-QAM TCM-scheme. Improved error performance has been observed with the following constraints:

- Key generation on the receiver side is an overhead on the decoding complexity to a lesser extent, at the same time performance improvement has been obtained. A trade off exists between the computational complexity and the performance gain obtained.
- The KBDA has been developed by providing a flexibility on the Key factor which decides the gain factor, and hence it needs proper tuning.
- The KBDA state-set interval is limited to certain range, beyond which significant performance improvement is possible with proper consideration of the variables.
- The Receiver Key generation is aided by the Key sent from the transmitter end, which requires additional hardware on the transmission side, to a smaller extent.
- The KBDA has the flexibility to implement it on any coded modulation scheme with suitable structural modifications.
- The KBDA is applicable for fading channels under suitable constraints.

7   Conclusions

We have proposed innovative decoding algorithm KBDA for Trellis coded modulation schemes transmission on bandlimited ISI channels. Simulation results are obtained for different channel conditions. Decoding is done using parallel decision feedback decoding as a case of reduced state estimation approach, and the results are used as reference for comparison. Also, decoding is done with the innovative KBDA. The KBDA has shown better decoding performance compared to reduced state estimation technique with little increase in computational complexity.

The KBDA results are obtained for one symbol as well as two symbols ISI effect plus AWGN effect, and the channel coefficients considered are: 1. h_0–0.6, h_1–0.3, h_2–0.1; and 2. h_0–0.6, h_1–0.4. Different state-set intervals are assumed.

The characteristic curves shown in Fig. 5 tells about coding gains obtained from 0.1dB to 0.5dB at an error rate of 10^{-4} as we move from the right most curve to the left most curve.

The KBDA results are obtained for the channel: h_0–0.6, h_1–0.4. Considering different state-set intervals a set of characteristic curves are obtained, and are plotted in Fig. 6. A coding gain of 0.6 dB has been observed for state-set interval 25 compared to zero state-set interval.

Better coding gain has been observed with the characteristics shown in Fig. 6 compared to Fig. 5. The KBDA can be extended to other decoding strategies with different equalization approaches and variable state-set interval. The KBDA can be tested for transmission over fading channels. The Key-code generation may vary from a simple code to any complexity as per the designer choice.

Acknowledgments. The author wishes to thank the reviewer of this paper. Also, the author would like to thank all those supported directly or indirectly in bringing out this paper.

References

1. Aljohani, A.J., Sun, H., Ng, S.X., Hanzo, L.: Joint source and turbo trellis coded hierarchical modulation for context-aware medical image transmission. In: IEEE 15th International Conference on e-Health Networking, Applications and Services (Healthcom 2013), Lisbon, pp. 1–5 (2013). https://doi.org/10.1109/HealthCom.2013.6720627
2. Haffane, A., Khelifi, A.H.M., Kadri, B.: Convergence analysis of the unpunctured turbo trellis-coded modulation (UTTCM). Ind. J. Electrical Eng. Comput. Sci. **13**, 447–452. https://doi.org/10.11591/ijeecs.v13.i2
3. Arora, K., Singh, J., Randhawa, Y.S.: A survey on channel coding techniques for 5G wireless networks. Telecommun. Syst. **73**(4), 637–663 (2019). https://doi.org/10.1007/s11235-019-00630-3
4. Li, B., Akbari, M., Liang, J., Wang, Y.: Deep learning-based image compression with trellis coded quantization, †arXiv:2001.09417v1 [eess. IV] 26 Jan 2020
5. Abdesselam, B., Mohammed, B., Abdelmalik, T.A., Abdelmounaim, M.L.: Application of 16-State TCM-UGM and TCM for improving the quality of compressed color image transmission. Int. J. Image Graph. Signal Process. **6**, 10–17. https://doi.org/10.5815/ijigsp.2014.10.02
6. Schlegel, C.: Trellis Coding. IEEE Press. ISBN 0-7803-1052-7
7. Vaithiyanathan, D., Nargis, J., Seshasayanan, R.: High performance ACS for Viterbi decoder using pipeline T-Algorithm. Alexandria Eng. J. (2015). https://doi.org/10.1016/j.aej.2015.04.007
8. Kaminsky, E.J., Deshpande, N.: TCM decoding using neural networks. Eng. Appl. Artif. Intell. **16**(5–6), 473–489 (2003)

9. Elhannachi, S.A., Benamrane, N., Abdelmalik, T.A.: Adaptive medical image compression based on lossy and lossless embedded zerotree methods. J. Inf. Process. Syst. **13**, 40–56 (2017). https://doi.org/10.3745/JIPS.02.0052

10. He, J., Wang, Z., Liu, H.: An efficient 4-D 8PSK TCM decoder architecture. IEEE Trans. Very Large Scale Integration (VLSI) Syst. **18**(5) (2010)

11. Rekkal, K., Abdesselam, B.: Improving the performance of trellis coded modulation over Rayleigh fading channel using locally rotated constellations. Int. J. Commun. Antenna propagation. https://doi.org/10.15866/irecap.v8i1.13625

12. Krishnamoorthy, N.R., Suriyakala, C.D.: Performance of underwater acoustic channel using modified TCM OFDM coding techniques. Ind. J. Geo Marine Sci. **46**(03), 629–637 (2017)

13. Mohamadi, P., Dadashzadeh, G.R., Naser-Moghadasi, M.: A new symmetric multimodal MIMO antenna with reduction of modal correlation coefficient using TCM. IETE J. Res. **66**(2), 150–159. https://doi.org/10.1080/03772063.2018.1481460

14. Tian, F., et al.: Probabilistic shaped trellis coded modulation with generalized frequency division multiplexing for data center optical networks. Opt Express. **27**(23), 33159–33169 (2019). https://doi.org/10.1364/OE.27.033159. PMID: 31878390

15. Le, V.H.S., Nour, C.A., Boutillon, E., Douillard, C.: Revisiting the Max-Log-Map algorithm with SOVA updates rules: new simplifications for high-radix SISO decoders. IEEE Trans. Commun. Inst. Electrical Electron Eng. **68**(4), 1991–2004 (2020). ff10.1109/TCOMM.2020.2966723ff. ffhal-02332503f

Optimizing the Performance of KNN Classifier for Human Activity Recognition

Ali Al-Taei[1]([⊠]), Mohammed Fadhil Ibrahim[2], and Nada Jasim Habeeb[2]

[1] Ministry of Higher Education and Scientific Research, Baghdad, Iraq
alitaei@tcm.mtu.edu.iq
[2] Middle Technical University (MTU), Baghdad, Iraq
{mfi,nadaj2013}@mtu.edu.iq

Abstract. Recently, computer and smartphone devices have been increasingly employed in various life aspects. Human Activity Recognition (HAR) is one of the most emerging fields that can be clearly noticed in terms of computer related researches. This study proposes a classification model to recognize human six basic activities based on smartphone embedded sensors. For feature selection, we utilized three optimization search algorithms which are: Particle Swarm Optimization (PSO), Greedy, and Genetic algorithm. K-nearest neighbors (KNN) classifier, with Manhattan distance function, is adopted to measure the models performance on two publicly available well-known datasets (i.e., WISDM and UCI-HAR). All experiments are achieved using 10-folds cross validation technique, and the results are evaluated and compared using different metrics, for instance, accuracy, root mean square error, f-measure, and number of features. The experimental results proved that PSO based KNN classifier presents the best performance compared to the other two algorithms, in addition to the baseline KNN classifier. Moreover, the proposed model outperforms those state-of-the-art models presented in related literature regarding the UCI-HAR dataset with performance rate of 98.89%, and close to the best performance achieved on WISDM dataset with performance rate of 91.07%.

Keywords: Human Activity Recognition (HAR) · Machine Learning · Accelerometer · Particle Swarm Optimization (PSO) · Genetic · Greedy · K-nearest neighbors (KNN)

1 Introduction

Recently, smartphone devices have been increasingly employed in various life aspects. Human Activity Recognition (HAR) is one of the most emerging fields that can be clearly noticed in terms of computer related researches [1–3]. HAR aims to identify the human physical activities such as walking, running, standing, sitting and many other actions that the human being can daily perform [4]. Such outstanding field presents several privileges that can enhance peoples' daily life quality within a verity of applications for instance: health, industries, environment, security and so on [3–5].

© Springer Nature Switzerland AG 2021
M. Singh et al. (Eds.): ICACDS 2021, CCIS 1440, pp. 373–385, 2021.
https://doi.org/10.1007/978-3-030-81462-5_34

Activity Daily Living (ADLs) recognition is also another type of human physical recognition which analyze human daily activities [6], it has been implemented prior to HAR. ADLs relies on ambient sensors surveillance cameras in order to register different actions of a particular individual [7, 8]. The main challenge that faces ADLs is the highly costed equipment [9, 10], in addition to the disturbance made by such technique make people less comfortable with it, because people think using such tools is a kind of intrusion of their life [11–13].

On the other side, HAR is considered as more acceptable by the people due to its less intrusive mode, where HAR relies on a type of sensors known as inertial sensors (i.e. accelerometer, and gyroscopes) [14–17].

Inertial sensors are normally activated by a human interference such as Adhoc wearable sensors, or they can be embedded with specific portable devices which is normally known as smartphone-based sensors [18, 19]. In spite of the aforementioned types of sensors present noteworthy privileges in terms of environment control, however, smartphone-based sensors are considered the most preferable sensors in terms of representing the ADLs recognition task, due to their popularity in daily use with various tasks [20, 21]. Additionally, smartphone-based sensors are less costly and easier to use if compared to other sensor types. Furthermore, a recent researches and studies stated that the data samples that provided by smartphone sensors were more reliable and accurate in terms of capturing and identifying human physical activities [22].

According to the previously mentioned characteristics, our research relies on the smartphone-based sensors in order to capture the required data. Several studies prefer conducting their classification experiments by using a data which are personally collected, such data are normally rarely used and difficult to be available online, hence it is difficult to conduct an objective comparison, specifically when a need of applying a new classification model. And to overcome such obstacle, we employ two of the most well-known datasets: 1) WISDM which stands for (Wireless Sensor Data Mining) [23], and 2) UCI-HAR [24]. The both datasets consist of an accelerometer-based samples which have been collected from different individuals, who have volunteered to perform a daily activity such as walking, standing, sitting, jogging, and climbing the stairs. The volunteers were performing the previous activities while holding their smartphones of Android operating systems.

Furthermore, in order to provide more precise results, a recent trend of studies raised toward performance optimization of the human activities by combining a variety of machine learning techniques, and select the best combination which outputs the best performance among the others (i.e. ensemble methods) [2, 25]. In the same context, there is also utilizing of deep learning techniques. In spite of that such techniques present a plausible result; however, they run with a noticeable difficulty related to the network architecture which require for multiple layer, and this increase the computational complications [26].

A few studies tried using optimization algorithms for HAR task such as [27, 28]. Therefore, the purpose of this paper is to enhance the performance of sensor-based HAR using optimization algorithms. In addition, we are presenting an evaluation and comparison of the performance of three optimization search algorithms (i.e., Genetic, Greedy, and PSO) for HAR using KNN algorithm with Manhattan function. Also, suggesting the best performance model and compare with state-of-the-art models.

The rest of this paper is organized as follows: Sect. 2 is dedicated to view the related works, Sect. 3 is devoted to illustrate the research methodology, Sect. 4 views the dataset characterization and preprocessing processes, while Sect. 5 is dedicated to display the experiments setup and classifications, and lastly Sect. 6 discusses the conclusions and future trends for the paper [27].

2 Related Work

As mentioned before, several studies have utilized human activity recognition (HAR) for both of the trends, Android smartphones sensors or other wearable sensors in order to perform different recognition tasks [29, 30]. in this section, we investigate the most recent studies that deal with smartphone-based datasets, WISDM and UCI-HAR, to achieve the activity recognition task.

In [23], a project presented involved a collection of raw data from (29) individuals who volunteered to perform a specific physical activity while carrying their Android smartphones, they have been asked to do six activities (walking, standing, climbing the stairs up and down, and sitting). For the recognition task, three classifiers are adopted (i.e. Multi-Layer Perceptron (MLP), Logistic Regression (LR), and Decision Tree (J48). The outcomes of the study stated that MLP classifier surpassed other classifiers with accuracy of (91.7%), while other classifiers came with accuracy of (78.1%) and (85.1%) regarding to (LR) and (J48) respectively. The experiment of the study showed that the stair climbing activity (up and down) was the most difficult, due to their rates similarity, so that the two activities were combined as a single activity (i.e. stairs). A number of experiments were conducted using J48, the outcomes of the study showed a noticeable enhancement with a rate of (90.04%) as overall performance. Nevertheless, the results still mediocre. In [31] a study has been presented to enhance the classification outcomes of the human activity recognition. The study preformed a feature extraction in order to reduce the obtained feature, for feature extraction, the study used Principal Component Analysis (PCA) to extract the most dominant features. The normalized features then tested by using two ML classifiers, Support Vector Machine (SVM), as well as Gaussian Discriminant Analysis (GDA). The study resulted a good rate of the performance accuracy with (92.22%) using the SVM classifier. Depending on the results presented in [23], a new study has been presented in [25], in which, a novel approach of accelerometer-based HAR is proposed. The study employed three classifiers as voting technique: J48, LR, and MLP. Based on different aggregation rules, the three classifiers were conducted. The results stated that the average of probabilities rule yielded highest level of recognition performance of (94.35%). Another study has been presented in [32], where three classifiers have been examined, Decision Tree (C4.5), LR, and MLP. The study presented an ensemble approach by using Adaptive Boosting (AdBoost) method to combine the classifiers. The outcomes of the study indicated a noticeable enhancement in the performance with an accuracy at (94.04%), it is also worth mentioning that the combined model performs much better than other models when implemented individually. In [33], a study presented an implementation of a set of ML algorithms, such as Nearest Neighbor, Naïve Bays

(NB), Rule Induction (JRip), Classification via Regression (CvR), J48, and Random Forest (RF). All of these classifiers have employed to perform a HAR task. The study used WISDM dataset, and utilized the AdBoost method with a conjunction of (J48) and (RF) classifiers respectively. The results showed a good performance with (94.31%) as overall accuracy for the combination of AdBoost and RF. [34] employed two ML methods Naïve Bayes Tree (NBTree) which is a combination of Decision Tree and Naïve Bayes classifiers, and MLP. The hybrid model aimed to recognize human daily activities. Firstly, the two models have been examined and evaluated separately, and then by combining them together using voting technique. The results stated that the hybrid model produced better accuracy with (97.56%). A Deep Learning approach has been presented in [26] by using Convolutional Neural Network (CNN) algorithm in order to classify human activities. The model implemented by using two types of datasets, WISDM and UCI-HAR, the results showed an accuracy of (93.32%) and (97.47%) for the WISDM and UCI-HAR datasets, respectively. However, due to the complexity of CNN model, the study adopted splitting technique for verification (i.e., the first 26 users for the training, and the other 10 users for the testing) rather than the leave-one-out validation technique that used in [35]. Nevertheless, the data splitting technique in contrast to 10-fold cross validation which is adopted in our study, does not ensure the same distribution of class labels. In [36], a research suggested a mixture of statistical approaches for the purpose of data preprocessing, such as, filtering, feature extraction, and noise reduction. In other words, temporal and spectral statistical characteristics, along with the reweighted Genetic Algorithm Classifier. In spite of the result of the study showed an accuracy performance with a rate of (94.02%), however, in addition to less sampling of the dataset, the approach used different window frames.

In [24], a study presented and used UCI-HAR dataset for activity recognition task. SVM classifier is implemented with 10-fold cross validation approach. The outcomes stated a high accuracy of (96%). A paper presented in [37], aimed to propose a new model of HAR task based on non-linear chaotic features. It focuses on how to measure the optimum delay time and how to calculate the embedding dimension of the reconstructed phase space. The experimental outcomes indicate that in contrast to existing methods, the chaotic function performed better with (96.80%) overall accuracy. In [38], a study presented to employ ML techniques along with HAR based on smartphone-based accelerometer sensors. The study evaluated the efficiency of the proposed approach using datasets of human activity recognition gathered and publicly accessible named UCI-HAR dataset. The study employed three algorithms in the context of classification which are: Reconstructed Phase Space (RPS), Gaussian Mixture Models (GMM), and Maximum Likelihood Classifier (MLE). The study assumed that it is possible to further study the implementation of the proposed device in wearable sensor-based activity recognition. The results of the study showed a good accuracy rate with (90%). A deep learning approach is presented in [39]. This paper designed an inertial accelerometer-based smartphone architecture for HAR. The smartphone captures the sensory sequence of data as participants conduct traditional everyday activities, extracts the features of high performance from the original data, Then, through many three-axis accelerometers, the user obtains physical behavior data. To extract useful feature vectors, the data is preprocessed by denoising, normalization and segmentation. In addition, CNN, LSTM,

BLSTM, MLP and SVM models are utilized in the context of classification on the UCI-HAR and Pamap2 datasets. Then the study discussed how to train methods of deep learning and show how the approach proposed outperforms others on both datasets. The results of the study showed a good accuracy level with a rate of (92.71%). In [40], A study suggested an effective group-based Context-aware classification system for recognizing human activity on smartphones, GCHAR is proposed, using the hierarchical group-based structure to enhance the efficiency of classification, instead of intensive computation, the classification error is minimized by context awareness, the method used various classifiers such as Random Tree, Bagging, J48, BayesNet, KNN and Decision Table, the results of the study presented a classification accuracy at (94.16%). In [41] proposed a wrapper filter approach based on Gradient-Boosted Tree (GBT) classifier for the recognition of human actions. The model achieved accuracy of (93.12%). In [42] suggested a model based on LSTM with CNN. The model was tested on three public datasets: UCI-HAR, WISDM, and OPPORTUNITY. Regarding UCI-HAR and WISDM datasets, the overall accuracy achieved was (95.78%) and (95.85%) respectively. In [27] presented a novel strategy to train Deep Neural Networks (DNN) using hybrid optimization model of Artificial Bee Colony (ABC) and Limited-memory BFGS. The proposed model was tested on various datasets, however, it achieved (96.98%) performance accuracy on UCI-HAR dataset. Moreover, in [28], a hybrid optimization model based on ABC and PSO is proposed and used along with Stacked Auto-Encoders (SAE) on WISDM and UCI-HAR datasets. The results showed a good classification rate for both datasets.

Combining optimization algorithms with various ML classifiers for HAR task is carried out by few previous studies. In specific, two original works [27] and [28] utilized optimization methods on WISDM and UCI-HAR datasets. Therefore, for the reason of coherent comparison, we considered the results of these papers to compare with our model.

3 Methodology

In this part of the paper, we present a detailed overview of our work methodology. In addition to illustrating an overview regarding the adopted datasets. Additionally, we demonstrate the preprocessing phase. Regarding to the feature selection process, we

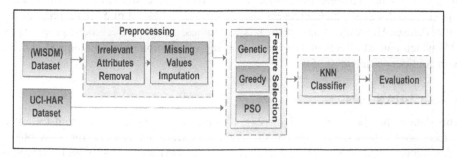

Fig. 1. Research methodology

employ three optimization methods: Genetic, Greedy, and PSO algorithms. All of the processing steps are illustrated in (Fig. 1).

4   Dataset Characterization and Preprocessing

As mentioned earlier, this paper adopts two datasets WISDM [43] and UCI-HAR [44] in order to implement the proposed model, a brief detail for each of the mentioned datasets is illustrates as below:

4.1   WISDM Dataset

This dataset composed of (5418) labeled accelerometer-based data, collected from (36) of the volunteers [23]. The collected data involved registering human physical activities for a particular person while holding his/her smartphone up in the front pant pocked. All members of the dataset were required to perform six common everyday activities (walking, jogging, climbing the stairs (up & down), sitting, and standing). The number of dataset attributes is (45), which is a continuous type of values. Table 1 shows the number and percentage of the examples per class.

Table 1. The number and percentage of examples per class

Class	Total	%
Walking	2081	38.41
Jogging	1625	29.99
Upstairs	632	11.66
Downstairs	528	9.75
Sitting	306	5.65
Standing	246	4.54

Before implementing the classification, and related experiments, we performed a number of preprocessing steps as follows: i) We eliminate the attributes (UNIQUE-ID, USER and X_{AVG}), since they have no meaning to be a part of the analysis, as explained in [41]. ii) There are a number of missing values in three successive attributes (i.e. X_{PEAK}, Y_{PEAK}, and Z_{PEAK}), which represent about 8.75% (i.e. 474 out of 5418 records) of the entire dataset. However, to impute the missing values, we adopt a technique proposed by [45] which named "Decision tree based Missing value Imputation techniques" (DMI), with confidence rate at (0.35).

4.2   UCI-HAR Dataset

This dataset has been proposed in [24], it consists of data recorded from (30) volunteers with (19- 48) ages. All of the volunteers were asked to perform six physical activities (WALKING, WALKING_UPSTAIRS, WALKING_DOWNSTAIRS, SITTING, STANDING, LAYING) while holding their Android smartphones mounted to their waists. The registered data from smartphone sensors is then mapped to produce a dataset contained (10299) labeled instances each with 562 attributes.

5 Experimental Setup and Results

For the feature selection phase, we adopted three different optimization algorithms which are (Genetic algorithm, Greedy backwards algorithm, and Particle Swarm Optimization algorithm), as will be stated in the upcomming sections, we also compare the performance for each algorithm. To classifiy the model, we adopt (KNN) classifier. Regarding the KNN classifier, a taxicab geometry, Manhattan distance function, is utilized for all experiments. The Manhattan distance between two points is the summation of the absolute differences between their coordinates. The Manhattan function can be computed using the following formula [46]:

$$D(x, y) = \sum_{i=1}^{n} |x_i - y_i| \qquad (1)$$

Where D is the distance between two points: x and y, and i is thier dimensional space.

All experiments are achieved using 10-fold cross validation. Furthermore, a combination of different performance metrics is utilized to evaluate the performance of the examined models. In specific, the performance metrics used are: accuracy, f-measure, number of features, and CPU time of the KNN.

Accuracy can be computed from the four confusion matrix terms [47]: true positive (TP) which is the number of positive classes predicted correctly, false positive (FP) which is the number of negative classes predicted as positive, true negative (TN) which is the number of negative classes that are predicted correctly as negative, and false negative (FN) which is the number of positive classes that are classified incorrectly as negative. Accordingly, accuracy can be calculated as follows:

$$Accuracy = \frac{TP + TN}{TP + FP + TN + FN} \qquad (2)$$

Precision is the percentage of relevant retrieved records, which can be calculated by:

$$Precision = TP/(TP + FP) \qquad (3)$$

and recall is the percentage of relevant returned records which can be calculated by:

$$Recall = TP/(TP + FN) \qquad (4)$$

Accordingly, F-measure can be calculated by [48]:

$$F - meaure = \frac{2 \times Recall \times Precision}{Recall + Precision} \qquad (5)$$

Where: TP represents the (True/Positive), TF represents the (True/Negative), FP represents the (False/Positive), FN represents the (False/Negative).

Regarding to the results of WISDM dataset presented in Table 2, it can be noticed that the behavior of both Greedy and PSO algorithms were identical and yielded higher performance than other algorithms in all metrics, in specific, the Greedy and PSO models' accuracy outperforms the baseline KNN model with about (1.8) percentage. Also, Genetic algorithm outperform the accuracy of the baseline model with about (1.53) percentage. In means of CPU time, Genetic algorithm performed slightly faster than Greedy and PSO models with about (0.4) seconds, and faster than the base learner with about (83) seconds.

The genetic model was maintained using population size and generations of (200), crossover probability is (0.6) and the probability of mutation is (0.033). Default value settings was used for both the Greedy backwards and PSO algorithms.

Table 2. Model performance for the WISDM dataset

	Baseline	Genetic	Greedy	PSO
Accuracy	89.28(1.99)	90.81(1.47)	91.07(1.92)	91.07(1.92)
RMSE	0.19(0.02)	0.17(0.01)	0.17(0.02)	0.17(0.02)
Weighted avg TP	0.89(0.02)	0.91(0.01)	0.91(0.02)	0.91(0.02)
Weighted avg FP	0.03(0.01)	0.03(0.01)	0.03(0.01)	0.03(0.01)
Weighted avg TN	0.97(0.01)	0.97(0.01)	0.97(0.01)	0.97(0.01)
Weighted avg FN	0.11(0.02)	0.09(0.01)	0.09(0.02)	0.09(0.02)
Weighted avg F-measure	0.89(0.02)	0.91(0.02)	0.91(0.02)	0.91(0.02)
CPU time in seconds	0.98(0.02)	0.15(0.01)	0.18(0.01)	0.19(0.01)
Number of features	42	7	8	8

Regarding to the results of UCI-HAR dataset illustrated in Table 3, it can be noticed that PSO model outperformed other models in terms of accuracy. In addition, Genetic model performed slightly better than the baseline KNN model with accuracy of (98.79) and (98.78), respectively. Greedy model's accuracy presented the lowest rate of (96.96) percentage. On the other hand, the CPU time for Greedy model is the lowest amongst other models with (2.6) second, while Genetic-based and PSO-based models required higher CPU time (6.56 and 7.39 s respectively). However, the KNN base classifier is the slowest with CPU time of (16.15) seconds. The genetic model was maintained using population size and generations of (200), crossover probability is (0.6) and the probability of mutation is (0.033). Defaults value settings was used for the Greedy backwards. For the PSO algorithm, the individual weight is (0.4), inertia and social weights are (0.3), mutation probability is (0.4) and seed is 2.

In general, from the experiments on both datasets (i.e. WISDM and UCI-HAR), PSO achieved higher performance among other models. Therefore, more detail results are presented in below, for only PSO based KNN performance metrics, for WISDM and UCI-HAR datasets (in Tables 4 and 5), respectively. In addition to that, the model perfroms better when dealing with UCI-HAR dataset in comparison with WISDM dataset.

A comparison with literature in term of accuracy is depicted in Table 6, which shows that the performance of our study outperforms the rates of the mentioned literature,

Table 3. Model performance for the UCI-HAR dataset

	Baseline	Genetic	Greedy	PSO
Accuracy	98.78(0.40)	98.79(0.38)	96.96(0.53)	98.89(0.23)
RMSE	0.06(0.01)	0.06(0.01)	0.06(0.01)	0.06(0.01)
Weighted avg TP	0.99(0.00)	0.99(0.00)	0.97(0.01)	0.99(0.00)
Weighted avg FP	0.00(0.00)	0.00(0.00)	0.01(0.00)	0.00(0.00)
Weighted avg TN	1.00(0.00)	1.00(0.00)	0.99(0.00)	1.00(0.00)
Weighted avg FN	0.01(0.00)	0.01(0.00)	0.03(0.01)	0.01(0.00)
Weighted avg F-measure	0.99(0.00)	0.99(0.00)	0.97(0.01)	0.99(0.00)
CPU time in seconds	16.15(1.11)	5.50(0.22)	2.60(0.13)	7.39(0.20)
Number of features	561	218	56	239

Table 4. WISDM dataset per class performance metrics for PSO based KNN

	Precision	Recall	F-measure	Accuracy
Walking	0.926	0.956	0.941	95.627
Jogging	0.976	0.975	0.975	97.476
Upstairs	0.733	0.733	0.733	73.259
Downstairs	0.773	0.689	0.729	68.939
Sitting	0.987	0.971	0.979	97.058
Standing	0.975	0.959	0.967	95.934

Table 5. UCI-HAR dataset per class performance metrics for PSO based KNN

	Precision	Recall	F-measure	Accuracy
Walking	0.999	1.000	1.000	100
Upstairs	0.999	0.999	0.999	99.935
Downstairs	0.999	0.999	0.999	99.857
Sitting	0.964	0.974	0.969	97.355
Standing	0.976	0.966	0.971	96.642
Laying	1.000	1.000	1.000	100

which indicates the positive impact on using PSO-based KNN classifier for human activity recognition task.

It noteworthy that the method used in [27] was tested by splitting the dataset into (70% for training and 30% for testing) and 10 independent runs, however, we followed

Table 6. Performance comparison with literature

Literature	WISDM	UCI-HAR
KNN [49]	89	
MLP [23]	91.7	
KNN [33]	85.93	
GBT [41]	93.12	
non-linear chaotic features [37]		96.80
Gaussian mixture model [38]		90
SVM [24]		96
CNN [39]		92.71
group-based scheme [40]		94.16
ABCPSO-based DNN [27]		96.98
CNN [26]	93.32	97.47
LSTM-CNN [42]	95.85	95.78
Ours	**91.07**	**98,89**

the same settings and our performance rate was at 98.33 (0.24), which is still greater than the others as stated in Table 6. In addition, and regarding to [28], our model outperforms the related works when running with UCI-HAR dataset, nevertheless, it presents close to the best results that deal with WISDM dataset. Furthermore, although our results on UCI-HAR dataset are higher than those presented in [42], the result of WISDM dataset are out of the comparison as they used a different dataset consisted of 16,690 instances which is also different than the standard WISDM dataset used in other works. To summarize, experimental test and result show that the model performs well when strongly correlated features are used for training/testing. The results of this model encourage various implications that might practically be adapted by many active applications such as healthcare, human computer interaction [41, 42], security [33], and game players and behavior analysis [50, 51].

6   Conclusion

This paper explored and analyzed the impact on performance on the use of three well-known optimization search algorithms for feature selection out from mobile-based data to recognize basic human activities. To ensure the stability and general applicability of the results, experiments are conducted on two publicly available benchmark datasets, specifically 1) WISDM which contains missing values of about 8.75% of the entire dataset, and 2) UCI-HAR which has a big number of features reaches to 560 in total. Results show that the use of PSO is more successful than other tested algorithms on both datasets. In addition, the results of the study are promising and outperform those presented in previous literature. Thus, for future work challenge, one might focus on using PSO to detect more complex human activities in real time. Another work challenge might be studying the quality of features related to every single smartphone sensor.

References

1. Micucci, D., Mobilio, M., Napoletano, P.: Unimib shar: a dataset for human activity recognition using acceleration data from smartphones. Appl. Sci. **7**(10), 1101 (2017)
2. Bayat, A., Pomplun, M., Tran, D.A.: A study on human activity recognition using accelerometer data from smartphones. Proc. Comput. Sci. **34**, 450–457 (2014)
3. Hassan, M.M., et al.: A robust human activity recognition system using smartphone sensors and deep learning. Futur. Gener. Comput. Syst. **81**, 307–313 (2018)
4. Sisodia, D.S., Yogi, A.K.: Performance Evaluation of Ensemble Learners on Smartphone Sensor Generated Human Activity Data Set Data, Engineering and Applications, pp. 277 284. Springer, Cham (2019). https://doi.org/10.1007/978-981-13-6351-1_22
5. Tian, Y., et al.: Inertial sensor-based human activity recognition via ensemble extreme learning machines optimized by quantum-behaved particle swarm. J. Intell. Fuzzy Syst. **38**(2), 1443–1453 (2020)
6. Gupta, P., Dallas, T.: Feature selection and activity recognition system using a single triaxial accelerometer. IEEE Trans. Biomed. Eng. **61**(6), 1780–1786 (2014)
7. Song, G., et al.: Multimodal similarity gaussian process latent variable model. IEEE Trans. Image Process. **26**(9), 4168–4181 (2017)
8. Song, G., et al.: Harmonized multimodal learning with Gaussian process latent variable models. IEEE Trans. Pattern Anal. Mach. Intell. (2019)
9. Zerrouki, N., et al.: Vision-based human action classification using adaptive boosting algorithm. IEEE Sens. J. **18**(12), 5115–5121 (2018)
10. Jegham, I., et al.: Vision-based human action recognition: an overview and real world challenges. Forensic Sci. Int. Digital Investigat. **32**, 200901 (2020)
11. Feldwieser, F., et al.: Acceptance of seniors towards automatic in home fall detection devices. J. Assistive Technol. (2016)
12. Jeffs, E., et al.: Wearable monitors for patients following discharge from an intensive care unit: practical lessons learnt from an observational study. J. Adv. Nurs. **72**(8), 1851–1862 (2016)
13. Pal, D., et al.: Internet-of-things and smart homes for elderly healthcare: an end user perspective. IEEE Access **6**, 10483–10496 (2018)
14. Attal, F., et al.: Physical human activity recognition using wearable sensors. Sensors **15**(12), 31314–31338 (2015)
15. Yusif, S., Soar, J., Hafeez-Baig, A.: Older people, assistive technologies, and the barriers to adoption: a systematic review. Int. J. Med. Inform. **94**, 112–116 (2016)
16. Reyes-Ortiz, J.-L., et al.: Transition-aware human activity recognition using smartphones. Neurocomputing **171**, 754–767 (2016)
17. Riboni, D., Murtas, M.: Sensor-based activity recognition: one picture is worth a thousand words. Futur. Gener. Comput. Syst. **101**, 709–722 (2019)
18. Kwon, Y., Kang, K., Bae, C.: Unsupervised learning for human activity recognition using smartphone sensors. Expert Syst. Appl. **41**(14), 6067–6074 (2014)
19. San-Segundo, R., et al.: Segmenting human activities based on HMMs using smartphone inertial sensors. Pervasive Mob. Comput. **30**, 84–96 (2016)
20. Vilarinho, T., et al.: A combined smartphone and smartwatch fall detection system. In: 2015 IEEE International Conference on Computer and Information Technology; Ubiquitous Computing and Communications; Dependable, Autonomic and Secure Computing; Pervasive Intelligence and Computing. IEEE (2015)

21. Luque, R., et al.: Comparison and characterization of android-based fall detection systems. Sensors **14**(10), 18543–18574 (2014)
22. Mourcou, Q., et al.: Performance evaluation of smartphone inertial sensors measurement for range of motion. Sensors **15**(9), 23168–23187 (2015)
23. Kwapisz, J.R., Weiss, G.M., Moore, S.A.: Activity recognition using cell phone accelerometers. ACM SIGKDD Explorations Newsl **12**(2), 74–82 (2011)
24. Anguita, D., et al.: A public domain dataset for human activity recognition using smartphones. In: Esann (2013)
25. Catal, C., et al.: On the use of ensemble of classifiers for accelerometer-based activity recognition. Appl. Soft Comput. **37**, 1018–1022 (2015)
26. Ignatov, A.: Real-time human activity recognition from accelerometer data using convolutional neural networks. Appl. Soft Comput. **62**, 915–922 (2018)
27. Badem, H., et al.: A new efficient training strategy for deep neural networks by hybridization of artificial bee colony and limited–memory BFGS optimization algorithms. Neurocomputing **266**, 506–526 (2017)
28. Ozcan, T., Basturk, A.: Human action recognition with deep learning and structural optimization using a hybrid heuristic algorithm. Clust. Comput. **23**(4), 2847–2860 (2020). https://doi.org/10.1007/s10586-020-03050-0
29. Raut, A.R., Khandait, S.: Review on data mining techniques in wireless sensor networks. In: 2015 2nd International Conference on Electronics and Communication Systems (ICECS). IEEE (2015)
30. Poon, C.C., et al.: Body sensor networks: In the era of big data and beyond. IEEE Rev. Biomed. Eng. **8**, 4–16 (2015)
31. Kuspa, K., Pratkanis, T.: Classification of mobile device accelerometer data for unique activity identification. Stanford Center for Professional Development, Palo Alto (2013)
32. Daghistani, T., Alshammari, R.: Improving accelerometer-based activity recognition by using ensemble of classifiers. Int. J. Adv. Comput. Sci. Appl **7**(5), 128–133 (2016)
33. Al-Taei, A.: A smartphone-based model for human activity recognition. Ibn AL-Haitham J. Pure Appl. Sci. **30**(3), 243–250 (2017)
34. Azmi, M.S.M., Sulaiman, M.N.: Accelerator-based human activity recognition using voting technique with NBTree and MLP classifiers. Int. J. Adv. Sci. Eng. Inf. Technol. **7**(1), 146–152 (2017)
35. Kolosnjaji, B., Eckert, C.: Neural network-based user-independent physical activity recognition for mobile devices. In: International Conference on Intelligent Data Engineering and Automated Learning. Springer, Cham (2015). https://doi.org/10.1007/978-3-319-24834-9_44
36. Quaid, M.A.K., Jalal, A.: Wearable sensors based human behavioral pattern recognition using statistical features and reweighted genetic algorithm. Multimedia Tools Appl. **79**(9–10), 6061–6083 (2019). https://doi.org/10.1007/s11042-019-08463-7
37. Tu, P., et al.: Non-linear chaotic features-based human activity recognition. Electronics **10**(2), 111 (2021)
38. Gani, M.O., et al.: A light weight smartphone based human activity recognition system with high accuracy. J. Netw. Comput. Appl. **141**, 59–72 (2019)
39. Wan, S., et al.: Deep learning models for real-time human activity recognition with smartphones. Mobile Netw. Appl. **25**(2), 743–755 (2020)
40. Cao, L., et al.: GCHAR: an efficient group-based context—aware human activity recognition on smartphone. J. Parallel Distribut. Comput. **118**, 67–80 (2018)
41. Al-Frady, L., Al-Taei, A.: Wrapper filter approach for accelerometer-based human activity recognition. Pattern Recognit Image Anal. **30**(4), 757–764 (2020). https://doi.org/10.1134/S1054661820040033
42. Xia, K., Huang, J., Wang, H.: LSTM-CNN architecture for human activity recognition. IEEE Access **8**, 56855–56866 (2020)

43. WISDM Dataset for Human Activity Recognition. [cited 2021 Jan 5]; https://www.cis.for dham.edu/wisdm/dataset.php
44. Anguita, D., et al.: Human Activity Recognition Using Smartphones Data Set (2012) [cited 2020 Dec 19]. https://archive.ics.uci.edu/ml/datasets/Human+Activity+Recognition+Using+ Smartphones
45. Rahman, G., Islam, Z.: A decision tree-based missing value imputation technique for data pre-processing. In: Proceedings of the Ninth Australasian Data Mining Conference, vol. 121 (2011)
46. Szabo, F.: The Linear Algebra Survival Guide: Illustrated with Mathematica. Academic Press, Cambridge (2015)
47. Deng, X., et al.: An improved method to construct basic probability assignment based on the confusion matrix for classification problem. Inf. Sci. **340**, 250–261 (2016)
48. Witton, I.H., Frank, E.: Data mining: practical machine learning tools and techniques with Java implementations. ACM SIGMOD Rec. **31**(1), 76–77 (2002)
49. Ferreira, P.J., Cardoso, J.M., Mendes-Moreira, J.: kNN prototyping schemes for embedded human activity recognition with online learning. Computers **9**(4), 96 (2020)
50. Al-Taei, A.: Automated classification of game players among the participant profiles in massive open online courses (2015)
51. Chen, K., et al.: Deep learning for sensor-based human activity recognition: overview, challenges and opportunities. arXiv preprint arXiv:2001.07416 (2020)

Face Recognition with Disguise and Makeup Variations Using Image Processing and Machine Learning

Farah Jawad Al-ghanim[✉] and Ali mohsin Al-juboori[✉]

College of Computer Science and Information Technology,
Al-Qadisiyah University, Al Diwaniyah, Iraq
{com.post07,Ali.mohsin}@qu.edu.iq

Abstract. Face recognition is a research challenge continuous and has seen colossal development during the last two decades. While coming algorithms keep on achieving improved the performance, a greater part of the face recognition systems are receptive to failure under disguise and makeup variations that is one of the common challenging covariates of facial recognition. In past researches, some algorithms show promising results on the existing disguise datasets, still, most of the disguise datasets include images with limited variations (oftentimes captured in controlled settings). This not simulate a real-world scenario, wherever both the intended/ unintended unconstrained disguises and makeup are encountered by a face recognition systems. In this paper, the disguised and makeup faces database (DMFD) is used. In order to handle this problem, One of simple, yet efficient ways for extracting face image features is (LBPH), Principal Component Analysis (PCA) that was majorly utilized in pattern recognition. Also, the technique of Linear Discriminant Analysis (LDA) employed for overcoming PCA limitations was efficiently used in face recognition. Further, classification is employed following the feature extraction. The Naïve Bayes, KNN and Random forest RF algorithms are used. The results paper show the effectiveness and generalization of the proposed system on the Disguise and makeup face database (DMFD) and the features which are extracted by means of (LDA) with (RF) provided the better results of (F-measure, Recall, and Precision).

Keywords: Face recognition · Disguised and makeup faces · Machine learning · Classification

1 Introduction

With fast advancement of the computer and networks technology, data security shows remarkable significance. Personality Identity is a fundamental essential to guarantee the security of the system. The precise recognizable is needed in the fields, national security, finance, e-commerce, justice, and so on. The personal ID system dependent on biometric acknowledgment innovation is getting the most attention for its superior security, validity, and reliability, and has begun to go on into all scopes of our lives. today and in the

© Springer Nature Switzerland AG 2021
M. Singh et al. (Eds.): ICACDS 2021, CCIS 1440, pp. 386–400, 2021.
https://doi.org/10.1007/978-3-030-81462-5_35

time of the Technologies evolve helped Humanity, image Processing has a broadrange of applications such as the video surveillance, crowd analyses, behavior analysis [1]. Face recognition is basically utilized in a security system, payment systems, public security systems, and so on. The specialty of perceiving the human face is very troublesome and critical challenges as it shows shifting attributes like expressions, illumination, pose, disguise, age, change in hairstyle, etc. [2] The human face also suffers irreversible changes consequent to aging. These factors make the operation of face recognition non-trivial and so hard. The conception of face recognition is to hand a computer system the ability to discover and recognizing human faces speedy and precisely in videos or images [3]. An image processing applications simultaneously with the machine learning approaches, in this paper PCA, LBPH and LDA algorithms are utilized to extract the face features. NB and KNN and RF algorithms are chosen for the face classification. The Experimental results on the Disguise and makeup database (DMFD) contain (2460) images of (410) subjects, most these images are of celebrities (movie/TV stars, politicians, athletes) in the disguise covariate and/or make-up facial with growth truth (beard, goggle, mustache, eye-glasses) as (Fig. 1 shows an example), gained under real environments. The first image of the all subjects is the frontal without disguise or low makeup. There at least 1 to 2 are clean (without makeups, disguises or pure facial) within each subject, and the rest have varied kinds of disguises as glasses, mustache and other face accessories [4]. All the results were tabulated and studied.

Fig. 1. Example image pair with different types of disguises or makeup

2   Literature Survey

Sabri N. et al. [5] present compare four varied ML classifiers (NB), (MLP), and (SVM) by mean distance metrics of the facial geometry to classify humans face. Using a webcam to capture (30) images of 8 (eight) person faces. Results from the experiments reveals superiority the (NB) on the MLP and SVM classification with the high precision. Putranto et al. [6] this study predicts Eigen-face is joint with (NB). Used Eigen-face for feature extraction and NB for dataset class predictions. Then, added the normalization (z-score) for sharping the accuracy. Used sample of image as a dataset and divided the 200 datasets into train and test data by applying (k = 10) in cross-validation method, to see the performance of proposed methodology. In the results, the proposed method can predicts the image face up to (70%). Over and above, adding normalization (Z-Score), the accuracy of prediction lifts up to (89.5%). Chen, Yong et al. [7] suggests a facial recognition systems established on the extended (LBP). LBP to extracted Features and to dimensions reduction used PCA. Lastly, used the sparse representation (SRC) classification of (L2) minimization to the identify face and verify. Experimental results to works can obtain a higher recognition rate in many databases ORL database, UMIST database, JAFFE database, CK database and VSS database. Zhao et al. [8] used LBP and GA for face recognition. The consideration illumination algorithm minimizes effect of illumination via gamma corrected, histogram equalization HE, logarithmic and exponential transform in capture YUV video image data. Finally, the system selected the approach on histogram equalizations HE and used for the classification, (NN) and (SVM). Sovitkar and Kawathekar [9] propose an automated attendance recording system to update the student's attendance records. This done by recognizing the physical presence to the students in classroom with the aid the techniques for detect face and recognition. Used Viola Jones to detect the face and two algorithms to feature selection (LDA, PCA) and then using both LDA and PCA with SVM for classification. In this work utilize various facial poses, expressions, and lighting for creating dataset. The average rate recognition of the system of the combination of all algorithms is 95%. When the persons to be identified increases there is a drop in the recognition rate. Kamlesh and Mayank [10] presented system to recognize facial-expression by using the hybrid method (Harris corner features and Gabor wavelet) then classifying by random forest algorithm. This work used the (JAFFE) Database. For training the feature vectors using the subspace is composed by the proposed algorithm. In performance evaluation, the random forest computes the similarity that provides enhanced results in recognition accuracy and the confusion matrix illustrates the random forest that proposed provides higher accuracy than the earlier research works.

3 Proposed Work

The architecture work for disguise and makeup recognition (Fig. 2 shows the proposed system). It describes by the following steps.

a) **Image Acquisition:** This step uses cropped image face from (DMFD) dataset from (Hong Kong Polytechnic) comes with cropped images created in 2016 [11].

b) **Image Preprocessing:** Image processing IP is fraction of various fields such as remote sensing, security systems, robot vision, etc., and it's an incredible and demanding area [12]. Image contrast enhancement is a significant objective in digital image processing. In image preprocessing, to enhance the input image quality used histogram equalization to dispose of or smother the impact from shadow (and, or) brightening varieties [13].

c) **Feature Extraction:** PCA, LBPH and LDA were utilized for feature extraction.

d) **Face Recognition:** Naïve Bayes, (k-NN) and (RF) methods used to classifying images for face recognition purpose.

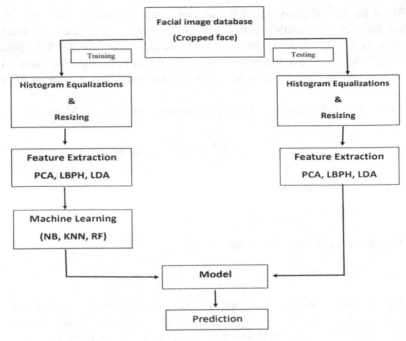

Fig. 2. Flowchart of the proposed system

4 Feature Extraction

Feature extraction have become a clear requirement in numerous processes that have frequently to do in computer vision, image processing, object detection, image retrieval, pattern recognition, bioinformatics, machine learning. Feature extraction is utilized to extract features that special existing in a dataset (text, image, voice) which are utilized to perform and describe the data [14]. The feature extraction step of the face recognition system comprises the decreasing of the number of resources that describe large data amounts [15]. Feature extraction from specific data was one of the critical issues for effective applications related to ML. The methods are vastly used to extract the face features are eigenface, (ICA), (LDA), (HOG), (SIFT), gabor filter, Haar wavelets, Fourier transforms, (LBP) and (LPQ) [16]. In this paper presented PCA (Eigenface), LBPH and LDA (Fisher face) were utilized as feature extraction approaches from original face images.

4.1 Principal Component Analysis (PCA)

PCA the main approach utilized in pattern recognition also compression and it is given as a dimensions reduction and feature extraction approach [17]. In 1991 Turk and Pentland submit the Eigenfaces method for face recognition [18]. PCA is a known as face recognition methods as well called the Eigen-face [19]. It is a statistical approach that utilizes a holistic method for discovering the patterns in the high-dimension data's. A goal of PCA has been possessed from the approach of information theory that is a fraction down facial images to little groups of the distinguishing feature images that indicate to as Eigen-faces used for representing current and new faces [20]. In PCA should be converted 2D facial image matrices to a 1D vector (might be a column or a row). Thus, a picture representation generates in high-dimensions space [21]. A stage of PCA are: [22]

1- Train a set of (M) images utilized for computing the Average Mean as:

$$Average = \frac{1}{M} \sum_{n=1}^{M} Training\ images(n) \tag{2}$$

2- Subtracting main image from Average Mean as:

$$SUB = Training\ images - Average \tag{2}$$

3- Then Covariance Matrix is computed as::

$$covariance = \sum_{n=1}^{M} Sub(n)Sub(n)^T \tag{3}$$

4- The Eigen-value and Eigen-vectors related to the covariance matrix are computing.

5- Sorted and chosen the best Eigen-values, then selected the highest Eigen-values belonging to the collection of Eigen-vectors, such as (M Eigen-vectors) are describing Eigenfaces. Eigenfaces might be re-account or updated because encountering new faces. Finally, projecting the training samples on Eigen-faces.

4.2 Linear Discriminant Analysis (LDA)

It is studied in machine learning, statistics, and pattern recognition in widely. LDA can be considered as a Fisher's linear discriminant (FLD), which is designed to find an optimal conversion to extract discriminant features that differentiate two or more classes [23]. Its application of LDA has been kept in a small image database that used to overcome the limitations of PCA, was accomplished by the projection of an image onto Eigen-face space over PCA, then perform pure LDA through it for classifying Eigen-face projected data (Fig. 3 illustrate PCA and LDA) [20]. LDA is seeking for vectors in implicit space which are superior discriminating between the classes. Over and above, the images of the LDA group that are linked to the same class and separating distinguishing class images. Mathematically, (SB) between class scatter and (SW) within class scatter matrixes are a particular measures been set [24]. For every class sample the SB and SW were specified as follow: [25]

1- Calculating (d)-dimensions mean vectors of classes from the data-set.
2- Calculating (SW and SB) scatter matrices.
3- To the scatter matrices, calculate the eigen-vectors (e1, e2, e3, e4, ..., ed) and identical eigen-value ($\lambda 1, \lambda 2, \lambda 3, \lambda 4, ..., \lambda d$).
4- Sorting the eigen-vectors via decrease eigen-values then pick out (k) eigen-vectors with biggest eigen-values for the shape a (d × k) dimensional matrix W (any column act an eigen-vector).

5- Employing the matrix ((d × k) eigen-vector) for converting a samples into (new) sub-space may be specified by multiplication the matrix: $Y = X \times W$, which X is an (n × d - dimension matrix) and (n = samples and y = converted (n × k - dimension) samples in new sub-space).

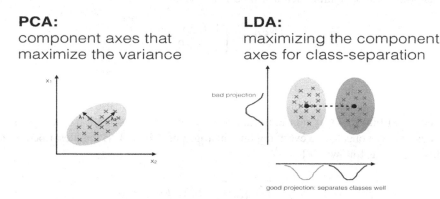

PCA:
component axes that maximize the variance

LDA:
maximizing the component axes for class-separation

bad projection

good projection: separates classes well

Fig. 3. PCA and LDA.

4.3 Local Binary Pattern Histogram (LBPH)

It is combination of (LBP) Local Binary Patterns and (HOG) Histograms of Oriented Gradients descriptor. Represent facial images with only a straightforward vector can simply do by applying the LBPH [26]. LBP was originally provided and developed for doing a texture analysis for gray images. The principal benefit of this method is that are coupling statistical and structural approaches end in growing texture analysis performance. Using the LBPH method for the represented and dimension reduced for a face image, The LBPH utilizes the histogram from the (LBP) distinctive spectrums as (feature-vector) for the classification. LBPH splits image into numerous sub-regions, thereafter extracts feature of (LBP) from every pixel of the sub-region, because that every sub-region can utilizing this histogram to describing the whole image into a numerous of components from the statistical histogram. The advantage is to decrease the error of an image that is not wholly aligned to a specific range. An operator is labeling the image pixels via thresholding 3x3-neighborhood regarding each one of the pixels for center value then assuming the output as a (binary number). Steps of LBPH are: [27].

1- Assuming an image with dimensions of N × M, it will be divided into regions of the same width and height leading to (m × m) dimension for each one of the regions.
2- A local binary operator was utilized in each one of the regions, LBP operator was specified as a window of 3 × 3.

$$LBP(x_c, y_c) = \sum_{P=0}^{7} S(i_p - i_c)2^P \qquad (4)$$

iC: central pixel values.

iP: neighbor pixel values.

3- With the use threshold (a median pixel value), it is compares the pixel to its 8 pixels (closest) with use of such, in the case when the value of neighbor is equal or more than the value of central, then it will be 1 or else 0.

$$S(t) = \begin{cases} 1 & t \geq 0 \\ 0 & t < 0 \end{cases}$$

4- Histogram used to reduce features of image from a decimal (256-dimensional) to a histogram (59- dimensional), that contains information on local patterns. We have (58) uniform patterns in 8-bit binary number; accordingly, we use (58 bins) to them and use (1 bin) for every non-uniform patterns. Global representation of image face is gained by concatenate every regional histograms. The LBPH can described in a histograms as follow: [28]

$$H(k) = \sum_{i=0}^{n} \sum_{j=1}^{m} f\left(LBP_{P,R}(i, j), k\right) \quad k \in [0, k] \tag{5}$$

P: the sampling points

R: the radius.

5 Machine Learning (ML)

It is a domain of computer science and subcategories of AI, which continues algorithms that allow systems to recognize. ML furnishes the PCs with the ability to thinks about and learns without programmed expressly. ML is utilized in different computational tasks to train the machine with the assistance of data given [29]. The data can be labeled in the status of supervised learning and unlabeled in the status of unsupervised learning to output better results for fixed problem. The essential focus is to cause computers to learn and gain from past experience [30, 31]. Image classification depends on various qualities reflected in the image data to recognize various sorts of targets. The feature classification phase of the facial recognition system includes the process that obtained face feature information is arranged into various and given the number classes relying upon a specific task [32]. There are multiple classification ML classification and prediction algorithm's, Naïve Bayes (NB), KNN algorithm and Random forest some of the Commonly used image classification algorithms.

5.1 Naïve Bayes (NB)

NB is one of the supervised algorithms and basic methodology that can perform in a way that better than some other classifiers. The thought behind this contemplation is that one ought to consistently attempt the simple way in the start. Attributes are overseen as they were however and independent in class's. This may in the end be better via remembering subsets of features for activity of decision settling on by settling on a complete selection of attributes that will be utilized [33]. The beauty of the NB approach, the estimation of (one) feature distribution is fully decoupled from the estimation of the others. Bayesian methods assign more likely class to a specific example describing on (feature's vector) show (Fig. 4 illustrate Naïve Bayes). The Naïve Bayesian works as follows: [34] Given (Cn) classes and every one of classes has its own probability (P(Cn)) evaluated from training dataset and showing the prior probability of classifying an attribute (xj) into (Cn). For the attribute value, (xj), so the classification used is to find this probability is illustrate in the following equation:

$$\frac{P(x_1 \wedge x_2 \ldots x_j \mid C_n)P(C_n)}{P(x_1 \wedge x_2 \ldots x_j)}$$

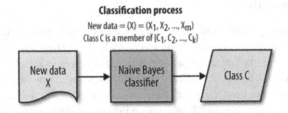

Fig. 4. Classification model of Naïve Bayes

5.2 K-Nearest Neighbor Algorithm

KNN is one of the easiest supervised machine learning algorithms it is established on analogy, as it analyzes training case that are like the case that are given to it. It utilizes (n) attributes to portray the training case [33]. Each case is a point in (n-dimension) space that produces an (n-dimensional) pattern space where each train tuple is saved. For a situation, unknown instance was assigned, a classifier searches the known qualities that are the more like the unknown case. The known as the k "nearest neighbors" to the case that is (unknown) [35] (Fig. 5 an example of KNN).

Generally, the distance function uses Euclidean distance or Manhattan distance:

$$D(X_q, X_i) = \sqrt{\sum nr} = 1(ar(X_q) - ar(X_i)^2 \tag{7}$$

$$D(X_q, X_i) = \sqrt{\sum nr} = 1(ar(X_q) - ar(X_i)^1 \tag{8}$$

Where (X) is input (n-dimensional) vector (a1, a2, ..., an), and query vector (Xq, i) is index vector in the training data, (ar) is number value of rth feature of the vector (X).

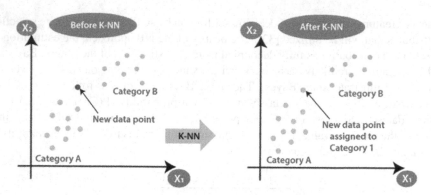

Fig. 5. An example of K-Nearest Neighbor algorithm approach

5.3 Random Forest (RF)

It is (ensemble learning) with random subspace methods and integration from bootstrap aggregating. Integration of many decision trees made the performance of the prediction is strong, but RF output is limited to set of output parameters values. It's widely used in medical image analysis to the classification of diseases and it was set to be effective in the brain stroke images classifying [36]. The weaknesses of the overfitting of decision tree calculations are decreased in RF by joining numerous decision trees to get the exact ultimate conclusion. The RF algorithm is a supervised technique where the multiple decision trees are utilized to make a forest and works better when the larger dataset. The forest is more powerful when a higher number of trees is utilized in the decision making process. The dataset is divided into two parts randomly with a similar structure. In stowing, (n bootstrap sets) are made by inspecting with substitution N trains examples from a trains set. The number of examples in the bootstrap train set minimal than the initial train set and is arbitrary. Then, each bootstrap set is utilized to build a decision

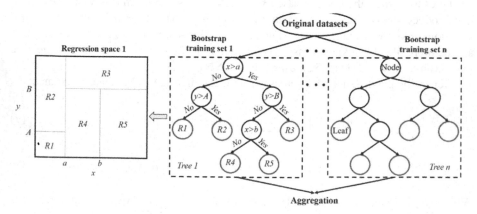

Fig. 6. Schematic view of (RF) algorithm

tree (Fig. 6 illustrate RF) [37], described works as:

$$Y = \frac{1}{n} \sum_{i=1}^{n} y_i(x) \tag{9}$$

yi(x) = output predicted for an (X) input vector tree.

6 K-Fold Cross Validation Algorithm

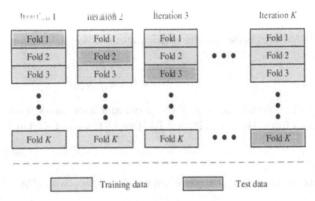

Fig. 7. K-fold cross validation algorithm diagrams

Which can entire process of building up the model of ML incorporates (3) stages (train, validate and test). A goal of validation is get better the robustness' of train models and avert over fitting and the trained model is accordingly most solid for testing set. Here, k-FCV method are widely used to validate model [38]. In k-FCV method, initial training set is split into (k) sub-datasets at random. The (K–1) sub datasets utilized to train models and a residual to validate models. Thus every sample has chance to train models and validated. In this study (K) set as (10). An overview of (k-FCV) process as the model evaluation seen in (Fig. 7 illustrate k-FCV).

7 Result and Discussion

After performing the preprocessing, image face enhancement, and resizing, the second part of this work is feature extraction using (PCA, LBPH and LDA), the input of this part is cropped facial image and the output is image features.

In the third part, a description is obtained about the results of applying both (Naïve Bayes, KNN and Random forest (RF)) machine learning algorithms for classifying the features obtained from the previously part described feature extraction algorithms.

The evaluation of the ML model utilized is the k-FCV algorithm. In cross validation CV, the dataset is split by k fold. each fold where in iteration, is utilized once as test data and the residual fold is utilized as training data and the process is repeated until all the data is evaluated.

The metrics that used to evaluation performance that extensively used metrics are (Precision, Recall, and F-measure). Precision determines the percentage of the results that are relevant, more the F-measure, better are the achieved results.

$$Precision = \frac{True\ Positive}{True\ Positive + False\ Positive} \tag{10}$$

Recall indicates the percentage of the total results that correctly classified.

$$Recall = \frac{True\ Positive}{True\ Positive + False\ Negative} \tag{11}$$

F1 score (F-measure) is the simpler metric that used both (Precision and Recall).

$$F1\ Score = 2 \cdot \frac{Precision \cdot Recall}{Precision + Recall} \tag{12}$$

The experimental results of implemented classification algorithms Naïve Bayes, KNN and Random forest (RF) with PCA, LBPH and LDA displayed in Tables (1, 2 and 3).

Table 1. The experiment results of implemented the Naïve Bayes with PCA, LBPH and LDA

Methods	Precision %	Recall %	F-measure %
Naïve Bayes + PCA	0.93	0.18	0.29
Naïve Bayes + LBPH	44.02	41.77	41.12
Naïve Bayes + LDA	72.9	65.9	64.9

Table 2. The experimental results of implemented the KNN with PCA, LBPH and LDA

Methods	Precision %	Recall %	F-measure %
KNN + PCA	0.067	0.18	0.05
KNN + LBPH	47.32	44.3	37.05
KNN + LDA	80.97	51.7	59.6

The given (Fig. 8 illustrate Performance Metrics Diagrams).

Table 3. The experimental results of implemented the RF with PCA, LBPH and LDA

Methods	Precision %	Recall %	F-measure %
RF + PCA	0.95	0.97	0.87
RF + LBPH	47.14	45.99	45.31
RF + LDA	**86.22**	**84.64**	**84.019**

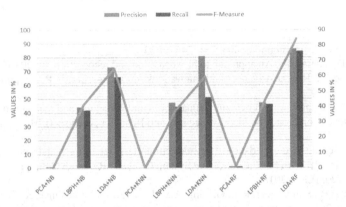

Fig. 8. Classifier metrics diagrams

8 Conclusion

The aim of this study is to implement a system for automatic facial recognition under disguised and makeup. In this paper used cropping faces in a certain database (Disguise and
makeup database (DMFD)). PCA (Eigenface), LBPH and LDA (Fisherface) is utilized
for feature extraction and dimension reduction. The study explores image processing
with machine learning algorithms that ensure high efficiency and accuracy Naive Bayes,
KNN and RF classification algorithms used with features extracted from PCA, LBPH
and LDA. It is concluded that LDA with Random forest classification algorithm gives
high results (Precision, Recall, and F-measure). Recognition time has been agreeable and
it took only a small number of seconds. Results display an increase in the (rates of recognition) in the case where increased the number of training images. It is our conviction that
the availability of the (DMFD) disguised and makeup faces dataset will further facilitate
the improvement of powerful face recognition system. The proposed system achieves
interesting results Because three different techniques for extracting features from facial
images were implemented and compared, three machine learning algorithms were used
for classification. Also under unfavorable conditions (illumination, facial expression),
the proposed system contributes to enhancing the system's accuracy. In future researches,
endeavors could be made to apply other classifiers of machine learning and combine any

of them to develop a more complex framework. They ought to have higher recognition precision.

Acknowledgments. My sincere thanks and gratitude to Dr. Ali mohsin Al-juboori for supervision and technical support during the project.

References

1. Hapani, S., Prabhu, N., Parakhiya, N., Paghdal, M.: Automated attendance system using image processing. In: 2018 4th International Conference on Computing Communication Control and Automation (ICCUBEA), pp. 1–5. IEEE (2018)
2. Jayaraman, U., Gupta, P., Gupta, S., Arora, G., Tiwari, K.: Recent development in face recognition. Neurocomputing **408**, 231–245 (2020)
3. Meena, D., Sharan, R.: An approach to face detection and recognition. In: 2016 International Conference on Recent Advances and Innovations in Engineering (ICRAIE), pp. 1–6. IEEE (2016)
4. Wang, T.Y., Kumar, A.: Recognizing human faces under disguise and makeup. In: 2016 IEEE International Conference on Identity, Security and Behavior Analysis (ISBA), pp. 1–7. IEEE (2016)
5. Sabri, N., et al.: A comparison of face detection classifier using facial geometry distance measure. In: 2018 9th IEEE Control and System Graduate Research Colloquium (ICSGRC), pp. 116–120. IEEE (2018)
6. Putranto, E.B., Situmorang, P.A., Girsang, A.S.: Face recognition using eigenface with naive Bayes. In: 2016 11th International Conference on Knowledge, Information and Creativity Support Systems (KICSS), pp. 1–4. IEEE (2016)
7. Chen, Y.-P., Chen, Q.-H., Chou, K.-Y., Wu, R.-H.: Low-cost face recognition system based on extended local binary pattern. In: 2016 International Automatic Control Conference (CACS), pp. 13–18. IEEE (2016)
8. Li-Hong, Z., Fei, L., Yong-Jun, W.: Face recognition based on LBP and genetic algorithm. In: 2016 Chinese Control and Decision Conference (CCDC), pp. 1582–1587. IEEE (2016)
9. Sovitkar, S.A., Kawathekar, S.S.: Comparative study of feature-based algorithms and classifiers in face recognition for automated attendance system. In: 2020 2nd International Conference on Innovative Mechanisms for Industry Applications (ICIMIA), pp. 195–200. IEEE (2020)
10. Tiwari, K., Patel, M.: Facial expression recognition using random forest classifier. In: Mathur, G., Sharma, H., Bundele, M., Dey, N., Paprzycki, M. (eds.) International Conference on Artificial Intelligence: Advances and Applications 2019. AIS, pp. 121–130. Springer, Singapore (2020). https://doi.org/10.1007/978-981-15-1059-5_15
11. Weblink for downloading The Hong Kong Polytechnic University Disguise and Makeup Faces Database described in this paper (2016). http://www.comp.polyu.edu.hk/~csajaykr/DMFaces.htm
12. Joseph, R.P., Singh, C.S., Manikandan, M.: Brain tumor MRI image segmentation and detection in image processing. Int. J. Res. Eng. Technol. **3**, 1–5 (2014)
13. Senthilkumaran, N., Thimmiaraja, J.: Histogram equalization for image enhancement using MRI brain images. In: Computing and Communication Technologies, WCCCT, pp. 80–83. IEEE (2014)
14. Salau, A.O., Jain, S.: Feature extraction: a survey of the types, techniques, applications. In: 2019 International Conference on Signal Processing and Communication (ICSC), pp. 158–164. IEEE (2019)

15. Oloyede, M.O., Hancke, G.P., Myburgh, H.C.: A review on face recognition systems: recent approaches and challenges. Multimedia Tools Appl. **79**(37–38), 27891–27922 (2020). https://doi.org/10.1007/s11042-020-09261-2
16. Kortli, Y., Jridi, M., Falou, A.A., Atri, M.: Face recognition systems: a survey. Sensors **20**(2), 342 (2020). https://doi.org/10.3390/s20020342
17. Kaushik, S., Dubey, R.B., Madan, A.: Study of face recognition techniques. Int. J. Adv. Comput. Res. **4**(4), 909 (2014)
18. Turk, M.A., Pentland, A.P.: Face recognition using eigenfaces. In: Proceedings of the 1991 IEEE Computer Society Conference on Computer Vision and Pattern Recognition, pp. 586–587. IEEE Computer Society (1991)
19. Pereira, J.F., Barreto, R.M., Cavalcanti, G.D.C., Tsang, R: A robust feature extraction algorithm based on class-modular image principal component analysis for face verification. In: 2011 IEEE International Conference on Acoustics, Speech and Signal Processing (ICASSP), pp. 1469–1472. IEEE (2011)
20. Singh, A., Singh, S.K., Tiwari, S.: Comparison of face recognition algorithms on dummy faces. Int. J. Multimedia Appl. **4**(4), 121 (2012)
21. Barnouti, N.H.N.: Face recognition using eigen-face implemented on Dsp Professor. Ph.D. Dissertation, School of Computer and Communication Engineering, Universiti Malaysia Perlis (2014)
22. Chen, J., Kenneth Jenkins, W.: Facial recognition with PCA and machine learning methods. In: 2017 IEEE 60th International Midwest Symposium on Circuits and Systems (MWSCAS), pp. 973–976. IEEE (2017)
23. Martinez, A.M., Kak, A.C.: PCA versus LDA. IEEE Trans. Pattern Anal. Mach. Intell. **23**(2), 228–233 (2001)
24. Bhattacharyya, S.K., Rahul, K.: Face recognition by linear discriminant analysis. Int. J. Commun. Netw. Secur. **2**(2), 31–35 (2013)
25. Patil, V., Narayan, A., Ausekar, V., Dinesh, A.: Automatic students attendance marking system using image processing and machine learning. In: 2020 International Conference on Smart Electronics and Communication (ICOSEC), pp. 542–546. IEEE (2020)
26. Deeba, F., Memon, H., Ali, F., Ahmed, A., Ghaffar, A.: LBPH-based enhanced real-time face recognition. Int. J. Adv. Comput. Sci. Appl. **10**(5), 274–280 (2019). https://doi.org/10.14569/IJACSA.2019.0100535
27. Ahmed, A., Guo, J., Ali, F., Deeba, F., Ahmed, A.: LBPH based improved face recognition at low resolution. In: 2018 International Conference on Artificial Intelligence and Big Data (ICAIBD), pp. 144–147. IEEE (2018)
28. Abuzneid, M.A., Mahmood, A.: Enhanced human face recognition using LBPH descriptor, multi-KNN, and back-propagation neural network. IEEE Access **6**, 20641–20651 (2018)
29. Bhavitha, B. K., Rodrigues, A.P., Chiplunkar, N.N.: Comparative study of machine learning techniques in sentimental analysis. In: 2017 International Conference on Inventive Communication and Computational Technologies (ICICCT), pp. 216–221. IEEE (2017)
30. Das, S., Dey, A., Pal, A., Roy, N.: Applications of artificial intelligence in machine learning: review and prospect. Int. J. Comput. Appl. **115**(9), 31–41 (2015). https://doi.org/10.5120/20182-2402
31. Dhall, D., Kaur, R., Juneja, M.: Machine learning: a review of the algorithms and its applications. Proc. ICRIC **2020**, 47–63 (2019)
32. Zhang, S., Wu, Y., Chang, J.: Survey of image recognition algorithms. In: 2020 IEEE 4th Information Technology, Networking, Electronic and Automation Control Conference (ITNEC), vol. 1, pp. 542–548. IEEE (2020)
33. Yaman, M., Subasi, A., Rattay, F.: Comparison of random subspace and voting ensemble machine learning methods for face recognition. Symmetry **10**(11), 651 (2018). https://doi.org/10.3390/sym10110651

34. Sen, P.C., Hajra, M., Ghosh, M.: Supervised classification algorithms in machine learning: a survey and review. In: Mandal, J.K., Bhattacharya, D. (eds.) Emerging Technology in Modelling and Graphics. AISC, vol. 937, pp. 99–111. Springer, Singapore (2020). https://doi.org/10.1007/978-981-13-7403-6_11

35. Han, J., Pei, J., Kamber, M.: Data Mining: Concepts and Techniques. Elsevier, Waltham (2011)

36. Subudhi, A., Dash, M., Sabut, S.: Automated segmentation and classification of brain stroke using expectation-maximization and random forest classifier. Biocybern. Biomed. Eng. **40**(1), 277–289 (2020)

37. Liaw, A., Wiener, M.: Classification and regression by randomForest. R News **2**(3), 18–22 (2002)

38. Zhang, P., Yin, Z.-Y., Jin, Y.-F., Chan, T.H.T., Gao, F.-P.: Intelligent modelling of clay compressibility using hybrid meta-heuristic and machine learning algorithms. Geosci. Front. **12**(1), 441–452 (2021). https://doi.org/10.1016/j.gsf.2020.02.014

Attention-Based Deep Fusion Network for Retinal Lesion Segmentation in Fundus Image

A. Mary Dayana$^{(\boxtimes)}$ and W. R. Sam Emmanuel

Department of Computer Science, Nesamony Memorial Christian College, Marthandam, Affiliated to Manonmaniam Sundaranar University, Tirunelveli, India
{mary_dayana_csa,sam_emmanuel}@nmcc.ac.in

Abstract. Segmentation of subtle lesions in fundus images has become a vital part of diagnosing ocular diseases such as Diabetic Retinopathy (DR). Diabetic eye disease is characterized by the scattered lesions in the retina. Detection of these lesions at the early stage is important as its progression leads to vision loss if proper treatment is not taken. The main objective of the work is to assist ophthalmologist in the effective diagnosis of eye disease providing timely treatment. This paper focuses on developing a deep learning-based Fusion Network (Fu-Net) with an attention mechanism for lesion segmentation in color fundus images. The network was developed based on the baseline U-Net model with trivial modification in the encoder and decoder part of the model. A multi-feature fusion block (MFuse) is integrated with the encoder of the network to extract the lesion features and a channel attention module is integrated with the decoder part to fuse the feature information effectively. Besides, a modified weighted focal loss function is introduced to mitigate the problem of class imbalance in the fundus image. The computational results obtained signifies the superior performance of the proposed method in the lesion segmentation task.

Keywords: Segmentation · Diabetic Retinopathy · Fusion network · Channel attention module

1 Introduction

Diabetic Retinopathy, a major consequence of Diabetes Mellitus is a microvascular disorder that affects the human eye in working-age adults. As per the report of the International Diabetes Federation 2019, at present 463 million adults have diabetes and it is estimated to rise 578 million by 2030 [1]. Furthermore, the prevalence of vision loss due to DR is likely to increase rapidly over the next 30 years. However, regular screening of DR possibly helps to prevent vision loss at the early stage. The screening process requires an experienced ophthalmologist to examine and analyze the fundus photographs for the presence of retinal pathologies like Microaneurysms (MAs), Hemorrhages (HEMs), Hard Exudates (HEs), and Soft Exudates (SEs).

© Springer Nature Switzerland AG 2021
M. Singh et al. (Eds.): ICACDS 2021, CCIS 1440, pp. 401–409, 2021.
https://doi.org/10.1007/978-3-030-81462-5_36

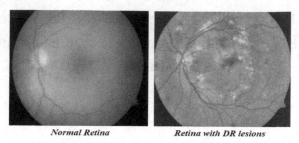

Normal Retina   *Retina with DR lesions*

Fig. 1. Sample Retinal images (Color figure online)

Microaneurysms are tiny dark red dots and they are only a few pixels wide. As the disease progress, MAs rupture, and the leakage of blood on the retinal vessels produce HEMs. Exudates are bright yellowish patches with varying sizes and intensities formed by the fluid substance that leaks due to vascular rupture [2]. Figure 1 depicts the sample retina without DR lesions and a retina with DR lesions. Researchers in the past have been using traditional image processing techniques such as region growing, clustering, morphological operations and threshold-based segmentation methods [3, 4]. These models attained poor performance, asserting that detection of subtle lesions in fundus images is a challenging and difficult task for the researchers. Deep Neural Networks has shown great potential in the analysis of medical images as it can automatically extract the most discriminant features and make accurate predictions. This paper presents a deep-learning-based fusion network called Fu-Net with encoder-decoder architecture based on the baseline U-Net model [5] for the segmentation of retinal lesions in fundus images.

The rest of the paper is structured as follows: Sect. 2 presents an overview of the deep-learning-based segmentation methods in the literature, Sect. 3 describes the proposed methodology, Sect. 4 reveals the experimental details, Sect. 5 demonstrates the results with discussion and Sect. 6 concludes the work and suggests a future direction for the research.

2   Related Work

In recent years, various deep learning-based segmentation methods have been evolved for analyzing the retinal fundus image [6–11]. In particular, U-Net proposed by [5] has been widely used for the segmentation of medical images. Jiang et al. [6] proposed a multipath recurrent U-Net architecture for the segmentation of optic disc and blood vessels and achieved good segmentation results combining Convolutional Neural Networks (CNNs) and Residual Neural Network (RNN). Zhang et al. [7] used an Attention guided network (AG-Net) for blood vessel and optic disc segmentation by incorporating guided filter into CNNs to learn the required detailed features for segmentation. A lightweight feature refinement network was introduced by Wang et al. [8] for retinal vessel segmentation by using spatial and semantic refinement path in the encoding and decoding layers. Also, a feature adaptive fusion block is introduced to combine different depth features and to improve the efficiency of the model. Hu et al. [9] designed a bridge-style U-Net model with a saliency mechanism for retinal vessel segmentation. Similarly, Wu et al. [10] proposed a U-Net-based dense convolutional network with an attention gate to segment the retinal blood vessels and this method was able to address the problem of vanishing gradient. A lightweight multi-feature fusion network was introduced by Guo et al. [2] with spatial and channel attention to detect the hard exudates more efficiently and achieved high accuracy. In [11], a fully convolutional network was developed with short and long skip connections for accurate segmentation of exudates and optic disc. However, this method fails to detect Microaneurysms and Hemorrhages. An ensemble deep CNN based on modified U-Net (MU-Net) was introduced by Zheng et al. [12] for exudate detection employing conditional Generative Adversarial Network (cGAN) to improve the generalization property of the network and to solve the class imbalance problem in fundus photographs. In [13], a deep learning-based encoder-decoder architecture was developed for the semantic segmentation of bright lesions called exudates. Sambyal et al. [14] proposed an U-Shaped deep architecture based on a pre-trained residual network (ResNet34) for semantic segmentation of Exudates and Microaneurysms in IDRiD and e-Optha dataset and achieved state-of-art results. A small object detection algorithm was developed by [15] with multilayer attention and spatial confidence for the detection of subtle MAs in fundus images. In [16], a multi-lesion segmentation network called L-Seg was designed for segmenting multiple lesions in fundus images using a multi-channel bin loss function. A weakly-supervised multitask learning architecture was developed by Playout et al. [17] for the joint segmentation of dark and bright lesions in fundus images and achieved better results in different databases. Yan et al. [18] adopted U-Net as the basic component of their model integrating a local and global network with a fusion module for lesion segmentation in the fundus image. The existing deep learning-based lesion segmentation methods [6–10] segment the blood vessels and optic disc or a single type of lesion is considered for segmentation [2, 11–13, 15]. The retinal image comprises four different kinds of lesions and it is necessary to detect the multiple lesions simultaneously for DR diagnosis. Most of the existing methods [13, 16, 17] could not alleviate the problem of misclassification, there are still misclassified pixels. To overcome these limitations, a deep-learning-based Fusion Network was developed with an attention mechanism for lesion segmentation in the fundus image.

3 Proposed Methodology

The proposed deep Fusion Network (Fu-Net) is framed as an encoder-decoder architecture with the aim of detecting the DR lesions in fundus image. Initially, preprocessing is done to enhance the contrast and then segmentation is carried out using the improved Fu-Net with a weighted focal loss function. The final segmented results are obtained with a postprocessing stage.

3.1 Preprocessing

The images captured by a fundus camera are influenced by uneven illumination, noise, and low contrast. This impacts the lesion segmentation performance during training. To address these issues, the fundus images are preprocessed to enhance the image quality.

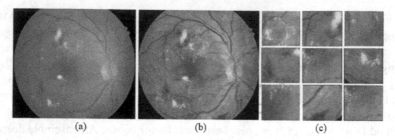

Fig. 2. (a) Input Image (b) Pre-processed Image (c) Sub images

At first, image denoising is performed using a Median filter and then image quality is upgraded using Contrast Limited Adaptive Histogram Equalization (CLAHE) [15]. Then the fundus images are normalized and are subjected to image dicing, in which a sliding window method with stride 64 is utilized to transform the fundus images with larger pixel dimension into sub-images with 256×256 pixels. Figure 2(a) represents a fundus image with DR lesions, Fig. 2(b) indicates the preprocessed image and Fig. 2(c) denotes the sub-images generated through the sliding window strategy.

3.2 Network Description

The proposed Fusion Net is the extension of the baseline U-Net model originally proposed by [5] for medical image segmentation. The Fusion Network architecture is designed as an encoder-decoder network with downsampling and upsampling layers followed by an activation function. As illustrated in Fig. 3, in the encoder part, a convolutional layer replaces the pooling layer for downsampling and reduces the loss of spatial information in the fundus image. An improved multiscale feature fusion (MFuse) block is embedded with the encoder network to extract the detailed lesion features of the fundus image at different scales. Furthermore, the MFuse block has a sequence of splitting operations with residual connections to extract the lesion features. A 3×3 dilated convolution [19] is followed by RReLU (Randomized Rectified Linear Unit) activation [20] and then for each step, the MFuse block performs a 3×3 and 5×5 convolution to divide the features. The features are then concatenated after three steps and are then fed into 1x1 convolution. In the decoder part, a channel attention module [21] is integrated to improve the concatenation procedure between the low-resolution decoder and the skip connection. A new weighted focal loss function is introduced during training to overcome the class imbalance problem and to enhance the discrimination ability of the model.

3.3 Loss Function

To optimize a model during the training process and to improve the network stability and segmentation accuracy the loss function is used. A modified weighted focal loss function is constructed by assigning different weights to different classes. The weighted focal loss function (F_{loss}) is represented as in Eq. (1)

$$F_{loss} = \sum_{k=1}^{m} - c\left(\left|z - P_k^z\right|\right)((1 - z)log(1 - P_k) + zlogP_k) \tag{1}$$

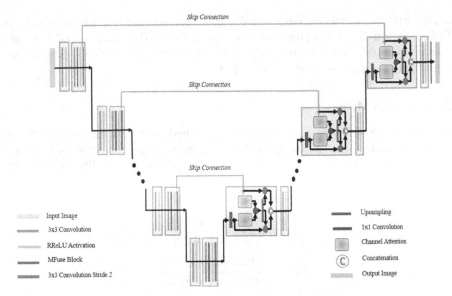

Fig. 3. Architecture of Fusion Network (Fu-Net) (Source: Xu et al. 2021, p. 3)

where m represents the total number pixels in the retinal fundus image and k signifies the kth sample. The pixels belonging to the normal unaffected area is set to a value 0 and the pixels from the lesion area is set to a value 1. The weight coefficient c is the ratio of abnormal pixels and the total pixels in all samples. The predicted probability of the model is defined as P_k and z is the tuning parameter.

3.4 Postprocessing

The segmentation output produced by the trained model has the same size as the input sub-image. Therefore, the sub-images are combined to obtain the final result of segmentation and by computing the average of the resultant sub-images the predicted label of a pixel is determined [21].

4 Experimental Setup

The proposed framework is implemented in Python using Keras with windows 10 OS and 8 GB memory. The batch size is taken as 64 and the network training is performed for 100 epochs with a initial learning rate 0.0002. Adam optimizer is adopted with $\beta_1 = 0.9$ and $\beta_2 = 0.999$ setting to optimize the model. Nonlinear activation RReLU is used in all the layers to reduce overfitting during training.

4.1 Dataset

The Indian Diabetic Retinopathy Image Dataset (IDRiD) [22] was used for the experiments conducted by the proposed deep Fusion Network. There are 81 color fundus images for pixel-level annotations which depict the distinct type of lesions such as MAs, HEMs, HEs, and SEs. Among these 54 images are considered for training and 27 images are taken for testing. The fundus images in the dataset are taken with 50 – degree field of view and have a resolution of 4288 × 2848.

4.2 Evaluation Metrics

To evaluate the experimental results and analyze the performance of the model in the DR lesion segmentation task, Sensitivity and F1-Score measures are computed. Sensitivity is a measure computed based on true positives (TP) and false negatives (FN) as on Eq. (2). F1-Score is the weighted harmonic mean of precision and recall that takes into account both false positives (FP) and false negatives (FN) as given in Eq. (3).

$$Sensitivity = \frac{TP}{TP + FN} \tag{2}$$

$$F1 = 2 \cdot \frac{Precision \cdot Recall}{Precision + Recall} \tag{3}$$

5   Results and Discussion

The experiments are carried out in the IDRiD dataset to assess and evaluate the performance of lesion segmentation in Diabetic Retinopathy images. The proposed Fusion Network model is trained and tested for different experiments. Initially, the model is evaluated with the original U-Net configuration and the results are computed. The next experiment is conducted on U-Net model replacing cross-entropy loss fuction with weighted focal loss (F_{loss}). The final experiment is conducted with the Fusion Network (Fu-Net) integrated with MFuse block and channel attention module. The segmentation results obtained for all the experiments in the IDRiD dataset are illustrated in Fig. 4.

The comparative analysis of the proposed Fu-Net model with the existing DR segmentation methods is tabulated in Table 1. The effectiveness of the Fu-Net model in the lesion segmentation task is evaluated by computing the metrics Sensitivity and F1-Score. From Fig. 4, it is observed that the segmentation results determined by the Fu-Net model are more nearby to the ground truth. The U-Net model and the U-Net with F_{loss} fails to identify the smaller lesions and are not able to produce clear lesion boundaries. Compared with the existing techniques [5, 13, 14, 21], Fu-Net method achieves 0.614%, 0.754%, 0.875%, and 0.815% sensitivity values and 0.737%, 0.864%, 0.913% and 0.895% F1-Score values for MAs, HEMs, HEs and SEs segmentation respectively. However, the sensitivity and F1-Score measure of Hard Exudate segmentation is the same for the Fu-Net model and the method developed by [21]. The Fu-Net method acheives 0.021%, 0.02%, and 0.022% improvement in the sensitivity measure and 0.019%, 0.019%, and 0.014% improvement in F1-Score measure for MAs, HEMs and SEs segmentation. The baseline U-Net model shows poor performance in segmenting all the four types of lesions. The method developed by [13] is suitable for the bright lesion detection but it was unable to segment the dark lesions. Furthermore, the modified U-Net designed by [14] is incapable of processing the MAs and Hemorrhages.

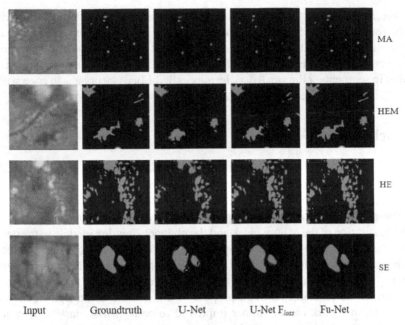

Input   Groundtruth   U-Net   U-Net F$_{loss}$ Fu-Net

Fig. 4. Segmented results of U-Net, U-Net F$_{loss}$ and Fu-Net

Table 1. Comparative results of the Fu-Net method and the existing methods on the IDRiD dataset

Methods	Sensitivity				F1-Score			
	MA	HEM	HE	SE	MA	HEM	HE	SE
Baseline U-Net [5]	0.482	0.636	0.781	0.673	0.619	0.775	0.863	0.802
Sambyal et al. [14]	0.564	0.698	0.832	0.756	0.665	0.806	0.890	0.856
Cristiana et al. [13]	0.581	0.712	0.840	0.788	0.685	0.822	0.904	0.877
Xu et al. [21]	0.593	0.734	0.875	0.793	0.718	0.845	0.913	0.881
Proposed Fu-Net	0.614	0.754	0.875	0.815	0.737	0.864	0.913	0.895

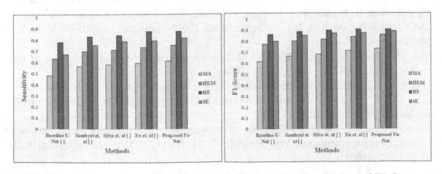

Fig. 5. Analysis of the comparative methods for metrics Sensitivity and F1-Score

Figure 5 depicts the performance analysis of the Fu-Net model and the existing methods for the metrics Sensitivity and F1-Score. Based on the performance analysis and the comparative results obtained, it is observed that the proposed deep fusion network outperforms the other methods in the lesion segmentation task. But still improvement is needed in segmenting the small dark lesions called Microaneurysms to cope with the real world clinical applications.

6 Conclusion

In this paper, a new deep fusion network called Fu-Net was developed for diabetic retinopathy lesion segmentation in fundus photographs. The network was developed based on the original U-Net with the max-pooling layer exchanged by a convolutional layer. The multiscale features of the input image are extracted in the encoder network with the integration of the MFuse block and the channel attention module in the decoder part fuse the information between the low-resolution decoder and the skip connection. The enhanced weighted focal loss function alleviates the misclassification problem and improves the network stability during training. The experimental results done on the IDRiD dataset show the efficacy of the proposed Fu-Net in the lesion segmentation task. In the future, this work can be further improved to perform DR classification and to make the model suitable for real-world application.

References

1. International Diabetes Federation: IDF Diabetes Atlas Ninth edition 2019 (2019)
2. Guo, X., Lu, X., Liu, Q., Che, X.: EMFN: Enhanced multi-feature fusion network for hard exudate detection in fundus images. IEEE Access 7, 176912–176920 (2019)
3. Kaur, J., Mittal, D.: A generalized method for the segmentation of exudates from pathological retinal fundus images. Biocybern. Biomed. Eng. 38(1), 27–53 (2018)
4. Saha, R., Chowdhury, A.R., Banerjee, S.: Diabetic retinopathy related lesions detection and classification using machine learning technology. In: Rutkowski, L., Korytkowski, M., Scherer, R., Tadeusiewicz, R., Zadeh, L.A., Zurada, J.M. (eds.) ICAISC 2016. LNCS (LNAI), vol. 9693, pp. 734–745. Springer, Cham (2016). https://doi.org/10.1007/978-3-319-39384-1_65
5. Ronneberger, O., Fischer, P., Brox, T.: U-Net: Convolutional networks for biomedical image segmentation. In: Navab, N., Hornegger, J., Wells, W.M., Frangi, A.F. (eds.) MICCAI 2015. LNCS, vol. 9351, pp. 234–241. Springer, Cham (2015). https://doi.org/10.1007/978-3-319-24574-4_28
6. Jiang, Y., Wang, F., Gao, J., Cao, S.: Multi-path recurrent U-Net segmentation of retinal fundus image. Appl. Sci. 10(11), 3777 (2020)
7. Zhang, S., et al.: Attention guided network for retinal image segmentation. In: Shen, D., et al. (eds.) MICCAI 2019. LNCS, vol. 11764, pp. 797–805. Springer, Cham (2019). https://doi.org/10.1007/978-3-030-32239-7_88
8. Wang, D., Hu, G., Lyu, C.: FRNet: An end-to-end feature refinement neural network for medical image segmentation. Vis. Comput. 37, 1101–1112 (2020). https://doi.org/10.1007/s00371-020-01855-z
9. Hu, J., et al.: S-UNet: A bridge-style U-net framework with a saliency mechanism for retinal vessel segmentation. IEEE Access 7, 174167–174177 (2019)

10. Wu, C., Zou, Y., Zhan, J.: DA-U-Net: Densely connected convolutional networks and decoder with attention gate for retinal vessel segmentation. IOP Conf. Ser. Mater. Sci. Eng. **533**, 012053 (2019)
11. Feng, Z., Yang, J., Yao, L., Qiao, Y., Yu, Q., Xu, X.: Deep retinal image segmentation: a fcn-based architecture with short and long skip connections for retinal image segmentation. In: Liu, D., Xie, S., Li, Y., Zhao, D., El-Alfy, E.S. (eds.) ICONIP 2017. LNCS, vol. 10637, pp. 713–722. Springer, Cham (2017). https://doi.org/10.1007/978-3-319-70093-9_76
12. Zheng, R., et al.: Detection of exudates in fundus photographs with imbalanced learning using conditional generative adversarial network. Biomed. Opt. Express **9**, 4863 (2018)
13. Silva, C., Colomer, A., Naranjo, V.: Deep learning-based approach for the semantic segmentation of bright retinal damage. In: Yin, H., Camacho, D., Novais, P., Tallón-Ballesteros, A.J. (eds.) IDEAL 2018. LNCS, vol. 11314, pp. 164–173. Springer, Cham (2018). https://doi.org/10.1007/978-3-030-03493-1_18
14. Sambyal, N., Saini, P., Syal, R., Gupta, V.: Modified U-Net architecture for semantic segmentation of diabetic retinopathy images. Biocybern. Biomed. Eng. **40**(3), 1094–1109 (2020)
15. Zhang, L., Feng, S., Duan, G., Li, Y., Liu, G.: Detection of microaneurysms in fundus images based on an attention mechanism. Genes (Basel) **10**(10), 817 (2019)
16. Guo, S., Li, T., Kang, H., Li, N., Zhang, Y., Wang, K.: L-Seg: An end-to-end unified framework for multi-lesion segmentation of fundus images. Neurocomputing **349**, 52–63 (2019)
17. Playout, C., Duval, R., Cheriet, F.: A novel weakly supervised multitask architecture for retinal lesions segmentation on fundus images. IEEE Trans. Med. Imaging **38**, 2434–2444 (2019)
18. Yan, Z., Han, X., Wang, C., Qiu, Y., Xiong, Z., Cui, S.: Learning mutually local-global u-nets for high-resolution retinal lesion segmentation in fundus images. In: Proceedings - International Symposium Biomedical Imaging, 2019-April, pp. 597–600 (2019)
19. Yu, F., Koltun, V., Funkhouser, T.: Dilated residual networks. In: Proceedings - 30th IEEE Conference on Computer Vision Pattern Recognition, CVPR 2017, 2017-January, pp. 636–644 (2017)
20. Xu, B., Wang, N., Chen, T., Li, M.: Empirical Evaluation of Rectified Activations in Convolutional Network (2015)
21. Xu, Y., Zhou, Z., Li, X., Zhang, N., Zhang, M., Wei, P.: FFU-Net: Feature fusion U-Net for lesion segmentation of diabetic retinopathy. Biomed Res. Int. **2021**, 1–12 (2021)
22. Porwal, P., et al.: IDRiD: Diabetic retinopathy - segmentation and grading challenge. Med. Image Anal. **59**, 101561 (2019)

Visibility Improvement in Hazy Conditions via a Deep Learning Based Image Fusion Approach

Satbir Singh[1], Asifa Mehraj Baba[2]([⊠]), Md. Imtiyaz Anwar[3], Ayaz Hussain Moon[2], and Arun Khosla[1]

[1] Dr B R Ambedkar National Institute of Technology, Jalandhar, India
khoslaak@nitj.ac.in
[2] Islamic University of Science and Technology, Kashmir, India
{asifa.baba,ayaz.moon}@islamicuniversity.edu.in
[3] B.R.A. Bihar University, Muzaffarpur, India

Abstract. Foggy weather prominently causes degradation in visibility due to the scattering of the atmospheric particles. Consequently, there arises a problem in identification of the precise object features by the human eye as well as the machine based computer vision systems. To encounter such situations, various adept mechanisms are required. The proposed scheme attempts to encompass the deep learning approach for the amalgamation of the RGB and Infra-red imaging in order to improve the vision quality of the hazy images. A fused image is obtained via intelligent conjunction of significant information from both the imaging schemes. Subsequently, the combined image is processed using a Dark Channel Prior algorithm and a bilateral filtering is used to maintain the edge information. Comparative results using various quality parameters including entropy, Standard Deviation, Similarity Index, and Peak Signal to Noise Ratio signifies that the proposed fusion scheme performs better than contemporary single image de-hazing algorithms.

Keywords: Image defogging · Fusion methods · Near-Infrared (NIR) · Evaluation metric · Deep Learning

1 Introduction

Image captured during poor weather conditions like fog and haze are degraded because the light reflected from the objects passing through atmospheric particles are partially attenuated. This degradation leads to the reduction in the colour and variance of the captured image and thus the objects within the image are difficult to identify by various computer vision systems. The image blurring and hence the distortion of a captured image leads to increase in road accidents every year. More than two and a half people die due to road accidents throughout the world from which 25% of such road accidents are due to the poor weather. The statistics from Ministry of Road Transport and Highways (MORTH) [1] shows the increase in number of accidents in India since last few years due to foggy weather and is depicted in a bar graph shown below (Fig. 1):

© Springer Nature Switzerland AG 2021
M. Singh et al. (Eds.): ICACDS 2021, CCIS 1440, pp. 410–419, 2021.
https://doi.org/10.1007/978-3-030-81462-5_37

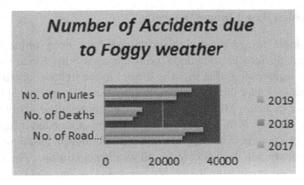

Fig. 1. Total Number of Accidents due to Foggy Weather in year 2017, 2018 & 2019

Apart from decreasing the number of road accidents there are several real world appli-cations like analysis of remote sensing imagery, video surveillance etc. where images with high quality and contrast are needed for data analysis [2, 3]. Therefore there is a need of an efficient fusion based algorithm for Image defogging for the removal of such unwanted weather degradations by various computer vision techniques [4].

Most of the computer vision algorithms are currently designed for use in clear weather and high visibility conditions. Since the beginning of transportation system, driving a vehicle is a challenge in adverse weather with poor visibility than in normal weather. Most of the accidents all over the world happen due to poor visibility and is increasing day by day. Vision enhancement process is just reducing these visual effects of bad weather for most of the outdoor applications. In bad weather the visibility is degraded and due to this poor visibility the driver can't see the vehicles/person/animal travelling towards or standing on the way. The degradation in visibility is mostly affected in fog during winters in north India and other similar regions of the world. It is estimated that more than two million people die per year of car accidents across the world out of which 24% of all crashes are due to bad weather. The main cause of road accidents is driver's inability to assimilate all visual information i.e. road signs or traffic signs while driving which are meant to assist the driver to reach the destination safely. Real-time vehicle recognition system is also helpful for intelligent system. Development of Road Sign Recognition (RSR) system in real time can be used to assist a driver in a vision based Driver Assistance System (DAS) to navigate the vehicle by keeping road signs information. These vision enhancement systems must be robust to any variation in weather conditions. A number of single image fog removal techniques have been reported in the literature. A single image fog removal techniques with visible light images fail in heavy fog with very less visibility if the edges of the objects in an image are not detected. Single image defogging techniques mostly enhances the contrast but fusion based defogging algorithm locates the position of the atmospheric light with high accuracy and then determines the enhancement of intensity of the image pixels. Thus thermal images provide better information than a normal visible light camera in foggy weather. A fusion technique [5] creates a single image is helpful to design a defogging algorithm that improves visibility.

1.1 Need of Fusion Based Image Defogging

A number of single image defogging techniques have been implemented by the researchers by taking into consideration visible images only. Since the visibility of details in such images captured at night is subject to the lighting condition, therefore the parts of the image under the poor lightning condition are usually displayed with low visibility and poor contrast. Moreover, the enhancement result of visible image would not clearly reveal all objects in the scene at night. On the other hand infrared images are used to differentiate targets from their background information based on the difference in the radiation pattern which works well in all weather conditions. Therefore there is a need of fusion process integrating the visible image and the infrared image together due to their complimentary characteristics, combining the advantages of Thermal information acquired from the infrared image and in-depth texture information extracted from the images captured in visible mode.

2 Literature Review

In the past, various image defogging algorithms based on fusion strategy have been proposed for visibility improvement; however there is a need for further enhancement in order to get the improved value of the quality parameters. Image defogging employing the fusion of visible and Infrared Image has gradually become a hot research field these days in order to provide better visual results. Many researchers have put forth different fusion strategies for combining the thermal image characteristics with the details extracted from the visual image.

In [5] a novel fusion based method has been employed for image defogging by decomposing the input images into constituent sub-images using Discrete Stationary wavelet transform. The Features obtained by applying DSWT to the individual images are applied with the Discrete Cosine Transform in order to centralize the Image components as per the energy of different frequencies. From the Discrete Cosine Transformed image, the regional detailed of the image are extracted by using the technique of Local Spatial Frequency (LSF). The features of the images extracted are then fused together by using the Contemporary fusion rule methods. The evaluation parameters obtained thus achieve better contrast enhancement. However, because of the smaller block size of chosen window, the average running time of the proposed system is more than existing fusion methods results in high computational cost. The Block size can be increased in order to decrease the computational time however that leads to the blocking artifacts in the reconstructed image.

Visible and NIR Image fusion technique put forth by researchers in [6] makes use of Weight map guided Laplacian-Guassian Pyramid for enhancing the visibility of the scene. In this proposed technique the luminance part of the RGB image is extracted and is fused with the NIR image. The luminance component after fusion is again converted from HSV colour space to RGB colour space. Further enhancement has been made by increasing the Colour and sharpness of fused image as a post-processing step. The only limitation with this proposed technique is that the Colour sharpness block produces more vivid image resulting in increased colour saturation in an image.

In [7] researchers have design a fusion algorithm based in which the base and the detail layers of both NIR image and the RGB image are obtained using Bilateral Filter (BF) and Weighted Least Square (WLS) filters. The fused detail layer of Infrared Image is then combined with the base layer of RGB image resulted in new luminance image. In the last step the new obtained luminance image is then combined with the chrominance part of RGB image to reconstruct the improved NIR-RGB fused image. Due to the fusion of images obtained from BF and WLS filters, the proposed technique can extract the fine details of an image resulting in better high frequency edges in the enhanced image However due to the use of bilateral filters several types of image artifacts like staircase effect and the appearance of false edges are introduced.

A novel fusion strategy combining the complementary characteristics of a guided filter and a two scale decomposition was implemented by the authors in [8]. In this proposed scheme both visible image and the infrared image are disintegrated into base layers and the detail layers with a two scale averaging filter. A phase congruency with guided filtering fusion rule is used to get the base layer and a large sum modified laplacian with guided filtering fusion rule is applied to get the detail layer. Both the base and the detail layer are fused together to obtain the resultant image with high contrast. This fusion method not only preserves the maximum details in a reconstructed image but can also remove the image artifacts successfully. However there is a need for improving the performance of the proposed technique which can be done by optimizing the parameters of a guided filter.

In [9] author has put forth a novel technique of visibility enhancement employing the fusion strategy using saliency detection and two scale decomposition. In this proposed scheme an averaging filter is used to convert both the images into base and detail layers to get the large scale and the small scale variations within the image. A saliency map detection algorithm is employed to get the more visual information interms of edges and lines from IR and Visual scenes for the fusion method thereby drags human attention towards the image quality. Finally, a new weight map construction method is employed to fuse the two images together in order to exploit their complimentary characteristics. The saliency features extracted by the proposed technique gives the accurate estimation of the edges and the lines present in the scene. Since the author has utilized the technique of 2-level decomposition, the proposed method is fast and efficient compared to other contemporary fusion techniques. Moreover, the method presented can be extended to the colour images by applying the fusion method on each colour channel individually. Although the technique presented by the author is best suitable for the removal of image artifacts and the noise present in the image there is still a scope of improvement as far as distortion in the output image is concerned.

In [10], a novel fusion based approach of combining visible and Infrared Images by Latent low-rank representation (LatLRR) has been put forth for visibility enhancement. Authors proposed a fusion technique in which low rank and the saliency parts are generated using LatLRR by disintegrating the Infrared and visible images into two scale representation. On the saliency parts of images, guided filter is applied for reducing the image artifacts, thereby making use of spatial consistency of the Image. Later Adaptive weights of the low rank parts are constructed using Fusion Global-Local Topology Particle Swarm Optimization (FGLT-PSO) in order to retrieve maximum information from the VI-NIR pair. The low rank and fused saliency part of an image are summed together in order to recover the resultant image. The presented scheme proved advantageous than several existing state of art fusion techniques though much execution time is required when number of iterations are increased.

A technique of hyper-spectral fusion method has been employed in [11] for Image dehazing by measuring the difference in intensity levels in certain areas of infrared-visible image pair. The method exploits the correlation of NIR image to red channel of color image by measuring the pixel irregularities. The probability of existence of the inconsistency of each pixel is measured and accordingly a mask is created for the modification. Finally a dehazzed image is obtained by incorporating the probability estimation into the fusion process. No halo effects and image artifacts have been seen in the dehazzed image preserving all the natural colors in the output image. However, the proposed method performs dehazing on luminance channel only and determining a threshold for classification of inconsistencies can be a challenging task.

3 Proposed Technique

In this proposed scheme the images in visible and thermal domain are converted into separate Base and detail Images as shown in Fig. 2. Since the individual base layers generally contain common features and redundant information therefore a less complex weighted average method is incorporated for the fusion of base images resulting in a fused based Image Fb. Further the detail images are enriched with varied information, so a deep learning framework known as VGG-19 Network is used to extract the multilayer deep features out of detail information of both IR-RGB pair of images. The VGG-19 is a Deep Convolutional Neural Network (CNN) with 19 fully connected layers mainly used to classify Images [16]. This Neural Network is used as a pre-processing model and has improved network depth than other traditional Convolutional Neural Networks. It consists of multiple Convolutional and non-linear activation layers, which is better than a single convolution Neural Network to extract image features in an improved manner, thus making this proposed scheme more beneficial in getting the enhanced results. Multilayer fusion strategy is employed for the fusion of detail information of each image using Deep Learning Artificial Neural Networks to get the Deep Fused Image Fd. The Fused base Image and the Deep Fused image are finally combined together to get the final fused output image as depicted in the block diagram given below. The fused image is then dehazzed using Dark Channel Prior which is the technique used to approximate the depth of haze directly and is used to recover the haze free images with high contrast and quality. Bilateral Filter is also used as an additional block to preserve the edges in the Dehazzed Image. The guided filter used proved to be efficient being naturally fast regardless of Kernel Size and the range of Intensity levels.

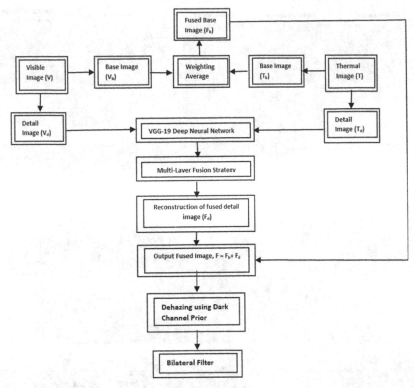

Fig. 2. Proposed Deep Learning based Dehazing Algorithm

4   Experimental Results and Discussion

The proposed algorithm has been simulated on HP Zbook with 1.65 GHz Intel Core i5, 16 GB Memory using MATLAB software in which eight different pairs of Visible-Infrared Images are used for qualitative and quantitative comparisons as shown in Fig. 3 and 4 respectively. The Image database from scene0 to scene 8 has been obtained from the source specified in [13] and [14]. The database was captured by the FLIR Duo R 640, 19 mm, 30 Hz Camera and was pre-processed by means of a FLIR tools Software by the authors in [13] and [14].

After the visual inspection of the images in Fig. 3, it can be easily observed that the edges remain preserved and clear in appearance due to the use of bilateral filters [12] compared with the images dehazzed by the scheme presented in [15]. Out of eight images under examination, it has been seen that few images like scene 3, 4 and 6 are vivid in colour and for other images, the performance still remains the challenge. Further, the results obtained by the proposed scheme are free from any halo effects, making the algorithm suitable for both outdoor and indoor applications.

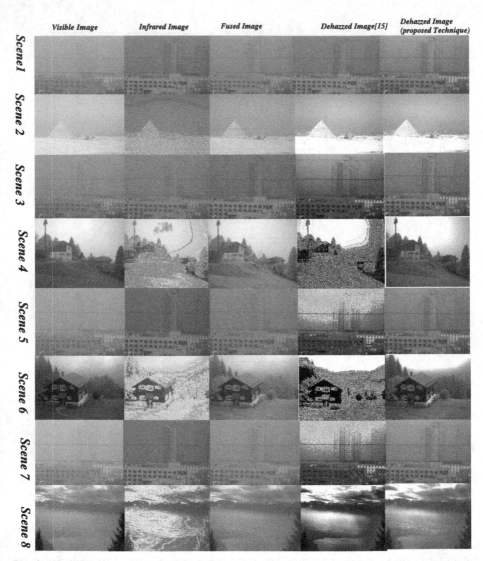

Fig. 3. Fusion performance of eight Infrared and Visible Image Pairs (Images Courtesy: Database ([13] (Scene 1, Scene 3, Scene 5, Scene 7)), ([14] (Scene 2, Scene 4, Scene 6, Scene 8))

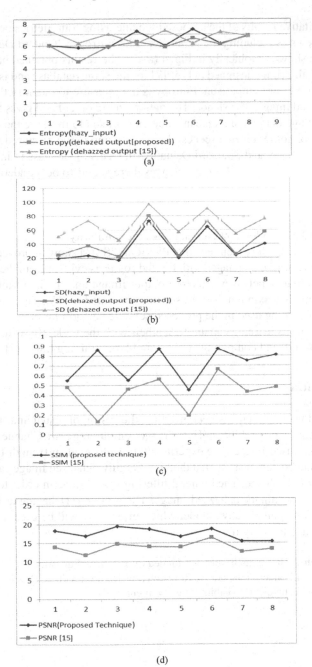

Fig. 4. Quantitative Comparisons of four metrics on eight visible-Infrared Image Pairs using the proposed scheme and Scheme presented in [15]; (Images Courtesy: Database [13, 14])

The quantitative results of the proposed scheme using four metrics on eight Visible-Infrared Image pairs including the parameters like Entropy, Standard Deviation, SSIM and PSNR are shown graphically in Fig. 4 (a–d). The results obtained have been compared with the algorithm designed in [15, 17] for the same database and is shown graphically in the figure for comparison. Entropy is the measure to quantify the information content in a fused image. The standard deviation on the other hand gives the measure of deviation of image data from its mean value and is used to measure the contrast in the fused image. For some selected scenes like scene 4 and 6, the proposed method shows advantages in Entropy and Standard Deviation in terms of richness in information and increase in contrast however, for other images there seems to be a tradeoff between the Entropy and PSNR.

The Peak Signal to Noise Ratio (PSNR) reflects the measure of distortion in a fused image and is equal to ratio of Peak Value Power to the Noise Power in the fused Image. Structural Similarity Index however, signifies the similarity between the source Image and the fused Image. It has been seen that PSNR and SSIM show improved performance in recovering the grayscale images with limited fog as is evident from the line graph (c) and (d), the similarity between dehazzed output image and the input image using the proposed scheme is maximum than the scheme presented in [15]. This is due to the fact that the Deep Learning used in the proposed scheme perform far better than traditional algorithms due to its learning capabilities, albeit with the trade-offs with respect to the requirement of high end computing systems and large training time [17].

5   Conclusion

A Deep Learning Approach has been used for the fusion of visual and Infrared Image intended to enhance the visibility of Hazy Images employing Dark Channel Prior method for dehazing the fused Image. The experimental results evident from different analysis parameters obtained demonstrate that the high Quality dehazzed Image can be obtained using the proposed scheme. The bilateral filtering is also done in order to maintain the edge information within the dehazzed Image. The proposed Dehazing technique can be simply changed into an effective model which effectively will be used in cars to avoid road accidents due to heavy fog.

Acknowledgment. The research work was funded by Technical Education Quality Improvement Program (TEQIP-III) under Collaborative Research Scheme project titled: Development of Fusion based Defogging Technique for visibility improvement.

References

1. Ministry of Road Transport & Highways, Government of India. https://morth.nic.in/. Accessed 22 Feb 2021
2. Fan, X., Wang, L.: Image defogging approach based on incident light frequency. Multimed. Tools Appl. **78**(13), 17653–17672 (2019). https://doi.org/10.1007/s11042-018-7103-1
3. Narasimhan, S.G., Nayar, S.K.: Vision and the atmosphere. Int. J. Comput. Vis. **48**, 233–254 (2002)

4. Nayar, S.K., Narasimhan, S.G.: Vision in bad weather. In: Proceedings of the IEEE International Conference on Computer Vision, pp. 820–827. IEEE (1999)
5. Jin, X., et al.: Infrared and visual image fusion method based on discrete cosine transform and local spatial frequency in discrete stationary wavelet transform domain. Infrared Phys. Technol. **88**, 1–12 (2018)
6. Vanmali, A.V., Gadre, V.M.: Visible and NIR image fusion using weight-map-guided Laplacian-Gaussian pyramid for improving scene visibility. Sadhana - Acad. Proc. Eng. Sci. **42**, 1063–1082 (2017)
7. Sharma, V., Hardeberg, J.Y., George, S.: RGB-NIR image enhancement by fusing bilateral and weighted least squares filters. J. Imaging Sci. Technol. **61**, 1–9 (2017)
8. Wang, X., Nie, R., Guo, X.: Two-scale image fusion of visible and infrared images using guided filter. In: ACM International Conference Proceeding Series, pp. 217–221. Association for Computing Machinery, New York (2018)
9. Bavirisetti, D.P., Dhuli, R.: Two-scale image fusion of visible and infrared images using saliency detection. Infrared Phys. Technol. **76**, 52–64 (2016)
10. Han, X., et al.: An adaptive two-scale image fusion of visible and infrared images. IEEE Access. **7**, 56341–56352 (2019)
11. Umbgen, F.D., El Helou, M., Gucevska, N., Usstrunk, S.S.: Near-Infrared Fusion for Photorealistic Image Dehazing (2018)
12. Li, S., Kang, X., Hu, J.: Image fusion with guided filtering. IEEE Trans. Image Process. **22**, 2864–2875 (2013)
13. Liang, J., Zhang, W., Ren, L., Ju, H., Qu, E.: Polarimetric dehazing method for visibility improvement based on visible and infrared image fusion. Appl. Opt. **55**, 8221 (2016)
14. Brown, M., Susstrunk, S.: Multi-spectral SIFT for scene category recognition. In: Proceedings of the IEEE Computer Society Conference on Computer Vision and Pattern Recognition, pp. 177–184. IEEE Computer Society (2011)
15. Anwar, M.I., Khosla, A.: Vision enhancement through single image fog removal. Eng. Sci. Technol. Int. J. **20**, 1075–1083 (2017)
16. Khan, A., Sohail, A., Zahoora, U., Qureshi, A.S.: A survey of the recent architectures of deep convolutional neural networks. Artif. Intell. Rev. **53**, 5455–5516 (2021). https://doi.org/10.1007/s10462-020-09825-6
17. Mathew, A., Amudha, P., Sivakumari, S.: Deep learning techniques: an overview. In: Hassanien, A.E., Bhatnagar, R., Darwish, A. (eds.) AMLTA 2020. AISC, vol. 1141, pp. 599–608. Springer, Singapore (2021). https://doi.org/10.1007/978-981-15-3383-9_54
18. Tyagi, V.: Understanding Digital Image Processing. CRC Press, Boca Raton (2018). https://doi.org/10.1201/9781315123905

Performance of Reinforcement Learning Simulation: x86 v/s ARM

Sameer Pawanekar[1(✉)] and Geetanjali Udgirkar[2]

[1] Indian Institute of Technology Guwahati, Guwahati, India
p.sameer@iitg.ac.in
[2] CMR Institute of Technology, Bengaluru, India
geethanjali.p@cmrit.ac.in

Abstract. In this paper, we develop a method for deep reinforcement of training simulation for Quadcopter and perform a comparison in the training of the neural network on x86 based machines and also on ARM-based machines. We consider the workload of training a quadcopter neural which is run on both the machines. We use a desktop for x86 machine, whereas, we consider a Raspberry pi based computer for simulation of RL algorithm. Both the machines are operating on Linux based systems. We observe that x86 based machine has a better performance than its ARM counterpart.

Keywords: CPU performance · Reinforcement learning · Quadcopter simulation · Network training

1 Introduction

There has been a large influence and work in progresss on Reinforcement learning since the DeepMind had their results published about using deep RL to play Atari games [1]. In traditional methods of reinforcement learning implementation, the Q-values are arranged in a table. Is such scenario, the model finds it difficult to extract the Q-values. In deep RL, the Q-values are associated with the neural network which outputs Q-values, which makes it convenient. Researcher can avoid creating large Q-table and obtain the Q-value from neural network. This change makes deep Q learning handle large input (example, image data). In reinforcement learning, the agent receives observation and reward from the unknown environment in return of the action it performs towards the goal. Reinforcement learning does not deal with large amount of data as compared to the deep learning methods. The reinforcement learning methods learn from the environment in a step by step manner. In Sect. 2 we discuss the works that present reinforcement learning methods. The Sect. 3 talks about Problem formulation for Reinforcement learning. This is followed by Experiments in Sect. 4 and then we Conclude in Sect. 5.

© Springer Nature Switzerland AG 2021
M. Singh et al. (Eds.): ICACDS 2021, CCIS 1440, pp. 420–430, 2021.
https://doi.org/10.1007/978-3-030-81462-5_38

2 Related Work

Deep reinforcement learning has been used in many applications and games. Examples of these games are Go [2] and first person shooter (FPS) game [3]. The research in [3] indicates that their agent plays single player and multiple player FPS better than humans. Besides playing games and robotics, the deep RL is also used in controlling quadcopter. There are multiple objectives when ying a quadcopter. Formost objective of control is making the quadcopter stable.

Besides stability, the quadcopter is also supposed to avoid collisions from the blockages in the path. The stability may be hindered by strong wind, whereas, the collision can occur due to blockages in the trajectory paths, The recent work in quadcopter stability has been presented in the paper [4]. Researchers aim to avoid quadcopter colliding intensively, or avoid potential collision early. This was the main topic of interest control robot and autopilot [5–7]. Notice that, robot and car can only move in horizontal, but quadcopter can y up and down which makes the quadcopter control different and difficult. There are some obstructions (such as a wall), where the vehicle can get stuck, but the quadcopter does not get stuck. There are some studies which present methods to avoid the quadcopter obstacle crash. [8] present a method to reduce the velocity of the quadcopter when it is about to collide an obstacle. In [9], Pawanekar et al. present a hardware architecture for reinforcement learning. Some of the latest works, such as, [10–16], worked on the problem of reinforcement learning. In this paper, we present a quadcopter agent for bypassing obstacles in 3D environment and perform the comparison of neural network training times on x86 and ARM machines.

3 Reinforcement Learning Problem Formulation

Deep Q-Network: is Reinforcement learning policy of learning for an agent which interacts with an unknown environment. The agent receives observation and reward from the unknown environment in return of the action it performs towards the goal. Reinforcement learning does not deal with large amount of data as compared to the deep learning methods. The reinforcement learning methods learn from the environment in a step by step manner. At every step, the agent receives an observation and present state from the environment which helps it to decide the action that it is going to take using a suitable policy, and achieves a reward R_t from the function of reward. The primary objective of an agent is to maximize the rewards R_t that it receives.

$$ R_t = \sum_{t'=t}^{T} \gamma^{t'-t} r_{t'} \tag{1} $$

where the terminal step is represented by T, γ is the discounted rate which affects the value of rewards in the future. In Q-learning, we will estimate the value of Q-value for every action with policy π and preserve the value of Q-value in table Q-table for querying.

$$Q^\pi(S, a) = E[R_t | S_t = s, a_t = a] \tag{2}$$

Generally, we select the action which has the largest value of Q, and we define Q*(S, a) as the maximum value for expected return which can be achieved by following a strategy.

$$Q(S, a) = \max_\pi E[R_t | S_t = s, a_t = a] \tag{3}$$

$$= \max_\pi Q^\pi(S, a) \tag{4}$$

However,if the highest Q-value is chosen, it may result in agent being not able to get new information about the environment. ϵ-greedy strategy can be used to overcome this problem. When the agent takes any action, there is a probability ϵ that any action is taken randomly instead of instead of doing the action with maximal Q-value. ϵ is not stable value, it gradually decreases with the increase in number of steps. After taking action, agent will obtain the reward for the action. With every reward received, agent updates the value of Q saved in Q-table, and which helps the agent to take a better action in each step.

$$Q^\pi(S, a) = (1 - \alpha) \cdot Q^\pi(S_t, a_t) + \alpha \cdot Q^*(S, a) \tag{5}$$

where, the value of α is the rate of learning. Deep Q-Network does replacement of Q-table to a neural network with θ as the weights of the network. Hence, the weights parameters θ also need to be updated. We need to adjust θ parameters in such a way that Q becomes closer to Q^*. Q-network can be trained by loss function

$$L_i(\theta_i) = E_{s_t, a, R, s_{t+1}}[(y_i - Q_\theta(S, a))^2] \tag{6}$$

then, we can update gradient with following equation:

$$\nabla_{\theta_i} L_i(\theta_i) = E_{s_t, a, R, s_{t+1}}[(y_i - Q_\theta(S, a))\nabla_{\theta_t} Q_{\theta_t}(S, a)] \tag{7}$$

The above formula could be approximated as:

$$\nabla_{\theta_i} L_i(\theta_i) \approx (y_i - Q_\theta(S, a))\nabla_{\theta_t} Q_{\theta_t}(S, a) \tag{8}$$

In general, Q-learning requires time to update its parameters. This update happens by experience of replay. Experience is recorded by the agent(including state, action, reward, next state). When enough steps are performed by the agent select mini-batch from experience randomly. Training the model by using mini-batch, and weights update is known as replay experience. We have designed a reinforcement learning system for quadcopter where it takes off from the ground, starting at required angle and velocity. We implement the system on two machines (x86 and ARM)

3.1 Quadcopter Navigation Function

is to obtain the shortest path which is a straight line from current position to the goal position. We then calculate the angle of departure of the quadcopter as well as its velocity. We use simulation using tensor ow in our experiments. The code is written in Python and simulated on Linux (x86 and ARM)

Collision avoidance function To avoid the obstacles on the path towards the goal, deep Q-network is used. The images obstacles are input into the program We have a queue for depth image. As we get new depth image, new one will enter queue and pop the processed one.

Convolutional and pooling layer Once the preprocessing of the depth image is performed, the image is then fed to the convolution and max pooling layer. The convolution layer sends the output to ReLU and connection is established with pooling layer. We have three convolution and pooling structures in out neural network model. We use stride of 2 and padding of 2. Fully connected layer At fully connected layer we attend the output of convolution and pooling layer. These attened nodes are then fed to FC layer to get 512 outputs. After sending 512 outputs to ReLU, we obtain 13 output nodes. These become the Q-value for 13 actions of the quadcopter.

4 Experiment

(See Fig. 1).

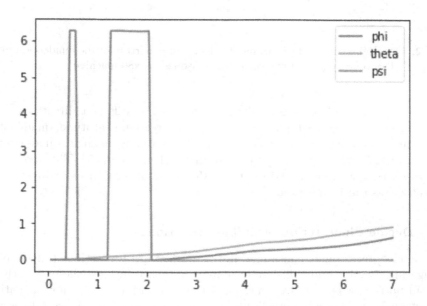

Fig. 1. This Figure shows the learning of angle of departure of the quadcopter as it takes off. The angle is with respect to the episodes for x86 machine

4.1   Exploration and Action Strategy

As we discussed in the previous section, Q-learning typically chooses the action considering ϵgreedy strategy. In our approach, we output 13 different agent actions related with quadcopter motion, and set the initial value of ϵ to 0.1. The value of ϵ gradually decreases with the increase in the steps. In the Final stage, ϵ decreases to 0.001.

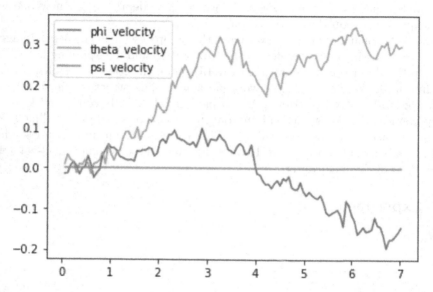

Fig. 2. This Figure shows the learning of velocity of departure of the quadcopter as it takes off. The velocity is with respect to the episodes for x86 machine

Using the 13 actions quadcopter is guided to y in different directions. The actions can be catagorized into two parts. First direction is altitude, quadcopter will y up, y down, y forward. Other movements are taking an angle, quad- copter will turn left 5°, turn right 15°, turn left 15°, and turn right 5°. The above two categories comprises of total 12 action. Movement in the forward direction is considered as the 13th action.

4.2   Decide Final Action with Two Functions

The quadcopter training considered in this project has two actions (quadcopter navigating function and collision avoidance function). In both these objectives, the output is a degree of movement. Thus, the degree of movement for navigation and obstacle avoidance iscontroller by the actions of quadcopter agent (Fig. 5).

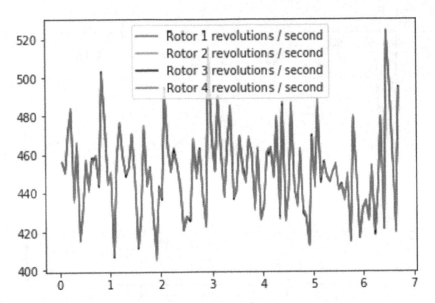

Fig. 3. This Figure shows the learning of choice of action by the agent at the time of departure of the quadcopter as it takes off. The choice of action is shown with respect to the episodes for x86 machine

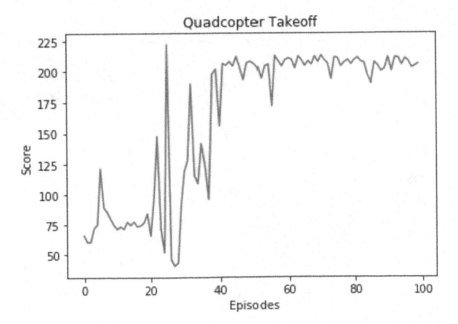

Fig. 4. This Figure shows the reward obtained by the agent after the departure of the quadcopter as it takes off. The obtained reward are with respect to the episodes for x86 machine

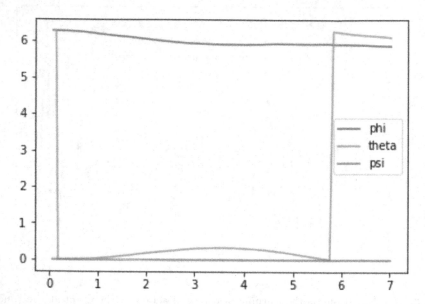

Fig. 5. This Figure shows the learning of angle of departure of the quadcopter as it takes off. The angle is with respect to the episodes for ARM machine

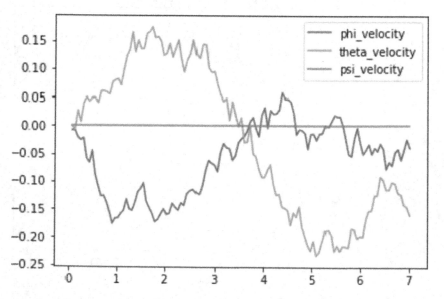

Fig. 6. This Figure shows the learning of velocity of departure of the quadcopter as it takes off. The velocity is with respect to the episodes for ARM machine

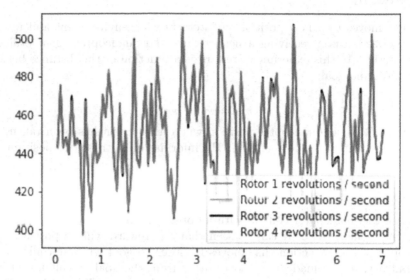

Fig. 7. This Figure shows the learning of choice of action by the agent at the time of departure of the quadcopter as it takes off. The choice of action is shown with respect to the episodes for ARM machine

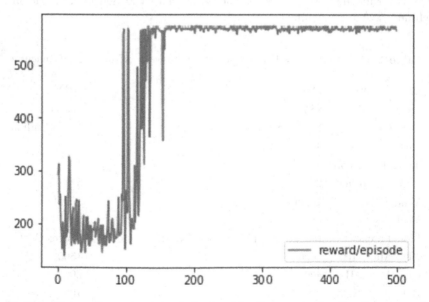

Fig. 8. This Figure shows the reward obtained by the agent after the departure of the quadcopter as it takes off. The obtained reward are with respect to the episodes for ARM machine

4.3   Reward

As agent moves to a new position, it interacts with environment and receives a new reward. Upon receiving a new reward, the quadcopter agent evaluates its last action. In this experiment, reward is a function of the distance between quadcopter and goal:

$$D = \|P_{move} - P_{goal}\|_2 \qquad (9)$$

where, P_{move} is quadcopter position where the action has been taken, and P_{goal} is the position of the goal. With the information of these two-positions, we are able to form reward function.

$$R = D_{now} - D_{last} - 100 \cdot Collision \qquad (10)$$

When the quadcopter collides with an obstacle, we say that the Collision is 1 else the Collision is 0. It can be seen that the reward will be positive when quadcopter takes an action that moves it closer to goal without collision. On the contrary, if the quadcopter moves away from the goal or collides with an obstacle, the reward becomes negative. Or quadcopter collide with anything. We have plotted the angle, velocity, action and reward for x86 and ARM in the Fig. 3, 2, 3, 4, 7, 6, 7 and 8 respectively. Table 1 present a comparison between the CPU characterstics of x86 and ARM. Table 2 presents comparison of runtime consumed by x86 and ARM machines against the number of episodes. Figure 9 shows the runtime comparison for the training of the neural network on x86 machine v/s ARM machine.

Table 1. Comparison of CPU characterstics between ARM and x86 machine

Characterstics	x86	ARM
Architecture	x86_64	armv7l
Byte Order	Little Endian	Little Endian
CPUs	4	4
Threads per core	2	1
Core per socket	2	a4
Vendor ID	GenuineIntel	ARM
Model	78	3
Model name	IntelR CoreTM i3-6006U CPU @ 2.00 GHz	Cortex A72
CPU Max MHz	2000	1500
CPU Min MHz	400	600
BOGOMIPS	3999.93	108

Table 2. Comparison of runtime consumed by ARM and x86 machines

Episodes	ARM runtime	x86 runtime
250	33 s	12 s
500	65 s	23 s
750	95 s	32 s
1000	134 s	46 s

5 Conclusion

This paper presents a deep reinforcement learning method for Quadcopter training simulation and performance a comparsion in the training of the neural network on x86 based machine and also on ARM based machine. We observe that the x86 machine of the given configuration is more suitable for training of the rein- forcement learning workload.

References

1. Mnih, V., et al.: Playing atari with deep reinforcement learning (2013). CoRR, vol. abs/1312.5602 http://arxiv.org/abs/1312.5602
2. Silver, D., et al.: Mastering the game of Go with deep neural networks and tree search. Nature **529**(7587), 484–489 (2016)
3. Lample, G., Chaplot, D.S.: Playing FPS games with deep reinforcement learning (2016). CoRR, vol. abs/1609.05521, http://arxiv.org/abs/1609.05521
4. Hwangbo, J., Sa, I., Siegwart, R., Hutter, M.: Control of a quadrotor with rein- forcement learning (2017). CoRR, vol. abs/1707.05110, http://arxiv.org/abs/1707.05110
5. Beitelspacher, J.: Applying reinforcement learning to obstacle avoidance (2005)
6. Azouaoui, O., Ouaaz, M., Farah, A.: Reinforcement learning (rl) based collision avoidance approach for multiple autonomous robotic systems (ars) (2001)
7. Long, P., Fan, T., Liao, X., Liu, W., Zhang, H., Pan, J.: Towards optimally decen- tralized multi-robot collision avoidance via deep reinforcement learning (2017). CoRR, vol. abs/1709.10082, http://arxiv.org/abs/1709.10082
8. Kahn, G., Villaflor, A., Pong, V., Abbeel, P., Levine, S.: Uncertainty-aware rein- forcement learning for collision avoidance (2017). CoRR, vol. abs/1702.01182, http://arxiv.org/abs/1702.01182
9. Pawanekar, S., Udgirkar, G.: Highly scalable processor architecture for reinforce- ment learning. In: 2020 Third International Conference on Smart Systems and Inventive Technology (ICSSIT), pp. 987–991 (2020)
10. Dooraki, A.R., Lee, D.J.: Reinforcement learning based flight controller capable of controlling a quadcopter with four, three and two working motors. In: 2020 20th International Conference on Control, Automation and Systems (ICCAS), p. 161–166 (2020)
11. Wu, T., Tseng, S., Lai, C., Ho, C., Lai, Y.: Navigating assistance system for quad- copter with deep reinforcement learning. In: 2018 1st International Cognitive Cities Conference (IC3), pp. 16–19 (2018)

12. Walvekar, A., Goel, Y., Jain, A., Chakrabarty, S., Kumar, A.: Vision based autonomous navigation of quadcopter using reinforcement learning. In: 2019 IEEE 2nd International Conference on Automation, Electronics and Electrical Engineering (AUTEEE), pp. 160–165 (2019)
13. Karthik, P.B., Kumar, K., Fernandes, V., Arya, K.: Reinforcement learning for altitude hold and path planning in a quadcopter. In: 2020 6th International Conference on Control, Automation and Robotics (ICCAR), pp. 463–467 (2020)
14. Ruan, X., Ren, D., Zhu, X., Huang, J.: Mobile robot navigation based on deep reinforcement learning. In: 2019 Chinese Control And Decision Conference (CCDC), pp. 6174–6178 (2019)
15. Al-Mahbashi, A., Schwartz, H., Lambadaris, I.: Machine learning approach for multiple coordinated aerial drones pursuit-evasion games. In: 2020 IEEE International Conference on Systems, Man, and Cybernetics (SMC), pp. 642–647 (2020)
16. Wiedemann, T., Vlaicu, C., Josifovski, J., Viseras, A.: Robotic information gathering with reinforcement learning assisted by domain knowledge: an application to gas source localization. IEEE Access 9, 13159–13172 (2021)

A Performance Study of Probabilistic Possibilistic Fuzzy C-Means Clustering Algorithm

J. Vijaya$^{(\boxtimes)}$ and Hussian Syed

VIT –AP University, Amaravati, Andrapradesh, India

Abstract. With the rapid proliferation of data across every stream makes raw data practically unusable. In this scenario, clustering has a major impact in grouping similar data into a dataset. This enhances the usability and meaningfulness of data, and further, the quantitative analysis can also be performed. In our existing research, a novel Probabilistic Possibilistic Fuzzy C-Means (PPFCM) clustering method is proposed. In this paper, the proposed PPFCM clustering technique is quantitatively evaluated based on several metrics and the accuracy of the clustering outcome as well as the execution output are investigated. A comparative study is made with the proposed PPFCM clustering with the traditional clustering methods, and the results are plotted. In this work, six benchmark datasets based on different application is used for evaluating the performance of PPFCM clustering method. To measure the productivity of the proposed clustering technique the Sum of Square Error (SSE) metric is used and it is found that the methodology mentioned above performs well for segmentation.

Keywords: Clustering · K-Means · K-Medoid · FCM · PCM · FPCM · PFCM · PPFCM · SSE

1 Introduction

Clustering is method of distributing a collection of data into many clusters and the data members in each cluster will be identical to each other and different from other clusters [1]. Unsupervised learning is a part machine learning algorithm which draws implications from different datasets with contains input data which are unlabeled. This is used to find hidden patterns and cluster the data by using exploratory learning algorithms and data analysis. There are various clustering methods available in machine learning which produces a different output in several patterns. The selection of a clustering method depends on the user requirement factors and according to the requirement of the system. Primarily in data mining applications in recent times, there is an explosion of data, e.g., 10 trillion records, which is an unimaginable number in a human perspective. The usage of computational resources is the most important factor. The amount of resources required for processing differs greatly between the various methods. For all but minimal data sets, certain approaches become impractical, e.g., hierarchical clustering

© Springer Nature Switzerland AG 2021
M. Singh et al. (Eds.): ICACDS 2021, CCIS 1440, pp. 431–442, 2021.
https://doi.org/10.1007/978-3-030-81462-5_39

techniques are in $O(N*)$ and $O(N3)$ whereas non-hierarchical techniques usually have $O(N)$, the total number of records in the data set is denoted by N [2]. Second, there are a few clustering strategies the output is stronger when it comes to distinguishing specific types of clusters. The central component of computing the clustering consistency depends on the similarity measure used and how it incorporates the different types of attributes present in the data. According to these characteristics, a definitive account of similarity measures is given for each model. The surveys in hybrid models (Clustering combined with classification) unsupervised learning methods has a prominent role in forecasting quality results. If the clustering algorithm is chosen right and if it performs properly, the correctness of prediction will touch the maximum [3]. For this function, customers are grouped together and evaluated for enhanced segmentation. Also, many of the application models proposed in literature proved that a single classification model does not yield satisfactory results compared to clustering combined with classification techniques [4]. So our main objective in this work is to design a novel clustering technique that serves to solve issues that underlie in clustering. In this work, we evaluate the PPFCM clustering model. The data is divided into U clusters using this model. To cluster the data, the PPFCM algorithm uses the typicality and membership matrices, as well as probability values which are not labeled [5]. The novel method of PPFCM is a benchmark to the traditional fuzzy based clustering like Fuzzy C-Means (FCM), Probabilistic C-Means (PCM), Fuzzy Probabilistic C-Means (FPCM), Probabilistic Fuzzy C-Means (PFCM), Balanced Iterative Reducing and Clustering using Hierarchies (BIRCH), Hierarchal clustering, Self-Organizing Map (SOM) and Partial based clustering (K-Means, K-Medoid, Weighted K-Means). Six proven datasets from the UCI machine learning repository are used in this study [6]. An analysis is made with the help of the dataset to evaluate the performance of PPFCM algorithm quantitatively. Also, the same dataset is evaluated using the other traditional clustering algorithms to make a comparative study of the performance. To measure the productivity of the proposed clustering technique the SSE metric is used.

2   Literature Survey

Clustering analysis is widely used in many research areas such as medical data segmentation, business customer segmentation, text segmentation, recommender system, crime analysis, human genetic clustering, fraud deduction and so on [7, 8]. The authors of paper [9] proposed a modified K-Means algorithm for segment the data. This algorithm clusters the telecommunication customers churn data. The near instance is determined by calculating the similarity between the instance under examination and the K number of clusters. Customer mixed data, also known as churn data, was predicted using the cluster with the most similar instances. Consider the case of the identified close cluster having two class labels which have mixed and distinct instances; the test data is passes to FOIL classification algorithm for calculating the close and non-close data. The authors of paper [10] used the credit card churn prediction dataset for constructing the hybrid model which utilizes an upgraded Rough K-Means algorithm for grouping and traditional classification techniques for classification. By performing this comparison, they justified that their proposed clustering combined classifier technique to be more productive than the performance of single classifier. In paper [11] proposed an enhanced FCM clustering approach to segment the MovieLens recommendation dataset. In addition, an improved algorithm known as the Modified Cuckoo Search (MCS) algorithm was proposed to make each cluster's data points as optimal as possible, which aids us in making better recommendations. In paper [12] identified the credit card fraud based on K-Means clustering algorithm, which helps the user by identifying whether the transaction performed is valid or fraud transaction(invalid). Clusters are formed to identify the intensity of fraud in credit card transaction which is low risk, high risk and very high risk. In paper [13] proposed foggy K-Means clustering technique for lung cancer segmentation; It is a segmentation clustering method that divides objects into K groups by first computing the mean and then computing the distance between each point from the cluster mean. In paper [14] presented a novel supervised neural approach which aids the user for text segmentation. In this work an attention-based bidirectional LSTM model where combination of sentences is captured using the neural network and it is based on contextual knowledge, the various segments that are captured are predicted. This model has the flexibility of automatically handling variable sized context information which can contain different contextual information. In paper [15] proposed plant leaf disease segmentation using a genetic algorithm which aids the farmers to identify the type of leaf disease and apply pesticides accordingly. This genetic algorithm uses image segmentation technique which helps for automatic identification and with the help of the image segmentation the classification of plant leaf diseases is done. In paper [16] focused the social network segmentation which is initiator for a healthy discussion which help to form categories that are widely accepted by everyone and it aids others to identify these categories an data longer run they set high level objectives for the development of future world. Clustering and analysis helps to identify areas wherein crimes are high and also to find the specific type of crime in the particular area. This clustering which helps us identify the specific crime that happens in specific areas those are areas can be blacklisted or identified as hotspot and law and force can be managed accordingly. In paper [17], they analyzed violent crime trends from the Communities and Crime Un normalized data collection from the University of California-Irvine archive, as well as real crime statistics from neighborhoodscout.com for the state of Mississippi.

3   Methodology: PPFCM Clustering Algorithm

FCM clustering algorithm [18] is the base model of the proposed PPFCM which is discussed in this work. The FCM algorithm is designed in such a way that it can handle large dataset and after clustering it. The proposed PPFCM is similar in operation but in a better fashion. The stepwise procedure of working of PPFCM clustering algorithm is given below.

- The primary consideration in PPFCM is number of clusters (C). There is a description given for each cluster. If the number of clusters (C) is three, it constructs three clusters.
- With the number of clusters (C) the fuzzy membership (M_{uv}) distances is calculated. M_{uv} is the distance between each cluster created and the corresponding samples or user. The probability matrix (P_{uv}) and the member-ship matrix (M_{uv}) are derived

from probability clustering and FCM clustering. Then proposed PPFCM membership matrix is calculated by the product of P_{uv} and M_{uv}. Now the membership function is calculated with the calculated function. The customer is attached to the cluster that has the highest membership value. The correctness is validated by noting whether the grand total of the membership values calculated for all clusters should sum to 1. The membership matrix calculation is done with the following equation Eq. (1)–Eq. (2).

$$New.M_{uv} = M_{uv} \times P_{uv} \tag{1}$$

$$DG = \begin{cases} 1 & 0 \leq D < K \\ e^{-\frac{(D-k)^2}{k^2}} & D > K \end{cases} \tag{2}$$

$\|x_u - s_v(y)\|_A$ and $d_j(x_u)$ is the Euclidean distance between u_{th} data point and v_{th} cluster center in terms of the number of attributes A is represented in Eq. (3–5)

$$New.M_{uv} = \frac{1}{\sum_{j=1}^{C} \left[\frac{\|x_u - s_v(y)\|_A}{\|x_u - s_j(y)\|_A} \right]^{\frac{2}{m-1}}} \times \frac{\prod_{j \neq v} d_j(x_u)}{\sum_{t=1}^{V} \prod_{j \neq t} d_j(x_u)} \tag{3}$$

$$d(x_u, s_v(y)) = \|x_u - s_v(y)\|_A, \ 1 \leq v \leq C, \ 1 \leq u \leq N \tag{4}$$

$$\|x_u - s_v(y)\|_A = \sqrt{\sum_{A=1}^{F} [(x_u)_A - (s_v(y))_A]^2} \tag{5}$$

The number of features is denoted by the letter F and N is the total number of data points. The number of clusters is denoted by the letter C, the degree of fuzziness of the algorithm is denoted as m. After created our clusters, we need to update the cluster centroid using Eq. (6).

$$s_v^l(y) = \frac{\sum_{u=1}^{N} (New.M_{uv} + T_{uv})x_u}{\sum_{u=1}^{N} (New.M_{uv} + T_{uv})}, \ 1 \leq v \geq C \tag{6}$$

Here T_{uv} is the Typicality matrix. It is defined in Eq. 7 by the PCM clustering algorithm [19].

$$T_{uv} = \frac{1}{1 + \left[\frac{\|x_u - s_v(y)\|_A}{\gamma_v} \right]^{\frac{1}{(m-1)}}}, \ 1 \leq v \leq C, \ 1 \leq u \leq N \tag{7}$$

Once the centroid for each cluster has been updated, the above steps are repeated with the new centroid, and the process continues until the update of the new centroid has been calculated. This process is repeated until the modified centroids of each cluster are identical in consecutive iterations or minimum objective function is reached based on the following Eq. (8)

$$J_M = \sum_{u=1}^{N} \sum_{v=1}^{C} \left(aNew.M_{uv}^m + bT_{uv}^m \times \|x_u - s_v(y)\|_A^2 + \sum_{v=1}^{C} \gamma_v \sum_{u=1}^{N} (1 - T_{uv})^{\eta} \right) \tag{8}$$

Subject to the constraints $\sum_{v=1}^{C} M_{uv} = 1$, $\forall u$, and $0 \leq T_{uv} \leq 1$. Here, $a > 0$, $b > 0$, $\eta > 1$, $m > 1$ and J_M is the objective performance.

4 Experiment Setup

The outcome of the proposed PPFCM clustering approach is discussed in this section.

4.1 Dataset

In this work, six benchmark datasets based on different application, which is in use from UCI's machine learning repository is used for evaluating the performance of PPFCM clustering method. The description of the data set is given below.

Table 1. The description of datasets

S.No	Dataset	Number of samples	Number of attributes (class label removed)
1	Australian	690	14
2	Breast	699	9
3	Credit	653	15
4	German	1000	24
5	Heart	270	13
6	Teaching	151	5

4.2 Dataset Preprocessing

Some of the data in the above data sets has missing values. As a result, pre-processing is critical for improving the system's consistency and accuracy. The mean and mode of the attribute values from the above-obtained data set are used to fill in the missing values of numerical and nominal features. All string values are translated to numeric values since the clustering algorithm only works for numerical values. The next move is to normalize the data using min-max normalization.

4.3 Evaluation Metrics

The sum of square error is used to assess the quality of clustering techniques. The primary aim of the clustering algorithm is to lower the SSE value. The cluster quality is good when the SSE function is minimized because the samples inside cluster are of similar nature. Equation (9) is used to measure SSE.

$$SSE = \sum_{u=1}^{C} \sum_{v=1}^{N} \| C_u - O_{uv} \|^2 \tag{9}$$

The number of clusters is C, and the number of information points in each cluster Cu is N. The Euclidean distance between any information point Ouv within a cluster and the cluster centre Cu is determined for all C clusters and then summarized.

5 Result Analysis

The proposed PPFCM clustering algorithm is contrasted to other fuzzy-based clustering algorithms like FCM, PCM, PFCM [20], BIRCH [21], SOM [22], Partial based clustering (K-Means, K-Medoid, Weighted K-Means) [11] and Hierarchal clustering (HL) [23] by means of SSE value. The proposed PPFCM clustering algorithm generates a variety of clustering solutions, as shown in Table 2 and prevailing algorithms based on the number of cluster C = 2.

Table 2. SSE value comparison (c = 2)

S.No	Dataset	PPFCM	PFCM	PCM	FCM	WKM	K-MED	K-MEANS	BIRCH	SOM	HL
1	Australian	1433400	1472685	1471854	1464274	1510900	1549206	1549600	1551464	1532227	1555454
2	Breast	3251.1	3791.7	3546.1	3684.2	3875.4	3748.8	3710.8	3458.4	3555.5	3696.2
3	Credit	759710	890170	789050	884230	1434300	1561200	1550300	1512700	1441800	1124100
4	German	19176	19390	20245	29308	19421	22122	25617	22094	19176	29990
5	Heart	9175.1	9723	10654	10697	9619.5	9530	9450	9334.4	9399.1	12208
6	Teaching	1289.1	1920.2	2042.4	2675.4	1938.1	1718.1	1971.8	1348.2	1372.5	1558.8

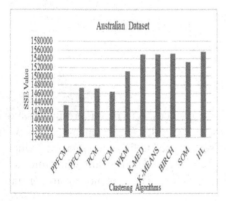

Fig. 1. SSE Comparison of considered benchmark datasets: SSE Comparison of Australian dataset

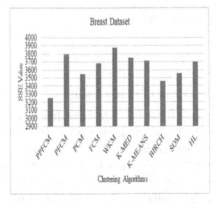

Fig. 2. SSE Comparison of considered benchmark datasets: SSE Comparison of Breast dataset

Figure 1, 2, 3, 4, 5 and 6 also shows a graphical representation of the outcome analysis. In this figure, the proposed algorithm and current clustering algorithms are represented on the horizontal axis, while the SSE value is represented on the vertical axis. The 10 clustering methods' SSE values are compared. From Fig. 1, 2, 3, 4, 5 and 6, it is clear that the proposed PPFCM method produces better clustering solutions than the ones obtained using existing methods.

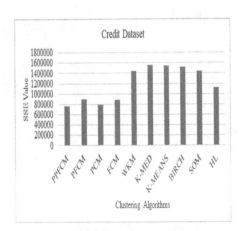

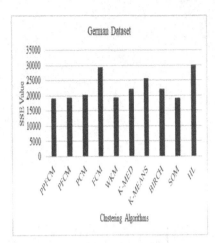

Fig. 3. SSE Comparison of considered benchmark datasets: SSE Comparison of Credit dataset

Fig. 4. SSE Comparison of considered benchmark datasets: SSE Comparison of German dataset

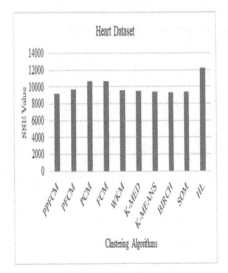

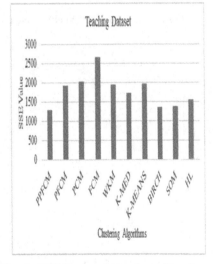

Fig. 5. SSE Comparison of considered benchmark datasets: SSE Comparison of Heart dataset

Fig. 6. SSE Comparison of considered benchmark datasets: SSE Comparison of Teaching dataset

Furthermore, the figure shows that the solution obtained using all fuzzy clustering has a lower SSE value than other clustering methods. The minimum SSE value revealed the clustering procedure's effectiveness. Figures 7, 8, 9, 10, 11 and 12 display the clustering results obtained by the parent FCM and the proposed PPFCM, respectively. The chosen 6 applications are envisioned as a binary classification problem, with all data samples presumed to belong to one of two distinct clusters (cluster1 and cluster2). Since the

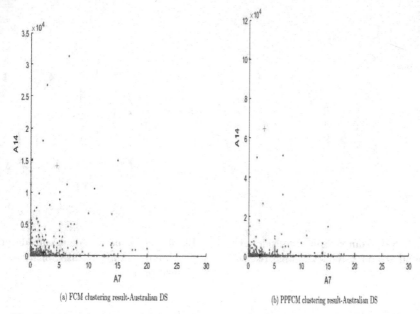

(a) FCM clustering result-Australian DS

(b) PPFCM clustering result-Australian DS

Fig. 7. FCM Vs PPFCM clustering comparison for considered benchmark datasets

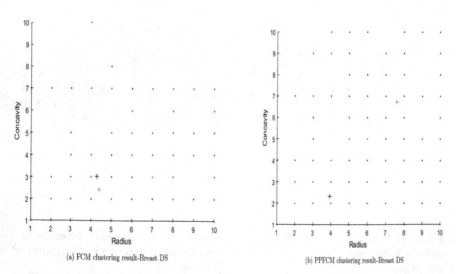

(a) FCM clustering result-Breast DS

(b) PPFCM clustering result-Breast DS

Fig. 8. FCM Vs PPFCM clustering comparison for considered benchmark datasets

datasets have more than two attributes, the information gain strategy is used to determine the two best features for each dataset in order to better reflect the quality of clustering. The X-axis and Y-axis reflect these two characteristics. The first cluster is shown in red, while the second cluster is shown in blue. The first cluster's centre is indicated by a green + symbol, while the second cluster's centre is indicated by a black + symbol. When

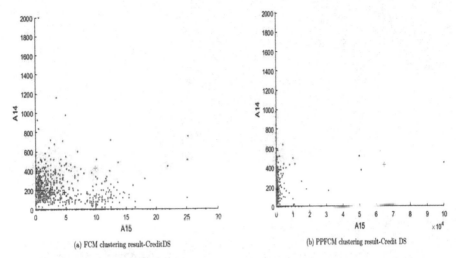

(a) FCM clustering result-CreditDS

(b) PPFCM clustering result-Credit DS

Fig. 9. FCM Vs PPFCM clustering comparison for considered benchmark datasets

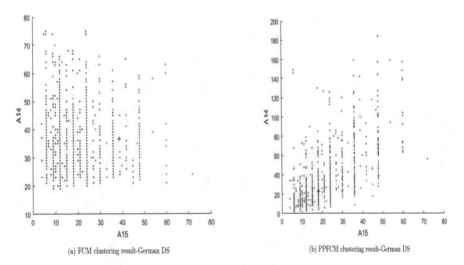

(a) FCM clustering result-German DS

(b) PPFCM clustering result-German DS

Fig. 10. FCM Vs PPFCM clustering comparison for considered benchmark datasets

compared to the current FCM (Fig. 7a, 8a, 9a, 10a, 11a and 12a), Fig. 7b, 8b, 9b, 10b, 11b and 12b shows high accuracy in data point separation.

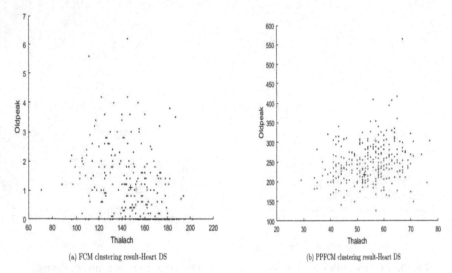

(a) FCM clustering result-Heart DS   (b) PPFCM clustering result-Heart DS

Fig. 11. FCM Vs PPFCM clustering comparison for considered benchmark dataset

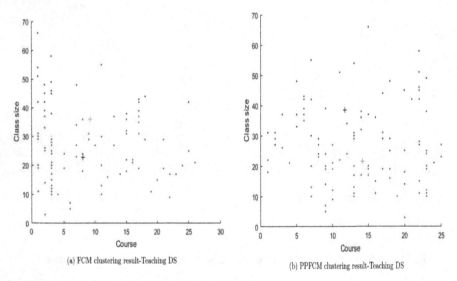

(a) FCM clustering result-Teaching DS   (b) PPFCM clustering result-Teaching DS

Fig. 12. FCM Vs PPFCM clustering comparison for considered benchmark datasets

6   Conclusions

In this paper, the proposed PPFCM clustering technique is quantitatively evaluated based on SSE value. A comparative study is made with the proposed PPFCM clustering with the traditional clustering methods, and the results are plotted. In this work, six benchmark datasets based on different application, which is taken from UCI's machine learning repository is used for evaluating the performance of PPFCM clustering method. To measure the productivity of the proposed clustering technique the SSE metric is used and it is found that the methodology mentioned above performs well for segmentation.

References

1. Van Ryzin, J., (ed.): Classification and Clustering: Proceedings of an Advanced Seminar Conducted by the Mathematics Research Center, the University of Wisconsin at Madison, 3–5 May 1976, no. 37. Elsevier (2014)
2. Liu, A., et al.: Hierarchical clustering multi-task learning for joint human action grouping and recognition. IEEE Trans. Pattern Anal. Mach. Intell. **39**(1), 102–114 (2017)
3. Vijaya, J., Sivasankar, E.: Improved churn prediction based on supervised and unsupervised hybrid data mining system. In: Mishra, D., Nayak, M., Joshi, A. (eds.) Information and Communication Technology for Sustainable Development. LNNS, vol. 9, pp. 485–499. Springer, Singapore (2018). https://doi.org/10.1007/978-981-10-3932-4_51
4. Bose, I., Chen, X.: Hybrid models using unsupervised clustering for prediction of customer churn. J. Organ. Comput. Electron. Commer. **19**(2), 133–151 (2009)
5. Sivasankar, E., Vijaya, J.: Hybrid PPFCM-ANN model: an efficient system for customer churn prediction through probabilistic possibilistic fuzzy clustering and artificial neural network. Neural Comput. Appl. **31**(11), 7181–7200 (2018). https://doi.org/10.1007/s00521-018-3548-4
6. Asuncion, A., Newman, D.: UCI machine learning repository (2007)
7. Sivasankar, E., Vijaya, J.: Customer segmentation by various clustering approaches and building an effective hybrid learning system on churn prediction dataset. In: Behera, H.S., Mohapatra, D.P. (eds.) Computational Intelligence in Data Mining. AISC, vol. 556, pp. 181–191. Springer, Singapore (2017). https://doi.org/10.1007/978-981-10-3874-7_18
8. Aggarwal, C.C., Reddy, C.K. (eds.): Data Clustering: Algorithms and Applications. CRC Press, Boca Raton (2013)
9. Huang, Y., Kechadi, T.: An effective hybrid learning system for telecommunication churns prediction. Expert Syst. Appl. **40**(14), 5635–5647 (2013)
10. Rajamohamed, R., Manokaran, J.: Improved credit card churn prediction based on rough clustering and supervised learning techniques. Cluster Comput. 1–13 (2017). https://doi.org/10.1007/s10586-017-0933-1
11. Selvi, C., Sivasankar, E.: A novel optimization algorithm for recommender system using modified fuzzy c-means clustering approach. Soft Comput. 1–16 (2017)
12. Tech, M.: Fraud detection in credit card by clustering approach
13. Yadav, A.K., Tomar, D., Agarwal, S.: Clustering of lung cancer data using foggy k-means. In: 2013 International Conference on Recent Trends in Information Technology (ICR-TIT). IEEE (2013)
14. Badjatiya, P., Kurisinkel, L.J., Gupta, M., Varma, V.: Attention-based neural text segmentation. In: Pasi, G., Piwowarski, B., Azzopardi, L., Hanbury, A. (eds.) ECIR 2018. LNCS, vol. 10772, pp. 180–193. Springer, Cham (2018). https://doi.org/10.1007/978-3-319-76941-7_14
15. Singh, V., Misra, A.K.: Detection of plant leaf diseases using image segmentation and soft computing techniques. Inf. Process. Agric. **4**(1), 41–49 (2017)
16. Perey, C.: Social Networking Segmentation: Celebrating Community Diversity in a Framework A W3C Workshop on the Future of Social Networking Position Paper (2008)
17. McClendon, L., Meghanathan, N.: Using machine learning algorithms to analyze crime data. Mach. Learn. Appl. Int. J. (MLAIJ) **2**(1), 1–12 (2015)
18. Bezdek, J.C., Ehrlich, R., Full, W.: FCM: the fuzzy c-means clustering algorithm. Comput. Geosci. **10**(2–3), 191–203 (1984)

19. Pal, N.R., et al.: A possibilistic fuzzy c-means clustering algorithm. IEEE Trans. Fuzzy Syst. **13**(4), 517–530 (2005)
20. Grover, N.: A study of various fuzzy clustering algorithms. Int. J. Eng. Res. (IJER) **3**(3), 177–181 (2014)
21. Du, H., Li, Y.: An improved BIRCH clustering algorithm and application in thermal power. In: 2010 International Conference on Web Information Systems and Mining. IEEE (2010)
22. Moya-Anegn, F., Herrero-Solana, V., Jimnez-Contreras, E.: A connectionist and multivariate approach to science maps: the SOM, clustering and MDS applied to library and information science research. J. Inf. Sci. **32**(1), 63–77 (2006)
23. Johnson, S.C.: Hierarchical clustering schemes. Psychometrika **32**(3), 241–254 (1967)

Optimized Random Forest Algorithm with Parameter Tuning for Predicting Heart Disease

Ajil D. S. Vins[1,2(✉)] and W. R. Sam Emmanuel[1,2]

[1] Nesamony Memorial Christian College, Marthandam, India
{ajil_csa,sam_emmanuel}@nmcc.ac.in
[2] Manonmaniam Sundaranar University, Tirunelveli, India

Abstract. The medical industry produces a large volume of data. Using this data, a disease can be detected, predicted and also cured. Heart diseases are one of the complex diseases and many people are suffering from this disease across the world. However, if the disease is detected early on, the mortality risk may be minimized. It is necessary to identify whether or not a person is at risk of heart disease in advance to reduce number of deaths occurring globally. This is a field in which researchers are working to help the medical practitioners to take decision regarding prediction of heart disease. To improve the accuracy for heart disease prediction is the primary goal of this research work. The cleveland dataset is used for our work which is obtained from the University of California, Irvine (UCI) Machine Learning Repository. There are 303 samples in the dataset, with 13 input features and one output feature. Here the train test ratio used is 90:10. Various classification algorithms are applied on the datasets to identify the most efficient algorithm but random forest (RF) algorithm has shown maximum accuracy in prediction. To further improve the accuracy of prediction system parameter tuning is done on the random forest algorithm.

Keywords: Heart disease · Random Forest (RF) · Parameter tuning · Disease prediction

1 Introduction

Heart diseases have become the main cause of death all over the world [1]. The number of people affected by heart disease is increasing rapidly irrespective of age in both men and women. Heart disease affect the heart's function and structure, and they come in a variety of forms. When advanced technology and medical experts are inaccessible, diagnosing and treating heart disease can become extremely difficult. [2]. The early detection of high-risk individuals probable of getting heart disease and the use of a predictive model to boost diagnosis has been proposed to lower the mortality rate [3, 4]. The development of exact models for predicting the occurrence of many diseases relies heavily on data analysis [5]. Many statistical methods are also used to build prediction models and understand important underlying risk factors. Nevertheless, artificial intelligence (AI),

© Springer Nature Switzerland AG 2021
M. Singh et al. (Eds.): ICACDS 2021, CCIS 1440, pp. 443–451, 2021.
https://doi.org/10.1007/978-3-030-81462-5_40

machine learning (ML) and big data are gaining more attention and are widely being used to build disease prediction models for a many diseases [6–11].

Big data analytics plays an important role in dealing with large volume of medical data and improving the quality of health-care services provided to patients. In this circumstance, one of the difficulties is choosing a classification algorithm for classifying the data, which depends on effective data mining and machine learning techniques. The improvement of technologies permits machine language to combine with data analytical techniques to deal with unstructured and dramatically growing data [12]. Random forest can be utilized to predict the heart disease. This algorithm has shown better accuracy than the classical algorithms such as Decision Tree (DT), Naïve Bayes (NB), K-Nearest Neighbor (KNN) and Support Vector Machine (SVM). The accuracy of prediction can also be further improved by using parameter tuning. In this work the author done parameter tuning in trial and error basis.

2 Related Works

A wide range of works are being done by different authors using different data mining algorithms to predict heart diseases. They have proposed efficient methods with high accuracy to identify the presence of heart diseases. Different datasets and algorithms, as well as their experimental effects, were investigated. More studies can be done in the future to get more detailed results. Iqra et al. [13] expressed that data mining has an extraordinary capacity to comprehend the unidentified patterns found in the data and therefore assisting in the prediction of diseases. Logistic Regression is used for classification. A 10-fold cross validation is used to ensure that the findings are accurate. Milan et al. [14] did a comparative study on data mining classification techniques like RIPPER classifier, Decision Tree, ANN and SVM. In that analysis an accuracy of 84.12% can be achieved using SVM. Miranda et al. [15] used Naive Bayes classification in order to predict cardiovascular diseases. The proposed method has gave 85% of accuracy, sensitivity and specificity. Otoom et al. [16] presented a method for detecting coronary artery disease that used the SVM technique to achieve an accuracy of 88.3%. The accuracy of SVM and Bayes net is 83.8% in the cross-validation test. After the use of FT the accuracy of 81.5% is achieved. Using the Best First algorithm for selection 7 best features are selected. 84.5% accuracy was achieved with Naïve Bayes, SVM predicted 85.1% accuracy, and FT predicted 84.5%. Vembandasamy et al. [17] proposed a heart disease detection method. Naive Bayes algorithm is used for disease detection offers 86.419% accuracy. Kuspraspta Mutijarsa et al. [18] provided an overview of single and hybrid data mining algorithms in order to decide the best algorithm for accurately predicting heart disease. The author [19] proposed a feedforward neural network based method. The proposed method is applied on Cleveland dataset consists of 300 patient's records and 80% accuracy is obtained. In another study, a neural network is combined with a multi-layer perceptron to detect the risk of heart disease in a patient using the patient's previous data [20]. Another study used multiple variables such as strain rate, hypertensive condition, and velocity to calculate a heart failure ratio with retained ejection fraction, obtained the accuracy of above 80% [21]. Men are more likely than women to suffer from cardiovascular disease. There are also children who are suffering from related health conditions

[22]. Gupta et al. [23] Using factor analysis of mixed data (FAMD) and RF-based MLA, a machine intelligence framework (MIFH) was developed. FAMD is used to find the right features, while RF is used to predict disease.

3   Discussion of Heart Disease Dataset

3.1   Dataset Description

The heart analysis dataset from the kaggle website was used in this research. It has 13 features, one target variable and 303 instances. The features and its description are mentioned in Table 1.

Table 1. Features and its Description

Feature	Feature description	Range
Age	Age (years)	29 to 79
Sex	1 means male; 0 means female	0, 1
CP	Chest pain type 1 = typical angina, 2 = atypical angina, 3 = non-anginal pain, 4 = asymptomatic	1, 2, 3,4
TRESTBPS	Resting blood pressure (in mm Hg)	94 to 200
CHOL	Serum cholesterol in mg/dl	126 to 564
FPS	If fasting blood sugar > 120 mg/dl then 1 else 0	0, 1
RESTECH	Resting electrocardiographic results	0, 1, 2
THALACH	Maximum heart rate achieved	71 to 202
EXANG	Exercise induced angina (yes means 1; no means 0)	0, 1
OLDPEAK	ST depression induced by exercise	1 to 3
SLOPE	The slope of the peak exercise ST segment	1, 2, 3
CA	Number of major vessels (0–3) colored by fluoroscopy	0 to 3
THAL	3 means normal; 6 means fixed defect; 7 means reversible defect	3,6, 7
TARGET	Class Attribute	0 or 1

3.2   Exploratory Data Analysis

In the dataset we have chosen has there are no null values and duplicates so our data is good for analysis. Out of 303 instances within the dataset, 138 people do not have a heart disease and 164 people were diagnosed to have heart disease. As we know, data in a classification problem should be balanced and this dataset has a almost balanced data. Now, we will analyze age with target and sex with target. The analysis of age with target is shown in Fig. 1 and the analysis of sex with target is shown in Fig. 2.

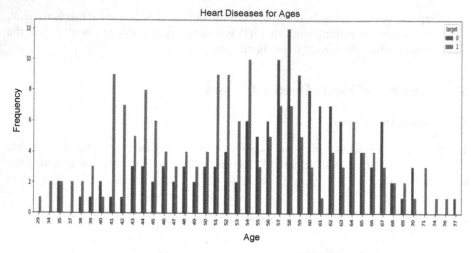

Fig. 1. Analysis of age with target

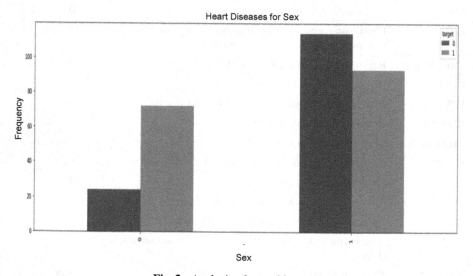

Fig. 2. Analysis of sex with target

4 Random Forest

Random forest algorithm works by forming a combination of prediction trees constructed using decision tree in such a way that all trees relies on the values of a random vector sampled separately within the same distribution for all trees in the forest. The intensity of the individual trees in the forest and the relation between them decide the generalization error of a random forest classifier. Random forest trees are more vigorous with respect to noise. It is a supervised classification algorithm used for prediction and is considered as the best due to the large number of trees in the forest thus giving improved accuracy than the decision trees. All the trees are trained separately and the prediction results of the trees are combined with techniques such as averaging and voting. Depending on the problem domain, the random forest algorithm can be used for both classification and regression. In random forest arbitrarily sampled attributes are used to split each node. Each node is split using the best split that is available among the predictors which are again chosen at random at the node. This method is different from the ones followed in decision trees where every node is split using the best split among all the attributes which are available in the dataset considered. Further, new values are predicted by combining the predictions of many decision trees constructed. Random forest represents an ensemble algorithm because it makes its final prediction from the results of a number of individual models. The individual models can be of either similar or different type. In the case of random forest algorithm, since decision trees are used, the individual models are chosen to be of the same type.

4.1 Parameter Tuning

Random forest algorithm can be tuned to give better accuracy than the traditional methods. We can alter and play with the features of algorithm in order to discover the combination of parameters that gives the best result for our problem. The values of the parameters can be changed randomly and the results can be studied. There are several parameters that can be tuned. The main parameters to adjust when tuning random forest algorithm are max_features and n_estimators. The max_features parameter is the size of the random subsets of features to be considered when a node is split. The lower value of max_feature, the greater is the reduction of variance. The number of trees in the random forest is denoted by the n_estimator parameter. A larger number of trees are better, but also it will take too long compute. The max_depth parameter denotes the maximum depth of the tree. If the max_depth is none, then nodes are expanded until all leaves are explored. After the base model for random forest has been created and evaluated, parameters can be tuned to increase some specific metrics like accuracy or F1 score of the model.

5 Performance Evaluation Methods

The performances obtained by tuning the parameters n_estimators, max_features and max_depth were evaluated on various quality measures such as accuracy, sensitivity or recall, precision and F1 score. Records that showed the occurrence of heart disease were rated as positive class. Negative class was those that did not show any signs of heart disease. The number of records correctly predicted is true positive. The number of records predicted incorrectly is true negative. The number of negative records that have been mistakenly identified as having heart disease is a false positive. False negative records are those in which a disease has been expected but mistakenly identified as a negative class. The quality measures are defined here. The accuracy can be evaluated using the Eq. (1).

$$accuracy = \frac{p1 + n1}{p1 + n1 + p2 + n2} \tag{1}$$

where p1 denotes true positivity, n1 denotes true negativity, p2 denotes false positivity, and n2 denotes false negativity. The proportion of patients with heart disease is determined by the sensitivity or recall is given in Eq. (2).

$$sensitivity\ or\ recall = \frac{p1}{(p1 + n2)} \quad (2)$$

The precision is the ratio of true positive with all instances classified as positive. This can be evaluated using the Eq. (3).

$$precision = \frac{p1}{(p1 + p2)} \quad (3)$$

The harmonic mean of precision and recall is the F1 score, which is defined in Eq. (4).

$$F1 = 2\left(\frac{precision * recall}{precision + recall}\right) \quad (4)$$

6 Result and Discussion

When the random forest algorithm is tuned with n_estimator = 10 the accuracy was 0.81. When the value of n_estimator is increased the accuracy also increased. The accuracy was 0.94 when n_estimator is 200. Values beyond 200 did not have any effect on the accuracy and other metrics. The values auto and log_2 for max_features gave the accuracy of 0.94 and with sqrt the accuracy was 0.91. The random forest algorithm was tuned with different values for max_depth. When max_depth is 1 the accuracy was 0.97 and with max_depth = 7 the accuracy was 0.94 and with max_depth = 20 the accuracy was 0.90. Lower values for max_depth gave better accuracy. The results obtained by tuning parameters with different values are shown in Table 2. And the performance metrics on the dataset is shown in Fig. 3. It has been noted that parameter tuning gave better result for heart disease prediction when compared with using the default values. As a result, parameter tuning can be performed to boost the prediction performance while using the random forest classifier algorithm for disease prediction.

Table 2. Performance comparison of results obtained by parameter tuning

| Performance metrics | Parameters tuned with different values | | | | | | | | |
| | n_estimator | | | max_features | | | max_depth | | |
	10	50	200	auto	sqrt	log₂	1	7	20
Precision	0.78	0.88	0.89	0.89	0.84	0.89	0.94	0.89	0.88
Recall	0.88	0.94	1	1	1	1	1	0.87	0.94
Accuracy	0.81	0.90	0.94	0.94	0.91	0.94	0.97	0.94	0.90
F1 score	0.82	0.91	0.94	0.94	0.90	0.94	0.97	0.93	0.91

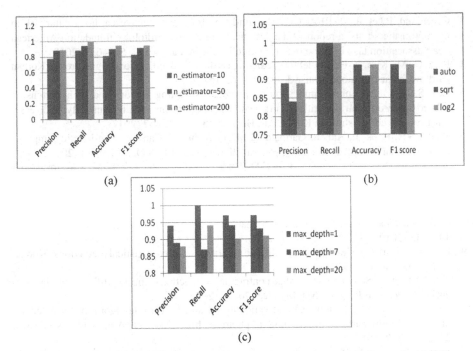

Fig. 3. Graphical representation of the performance metrics (a) Parameters tuned with different values of n_estimator (b) Parameters tuned with different values of max_features (c) Parameters tuned with different values of max_depth

7 Conclusion

The ratio of death due to heart failure is increasing day by day. So, a machine learning algorithm that can classify data and predict the probability of heart disease occurrence is needed. An efficient prediction algorithm can help to identify persons with high risk of getting heart diseases and reduce the number of deaths that occur globally. The aim of this research work is to improve the random forest algorithm's efficiency in heart disease prediction by tuning the parameters. Better results were obtained by tuning the parameter values when compared with the random forest algorithm using the default values. Hence tuning can be used in cases where feature selection does not give the expected accuracy. Future research with high accuracy in predicting heart disease will be of great use to the human society.

References

1. Pu, L.N., Zhao, Z., Zhang, Y.T.: Investigation on cardiovascular risk prediction using genetic information. IEEE Trans. Inf. Technol. Biomed. **16**, 795–808 (2012)
2. Ghwanmeh, S., Mohammad, A., Al-Ibrahim, A.: Innovative artificial neural networks-based decision support system for heart diseases diagnosis. J. Intell. Learn. Syst. Appl. **05**, 176–183 (2013)

3. Greenland, P., et al.: 2010 ACCF/AHA guideline for assessment of cardiovascular risk in asymptomatic adults: a report of the American college of cardiology foundation/American heart association task force on practice guidelines. Circulation **122**, 584–637 (2010)
4. Park, G.-M., Kim, Y.-H.: Model for predicting cardiovascular disease: insights from a Korean cardiovascular risk model. Pulse **3**, 153–157 (2015)
5. Lee, S.S., et al.: Clinical implication of an impaired fasting glucose and prehypertension related to new onset atrial fibrillation in a healthy Asian population without underlying disease: a nationwide cohort study in Korea. Eur. Heart J. **38**, 2599–2607 (2017)
6. Weng, S.F., Reps, J., Kai, J., Garibaldi, J.M., Qureshi, N.: Can machine-learning improve cardiovascular risk prediction using routine clinical data? PLoS One **12**, 1–14 (2017)
7. Alaa, A.M., Bolton, T., Angelantonio, E.D., Rudd, J.H.F., van der Schaar, M.: Cardiovascular disease risk prediction using automated machine learning: a prospective study of 423,604 UK Biobank participants. PLoS One **14**, 1–17 (2019)
8. Dinh, A., Miertschin, S., Young, A., Mohanty, S.D.: A data-driven approach to predicting diabetes and cardiovascular disease with machine learning. BMC Med. Inform. Decis. Mak. **19**, 1–15 (2019)
9. Park, J.: Can artificial intelligence prediction algorithms exceed statistical predictions? Korean Circ. J. **49**, 640–641 (2019)
10. Attia, Z.I., et al.: Screening for cardiac contractile dysfunction using an artificial intelligence–enabled electrocardiogram. Nat. Med. **25**, 70–74 (2019)
11. Gulshan, V., et al.: Development and validation of a deep learning algorithm for detection of diabetic retinopathy in retinal fundus photographs. JAMA - J. Am. Med. Assoc. **316**, 2402–2410 (2016)
12. Suguna, S., Sakunthala, S., Sanjana, S.: A survey on prediction of heart diseases using big data algorithms. Int. J. Adv. Res. Comput. Eng. Technol. **6**, 371–378 (2017)
13. Basharat, I., Raza Anjum, A., Fatima, M., Qamar, U., Ahmed Khan, S.: A framework for classifying unstructured data of cardiac patients: a supervised learning approach former specialist-business analysis and planning Mobilink Islamabad, Pakistan. (IJACSA) Int. J. Adv. Comput. Sci. Appl. **7**, 133–141 (2016)
14. Kumari, M., Godara, S.: Comparative study of data mining classification methods in cardiovascular disease prediction. Int. J. Comput. Sci. Trends Technol. **2**, 304–308 (2011)
15. Miranda, E., Irwansyah, E., Amelga, A.Y., Maribondang, M.M., Salim, M.: Detection of cardiovascular disease risk's level for adults using Naive Bayes classifier. Healthc. Inform. Res. **22**, 196–205 (2016)
16. Otoom, A.F., Abdallah, E.E., Kilani, Y., Kefaye, A., Ashour, M.: Effective diagnosis and monitoring of heart disease. Int. J. Softw. Eng. Appl. **9**, 143–156 (2015)
17. Vembandasamy, K., Sasipriya, R., Deepa, E.: Heart diseases detection using Naive Bayes algorithm. Int. J. Innov. Sci. Eng. Technol. **2**, 441–444 (2015)
18. Mutijarsa, K., Ichwan, M., Utami, D.B.: Heart rate prediction based on cycling cadence using feedforward neural network. In: Proceedings - 2016 International Conference on Computer Control, Informatics its Applications. Recent Progamming in Computer Control. Informatics Data Science, IC3INA 2016, pp. 72–76 (2017)
19. Ismaeel, S., Miri, A., Chourishi, D.: Using the Extreme Learning Machine (ELM) technique for heart disease diagnosis. In: 2015 IEEE Canadian International Conference on Humanities, IHTC 2015, pp. 1–3 (2015)
20. Gavhane, A., Kokkula, G., Pandya, I., Kailas, D.: Prediction of heart disease using machine learning. In: Proceedings of the 2nd International Conference on Electronics, Communication and Aerospace Technology, ICECA 2018, pp. 1275–1278 (2018)
21. Tabassian, M., et al.: Diagnosis of heart failure with preserved ejection fraction: machine learning of spatiotemporal variations in left ventricular deformation. J. Am. Soc. Echocardiogr. **31**, 1272–1284.e9 (2018)

22. Yadav, D.C., Pal, S.: Prediction of heart disease using feature selection and random forest ensemble method. Int. J. Pharm. Res. **12**, 56–66 (2020)
23. Gupta, A., Kumar, R., Singh Arora, H., Raman, B.: MIFH: a machine intelligence framework for heart disease diagnosis. IEEE Access **8**, 14659–14674 (2020)

Machine Learning Based Techniques for Detection of Renal Calculi in Ultrasound Images

Harsha Herle[(✉)] and K. V. Padmaja

RV College of Engineering, Bengaluru, Karnataka, India
{harsha,padmajakv}@rvce.edu.in

Abstract. The Ultrasound imaging is a non-invasive procedural technique which is used for detection of kidney diseases in the medical/clinical practice. This work emphasizes on different preprocessing methods to remove the speckle noise, embedded in kidney Ultrasound images. Preprocessing filters like, adaptive median and wiener are applied to both normal and renal calculi US images, evaluated for noise variance ranging from 0.01 and 0.08 against the parameters like Signal to Noise Ratio (SNR), Peak Signal to Noise Ratio (PSNR), Root Mean Square error (RMSE), Mean Squared Error (MAE), to determine the optimum noise variance value to be considered in preprocessing the Ultrasound images. The work also recommends adaptive median filter applied for kidney Ultrasound images with experimental results indicates increase in Peak Signal to Noise Ratio up to 33.51 dB, compared to Weiner filter. The next step is to select the best classifiers like Support Vector Machine with kernels, Multi-Layer Perceptron to preprocessed Ultrasound kidney images to estimate the accuracy, Recall, F1 score and precision. The experimental results obtained by Support Vector machine with poly kernel reaches an accuracy of 81.1% and are compared with results obtained from similar works.

Keywords: Ultrasound images (US) · Support Vector Machine (SVM) · Multilayer Perceptron Algorithm (MLP) · GLCM (Gray Level Co-occurrence Matrices) · Noise variance (NV)

1 Introduction

Medical imaging plays vital role in finding abnormalities related to human subjects, of which, medical image processing also have wide range of scope, thereby researchers and scientists have contributed significantly over decades [1]. The Ultra Sound(US) imaging is a leading imaging modality for examination of renal diseases due to superior advantages, cost reduction, portability, free from radiation and real time imaging capability. They have been extensively used to aid diagnosis and forecast of acute and chronic kidney diseases. US images correlate with speckle noise, resulting in image uncertainty, reduces the image resolution and contrast, and hard to interpret the actual information present in US image. The first step in image preprocessing is, despeckling

M. Singh et al. (Eds.): ICACDS 2021, CCIS 1440, pp. 452–462, 2021.
https://doi.org/10.1007/978-3-030-81462-5_41

of the US image to improve the image quality and also increase the accuracy of image segmentation and classification.

Mathematically, in US imaging, the speckle noise is modeled with multiplicative noise shown in Eq. 1:

$$b(m, m) = a(m, n) * n(m, n) \tag{1}$$

Here, original image a(m, n) is multiplied with n(m, n), the non-gaussian noise to obtain the speckle image b(m, n). Where, m and n represents the spatial position of the image. Primarily most of the despeckling algorithms are used for only removal of additive noise [2], and requires converting for the multiplicative noise into additive noise by applying logarithmic function to Eq. 2,

$$\log b(m, n) = \log a(m, n) + \log n(m, n) \tag{2}$$

The n(m, n) is an additive noise component with zero mean.

In preprocessing of US images, speckle filters like, spatial and wavelet filters have proved its effectiveness in terms of statistical parameters like increased Signal to noise ratio (SNR), Peak Signal to Noise Ratio (PSNR), decrease in Mean Squared Error (MSE), Mean Absolute Error (MAE) [3, 4].

Moreover, in US image, the Region of Interest (RoI) is retained, and other portions are removed by using segmentation techniques, to partition an image into multiple parts or regions, thus its features are likely to locate the exact region of renal calculi. Feature extraction technique aims at dropping the input data by discovering the features from several input patterns, resulting in an input vector consisting of appropriate image properties, which will be given to classifier techniques for predicting the new data sets.

1.1 Related Work

Ultrasound (US) technique is one of the widely used imaging techniques to detect the renal calculi, cysts in human subjects. The US images are embedded with speckle noise, of which two most common models, that are used to detect the speckle are, statistical and classification model. The main objective in statistical modeling is to remove the noisy images by concentrating only on statistical features from training data and later obtain the parameters of interest. Statistical filters such as Weiner filter [5], adopted filtering in spectral domains, finds applicable only for additive noise [6]. In order to deal with the multiplicative noise, Jain model, is applied [6], wherein, the logarithmic of the image is obtained, converts multiplicative noise to additive noise, and then the Weiner filter is applied but this process is tedious, as it takes more time for conversion and depends on the size of the image. A new adaptive filter was proposed and helps in significant improvement in sharpness of edges in US images with better smoothing [7–9]. A region-based segmentation method is applied to kidney along with Gabor filter, resulting in removing of speckle noise, smoothen the image with histogram method to improve the quality of the image [10, 11]. Other prominent method in removal of speckle noise is thresholding method with two level decomposition where the B mode scanned images are used in the experimentation to the accesses the parameters like SNR, PSNR, MSE and QI [12, 13]. A multiscale imaging technique can also be applied to remove the speckle noise, with increased accuracy [14] and 80% reduction in speckle noise with loss of image features in the image. The next step towards processing of US image includes segmentation and classification.

In classifying the kidney abnormalities, statistical methods like Grey level Co-occurrence matrix (GLCM) or Run Length Matrix (RLM) are used along with Support Vector Machine (SVM) to reach an accuracy of 85.8%. [15]. A two-level set segmentation method applied to US images increased the classification accuracy up to 98.8% by using with Artificial Neural Network (ANN) as classifier [16]. Recent advancements in US imaging have revealed the way to increased interest in removal of the speckle noise from the medical images using different algorithms, without reducing much of the diagnostic information [17]. A Meta-heuristic classifier finds application in US images, with the novel algorithm developed helps the researchers to achieve the accuracy up to 98.8% with high False Rejection Rate of 3.3. A convolution network classifier was initially applied to fetal US images but can be applied to other image datasets [18, 19]. The experimental result reached the segmentation accuracy up to 96.9%. A 4-fold multiple SVM classifier technique was applied to US images reaching an average accuracy (ACC) of 89.53% of all the classifier methods applied to US images; it is requested to estimate best classifier.

The current works aims to find the best preprocessing filter for detecting renal calculi in US images considering the parameter like SNR, PSNR, RMSE, MAE. Moreover, this work also estimates the best noise value to be well thought, when using different preprocessing filters like, Adaptive median, Weiner, by considering the above statistical parameters like SNR, PSNR, RMSE, MAE, in removal of speckle noise.

The next step is to apply segmentation method like, threshold and Region of Interest method with morphological operation to detect the renal calculi for further classification. A comparative analysis SVM based kernel functions and MLP are tabulated along with experimental results obtained from similar works.

2 Research Method

2.1 Proposed Design

Figure 1 depicts the system block diagram to detect renal calculi detection system from US images. At first, the US image (normal/renal calculi) is obtained, redundant information like anatomical information, removed by using binary threshold method. The noisy image filtered by using different filters like adaptive median and Weiner filter. The preprocessed image is segmented by threshold method and morphological operation [20].

Furthermore, the statistical features like GLCM and Histogram features are extracted from normal and abnormal images to classify the kidney abnormalities using classifiers like SVM and MLP. The detected abnormalities the best classifier for kidney US Images to detect abnormalities with increased accuracy, provide information for medical doctors, to take the medication to next level. All the preprocessing methods, classification and detection of area are carried out in MATLAB simulation environment.

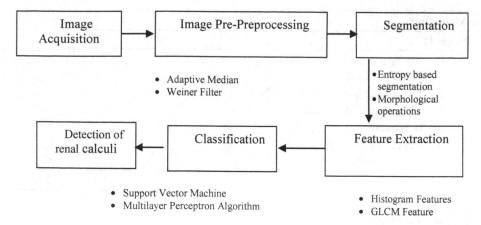

Fig. 1. Block diagram of renal calculi detection using US images.

2.2 Preprocessing Filters

This section will give the details of preprocessing filters, namely Adaptive Median and Weiner filters. This preproceesing stage will surpress the distortions in the image and enhance the image features and futher give to classify US image into nomal and abnormal.

2.2.1 Adaptive Median Filters

The adaptive median filter works with the three step methodology of which the primary step is to detect the noise in US Images, secondly, adaptively estimate the window size based on the number of noise pixels in the window. Lastly, determine the weight of each non-noise point in filtering window and remove noise points by means of weighted median filtering algorithm. This method is advantageous compared to other preprocessing filters, as it preserves the edges of its high frequency parts of an image.

Algorithm:

Stage 1:

1. $S_1 = X_{med} - X_{min}$;
2. $S2 = X_{med} - X_{max}$;
3. If $S_1 > 0$ and $S_2 < 0$ Goto Stage 2
4. Else increase the window size
5. If Window size $< S_{max}$
6. Repeat the Stage 1
7. Else output X_{med}.

Stage 2:

1. $R1 = X_{xy} - X_{min}$;
2. $R2 = X_{xy} - X_{max}$;
3. If $R1 > 0$ and $R2 < 0$
4. Output X_{xy}
5. Goto Stage 2
6. Else output X_{med}.
7. Stage 1 will determine the output is impulse or not (Black or white), next increase the window size. Ifit is not impulse, go to Stage 2.
8. If it is an impulse, the window size is increased until it reached S_{max} or Z_{med} is not an impulse. Where,

   S_{xy} is the support of the filter centered at (x,y),

   Z_{min} is the minimum grey level in S_{xy},

   Z_{max} is the minimum grey level in S_{xy}

   Z_{med} is the median grey level in S_{xy},

   Z_{xy} is the grey level at coordinates (x,y)

   S_{xy} is maximum allowed size of S_{xy}

2.2.2 Weiner Filter

Another prominent preprocessing filter that finds application in US image is Weiner filter that works on the principle of statistical features of US Images and its boundary of region finds applications to detect degradation of images. They are advantageous that compared to other filters due to, efficient in removal of speckle noise, but over smoothens the boundaries on vital image features, as compared to other methods. The frequency domain equation of Weiner filter is characterized by the Eq. 3 given below,

$$W(f1, f2) = \frac{H * (f1, f2)Sxx(f1, f2)}{|H(f1, f2)|^2 Sxx(f1, f2) + S_{\beta\beta(f1,f2)}} \tag{3}$$

Where $Sxx(f1, f2)$, and $S_{\beta\beta(f1,f2)}$ are the power spectra of original and denoising image. $H(f1, f2)$, is the blurring filter.

2.3   Classifier

2.3.1   Support Vector Machine (SVM)

The SVM is one of the most popular machine learning algorithms which is mainly used for classification and regression problems. The main aim of SVM algorithm is to build the best line or decision boundary that can separate n- dimensional space into class, so that future data point put in the corrected category. There are different types of SVM namely linear and nonlinear SVM [21–23, 25].

The training datasets of n points of the form $(x_1, y_1)....(x_n, y_n)$, Where y_n are either 1 or -1 indicating that class to which x_n belongs, x_n is input real vector. The hyper plane or decision boundary is given as

$w^T - x - b = 0$; where w is weight factor, b is biasing factor.

Algorithm:

1   Import the US Kidney Image data base
2   Pre-process the Image Data base. uses the Weiner filter
3   Discover the Histogram and GLCM feature from the database
4   Split the data into feature attributes and labels as renal calculi present or not.
5   Divide the Image data base into training and testing sets
6   Train the data using SVM algorithm.
7   Validate the datasets with new image from the database

2.3.2   Multilayer Perceptron Algorithm (MLP)

Another significant class of classifier algorithm is MLP, consists of an input layer, one or more hidden layers, and the output layer, of which the output in different stage is activated either using linear or non-liner activation function. In MLP model, there are three training steps involved three processing, like Forward phase, Error estimation phase and backward pass. In the first phase, input data is trained using the model into model and multiply the input into corresponding weights and add bias at every layer to determine the output of the model. In the next stage error estimation is obtained by taking the difference between actual outputs and desired output. Lastly, the backward phase, minimize the error by updating the weights at every stage as compared to forward stage, where the output is weighted average of input and the bias. Mathematically, this can be derived as,

$$W_{new} = W_{old} + \rho^*(d - y)^*x \tag{4}$$

W_{new}, W_{old} new and old weights, d- desired output, y-actual output, x-input,
ρ-learning rate where

$$y = £((W_T * x + b)) \text{ where } b - \text{bias [24]}. \tag{5}$$

The algorithm steps:

1.   Initialize the weights with random value.
2.   Calculating the actual output layer by layer in forward pass by using Eq. 5.

3. At the last layer finding the error by using formulae e = d–y.
4. Next minimizing the error by updating weight using backward stage with minimum learning rate.

3 Results and Discussion

The 100 normal and 125 abnormal kidney US images were obtained from radiologist, and publically available database comprising of normal and renal calculi images in form of DICOM format, further processed using in image acquisition toolbox in MATLAB® 2019a version. The images are data augmented to increase the number of abnormal US Image samples from 125 to 250. During processing, the associated labels with the US images were removed for better analysis and un-masked as shown in Fig. 2.

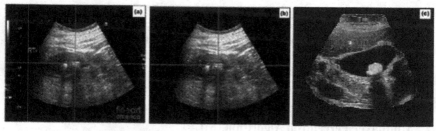

Fig. 2. (a) normal kidney image. (b) Removal of background and label of kidney images, (c) Contrasted image and sharpened image

In the next step as shown in Fig. 2, the resultant US image is enhanced and preprocessed to determine the statistical analysis like SNR, PSNR, RMSE, MAE for adaptive median and wiener filters discussed in earlier section. Furthermore, any NV greater than 0.08, there is indefinite decrease in SNR, and PSNR causing complexity in diagnosis of US Images [2]. The next step is to use apply the preprocessing filters for speckle noise reduction, for both normal and renal calculi US images. The spatial filters like, Adaptive median, Weiner are used, in estimating statistical parameters [26]. The performance analysis of US images is estimated by subjecting to different NV values starting from 0.01 to 0.08, against PSNR, SNR, RMSE and MAE as shown in Fig. 3 (Table 1).

From the Fig. 4, it is evident that the speckle noise is reduced relatively well using Adaptive filter, maximum speckle noise reduction is observed, and the edges features are well preserved. In Weiner filter, image is subjected to over enhancement, resulting in diagnosis problems and reduction in speckle noise. After the preprocessing of US images, it is required to classify the US Images to detect the presence of renal calculi or not. Two such algorithms, SVM with kernels like Linear, Poly, RBF, Sigmoid and MLP. From Table 2 it is observed that, the SVM algorithm yields better results of 84% precision,

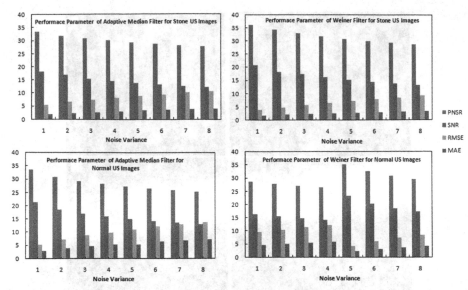

Fig. 3. Performance Parameter of Adaptive Median, Weiner filter for renal calculi and normal US images (X-Axis noise variance value = 0.01 x noise variance (NV))

Table 1. Statistical analysis of NV against different preprocessing filters

Renal Calculi US image					
Filter	NV	PSNR in dB	SNR in dB	RME	MAE
Adaptive Median	**0.01**	**33.51**	**18.01**	**5.38**	**1.70**
Weiner	0.04	31.85	16.35	6.51	2.58
Normal Kidney US image					
Adaptive Median	0.01	33.70	21.36	5.26	2.76
Weiner	**0.05**	**35.29**	**23.29**	**4.21**	**2.24**

and 81% accuracy against other SVM Kernels and MLP. Among the SVM classifier poly-kernel reach maximum accuracy as compared to other SVM Kernel functions.

The Table 3 shows the comparative analysis of experimental results obtained from similar works, it is observed that he SVM with poly kernel reached a maximum accuracy of 81.1% with data augmentation methods employed.

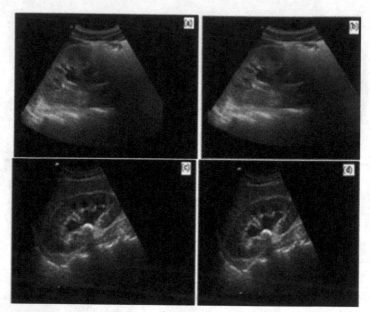

Fig. 4. (a–b) results of, adaptive median, Weiner filter for normal US images (c–d) results of adaptive median, Weiner filter for renal calculi US images

Table 2. Performance evaluation for US kidney images.

Performance evaluation

Classifier	Precision	Recall	F1 Score	Accuracy
SVM-Linear	0.65	0.59	0.55	0.6
SVM-Poly	**0.84**	**0.75**	**0.8**	**0.81**
SVM-RBF	0.69	0.64	0.62	0.64
SVM-Sigmoid	0.51	0.5	0.68	0.51
Multi-Layer Perceptron	0.5	0.57	0.56	0.57

Table 3. Comparative analysis of proposed method

Classifier method	Accuracy (%)
This work	**81**
Subramanya et al.	86
S. Selvaran et al.	97
Jyoti Verma et al.	87
D Pallavi et. al.	91
Prema et al.	84

4 Conclusion

The main focus of this work is to characterize the different value of noise parameter varying from 0.01 to 0.08 against the statistical parameters like SNR, PSNR, RMSE and MAE for different preprocessing filters like spatial and wavelet filters for US images. From preprocessing of US image with different filters, it observed with experimental results that, Adaptive Median filter, provide increased SNR (18.01 dB), PSNR (33.51 dB), with reduction in RMSE (5.38) and MAE (1.7) as compared other filtering methods. Further, morphological operations like erosion and dilation were applied to segment the filtered US image, as obtained filtered image. By applying entropy-based segmentation and morphological operations, feature extraction, RoI and exact area of renal calculi were located for in normal/ kidney renal calculi US image

The different classifiers such as SVM with different Kernel, MLP are used. The SVM reaches good accuracy up-to 81% for data sets against other techniques discussed above. The reduction in accuracy is due to data augmentation techniques used for US images, to increase the data samples sets and also the accuracy depends on quantity of the data sets used. As the images vary between hospitals, a direct comparison is beyond scope of this work considering other statistical parameters. In order to assist the medical doctor for further treatment, a GUI was developed, and that helps in detection of renal calculi and its area with minimum effort.

References

1. Couser, W.G., Remuzzi, G., Mendis, S., Tonelli, M.: The contribution of chronic kidney disease to the global burden of major non communicable diseases. Kidney Int. **80**(12). https://doi.org/10.1038/ki.2011.368
2. Kaur, R., Girdhar, A., Kaur, J.: A New Thresholding Technique for Despeckling of Medical Ultrasound Images, pp. 84–88. IEEE Conference Publications (2014)
3. Harsha, Padmaja, K.V.: Performance assessment of ultrasound kidney images using despeckling algorithms. Indian J. Comput. Sci. Eng. (IJCSE) **11**(6), 880–891 (2020)
4. Wagner, R.F., et al.: Statistics of speckle in ultrasound B-scans. IEEE Trans. Sonics Ultrasonic **30**(3), 156–163 (1983)
5. Hillery, A.D., Chin, R.T.: Iterative Wiener filters for image restoration. IEEE Trans. Sig. Process. **39**(8), 1892–1899 (1991)
6. Andria, G.: A suitable threshold for speckle reduction in ultrasound images. IEEE Trans. Instrum. Meas. **62**(8), 2270–2279 (2013)
7. Jain, L., Singh, P.: A novel wavelet thresholding rule for speckle reduction from ultrasound images. J. King Saud Univ. – Comput. Inf. Sci. 1–11 (2020)
8. Dutt, V., Greenleaf, J.F.: Adaptive speckle reduction filter for log compressed B-scan images. IEEE Trans. Med. Imaging **15**(6), 802–813 (1996)
9. Zong, X., Laine, A.F., Geiser, E.A.: Speckle reduction and contrast enhancement of echocardiograms via multiscale nonlinear processing. IEEE Trans. Med. Imag. **17**(4), 532–540 (1998)
10. Kuan, D., et al.: Adaptive noise smoothing filters for signal dependent noise. IEEE Trans. Pattern Anal. Mach. Intell. **PAMI-7**(2), 165–177 (1985)
11. Haralick, R.M., Shanmugam, K., Dinstein: Textural features for image classification. IEEE Trans. Syst. Man Cybern. **6**, 610–621 (1973)

12. Subramanya, M.B., Kumar, V., Mukherjee, S., Saini, M.: SVM-based CAC system for B-mode kidney ultrasound images. J. Digit. Imaging **28**, 448–458 (2015)
13. Garg, A., Goal, J., Malik, S., Choudhary, K., Deepika: De-speckling of medical ultrasound images using Wiener filter and wavelet transform. IJECT **2**(3), 21–24 (2011)
14. Kalaivani Narayanan, S., Wahidabanu, R.S.D.: A view on despeckling in ultrasound imaging. Int. J. Sig. Process. Image Process. Pattern Recogn. **2**(3), 85–98 (2009)
15. Chan, V., Perlas, A.: Basics of ultrasound imaging. In: Narouze, S. (eds.) Atlas of Ultrasound-Guided Procedures in Interventional Pain Management, pp. 13–19. Springer, New York (2010). https://doi.org/10.1007/978-1-4419-1681-5_2
16. Dhrumil, S., Shah, S.: Ultrasound image segmentation techniques for renal calculi - a review. Eur. J. Acad. Essays **1**(10), 51–55 (2014)
17. Kang, J., et al.: A new feature-enhanced speckle reduction method based on multiscale analysis for ultrasound B mode imaging. IEEE Trans. Biomed. Eng. **63**(6), 1178–1191 (2016)
18. Hafizah, W.M., Supriyanto, E., Yunus, J.: Feature Extraction of Kidney Ultrasound Images based on Intensity Histogram and Gray Level Co-occurrence Matrix, pp. 115–120. IEEE Conference Publications (2012)
19. Viswanath, K., Gunasundari, R.: Modified distance regularized level set segmentation based analysis for kidney stone detection analysis. Int. J. Rough Sets Data Anal. **2**(2), 22–39 (2015)
20. Loupas, T., McDicken, W.N., Allan, P.L.: An adaptive weighted median filter for speckle suppression in medical ultrasonic images. IEEE Trans. Circuits Syst. **36**(1), 129–135 (1989)
21. Krishna, K.D., Akkala, V., Bharath, R., Rajalakshmi, P., Mohammed, A.M.: FPGA based preliminary CAD for kidney on IoT enabled portable ultrasound imaging system. In: 2014 IEEE 16th International Conference on e-Health Networking, Applications and Services (Healthcom), Natal, pp. 257–261 (2014)
22. Verma, J., Nath, M., Tripathi, P., Saini, K. K.: Analysis and identification of kidney stone using kth nearest neighbour (KNN) and support vector machine (SVM) classification techniques. Pattern Recogn. Image Anal. **27**(3), 574–580 (2017)
23. Akkasaligar, P.T., Biradar, S.: Diagnosis of renal calculus disease in medical ultrasound images. In: IEEE International Conference on Computational Intelligence and Computing Research (2016). 978-1-5090-0612-0/16
24. Selvarani, S., Rajendran, P.: Detection of renal calculi in ultrasound image using meta-heuristic support vector machine. J. Med. Syst. **43**(9), 1–9 (2019). https://doi.org/10.1007/s10916-019-1407-1
25. Tyagi, V.: Understanding Digital Image Processing. CRC Press, Boca Raton (2018). https://doi.org/10.1201/9781315123905
26. Viswanath, K., Gunasundari, R.: VLSI implementation and analysis of kidney stone detection by level set segmentation and ANN classification. In: International Conference on Intelligent Computing, Communication, Convergence, vol. 48, pp. 612–622. Elsevier (2015)

Unsupervised Change Detection in Remote Sensing Images Using CNN Based Transfer Learning

Josephina Paul[1]($\boxtimes$), B. Uma Shankar[2], Balaram Bhattacharyya[1], and Alak Kumar Datta[1]

[1] Department of Computer and System Sciences, Visva-Bharati University, Santiniketan 731 235, West Bengal, India
[2] Machine Intelligence Unit, Indian Statistical Institute, 203 B.T. Road, Kolkata 700 108, India

Abstract. Change detection (CD) using remote sensing images have gained much attention in recent past due to its diverse applications. Devising reliable CD techniques that integrate huge topographical information is highly challenging. Researches in deep learning paradigm, particularly with Convolutional neural networks (CNN), have proven that CNN are efficient in abstracting knowledge from mul- tiple spectral bands, easy to be trained, and capable of deriving inference from unseen datasets. However, gathering training patterns are difficult in many real life problems and therefore, the pre-trained CNN models can be applied effectively. Hence, we consider three CNN models, VGG19, InceptionV3 and ResNet50 for feature extraction using transfer learning, followed by KMeans and Fuzzy C-Means(FCM) clustering algorithms for generating change maps. The proposed methods have been tested on two representative datasets of different land cover dynamics and have exhibited promising results with high overall accuracy and Kappa statistic (95.09 & 0.8173 respectively on Dubai city dataset and 97.12 & 0.8970 respectively on Texas dataset for Resnet+FCM) as well as superior to the state-of-the-art methods compared.

Keywords: Change detection · Convolutional neural networks · Transfer learning · Feature extraction · Clustering

1 Introduction

Change detection (CD) using images is aimed at understanding the changes occurred in the same area during an interval of time. Over the past decades the significance of change detection studies has been increasing due to its diverse applications such as medical diagnosis, video surveillance, land use change monitoring, crop stress detection, urbanization studies and, forest fire and infestations detection [1–5]. A wide array of CD techniques using remote sensing images are

© Springer Nature Switzerland AG 2021
M. Singh et al. (Eds.): ICACDS 2021, CCIS 1440, pp. 463–474, 2021.
https://doi.org/10.1007/978-3-030-81462-5_42

seen in the literature [1–8] that ranges from simple methods to complex algorithms. Nevertheless, most of them fail in mapping the changes occurred when the images are noisy or of low resolution. Further, many techniques are application specific and cannot be adapted to other domains. Moreover, they are inefficient in exploiting the embedded information from multiple spectral bands, unless combinations of methods used [6]. The much resorted solution is deep learning architectures that can delineate the pixels in better way by learning from multichannel images [6].

The power of deep neural network has been explored by several research applications reported in the literature. Autoencoders [9,10] and Deep Belief networks [11] are some successful DNN architectures. However, the DNN needs extensive training and are computationally intensive. Nevertheless, the paradigm shift in supervised learning to unsupervised learning with deep convolution neural networks (CNN) have made the CD research into new trends in the present decade. The CNN are powerful tools to extract distinct features of images, based on the convolution operation [8,12,13]. Following the success of Lenet, since 2012, a large number of deep convolutional neural network architectures have emerged which includes Alexnet, Caffenet, GoogLenet, ResNet, VGGnet and so forth [12–15]. These networks were actually designed for image classification and object detection that contains many layers of diverse functions [9]. But, utilizing their unique characteristic viz. convolution, they can be customized for semantic and pixel based segmentation, which is a crucial procedure in change detection. While primary convolution layers of the CNN models extract simple features like lines and edges, the deeper layers are capable of deriving complex features such as patterns and shapes. These feature maps form a compressed representation of the input images [8]. The down sampling operation in the Maxpooling layer, next to the convolution layer reduces the size of the feature maps further and, the activation function, ReLU (Rectified Linear Unit), efficiently manages the gradient vanishing problem.

The performance of these convolution models relies on two factors, the volume of training patterns which are often difficult to gather and requirement of high-end computing resources for training. The transfer learning technique provides a solution to overcome both of these problems, therefore, they are widely adopted in many applications [16,17]. Transfer learning is a technique which uses pre-trained DNN models for classification or feature extraction tasks on a new but, similar problem [18]. Here, the knowledge learnt in the source domain is transferred to the target domain and therefore, further training can be reduced or bypassed. Transfer learning is implemented in two different ways (i) using pre-trained networks without adjust the weights (as feature extractors) and (ii) with fine tuning the network weights by training on new dataset (as classifiers).

There has been a boom of research outcomes in this area in recent years. Fine tuning of three pretrained models, VGG16, Nasnet and Resnet, for classifying the changed pixels is investigated in [19]. It process two temporal patches in parallel with shared weights and then a FCN classifier is trained to get the change map from it [19]. In [20] fine tuning of parameters of Resnet −101 by using a dilated convolutional neural network (DCNN) is proposed for change detection. Zhang et al. proposed a CD method for remote sensing images that involves a pair of Siamese network being trained by improved triplet loss functions [21]. Yang et al. used a U-net architecture for supervised CD coupled with a reconstruction network in the target domain in which the final layers are trained while the lower layers are not [22]. VGG19 based prediction of tropical cyclones by using Infrared satellite images is implemented in [10]. Some other transfer learning researches includes Caffenet based feature extraction [18], U-net based change detection of remote sensing images [23], wetland classification by fine tuning VGG, Xception and Inception networks [24] and high resolution remote sensing image change detection with various CNN models [25].

The present work is inspired by the studies in [8,18] and [23]. Here, we propose a new approach to change detection in the unsupervised framework, by using transfer learning technique based on three different CNN models. The major contributions of this paper are:

(i) Feature extraction from the bitemporal images by transfer learning with VGG19, InceptionV3 and ResNet50 models.
(ii) Generate a change map from the extracted features, by employing two clustering algorithms - KMeans(KM) and FCM algorithm.
(iii) The proposed methods are compared with few relevant algorithms qualitatively and quantitatively.

The remaining sections of this paper are organized as follows. Section 2 discusses the state-of-the-art CNN models and the methodology of the proposed technique in detail along with the datasets used. Section 3 details the experimental set up, results and their analysis. Finally, the conclusion and future works are discoursed in Sect. 4.

2   Methodology

Problem Definition. Our problem is to generate a change map from two images I1 and I2 having b bands (b=R, G, B) of size m × n, of the same scene at different dates. The problem can be represented as

$$G(x) = \begin{cases} 0 \text{ if } x \text{ belongs to changed region,} \\ 1 \text{ if } x \text{ belongs to unchanged region.} \end{cases} \tag{1}$$

and for that, we have proposed three Transfer learning models by utilizing the state-of-the-art CNN architectures, VGG19, InceptionV3 and ResNet50.

2.1   Feature Extraction and Clustering

The transfer learning method is suitable when the training samples are limited and here, due to limited number of samples, we use the same. The connected layers of the models are thrown out to adapt with our requirements. The flow diagram of the proposed method is given in Fig. 1. Image patches are extracted from the bitemporal images I1 and I2 and input to the CNN models in the required formats. For example, for VGG19 and ResNet50, the patches are input as $224 \times 224 \times 3$ format and for the InceptionV3, in $299 \times 299 \times 3$ format. The features are extracted from different convolution layers since the features at various scales and abstraction levels can better delineate the pixels. The extracted hyper features at different layers are of different scales and coarseness. Hence, we need to make all the feature maps into the same dimension as that of the input patch, so as to have a one-to-one correspondence of feature map and the input pixels. Therefore, the hyper features are upsampled by bilinear interpolation to the size of the input image patch. On upsampling, the feature values at $(i, j)^{th}$ position corresponds to the $(i, j)^{th}$ pixel of the input patch. For CD problems, this is so important that the embedded shapes and structures in the images remain undistorted in the new representation. Further, the feature vectors of all the layers are concatenated along the depth dimension and produce K feature maps from each input patch of I1 and I2. By applying a dissimilarity metric, we can compare the hyper features. By using Euclidean distance as the dissimilarity metric, the difference between the feature vectors is computed as

$$DI^k = (f_{1l}^k - f_{2l}^k)^2 \qquad (2)$$

where f_{1l}^k is the feature map obtained by k^{th} filter at layer l of the patch of image I1 and f_{2l}^k is the feature map obtained by k^{th} filter at layer l of patch of image I2.

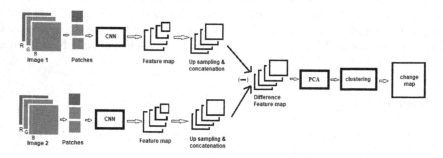

Fig. 1. Flow diagram of the proposed method

If we consider the complete set of resultant hyper features, it is computationally intensive and time consuming. For example, a total of 5504 feature maps are produced by VGG19 from its 16 convolution layers. In the case of ResNet50 and InceptionV3 these features are even more as the number of convolution layers are more in them. However, not all the feature vectors are carrying information equally that are useful for deriving the changed and unchanged pixels. Hence, for dimensionality reduction of the feature maps, we apply Principal Component Analysis(PCA) on the DI (difference feature map) and generate P principal components. These P principal component vectors are good enough to detect the changes occurred as they carry most of the relevant information.

By employing two unsupervised methods, Kmeans and FCM clustering on the principal components, the changed and unchanged pixels are delineated, from which we generate a binary change map. Kmeans is a popular partitioning algorithm that generates clusters based on the Euclidean distance between the data points and the centroid of clusters. The cluster centres are computed iteratively aiming to minimize the sum of square error for all objects in the data set, i.e., to make the intra-class distance minimum and interclass distance maximum. The FCM works on fuzzy approximation. This property is suitable for remote sensing problems as a single pixel in remotely sensed image may carry information from overlapping land cover classes. The objective of FCM is to minimize the sum of least square errors of all the clusters in the problem space. Rather than crisp membership as in Kmeans, FCM assigns the pixels with some degree of membership to every cluster [3]. Finally, the pixel is assigned to the cluster which is having maximum membership. The algorithm is detailed as below.

Algorithm 1: *Feature extraction and change map generation*
Input: RGB channels of bitemporal image patches
Output: Binary change map.
1. Extract patches from two source images I1 and I2 in the required format and input to the model (Eg: 224x224x3 for VGG19).
2. Extract features from intermediate convolution layers for each patch, say, f_{1l}^{k} and f_{2l}^{k}, the kth feature map at l^{th} layer of patch of I1 and I2.
3. Up sample the feature maps to the input image patch size by bilinear interpolation.
4. Concatenate the feature maps from all the layers along the depth dimension(pixelwise) to form two sets of feature maps (one set for patches of I1 and one set for patches of I2 respectively).
5. Compute the difference of feature maps of corresponding layers of I1 and I2 and produce DI using eqn. 2 (pixelwise)
6. Apply PCA for dimensionality reduction and select the P principal components of features.
7. Cluster the resultant principal components using Kmeans and FCM clustering algorithms.
8. From the resultant clusters, generate change map by assigning 1 for unchanged pixels (one cluster) and 0 for changed pixels (other cluster)

2.2 Data Sets

Data set 1: The satellite images of Dubai city of year 2001 and year 2010 captured by Landsat 7 ETM+ are selected for the study. Developmental activities such as built ups and roads have changed the topography of the location from 2001 to 2010 considerably. Three bands in the wavelength range $(0.45$–$0.52\ \mu m)$, $(0.52$–$0.60\ \mu m)$ and $(0.63$–$0.69\ \mu m)$ are considered for this study. A portion of 1000×1000 pixels were cropped from the co-registered images for the present experiments, the same along with the corresponding reference map are shown in Fig. 2.

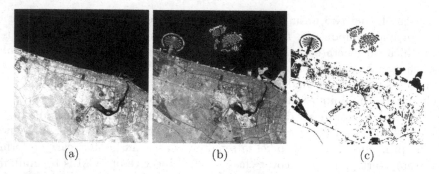

(a)   (b)   (c)

Fig. 2. Image of spectral band 2 of the multitemporal images of Dubai city (a) year 2001 (b) year 2010 (c) Reference Change Map (Ground truth)

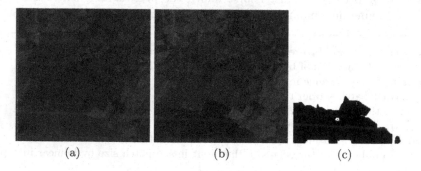

(a)   (b)   (c)

Fig. 3. Image of spectral band 2 of the multitemporal images of Central Texas, USA (a) April 2, 2001 (b) April 18, 2011 (c) Reference Change Map (Ground truth)

Data set 2: The images acquired by Landsat 5 Thematic Mapper (TM) on April 2, 2011 and April 18, 2011 are selected for the study to detect the changes due to the fire occurred in the region of central Texas, USA, on April 9, 2011. The area lies between longitude $100°28.31E$ and $100°34.16E$ and latitude $32°5.38N$ and $32°59.53N$. A portion of 300×300 pixels were cropped from the co-registered images and considered three bands, as selected in the first dataset. Figure 3 shows the portion 300×300 pixels of the two images along with the reference map.

2.3 Experiments and Evaluation Metrics

The bitemporal images were co-registered to the same co-ordinates by using ArcGis software. Image patches of $224 \times 224 \times 3$ were generated and input to the models VGG19 and ResNet50 and patches of $299 \times 299 \times 3$ were generated and input to the InceptionV3 model. Change maps were generated by employing the methodology in Sect. 2 and evaluated empirically. The metrics used are Precision, Recall, F1 score, Overall accuracy (OA) and Cohen's Kappa coefficient to measure the performance of the methods.

3 Results and Discussion

We analyse the results quantitatively and qualitatively. The detailed analysis of results are given in the following sections.

3.1 Quantitative Analysis of Results

Tables 1 and 2 demonstrate the quantitative results obtained for the Dubai city dataset and Texas, USA dataset respectively. The first column of the table displays the method and the remaining columns show the accuracy metrics OA, Kappa, precision, recall and F1 score. We choose the first three principal components for clustering with kMeans and FCM. In order to compare the results of two clustering methods, we have chosen the Otsu thresholding method. Otsu is a well established thresholding algorithm for segmentation. The Otsu algorithm was employed as follows: The sum of squared difference for every pixels of the first three principal components of the feature vector are taken and scaled between 0–255 as it corresponds to the range of pixel values and constructed the difference image for segmentation. The threshold is calculated from the histogram and probabilities of each intensity level iteratively. By employing the threshold, the changed and unchanged pixels are clustered into two.

While Kmeans algorithm obtained slightly better results compared to Otsu on VGG19 model on the Dubai city dataset, FCM has performed much better than Otsu and Kmeans. The same is true with Texas dataset also, which is evident from Table 2. For InceptionV3 with the two clustering methods and Otsu, we get better accuracies than that obtained for VGG19 on both datasets. In the case of ResNet50, we can notice that the highest values are obtained with FCM for various accuracy metrics- the Kappa, Precision, recall and F1 score, on the two datasets, compared to Otsu and Kmeans. From the tables, one can also observe that the accuracy also depends on the architecture of the CNN models

Table 1. Change detection results on Dubai City data set (OA-Over all accuracy, Kappa- Kappa statistic, Pr-precision, R-recall, F1-F1 score)

Method	OA	Kappa	Pr	R	F1
VGG19+Otsu	92.22	0.5366	0.66	0.51	0.55
VGG19+KM	92.27	0.5371	0.66	0.52	0.58
VGG19+FCM	93.40	0.5700	0.90	0.45	0.60
InceptionV3+Otsu	93.30	0.6321	0.66	0.67	0.67
InceptionV3+KM	93.34	0.6342	0.67	0.67	0.67
InceptionV3+FCM	94.45	0.7604	0.96	0.68	0.79
ResNet50+Otsu	93.26	0.7578	0.96	0.68	0.80
ResNet50+KM	94.45	0.7654	0.96	0.68	0.80
ResNet50+FCM	**95.09**	**0.8173**	**0.98**	**0.75**	**0.85**

Table 2. Change detection results on Texas, USA dataset (OA-Overall accuracy, Kappa- Kappa statistic, Pr-precision, R-recall, F1-F1 score)

Method	OA	Kappa	Pr	R	F1
VGG19+Otsu	90.90	0.7284	0.67	0.93	0.78
VGG19+KM	91.66	0.7487	0.69	0.94	0.79
VGG19+FCM	94.82	0.8236	0.85	0.86	0.86
InceptionV3+Otsu	95.46	0.8300	0.95	0.78	0.86
InceptionV3+KM	95.55	0.8388	0.93	0.81	0.86
InceptionV3+FCM	96.46	0.8717	0.96	0.83	0.89
ResNet50+Otsu	95.64	0.8428	0.93	0.82	0.87
ResNet50+KM	95.63	0.8444	0.91	0.83	0.87
ResNet50+FCM	**97.12**	**0.8970**	**0.97**	**0.87**	**0.91**

chosen. The deeper models perform better. In this study, the deepest model, ResNet50 achieved the best accuracy followed by InceptionV3 and the least is for VGG19. It can be noticed from the tables that the proposed methods based on FCM clustering have come up with the highest values for various accuracy metrics especially Kappa and F1 score. The vaues of OA, Kappa coefficient, Precision, recall and F1 score for the Dubai city dataset (95.09, 0.8173, 0.98, 0.75, 0.85) and for the Texas dataset (97.12, 0.8970, 0.97, 0.87, 0.91) confirm the superiority of the ResNet50 model with FCM algorithm.

The second comparison is with five popular methods - Gaussian Mixture model (GMM) clustering, FCM and standard Particle Swarm Optimization (SPSO) on normalized difference image, change vector analysis (CVA) with otsu and Kmeans with VGG16 features. The GMM is a soft clustering method that models the probability of membership of each data point from a mixture of Gaussian distributions. It can derive arbitrary shaped clusters. So, it is appropraite for remote sensing problems. SPSO is an population based algorithm inspired by the social behaviour of swarm of insects. It uses the certain characteristics of these swarm in optimizing problems. In CVA, the change of intensity values of multiple spectral bands is computed. Thus it is good in drawing information from multiple bands. The Kappa coefficient and OA obtained for these methods on the two datasets are shown in Table 3. From the table it is evident that the accuracy of the proposed methods outperform the state-of-the-art CD algorithms on the Dubai dataset and the Texas datasets in all cases.

Table 3. Comparison of OA and Kappa statistic for various methods(OA-Overall accuracy, Kappa- Kappa statistic, GMM-Gaussian Mixture model, NDI-Normalized Difference image)

Method	Dubai city dataset		Texas dataset	
	OA	Kappa	OA	Kappa
DI+GMM	86.49	0.4020	88.41	0.6785
NDI+FCM [7]	88.73	0.5370	88.77	0.6890
NDI+SPSO	89.34	0.5243	89.99	0.6792
CVA+Otsu	91.13	0.5295	90.34	0.6845
VGG16+KM	86.95	0.3919	89.15	0.6895
VGG19+Otsu	92.22	0.5366	90.90	0.7284
VGG19+KM	92.27	0.5371	91.66	0.7487
VGG19+FCM	93.40	0.5700	94.82	0.8236
IncepV3+Otsu	93.30	0.6321	95.46	0.8300
IncepV3+KM	93.34	0.6342	95.55	0.8388
IncepV3+FCM	94.45	0.7604	96.46	0.8717
ResNet50+Otsu	93.26	0.7578	95.64	0.8428
ResNet50+KM	94.45	0.7654	95.63	0.8444
ResNet50+FCM	**95.09**	**0.8173**	**97.12**	**0.8970**

3.2   Qualitative Analysis of Results

For the visual analysis, the change maps generated for various methods are displayed in Figs. 4 and 5. The proposed methods show better quality change map compared to that obtained by DI+Gaussian mixture model, NDI+FCM and VGG16+KM. While VGG19 with Otsu and KM generate change maps with thick boundaries indicating more false alarms, FCM map shows less false hits. InceptionV3 produces high quality images with less noisy spots, but a number of missed alarms can be noticed. With ResNet50 model, Otsu and KM generate good change maps but have a few false alarms, whereas FCM produces high quality change map. For both datasets, we observe the same trend which reveals that the deeper the networks, better the quality of change maps and, the clustering technique FCM outperform both Kmeans and Otsu on all CNN models.

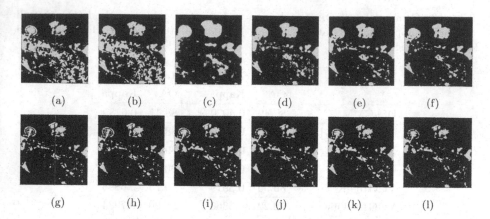

(a)   (b)   (c)   (d)   (e)   (f)

(g)   (h)   (i)   (j)   (k)   (l)

Fig. 4. Change map of Dubai city dataset by (a) DI+GMM (b) NDI+FCM (c) VGG16+KM (d) VGG19+Otsu (e) VGG19+KM (f) VGG19+FCM (g) InceptionV3+Otsu (h) InceptionV3+KM (i) InceptionV3+FCM (j) ResNet50+Otsu (k) ResNet50+KM (l) ResNet50+FCM

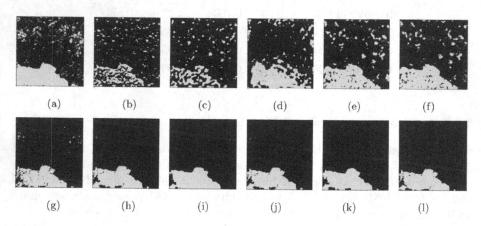

(a)   (b)   (c)   (d)   (e)   (f)

(g)   (h)   (i)   (j)   (k)   (l)

Fig. 5. Change map of Texas dataset by (a) DI+GMM (b) NDI+FCM (c) VGG16+KM (d) VGG19+Otsu (e) VGG19+KM (f) VGG19+FCM (g) InceptionV3+Otsu (h) InceptionV3+KM (i) InceptionV3+FCM (j) ResNet50+Otsu (k) ResNet50+KM (l) ResNet50+FCM

4   Conclusion and Future Work

A novel method for change detection of remote sensing images in the transfer learning framework is proposed. As CNN better are efficient in extracting deeper features from the images, we use three pre-trained CNN architectures to extract features from multiple abstraction layers. Later, these hyper features are used to compute difference image and employed PCA for the dimensionality reduction. Further, the change map is generated from it, by applying K-means and FCM clustering. On comparing results, it is noticed that the deeper networks perform better compared to less deep ones, and the clustering algorithm also act as a deciding factor for the quantitative and qualitative accuracy of change maps. The methods were compared with some relevant algorithms and found the results with higher accuracy values. As the feature extraction in the transfer learning framework requires no training patterns, this method is suitable when the training patterns are limited. The weakness of the method is that as the weights are not adjusted in the model, the accuracy is not much improved. Our next study aims at fine tuning of deep CNN architectures that involves weight adjustments also.

References

1. Gong, M., Zhou, Z., Ma, J.: Change detection in synthetic aperture radar images based on image fusion and fuzzy clustering. IEEE Trans. Image Process. **21**(4), 2141–2151 (2012)
2. Miller, J.D., Yool, S.R.: Mapping forest post-fire canopy consumption in several overstory types using multi-temporal landsat TM and ETM data. Rem. Sens. Enviro. **82**(2), 481–496 (2002)
3. Ghosh, A., Mishra, N.S., Ghosh, S.: Fuzzy clustering algorithms for unsupervised change detection in remote sensing images. Inf. Sci. **181**(4), 699–715 (2011)
4. Anees, A., Aryal, J., O'Reilly, M.M., Gale, T.J., Tim, W.: A robust multi-kernel change detection framework for detecting leaf beetle defoliation using Landsat 7 ETM+ data. ISPRS J. Photogram. Remote Sens. **122**, 167–178 (2016)
5. Hussain, M., Chen, D., Cheng, A., Wei, H., Stanley, D.: Change detection from remotely sensed images: from pixel-based to object-based approaches. ISPRS J. Photogram. Remote Sens. **80**, 91–106 (2013)
6. Saha, S., Solano-Correa, Y.T., Bovolo, F., Bruzzone, L.: Unsupervised deep learning based change detection in sentinel-2 images. In: 2019 10th International Workshop on the Analysis of Multitemporal Remote Sensing Images (MultiTemp), pp. 91–106. IEEE Xplore (2019)
7. Paul, J., Uma Shankar, B., Bhattacharyya, B.: Change detection in multispectral remote sensing images with leader intelligence PSO and NSCT feature fusion. ISPRS Int. J. Geo-Inf. **9**(462), 1–23 (2020)
8. de Jong, K.L., Bosman, S.A.: Unsupervised change detection in satellite images using convolutional neural networks. In: 2019 International Joint Conference on Neural Networks (IJCNN), Budapest, Hungary, pp. 1–8 (2019)
9. Alhassan, V., Henry, C., Ramanna, S., Storie, C.: A deep learning framework for land-use/land-cover mapping and analysis using multispectral satellite imagery. Neural Comput. Appl. **32**(12), 8529–8544 (2019). https://doi.org/10.1007/s00521-019-04349-9
10. Sublime, J., Kalinicheva, E.: Automatic post-disaster damage mapping using deep-learning techniques for change detection: case study of the tohoku tsunami. Remote Sens. **11**, 1123 (2019)
11. Ayhan, B., Kwan, C.: Application of deep belief network to land cover classification using hyperspectral images. In: Cong, F., Leung, A., Wei, Q. (eds.) ISNN 2017. LNCS, vol. 10261, pp. 269–276. Springer, Cham (2017). https://doi.org/10.1007/978-3-319-59072-1_32
12. Krizhevsky, A., Sutskever, I., Hinton, G.E.: Imagenet classification with deep convolutional neural networks. Commun. ACM **60**(6), 84–90 (2017)

13. Simonyan, K., Zisserman, A.: Very deep convolutional networks for large-scale image recognition. In: Published as a Conference Paper at ICLR 2015, pp. 1–14 (2015). arXiv:1409.1556v6
14. Lateef, F., Ruichek, Y.: Survey on semantic segmentation using deep learning techniques. Neurocomputing **338**, 321–348 (2019)
15. Goodfellow, I.J., Bengio, Y., Courville, A.: Deep Learning. MIT Press, Cambridge (2016)
16. Pan, S.J., Yang, Q.: A survey on transfer learning. IEEE Trans. Knowl. Data Eng. **22**(10), 1345–1359 (2010)
17. Khoshgoftaar, T.M., Weiss, K., Wang, D.D.: A survey on transfer learning. J. Big Data **3**(9), 1–40 (2016)
18. Amin, A.M.E., Qingjie, L., Yunhong, W.: Convolutional neural network features based change detection in satellite images. In: First International Workshop on Pattern Recognition, vol. 10011, p. 100110W (2016)
19. Larabi, M.E.A., Chaib, S., Bakhti, K., Karoui, M.S.: Transfer learning for changes detection in optical remote sensing imagery. In: IGARSS 2019–2019 IEEE International Geoscience and Remote Sensing Symposium, pp. 1582–1585 (2019)
20. Venugopal, N.: Sample selection based change detection with dilated network learning in remote sensing images. Sens. Imaging **20**(12), 20–31 (2019)
21. Zhang, M., Xu, G., Chen, K., Yan, M., Sun, X.: Triplet-based semantic relation learning for aerial remote sensing image change detection. IEEE Geosci. Remote Sens. Lett. **16**(2), 266–270 (2019)
22. Yang, M., Jiao, L., Liu, F., Hou, B., Yang, S.: Transferred deep learning-based change detection in remote sensing images. IEEE Trans. Geosci. Remote Sens **57**(9), 6960–6973 (2019)
23. Zhang, C., Wei, S., Ji, S., Lu, M.: Detecting large-scale urban land cover changes from very high resolution remote sensing images using cnn-based classification. ISPRS Int. J. Geo-Inf. **8**(189), 1–16 (2019)
24. Mahdianpari, M., Salehi, B., Rezaee, M., Mohammadimanesh, F., Zhang, Y.: Very deep convolutional neural networks for complex land cover mapping using multispectral remote sensing imagery. Remote Sens. **10**(7), 1119 (2018)
25. Larabi, M.E.A., Chaib, S., Bakhti, K., Hasni, K., Bouhlala, M.A.: High-resolution optical remote sensing imagery change detection through deep transfer learning. J. Appl. Remote Sens. **13**(4), 1–18 (2019)

Biological Sequence Embedding Based Classification for MERS and SARS

Shamika Ganesan, S. Sachin Kumar$^{(\boxtimes)}$, and K. P. Soman

Centre for Computational Engineering and Networking (CEN), Amrita School of Engineering, Coimbatore, Amrita Vishwa Vidyapeetham, Coimbatore, India
s_sachinkumar@cb.amrita.edu, kp_soman@amrita.edu

Abstract. Biological sequence comparison is one of the key tasks in finding similarities between different species. The primary task involved in computing such biological sequences is to produce embeddings in vector space which can capture the most meaningful information for the original sequences. Several methods such as one-hot encoding, Word2Vec models, etc.. have been explored for sequence embeddings. But these methods either fail to capture similarity information between k-mers or face the challenge of handling Out-of-Vocabulary (OOV) k-mers. In this paper, we aim at conducting an in-depth analysis of sequence embeddings using Global Vectors (GloVe) model and FastText n-gram representation. We thereby evaluate its performance using classical Machine Learning algorithms and Deep Learning methods. We compare our results with an existing Word2Vec approach. Results show that FastText n-gram based sequence embeddings provide the most meaningful sequences based on classification accuracy and visualization plots.

Keywords: Biosequence · NLP · GloVe · FastText · MERS · SARS

1 Introduction

Coronaviruses are a part of single-stranded RNA virus families. The Coronaviridae family consists of four genera viz., α-Cov, β-Cov, γ-Cov and δ-Cov. Among these genera, SARS-COV-2 belongs to the β-Cov genus. The Middle East Respiratory Syndrome (MERS) and the Severe Acute Respiratory Syndrome (SARS) also belong to the Coronaviridae family's β-Cov genus, but they differ in their subgenus. While SARS-COV-2 belongs to the Sarbecovirus subgenus, MERS and SARS belong to the Merbecovirus subgenus. According to a study conducted by the National Institute of Allergy and Infectious Diseases (NIAID) [1], SARS tracks back to November 2002 affecting more than 8,000 people and causing around 744 deaths. While MERS was first reported in September 2012, spreading to over 2,500 people worldwide causing around 866 deaths.

RNA sequences consist of four bases, Adenine (A), Guanine (G), Cytosine (C), Uracil (U). In DNA sequences, Uracil is replaced by Thymine (T). These bases combine with each other using Hydrogen bonds. The viruses, MERS and

© Springer Nature Switzerland AG 2021
M. Singh et al. (Eds.): ICACDS 2021, CCIS 1440, pp. 475–487, 2021.
https://doi.org/10.1007/978-3-030-81462-5_43

SARS, belonging to the Coronaviridae family, though closely related, differ in their effects on the human body as well as the course of medical treatment [1]. With the increasing capabilities of Computational Biology, several attempts have been made to acquire deeper insights into the characteristics and patterns showcased by gene sequences of various viruses. Conventional sequence analysis date back to Dynamic Programming approaches such as Needleman–Wunsch algorithm and Smith–Waterman algorithm to find the best possible alignment between sequences [3]. In this paper, we aim to find an efficient method for sequence representation of MERS and SARS using Global Vectors (GloVe) [4,5] and FastText n-gram embeddings [6] which are capable of capturing the characteristics of each of the virus sequences and provide meaningful embeddings. The remaining sections of the paper have the following organization of content. In Sect. 2, a description of existing works is provided. Section 3 contains the details regarding the methodology used in our work and Sect. 4 describes the proposed approach. Section 5 describes the experimental setup for the work, while Sect. 6 discusses the obtained results with a visual representation and analysis of the sequences using t-Distributed Stochastic Neighbour Embedding (t-SNE) plots. Section 7 provides an overall summary of the paper along with a discussion on future scope.

2   Related Works

The attempt to represent biosequences in meaningful ways has been a crucial task in computational biology. The approaches include probabilistic, distributive and similarity-based techniques [7,8]. Initial representation was based on numerical mapping where each of the bases A,T,G,C were numbered 0,1,2 or 3. However, this approach implied that the numerical value of one base was greater than the other, which did not present the structural meaning appropriately. In [8], each nucleotide base was encoded as 3-digit binary number using exclusive OR operation for expressing the variation in the different bond strengths. Though it is a computationally simple method, it could not capture contextual information of the sequences. Among other approaches, one-hot encoding of each of the individual bases or its k-mers had been widely used [9]. This approach faced a drawback due to its inability to capture the similarity relationships between k-mers. The bag-of-words approach in [10] does not take into account the position of k-mers in the original sequence. In [11], the authors took an approach of representing sequences using the set of four points, A,T,G,C, and projecting them on to the Fermat spiral.

With the advances in the Natural Language Processing (NLP) community, researchers found the commonality between k-mers in biosequences with the tokens in sentences. This instigated the advent of word embedding methods for representing sequences. In [12], the authors used the Word2Vec framework for generating sequence embeddings. The proposed Artificial Neural Network (ANN) based architecture, known as ProtVec, embeds the sequence into an n-dimensional vector representation. Authors of [13] and [14] use a similar approach to design seq2vec and dna2vec representations respectively for sequence embeddings. For classifying the promoter in a DNA, [15] took an approach using a combination of continuous FastText n-grams and thus feeding the embeddings to a Convolutional Neural Network (CNN) model for classification. For modelling a biosequence, along with obtaining a uniform representation, it is essential to take global contextual information into account [16]. There have been various attempts to classify sequences using methods that capture contextual information such as Long Short Term Memory (LSTM), Recurrent Neural Network (RNN) and Bidirectional Encoder Representation (BERT) [17,18].

3   Methodology

Sequence embeddings are a way of representing gene sequences in a numerical fashion. Representation that allows sequences with similar meaning to have similar embeddings contain more meaning and capture better relationship information. In this work, each sequence is split into tokens similar to what is followed in NLP. Here, k-mers are considered as tokens which are mapped to a vector. The key advantage of using a sequence embedding representation of a sequence is that along with capturing the similarity information, it also gives a fixed length representation for all words which is extremely important in classification tasks using Machine Learning or Deep Learning methods.

3.1   FastText

Word or Sequence Embeddings can be generated using various methods such as Word2Vec which uses Skip gram and CBOW methods [14]. But in these methods one of the major setbacks faced is the ability to handle Out-of-Vocabulary (OOV) words. Word2Vec model is trained on a certain corpus where it generates embeddings for each of the words. However, if a new word is received during the testing phase, it shows an error. In case of gene sequences, several mutations of the same virus might exist in different geographical locations. During such scenarios, if a particular k-mer did not exist in the pre-trained model vocabulary, then such a model would fail to generate embeddings for the sequence. Thus, FastText brought in a method to incorporate OOV words. It treats each word as composed of n-grams. Thus, if FastText has to generate the embeddings for an OOV k-mer, say 'ATGCCTTTAT' it would split the words as *'ATG'* + *'CCT'* + *'TTAT'* which might individually have word embeddings in the model, and sum up the embeddings for those individual chunks to generate the embedding for the OOV k-mer.

3.2   Global Vectors (GloVe)

GloVe is an unsupervised learning algorithm for obtaining vector representations for words or tokens. Training is performed on aggregated global word-word co-occurrence statistics from a corpus, and the resulting representations showcase interesting linear substructures of the word vector space [15].

There are two categories of word representation - Count based and Prediction based. Count based representation signifies the frequencies of words in a given document or a corpus. It does not provide the context information or semantic details of the words. Prediction based representation signifies the contextual representation of words. GloVe combines both these representations to give meaningful embeddings. This method is primarily essential in terms of computational biology since both frequency of k-mers and contextual similarity between them is essential to understand a sequence.

4   Proposed Approach

4.1   Dataset Description

The dataset used for the experiments was collected from the National Center for Biotechnology Information (NCBI) Virus databank [19], where the sequences belonged to SARS and MERS. The files were obtained in FASTA format with each sequence containing details pertaining to Accession ID, Sequence and Remarks. The FASTA files were pre-processed into Comma Separated Values (CSV) format to obtain a total of 804 sequences for MERS and 1691 sequences for SARS. From the obtained dataset, different k-mers were generated for carrying out experiments to find the optimal 'k' value for the k-mers and the embedding dimensions which represent the k-mers. The k-mers were created based on non-overlapping non-repetitive sliding windows to capture contextual information between k-mers [20].

4.2   Proposed Experimental Design

In this paper, we focus our study on MERS and SARS sequences by performing a detailed study on sequence embeddings and there-by conducting a binary classification of the obtained embeddings. The sequences were split into k-mers of 'k' sizes equal to 10, 20, 30 and 40. The k-mers thus obtained were used to generate sequence embeddings using the FastText n-grams approach and the GloVe models with various embedding dimensions. In order to evaluate the efficiency of the sequences, the sequence embeddings obtained from each of the methods were classified using various Machine Learning and Deep Learning classifiers. Figure 1 represents the block diagram for the proposed architecture.

5   Experimental Setup

5.1   Sequence Embeddings

The first step towards sequence classification is to obtain meaningful embeddings for the gene sequences. In this paper, we explore three different methods for sequence embeddings- two methods using the FastText open-source library[1]

---

[1] https://fasttext.cc/.

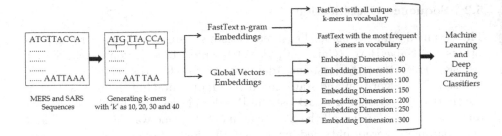

Fig. 1. Block diagram for proposed architecture

and another using the Global Vector (GloVe) representation[2]. To ascertain the optimal value of 'k' for the k-mer representation, we experimented with different values. Each of the k-mers were embedded using the embedding techniques as follows. The FastText embeddings were used to experiment with two methods of embeddings - 1. embeddings obtained from a model trained over the entire set of k-mer vocabulary 2. embeddings obtained from a model trained over the 100 most frequent k-mers obtained from the vocabulary. The embeddings, thus obtained, had an embedding dimension of 100. The third method of sequence embedding was GloVe, where the embedding dimensions were experimented with, to determine which dimension best captures the characteristics of the sequence. All the three embeddings of FastText and GloVe were trained for 500 epochs. The FastText embeddings were trained with a minimum n-gram count as 1, since biosequences exist in various lengths thereby leaving a possibility of unigrams during k-mer generation. GloVe embeddings were trained using a learning rate of 0.05 for 500 epochs. For all the k-mer embeddings generated, the sequence embeddings were calculated by averaging the k-mer embeddings.

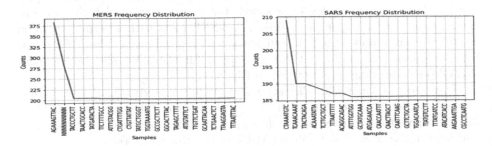

Fig. 2. Frequency distribution plots for 10-mers

---

[2] http://nlp.stanford.edu/projects/glove/.

5.2   Sequence Classification

The obtained sequence embeddings were classified using Machine Learning algorithms viz. Naive Bayes (Gaussian), SVM Classifier (Linear, Polynomial, RBF), Decision Tree, Random Forest and Adaboost with 5-fold cross validation. For the Decision Tree algorithm and Random Forest algorithm, Gini Index was used as the splitting criterion. For AdaBoost and Random Forest, the number of estimators were set to 100. For all the SVM classifiers, C value was taken as 1. These hyperparameters were obtained experimentally for obtaining the best results. For further verification of the results, the dataset was split into training and testing sets with an 80-20 split using the scikit-learn Machine Learning Library[3] to ensure that the test data was completely unseen for the trained model and the classification task was carried out using 1-D CNN [17] and CNN-BiLSTM [18]. The results for the classification task are discussed in Table 1 and 2. The embeddings are compared using dna2vec [14] embeddings in Table 3.

6   Results and Discussion

The experiments were conducted using the Google Colaboratory platform[4] with a hosted runtime of 12GB RAM support. To analyse and understand the FastText embeddings generated with the most frequent k-mers in both SARS and MERS datasets over the range of 'k' values, we generated frequency distributions using the Natural Language Toolkit (NLTK)[5]. The distributions were visually represented using the FreqDist module of NLTK which enables the creation of a graphical frequency distribution plot. Figure 2 gives the frequency distribution plots for 10-mers which indicate a steep slope. Similar patterns were shown for 20-mers, 30-mers and 40-mers which indicate patterns in the MERS and SARS sequences.

The t-Distributed Stochastic Neighbor Embedding (t-SNE) uses an unsupervised technique that enables lower dimensional visualization of higher dimensional data. For our experiments, we used t-SNE plots with a perplexity value of 30 to visually understand the efficiency and separability of the MERS and SARS sequence embeddings. The t-SNE plots for the most efficient FastText and GloVe embeddings are provided in Fig. 3.

The Receiver Operating Characteristic (ROC) is a visualization tool that plots the False Positive Rate with the True Positive Rate using thresholding method where the predicted probabilities play a major role in deciding the curve and in turn the Area Under Curve (AUC) value (Refer Fig. 4).

---

[3] https://scikit-learn.org/stable/.

[4] https://colab.research.google.com/notebooks/intro.ipynb.

[5] https://www.nltk.org/.

Table 1. Cross-validation scores for Machine Learning classifiers and Accuracy for CNN [17] and CNN-BiLSTM [18] with FastText n-gram embeddings

(a) Naive Bayes				
k	10	20	30	40
FastText	0.745	0.847	0.837	0.967
FastText with most frequent k-mers	1.000	1.000	1.000	1.000
(b) Decision Tree				
FastText	1.000	1.000	1.000	1.000
FastText with most frequent k-mers	1.000	1.000	1.000	1.000
(c) Support Vector Machine Classifier (Linear Kernel)				
FastText	0.994	0.998	0.997	0.999
FastText with most frequent k-mers	1.000	1.000	1.000	1.000
(d) Support Vector Machine Classifier (Polynomial Kernel)				
FastText	0.676	0.678	0.726	0.920
FastText with most frequent k-mers	1.000	1.000	1.000	1.000
(e) Support Vector Machine Classifier (RBF Kernel)				
FastText	0.917	0.908	1.000	1.000
FastText with most frequent k-mers	1.000	1.000	1.000	1.000
(f) Linear Support Vector Machine Classifier				
FastText	0.917	0.908	1.000	1.000
FastText with most frequent k-mers	1.000	1.000	1.000	1.000
(g) Random Forest				
FastText	1.000	1.000	1.000	1.000
FastText with most frequent k-mers	1.000	1.000	1.000	1.000
(h) AdaBoost				
FastText	1.000	1.000	1.000	1.000
FastText with most frequent k-mers	1.000	1.000	1.000	1.000
(i) CNN [17]				
FastText	1.000	0.997	1.000	1.000
FastText with most frequent k-mers	1.000	1.000	1.000	1.000
(j) CNN-BiLSTM [18]				
FastText	0.997	0.991	0.997	1.000
FastText with most frequent k-mers	1.000	1.000	1.000	1.000

Table 2. Cross-validation scores for Machine Learning classifiers and Accuracy for CNN [17] and CNN-BiLSTM [18] with GloVe embeddings over the embedding dimensions 40, 50, 100, 150, 200, 250 and 300

(a) Naive Bayes				
k	10	20	30	40
40	0.597	0.800	0.818	0.824
50	0.592	0.812	0.791	0.819
100	0.592	0.807	0.830	0.829
150	0.614	0.826	0.826	0.838
200	0.606	0.813	0.809	0.832
250	0.607	0.826	0.820	0.830
300	0.657	0.814	0.808	0.830
(b) Decision Tree				
40	0.983	0.981	0.963	0.958
50	0.992	0.959	0.963	0.965
100	0.990	0.988	0.965	0.963
150	0.997	0.987	0.992	0.978
200	0.999	0.995	0.985	0.961
250	0.999	1.000	0.996	0.968
300	0.997	0.992	0.971	0.979
(c) Support Vector Machine Classifier (Linear Kernel)				
40	0.682	0.726	0.805	0.787
50	0.658	0.716	0.772	0.766
100	0.793	0.734	0.813	0.805
150	0.820	0.814	0.781	0.678
200	0.817	0.807	0.818	0.716
250	0.808	0.790	0.816	0.800
300	0.811	0.818	0.787	0.817
(d) Support Vector Machine Classifier (Polynomial Kernel)				
40	0.693	0.680	0.691	0.736
50	0.678	0.678	0.765	0.721
100	0.687	0.696	0.729	0.725
150	0.684	0.731	0.715	0.722
200	0.678	0.682	0.689	0.721
250	0.683	0.691	0.680	0.682
300	0.688	0.731	0.689	0.684
(e) Support Vector Machine Classifier (RBF Kernel)				
40	0.544	0.636	0.662	0.671
50	0.636	0.642	0.670	0.675
100	0.634	0.658	0.677	0.680
150	0.643	0.660	0.684	0.670
200	0.648	0.670	0.684	0.702
250	0.646	0.658	0.702	0.702
300	0.642	0.663	0.698	0.697

(*continued*)

Table 2. (*continued*)

(f) Linear Support Vector Machine Classifier			
40	0.823	0.848	0.841	0.852
50	0.842	0.844	0.836	0.844
100	0.844	0.840	0.840	0.852
150	0.843	0.847	0.840	0.845
200	0.850	0.845	0.834	0.845
250	0.844	0.845	0.838	0.844
300	0.845	0.843	0.838	0.844
(g) Random Forest			
40	1.000	1.000	1.000	1.000
50	1.000	1.000	1.000	1.000
100	1.000	1.000	1.000	1.000
150	1.000	1.000	1.000	1.000
200	1.000	1.000	1.000	1.000
250	1.000	1.000	1.000	1.000
300	1.000	1.000	1.000	1.000
(h) AdaBoost			
40	0.949	0.946	0.939	0.941
50	0.959	0.946	0.941	0.936
100	0.973	0.967	0.965	0.961
150	0.981	0.974	0.976	0.963
200	0.988	0.977	0.971	0.975
250	0.986	0.977	0.984	0.978
300	0.993	0.987	0.977	0.983
(i) CNN [17]			
40	0.945	0.913	0.921	0.907
50	0.949	0.915	0.913	0.921
100	0.941	0.933	0.923	0.927
150	0.943	0.927	0.917	0.903
200	0.967	0.925	0.921	0.913
250	0.955	0.927	0.925	0.913
300	0.943	0.923	0.903	0.909
(j) CNN-BiLSTM [18]			
40	0.937	0.925	0.911	0.905
50	0.931	0.929	0.905	0.911
100	0.955	0.921	0.919	0.911
150	0.941	0.925	0.921	0.907
200	0.953	0.917	0.919	0.917
250	0.943	0.915	0.927	0.907
300	0.939	0.925	0.933	0.917

Table 3. Cross-validation scores for Machine Learning classifiers and Accuracy for CNN [17] and CNN-BiLSTM [18] with dna2vec embeddings [14]

(a) Classical Machine Learning Classifiers	
Classifier	Cross validation score
Naive Bayes	0.603
Decision Tree	0.890
SVC (Linear Kernel)	0.677
SVC (Polynomial Kernel)	0.676
SVC (RBF Kernel)	0.577
Linear SVC	0.842
Random Forest	0.890
AdaBoost	0.890
(b) CNN [17] and CNN-BiLSTM [18]	
Classifier	Accuracy
CNN [17]	0.881
CNN-BiLSTM [18]	0.879

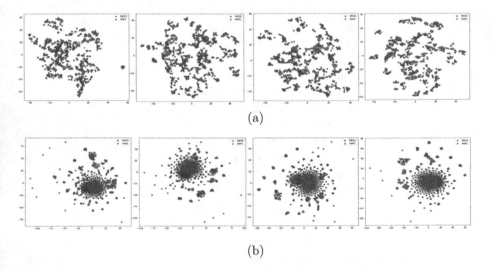

(a)

(b)

Fig. 3. t-SNE plots showing variation in MERS and SARS sequences for k-mers with value of 'k' as 10, 20, 30, 40 based on the following embeddings (a) FastText with frequent k-mers (b) GloVe embedding dimension 100

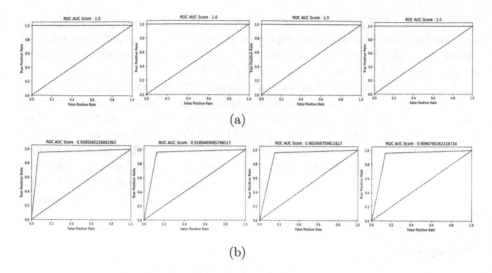

(a)

(b)

Fig. 4. ROC for CNN [13] classifier with value of 'k' as 10, 20, 30 and 40 using the following embeddings (a) FastText with frequent k-mers (b) GloVe embedding dimension 100

7 Conclusion and Future Works

This paper aims at the following three objectives - 1. performance evaluation of context information based k-mer generation method i.e. non-overlapping non-repetitive k-mer generation from sequences, 2. conducting detailed analysis of GloVe and FastText n-gram representation for MERS and SARS sequences which belong to the same genus and subgenus of the Coronaviridae family, 3. studying the performance of the embeddings over various 'k' values to interpret the effect of k-mer size and context information capturing for the sequence embeddings. The dataset collected for the work consists of MERS and SARS sequences of varying lengths in the form of FASTA files which were pre-processed into k-mers of various 'k' values. The effect of GloVe and FastText n-gram embeddings were evaluated based on their performances using various classifiers. The classifiers included classical Machine Learning algorithms viz. Naive Bayes, Decision Trees, Support Vector Machine Classifiers, Random Forest and AdaBoost, and Deep Learning methods viz. Convolutional Neural Networks (CNN) [17] and CNN with Bidirectional LSTM [18]. To further evaluate the performance of the embedding techniques, we deployed a visualization tool viz. t-Distributed Stochastic Neighbor Embedding (t-SNE) plot which depicts the variations of MERS and SARS sequence embeddings. To get a better understanding of the most frequent k-mers which were also used for creating k-mer embeddings using FastText n-gram method, we visualized the Frequency Distribution plots of k-mers over a range of 'k' values. In order to provide a deep dive into the sequence embedding performances using CNN classifier [17] we plotted the ROC graphs and calculated the AUC scores for the same. From the t-SNE plots and the ROC graphs it was evident that the FastText n-gram representations provide the most meaningful embeddings to the sequences when the most frequent k-mers were chosen to train the model. For a comparative study, we conducted the above experiments with the existing dna2vec [14] model.

As a future work, we aim to experiment with Transformers for sequence embeddings as well as for classification. Transformers have self attention units which focus on regions of primary importance in the biosequence [21]. This work can also be extended to analyse SARS-COV-2 sequences to understand the characteristics and similarities of the virus with MERS and SARS which belong to the same family but differ in the subgenus. In our current work, since using the most frequent k-mers could provide significant embeddings, we also wish to analyse the possibility of finding the most important k-mers to represent each sequence, which might contribute to reducing the computations.

References

1. National Institute of Allergy and Infectious Diseases. COVID-19, MERS and SARS (2020). https://www.niaid.nih.gov/diseases-conditions/covid-19
2. Lan, T.C.T., et al.: Structure of the full SARS-CoV-2 RNA genome in infected cells (2021). https://doi.org/10.1101/2020.06.29.178343
3. Saeed, U., Zainab, U.: Biological Sequence Analysis (2019). https://doi.org/10.15586/computationalbiology.2019.ch4
4. George, A., Ganesh, H.B.B., Soman, K.P.: Teamcen at semeval-2018 task 1: global vectors representation in emotion detection. In: Proceedings of the 12th International Workshop on Semantic Evaluation (2018)
5. George, A., Ganesh, H.B.B., Kumar, M., Soman, K.P.: Significance of Global Vectors Representation in Protein Sequences Analysis (2019). https://doi.org/10.1007/978-3-030-04061-1_27
6. Naveen, R., Hariharan, V., Ganesh, H.B.B., Kumar, M., Soman, K.P.: CENNLP at SemEval-2018 Task 2: enhanced distributed representation of text using target classes for emoji prediction representation, pp. 486–490 (2018). https://doi.org/10.18653/v1/S18-1078
7. Mikolov, T., Corrado, G., Chen, K., Dean, J.: Efficient estimation of word representations in vector space. In: Proceedings of the International Conference on Learning Representations (ICLR 2013), pp. 1–12 (2013)
8. Kwan, H., Arniker, S.: Numerical representation of DNA sequences, pp. 307–310 (2009). https://doi.org/10.1109/EIT.2009.5189632
9. Nguyen, N., et al.: DNA sequence classification by convolutional neural network. J. Biomed. Sci. Eng. **9**, 280–286 (2016). https://doi.org/10.4236/jbise.2016.95021
10. Rizzo, R., Fiannaca, A., La Rosa, M., Urso, A.: A deep learning approach to DNA sequence classification. In: Angelini, C., Rancoita, P.M.V., Rovetta, S. (eds.) CIBB 2015. LNCS, vol. 9874, pp. 129–140. Springer, Cham (2016). https://doi.org/10.1007/978-3-319-44332-4_10
11. Mo, Z., et al.: One novel representation of DNA sequence based on the global and local position information. Sci. Rep. **8** (2018). https://doi.org/10.1038/s41598-018-26005-3
12. Asgari, E., Mofrad, M.: Continuous distributed representation of biological sequences for deep proteomics and genomics. PLoS ONE **10**, e0141287 (2015). https://doi.org/10.1371/journal.pone.0141287

13. Kimothi, D., et al.: Distributed representations for biological sequence analysis (2016). ArXiv abs/1608.05949
14. Ng, P.: dna2vec: consistent vector representations of variable-length k-mers (2017) arXiv preprint. arXiv:1701.06279
15. Lee, K., Yapp, E., Nagasundaram, N., Yeh, I.H.-Y.: Classifying promoters by interpreting the hidden information of DNA sequences via deep learning and combination of continuous FastText N-Grams. Front. Bioeng. Biotechnol. **7**, 305 (2019). https://doi.org/10.3389/fbioe.2019.00305
16. Vazhayil, A., Soman, K.P.: DeepProteomics: protein family classification using shallow and deep networks (2018). arXiv preprint arXiv:1809.04461
17. Lopez-Rincon, A., et al.: Accurate identification of SARS-CoV-2 from viral genome sequences using deep learning (2020). https://doi.org/10.1101/2020.03.13.990242
18. Zhang, J., Chen, Q., Liu, B.: DeepDRBP-2L: a new genome annotation predictor for identifying DNA binding proteins and RNA binding proteins using Convolutional Neural Network and Long Short-Term Memory. IEEE/ACM Trans. Comput. Biol. Bioinf. (2019).https://doi.org/10.1109/TCBB.2019.2952338
19. NCBI Virus. https://www.ncbi.nlm.nih.gov/labs/virus/vssi
20. Min, X., Zeng, W., Chen, N., Chen, T., Jiang, R.: Chromatin accessibility prediction via convolutional long short-term memory networks with k-mer embedding. Bioinformatics **33**(14), i92–i101 (2017). https://doi.org/10.1093/bioinformatics/btx234
21. Ji, Y., Zhou, Z., Liu, H., Davuluri, R.: DNABERT: pre-trained Bidirectional Encoder Representations from Transformers model for DNA-language in genome (2021). https://doi.org/10.1101/2020.09.17.301879

Supply Path Optimization in Video Advertising Landscape

Ujwala Musku[(✉)] and Prakhar Yadav

Research and Development, MiQ Digital India, Bangalore 560001, KA, India
{ujwala,prakhar.yadav}@miqdigital.com
http://www.wearemiq.com

Abstract. It is quite intricate for a buyer to reach the publisher's advertising slot with many market players in the programmatic era. Auction Duplication, internal deals between Demand & Supply side platforms, and rife fraudulent activities are complicating the existing complex process - leading to a single impression being sold through multiple routes by multiple sellers at multiple prices. The dilemma: Which path should the buyer choose, and what should be the fair price to pay? has been staying put for years. The framework suggested in this paper solves the problem of choosing the best path at the right price in the Video Advertising Landscape, a significant contributor compared to other advertising channels. This framework embraces two techniques named Data Envelopment Analysis, where an unsupervised data set is ranked by estimating the relative efficiencies, and a statistical and machine learning hybrid scoring method based on Classification Modeling to help us decide the path worth bidding. These models' results are compared with each other to choose the best one based on campaign KPI, i.e., CPM (Cost per 1000 impressions) and VCR (Video Completion rate of the video ad). An average of 6%- 12% reduction in CPM and 1% - 4% increment in VCR is observed across 10 live video ad campaigns. The zenith improvements in CPM reduction give rise to a better return on investment(ROI) than the heuristic approach.

Keywords: Supply path optimization · Programmatic advertising ·
Video advertising · TV advertising · Real time bidding · CPM · VCR ·
DSP · SSP · Artificial Intelligence · Digital marketing · DEA

1 Introduction

Digital Advertising has emerged to be one of the most critical elements in the ecosystem that makes the whole internet sustainable. The Covid -19 outbreak has caused a sudden surge in the number of ad tech campaigns, thus creating a new need for optimizations to help companies maximize their Return On Investment(ROI). As opposed to the classic display ads, which have been in the market for years, the growth of Video Ads is outpacing progressively. The rise in Video ad share in the present market is threefold as compared to Display.

© Springer Nature Switzerland AG 2021
M. Singh et al. (Eds.): ICACDS 2021, CCIS 1440, pp. 488–499, 2021.
https://doi.org/10.1007/978-3-030-81462-5_44

Emerging OTT platforms, Connected TV, and video websites alluring the target audience have created a perfect platform for companies to target a specific set of users with video ads, leveraging the much-unexplored potential of video advertising. Because of better engagement with the audience and higher inventory costs(thus higher revenues), video advertising is gaining massive popularity, and various reports suggest video advertising dominating the market by the next decade. According to a recent survey, 35% [1] of the users were frustrated with video ads because of factors like irrelevancy and long video ads. Such factors make video advertising a high-risk domain, but with optimizations by the side can yield great ROI. The optimizations need to be real-time and data-driven since the real-time bidding [2] occurs in few milliseconds.

2 Video Advertising Ecosystem

2.1 Market Players

Due to an increased demand for video advertising in the digital marketing space these days, the convolution in terms of players and process has evolved invariably. It involves multiple steps in the process of selling-buying in between these multiple players. The most crucial players in the video advertising landscape are DSP, SSP, Publishers, and ad exchanges [3].

- Advertiser: The brand which/who wants to market their product via digital advertising.
- Publishers: Organizations with owned video media who sell inventory on their own websites.
- DSP: Helps the advertiser to buy and access the available video inventory
- SSP: Helps the publisher to sell the video inventory they own
- AdExchange: A marketplace where buying and selling of video ad slots occurs.

2.2 Key Performance Indicators

In Real-time bidding [2], the advertiser buys the ad slot via DSP from the seller (SSP) in the auction process so that they can show the video ad of their product on the ad slot purchased. Once the ad is launched, the user might choose to pause the video right at the start, or sometime in the middle, or watch the video till the end depending on the individual's interest. These particular criteria measure the performance of the video ad. A user who completed watching the full video ad is considered an acquired user compared to a user who watched half of the video. The video campaigns' key performance indicators are judged usually using VCR (Video Completion rate) and CPM(Cost per 1000 impressions).

- Impressions: The number of video ads delivered on an ad slot at a defined level.
- Completions: The number of video ads that were completely watched.
- Video Completion Rate (VCR): The ratio between the number of video view completions to the number of impressions delivered at a defined level.
- Cost per 1000 impressions (CPM): The ratio of overall media cost spent to deliver impressions with the number of impressions delivered.

2.3   Devices

Video campaigns are also mainly targeted via four common channels - Desktop, Mobile, Tablets, and CTV. The factors like better engagement with users are alluring advertisers to drive more spends in terms of media cost in CTV, mobile, and desktop, thus making these necessary devices to target ads. Digital video advertising across mobile and desktop is cited as a significant contributor, increasing 33.5% ($5.45 billion) to a total of $21.72 billion from 2018 [4]. Due to COVID, among all the paused and canceled campaigns, digital video remains relatively stable, driven by Connected TV (CTV). According to IAB's U.S. 2020 Digital Video Advertising Spend Report: Putting COVID in Context, CTV is largely responsible for the sector's resilience and was least impacted by COVID-19. The average CTV spend for 2020 is expected to reach $16MM per advertiser (+8% y/y) [5].

3   Literature Review

It was the problem of auction duplication which mainly led to the rise of Supply Path Optimization in Programmatic. In a white paper published by PubMatic, it is stated that the normal waterfall method [6] limited the publishers to certain SSPs. This led to the inventory being limited via top SSPs in waterfall. After introduction of header-bidding [7] this was reduced by some extent. GroupM also introduced Supply Path Optimization to make the programmatic more transparent thus making it more efficient. However, it again relied on the header-bidding framework for this. Not much of research is there on SPO and while most of the market players rely on header-bidding to ensure optimized supply paths, there is a lack of data driven technique to ensure good paths from advertiser to respective inventory slot. In this paper we introduce a machine learning based novel approach which ensures the selection of only the effective optimized paths.

4   Experimental Setup

4.1   Data Preparation

The activity data set of campaigns stored by the DSP is the primary data source. This data comprises the following features of campaigns such as site domain, timestamp, seller, publisher, country, etc., along with the metrics such as impressions, media cost spent, video completes, etc. A supply path is constituted by putting all the above features together, and key performance metrics are calculated at this supply path level. The historical performance data of a campaign within a fixed time span is stored at a supply path level with metrics. Sample data looks like the following (Table 1).

Table 1. Sample data

Path	Domain	Seller	Publisher	Country	Imps	VCR (in %)	cpm
p1	ebay.com	Appnexus	3576	US	1289	95	0.06
p2	ebay.com	Google	1137	US	7890	91	0.02
p3	yahoo.com	One video	8901	CA	4680	98	0.04
p4	nytimes.com	Openx	2609	US	9089	86	0.04
p5	zillow.com	Telaria	85001	GB	3467	89	0.03

4.2 Heuristic Approach

As a baseline model, clustering based on RFM [8] principle is implemented. A set of three key performance indicators such as VCR, CPM, and impressions in Video Advertising are used instead of recency, frequency, and monetary variables. Each supply path is segmented into four quartiles in the dataset based on each key performance indicator. The cluster with high performance (VCR), low cost(CPM), and high delivery(Impressions) is used for targeting. The testing (see Sect. 5) with the clustering technique yielded around a 3% reduction in cost spent. This is considered the baseline performance to evaluate the finalized techniques explained in upcoming sections.

4.3 Step 1 - Data Envelopment Analysis

Data Envelopment Analysis [9], commonly known as DEA, is an algorithm based on linear programming techniques to measure individual units' efficiency referred to as DMUs. For this use-case, each unique combination of site domain, supply vendor(supply path) with respective metrics(media cost, impressions) is termed as a Decision-Making Unit or more commonly known as DMU. DEA measures the relative performance of individual DMUs and ranks them by taking into account the multiple inputs and outputs. It is instrumental in cases where we have unlabelled data(like here) and want to rank them according to a specific feature set.The mathematical formulation of DEA is as below. Basic efficiency is calculated using the formula :

$$Efficiency = \frac{Output}{Input} \tag{1}$$

Efficiency for k_{th} DMU Θ_k, where O_i is i^{th} output, u_i the weight of that particular i^{th} output, I_j is the jth input and v_j is the weight of jth input is given by :

$$\Theta_k = \frac{\sum_{i=0}^{n} O_i * u_i}{\sum_{j=0}^{m} I_j * v_j} \tag{2}$$

where n = Number of Outputs, m = Number of Inputs and k $\in$ (0,1)

We aim to

$$Maximize(\Theta)$$

for the current DMU k(out of total l DMU's) given the constraints,

$$\Theta_k <= 1 \quad \forall x \in l$$
$$u_i, v_j > = 0 \quad \forall i, j \in (0, n)\,(0, m)$$

Once we get the optimized weights, DEA categorizes the points according to their efficiency. This can be visualized as an outer frontier line, which acts as an efficient front(P1, P2, P3, P4), encapsulating all the less efficient points(P5, P6). The frontier is created on the graph, and efficient points lie on the frontier. For all the other inefficient points, efficiency values are calculated based on how far the point is from the outer frontier line. The farther it is, the lower is the efficiency (Fig. 1).

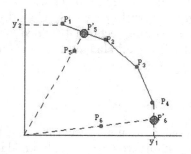

Fig. 1. Graphical visualization of DEA

Although DEA can be used in cases where we have multiple outputs to optimize upon but after experimenting, we decided to make a singular output as a combination of all outputs we needed to optimize on(hereby called redefined KPI) and ran the algorithm to optimize on this KPI instead of multiples KPI's.

$$rKPI = f(x) \ where \ x \in [CPM, VCR, Impressions] \tag{3}$$

Thus, maximizing the rKPI is the primary goal of DEA and paths are allotted with efficiency values according the rKPI values.

DEA considers only the numerical features of a DMU, however with constrained values of metrics like VCR and CPM, DEA cannot differentiate much amongst the DMU's leading to close efficiency values (Table 2). Also, the inventory features like site domain and supply vendor hold higher significance in supply path optimization. Due to the DEA's constraint of not handling categorical features, these features were not considered in the technique.

Table 2. Efficiency

Path	Imps	Cost(in $)	VCR(in %)	CPM	Efficiency
p1	1289	78	95	0.06	0.51
p2	7890	239	91	0.02	1.0
p3	4680	209	98	0.04	0.63
p4	9089	378	86	0.04	0.6 3
p5	3467	123	89	0.03	0.63

In the next section, we have introduced an algorithm that assesses the importance of categorical features and requires labeled data. However, we have unlabelled data, and there was a need to come up with a classifying technique to
label the data(good/bad in our case). DEA hence acts as a technique that helps
transform the unsupervised dataset to a supervised one. DEA's efficiency values
are studied in the context of digital advertising and are used to come up with a
threshold value. This threshold value helps us differentiate the path from good
to bad. All the paths with efficiency values above the threshold are marked as
good paths and vice versa. Labeled data is passed on to the next algorithm,
which considers the importance of all categorical features.

4.4 Step 2 - Scoring Methodology

Scoring methodology [10] accommodates each entity's importance within the
supply path such as publisher, site domain, seller, etc. This technique fundamentally calculates the proxy score signifying the importance of each entity in
the supply path. The flagged output data from DEA that will act as an input in
the scoring methodology looks as below. For the sake of simplicity, let's assume
the entities in the supply path are site domain, seller, and publisher. The following steps are to be followed in an order to obtain the score.

1. Binning
   Entities within each path in the input data (Table 2) need to be pre-processed
   to be divided into intervals named bins to avoid overfitting and minimizing the
   observational errors. Quantile - based [11] discretization function is used to bin
   using percentiles based on the data's distribution. Continuous and categorical
   variables are binned in a range format and discrete values respectively. For
   example, continuous variables such as video completion rate are allotted as
   (21%–42%) and categorical variables such as site domain will be split as
   "cnn.com" etc.

2. WOE and IV Calculation
   Once we have the binned features, the event and non-event rates are calculated using the flag values. Let the total number of entities in a supply path be x. We denote flag value as f and each row as n. $f_{n,i}$ basically denotes the flag value of an entity i of a particular row n. For each entity i,

$$Event(e_i) = \sum_{n=1}^{N} [\![f_{n,i} = 1]\!] \quad where \ n \in [1, N] \tag{4}$$

$$Non \ Event(ne_i) = \sum_{n=1}^{N} [\![f_{n,i} = 0]\!] \quad where \ n \in [1, N] \tag{5}$$

$$Distribution Event(d_i) = \frac{\sum_{n=1}^{N} [\![f_{n,i} = 1]\!]}{\sum_{i=1}^{x} \sum_{n=1}^{N} [\![f_{n,i} = 1]\!]} \tag{6}$$

$$Non \ Distribution Event(nd_i) = \frac{\sum_{n=1}^{N} [\![f_{n,i} = 0]\!]}{\sum_{i=1}^{x} \sum_{n=1}^{N} [\![f_{n,i} = 0]\!]} \tag{7}$$

A logarithmic value of the ratio between event distribution to non-event distribution is calculated, and this value is labeled as the weight of evidence(WOE). WOE fundamentally helps us measure the credibility of each bin. Information value is calculated as a product of the weight of evidence with a distribution event ratio. Information value (IV) signifies and points us to the best features that should be considered by measuring how well a feature distinguishes the target variable (here flag) from bad to good. As per business standards, the threshold of information value is typically 0.02 [12]. All the features with information value less than 0.02 are considered weak predictors and hence, abandoned out of the modeling process. Therefore, x is further reduced to y.

$$WOE_i = \log(\frac{d_i}{nd_i}) \tag{8}$$

$$IV_i = (d_i - nd_i) * \log(\frac{d_i}{nd_i}) \tag{9}$$

3. Classification modeling using Logistic regression
   The data, with features securing information value greater than 0.02 (from the above step) as inputs and flags as outputs, is fed to the Logistic regression [13] model to obtain each entity's coefficients. The logistic regression algorithm's coefficients are estimated from the historical data with the inclusion of only using maximum-likelihood estimation [14]. The robust coefficients would result in a model that would predict a flag value, i.e., 0 or 1 precisely, by minimizing the model's probability error–the coefficient of an entity i is denoted as c_i and overall intercept value is denoted as I.
4. Logit, Odds and Probability calculations
   Logit values are then obtained for each bin by taking a product of WOE of

the bin with a specific feature's coefficient. Using logit values, we come up with odds and probabilities for each bin. For an entity i,

$$Logit(L_i) = c_i * WOE_i = c_i * \log(\frac{d_i}{nd_i}) \tag{10}$$

$$Odds(O_i) = \exp(Li) = \exp(c_i) * \frac{d_i}{nd_i} \tag{11}$$

$$Probability(P_i) = \frac{O_i}{O_i + 1} = \frac{\exp(c_i) * d_i}{\exp(c_i) * d_i + nd_i} \tag{12}$$

5. Scoring and Bid Modifier logic

To set up score logic, we fix the range of scores, factor (f) and offset (θ) values to come up with a score using the following formula.

$$Score(S_i) = f * (L_i + \frac{I}{y}) + \frac{\theta}{y} \tag{13}$$

The score values are transformed into a functioning format that fits into the DSP feature targeting syntactic structure. Using Min-Max feature scaling [15], the bid modifier value is scaled between 0.5 and 1 according to the scores obtained. The boundaries 0.5 and 1 are fixed according to the domain context, which can vary as per the use case (Table 3).

Table 3. Score and Bid modifier

Path	Imps	Cost (in \$)	VCR (in %)	CPM	Efficiency	Score	Bid modifier
p1	1289	78	95	0.06	0.51	254	0.5
p2	7890	239	91	0.02	1.0	550	1
p3	4680	209	98	0.04	0.63	369	0.69
p4	9089	378	86	0.04	0.63	278	0.54
p5	3467	123	89	0.03	0.63	410	0.76

5   Testing

A/B testing is carried out by cloning the control campaign group in a DSP into a testing campaign group. This kind of testing helps us measure the performance of the supply paths with specified bid value while sustaining the rest of the environment in both the campaign groups. The list of supply paths (bid list) with the specified bid values from the scoring methodology is attached to the testing campaign group. The testing campaign group ensures the impressions delivery solely from the specified paths. No list of supply paths is attached to the control group, and hence the back end logic of this campaign group delivers based on whatever DSP decides. Various other settings, such as the geographical, ad environment, campaign flight, creatives, and device targeting, will be identical in both campaign groups. Once the bid list is attached and both the campaign groups are set live, the delivery and bidding process start accordingly. This data is stored in the back end and monitored daily to check the testing campaign group's performance with the control campaign group (Fig. 2).

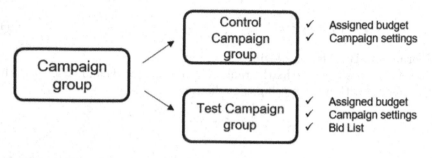

Fig. 2. Testing setup

6   Results

The results of A/B testing of multiple campaigns are observed during the campaign cycle. Exceptionally differentiating KPI performance was observed between control and test campaign groups. An average of 6%–12% CPM drop and 1%–4% VCR rise was observed in 10 testing campaigns.

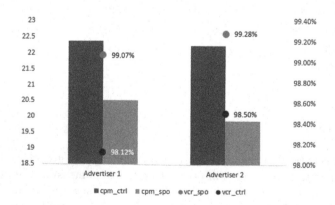

Fig. 3. Performance comparison

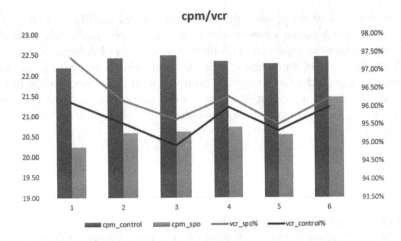

Fig. 4. DOD performance comparison of advertiser 1

From Fig. 3, it is evident that the SPO campaign group outperforms the control campaign group in both the advertisers showing a reduction of 9% in terms of cpm and an increment of 1% in terms of VCR. Figure 4 shows how the performance pans out for the control campaign group and test campaign group for Advertiser 1. Both CPM and VCR values are better than the respective control campaign group each day, which depicts how SPO can be leveraged to get better performance for the advertisers at reduced costs.

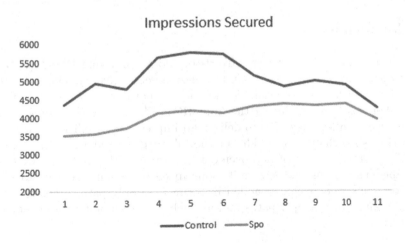

Fig. 5. Gap in delivery

Though the performance is great in the testing campaigns, the number of impressions delivered through test campaigns have slightly plummeted. Restricting the impression delivery through specified supply paths led to this delivery drop. This behaviour is expected due to decreased bid modifiers in the bid list, leading to reduced media costs in the targeted supply paths. Figure 5 above shows that the test campaign group impressions delivery of Advertiser 1 is always lower than the control campaign group.This behavior is acceptable as long as the delivery drop in the test campaign group is within the permissible range.

Out of 10 live testing campaigns, the overall success rate is 80% considering both the KPI's (i.e., Video Completion Rate and Cost spent per 1000 impressions) into account.

7   Limitations and Future Scope

As mentioned in the results, the minor delivery drop is inevitable. This is tackled by adding relevant extra paths to make up for delivery drop. Though advertisers see the future of ad-tech in video channel but the current market share of display channel is also quite huge; expanding SPO to display will result in agencies spending money much more efficiently and ultimately saving up a lot.

Apart from this, a path from advertiser to publisher involves many intermediating players. Therefore a path can be considered as a combination of any of those mediating players. For example, paths can be as long as DSP exchange publisher domain deal placement ID. Involving more players in paths instead of just 2(which we are doing currently) can help us take more control in how the impressions are delivered in the ad tech ecosystem, much of which currently is similar to a black box. This increased control in path delivery is saving up huge on cost and thus maximizing ROI.

8   Conclusion

In the current ad-tech ecosystem, advertisers do not have much control on how and through which paths/players an impression gets delivered. This leads to an impression, which was also available via another path at a lower cost, delivered at a higher cost, thus reducing the final ROI. The proposed approach shows how agencies can leverage SPO to deliver an impression via efficient paths, thus eradicating superfluous costs. The methodology discussed was tested on live campaigns, and a significant improvement in terms of CPM and VCR were seen. The paper also discusses a few limitations in terms of some delivery drop and how more constituents can be added in terms of supply path to create a more efficient control over supply paths and a much better scope for improvement in the future.

References

1. The Complete Guide to Online Video Advertising. https://www.outbrain.com/blog/online-video-advertising-guide/, Accessed 24 Mar 2021

Stack-Based CNN Approach to Covid-19 Detection

V. S. Suryaa$^{(\boxtimes)}$ iD and Z. Sayf Hussain iD

College of Engineering, Guindy, Anna University, Chennai, India

Abstract. Covid-19, declared as a pandemic by the World Health Organization (WHO), has infected more than 113 million globally across 221 countries. In this work, we propose a method for automatic detection of coronavirus based on analyzing the Chest X-ray images. The dataset used for the study composes of 1200 Covid-19 infected, 1,345 Viral Pneumonia infected and 1,341 healthy patient X-ray images. We use different CNN architectures pretrained on ImageNet dataset, fine tune them to adapt the dataset and use it as feature extractors. We determine the best feature extractor among them, stack them with fully connected layers and employ different classification approaches such as softmax, XGBoost and Support Vector Machines (SVM). The results show that the stacked CNN model with DenseNet169, fully connected layers and XGBoost achieves an accuracy, recall and F1-score of 99.679% and precision of 99.683%. Hence, the proposed model showcases potential to assist physicians and make the diagnosis process more accurate and efficient.

Keywords: Deep learning · Covid-19 · Convolutional Neural Networks · Chest X-Ray

1 Introduction

The Coronavirus pandemic which began with a few cases reported in the city of Wuhan, China has quickly spread all over the world. As of 13th January 2021, a year since the first case was reported, over 91 million people have tested positive for the virus. The virus by itself is not life threatening to most people, but people over the age of 65 and those with existing medical conditions are more likely to develop serious illnesses. More than 2 million people have lost their lives to the virus [1].

Chest X-Ray (CXR) is used for the preliminary analysis of various respiratory abnormalities. Studies have shown that patients infected with Covid-19 develop infectious lesions in their chest region [2]. Radiologists at laboratories look for these infectious lesions in the chest region as an initial screening procedure to diagnose Covid-19. However, the manual process to look for these infectious lesions is challenging and can be done only by expert radiologists. The limited number of experienced radiologists and the constant spike in the number of suspected cases makes it difficult for the radiologist to conduct the screening both accurately and in a timely manner. Using a computerized system to identify these infectious lesions based on various spatial features makes the whole process more precise, faster and also easier for the radiologists.

© Springer Nature Switzerland AG 2021
M. Singh et al. (Eds.): ICACDS 2021, CCIS 1440, pp. 500–511, 2021.
https://doi.org/10.1007/978-3-030-81462-5_45

2. Adikari, S., Dutta, K.: Real time bidding in online digital advertisement. In: Donnellan, B., Helfert, M., Kenneally, J., VanderMeer, D., Rothenberger, M., Winter, R. (eds.) DESRIST 2015. LNCS, vol. 9073, pp. 19–38. Springer, Cham (2015). https://doi.org/10.1007/978-3-319-18714-3_2

3. Lee, K.-C., Orten, B., Dasdan, A., Li, W.: Estimating conversion rate in display advertising from past performance data. In: ACM SIGKDD Conference on Knowledge Discovery and Data Mining, pp. 768–776 (2012)

4. Chelsea Fine. First-Quarter Ad Revenue Growth Despite COVID-19 (2020). https://geniusmonkey.com/blog/first-quarter-ad-revenue-growth-despite-covid-19/, Accessed 26 Feb 2021

5. IAB U.S.2020 Digital Video Advertising Spend Report: Putting COVID in Context, https://www.iab.com/insights/iab-us-2020-digital-video-advertising-spend-report/, Accessed 26 Feb 2021

6. PubMatic Understanding Auction Dynamics. https://pubmatic.com/wp-content/uploads/2017/08/PubMatic-UnderstandingAuctionDynamics.pdf, Accessed 24 Mar 2021

7. Pachilakis, M., Papadopoulos, P., Markatos, E.P., Kourtellis, N.: No more chasing waterfalls: a measurement study of the header bidding ad-ecosystem. In: Proceedings of the Internet Measurement Conference (IMC '19), pp. 280–293 (2019). Association for Computing Machinery, New York. https://doi.org/10.1145/3355369.3355582

8. Dogan, O., Ayçin, E., Bulut, Z.: Customer Segmentation by using RFM model and clustering methods: a case stufy in retail industry. Int. J. Contemp. Econ. Adm. Sci. **8**, 1–19 (2018)

9. Zhu, J.: Data Envelopment Analysis: A Handbook of Models and Methods (2015). https://doi.org/10.1007/978-1-4899-7553-9.

10. Thomas, L., Edelman, D., Crook, J.: Credit Scoring and its Applications (2002)

11. Nair, N., Paduthol, S., Balakrishnan, N.: Quantile-Based Reliability Analysis (2013). https://doi.org/10.1007/978-0-8176-8361-0

12. Siddiqi, N.: Credit Risk Scorecards: Developing and Implementing Intelligent Credit Scoring. SAS Publishing, Cary (2005)

13. Alzen, J.L., Langdon, L.S., Otero, V.K.: A logistic regression investigation of the relationship between the Learning Assistant model and failure rates in introductory STEM courses. Int. J. STEM Educ. **5**(1), 1–12 (2018). https://doi.org/10.1186/s40594-018-0152-1

14. Miura, K.: An introduction to maximum likelihood estimation and information geometry. Interdisc. Inf. Sci. (IIS) **17** (2011). https://doi.org/10.4036/iis.2011.155

15. Raschka, S.: About Feature Scaling and Normalization and the effect of standardization for machine learning algorithms (2014). https://bit.ly/2U0DJIG, Accessed 26 Feb 2021

Deep Learning has been successfully used to perform several image processing tasks in different domains with a very high accuracy and small processing time. In the domain of medical science, Deep Learning models have been used to accurately detect malignant brain tumor [3], breast cancer [4], lung tumor [5] and various other deadly diseases. In Deep Learning, Convolutional Neural Network (CNN) eliminates the process of manual feature extraction as the features are learned by the model during training from the provided visual data. The recent advancements in Deep Learning frameworks along with increased CPU and GPU processing capabilities have aided faster and more accurate detection.

In this paper we consider the state-of-the-art Convolutional Neural Network architectures proposed in the last few years for feature extraction and assess their performance to classify CXR images on various parameters such as accuracy, precision, recall and F-1 score. Further the feature extractor with the best performance is considered and a stacked CNN approach is proposed. The proposed stacked CNN model consists of the best performing feature extractor followed by two fully connected layers and finally an XGBoost layer for classification. The stacked model was proposed after analyzing the performance of the combination of different CNN architectures with different machine learning algorithms for the classification task.

2   Related Works

Recently, significant research has been done to automate Covid-19 detection by using several Deep Learning methods. The type and size of data used by studies changes as some studies use raw data while some studies use X-Ray images or CT scans to diagnose Covid-19. Among the proposed studies, the most preferred method to detect Covid-19 is by use the of Convolutional Neural Networks.

Tao Ai et al. performed a comprehensive study with over 1000 patients from Wuhan, China to validate the consistency of chest CT scans in detecting Covid-19 as compared to RT-PCR tests [6]. During the study, the patients underwent both CT scans and RT-PCR tests for a period of 1 month. The study concluded that compared to RT-PCR tests, CT scans are a more reliable, practical and rapid method for early diagnosis of Covid-19. Singh et al. performed two-way classification to classify chest CT scans as Covid-19 positive or negative by using CNN, the initial parameters were trained using Multi-Objective Differential Evaluation (MODE) [7]. The proposed model outperformed traditional CNN models with an F measure of 2.0928%.

Ioannis et al. used the popular transfer learning strategy for three-way classification between Covid-19, bacterial pneumonia and healthy patients' X-Ray images on state-of-the-art CNN architectures like VGG19, MobileNetv2, Inception, Xception and Inception Resnet v2 [8]. The model with the highest accuracy proposed by them was the VGG19 with three-way accuracy of 93.48%. Khan et al. proposed a model CoroNet for four-way classification of X-Ray images between Covid-19, viral pneumonia, bacterial pneumonia and normal patients [9]. The model prepared by them is based on the Xception architecture. Ningwei Wang et al. proposed an integrated pretrained ResNet101 and ResNet152 model, the model achieved 96.1% accuracy on the test set [10]. Asmaa Abbas et al. proposed a Decompose, Transfer, and Compose (DeTraC) approach for the classification of COVID-19 chest X-Ray images and the observations reported show that pretrained VGG19 using DeTrac approach achieves higher accuracy than other models [11].

3   Transfer Learning

Transfer learning is a popular strategy used in computer vision and arduous deep learning tasks to overcome the isolated learning paradigm [12]. Currently transfer learning is very popular in the field of deep learning as it can be used to train deep neural networks with comparatively lesser data. Transfer learning is a strategy where a model developed for a task trained on a specific dataset is used as the starting point to solve a related problem. When transfer learning is used, the models do not require huge modifications to better adapt to the problem in hand.

3.1   VGG

The VGG architecture differs from earlier CNN architectures such as AlexNet; instead of using large receptive filters (11×11 with a stride of 4) VGG uses multiple small receptive filters (3×3 with a stride of 1) [13]. The 3×3 is the smallest size filter which can capture the notion of left or right, up or down and center. VGG also incorporates 1×1 convolutions that make the decision function more nonlinear without changing the size of the feature map.

3.2   Inception

Salient features used to classify images may have extremely large variation in size. A large kernel is preferred for information that is distributed more globally and a small kernel is preferred for information that is distributed more locally. Inception networks allow us to use kernels of multiple sizes simultaneously, thereby making the network wider rather than deeper [14]. InceptionV3 made convolutions more efficient in terms of computational capacity by using smart factorization methods. The Inception network also introduced techniques to improve computational speed and used label smoothing to prevent overfitting.

3.3   Resnet

Researchers have found that while training very deep neural networks, we run into problems of exploding or vanishing gradients. The problem of training very deep neural network layers can be alleviated by the ResNet architecture [15]. ResNet introduced a shortcut or a skip connection that skips one or more layers. The architecture says that if the deeper layers are found to hurt the model's performance then it is always easy to push the residuals to 0 ($F(x) = 0$). If the deeper layers are found to be useful, then the weights of the layers would be non-zero values and the model performance could increase slightly.

3.4 DenseNet

Another way to solve the exploding/vanishing gradient problem of convolutional networks is by adding connections between layers close to the input layers and those that are close to the final output. In the DenseNet architecture, the output of each layer is connected to every subsequent layer in the model [16]. For each layer the outputs or the feature maps of all preceding layers are used as inputs. The various inputs to each layer are concatenated to create a new feature map as the input to that layer. This concatenation of feature maps learned by different blocks increases the variation in the input of subsequent blocks and improves efficiency.

3.5 Xception

The Xception network is also called the extreme version of the Inception network [17]. The Xception network aims to reduce the number of parameters required for the entire neural network, the number of operations and the training time. The Xception network relies heavily on Depthwise Separable Convolution and shortcuts between Convolutional blocks (Fig. 1).

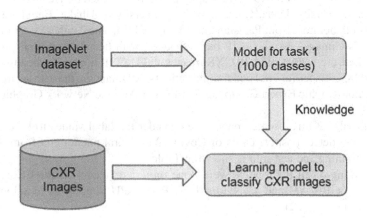

Fig. 1. Transfer learning method

4   Classification Approaches

4.1 SVM

Support Vector Machines, more commonly known as SVM, is a supervised learning model used most often for classification, as done in this project [18]. In an SVM, a Kernel function is used to increase the dimensions of the data, thereby making the data separable. The main idea of SVM lies in choosing an optimal hyperplane instead of just a hyperplane that classifies. To do so, two new concepts of margins and support vectors are introduced. Margin is the largest region that can separate the classes without any data points lying within it. They are formed by two lines running parallel to the decision boundary hyperplane. Support Vectors are those datapoints that lie closest to the hyperplane. They are the most useful datapoints since they form the margin.

4.2 XGBoost (eXtreme Gradient Boosting)

XGBoost is an algorithm that has recently dominated machine learning and is used for many classification tasks [19]. XGBoost is a decision tree-based ensemble learning algorithm that uses gradient boosting to improve performance. Being based on decision trees, it has the potential of being one of the more powerful supervised classification algorithms. In this algorithm, multiple weak classifiers are combined just as in Adaptive Boosting. However, performance is further improved with the help of Extreme Gradient Boosting. It acts as a special case of boosting where errors are minimized by gradient descent algorithm which helps in pruning and weeding out less qualified candidates.

5 Dataset

The dataset used is "COVID-19 Radiography Database" curated by a team of researchers from Qatar University, Doha, Qatar, and the University of Dhaka, Bangladesh along with their collaborators from Pakistan and Malaysia [20]. The database contains a total of 3,886 CXR images for Covid-19 patients along with healthy and viral pneumonia patients. There are 1,200 Covid-19 CXR images, 1,341 CXR images for healthy patients and 1,345 CXR images for viral pneumonia patients. All images in the dataset are either in the Joint Photographic Expert Group (JPG/JPEG) or Portable Network Graphics (PNG) format.

For the sake of this research paper, we consider the label values to be the ground truth values to detect positive cases of Covid-19 and viral pneumonia from the given input. The training and test split is shown in Table 1.

The training of all the models were done on a computer with Intel Core i5 10th generation, 16GB RAM and an Nvidia Tesla P100 GPU running on Windows 64-bit operating system (Fig. 2).

Table 1. Covid-19 CXR dataset

	Training images	Test images	Total
Covid-19	1,096	104	1,200
Viral pneumonia	1,241	104	1,345
Healthy	1,237	104	1,341
Total	3,574	312	3,886

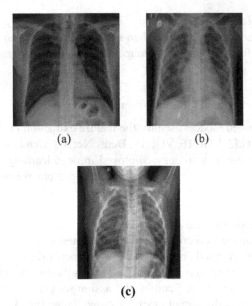

(a)   (b)

(c)

Fig. 2. Samples from the dataset used in the study (a) CXR of a Covid-19 patient (b) CXR of a viral pneumonia patient (c) CXR of a healthy patient

6 Metrics

Several performance measures are defined to evaluate a classifier. Classification accuracy is the most widely used measure to evaluate a model. Classification accuracy is defined as the ratio of the number of correctly classified samples to the total number data samples.

To calculate the accuracy of the model we will need a) Correctly identified diseased cases (True Positives, TP) b) Correctly identified healthy cases (True Negatives, TN) c) Incorrectly identified diseased cases (False Negatives, FN) and d) Incorrectly identified healthy cases (False Positives, FP).

Another widely used performance metric is precision. The precision for a model is calculated as the ratio between the number of positive cases correctly classified to the total number of cases classified as positive (correctly or incorrectly). The precision for a model is high when the model predicts many correct positive cases (TP) or if the model predicts fewer incorrect positive predictions (FP). Recall, another popular performance metric, is defined as the ratio between the number of positive cases correctly classified as positive to the total number of positive cases. F1-score is a performance metric which combines precision and recall. It is defined as the harmonic mean of precision and recall.

7 Methodology

7.1 Pre-processing

Before feeding the obtained CXR images as input to the models, the images were rescaled to 224 × 224 resolution with RGB channels. Data Normalization techniques were then used to improve computational efficiency. This was achieved by scaling down each individual pixel by a factor of 255 and then setting the input mean to 0 over the entire dataset for different features. The input images are also randomly zoomed in by a factor of 0.07.

7.2 Feature Extraction and Classification

The ImageNet pretrained models used for the feature extraction in the experiment were - ResNet101, ResNet152, VGG16, VGG19, DenseNet169, DenseNet201, Xception and InceptionV3. Fine tuning is a strategy employed in deep learning to allow the convolutional model to better fit the dataset and extract more precise information from the last convolution layers by retraining them. We adopt this strategy in our experiment and unfreeze the last few layers in all the models.

The pretrained models were used only as a feature extractor and the classification task on the extracted features was accomplished by using new fully connected networks and a softmax layer that were fitted on top of the existing network. The two fully connected networks FC1 and FC2 were used as normal feed forward networks with ReLU activation function to analyze the extracted features and assign weights. Further on, the softmax layer was used to classify the output labels as shown in the Fig. 3.

For the CNN, the learning rate was set to 0.01 with batch size as 32, optimization function as Stochastic Gradient Descent (SGD) and loss function as Sparse Categorical Cross-Entropy. The model was trained over 5 epochs. The performances of all of the models have been measured on parameters like accuracy, precision, recall, and F1-score. From Table 2, we can observe that DenseNet169 showcases better results than all other models considering the different evaluation metrics used.

DenseNet169 shows an impressive accuracy, recall and F1-score of 99.359% and precision of 99.371%. ResNet152 also shows promising results with an accuracy and F1-score of 99.038% with 309 correctly classified images out of 312 test images. The other configurations in the ResNet and DenseNet architecture - ResNet101 and DenseNet201, both scored more than 98% in all of the metrics. VGG architectures used in the experiments show comparatively lower performances than other models. InceptionV3 produced 97.436% accuracy and 97.434% F1-score on the testing data which was slightly more than Xception network which produced 97.115% accuracy and 97.112% F1-score.

DenseNet169 has 14 million parameters in the network which is lesser than all of the other state-of-the-art CNNs considered for the experiment. So, we further use the DenseNet169 convolutional neural network for feature extraction.

Table 2. Result description of Architecture + FCN + Softmax.

Architecture + FCN + Softmax	Accuracy	Precision	Recall	F1-score	Number of parameters
VGG16	96.154	96.251	96.154	96.117	138,357,544
VGG19	97.436	97.496	97.436	97.419	143,667,240

(*continued*)

Table 2. (*continued*)

Architecture + FCN + Softmax	Accuracy	Precision	Recall	F1-score	Number of parameters
ResNet101	98.718	98.748	98.718	98.713	44,675,560
ResNet152	99.038	99.065	99.038	99.038	60,380,648
DenseNet169	99.359	99.371	99.359	99.359	14,307,880
DenseNet201	98.397	98.448	98.397	98.391	20,242,984
Xception	97.115	97.275	97.115	97.112	22,910,480
InceptionV3	97.436	97.558	97.436	97.434	23,851,784

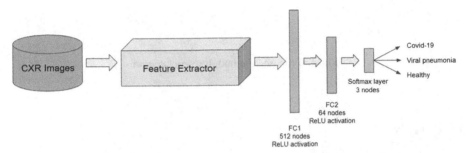

Fig. 3. Methodology adopted to find the suitable feature extractor

8 Stack Based CNN

From Table 2, it is observed that DenseNet169 has both the best performance metrics and the fewest number of parameters. The DenseNet architecture solves the problem of exploding and vanishing gradients by introducing connections from layers close to the input layer to layers that are close to the output of the feature extractor. The various inputs to each block are concatenated to create a new feature map as the input to that block. That is for each layer, the output of all layers preceding it are used as inputs. This improves feature propagation in both forward and backward directions and also increases feature reuse, thus allowing the model to learn better and faster.

In DenseNet169 the input image is first passed through a 7×7 convolutional layer with a stride of 2 and then through a 3×3 max pooling layer. DenseNet169 consists of 4 blocks with varying number of layers in each block. The first block of DenseNet169 consists of 6 sets of 1×1 convolutions followed by 3×3 convolutions. The second block consists of 12 sets of 1×1 convolutions followed by 3×3 convolutions. The third block consists of 32 sets of 1×1 convolutional layers followed by 3×3 convolutions. The final block consists of 32 sets of 1×1 convolutional layers followed by 3×3 convolutions. Each block is separated by a 1×1 convolutional layer followed by a 2×2 average pooling layer with a stride of 2.

A new stack-based CNN approach is proposed where DenseNet169 is used for feature extraction followed by two fully connected layers and an XGBoost layer for classification as shown in Fig. 4. XGBoost is a decision tree-based ensemble learning algorithm in which multiple weaker classifiers are combined. The XGBoost objective is softmax and the learning rate was set to 0.01.

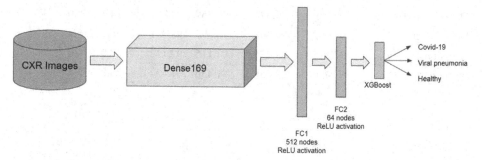

Fig. 4. Proposed methodology architecture

9   Results

For comparison with the proposed model of stack-based CNN with DenseNet169, fully connected layer and XGBoost, support vector machines with radial basis function kernel and softmax were used for classification instead of XGBoost. The performances of the different models considered have been recorded in Table 3 and we can infer that our proposed model outperforms the other configurations on all the performance metrics considered. From the confusion matrix in Fig. 5(c), we can observe that our model has correctly classified 311 images out of the total test images of 312 and achieved an accuracy, recall and F1-score of 99.679% and precision of 99.683%.

Table 3. Result data for different stack model used in the experiment

Stack model	Accuracy	Precision	Recall	F1-score
DenseNet169 + FCN + softmax	99.359	99.371	99.359	99.359
DenseNet169 + FCN + XGBoost	99.679	99.683	99.679	99.679
DenseNet169 + FCN + SVM	98.397	98.471	98.397	98.397

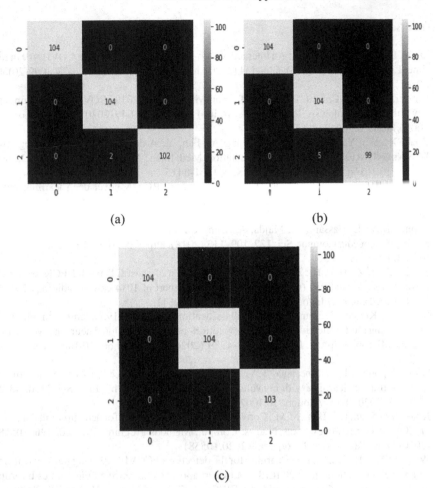

Fig. 5. Confusion Matrix (a) DenseNet169 + FCN + softmax (b) DenseNet169 + FCN + SVM (c) DenseNet169 + FCN + XGBoost

10 Conclusion

Early detection of Covid-19 is essential to prevent the quick spread of the virus. The study presents a transfer learning based stacked CNN approach with fully connected layers and XGBoost to automate the detection of Covid-19 from CXR images. Different state-of-the-art fine-tuned CNN architectures were considered and the results were analyzed to finally propose a stacked CNN approach. XGBoost was used for classifying CXR images due to its advanced features for module tuning, algorithm enhancement and execution speed. The proposed model has an accuracy, recall and F1-score of 99.679% and precision of 99.683%.

The proposed method has not undergone a medical study. Thus, it cannot replace radiologists, but can instead be used as an assistive tool to detect Covid-19 from CXR images efficiently.

References

1. Dong, E., Du, H., Gardner, L.: An interactive web-based dashboard to track COVID-19 in real time. Lancet Infect. Dis. **20**(5), 533–534 (2020). https://doi.org/10.1016/S1473-3099(20)301 20-1
2. Carotti, M., et al.: Chest CT features of coronavirus disease 2019 (COVID-19) pneumonia: key points for radiologists. Radiol. Med. (Torino) **125**(7), 636–646 (2020). https://doi.org/10. 1007/s11547-020-01237-4
3. Zhao, X., Wu, Y., Song, G., Li, Z., Zhang, Y., Fan, Y.: A deep learning model integrating FCNNs and CRFs for brain tumor segmentation. Med. Image Anal. **43**, 98–111 (2018). https:// doi.org/10.1016/j.media.2017.10.002. ISSN 1361-8415
4. Khan, S., Islam, N., Jan, Z., Din, I.U., Rodrigues, J.J.P.C.: A novel deep learning based framework for the detection and classification of breast cancer using transfer learning. Pattern Recogn. Lett. **125**, 1–6 (2019). https://doi.org/10.1016/j.patrec.2019.03.022. ISSN 0167-8655
5. Skourt, B.A., El Hassani, A., Majda, A.: Lung CT image segmentation using deep neural networks. Procedia Comput. Sci. **127**, 109–113 (2018). https://doi.org/10.1016/j.procs.2018. 01.104. ISSN 1877-0509
6. Ai, T., Yang, Z., Hou, H., Zhan, C., et al.: Correlation of chest CT and RT-PCR testing for coronavirus disease 2019 (COVID-19) in China: a report of 1014 cases. Radiology **296**(2), 32–40 (2020). https://doi.org/10.1148/radiol.2020200642
7. Singh, D., Kumar, V., Kaur, M., et al.: Classification of COVID-19 patients from chest CT images using multi-objective differential evolution–based convolutional neural networks. Eur. J. Clin. Microbiol. Infect. Dis. **39**, 1379–1389 (2020). https://doi.org/10.1007/s10096-020-03901-z
8. Apostolopoulos, I.D., Mpesiana, T.A.: Covid-19: automatic detection from X-ray images utilizing transfer learning with convolutional neural networks. Phy. Eng. Sci. Med. **43**(2), 635–640 (2020). https://doi.org/10.1007/s13246-020-00865-4
9. Khan, A.I., Shah, J.L., Bhat, M.M.: CoroNet: a deep neural network for detection and diagnosis of COVID-19 from chest x-ray images. Comput. Methods Programs Biomed. **196**, 105581 (2020). https://doi.org/10.1016/j.cmpb.2020.105581
10. Wang, N., Liu, H., Xu, C.: Deep learning for the detection of COVID-19 using transfer learning and model integration. In: 2020 IEEE 10th International Conference on Electronics Information and Emergency Communication (ICEIEC), Beijing, China, pp. 281–284 (2020). https:// doi.org/10.1109/ICEIEC49280.2020.9152329
11. Abbas, A., Abdelsamea, M.M., Gaber, M.M.: Classification of COVID-19 in chest X-ray images using DeTraC deep convolutional neural network. Appl. Intell. **51**(2), 854–864 (2020). https://doi.org/10.1007/s10489-020-01829-7
12. Pan, S.J., Yang, Q.: A survey on transfer learning. IEEE Trans. Knowl. Data Eng. **22**(10), 1345–1359 (2010). https://doi.org/10.1109/TKDE.2009.191
13. Simonyan, K., Zisserman, A.: Very deep convolutional networks for large-scale image recognition. In: International Conference on Learning Representations (2015). arXiv:1409. 1556v6
14. Szegedy, C., et al. Rethinking the inception architecture for computer vision. In: Proceedings of the IEEE Conference on Computer Vision and Pattern Recognition (2016). arXiv:1512. 00567v3
15. He, K., Zhnag, X., Ren, S., Sun, J.: Deep residual learning for image recognition. In: 2016 IEEE Conference on Computer Vision and Pattern Recognition (CVPR), Las Vegas, NV, pp. 770–778 (2016). https://doi.org/10.1109/CVPR.2016.90
16. Huang, G., Liu, Z., Van Der Maaten, L., Weinberger, K.Q.: Densely connected convolutional networks. In: 2017 IEEE Conference on Computer Vision and Pattern Recognition (CVPR), Honolulu, HI, USA, pp. 2261–2269 (2017). https://doi.org/10.1109/CVPR.2017.243.

17. Chollet, F.: Xception: Deep learning with depthwise separable convolutions. In: Proceedings of the IEEE Conference on Computer Vision and Pattern Recognition (2017). arXiv:1610.02357v3
18. Hearst, M.A., Dumais, S.T., Osuna, E., Platt, J., Scholkopf, B.: Support vector machines. IEEE Intell. Syst. Appl. **13**(4), 18–28 (1998). https://doi.org/10.1109/5254.708428
19. Chen, T., Guestrin, C.: Xgboost: a scalable tree boosting system. In: Proceedings of the 22nd ACM SIGKDD International Conference on Knowledge Discovery and Data Mining (2016). https://doi.org/10.1145/2939672.2939785.
20. Chowdhury, M.E.H., et al.: Can AI help in screening viral and COVID-19 pneumonia? IEEE Access **8**, 132665–132676 (2020)

Performance Analysis of Various Classifiers for Social Intimidating Activities Detection

Mansi Mahendru[✉] and Sanjay Kumar Dubey[✉]

Department of Computer Science and Engineering, Amity University Uttar Pradesh, Sec-125, Noida, UP, India
skdubey1@amity.edu

Abstract. The emergence of social networks is at a great boom today. Every big news before telecasting on television comes to these forums, therefore raises many dilemmas due to misinterpretation regarding the freedom of speaking. One of this trouble is social intimidation that is very disturbing misbehavior that can cause troubling consequences for the victim. Existing works of social intimidation focuses on only one or two topics of harassment. The main aim of this study is to analyze the hub of social intimidation i.e. twitter, consisting of 25,000 tweets covering five topics of harassment i.e. sexism, racism, appearance related, political and intellectual. Moreover, five machine learning and four deep learning techniques were used namely sequential minimal optimization (SMO), random forest, multinomial naïve bayes, logistic regression (LR), decision tree J48, CNN-CB, CNN-GRU, CNN-LRCN and CNN-Bi-LSTM. Each of the classifiers are evaluated using accuracy, precision, recall and f-measure as a performance metric on the dataset. Results indicate the dominance of CNN-Bi-LSTM and logistic regression among all classifiers used.

Keywords: Social intimidation · SMO · Multinomial NB · Logistic regression · J48 · CNN-LRCN · CNN-GRU · CNN-CB · CNN-Bi-LSTM

1 Introduction

COVID-19 has converted the whole world into virtual reality. Today everyone from small kids to old people is doing all their work from home by using laptops, mobile phones etc. This pandemic has changed everything in ways we have never thought before. COVID-19 is the transformation button that has impacted our whole life i.e. how we live, how we work, how we adapted this new change and finally how we use different technologies to live our life. Apart from many advantages there are many disadvantages too. During lockdown people from all ages start spending their time on social media networks as it is the exclusive means through which they can communicate with their dearest ones who are sitting at a far distance from them. Some people have adapted social networks to interact, share their experience with their loved ones, on the other hand some people have used it in negative terms i.e. cybercrimes, social harassment, phishing etc. Through social media people can discuss, share various types of details in the form of multiple ways i.e. textual comments, audio clips, video, memes etc. The user can create their profile irrespective of revealing their personal information and location from where they are sending this information. Social intimidation is a very disturbing misbehavior that can have a threatening consequence for someone whom we are harassing just for our fun. According to the "Global Threat Report 2020" online harassment takes place in various forms i.e. 35% bully people by sharing the screenshot of something and by sharing audio and video clips to harass that person, 25% teenagers trolled someone while playing online games in a large group, 17% people share something on social network that mock some other person who is indirectly affiliated to that content, 12–13% persons send disgusting, unkind messages to annoy someone and 5% people create their fake profile just to irritate and upset others. Apart from these forms of offensive behavior there are many other questions that are to be resolved i.e. various factors on which peoples are harassed on social media. According to cyberbullying research Centre data 61% people bullied due to their appearance 61%, Race 17%, Sex 15%, financial status 15%, religion 12% and other 10% etc. [1, 2]. However social media content is vast and difficult to control, therefore social intimidation detection should pay more awareness to protect children and society from its consequences. Existing works related to social intimidation detection only focuses on one or two topics of harassment. The main objective behind this comparative analysis is to review the quality papers to determine several machine learning and deep learning methods mostly used in the past studies. The classifiers attaining high accuracy are investigated on a sizable dataset to test their usability in more effective terms i.e. high accuracy and less processing time. Accordingly we prepared and aimed to answer the following research questions from this study.

RQ1 Which types of machine learning and deep learning classifiers are actively used to detect social intimidation from social media?

RQ2 How can the social intimidation detection model developed on various topics with high accuracy and less processing time?

2   Related Work

For the past multiple years many researchers are trying to detect the online intimidation activities at the earliest stage. Most of the existing research have approached this issue sing machine learning classifiers and majority of them focuses on one topic of social intimidation i.e. racism and sexism. In recent researches deep learning classifiers have claim that they can overcome the problems of machine learning algorithms and can improve the performance. In this section the overview of various studies carried out to resolve the problem definition by applying machine learning and deep learning technique is illustrated along with the dataset adopted and the key analysis from the research is shown in Table 1. The findings from this related work is validated on different dataset having five categories of harassment. Our findings shows that deep learning classifiers outperform machine learning classifiers on same dataset. Detailed analysis based on performance measures is discussed in results section.

Table 1. Comparison of various classifiers

Objective	Method or technique used	Dataset used	Analysis	Ref.
Supervised Machine Learning for the Detection of Troll Profiles in Twitter Social Network: Application to a Real Case of Cyberbullying	Random Forest, J48,KNN, Sequential Minimal Optimization	Twitter Text Dataset	In this study researchers proposed a methodology that will try to extract the real user behind the fake profile and then try to keep track of parallel activities while trolling someone. They have applied four supervised machine learning algorithms and found Decision trees and SMO with poly kernel algorithms achieve a best AUC of 0.96 among all other algorithms	[4]
Cyberbully Detection Using Intelligent Techniques	Gen leven and Fuzzy Classifier Algorithm	Formspring.me Dataset	In this paper researchers have developed a system that will detect the cyberbully words from the text using genetic algorithm and then classified using fuzzy classifier under four categories harassment, insult, terrorism and flaming. They have also compared their technique with existing method DCLANC and achieve better accuracy between 85–90%	[5]

(*continued*)

Table 1. (*continued*)

Objective	Method or technique used	Dataset used	Analysis	Ref.
Improving cyberbullying detection using Twitter users' psychological features and machine learning	Naïve Bayes, Random forest and J48 algorithm	Twitter Dataset #Game gate	In this study they have used the sentiments and psychological features for detection of abusive comments from tweets and classify them under various categories. The execution of the J48 algorithm has achieved the overall accuracy of 91.88% and improves the overall results when sentiments and personality features are combined with each other	[6]
A Framework to Predict Social Crime through Twitter Tweets by Using Machine Learning	Multinomial Naive Bayes, KNN and SVM Algorithm	Twitter Dataset	In this study researchers have developed the machine learning framework that will classify the data in four categories of crimes. Further n-gram is used with machine learning algorithms to estimate the accuracy of the model. Results show that with SVM the overall accuracy is 92% which is slightly better than other algorithms	[7]

(*continued*)

Table 1. (*continued*)

Objective	Method or technique used	Dataset used	Analysis	Ref.
Machine Learning and Semantic Sentiment Analysis based Algorithms for Suicide Sentiment Prediction in Social Networks	IBI, J48, Cart, SMO and Naïve Bayes Algorithm	Twitter j4 Library to extract tweets	In this study authors have develop the methodology that will try to detect one threating consequence of harassment i.e. suicide. They have used WEKA tool along with some machine learning algorithms to extract the sentiment of suicide behind the message and found that 33% of messages are suspect of risk	[8]
Machine Learning Solutions for controlling Cyberbullying and Cyber stalking	SVM and Neural Networks	Enron 1 email and SMS collection Dataset	In this paper researchers have developed the automatic system that will detect the email cybercrime and document the evidence to protect the user from bullying content and spam mails	[9]

(*continued*)

Table 1. (*continued*)

Objective	Method or technique used	Dataset used	Analysis	Ref.
Detecting misogyny in Spanish tweets. An approach based on linguistics features and word embedding's	Random forest, Sequential minimal optimization (SMO) and Linear support vector machine algorithms	VARW, SELA and DDSS Datasets	In this paper authors try to find the misogyny from Spanish tweets on the most frequently targeted group i.e. women's. Here researchers evaluated their model on three different machine learning approaches and classified all the tweets in different categories. The model obtained 87.15% accuracy with the SMO algorithm which is considered as the baseline model	[10]
Developing an online hate classifier for multiple social media platforms	Logistic Regression, SVM, Naïve Bayes and XGBoost Algorithm	Reddit, Wikipedia, YouTube and Twitter Dataset	Here researchers develop the online hate comments detection for multiple social media platforms on various machine learning algorithms and find XG boost with BERT is the best combination and achieve the best ROC-AUC value	[11]

(*continued*)

Table 1. (*continued*)

Objective	Method or technique used	Dataset used	Analysis	Ref.
Cyberbully Image and Text Detection using Convolutional Neural Networks	Bag of words, Naïve Bayes and CNN Algorithm	Instagram Dataset	Here authors have developed the system that will detect harassing content from both the visuals and text. The sequence and visuals are processed using a bag of words and CNN algorithm and then labelled into four categories as insulting, harassing, trolling, and Threatening. They have achieved the accuracy of nearly 45% on 50 input samples	[12]
Using Convolutional Neural Networks to Classify Hate-Speech	CNN Algorithm	English Twitter Hate speech Dataset	Here Researchers have develop the online hate speech classification using CNN algorithm using four nlp techniques i.e. random vec, word 2 vec, character n gram and word 2 vec + character n gram and find best accuracy with CNN + word 2 vec + character n gram of 0.7389	[13]

(*continued*)

Table 1. (*continued*)

Objective	Method or technique used	Dataset used	Analysis	Ref.
A Hybrid Deep Learning System of CNN and LRCN to Detect Cyberbullying from SNS Comments	CNN and LRCN Algorithm	Cyberbullying comments on twitter dataset from kaggle released in 2012	In this paper researchers have develop the hybrid approach by combining two deep learning algorithm i.e. CNN as word approach and LRCN as character approach. They observe that by using CNN they get 83% accuracy and LRCN achieve 86% accuracy while on the other side by combining both they get 88% accuracy	[14]
Detecting Hate Speech on Twitter Using a Convolution-GRU Based Deep Neural Network	CNN-GRU Algorithm	Hate speech detection dataset	Here authors have developed a system that will detect the hate speech using CNN-GRU architecture on twitter dataset and achieve 0.92 accuracy. They have also compared their model with SVM, CNN on seven different datasets and observe that CNN-GRU has improved the overall performance	[15]

(*continued*)

Table 1. (*continued*)

Objective	Method or technique used	Dataset used	Analysis	Ref.
Aggression detection through deep neural model on Twitter	CNN with LSTM and Bi-LSTM	Cyber Trolls Dataset	Here Authors have developed the system that will detect the trolling comments from twitter using deep neural networks. They achieve accuracy of 86% with CNN LSTM and 87% with CNN-BiLSTM	[16]
Deep Learning Algorithm for Cyberbullying Detection	CNN-CB Algorithm	Dataset from Twitter API	In this study authors have used CNN-CB algorithm to detect online abuse and test it for 10 epoch. They attain 65% accuracy at first epoch and it raises to 95% in 10th one	[17]
Unsupervised Cyberbullying Detection via Time-Informed Gaussian Mixture Model	Gaussian mixture model	Instagram and vine Dataset	In this paper researchers develop the UCD model on two datasets and compare them with many supervised and unsupervised algorithms by extracting multimodal features i.e. text, time and graph. They achieve 0.73 AUROC on Instagram dataset and 0.71 AUROC on vine dataset	[18]

(*continued*)

Table 1. (*continued*)

Objective	Method or technique used	Dataset used	Analysis	Ref.
Antisocial online behavior detection using deep learning	Deep machine learning and NLP techniques	Toxic comment classification dataset	In this paper authors examine the various types of social abusive behavior in order to obtain detection and prevention from those messages. In their study they have compared their models with many other NLP and DL based text classifiers and comparison in terms of precision and AUC is also given in order to have complete understanding	[19]

3 Research Methodology

This segment outlines the dataset used for evaluation between various ML/DL classifiers and the proposed methodology done on the chosen dataset.

3.1 Bench-Mark Dataset

Social intimidation detection at an early stage is a very challenging problem. Many researchers are still trying to detect the harassing post from the social network through abusive words. However, choosing the right classifier, achieving high accuracy and less processing time is still a critical challenge. Previous research has mainly focused on one or two topics of harassment i.e. racism and sexism. In this study five machine learning and four deep learning algorithms that are frequently used by researchers are analyzed for more contents of social harassment. In this study a dataset of 25,000 tweets [20] are applied to evaluate all the techniques. This dataset covers five contents of social intimidation i.e. sexism, racism, appearance related, political and intellectual.

3.2 Model Outlook

This section represent the proposed structure for online intimidation detection where it has three states i.e. Pre-processing state, feature analysis state, classification state as shown in Fig. 1. Firstly, data moved for pre-processing where data is cleaned in order to enhance the accuracy. Pre-processing state consist of several steps i.e. Tokenization in which the whole paragraph of post is converted into tokens in order to handle every word individually, Lowering of text, stop words and redundant words removal and finally handling of missing data [21] by filling the best word fits in the sentence. Once the data is cleaned then it moves for feature analysis. Feature analysis is further divided into two steps i.e. feature extraction and feature selection. Feature extractions consist of extracting all nouns, pronouns and adjectives from the sequence and make it suitable to feed in a classifier for testing. Feature eradication is executed using TF-IDF-NCF [22] and kept inside the feature list for further selection. Once the feature list of intimidating words is prepared then the sentence is checked for the type of harassment content i.e. gender, disability, racism, sexism etc. The last step in the proposed framework is classification phase in which the extracted characteristics are fed into the classifiers for training and testing. In this study analysis on various techniques is adapted namely machine learning classifier i.e. random forest [11] is an ensemble algorithm that uses multiple decision trees randomly to choose variable for the classifier data, multinomial naïve bayes [23] is a learning algorithm that work on bayes theorem and treat every value independent and assign the probability to each word, decision tree J48 [24] is an enhanced version of decision tree classifier in which data is divided by processing it through decision tree and splitting is done based on the selection criteria at particular step. Thus features enhance accuracy while splitting the best attribute. Logistic regression [25] is a well-known technique that works when the output needs to be divided into binary classification put together on the probability based class suitable to the input values. LR goes well when the size of the data increases. SMO [4] is the modern version of the SVM algorithm that is designed to solve the training problem of SVM by designing the heuristics. It splits the training data into little subsets and carries out analysis separately in order to improve the accuracy. Deep learning classifier i.e. CNN-LRCN [14] is applied to enhance the outcomes of classification by introducing the word level embedding that will check

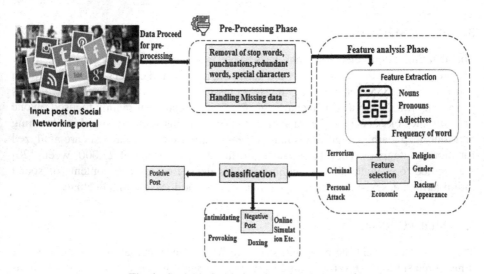

Fig. 1. Social intimidation detection framework

Table 2. Performance measures of different classifiers

Classifier	Accuracy	Precision	Recall	F-measure	Prediction time
SMO	91.24%	89.50%	89.11%	89.30%	0.1020
Random Forest	87.74%	93.36%	84.69%	88.81%	3.4943
Multinomial NB	78.41%	79.52%	88.44%	83.74%	1.5429
Decision Tree J48	88.05%	81.20%	84.20%	82.60%	2.5950
Logistic regression	91.31%	91.18%	90.53%	90.85%	0.0120
CNN-LRCN	88.00%	80.49%	85.45%	82.89%	2.4560
CNN-GRU	92.05%	85.00%	82.05%	83.49%	1.6732
CNN-CB	91.04%	79.00%	81.67%	79.98%	2.8956
CNN-Bi-LSTM	95.00%	93.89%	91.05%	91.98%	0.0010

semantics of the sentence instead of syntax, CNN-GRU [15] works in a similar manner as LSTM but the only difference is the number of gates. LSTM has three gates i.e. input, output and forget gate and GRU has two gates i.e. reset and update. CNN-GRU works well with small dataset and for large dataset LSTM is preferred, CNN-CB [17] is developed by making some modifications in other traditional classifiers as it make detection without any feature extraction by transforming text into word embedding and put it in CNN for classification, CNN-Bi-LSTM [16] is the latest version of LSTM as it checks the presence of every word in the sequence from ahead as well as from reverse and it is very much effective in handling missing statistics by forecasting the correct sequence.

4   Experimental Results and Analysis

In this segment the outcomes of the experiments conducted on each classifier is shown in Table 2. The evaluation of each classifier is done on the basis of accuracy, precision, recall and f-measure. Secondly the time complexity of every classifier in terms of prediction time is displayed in Table 3.

4.1   Performance Metrics of Different Classifiers

The efficacy of the proposed model is measured using several performance metrics to analyze how much one classifier can identify social intimidation than other. In this comparative analysis nine classifiers i.e. five machine learning and four deep learning techniques namely SMO, Random Forest, Multinomial NB, Decision Tree J48, Logistic Regression, CNN-LRCN, CNN-GRU, CNN-CB, CNN-Bi-LSTM are used. It is very necessary to do analyses using standard performance metrics to understand the conflicting classifiers. Criteria to evaluate the classifiers is described below [26].

- Accuracy tells the proportion of the correct results among the total cases analysed.

$$\text{Accuracy} = (TP + FP)/T \tag{1}$$

- Precision measures how much of predicted positives values is actually true.

$$\text{Precision} = TP/(TP + FP) \tag{2}$$

- Recall tells how much actual positives are truly classified.

$$\text{Recall} = TP/(TP + FN) \tag{3}$$

- F-measure is the foremost estimation metric that notify the harmonic mean of precision and recall value.

$$F - \text{measure} = (2 * \text{precision} * \text{recall})/(\text{precision recall}) \tag{4}$$

Performance visualization of different classifiers is shown in Fig. 2. From the graph it is clear that deep learning classifier CNN-Bi-LSTM achieves highest accuracy and F-measure on the dataset used whereas F-measure and classification accuracy are 91.98% and 95.00% respectively. Meanwhile Logistic regression among all machine learning classifiers achieve the highest accuracy and f-measure of 91.31% and 90.85%. There is a slight difference between the accuracy of LR and SMO. LR achieves more F-measure than SMO, decision tree J48, random forest and multinomial NB. Therefore, logistic regression performs finer than remaining machine learning classifiers on the dataset. CNN-CB and Multinomial NB achieves the lowest F-measure on the dataset. Nevertheless, multinomial NB has achieved fourth largest recall value among other classifiers, but the accuracy is lowest. Many studies showed that multinomial NB has achieved highest accuracy, however the accuracy decreased when the size of the dataset increased. F-measure is the most important feature to contrast the performance of the quantifiers. Multinomial NB presumes that every value is independent of each other, but

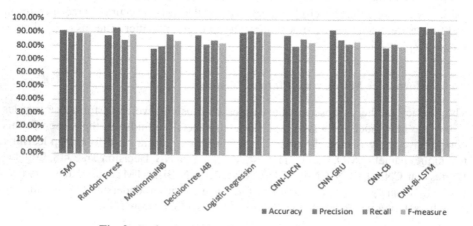

Fig. 2. Performance visualization of different classifiers

it is not correct in every problem. Due to this purpose multinomial NB is not performing well on this research as well. On the other hand, LR treats every tweet in binary form also and works better when the size of data increases. CNN-Bi-LSTM has achieved highest f-measure among all nine classifiers as CNN gives good accuracy in classification and Bi-LSTM checks the presence of every word from forward and backward side, therefore increase the accuracy and F-measure.

4.2 Time Complexity of Classifiers

Table 3 displays the time complexity of classifiers in terms of time taken to predict the outcome. The result indicates that CNN-Bi-LSTM takes less prediction time among all classifiers and Random forest takes the worst time to forecast the output among all classifiers. However there is the slight difference between LR and SMO algorithms. Multinomial NB and CNN-GRU takes nearly equal time to predict. There is a slight difference between J48 and CNN-CB quantifiers as shown in Table 2.

Table 3. Time complexity of classifiers in terms of prediction time

Specification	Classifier	Prediction time
Best prediction time	CNN-Bi-LSTM	0.0010
Worst prediction time	Random Forest	3.4943

5 Conclusion and Future Scope

Social harassment has been increasing in our society day by day. People from all ages are actively using these internet forums, therefore some people are taking false advantage of this freedom by harassing someone without revealing their true identity just for their fun. This paper proposed a methodology in which several machine learning and deep learning classifiers that are most frequently used in past studies are analyzed on the twitter dataset. The experimental outcomes show the superiority of CNN-Bi-LSTM and logistic regression from all classifiers in terms of f-measure and accuracy. Apart from these performance metrics prediction time of every classifier is also calculated to record the time complexity of each algorithm. Best prediction time is of CNN-Bi-LSTM and worst prediction time is of random forest classifier. Other than CNN-Bi-LSTM and logistic regression, random forest achieve the greatest precision even more than logistic regression and SMO attain more recall value. This time study is done on the textual post but another direction of this research is to have online intimidation detection from multiple types of posts i.e. audio, video, visuals for various social media networks and for multiple languages.

References

1. Patchin, J.W., Hinduja, S.: It is time to teach safe sexting. J. Adolesc. Health **66**(2), 140–143 (2020)
2. Patchin, J.W., Hinduja, S.: Sextortion among adolescents: results from a national survey of US youth. Sex. Abuse **32**(1), 30–54 (2020)
3. Samghabadi, N.S., Monroy, A.P.L., Solorio, T.: Detecting early signs of cyberbullying in social media. In: Proceedings of the Second Workshop on Trolling, Aggression and Cyberbullying, pp. 144–149 (2020)
4. Galán-García, P., Puerta, J.G.D.L., Gómez, C.L., Santos, I., Bringas, P.G.: Supervised machine learning for the detection of troll profiles in twitter social network: application to a real case of cyberbullying. Log. J. IGPL **24**(1), 42–53 (2016)
5. Sheeba, J.I., Devaneyan, S.P.: Cyberbully detection using intelligent techniques. Int. J. Data Min. Emerg. Technol. **2**(2), 86–94 (2016)
6. Balakrishnan, V., Khan, S., Arabnia, H.R.: Improving cyberbullying detection using Twitter users' psychological features and machine learning. Comput. Secur. **90**, 101710 (2020)
7. Abbass, Z., Ali, Z., Ali, M., Akbar, B., Saleem, A.: A framework to predict social crime through Twitter tweets by using machine learning. In: 2020 IEEE 14th International Conference on Semantic Computing (ICSC), pp. 363–368. IEEE (2020)
8. Birjali, M., Beni-Hssane, A., Erritali, M.: Machine learning and semantic sentiment analysis based algorithms for suicide sentiment prediction in social networks. Procedia Comput. Sci. **113**, 65–72 (2017)
9. Frommholz, I., Al-Khateeb, H.M., Potthast, M., Ghasem, Z., Shukla, M., Short, E.: On textual analysis and machine learning for cyberstalking detection. Datenbank-Spektrum **16**(2), 127–135 (2016)
10. García-Díaz, J.A., Cánovas-García, M., Colomo-Palacios, R., Valencia-García, R.: Detecting misogyny in Spanish tweets. An approach based on linguistics features and word embeddings. Future Gen. Comput. Syst. **114**, 506–518 (2021)
11. Salminen, J., Hopf, M., Chowdhury, S.A., Jung, S.-G., Almerekhi, H., Jansen, B.J.: Developing an online hate classifier for multiple social media platforms. HCIS **10**(1), 1–34 (2020). https://doi.org/10.1186/s13673-019-0205-6
12. Drishya, S.V., Saranya, S., Sheeba, J.I., Devaneyan, S.P.: Cyberbully image and text detection using convolutional neural networks. CiiT Int. J. Fuzzy Syst. **11**(2), 25–30 (2019)
13. Gambäck, B., Sikdar, U.K.: Using convolutional neural networks to classify hate-speech. In: Proceedings of the First Workshop on Abusive Language Online, pp. 85–90 (2017)
14. Bu, S.J., Cho, S.B.: A hybrid deep learning system of CNN and LRCN to detect cyberbullying from SNS comments. In: de Cos Juez, F. et al. (eds.) Hybrid Artificial Intelligent Systems. HAIS 2018. LNCS, vol.10870, pp. 561–572. Springer, Cham (2018). https://doi.org/10.1007/978-3-319-92639-1_47
15. Zhang, Z., Robinson, D., Tepper, J.: Detecting hate speech on twitter using a convolution-gru based deep neural network. In: Gangemi, A., et al. (eds.) ESWC 2018. LNCS, vol. 10843, pp. 745–760. Springer, Cham (2018). https://doi.org/10.1007/978-3-319-93417-4_48
16. Sadiq, S., Mehmood, A., Ullah, S., Ahmad, M., Choi, G.S., On, B.W.: Aggression detection through deep neural models on twitter. Futur. Gener. Comput. Syst. **114**, 120–129 (2021)
17. Al-Ajlan, M.A., Ykhlef, M.: Deep learning algorithm for cyberbullying detection. Int. J. Adv. Comput. Sci. Appl. **9**(9), 199–205 (2018)
18. Cheng, L., Shu, K., Wu, S., Silva, Y.N., Hall, D.L., Liu, H.: Unsupervised cyberbullying detection via time-informed gaussian mixture model. ArXiv preprint arXiv: 2008.02642 (2020)

19. Zinovyeva, E., Härdle, W.K., Lessmann, S.: Antisocial online behavior detection using deep learning. Decis. Support Syst. **138**, 113362 (2020)
20. Rezvan, M., Shekarpour, S., Balasuriya, L., Thirunarayan, K., Shalin, V.L., Sheth, A.: A quality type-aware annotated corpus and lexicon for harassment research. In: Proceedings of the 10th ACM Conference on Web Science, pp. 33–36 (2018)
21. Garg, A., Gupta, A., Gowda, D., Singh, S., Kim, C.: Hierarchical multi-stage word-to-grapheme named entity corrector for automatic speech recognition. In: Proceedings of Interspeech, vol. 2020, pp. 1793–1797 (2020)
22. Mageshwari, V., Aroquiaraj, L.: An efficient feature extraction method for mining social media. Int. J. Sci. Technol. Res. **8**(11), 640–643 (2019)
23. Hinduja, S., Patchin, J.W.: Cyberbullying: an exploratory analysis of factors related to offending and victimization. Deviant Behav. **29**, 129–156 (2008)
24. Altay, E.V., Alatas, B.: Detection of cyberbullying in social networks using machine learning methods. In: 2018 International Congress on Big Data, Deep Learning and Fighting Cyber Terrorism (IBIGDELFT), pp. 87–91. IEEE (2018)
25. Raza, M.O., Memon, M., Bhatti, S., Bux, R.: Detecting cyberbullying in social commentary using supervised machine learning. In: Arai, K., Kapoor, S., Bhatia, R. (eds.) FICC 2020. AISC, vol. 1130, pp. 621–630. Springer, Cham (2020). https://doi.org/10.1007/978-3-030-39442-4_45
26. Pawar, R., Raje, R.R.: Multilingual cyberbullying detection system. In: 2019 IEEE International Conference on Electro Information Technology (EIT), pp. 040–044. IEEE (2019)

Technique for Enhancing the Efficiency and Security of Lightweight IoT Devices

Santosh P. Jadhav, Georgi Balabanov[✉], and Vladimir Poulkov[✉]

Faculty of Telecommunications, Technical University of Sofia, Sofia, Bulgaria
{grb,vkp}@tu-sofia.bg

Abstract. Internet of Things (IoT) has become the integral part of our everyday life. IoT devices are delegated for observing our wellbeing, houses, workplaces, industrial facilities, etc., so they need to comply with necessary security requirements. Moreover, such devices should be handled with extra care as they have extraordinary significance and handles vital information that might be identified with one's life. Their effectiveness matters when these devices are considered as resource constraint devices. Numerous procedures are utilized to make these devices profoundly secure and productive. In the proposed system, we have applied hyper elliptic curve (HEEC) for key generation, considering the fact that such approach is more appropriate for lightweight devices due to its smaller key size. We implemented signcryption approach, where digital signature and encryption should be possible in single logical step. By applying HEEC-based signcryption, we have achieved significant improvement in the security and efficiency for resource constraint devices in IoT .

Keywords: Resource constraint devices · Signcryption · Hyper elliptic curve · Sensing unit

1 Introduction

It is being observed that IoT has changed the scenario in our surrounding environment and way of life. The four major areas that have shown major contribution of IoT recently are [1]. Wearable Devices: Applications such as heart rate tracking, health and fitness tracker, tracking user habit and continuous monitoring are all done by the wearable devices.

- Smart Home: due to up gradation of wireless internet services it was noticed that more and more home automaton devices were consumed recently.
- Smart Cities: Municipal services are getting safer and people are adapting the change in service provision such as optimized energy consumption.
- Smart traffic surveillances. Smart Industries: real time data fetching from machine sensors which optimizes various operations in industries which prevent various financial disaster.

© Springer Nature Switzerland AG 2021
M. Singh et al. (Eds.): ICACDS 2021, CCIS 1440, pp. 528–537, 2021.
https://doi.org/10.1007/978-3-030-81462-5_47

These several areas of IoT application generate very valuable data which may be health related data or money transaction data that are useful for various applications and decisions. Data can be confidential and sensitive and must be taken care of. Because of its rapid growth in IoT, it had become more prone to various cyber-attacks. Hence, it becomes of great importance to look into security measures, taking care that it should not affect the efficiency of IoT devices.

In this paper, we propose a technique for making IoT devices more secure and efficient. The technique includes integration of various standard algorithms which have been thought to be implemented in IoT scenarios where there are limitations of resources, called IoT with resource constraint devices. IoT devices commonly have low handling abilities, restricted memory and capacity and insignificant organization convention uphold. It is a huge test for IoT device producers and programming engineers to plan intricate and thorough safety efforts. They need to keep the plan straightforward and abstain from adding pointless highlights while additionally leaving enough space for security programming controls to shield against security dangers. Like some other security techniques, encryption and unscrambling are asset serious undertakings. They require huge preparing and capacity limits that IoT devices need. Besides, for storing and transmitting information after encryption-decryption process and putting away data limit prerequisites, making it significantly harder for IoT devices to deal with.

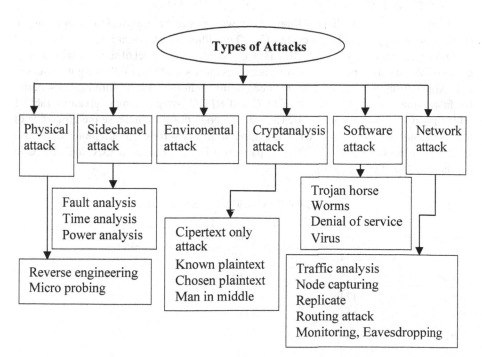

Fig. 1. Different attacks on IoT devices

Producers of IoT devices should utilize lightweight calculations that are reasonable for asset obliged conditions. These incorporate calculations which are quick and responsive are more energy and capacity effective than ordinary encryption and unscrambling calculations.

To secure the connected devices, as for example in a smart home, owners normally lack in expertise, thus neglecting to change password periodically. Thus the latter can become open to attacks and in such cases be difficult to secure users sensitive information. Several different types of attacks are possible on IoT devices.

The Fig. 1 shows the various types of attacks that can be possible on any IoT device. Various techniques are in thought to avoid such attacks as the data captured by these devices is sensitive and important. While securing this, care should be taken that the efficiency of the devices must not be hampered. Yan Naung Soe et al. [2] proposed a lightweight detection system which requires very less feature to identify the corresponding attacks. Also, this detection system is well suited for the lightweight devices in terms of efficiency of the system.

2   Proposed Scheme

2.1   Enhancing Security and Efficiency by Using Hyper Elliptic Curve Cryptography

It is observed that use of hyper elliptic curve with genus 2 is lightweight than its origin elliptic curve cryptography algorithm [3]. The arithmetic of operations gets more and more complex as genus increases. But, HECC provides same level of security with shorter keys and also its operand size is small it becomes more suitable for lightweight devices in IoT. Applications such as ecommerce need lightweight algorithms with higher security for faster transactions [4]. Result of ECC and HECC comparisons is given in Table 1 which precisely says that use of HECC can be good option to increase the efficiency of resource constraint devices in IoT. The computational cost for cryptographic mechanism for HECC is lesser then ECC. The comparisons of the key sizes of ECC and HECC in bits is shown in Table 1 [5, 6].

Table 1. ECC and HECC key comparisons

Security level	Elliptic curve	HECC Genus 2	HECC Genus 3
256	94	47	32
512	128	64	43
1024	174	87	58
2048	234	117	78
4096	313	157	105
8192	417	209	139

The HEEC of genus $g > 1$ is an algebraic curve given by an equation

$$y^2 + h(x)y = f(x) \tag{1}$$

Where,

f(x) is a polynomial of degree n $= 2g + 1 > 4$ or with n $= (2g + 2) > 4$ different roots and h(x) is a polynomial of degree $< g + 2$ (if the characteristic of the ground field is not equal to 2, we can take h(x) $= 0$).

Graph C $: y^2 = f(x)$.

Where, $f(x) = x^5 - 2x^4 - 7x^3 + 8x^2 + 12$ the curve is represented as fallow.

HECC is an asymmetric type of public-key cryptography technique where key generation process is involved for private and public keys. Every member in the process have private public key pairs where, private key is used for decryption and digital signature generation and public key is used for encryption and verification of signature (Fig. 2).

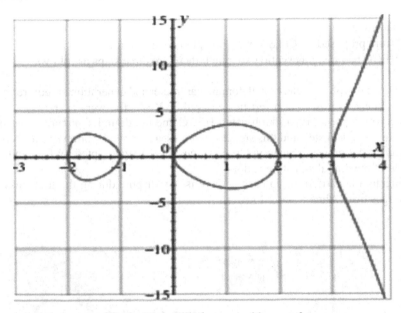

Fig. 2. Hyper elliptic curve with genus 2

2.2 Use of HECC with Signcryption Approach

Signcryption mechanism is one of the techniques to increase the efficiency of IoT devices. In the normal scenario of the cryptography encryption then assigning signature is performed in any mechanism but, this approach requires two different logical steps which also increases the computational time of the processor. Signcryption techniques performs the encryption and signature in single logical step therefore use of this technique along with HECC becomes most suitable for making the security mechanism overhead lightweight which can be the recommendation to increase the security and efficiency of resource constraint devices in IoT.

In 1997, Zheng et al. [7] presented a signcryption plot which incorporates an elliptic bend based sighncryption approach which saves 40% of correspondence cost and

58% of the computational expense when contrasted with the conventional methodology of encryption then signature approach. This methodology of Zheng made specialists consider how signcryption can be valuable to build the productivity of any framework. According to Deng and Bao [8], signcryption conspire attempts to lessen the computational expense by 16% and correspondence cost up to 85%. Gamage et al. [9] proposed the improved plan compared to Deng and Bao which give verification to get messages. Sharma et al. [10] proposed the plan which needs open unquestionable status which depended on personality based signcryption. In 2005, Hwang RJ et al. [11] Proposed an elliptic bend based signcryption conspire which was discovered to be appropriate for lightweight gadgets. These plans were discovered to be secure in any event, when the sender's private key was undermined. Similarly, Nizamuddin et al. [12] worked on distinctive signcryption plans which additionally diminish the correspondence and computational time yet they need public obviousness. Public certain plans dependent on HECC were proposed by Ch. SA et al. this plan need sending mystery.

In HECC based sighncryption approach there are three steps as follows:

- Initiate: In this phase, selection of domain parameter and generation of required private and public keys as well as a certificate of public Keyes for every user.
- Signcryption phase: Encryption using HECC and the digital signature is performed in a single logical step and this signcrypted message is send from Alice to Bob.
- Unsigncryption phase: Signcrypted messages are un-signcrypted by Bob and verification of the digital signature is done.
- Verification is carried out in case if there is any dispute during the transmission of messages from Alice to Bob (Fig. 3).

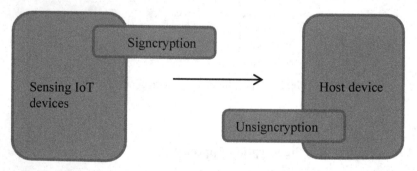

Fig. 3. Signcryption and unsigncryption approach

HECC overcomes ECC when it could be thought in the environment where there is limitation of resources. Such lightweight techniques can be performed to secure lightweight devices in IoT.

2.3 Signcryption Algorithm

Signcryption based on hyper elliptic curve cryptography requires one encryption algorithm and digital signature algorithm. Based on Hyper Elliptic Curve Discrete Logarithm

Problem (HEECDLP), PU = PR.G is called hyper elliptic curve discrete logarithmic problem. Private Key generator generates the keys for sensing unit and the mobile unit. The distribution of this key i.e. PR_{sensor}, PU_{sensor}, PR_{host}, PU_{host} is done in secure way in the initialization phase. When these keys are shared securely then devices store their private keys and are the public keys with one another.

An HEEC over the finite field Fp is denoted by E(Fp) with a base point $G \in Fp$ of order q, where G is taken indiscriminately from set of points on E(Fp). the parameter P is a prime number identifying the finite field Fp. Private keys are generated by choosing a number from the set of large prime number. Three parameters required for signcryption are.

- $PU = PR.G$ (Where PU is public key and PR is private key)
- Digital signature (HECDSA)
- Encryption (AES)

Key pairs for sensing unit and host unit are given by PR_{sensor}, PU_{sensor} $PR_{,host}$ PU_{host} respectively. In Signcryption algorithm we choose the random variable $P \in (1, 2, \ldots, q - 1)$

$$k1 = \text{hash}(P * G) \tag{2}$$

$$k2 = hash(P * PU_{host}) \tag{3}$$

$$c = Encrypt k2(data\ frame) \tag{4}$$

$$r = hash(c, k1) \tag{5}$$

$$s = \frac{P}{(r + PR_{sensor})} mod q \tag{6}$$

$$R = r.G \tag{7}$$

Therefore, signcryption output is given by these (c, R, s) signcrypted message is then transferred to the host device where un-signcryption is performed. Sensing unit authenticate the signcrypted data by its public key and confirms its authenticity if r.G = R.

2.4 Un-signcryption Algorithm

Unsigncryption is performed when the signcrypted message is received at the mobile or host device

$$k1 = \text{hash}(s(R + PU_{sensor})) \tag{8}$$

$$r = hash(c, k1) \tag{9}$$

$$k2 = \text{hash}(\text{PR}_{\text{host}}(s(R + \text{PU}_{\text{sensor}}))) \qquad (10)$$

$$\text{dataframe} = \text{Decrypt}k2(c) \qquad (11)$$

$$r.G = R \qquad (12)$$

3 Result Analysis

To perform the demonstration, we established a setup where we choose Arduino Uno board which is associated with its libraries called Arduino pack. Arduino is open source microcontroller board which creates an environment where sensing unit is able to transfer data frames to the host device i.e. mobile application. Arduino IDE i.e. integrated development environment is a cross-stage application for instance made from C and C++ It is used to create and shift different task to Arduino board. Its usage is reliant on C language the undertakings are as of now done to execute ECC with distorted Montgomery on Arduino Uno [13]. The source code for Arduino IDE is conveyed under the general public licenses. In our scheme, we preload the hyper elliptic curve used and its limits on the device memory. Hyper elliptic curve are suitable for cryptography for lightweight devices in IoT. Each hyper elliptic curve can't be used. For our inspiration, we need to pick them zeroing in on their family, the solicitation for the get-together and their field of definition. The HECC and DSA mixing can handle the issue of checking decency of report and imprint Id in electronic circumstance where id endorsement is necessary [14].

The Arduino Uno is most reassuring open-source equipment stage and it is ordinarily used as controller for the parts in IoT things. HECC is executed on Arduino Uno, which is a microcontroller board subject to the ATmega328P which is an Atmel 8-digit AVR microcontroller. Program codes for Arduino are written in Arduino language dependent on inserted C with Arduino IDE.

The particulars of the devices in hand are ATmega328P Microcontroller, clock speed 16 MHz, 32 kb memory, SRAM 2 kb, flash memory 32 kb. It computes the computational time of the various data frames received at the sensing unit shown in Table 2. The data frames are stored in Base-32 encoding to enable AES encryption during the signcryption process. HECC key pairs are smaller in size as compare to ECC and RSA. The signcryption based on HECC found to be more efficient with the same level of security with smaller keys sizes. Therefore, the computational time required for HECC based signcryption on the data frame received is less than the ECC based signcryption. The addition and multiplication operations in HECC require less time as compared to ECC [15]. While demonstrating we take the data frames of varying sizes and observe the computational cost for HECC based sighncryption and ECC based sighncryption as shown in Table 2.

Table 2. Computational time in milliseconds

Block size kb	ECC based signcryption (ms)	HECC based signcryption (ms)
50	376	270
103	496	456
298	530	492

Because of the shortest key size and the single-step usage of digital signature and encryption, HEEC-based signcryption has possible focal points over existing works. It was shown that a same level security level can be achieved by HEEC utilizing a shortest key length consequently, the execution of HEEC put together signcryption with respect to the data frames requires less computational expenses as shown in Fig. 4. Where, computational time is given in milliseconds. The addition and multiplication activities of HEEC are the most tedious operations of signcryption and unsigncryption measures. Nonetheless, it is worth seeing that these parts should be executed just a single time toward the start of the signcryption interaction and from that point forward, just encryption or unscrambling operator impacts the execution time. An equipment quickening agents for hash capacity, AESand HEEC on the sensing devices can improve the computational effectiveness.

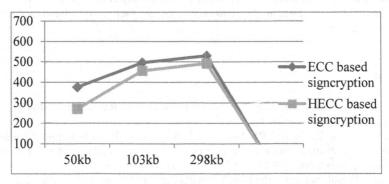

Fig. 4. Comparative analysis of HECC and ECC

3.1 General Performance Metric for Lightweight Cryptography Algorithms.

General performance metrics for any lightweight cryptographic algorithm is given by

$$\text{General Metric}\left(A^\alpha, T_B^\beta, E_B^\lambda, C_B^\tau, N_B^\mu\right) = \frac{A^\alpha, T_B^\beta, E_B^\lambda, C_B^\tau}{N_B^\mu}, \tag{13}$$

Where, A is the area; TB the time to encrypt one block; E is the energy; CB is the number of cycles to encrypt one block; NB is the block size; α β λ τ and μ are power coefficients. By using this metrics and the values of coefficient as mentioned in [16]. Therefore considering the computational time for 50 kb block according to the HEEC based signcryption approach is 270 ms. We can calculate the throughput of any software or hardware system as.

$$\text{Throughput} = \frac{T_B^\beta C_B^\tau}{N_B^\mu} = \frac{270^{-1}}{50^{-1}} = 0.18 \text{ block/ms}.$$

Thus, by knowing the throughput of the algorithms we can determine efficiency of the algorithms and can check the feasibility of using it considering the application.

4   Conclusion

In the proposed system, HEEC-based signcryption has been utilized to ensure information caught by Sensing unit for occasion set off observing in IoT applications. We initially recognized the expected dangers for such applications and afterward examined chosen security issues. The proposed signcryption, which is executed on the detecting unit, gives countermeasures to the potential dangers and empowers the legitimacy of encoded pictures on the untrusted sensing devices have part without bargaining its classification. The result shows that HEEC based signcryption approach is suitable for the resource constraint devices that normally found in IoT. In future work, block chain technology can be thought to use for IoT through some specialized intermediaries and can enhance the security of the devices.

Acknowledgments. This work was supported by research project KP06-N27/3 "Resource self-configuration and management in ultra-dense networks with user centric wireless access" of the Bulgarian Science Fund.

References

1. https://www.weforum.org/agenda/2021/03/ai-is-fusing-with-the-internet-of-things-to-cre ate-new-technology-innovations
2. Soe, Y.N., Feng, Y., Santosa, P.I., Hartanto, R., Sakurai, K.: Towards a lightweight detection system for cyber attacks in the IoT environment using corresponding features. Electronics **9**, 144 (2020). https://doi.org/10.3390/electronics9010144
3. Jadhav, S.P., Balabanov, G., Poulkov, V., Shaikh, J.R.: Enhancing the security and efficiency of resource constraint devices in IoT. In: 2 e 020 International Conference on Industry 4.0 Technology (I4Tech), Pune, India, pp. 163–166 (2020). https://doi.org/10.1109/48345.2020. 9102639. Signcryption or how to achieve cost (signature & encryption I4Tech)
4. Shaikh, J.R., et al.: Enhancing e-commerce security using elliptic curve cryptography. Int. J. Curr. Adv. Res. **06**(08), 5338–5342 (2017). https://doi.org/10.24327/ijcar.2017.5342.0701
5. Alimoradi, R.: A study of hyperelliptic curves in cryptography. I. J. Comput. Netw. Inf. Secur. **8**, 67–72 (2016)
6. Roman, R., Alcaraz, C., Lopez, J.: A survey of cryptographic primitives and implementations for hardware-constrained sensor network nodes. J. Mob. Netw. Appl. **12**(4), 231–244 (2007)
7. Zheng, Y.: Digital signcryption or how to achieve cost(signature & encryption) ≪ cost(signature) + cost(encryption). In: Kaliski, B.S. (ed.) Advances in Cryptology — CRYPTO'97, CRYPTO 1997. LNCS, vol. 1294, pp. 165–179. Springer, Heidelberg (1997). https://doi.org/10.1007/BFb0052234

8. Bao, F., Deng, R.H.: A signcryption scheme with signature directly verifiable by public key. In: Imai, H., Zheng, Y. (eds.) PKC 1998. LNCS, vol. 1431, pp. 55–59. Springer, Heidelberg (1998). https://doi.org/10.1007/BFb0054014
9. Gamage, C., Leiwo, J., Zheng, Y.: Encrypted message authentication by firewalls. In: Imai, H., Zheng, Y. (eds.) PKC99. LNCS, vol. 1560, pp. 69–81. Springer, Cham (1999). https://doi.org/10.1007/3-540-49162-7_6
10. Sharma, G., Bala Verma A.K.: An identity-based ring signcryption scheme. In: Kim, K., Chung, K.Y. (eds.) IT Convergence and Security 2012. LNEE, vol. 215, pp. 151–157. Springer, Dordrecht (2013)https://doi.org/10.1007/978-94-007-5860-5_18
11. Hwang, R.J., Lai, C.H., Su, F.F.: An efficient signcryption scheme with forwarding secrecy based on an elliptic curve. Appl. Math. Comput. 167(2), 870–888 (2005)
12. Nizamuddin, S., Ch, S.A., Nasar, W., Javaid, Q.: Efficient signcryption schemes based on hyperelliptic curve cryptosystem. In: 7th International Conference on Emerging Technologies (ICET), pp 1–4 (2011)
13. Hashimoto, Y., et al.: An Implementation of ECC with Twisted Montgomery Curveover 32nd Degree Tower Field on Arduino Uno. Int. J. Netw. Comput. 8(2), 341–350 (2018). www.ijnc.org. ISSN 2185-2847
14. Jian-zhi, D., Xiao-hui, C., Qiong, G.: Design of hyper elliptic curve digital signature. In: International Conference on Information Technology and Computer Science (2009)
15. Wankhede-Barsgade, M.T., Meshram, S.A.: Comparative study of elliptic and hyper elliptic curve cryptography in discrete logarithm problem. IOSR J. Math. (IOSR-JM) 10(2), 61–63 (2014). www.iosrjournals.org. e-ISSN: 2278-5728, p-ISSN: 2319-765X
16. Jadhav, S.P.: Towards light weight cryptography schemes for resource constraint devices in IoT. J. Mob. Multimedia 15, 91–110 (2020). https://doi.org/10.13052/jmm1550-4646.15125

Performance Improvement in Deep Learning Architecture for Phonocardiogram Signal Classification Using Spectrogram

R. Sai Kesav, M. Bhanu Prakash[✉], Krishanth Kumar, V. Sowmya, and K. P. Soman

Centre for Computational Engineering and Networks (CEN), Amrita School of Engineering, Amrita Vishwa Vidyapeetham, Coimbatore, Tamil Nadu, India
v_sowmya@cb.amrita.edu,kp_soman@amrita.edu

Abstract. Phonocardiogram (PCG) assumes a critical part in the early determination of heart irregularities. Phono-cardiogram can be utilized as an underlying diagnostics apparatus in far-off applications because of its effortlessness and cost-adequacy. The proposed work targets utilising a CNN architecture, with multiple preprocessing strategies like converting to Spectrogram or normalizing the signals, which analyze various cardiovascular anomalies from PCG signals gathered from different sources. Our study shows the viability of utilising Spectrogram and Normalization of signals in cardiac abnormalities identification. This work avoids feature extraction and trivial pre-processing mechanisms, and we have achieved promising results.

Keywords: Phonocardiogram (PCG) · Spectrogram · Deep learning

1 Introduction

The Phonocardiogram (PCG) is the graphical portrayal of the sound delivered from the cardiac muscles' cardiovascular activity inside the heart. During the heart's cardiovascular movement, murmur sounds are generated, which are unidentified by an Electrocardiogram [1] but can be recognized by a Phonocardiogram and gives an assistive analysis in the early location of cardiovascular illnesses [1]. The PCG signal comprises s1 and s2, the two significant heart sounds. s1, the low pitch sound (lub) is generated because of atrioventricular valve closure and when the blood moves from the heart chamber to the ventricle. s2, the high pitch sound (dub) is generated when blood flows through the vessels from the heart to the lung, for a limited duration. A systolic period is defined from the early s1 to s2, and a diastolic period is defined from the early s2 to s1. The systolic and diastolic periods together constitute a single heart cycle. In addition to the fundamental sound segments, there are multiple and unusual sounds like a murmur, extrasystole delivered in case of any cardiovascular abnormality [11].

© Springer Nature Switzerland AG 2021
M. Singh et al. (Eds.): ICACDS 2021, CCIS 1440, pp. 538–549, 2021.
https://doi.org/10.1007/978-3-030-81462-5_48

Phonocardiogram is generally recorded in a clinical climate utilising an innovatively advanced stethoscope. The Phonocardiogram is additionally recorded inside a non-clinical arrangement. The recorded Phonocardiogram signal consists of different noises, making the classification of the systolic and diastolic beats generated by the heart's mechanical activity challenging to be classified.

Spectrogram has been used in this work, an electronic or visual representation of the spectrum of frequencies for a signal, which varies with time generated through a Fourier transform where frequency and time are visually represented. Different colours are used to show the spectrum's magnitude. Further, we have trained the standard CNN architecture citeb4 using the Spectrogram. Our objective is to show the use of simple preprocessing techniques like spectrogram, and normalization on the signals can give better classification results.

2   Related Works

In the literature, a study about the automatic classification of sounds generated by the heart's mechanical activity is described [1,2]. The localization algorithm is used to localize the peaks from the input signal and constructing windows from those peaks, which are further used to classify the extracted features. Further study shows the analysis of the PCG signal's time and frequency analysis, which overcomes the frequency and time domains' anomalies. The time-frequency methods include wavelets and Empirical Mode Decomposition. In Sujadevi et al. [9] Variational Mode Decomposition (VMD) based denoising was used to remove noise from the heart sounds and using the same technique to denoise the signal and visually displaying the waveforms. In Sujit et al. [8] the abnormality produced in the heart sounds were detected by a back-end classifier that extracts the time and frequency from the PCG signal using the Ada-Boost technique and SMOTE. In Schimdt et al. [5], a diagnosis system is proposed which utilizes Support Vector Machines (SVM) to classify the diseases developed in heart valves. In Sujadevi et al. [7], different algorithms RNN, B-RNN, LSTM, B-LSTM, CNN, and GRU were used for PCG classification. 80% accuracy was achieved for CNN.

In Sujadevi et al. [4], without any denoising and trivial pre-processing techniques, a convolutional neural network (CNN) on the physionet datasets using a raw signal achieved better results. Similarly, in [3], CNN was declared by them as the better network to classify Phonocardiogram signals. Both the Physionet datasets and the AISTATS 2012 datasets were collected from multiple sources.

3   Methodology

All the existing works utilize the input signal based on the time domain used for the PCG Signal classification using deep learning architectures. The current work is based on a spectrogram that converts the input signal from the time domain to the frequency domain through a visual portrayal of the spectrum of frequencies of a signal as it varies with time. A spectrogram is generated by

Fourier transform using the time and frequency as x-axis and y-axis respectively and different colours to show the magnitude of the spectrum. The spectrogram images are fed as input to the CNN architecture, specified in [4], where training and classification are done accordingly for the different datasets used. The CNN architecture used for this work is shown in Fig. 7.

Before converting the datasets into spectrograms, the signals should be fixed to a particular length (here 5 s) and then normalized. Normalization was done because we found that datasets 1 and 3 are noisy signals compared to dataset 2. Which are shown in Fig. 1, Fig. 2, and Fig. 3

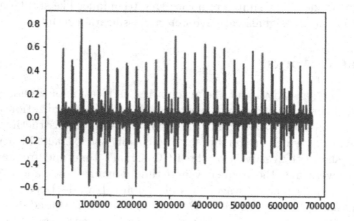

Fig. 1. Normal Signal sample was taken from dataset 1

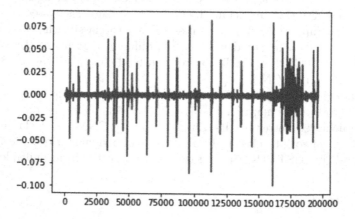

Fig. 2. Normal signal sample was taken from dataset 2

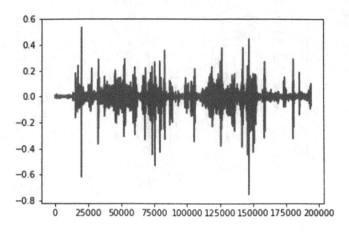

Fig. 3. Normal signal sample was taken from dataset 3

As observed from Fig. 1, Fig. 2, Fig. 3, there is very little noise in dataset 2. So the normalization is done for datasets 1 and 3 using the min-max normalization(for which the pseudo-code is shown below).

Algorithm 1. Psuedo Code for Normalization
___
Input: Signal;
Output: Normalized Signal;
def audio-norm(signal):
max_data = max(signal)
min_data = min(signal)
norm_signal = (data- min_data)/(max_data - min_data + 0.0001)
return norm_signal-0.5

All the signals are trimmed to 5 s, but if a signal is less than 5 s, then zero paddings are added, which is shown in Fig. 4.

From Fig. 4, it has been observed that there is zero-padding added to the signal. When we compare the original signal's corresponding amplitude values with the normalized signal, the amplitude has been normalized, giving better classification results. Amplitude for comparison is shown below:

Original amplitude values: [−0.00806781 −0.00920942 −0.00991806 ... −0.03190301 −0.02225797]

Normalized amplitude values: [0.08240861 0.08149707 0.08093125 ... 0.08885056 0.08885056 0.08885056]

When compared to the previously achieved output in [3] and the metrics which was achieved here for dataset-3, as seen in the results in Table 6, an improvement in output is observed when the proposed method of fixing to a particular length and normalization is used.

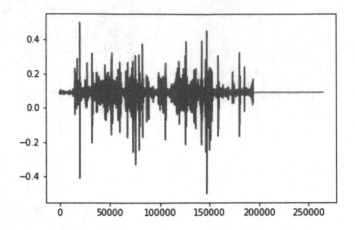

Fig. 4. Dataset 3 normalized normal signal

3.1   Input Description

In our current work, the Phonocardiogram signal(PCG) is gathered from different sources, and a total of three datasets are used for this work. Dataset-1, collected from clinical and non-clinical conditions, is accessible from physionet challenge 2016 and consisting of classes: normal and abnormal. Dataset-2 and Dataset-3 are accessible from AISTATS 2012 test. Dataset-2 was gathered utilising the iStethoscope Pro iPhone application and consisted of four classes: extrasystole, murmur, artifact, and normal. Dataset-3 was gathered using a digital stethoscope – DigiScope application and consisted of three classes: normal, murmur, extrasystole.

3.2   Spectrogram

A spectrogram is generated through a Fourier transform which visually represents the spectrum of frequencies of a given signal as it varies with time. Frequency and time are horizontal and verticals in a formed visual representation and different colours show the spectrum's magnitude [7].

For a given signal of x with a length of N, there are consecutive segments of the signal of m, where m $\leq\leq$ N, and the x $\in$

$$R^{m \times (N-m+1)}$$

where, in the formed matrix, rows and columns of x are indexed by time.

$\dot{x} = F * x$ and $x = (1/m) * F * \dot{x}$, of size m and matrix F, which are DFT columns of x, and F is the Fourier matrix with Fi, being its complex conjugate.

$$\begin{bmatrix} 1 & 1 & 1 & .. & 1 \\ 1 & e^{i2\pi} & e^{i\frac{4\pi}{N}} & .. & e^{i2\pi\frac{(N-1)}{N}} \\ 1 & e^{i\frac{4\pi}{N}} & e^{i\frac{8\pi}{N}} & .. & e^{i2\pi\frac{2(N-1)}{N}} \\ \vdots & \vdots & \vdots & \vdots & \vdots \\ 1 & e^{i2\pi\frac{(N-1)}{N}} & e^{i2\pi\frac{2(N-1)}{N}} & .. & e^{i2\pi\frac{(N-1)^2}{N}} \end{bmatrix} \tag{1}$$

Rows and columns of $\dot{x}$ are indexed by the frequency and time, respectively, and their location corresponds to the point in frequency and time. The spectrogram is a visualised matrix where the matrix image with the ith and jth entry in the matrix, corresponds to the intensity or colour of the ith, and jth pixel in the visually represented image general the bright colours denote the strong frequencies in a spectrogram. Figure 5 and Fig. 6 portray the spectrographic images of datasets 1 and 2.

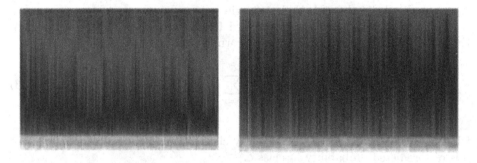

Fig. 5. Spectrographic images of dataset-1 showing normal and abnormal classes

Fig. 6. Spectrographic images of dataset-2 showing normal, murmur, extrahls and artifact classes

3.3 Deep Learning Architecture

In [4], a distinguished classification for the PCG model was implemented. In this work, similar topologies and hyperparameters for crude PCG Signal classification are gathered from different clinical and non-clinical conditions. The experiments are done using the benchmark architecture of CNN and using the three datasets (Dataset 1,2, and 3). The architecture of CNN used can be seen in Fig. 7

Convolutional Neural Network (CNN): The CNN architecture used here comprises of four Convolution layers stacked along with 64 filters, each of filter size 3, and followed by an average pooling layer and a ReLu activation function. The average pooling layer reduces the size of the feature without losing any information and a flattening layer after the 4th convolution layer followed by five dense layers with the softmax activation function. The architecture details are mentioned in Table 1.

The loss function and optimizer used in this architecture are Logcosh and ADAM optimizer. Whose formula can be seen as:

$$L\left(y, y^{P}\right) = \sum_{i=1}^{n} \log\left(\cosh\left(y_i^P - y_i\right)\right) \tag{2}$$

where y_i^P is the predicted values, and y_i are the original values.

As we have achieved less accuracy for dataset 3 using spectrogram, we implemented 1D convolution layers instead of using spectrogram images and have used the normalized signal for their classification.

The loss function and optimizer used for dataset 3 are Categorical Cross-Entropy and ADAM optimizer. Whose formula can be seen as

$$CCE(p, t) = -\sum_{c=1}^{C} t_{0,c} \log\left(p_{0,c}\right) \tag{3}$$

C is no. of classes
t is the binary indicator for the correctly classified observation o.
p is o's predicted probability in class c.

4   Experimental Results

4.1   Dataset Description

Phonocardiogram (PCG) Data Gathered from both Clinical and Non-Clinical Environments (Dataset 1): This information source is a segment from physionet challenge 2016, which is gathered from healthy and unhealthy patients worldwide in clinical and non-clinical conditions. Detailed information regarding datasets is clarified underneath in Table 2, which sums up the total number of samples considered from each class.It is purposefully used as training and testing input for the model from 665 abnormal and 2575 normal signals.

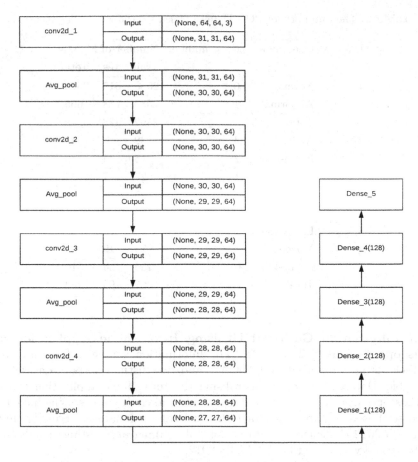

Fig. 7. Architecture of CNN used for classification

Table 1. The Architectural details of the PCG Signal Classification Collected from multiple sources with distinct Cardiac Abnormalities

Architecture details		CNN
No. of neurons in the stacked layers		64
Zeropaddings (Border mode)		Yes
Filter size		3
Pool length		2
Number of neurons in dense layers		128
Output layer	Dataset 1	2
	Dataset 2	4
	Dataset 3	3

Table 2. The Summary of PCG Datasets used for the PCG Classification

Dataset	Class name	Class number	Number of Signals		
			Train	Test	Total
1	Normal	0	2060	515	2575
	Abnormal	1	532	133	665
	Total		2592	828	3240
2	Artifact	3	32	8	40
	Extrasystole	2	15	4	19
	Murmur	1	27	7	34
	Normal	0	25	6	31
	Total		99	25	124
3	Extrasystole	2	36	10	46
	Murmur	1	76	19	95
	Normal	0	255	64	319
	Total		367	93	460

PCG Information Gathered Utilising Ipro Phone Application and Digiscope (Dataset 2 and 3): The information source is from an event, AISTATS 2012, supported by PASCAL, and two distinct sorts of datasets are accessible. Dataset 2 is gathered utilising the Ipro phone application utilising iStethoscope. Dataset 3 is gathered from a clinical environment using a computerized stethoscope. Table 2 portrays the outline of the dataset accessible in the AISTATS 2012 challenge. A dataset split of training and testing signals based on the strategy proposed by AISTATS 2012.

Table 3. Hyper-parameter set for the CNN

Hyper-parameters	CNN
Batch–size	32
Learning rate	0.1
No. of hidden layers	8
Optimizer	Adam
No. of epochs	50

Table 4. The summary of the results obtained using the above Architecture and Spectrogram images

Metrics/Datasets	Precision	Accuracy	Recall	f1-score
Dataset I	0.66	0.82	0.82	0.73
Dataset II	0.85	0.85	0.85	0.84
Dataset III	0.58	0.67	0.67	0.62

Table 5. The Summary of the results obtained for Dataset III using 1D Convolution without Spectrogram

Metrics/Datasets	Precision	Accuracy	Recall	f1-score
Dataset III	0.81	0.83	0.83	0.80

4.2 Result Analysis

In Sujadevi et al. [6], for dataset 1, when trained with raw signals promising results were obtained using CNN. They have concluded that the architecture, as seen in Fig. 7, gave better results. For benchmarking the optimum value, multiple experiments were done with various configurations. In this work, the same hyper-parameters used in [6] are considered. The learning rate and batch size are fixed at 0.1 and 32 for the deep learning architecture. All the parameters used are shown in Table 3.

The model's performance was evaluated using precision, recall, F1–score, and accuracy. The results obtained when spectrogram is used shown in Table 4. And the results of dataset 3 without using spectrogram are shown in Table 5.

The comparison of our results and previous results obtained in [3] are displayed in Table 6. Here, an improvement is observed in the classification performance of dataset 2 using spectrogram when compared with the results of the existing methodology used.

Utilising dataset 1, work done has been almost a replica of the previous existing classification performance accuracy of 82%. For dataset 2, the proposed work improved resulted in achieving better classification performance than the previously existing one, from accuracy 82% to 85%. In the case of dataset 3, we have achieved better results than the previously mentioned performance. We have normalized the signal and used 1D convolutions in place of Fast Fourier Transform.

Table 6. Performance comparison of the PCG Classification using CNN for Existing work and Proposed work

Data	Metrics	Existing	Proposed
1	Precision	0.83	0.66
	Accuracy	0.82	0.82
	Recall	0.82	0.82
	F1-score	0.83	0.73
2	Precision	0.85	0.85
	Accuracy	0.80	0.85
	Recall	0.79	0.85
	F1-score	0.82	0.84
3	Precision	0.81	0.81
	Accuracy	0.75	0.83
	Recall	0.76	0.83
	F1-score	0.79	0.80

5 Conclusion

In this work, it was found that data preprocessing is an essential task for training. As seen in dataset 2, it has given a very good output. And class imbalance problems can lead to overfitting of data of the majority class, as seen in dataset 1. Therefore downsampling was used to overcome the imbalance. As it was noticed that the challenge in which dataset 3 was given is for feature extraction, we used fixed signal length and normalized the signals, and performed 1D convolutions with them, which gave better classification results than that of FFT results. The current work can be extended to make further advancements in cardiac diseases classification.

References

1. Amit, G., Gavriely, N., Intrator, N.: Cluster analysis and classification of heart sounds. Biomed. Signal Process. Control 4(1), 26–36 (2009)
2. Maglogiannis, I., Loukis, E., Zafiropoulos, E., Stasis, A.: Support vectors machine-based identification of heart valve diseases using heart sounds. Comput. Methods Prog. Biomed. 95(1), 47–61 (2009)
3. Gopika, P., Sowmya, V., Gopalakrishnan, E.A., Soman, K.P.: Performance improvement of deep learning architectures for phonocardiogram signal classification using fast fourier transform. In: International Conference on Intelligent Computing and Communication Technologies, (ICICCT-2019). Springer, Heidelberg (2019)

4. Sujadevi, V.G., Soman, K.P., Vinayakumar, R., Prem Sankar, A.U.: Anomaly detection in phonocardiogram employing deep learning. In: Behera, H.S., Nayak, J., Naik, B., Abraham, A. (eds.) Computational Intelligence in Data Mining. AISC, vol. 711, pp. 525–534. Springer, Singapore (2019). https://doi.org/10.1007/978-981-10-8055-5_47

5. Schmidt, S.E., Holst-Hansen, C., Graff, C., Toft, E., Struijk, J.J.: Segmentation of heart sound recordings by a duration-dependent hidden markov model. Physiol. Meas. **31**(4), 513 (2010)

6. Sujadevi, V., Soman, K.P., Vinayakumar, R., Sankar, A.P.: Deep models for phono-cardiography (PCG) classification. In: The 2017 International Conference on Intelligent Communication and Computational Techniques (ICCT), pp. 211–216. IEEE (2017)

7. Sujadevi, V.G., Soman, K.P., Vinayakumar, R., Sankar, A.U.P.: Anomaly detection in phonocardiogram employing deep learning. In: Advances in Intelligent Systems and Computing, vol. 711, pp. 525–534 (2019)

8. Sujit, N.R., Kumar C.S., Rajesh, C.B.: Improving the performance of cardiac abnormality detection from PCG signal. In: AIP Conference Proceedings, vol. 1715 (2016)

9. Sujadevi, V.G., Soman, K.P., Kumar, S.S., Mohan, N., Arunjith, A.S.: Denoising of phonocardiogram signals using variational mode decomposition. In: 2017 International Conference on Advances in Computing, Communications and Informatics (ICACCI), Udupi, pp. 1443–1446 (2017). https://doi.org/10.1109/ICACCI.2017.8126043

10. https://www.princeton.edu/~cuff/ele201/files/spectrogram.pdf

11. https://www.ncbi.nlm.nih.gov/books/NBK333/

Performance Analysis of Machine Learning Techniques in Device Free Localization in Indoor Environment

K. S. Anusha[✉], R. Ramanathan, and M. Jayakumar

Department of Electronics and Communication Engineering, Amrita School of Engineering,
Amrita Vishwa Vidyapeetham, Coimbatore, India
{ks_anusha,r_ramanathan,m_jayakumar}@cb.amrita.edu

Abstract. The potential benefits of device free localization technique has accelerated the research in target detection domain in indoor environment. Received signal strength based target localization technique is commonly used as it is easily equipped in commercially available modules. Machine learning models that understand the variation in received signal strength due to the presence of target and assessing the position of the target is more reliable in a noisy and changing environment compared to conventional models. In this work, two popularly used machine learning models, deep neural network (DNN) based deep learning model and support vector machine models are used to assess the location of the target in an indoor. The performance analysis of the models is assessed with and without measurement error variance condition in four different node configuration setup using mean localization error parameter. The performance of the deep learning model is found to be better in higher node configuration scenario.

Keywords: Device free localization · Received signal strength · Deep neural network · Multi-output support vector regression

1 Introduction

Target location estimation is an upcoming area of research particularly in academics, military, industrial applications [1, 2]. In indoor localization scenario, the popular techniques available for target localization is broadly classified into device based localization (DBL) techniques and device free localization (DFL) techniques. In device free localization, no device is required to be equipped by the target for getting its location where as in device based localization technique, a device need to be held by the target for getting its location. Thus getting the information of the location of the target using DFL technique based on wireless sensor networks is gaining attention as the technique helps in context aware applications especially in case of non-cooperative targets [3]. The non-co-operative targets include elderly person, invader, differently abled person etc.

The DFL over wireless senor network measures received signal strength (RSS) between the link connecting a pair of nodes [4]. In a given predefined network with the availability of a target, the received signal strength gets attenuated for the links

© Springer Nature Switzerland AG 2021
M. Singh et al. (Eds.): ICACDS 2021, CCIS 1440, pp. 550–560, 2021.
https://doi.org/10.1007/978-3-030-81462-5_49

passing through the target. The indoor localization is highly affected due to multipath interference, reflection, scattering etc. Many researchers have carried out their studies on DFL technique categories, comparison made, their shortcomings and future scope [5]. Out of which machine learning technique is found to be efficient in changing environmental scenarios due to their adaptability to such conditions easily without much redesign but only retraining. Machine learning techniques involve two phases; a training and a testing phase. During the training process, with the training dataset the machine learning model studies the output for different inputs. During testing phase, the model will be able to predict the output for a new input.

The machine learning models used in our research is multi-layer perceptron based deep learning model and multi-output support vector regression (M-SVR) model. The support vector machine models due to its acquaintance ability in multi-class scenario and sparse data has made it popular in versatile applications [6, 7]. The Deep learning (DL) models are gaining attention nowadays due to its capability to analyze noisy big data and in providing exact input-output mapping through automatically learning features [8, 9]. The deep models have started putting its imprints in various applications like non cooperative target detection. Because of the potential benefits of the mentioned machine learning models in versatile applications, deep neural network (DNN) and and multi-output support vector regression (M-SVR) model is chosen for performance analysis in indoor.The performance of both the machine learning models were analyzed within a room setup with different number of nodes and measurement error variance conditions. The model performance is assessed with minimum assumptions.

Nowadays DFL technique has emerged as a promising topic of interest in the field of security, health care, smart home, military applications etc. Various machine learning algorithms has been developed in locating the position of a target which include random forest, support vector machines (SVM), linear discriminant analyzers (LDA) [10]. SVM's are one of the popular technique used to address classification and regression problems [6, 11, 12]. Nowadays the popularity and effectiveness of deep learning models have gained attention in comparison to the conventional machine learning models [13]. Thus many deep learning architectures are developed in the field of DFL. The widely used parameter in finding the location of the target is RSS between the links connecting sensor nodes within a network [3]. Fengjun Shang et al. [14] conducted a study on RSS based target localization based on experimental data.

The main challenges faced by many of the researchers in addressing DFL problem is with a single variable received signal strength and its variation apart from target position within the network.In this work, authors has address the DFL problem using a single variable RSS value. Performance of two popular machine learning algorithm based device free localization techniques are compared to assess the suitability of algorithm in different scenarios.

The remaining part of the paper is structured as follows: Sect. 2 gives the problem statement of the proposed work. Section 3 elaborates on the machine learning models used in this work along with methodology used. Section 4 elaborates on results and discussion of the work done. Section 5 give conclusive remarks on the work done along with future scope.

2 Problem Statement

In a wireless sensor network based DFL technique, the links connecting the nodes within the network changes that are affected by target position. Suppose there are N number of nodes within a network, number of links will be $(N \times N) - N$. The received signal power obtained from all nodes can be used to identify the position of the target using an appropriate machine learning technique. If P_{11} is the measured received power between transmitter T_1 and R_1 where $T_1 \neq R_1$, an $N \times N$ received power matrix can be generated with received power values from P_{11} to P_{nn}.. When a target is positioned within a given network the received power of the links affected by the target also changes and the difference matrix can be obtained. In a difference matrix each element is the difference in received power obtained with and without target condition. The sample difference matrix is shown below where target has affected link between a and b.

$$\begin{bmatrix} 0 & \cdots & & \cdots & \cdots & 0 \\ 0 & \cdots & & \cdots & \cdots & 0 \\ \cdots & \cdots & \Delta P_{ab} & \cdots & \cdots \\ \cdots & \Delta P_{ba} & \cdots & \cdots & \cdots \\ 0 & \cdots & & \cdots & \cdots & 0 \end{bmatrix}$$

Thus a difference matrix is a matrix with difference in received power along all the links within the network as its elements. Let the difference in received power is denoted as ΔP. So ΔP_{ab} denotes the difference in received signal power in the link from transmitter a to receiver b. Similarly, ΔP_{ba} is the difference in received power in link between transmitter b and receiver a. Thus if ΔP is a non-zero value then it is understood that the link connecting the corresponding transmitter and receiver is affected by the target location. If this difference matrix is given as an input to a machine learning model, the model should be able to assess the target location. One of the main challenge faced by most of the researchers are to develop an accurate RSS based DFL technique due to influence of dynamic variations of RSS values.

3 Methodology

Consider a network with N number of sensor nodes in which the link connecting any pair of nodes that act as transmitter and receiver measures RSS between the link connecting that node pair. To generate a ΔP matrix, each element is calculated by link distance based device free localization approach. [7]. In this approach the difference in the received power of the target affected links are obtained by:

$$\Delta P = \frac{kP_t}{d^2} - \frac{L_1 k^2 P_t}{d_1^2 (d - d_1)^2} \tag{1}$$

Where ΔP is the difference in received signal power matrix, d_1 is the distance of the target from transmitter, d is the separation from transmitter and receiver modules, P_t is the transmitted power in dBm, L_1 is the loss function w.r.t. nature of the target, k is the constant associated with Friis transmission formulae in free space given by:

$$k = G_t G_r \left(\frac{\lambda}{4\pi} \right)^2 \tag{2}$$

Where G_t is the transmitter gain, G_r is the receiver gain and λ is transmitted signal wavelength in meters.

The difference in received power matrix or ΔP matrix is given as input to a machine learning model in order to predict the target location. In this work, support vector regression model and deep neural network regression model is used to predict the exact coordinate of the target.

3.1 Multi-output Support Vector Regression

In general, SVM model is used to give a single output prediction based on an appropriate relationship between input $\times$ and single output y_i from a given dataset S_i. The training set comprises a set of observations of x with a corresponding observation of y values. The aim of the model is to assess the value of y for a new input x, This is obtained by a regression parameter $\omega \in R^{m \times 1}$ and the bias term b $\in$ R that reduces the function as given in Eq. 3 [10]:

$$\frac{1}{2}\left(\|\omega\|^2\right) + \chi \sum_{l=1}^{N} L(y^{(l)} - (\varphi(x^{(l)})^T \omega + b)) \tag{3}$$

Where $\Phi(.)$ is a non-linear function that transforms to a high dimensional feature space. Further, the parameter χ is used to obtain a balance between regularization term and error reduction.

In Eq. 4, L is denoted as ε based insensitive loss function which is equal to:

$$0 \text{ for } \left|(y^{(l)} - (\varphi(x^{(l)})^T \omega + b)\right| < \varepsilon$$
$$\left|(y^{(l)} - (\varphi(x^{(l)})^T \omega + b)\right| - \varepsilon \text{ for} \left|(y^{(l)} - (\varphi(x^{(l)})^T \omega + b)\right| \geq \varepsilon \tag{4}$$

The probable solution (ω and b) is obtained by the linear combination of the trained dataset onto transformed vector space with an absolute error $\geq \varepsilon$. The architecture of M-SVR model is shown in Fig. 1. Where R1, R2 and R3 are individual SVR model that are combined to give target coordinate.

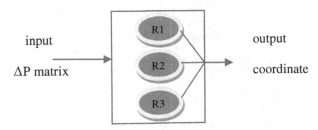

Fig. 1. Multi-output support vector regression architecture.

To develop a multi output regression model, the single output model is applied to each output independently [11]. This is done with a wrapper which will have one SVR model corresponding to each output. Same input is given to each SVR model during training and testing phase. Thus, an n output SVR model is developed by combining n individual SVR output models which are not depending on each other.

3.2 Deep Neural Network Regression (DNN-R)

The architecture of the DNN-R model comprising all layer details is shown in Fig. 2.

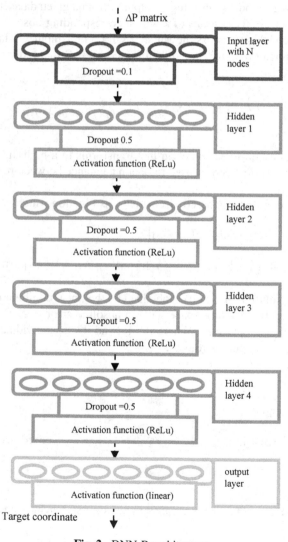

Fig. 2. DNN-R architecture

Estimation of target coordinates using deep learning models are becoming popular due to the potential benefits of the deep networks. It is possible to model the dependency of target location coordinates and ΔP matrix with DNN via regression approach. Thus when ΔP matrix is given as input to the DNN regression model for an unknown target location, the model will assess the exact target location. The DNN model is multilayer perceptron (MLP) based with one input layer, one output layer and four hidden layers. A dropout of 0.1 is applied in input layer and 0.5 applied in hidden layers. The dropout regularization method enables only certain number of nodes update their weights during different epochs of training and thus making the model robust to inputs. The activation function used in the hidden layers of DNN regression model is ReLu and linear for output layer.

4   Results and Discussions

4.1   Simulation Setup

The setup used for simulation is within a 3 m × 3 m × 3 m room dimension with sensor nodes randomly positioned but distributed equally over the four walls in layers of two. For a given length D of the room, the two layers are at 1.5 m and 0.5 m from ceiling. The performance of the model is assessed using this simulation setup with 16 nodes, 32 nodes, 48 nodes and 64 node configurations. In each node configuration, during training one target is placed randomly and ΔP matrix training samples are generated. In this work target is considered as spherical in shape with a radius of 15 cm. The sensor nodes used in the setup is working in 2.4 GHz band. During testing phase, when a target enters the room setup, due to which a non-zero ΔP matrix is developed and is given as input to the machine learning model to assess the exact location of the target. For assessing the effect of measurement error, a received power error variance in the range of −10 dB to +10 dB is added in steps of two during testing. The training samples used for evaluation 4000 samples and 5000 samples each for testdata for each error variance condition. The performance of the DNN-R and M-SVR model is assessed using mean localization error. The simulations are done in python in google Colab platform.

4.2   Performance of the DNN-R and M-SVR in 16 Node Configuration

The Fig. 3. Depicts the comparison of mean localization error of the DNN-R and M-SVR models within a 16 node room configuration in different error variance scenario. It is very much evident that DNN-R is performing good in an error variance condition compared to M-SVR. In without measurement error variance condition mean localization error is less than 0.2 m for M-SVR and nearly 0.3 m for DNN-R. No var. indicates no variance condition. DNN-R indicates a decreasing trend in error from −10 dB to +10 dB variance band. In case of M-SVR, the mean localization error is comparatively higher from −6 dB to 10 dB band except for 2 bands (−10 dB and −8 dB) the performance of both the models are almost same.

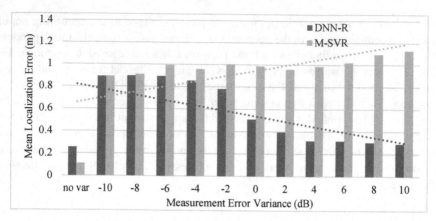

Fig. 3. Performance analysis in 16 node configuration

4.3 Performance of the DNN-R and M-SVR in 32 Node Configuration

The Fig. 4. Depicts the comparison of mean localization error of the DNN-R and M-SVR models within a 32 node room configuration in different error variance scenario. From the plot it is understood that for most of the variance condition scenario DNN is performing better except for −8 dB and −6 dB for which localization error has crossed unity. The performance of DNN-R is comparatively good and is projected with low localization error when the measurement variance is changed from −10 dB to 10 dB. In no variance condition, error of DNN-R has gone less than 0.2 m.

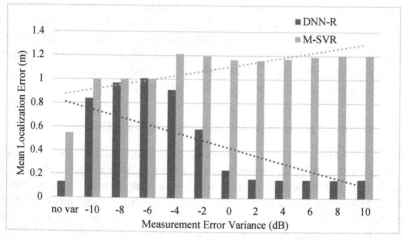

Fig. 4. Performance analysis in 32 node configuration

4.4 Performance of the DNN-R and M-SVR in 48 Node Configuration

In the 48 node configuration given in Fig. 5, the performance of DNN-R has improved and for all measurement error conditions, mean localization error has reduced to less than 1 m. In contrast, the variance band has no considerable effect on M-SVR performance. In no variance condition, localization error of DNN-R is nearly 0.2m and of M-SVR is more than 1m. In the error variance scenario, the M-SVR performance was nearly same without any much improvement throughout the variance band.

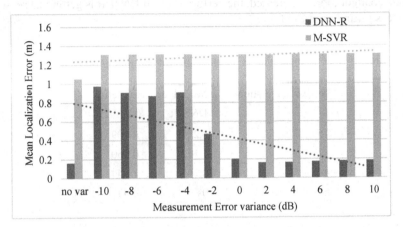

Fig. 5. Performance analysis in 48 node configuration

4.5 Performance of the DNN-R and M-SVR in 64 Node Configuration

In the 64 node configuration scenario given in Fig. 6 also, the performance of DNN-R is good compared to M-SVR model with measurement error variance. Except for the error variance band from −10 dB to 0 dB the performance of DNN-R is excellent with mean localization error zero. The mean localization error of the M-SVR in no variance condition is in this node configuration is nearly 1m and that of DNN-R is excellent with mean localization error to be zero.

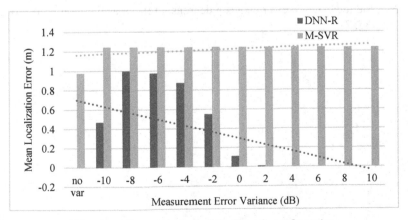

Fig. 6. Performance analysis in 64 node configuration

4.6 Node Configuration vs Mean Localization Error in no Measurement Error Variance Condition

In no measurement error variance scenario, the performance of DNN-R is found to promising when compared to M-SVR when higher configurations are imparted. This is very much evident from Table 1. For 64 node configuration the mean localization error of DNN-R is zero. The mean localization error is 25.5 cm for DNN-R in no measurement error variance condition and for M-SVR it is only 10.9 cm for 16 node configuration. As node configuration is increased, the performance of DNN-R is getting increased and M-SVR is getting decreased.

Table 1. Node configuration vs mean localization error in no variance condition

Node configuration	Mean localization error (m)	
	DNN-R	M-SVRs
16	0.255	0.109
32	0.133	0.550
48	0.161	1.054
64	0	0.977

4.7 Node Configuration vs Mean Localization Error in Maximum Error Variance Condition

The performance of the DNN-R and M-SVR model is evaluated in maximum error variance condition which is with -10 dB error variance condition. The performance of DNN-R is found to be good compared to M-SVR in higher node configurations. For all the node configuration setup, the mean localization error is less than 1m for DNN-R whereas for M-SVR it is more than unity. The localization error is found to be showing an increasing trend as node configuration is increased in case of M-SVR. But in case of DNN-R the localization error shows a decreasing trend as the number of nodes are increased (Fig. 7).

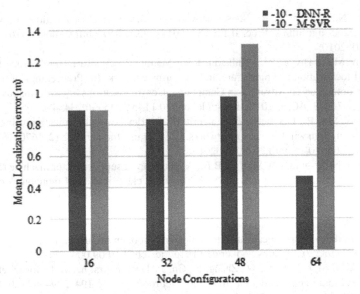

Fig. 7. Node configuration vs mean localization error in maximum variance condition (−10 dB)

5   Conclusion

The performance of popular machine learning models DNN-R and M-SVR are analyzed by assessing the location of a target in a room setup with different number of node configurations. The received signal measurement error condition is also imparted to know the better performing model. For 16 node configuration, in no measurement variance condition, mean localization error of M-SVR model is found to be lesser compared to DNN-R model. Also in maximum error variance scenario, mean localization error of M-SVR and DNN-R in 16 node configuration is comparable. The DNN-R model is found to be performing better with the measurement error variance condition. For higher node configurations, DNN-R seems to be better performing due to the large data handling capability of DNN-R. The M-SVR model is found to be underperforming for higher node configurations as large number of data points in higher node configuration increases the complexity of the hyperplane. For lower node configuration M-SVR performance is good compared to higher node configurations. If training samples are increased the performance of DNN-R can be improved further. Thus the analysis helps to choose appropriate machine learning models effectively for different node configuration setups.

References

1. Anusha, K.S., Ramanathan, R., Jayakumar, M.: Device free localisation techniques in indoor environments. Def. Sci. J. **69**, 378–388 (2019). https://doi.org/10.14429/dsj.69.13214
2. Savazzi, S., Rampa, V., Vicentini, F., Giussani, M.: Device-free human sensing and localization in collaborative human-robot workspaces: a case study. IEEE Sens. J. **16**, 1253–1264 (2016). https://doi.org/10.1109/jsen.2015.2500121

3. Patwari, N., Wilson, J.: RF Sensor networks for device-free localization: measurements, models, and algorithms. Proc. IEEE **98**, 1961–1973 (2010). https://doi.org/10.1109/jproc. 2010.2052010
4. Ramesh, M.V., Divya, P.L., Kulkarni, R.V., Manoj, R.: A swarm intelligence based distributed localization technique for wireless sensor network. In: Proceedings of the International Conference on Advances in Computing, Communications and Informatics - ICACCI 12, pp. 367–373. ACM (2012). https://doi.org/10.1145/2345396.2345457
5. Talampas, M.C.R.: Geometric filter algorithms for device-free localization using received-signal strength in wireless sensor networks. IEEE Trans. Industr. Inf. **12**, 1670–1678 (2016). https://doi.org/10.1109/tii.2015.2433211
6. Sakthivel, N., Sugumaran, V., Nair, B.B.: Application of support vector machine (SVM) and proximal support vector machine (PSVM) for fault classification of monoblock centrifugal pump. Int. J. Data Anal. Tech. Strat. **2**, 38 (2010). https://doi.org/10.1504/IJDATS.2010. 030010
7. Anusha, K., Ramanathan, R., Jayakumar, M.: Link distance-support vector regression (LD-SVR) based device free localization technique in indoor environment. Eng. Sci. Technol. Int. J. **23**, 483–493 (2020). https://doi.org/10.1016/j.jestch.2019.09.004
8. Zhang, W., Liu, K., Zhang, W., Zhang, Y., Gu, J.: Deep neural networks for wireless localization in indoor and outdoor environments. Neurocomputing **194**, 279–287 (2016). https:// doi.org/10.1016/j.neucom.2016.02.055
9. Sukor, A.S.A., Kamarudin, L.M., Zakaria, A., Rahim, N.A., Sudin, S., Nishizaki, H.: RSSI-based for device-free localization using deep learning technique. Smart Cities **3**, 444–455 (2020). https://doi.org/10.3390/smartcities3020024
10. Mager, B., Lundrigan, P., Patwari, N.: Fingerprint-based device-free localization performance in changing environments. IEEE J. Sel. Areas Commun. **33**, 2429–2438 (2015). https://doi. org/10.1109/JSAC.2015.2430515
11. Zhang, W., Liu, X., Ding, Y., Shi, D.: Multi-output LS-SVR machine in extended feature space. In: 2012 IEEE International Conference on Computational Intelligence for Measurement Systems and Applications (CIMSA) Proceedings, pp. 130–134 (2012). https://doi.org/10. 1109/cimsa.2012.6269600
12. Borchani, H., Varando, G., Bielza, C., Larrañaga, P.: A survey on multi-output regression. Wiley Interdisc. Rev. Data Min. Knowl. Discov. **5**, 216–233 (2015). https://doi.org/10.1002/ widm.1157
13. Zhou, R., Hao, M., Lu, X., Tang, M., Fu, Y.: Device-free localization based on CSI fingerprints and deep neural networks. In: 2018 15th Annual IEEE International Conference on Sensing, Communication, and Networking (SECON), pp. 1–9 (2018). https://doi.org/10. 1109/SAHCN.2018.8397121.
14. Shang, F., Su, W., Wang, Q., Gao, H., Fu, Q.: A location estimation algorithm based on RSSI vector similarity degree. Int. J. Distrib. Sens. Netw. **10**, 371350 (2014). https://doi.org/10. 1155/2014/371350

D-Leach: An Energy Optimized Deterministic Sub-clustering and Multi-hop Routing Protocol for Wireless Sensor Networks

Subhash Chandra Gupta[✉] and Mohammad Amjad

Computer Engineering, Jamia Millia Islamia, New Delhi, India
mamjad@jmi.ac.in

Abstract. The The introduction of Wireless Sensor Networks (WSN) has brought comfort of uninterrupted wireless networks into many lives. The data transmission nodes are a group of heterogeneous sensor node (SN) fitted with a battery, which are deployed randomly for monitoring of surrounding. Clustering algorithms with effective routing protocols are used to handle the random deployment of nodes. This results in redundant data packets being aggregated and dropped, enabling sound data transfer from sensor node to the Base Station (BS) through Cluster Head (CH). Different Energy optimization Routing Protocols have been introduced in previous years but have not been able to examine protocol behavior in different environments. An adaptive, sub-clustering and multi-hop routing protocol is proposed in this paper with experimental based analysis, taking into consideration energy and distance. That create smooth and simple route from the cluster nodes, cluster head, and sub-cluster head to the base station. Experimental studies show a substantial increase in network lifetime efficiency by comparing the proposed method with the present situation. Proposed protocol behaviour shows deterministic, so it is called Deterministic-LEACH (D-LEACH).

Keywords: Multi-hop routing protocol · Network life-time · Wireless sensor network · Energy efficiency

1 Introduction

Wireless Sensor Networks are networks of sensor nodes that track corporal and environmental condition such as sound, vibration, heat, motion, and pressure. Sensor nodes (SNs) with adequate processing power but insufficient power storage make up WSNs [1, 4]. A sensor node contains three main sections: a processing section for data aggregation and repository, a transmission section for data receiving and sending data, and a sensing section for finding of data from the surrounding [10].

SNs are typically placed at random in areas where humans are unable to track weather conditions. Nodes collect data and relay it to base station (BS), that uses battery power [9]. Continuous battery use causes battery power loss and sensing failure [12]. When nodes are installed in dangerous conditions such as volcanoes, battlefields [13], and so on, replacing SN batteries is virtually impossible. As a result, the network needs a longer life in order to prolong lifetime of networks [15] (Fig. 1).

© Springer Nature Switzerland AG 2021
M. Singh et al. (Eds.): ICACDS 2021, CCIS 1440, pp. 561–570, 2021.
https://doi.org/10.1007/978-3-030-81462-5_50

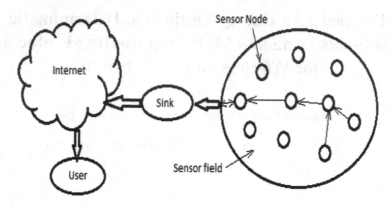

Fig. 1. Typical structure of WSN

Simple structure of WSN shown above. The message or information is routed via the sink node [2], from sensor (source) node, to the base station (destination), where end user can get these data through the internet [14]. When designing clustering protocols, factors such as fault tolerance, power consumption, data aggregation, load balance of nodes, QoS of sensor nodes, node deployment, and data latency are consider into account [4].

Routing protocols based on cluster, which split a big network (cluster) in to small and easily manageable multiple clusters, provide a cost-effective solution to the problem. Clusters and BS can communicate more effectively within this protocols and formation of a simple multi-hop routing path. So that a minimum energy consumption is accomplished by combining the obtained packets from the cluster., In the end, Cluster load balancing as well as network's lifespan also increases.

2 Literature Review

Energy Efficient Multi-hop Routing Protocol proposed by Khanoucheet al. [1] that is based on Clusters Re-organization, which includes different phases of structural type: cluster creation, sub-zone division, and data transmission through multiple path in inter-cluster route. The proposed idea ignores the distance between sensor nodes and sink node.

Cengiz et al. [2] suggested starting with a fixed number of cluster heads and then selecting new cluster heads based on threshold values or energy consumed after a few rounds. Inter-clusters and intra-clusters transmit data through number of relay and that is used to measure the energy change in overall network utilisation. By adding EAMR, they were able to reduce LEACH's unnecessary overhead by implementing a fixed cluster with a low number of cluster head changes.

Because of the battery constraint, Nam et al. [3] suggested the Energy-efficient data transmission that WSN needs. By creating a local cluster head and developing a formula for the number of packets to forward to the sink node, they suggest the optimal number of Cluster Head (CH).

Liu et al. [4] suggested HCNM, a hierarchical clustering method for managing sensor nodes. The networks calculates the distance between every node to disperse the equivalent number of nodes. The proposed model prevents over-fitting as well as under-fitting cluster head in a network by conducting subsequent clustering.

Yang et al. [5] suggested a CH selection algorithm that took into account the effect of the distance between the cluster-head and the base-station, as well as the WSN routing algorithm based on the enhanced LEACH protocol. When comparing simulation results to the LEACH algorithm, it is discovered that there is a gap in node die-time, an increase in its continued existence rate, and a dispersed existence in the position of a dead node. Its average battry consumption has also been enhance, and its lifespan has been extended.

Kumar et al. [6] divides the entire network into smaller network that is based on distance and suggested a good routing algorithm, more structured clusters, each cluster having their cluster head to take care of data transfer. By maximising the battery power of nodes, the proposed framework extends the lifespan of wireless networks. Move the sensed data to the base station after it has been collected.

By selecting cluster-heads with high residual energy through local radio frequency, M. Ye, C. Li, G. Chen, and J. Wu [7] proposed the Energy Efficient Clustering Scheme (EECS) protocol. The competition method achieves uniformity among all cluster and cluster heads without requiring iteration. Cluster foundation phase distributes data transfer load among all cluster head, with CHs handling packet routing to the sink.

Abdellatief et al. [9] find that distribution of nodes randomly is the main reason of battery drain in the network. The problem of different levels of energy at different levels of the region makes the network unbalance. To avoid this problem author proposed a distributed density-based clustering techniques based on sub-regions according to the density of nodes in that region.

Chan et al. [10] proposed a comprehensive survey on hierarchical routing protocols for WSN. It gives a vast idea to upcoming research in this area.

Arunmugam, G.S., Ponnuchamy, T. [11] suggested an energy-efficient LEACH protocol for data aggregation. It proposed an energy-efficient protocol in WSN based on the sorting algorithm residual energy while aggregating nodes.

3   Proposed Model

Since the advent of WSN, various energy-efficient routing protocols have been proposed to date. In most of the existing protocols, researchers did not investigate the protocol's behavior in various environments. Usually, CHs in standard field size are selected as proposed at the time of LEACH [8]. Existing models also did not perform in-depth analysis on dense [9] and sparse [10] environments of WSN with various field sizes and various numbers of nodes.

In this paper, we performed an in-depth performance analysis of existing protocols [7] based on the number of sensor nodes corresponding to the variable network size. Here, the numbers of heterogeneous sensor nodes are increased and decreased according to a sensor network's size. This also changes the number of cluster heads, Sub-CHs, cluster nodes, and the multi-hop routing path formation.

Table 1. Parameters of the suggested model

Parameters	Description	Value
Topography	Dimensions of Field	100 m x 100 m, 300m x 300m 500m x 500m
N	No of Nodes	100, 300 and 500
Rounds	Max no of Rounds	6000
Eo	Initial energy of each node	0.5 J
ETx	Transmission energy of node	50 x 0.000000001 J
ERx	Receiving energy of node	50 x 0.000000001 J
EDA	Data aggregation energy	5 x 0.000000001 J
Efs	Energy dissipation for free space	10 x 0.000000000001 J
Emp	Energy dissipation for multi-path delay	0.0013 x 0.000000000001 J
Packet	Packet size	4000

Table 1 summarizes the initial parameters to implement the proposed model for both scenarios: sparse (field size 100×100 with 100 nodes), medium (field size 300×300 with 300 nodes) and dense (field size 500×500 with 500 nodes) environment. Apart from several topography and node everything is common.

Our proposed protocol is based on a dynamic selection of cluster head (CHs) and subcluster head (SCH) in multi-hop route. In intra cluster formation of sub-route between the cluster nodes to sub–cluster head to cluster head, and to the base station makes proposed protocol scalable andreliable energy efficient tools to face the in real challenges [2].

4 Flowchart and Algorithm

The working algorithm and flow methodology are shown in Fig. 2a and Fig. 2b, whenever data sensed by any node, it transfer data to a nearest node, that forming a link of all intermediate node until the data received by the destination node or cluster head. The main cluster head forms the Cluster head chain, where transmissions allow in the energy-efficient way of the data to the sink.This techniques makes the network more reliable for real-time issues and make it more scalable also.

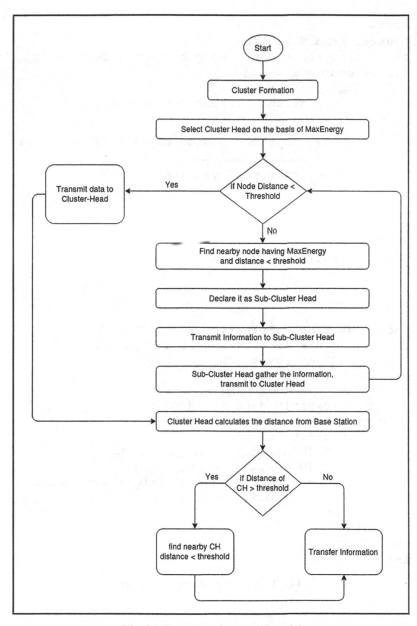

Fig. 2. Flowchart of proposed model

- **Start**
- **Cluster Head Selection**

```
Initialize the cluster neighbor node, Set -> S_nbr-CH
For (eachnode ∈ S_nbr-CH )
    If (S_nbr-CH = φ):
            Select the sink as the next hop
    else:

            Compute the distance of each CH of S_nbr-CH by :
```

$$D_i = \sqrt{(X2-X1)^2 + (Y2-Y1)^2}$$

```
            If(D_i [S_nbr-CH] < D_{i+1} [S_nbr-CH]) :
                    Set Nexthop <- D_i [S_nbr-CH ]
            else :
                    Set Nexthop <- D_{i+1} [S_nbr-CH ]
            endif

    endif

endForLoop
```

- **Sub-cluster Head Selection**

```
Initialize distance threshold as D_threshold
ClusterNode energy as CN_energy
For (each CN ∈ Cluster)
    If (CN_Distance < D_threshold ):
            Transmit data to CH/BS
    else :
            Select Sub_CH
            If (Node_distance > D_threshold ) AND
                (Node_energy = Node_MaxEnergy ):
                    Declare Node as Sub_CH and gather
                    information
            endif
    endif
endForLoop
```

- **End**

Fig. 3. Algorithm of proposed model

Furthermore, we will make intensive comparisons of proposed energy-efficient protocols with the existing researches proposed in the literature. In this paper, parametric analysis of energy efficiency, data aggregations, and throughput of the network are focused on in the next section.

5 Result and Discussion

As compared through simulation result of D-LEACH with other routing protocols LEACH and EE-LEACH in sparse, medium and dense environment considering all condition and standard parameters.

Figure 3 shows the representation of the performance analysis with the network lifetime of the node. In terms of network lifetime, the proposed protocol is outperformed the other two protocols.

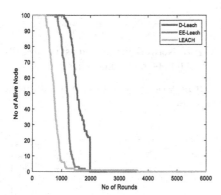

Fig. 4. Alive nodes vs. round for 100 nodes and field size (100 × 100).

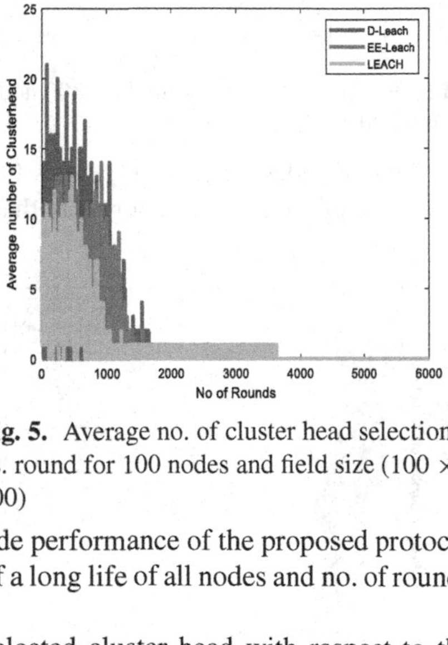

Fig. 5. Average no. of cluster head selection vs. round for 100 nodes and field size (100 × 100)

Figure 3 shows that the number of dead node performance of the proposed protocol is better than the other two protocols in terms of a long life of all nodes and no. of rounds to the beginning of the first dead node.

Figure 4 shows the average number of selected cluster head with respect to the number of rounds, keeping the selection probability the same for all protocols p = 0.05. The proposed protocol can exploit it fully and maintain the maximum cluster selection since the beginning. Cluster heads quantity goes down as the number of surviving nodes reduced. From the analysis of the small scenario network 100 m × 100 m with 100 nodes, that shows our proposed D-LEACH protocol is the most stable network performance.

After satisfaction result in sparse environment of proposed protocol, let us see the simulation result in medium (300 × 300) and dense (500 × 500) environment with 300 and 500 nodes respectively.

Figure 5 shows the comparison simulation result of the proposed protocol with LEACH and EE-LEACH, in terms of network lifetime. We can see from the graph that LEACH and EE-Leach's performance decreases as we increased the field size. For LEACH, 90% of nodes die before 800 rounds, unlike the smaller field size where they survive for 1000 rounds. Same downgrade the performance of EE-Leach as field size increases. Moving toward our proposed protocol D-Leach. We see that protocols show not only scalable behavior but also their performance increase in terms of network lifetime.

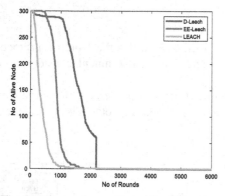

Fig. 6. Alive nodes vs. Round for 300 nodes and field size (300 × 300)

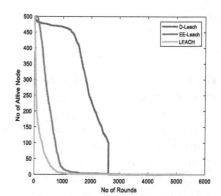

Fig. 7. Network throughput vs. number of round for 300 nodes and field size (300 × 300)

90% of nodes die in 2000 rounds for smaller network size, while for medium field size, 90% of nodes survive for more than 2200 rounds.

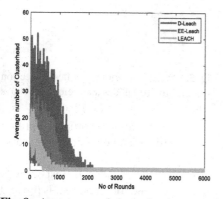

Fig. 8. Average no. of cluster head vs. round for 300 nodes and field size (300 × 300)

Fig. 9. Alive nodes vs. round for 500 nodes and field size (500 × 500)

Cluster selection in each round is showing the same performance as all protocols show for smaller field sizes. The simulation performance of network throughput for our proposed protocol is also significantly high, as shown in Fig. 6 and 7.

In the analysis for larger field size (500 × 500) and node count 500, our proposed protocol shows exciting and significant improvement in network lifetime and network throughput, as shown in Fig. 8 and 9 respectively.

Figure 8 shows the performance graph of the proposed protocol with LEACH and EE-Leach, in terms of network lifetime. We can see from the graph that LEACH and EE-Leach's performance decreases as we increased the field size. For LEACH, 90% of nodes die before 500 rounds. Unlike the smaller (100 × 100) field size where they survive for 1000 rounds, performance remains only 50% as we have increased the network size 5-times and the performance of EE-Leach remain 50% as field size increase. In contrast, our proposed protocol shows a 25% increase in network lifetime performance as we increased the network field size 5-times.

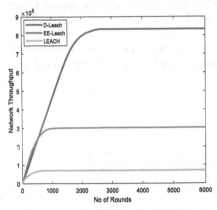
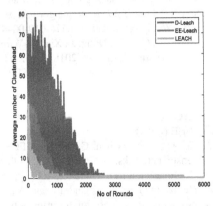

Fig. 10. Network throughput vs. Round for 500 nodes and field size (500 × 500)

Fig. 11. Average no. of cluster head selection round for 500 nodes and field size (500 × 500)

The network throughput is increased two times from the large field size network, and the average number of cluster head performances shown in Fig. 9 and Fig. 10 respectively. The cluster head selection for each round remains constant, showing almost similar performance for all field sizes (Fig. 11).

6   Conclusion and Future Work

The simulation analysis of a current energy optimized protocol with different environment and different number of nodes was the starting point for this paper. The simulation depicts the performance of existing protocols, which indicates a substantial reduction in performance due to the fact that none of them are flexible and reliable to sustain in various scenario and are statically designed for a static field size with a static number of nodes. We developed a protocol for a dynamic network based on multi-hop sub-clustering and clustering routing path for transmission of data to a base station, taking into account all of these limitations. Simulation and experimental results represent that the suggested protocol well performed in all field sizes, with performance gradually increasing as we transition to dense environment, unlike other protocols, which perform worse as field sizes grow larger. The parameters we considered in this study were a longer lifespan and a higher throughput. The protocol's end-to-end delay and protection will be investigated further.

References

1. Khanouche, F., Maouche, L., Mir, F., Khanouche, M.E.: Energy efficient Multi-hops routing protocol based on clusters reorganization for wireless sensor networks. In: ICFNDS'19:

Proceedings of the 3rd International Conference on Future Network Distributed Systems, pp. 1–10 (2019)
2. Cengiz, K., Dag, T.: Energy aware multi-hop routing protocol for WSNs. IEEE Access **6**, 2622–2633 (2017)
3. Nam, C.S., Han, Y.S., Shin, D.R.: Multi-hop routing-based optimization of the number of cluster-heads in wireless sensor networks. Sensors **11**(3), 2875–2884 (2011)
4. Liu, S.: Energy-saving optimization and matlab simulation of wireless networks based on clustered multi-hop routing algorithm. Int. J. Wireless Inf. Networks **27**(2), 280–288 (2019). https://doi.org/10.1007/s10776-019-00448-5
5. Yang, T., Guo, Y., Dong, J., Xia, M: Wireless routing clustering protocol based on improved LEACH algorithm. In: 2018 IEEE International Conference on RFID Technology and Application (RFID-TA), pp. 1–6 IEEE (2018)
6. Kumar, N., Kaur, S.: Performance evaluation of Distance based Angular Clustering Algorithm (DACA) using data aggregation for heterogeneous WSN. In: 2016 International Conference on Computation of Power, Energy Information and Commuincation (ICCPEIC), pp. 097–101 IEEE (2016)
7. Ye, M., Li, C., Chen, G., Wu, J.: EECS: an energy efficient clustering scheme in wireless sensor networks. In: PCCC 2005. 24th IEEE International Performance, Computing, and Communications Conference, 2005, pp. 535–540. IEEE (2005)
8. Heinzelman, W.R., Chandrakasan, A., Balakrishnan, H.: Energy-efficient communication protocol for wireless microsensor networks. In: Proceedings of the 33rd Annual Hawaii International Conference on System Sciences, pp. 10-pp. IEEE (2000)
9. Abdellatief, W., Youness, O., Abdelkader, H., Hadhoud, M.: Balanced density-based clustering technique based on distributed spatial analysis in wireless sensor network. Int. J. Wireless Inf. Networks **26**(2), 96–112 (2019). https://doi.org/10.1007/s10776-019-00425-y
10. Chan, L., Gomez Chavez, K., Rudolph, H., Hourani, A.: Hierarchical routing protocols for wireless sensor network: a compressive survey. Wireless Netw. **26**(5), 3291–3314 (2020). https://doi.org/10.1007/s11276-020-02260-z
11. Arumugam, G.S.; Ponnuchamy, T.: EE-Leach: Development of energy-efficient LEACH protocol for data gathering in WSN. EURASIP J. Wirel. Commun. Netw. **76**, (2015). https://doi.org/10.1186/s13638-015-0306-5
12. Akyildiz, I.F., Su, W., Sankarasubramaniam, Y., Cayirci, E.: Wireless sensor networks: a survey. Comput. Netw. **38**(4), 393–422 (2002)
13. Bekmezci, I., Alagz, F.: Energy efficient, delay sensitive, fault tolerant wireless sensor network for military monitoring. Int. J. Distrib. Sens. Netw **5**(6), 729–747 (2009)
14. Mainwaring, A., Culler, D., Polastre, J., Szewezyk, R., Anderson, J.: Wireless sensor networks for habitat monitoring. In: Proceedings of the 1st ACM International Workshop on Wireless Sensor Networks and Applications, pp.88–97. NewYork, NY, USA (2002)
15. Tan, H.Ö., Ibrahim, K.: Power efficient data gathering and aggregation in wireless sensor networks. ACM Sigmod Rec. **32.4**, 66–71 (2003)

Robust Image Watermarking Using Support Vector Machine and Multi-objective Particle Swarm Optimization

Kapil Jain$^{(\boxtimes)}$ and Parmalik Kumar

Madhyanchal Professional University, Bhopal, India

Abstract. Multimedia data security is paramount due to the high traffic on the internet. The internet traffic compromised with security threats such as inten tional and time differential attack of multimedia data. This paper proposed an image watermarking method based on support vector machine and particle swarm optimization. The proposed algorithm works in dual mode as a selection of the embedding location of the watermark and increases the randomness factor of the watermark image. The increase randomness factor increases the value of imperceptibility and robustness factor. The support vector machine classifies the pattern of the host image and the watermark image. The processing of feature components of an image extracted by stationary wavelet transform (SWT). The stationary wavelet transform is also called redundant wavelet transform; these transforms overcome the discrete wavelet transform's limitation. The extraction of watermark image applies various attacks and measures standard parameters such as SSIM (similarity index matrix) and NC (number of correlation). The art of analysis of the proposed algorithm compares with SVM and DWT based watermarking methods. The proposed algorithm simulated in MATLAB software and tested with a reputed image dataset such as cameraman, peppers, baboon and Lena. The support vector machine applied the non-linear kernel function. The proposed algorithm decreases the 5–8% risk of geometrical attacks.

Keywords: Watermarking · SWT · MPSO · SVM · MATLAB · Geometrical attacks

1 Introduction

The rapid growth of information technology and easy accessibility needs the concept of data hiding and security. The security of multimedia data has various approach such as steganography, encryption and digital watermarking [1–3]. The contribution of digital watermarking is very high concerning other methods of digital data security and copyright protection [4]. The process of digital watermarking methods proceeds in two different ways, such as spatial domain and frequency-based methods of digital image watermarking [5, 6]. The spatial based watermarking methods operate the shifting of pixel to the left to right and vice-versa. The very famous algorithm of the spatial domain is LSB and RSB [7, 8]. The strength of these watermarking methods is week and easily

© Springer Nature Switzerland AG 2021
M. Singh et al. (Eds.): ICACDS 2021, CCIS 1440, pp. 571–591, 2021.
https://doi.org/10.1007/978-3-030-81462-5_51

compromised with the geometrical attack. Despite the spatial image digital watermarking algorithm, the transform-based watermarking methods are very efficient and more secured [9–11]. The multiple derivatives of transform are applied for the processing of digital image watermarking. The major contribution of wavelet transforms, the wavelet transform has multiple variants that change the characteristics of digital image watermarking [12–14]: the concept of multi-layer decomposition of transform in the form of high frequency and low frequency [15]. The processing of low-frequency watermark embedding moves to the next level of security. The transform DWT and WPT have certain bottleneck issue, such as translation invariance is not supported by DWT [16]. It also faces a problem of the input image's dyadic size and many other issues such as impulsive noise. These limitations overcome with stationary wavelet transform (SWT) [17], also called redundant transform. The application of transform in digital image watermarking direct embedding process and feature-based watermarking. Recently various authors and research scholar reported the feature-based watermarking algorithm is efficient instead of the direct embedding of transform. The feature-based image watermark methods derived the process of pattern-based template embedding [18, 19]. The pattern-based embedding of watermarking enhances the robustness factor of watermarking methods. For the generation and formation of patten, various machine learning algorithm applied in digital image watermarking [20, 21]. Machine learning provides various supervised algorithms such as decision tree, KNN and supports vector machine. Various authors reported support vector machine is very efficient for digital image watermarking: the pattern-based digital image watermarking compromised with two parameters, such as feature selection and feature optimization [22–24]. The selection and optimization process call a meta-heuristic function such as ant colony optimization (ACO), particle swarm optimization (PSO), genetic algorithm (GA), glow-worm swarm optimization (GSO) and firefly algorithm. The participate of the swarm intelligence algorithm moves to the next level of the digital image watermarking algorithm [23, 25]. The evaluation condition of digital image watermarking based on robustness and imperceptibility. The trade-off between robustness and imperceptibility in terms of visual quality and security of the watermark image. This trade-off reduces with optimization algorithms [7, 9]; various authors applied a swarm-based optimization algorithm and increased the watermark image's quality index and the watermark image's security correlation [19, 20]. This paper poses a support vector machine-based digital image watermarking. The support vector machine generates the dynamic patterns of transform features of SWT [3, 7]. The multi-objective particle swarm optimization hands the process of embedding and extraction of digital image watermarking. The major contribution of this paper (1) enhances the security strength of digital image watermarking. (2) validate the algorithm with a geometrical attack (3) art of investigation with SVM and DWT methods. The rest of the paper describes as in Sect. 2. Methods description in Sect. 3—proposed methodology, in Sect. 4. Experimental analysis and finally conclude in Sect. 5.

2 Methods Description

This section describe the process of feature extraction based on stationary wavelet transform (SWT) [3], multi-objective particle swarm optimization and support vector machine. These methods applied for the process of watermark embedding.

a. Stationary Wavelet Transforms

Stationary wavelet transform is overcome the limitation of discrete wavelet transform and preserved translation invariance. The process of SWT transform cannot translation of wavelet coefficient. Despite of sampling, SWT applied recursive dilated filters [1, 3]. The scaling factor of transform is 2^j the filter dilated by inserting 2^{j-1}. The process of transform reduces the bandwidth of one level to another level.

$$F1_k^{(j)} = \begin{cases} F1_{\frac{k}{2^j}}, & , k = 2^j m \ if \ m \epsilon Z \\ 0 & else \end{cases} \tag{1}$$

$$F2_k^{(j)} = \begin{cases} F2_{\frac{k}{2^j}}, k = 2^j m \ if \ m \epsilon Z \\ 0 \ else \end{cases} \tag{2}$$

Where F1 and F2 are low pass and high pass filter. For processing of image rows and column filtered separately. The process of filtering drive at level J + 1 from level J are the following (where (p, q) is the pixel position)

$$D_{X_{j+1}(p,q)} = \sum_{k,m} h_k^{(j)} h_m^{(j)} D_{X,J}(P+K, q+m) \tag{3}$$

$$C_{X,J+1}^h(p, q) = \sum_{k.m} F1_k^{(j)} h_m^{(j)} D_{X,j}(p+k, q+m) \tag{4}$$

$$C_{X,J+1}^v(p, q) = \sum_{k.m} F2_k^{(j)} h_m^{(j)} D_{X,j}(p+k, q+m) \tag{5}$$

$$C_{X,J+1}^d(p, q) = \sum_{k.m} F1_k^{(j)} F1_m^{(j)} D_{X,j}(p+k, q+m) \tag{6}$$

Where Dx, j is coefficient of approximate of original image at the scale 2^j finds the lower content of image and correspond C finds the value of horizontal, vertical and diagonal of SWT transform of source image.

b. Multi-Objective Particle Swarm Optimization (MPSO)

The process of multi-objective optimization deals at least two objective function to be optimized. The characteristics of objective function minimized the constraints of features components of SWT transform function. Another objective function to estimate optimal location of pattern to embedded watermarking. The process of objective defines as

$$Minimize \ f(x) := [f1(x), f2(x), \ldots \ldots \ldots fk(x)] \tag{7}$$

Realization to

$$gi(x) \leq 0 \ i = 1, 2, \ldots \ldots \ldots, m, \tag{8}$$

$$hj(x) = 0 \ j = 1, 2, \ldots \ldots \ldots .p \tag{9}$$

where the variable [x1, x2,............, xn] is decision factor of constraints and gi and hj are equality equation of desired function of minimized value.

The particle swarm optimization derived by natural intelligence algorithm. The working concept of particle swarm optimization is bird of fork. The MPSO algorithm applied the concept of pareto dominance. The process of algorithm describes here.

1. Define the population, X^j for $j = 1, 2, ., n$ where n is population size.
2. Define the speed, Ve^j for each particle $ve^j = 0$
3. Estimate each particle
4. Estimated non-dominated selection in memory, Me
5. Define search space
6. Define the memory of each particle and found new position of particle
7. $BP^j = X^j$
8. measure speed of particle

$$Ve^j = WX \, Ve^j + R1X \left(BP^j - X^j \right) + R2X \left(Me - X^j \right)$$

here w is wieght value is 0.5 R1 and R2 is random number in the range[0..1]
9. Estimate next position of each particle

$$X^j = X^j + Ve^j$$

10. Applied mutation of particle
11. Update memory Me of particle
12. If maximum iteration is archive terminate otherwise go to step 8.

c. Support Vector Machine

SVM (Support vector machine) is machine learning algorithm derived by Vipin in 1990. The support vector machine applied in various filed of image classification and pattern recognition. The nature of support vector machine is linear, non-linear and sigmoid. The non-linear support vector machine mapping the feature data with respect to one plane to another plan [16]. The separation of data plan is non-linear and decision factor correlate with margin function of support vector. The hyperplane of equation is derived as

$$WD.\, xi + b \geq 1 \; if \; yi = 1 \tag{10}$$

$$WD.\, xi + b \leq -1 \; if \; yi = -1$$

Here W is weight vector, x is input vector yi label o class and b is bias (Fig. 1). The minimization formulation of support vector

$$Minimize \; \frac{1}{2} ||w||2 + C \sum_{i=1}^{n} \varepsilon i, i = 1, 2, \ldots \ldots, n$$

$$subject \; to \; y_i \left(w^T D.x1 + b \right) \geq 1 - \varepsilon 1$$

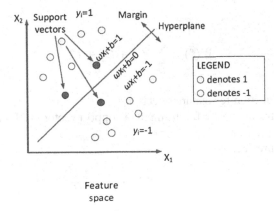

Fig. 1. Process block diagram of support vector machine.

$$\varepsilon_i \geq 0 \ i = 1, 2, \ldots\ldots, n \tag{11}$$

Here C is constant, n is number of observation and $\varepsilon 1$ is slack variable. The rule of decision function is

$$f(x) = \sum_{i=1}^{n} yi\alpha iK(xi, xj) + b \tag{12}$$

3 Proposed Methodology

The proposed digital image watermarking algorithm based on stationary wavelet transform, multi-objective particle swarm optimization and non-linear support vector machine (NSVM). The SWT transform applied for the extraction of feature components of host image and watermark image. The extraction of features components passes through the optimization of features of one objective function of particle swarm optimization. The other objective function of particle swarm optimization searches the position of P (xi, yj) to embedded the watermark. The non-linear support vector machine trains the watermark image for the generation of patterns and embedding with host image. The processing of algorithm describe here.

1. Let us consider I = g(I, j) and I' = g(I, j) is host image and watermark image
2. Applied SWT transform of both images (host, watermark)
3. Extraction of feature components as horizonal, diagonal and vertical as derived by equation of SWT transform as C_x^h, C_x^v, C_x^d
4. Compute the feature matrix of both images as

$$H(f) = \sum_{i=1}^{n} C_x^h + C_x^v + C_x^d) \tag{13}$$

$$W(f) = \sum_{i=1}^{n} C_x^h + C_x^v + C_x^d \qquad (14)$$

5. Call MPSO function for feature selection
6. Estimate the feature vector for the input of support vector machine as (ik, jk) where k = 1, 2, 3,.........,n

   Now feature vector

$$Fd = \sum_{m=1}^{n} H(f)XW(f) \qquad (15)$$

7. Train the feature vector for pattern generation

   Let $D_{kk=1,2,......,n}$ and Sk k = 1, 2 ,3 ,.......n be the vector of class Ok = { Fd(ik.jk) k = 1}.

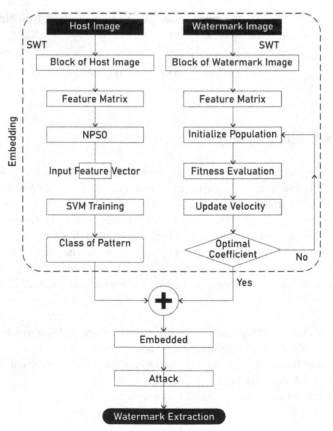

Fig. 2. Process model of proposed model of watermark embedding and extraction process with applied different attacks.

8. Call second objective function of MPSO by equation and search the position of I(xi, yj) for the position of watermark
9. Selection of sub-blocks with position with pattern (I, j)
10. Watermark embedding process is done
11. Applied geometrical attack
12. Extract watermark
13. Measure parameters
14. Process terminated
15. Exit (Fig 2)

4 Experimental Analysis

To evaluate the validation of proposed digital image watermarking algorithm based on support vector machine and SWT. The major concern regarding robustness and imperceptibility of digital watermark. All simulation process run over a personal computer I7 processor, 2.84 GHZ, 16 GB RAM and using MATLAB software. The used host image is Leena, Baboon, cameraman, Peppers. The symbol image shown in figure [18, 21, 22, 25]. The size of host image is 512×512 and symbol image 128×128. The following parameters are estimated.

$$\text{PSNR (peak to signal noise ratio)} = 10 \log 10 \frac{M \, X \, N}{\sum_{m=1}^{M} \sum_{n=1}^{N} \cdot [I(m,n) - I^*(m,n)]} \dots \dots \dots \quad (16)$$

Hera A and A* shows the host image and symbol image for watermarking M and N is size of matrix.

$$\text{Now } SSIM\,(I, I*) = \frac{(UAUI * + c1)(2\sigma II * + c2)}{(U_I^2 + U_{I*}^2 + c1)(\sigma_I^2 + \sigma_{I*}^2 + c2)} \quad (17)$$

Where UI and UI* is average of host image and symbol image respectively. σI and $\sigma I*$ is variation and error of host image and symbol image. C1 and C2 is MPSO update value.

$$NC(I, I*) = \frac{\sum_{m=1}^{M} \sum_{n=1}^{N} A(m, n)A * (m, n)}{\sqrt{\sum_{m=1}^{M} \sum_{n=1}^{N} [I(m, n)]^2} \sqrt{\sum_{m=1}^{M} \sum_{n=1}^{N} [I * (m, n)]^2}} \quad (18)$$

$$BER = \frac{R(Xi)}{\sum_{J=1}^{M} R|(Xj)} \quad (19)$$

Where R is index coefficient of watermark image.

The description of attacks applied on the process of watermark extraction. This attack deformed the process of digital image watermarking methods.

Attacks Abbreviations.

- Median filter attack-MFA
- Rescaling attack-RA
- Salt and pepper attack-SPNA
- Gaussian attack-GNA
- JPEG compression attack-JCA (Tables 1, 2, 3 and 4)

Table 1. The comparative analysis of SVM, DWT and Proposed techniques with image baboon host image using BER, NC and SSIM parameters.

Attacks	Svm [16]	Dwt [1]	Proposed	Parameter	Host Image: Baboon
MFA	0.395	0.436	0.392	BER	
RA	0.991	0.993	0.990		
SPNA	0.986	0.988	0.981		
GNA	0.994	0.995	0.990		
JCA	0.998	0.998	0.996		
MFA	0.004	0.0015	0.006	NC	
RA	0.009	0.0018	0.016		
SPNA	0.008	0.0058	0.032		Watermark Image
GNA	0.0007	0.0003	0.007		
JCA	0.045	0.009	0.078		
MFA	0.90	0.94	0.95	SSIM	
RA	0.94	0.98	0.99		
SPNA	0.54	0.51	0.67		
GNA	0.90	0.92	0.95		
JCA	0.95	0.96	0.98		

Table 2. The comparative analysis of SVM, DWT and Proposed techniques with image cameraman host image using BER, NC and SSIM parameters.

Attacks	Svm [16]	Dwt [1]	Proposed	PARAMET ER	Host Image: Cameraman
MFA	0.990	0.984	0.980	BER	
RA	0.995	0.996	0.991		
SPNA	0.994	0.995	0.993		
GNA	0.999	0.998	0.994		
JCA	0.541	0.667	0.519		
MFA	0.010	0.008	0.015	NC	
RA	0.0001	0.0002	0.0005		
SPNA	0.042	0.070	0.070		Watermark Image
GNA	0.015	0.021	0.153		
JCA	0.006	0.002	0.008		
MFA	0.57	0.59	0.63	SSIM	
RA	0.78	0.79	0.82		
SPNA	0.88	0.87	0.88		
GNA	0.86	0.84	0.91		
JCA	0.79	0.76	0.88		

Table 3. The comparative analysis of SVM, DWT and Proposed techniques with image leena host image using BER, NC and SSIM parameters.

Attacks	SVM [16]	DWT [1]	Proposed	Parameter	Host Image: Leena
MFA	0.59	0.53	0.49	BER	
RA	0.79	0.89	0.69		
SPNA	0.58	0.58	0.57		
GNA	0.79	0.76	0.75		
JCA	0.68	0.69	0.66		
MFA	0.005	0.006	0.008	NC	
RA	0.008	0.007	0.008		
SPNA	0.003	0.001	0.003		Watermark Image
GNA	0.004	0.004	0.005		
JCA	0.045	0.039	0.078		
MFA	0.80	0.84	0.88	SSIM	
RA	0.92	0.93	0.96		
SPNA	0.74	0.71	0.77		
GNA	0.58	0.50	0.58		
JCA	0.89	0.86	0.89		

Table 4. The comparative analysis of SVM, DWT and Proposed techniques with image pepper host image using BER, NC and SSIM parameters.

Attacks	SVM [16]	DWT [1]	Proposed	Parameter	Host Image: Pepper
MFA	0.88	0.89	0.78	BER	
RA	0.89	0.90	0.89		
SPNA	0.79	0.82	0.69		
GNA	0.82	0.87	0.79		
JCA	0.81	0.86	0.80		
MFA	0.016	0.010	0.018	NC	
RA	0.002	0.001	0.003		
SPNA	0.042	0.070	0.070		Watermark Image
GNA	0.022	0.012	0.053		
JCA	0.0016	0.0011	0.0018		
MFA	0.88	0.91	0.92	SSIM	
RA	0.93	0.94	0.95		
SPNA	0.88	0.88	0.91		
GNA	0.72	0.75	0.78		
JCA	0.63	0.68	0.69		

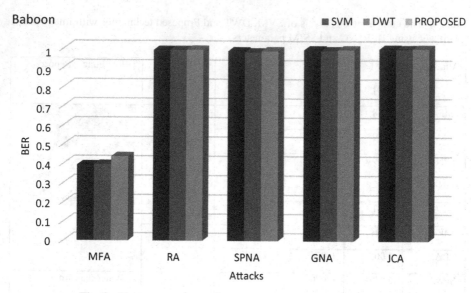

Fig. 3. The comparative performance of BER with baboon image.

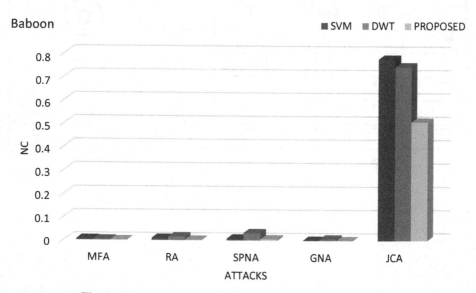

Fig. 4. The comparative performance of NC with baboon image.

Here using SVM [16], DWT [1] and proposed techniques with all given attacks Median filter attack, Rescaling attack, Salt and pepper attack, Gaussian attack, JPEG compression attack using baboon host image and flower symbol image. Here we observe the performance of Proposed technique shown better performance with the comparison of other two SVM and DWT techniques (Figs. 3 and 4).

Here using SVM [16], DWT [1] and proposed techniques with all given attacks Median filter attack, Rescaling attack, Salt and pepper attack, Gaussian attack, JPEG compression attack using baboon host image and flower symbol image. Here we observe the performance of Proposed technique shown better performance with the comparison of other two SVM and DWT techniques (Fig. 5).

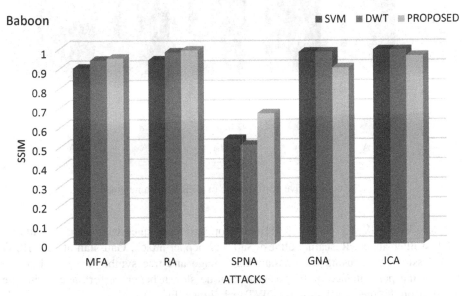

Fig. 5. The comparative performance of SSIM with baboon image.

Here using SVM [16], DWT [1] and proposed techniques with all given attacks Median filter attack, Rescaling attack, Salt and pepper attack, Gaussian attack, JPEG compression attack using baboon host image and flower symbol image. Here we observe the performance of Proposed technique shown better performance with the comparison of other two SVM and DWT techniques (Fig. 6).

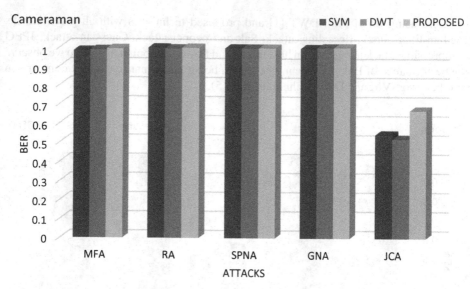

Fig. 6. The comparative performance of BER with cameraman image.

Here using SVM [16], DWT [1] and proposed techniques with all given attacks Median filter attack, Rescaling attack, Salt and pepper attack, Gaussian attack, JPEG compression attack using cameraman host image and tree symbol image. Here we observe the performance of Proposed technique shown better performance with the comparison of other two SVM and DWT techniques (Fig. 7).

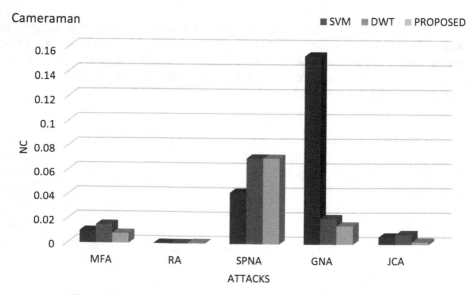

Fig. 7. The comparative performance of NC with cameraman image.

Here using SVM [16], DWT [1] and proposed techniques with all given attacks Median filter attack, Rescaling attack, Salt and pepper attack, Gaussian attack, JPEG compression attack using cameraman host image and tree symbol image. Here we observe the performance of Proposed technique shown better performance with the comparison of other two SVM and DWT techniques (Fig. 8).

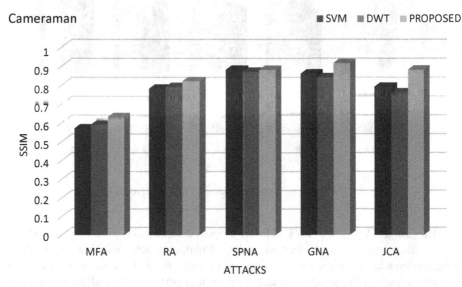

Fig. 8. The comparative performance of SSIM with cameraman image.

Here using SVM [16], DWT [1] and proposed techniques with all given attacks Median filter attack, Rescaling attack, Salt and pepper attack, Gaussian attack, JPEG compression attack using cameraman host image and tree symbol image. Here we observe the performance of Proposed technique shown better performance with the comparison of other two SVM and DWT techniques (Fig. 9).

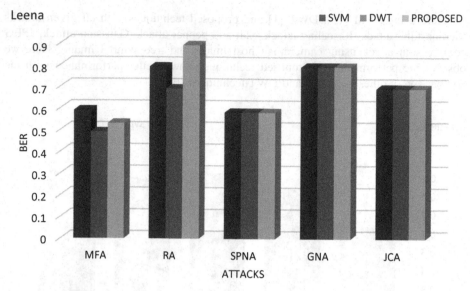

Fig. 9. The comparative performance of BER with leena image.

Here using SVM [16], DWT [1] and proposed techniques with all given attacks Median filter attack, Rescaling attack, Salt and pepper attack, Gaussian attack, JPEG compression attack using leena host image and infinity symbol image. Here we observe the performance of Proposed technique shown better performance with the comparison of other two SVM and DWT techniques (Fig. 10).

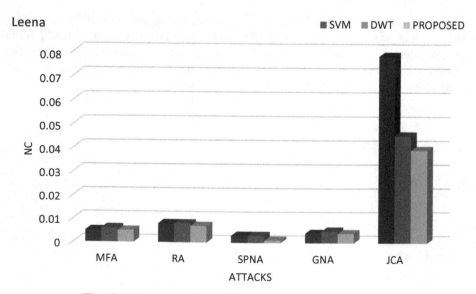

Fig. 10. The comparative performance of NC with leena image.

Here using SVM [16], DWT [1] and proposed techniques with all given attacks Median filter attack, Rescaling attack, Salt and pepper attack, Gaussian attack, JPEG compression attack using leena host image and infinity symbol image. Here we observe the performance of Proposed technique shown better performance with the comparison of other two SVM and DWT techniques (Fig. 11).

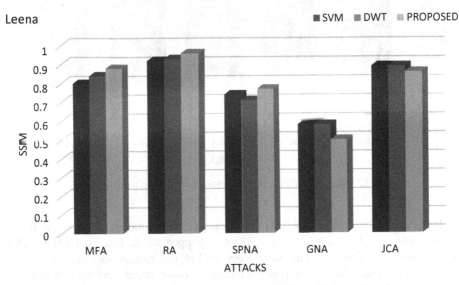

Fig. 11. The comparative performance of SSIM with leena image.

Here using SVM [16], DWT [1] and proposed techniques with all given attacks Median filter attack, Rescaling attack, Salt and pepper attack, Gaussian attack, JPEG compression attack using leena host image and infinity symbol image. Here we observe the performance of Proposed technique shown better performance with the comparison of other two SVM and DWT techniques (Fig. 12).

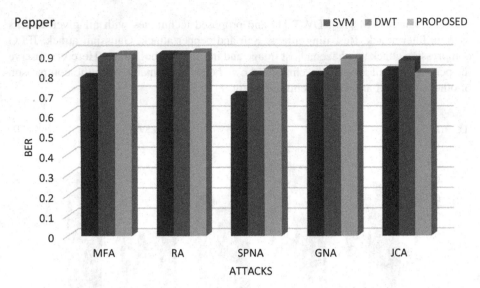

Fig. 12. The comparative performance of BER with pepper image.

Here using SVM [16], DWT [1] and proposed techniques with all given attacks Median filter attack, Rescaling attack, Salt and pepper attack, Gaussian attack, JPEG compression attack using pepper host image and at_the_rate symbol image. Here we observe the performance of Proposed technique shown better performance with the comparison of other two SVM and DWT techniques (Fig. 13).

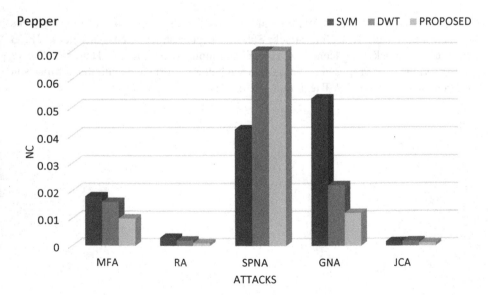

Fig. 13. The comparative performance of NC with pepper image.

Here using SVM [16], DWT [1] and proposed techniques with all given attacks Median filter attack, Rescaling attack, Salt and pepper attack, Gaussian attack, JPEG compression attack using pepper host image and at_the_rate symbol image. Here we observe the performance of Proposed technique shown better performance with the comparison of other two SVM and DWT techniques (Fig. 14).

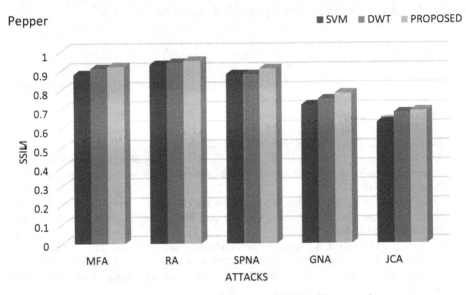

Fig. 14. The comparative performance of SSIM with pepper image.

Here using SVM [16], DWT [1] and proposed techniques with all given attacks Median filter attack, Rescaling attack, Salt and pepper attack, Gaussian attack, JPEG compression attack using pepper host image and at_the_rate symbol image. Here we observe the performance of Proposed technique shown better performance with the comparison of other two SVM and DWT techniques.

5   Conclusion and Future Work

This paper proposed robust digital image watermarking method based on support vector machine and stationary wavelet transform. The proposed algorithm derives from the non-linear support vector machine and multi-objective particle swarm optimization. The MPSO (multi-objective swarm optimization) algorithm define two objective function for the minimization and selection of feature components of feature input vector. The process of support vector machine learning model reduces the boundary value of feature vector. The proposed algorithm is very efficient in concern of imperceptibility and robustness factor. The increased value of NC, increase the robustness of digital image watermark. BER's minimized value shows that the processing of the transform function is fair in terms of scaling. The proposed algorithm compares with DWT and SVM algorithms. The art of analysis of results indicates that the proposed algorithm is efficient. The scaling factor controlled with MPSO position value updates. The controlled population of the position of watermark embedding mark as robust embedding. The strength of the algorithm measure with geometrical applied attack. The applied attack decreases the value of NC. However, in the case of the proposed algorithm, the value of NC is remained constant and show the strength of digital image watermarking algorithm. The iteration process of transform and MPSO increases the algorithm's time complexity, minimizing the number of iteration and increasing the efficiency of the algorithm.

References

1. Araghi, T.K., Manaf, A.: An enhanced hybrid image watermarking scheme for security of medical and non-medical images based on DWT and 2-D SVD. Future Gener. Comput. Syst. **101**, 1223–1246 (2019). https://doi.org/10.1016/j.future.2019.07.064
2. Jiao, S., Zhou, C., Shi, Y., Zou, W., Li, X.: Review on optical image hiding and watermarking techniques. Opt. Laser Technol. **109**, 370–380 (2019)
3. Xia, Z., Wang, X., Zhou, W., Li, R., Wang, C., Zhang, C.: Color medical image lossless watermarking using chaotic system and accurate quaternion polar harmonic transforms. Signal Process. **157**, 108–118 (2019)
4. Ernawan, F., Kabir, M.N.: A block-based RDWT-SVD image watermarking method using human visual system characteristics. Vis. Comput. **36**(1), 19–37 (2018). https://doi.org/10.1007/s00371-018-1567-x
5. Swaraja, K., Meenakshi, K., Kora, P.: An optimized blind dual medical image watermarking framework for tamper localization and content authentication in secured telemedicine. Biomed. Sig. Process. Control **55**, 101665 (2020). https://doi.org/10.1016/j.bspc.2019.101665
6. Chandan, A.K., Singh, P. K., Singh, R., Singh, S.: SPIHT-based multiple image watermarking in NSCT domain. Concurr. Comput.: Pract. Exper. **32**(1), e4912 (2020). https://doi.org/10.1002/cpe.4912
7. Kumar, C., Singh, A.K., Kumar, P.: Improved wavelet-based image watermarking through SPIHT. Multimedia Tools Appl. **79**(15–16), 11069–11082 (2018). https://doi.org/10.1007/s11042-018-6177-0
8. Sun, L., Jiucheng, X., Liu, S., Zhang, S., Li, Y., Chang'an, S.: A robust image watermarking scheme using Arnold transform and BP neural network. Neural Comput. Appl. **30**(8), 2425–2440 (2017). https://doi.org/10.1007/s00521-016-2788-4
9. Hosny, K.M., Darwish, M.M.: Robust color image watermarking using invariant quaternion Legendre-Fourier moments. Multimedia Tools Appl. **77**(19), 24727–24750 (2018). https://doi.org/10.1007/s11042-018-5670-9
10. Loan, N.A., Hurrah, N. , Parah, S.A., Lee, J.W., Sheikh, J.A., Mohiuddin, G.: Secure and robust digital image watermarking using coefficient differencing and chaotic encryption. IEEE Access **6**, 19876–19897 (2018). https://doi.org/10.1109/ACCESS.2018.2808172
11. Ernawan, F., Kabir, M. N.: A robust image watermarking technique with an optimal DCT-psychovisual threshold. IEEE Access **6**, 20464–20480 (2018). https://doi.org/10.1109/ACCESS.2018.2819424
12. Kumar, C., Singh, A.K., Kumar, P.: A recent survey on image watermarking techniques and its application in e-governance. Multimedia Tools Appl **77**(3), 3597–3622 (2017). https://doi.org/10.1007/s11042-017-5222-8
13. Wang, C., Zhang, Y., Zhou, X.: Robust image watermarking algorithm based on ASIFT against geometric attacks. Appl. Sci. **8**(3), 410 (2018)

14. Etemad, E., et al.: Robust image watermarking scheme using bit-plane of hadamard coefficients. Multimedia Tools Appl 77(2), 2033–2055 (2017). https://doi.org/10.1007/s11042-016-4278-1
15. Mehta, R., Gupta, K., Yadav, A.K.: An adaptive framework to image watermarking based on the twin support vector regression and genetic algorithm in lifting wavelet transform domain. Multimedia Tools Appl 79(25–26), 18657–18678 (2020). https://doi.org/10.1007/s11042-020-08634-x
16. Parmalik, K., Sharma, A.K.: A robust digital image watermarking technique against geometrical attacks using support vector machine and glowworm optimization. In: Jude , D., Shakya, S., Baig, Z. (eds.) ICICI 2019. LNDECT, vol. 38, pp. 733–747. Springer, Cham (2020). https://doi.org/10.1007/978-3-030-34080-3_82
17. Susanto, A., Christy, A.S., Eko, H.R.: Hybrid method using HWT-DCT for image watermarking. In: 2017 5th International Conference on Cyber and IT Service Management (CITSM), pp. 1–5. IEEE (2017)
18. Thakkar, F.N., Srivastava, V.K.: A blind medical image watermarking: DWT-SVD based robust and secure approach for telemedicine applications. Multimedia Tools Appl. 76(3), 3669–3697 (2016). https://doi.org/10.1007/s11042-016-3928-7
19. Muhammad, N., Bibi, N., Qasim, I., Jahangir, A., Mahmood, Z.: Digital watermarking using Hall property image decomposition method. Pattern Anal. Appl. 21(4), 997–1012 (2017). https://doi.org/10.1007/s10044-017-0613-z
20. Dey, N., et al.: Watermarking in biomedical signal processing. In: Nilanjan, V., Santhi (eds.) Intelligent techniques in signal processing for multimedia security. SCI, vol. 660, pp. 345–369. Springer, Cham (2017). https://doi.org/10.1007/978-3-319-44790-2_16
21. Abraham, J., Paul, V.: An imperceptible spatial domain color image watermarking scheme. J. King Saud Univ.-Comput. Inf. Sci. 31(1), 125–133 (2019)
22. Zhou, R.-G., Hu, W., Fan, P.: Quantum watermarking scheme through Arnold scrambling and LSB steganography. Quantum Inf. Process. 16(9), 1–21 (2017). https://doi.org/10.1007/s11128-017-1640-9
23. Thakur, S., Singh, A.K., Ghrera, S.P., Elhoseny, M.: Multi-layer security of medical data through watermarking and chaotic encryption for tele-health applications. Multimedia Tools Appl. 78(3), 3457–3470 (2018). https://doi.org/10.1007/s11042-018-6263-3
24. Sreenivas, K., Kamkshi, V.: Fragile watermarking schemes for image authentication: a survey. Int. J. Mach. Learn. Cybern. 9(7), 1193–1218 (2017). https://doi.org/10.1007/s13042-017-0641-4
25. Singh, S., Rathore, V.S., Singh, R.: Hybrid NSCT domain multiple watermarking for medical images. Multimedia Tools Appl. 76(3), 3557–3575 (2016). https://doi.org/10.1007/s11042-016-3885-1

Generalized Intuitionistic Fuzzy Entropy on IF-MARCOS Technique in Multi-criteria Decision Making

Rishikesh Chaurasiya[✉] and Divya Jain

Department of Mathematics, Jaypee University of Engineering and Technology,
Guna, Madhya Pradesh, India
divya.jain@juet.ac.in

Abstract. In the present era the use of mobile phones in doing an enormous progress. Discovering the best service providers on the bases of functioning constraint can assist to attain the objective of mobile facility distributor selection (MFDs). To solve such multi-criterion decision making problem, an Intuitionistic-Fuzzy Measurement Alternatives and Ranking according to the Compromise Solution (IF-MARCOS) technique is applied in the current paper to estimate the optimal selection. In solving the MCDM problems weight function plays a very vital role, here in this paper the weight function is evaluated on the bases of generalized intuitionistic fuzzy entropy measures and their corresponding operations is discussed. The results attained by the generalized technique is associated with the already prevailing ones, and it reveals that the generalized technique is in synchronization with the already existing ones.

Keywords: Intuitionistic fuzzy sets · IF-MARCOS · MFDs · MCDM

1 Introduction

Fuzzy sets (FS) have important applications in various fields of study Zadeh [1] The value of membership of an object in a FS is the value among zero and one, though, the value of nonmembership may not always be an entity in the fuzzy set, as it may be somewhat hesitant. Meanwhile, intuitionistic fuzzy sets, presented by Atanassov [2, 3] offers a capable framework to cope, along with the occurrence of haziness, Hesitant to start with information. Intuitionistic fuzzy sets are defined as relating two distinctive purposes exposing the value of membership and non-membership of objects of the space to IFS. In this set, each element is represented by $\mu + \vartheta < 1$. Since its evolution, IFS has been successfully applied by several researchers in various fields to solve decision-making problems.

Atanasov [2] studied their properties. Later, Atanasov [3] commenced operations at @ and $ IFS. De et al. [4] defined concentration, expansion, and generalization of IFS to deal with linguistic variables. Ejegwa and Akowe [5] presents some brief descriptions of IFSs, some definitions, processes, algebra and modal operators based on IFSs. For

© Springer Nature Switzerland AG 2021
M. Singh et al. (Eds.): ICACDS 2021, CCIS 1440, pp. 592–603, 2021.
https://doi.org/10.1007/978-3-030-81462-5_52

example, Xu [6], Xu and Yager [7] first proposed a weighted average and geometric aggregation operator. Subsequently, Liu and Tang [8] developed some novel Einstein operations of intuitionistic fuzzy set environment. Balaman and Cali [9] presented the ELECTRE and VIKOR method for solve the MCGDM. Mishra et al. [10] presented the combined SWARA-COPRAS method under the multi-criteria intuitionistic fuzzy entropy. Rani and Mishra [11] studied Information measures on the TOPSIS method. Kozae et al. [12] applied the Application to study the Corona COVID-19 for IFS. Also, Szmidt and Kacprzyk [13] presented distance measures between IFS.

In current times, IFSs predict with FS in handling real-life problems and uncertainties due to the rise difficulty of the socio-economic atmosphere. IFS has been broadly applied by researchers to investigate problems with uncertainties, one of them may mention to MCDM and MADM problems. Purpose of it is to appraise the best choice under a set of diverse objects. Weighted and ordered weighted aggregation operators for alternative options to accumulate all performance of criteria. So that, abundant MCDM methods such as, MARCOS, ARAS, TODIM, MABAC, are generalized under ambiguous scenario with various weight determining methods launch many authors.

In this paper, we generalized an entropy measure on IFS. Here in the current paper, an outstanding technique, intuitionistic-fuzzy MARCOS for MCDM is applied on the generalized entropy measure. As it is well known that in the multi-criteria decision making, the criterion weights evaluation is a vibrant matter for accuracy of estimation outcomes. In this procedure, it is calculated using the generalized entropy measure for IFSs to overwhelmed the inadequacy of the usual ones. As the vagueness is an inevitable trait of multi-criteria decision making, the generalized technique can be an esteemed weapon for policymaking in an ambiguous environment to calculate the order ranking. To signify the strength and rationality of the generalized technique in the everyday life MCDM problems, a MFDs problem is executed. Applying the generalized skeleton, a decision architect formulates a policy to improve the performance by altering the working parameters. Finally, a comparison between the results of the generalized technique and the existing ones in order to validation of the generalized technique is discussed. This study displays that the generalized technique is efficient and well reliable with the already existing techniques.

The present paper is arranged as follows: Sect. 2, depict the basic Fundamental and operators of IFSs are discussed. Next Sect. 3, we generalized the entropy measure of intuitionistic fuzzy sets. In Sect. 4, an MCDM problem of MFDs is solved using the intuitionistic fuzzy MARCOS technique and weights computed on the bases of generalized measure entropy. Thereafter, we compare generalized intuitionistic-fuzzy entropy with the already existing entropies in Sect. 5.

2 Preliminaries

The basic definitions and operations related to IFSs are given as:

Definitions 1. Zadeh [1] Let $T = \{t_1, t_2, \ldots, t_n\}$ be a fixed set, then the FSs $A \subset T$ is defined by:

$$A = \{(t_i, \ \mu_A(t_i))| \ t_i \in T\}.$$

Where $\mu_A(t_i) : T \rightarrow [0, 1]$ is membership function of A. $\mu_A(t_i)$ is the membership degrees.

Definitions 2. Atanassov [2] An intuitionistic fuzzy set $A \subset T$ in a fixed set let $T = \{t_1, t_2, \ldots, t_n\}$ is defined by.

$$A = \{(t_i, \ \mu_A(t_i), \ \vartheta_A(t_i) \ )|t_i \in T\}$$

Where, $\mu_A : T \rightarrow [0, 1]$ and $\vartheta_A : T \rightarrow [0, 1]$, are the MF and NMF, $t_i \in T$ to the set A, which is a subset of T, with the condition $0 \leq \mu_A(t_i) + \vartheta_A(t_i) \leq 1$. For IFS $A \subset T$, $\pi_A(t_i) = 1 - \mu_A(t_i) - \vartheta_A(t_i)$ is said to be the hesitancy degree of t_i in A. $\pi_A(t_i)$ is the indeterminacy degree of $\forall t_i \in T$ to the condition is defined as: $0 \leq \pi_A(t_i) \leq 1$.

Definitions 3. Xu et al. [6] Consider (μ_ξ, ϑ_ξ) is called two intuitionistic fuzzy number (IFN) is represented by $\xi = (\mu_\xi, \vartheta_\xi)$, such that $\mu_\xi, \vartheta_\xi \in [0, 1]$ and $0 \leq \mu_\xi + \vartheta_\xi \leq 1$.

Definitions 4. Xu et al. [6] Assume $\xi_j = (\mu_j, \vartheta_j)$, is called IFNs. Then, the score & accuracy function are given as:

$$\Psi(\xi_j) = (\mu_j - \vartheta_j), \hbar(\xi_j) = (\mu_j + \vartheta_j); \text{ For } \Psi(\xi_j) \in [-1, 1], \hbar(\xi_j) \in [0, 1].$$

Definitions 5. Xu et al. [14] Let $\xi_j = (\mu_j, \vartheta_j)$, is called IFN. certainly, the normalized score and accuracy values are given by.

$$\Psi^*(\xi_j) = \frac{1}{2}(\Psi(\xi_j) + 1), \hbar^*(\xi_j) = \frac{1}{2}(\mu_j + \vartheta_j) \text{ For } \Psi^*(\xi_j), \hbar^*(\xi_j) \in [0, 1]$$

Definitions 6. Xu et al. [7] Consider $\xi_j = (\mu_j, \vartheta_j)$, is called an intuitionistic fuzzy number (IFNs). The operator

$$IFWA_w(\xi_1, \xi_2, \ldots, \xi_n) = \left(1 - \prod_{j=1}^{n} (1 - \mu_j)^{w_j}, \prod_{j=1}^{n} (\vartheta_j)^{w_j}\right) \tag{1}$$

$$IFWG_w(\xi_1, \xi_2, \ldots, \xi_n) = \left(\prod_{j=1}^{n} (\mu_j)^{w_j}, 1 - \prod_{j=1}^{n} (1 - \vartheta_j)^{w_j}\right) \tag{2}$$

Where IFWA means intuitionistic fuzzy weighted average and IFWG- intuitionistic fuzzy weighted geometric.

Definition 7. Szmidt and Kacprzyk [13] A mapping E : IFS(T) $\rightarrow$ [0, 1] is said to be an entropy measure for IFSs, if it fulfills the postulates as given below:

(P$_1$) $E(A) = 0$, iff A is crisp set.
(P$_2$) $E(A) = 1$ iff $\mu_A(t_i) = \vartheta_A(t_i), \forall t_i \in T$
        $E(A) \leq E(B)$ if A subset of B

(P$_3$)   i.e.$\mu_A(t_i) \leq \mu_B(t_i)$ and $\vartheta_A(t_i) \geq \vartheta_B(t_i)$ for $\mu_B(t_i) \leq \vartheta_B(t_i)$
          or $\mu_A(t_i) \geq \mu_B(t_i)$ and $\vartheta_A(t_i) \leq \vartheta_B(t_i)$ for $\mu_B(t_i) \geq \vartheta_B(t_i)$,
    for each $t_i \in T$
(P$_4$) $E(A) = E(A^c)$.

3 Entropy Measure on Fuzzy and Intuitionistic-Fuzzy Sets

Harvda and Charvat [15] characterized entropy of a discrete probability distribution as:

$$H^\beta(A) = \frac{1}{2^{1-\beta}-1}\left[\sum_{i=1}^{n}\left(P_i^\beta\right) - 1\right]; \beta > 0, \beta \neq 1 \tag{3}$$

Corresponding to above entropy Kapur [16] proposed and characterized the following fuzzy information measure:

$$H^\beta(A) = \frac{1}{1-\beta}\sum_{i=1}^{n}\left[\mu_A^\beta(t_i) + (1 - \mu_A(t_i))^\beta - 1\right]; \beta > 0, \beta \neq 1 \tag{4}$$

Corresponding to Kapur entropy, we have generalized the following entropy measure for IFSs as follows:

$$E^\beta(A) = \frac{1}{n(2^{1-\beta}-1)}\sum_{i=1}^{n}\left[\left|\frac{\mu_A(t_i)+1-\vartheta_A(t_i)}{2}\right|^\beta + \left(1 - \left|\frac{\mu_A(t_i)+1-\vartheta_A(t_i)}{2}\right|\right)^\beta - 1\right] \tag{5}$$

Theorem 1. The function E(A) given in (5) be an entropy measure on IFS(T), satisfies the postulates [13]:

Proof. To prove fulfil the axioms (P_1)–(P_4), these theorem, Eq. (5).

P_1: Let A be a crisp set; $\mu_A(t_i) = 0, \vartheta_A(t_i) = 1$ or $\mu_A(t_i) = 1, \vartheta_A(t_i) = 0 \; \forall t_i \in T$, then $E(A) = 0$. Hence P_1 is satisfied.

P_2: For $\mu_A(t_i) = \vartheta_A(t_i); \forall t_i \in T$ it implies.

$$E(A) = \frac{1}{n(2^{1-\beta}-1)}\sum_{i=1}^{n}\left[\left(\frac{1}{2}\right)^\beta + \left(1 - \frac{1}{2}\right)^\beta - 1\right]$$

$$E(A) = \frac{1}{n(2^{(1-\beta)}-1)}n(2^{(1-\beta)} - 1); \; i.e., \; E(A) = 1$$

Conversely if $E(A) = 1$ then we have

$$E(A) = \frac{1}{n(2^{1-\beta}-1)}\sum_{i=1}^{n}\left[\frac{\mu_A(t_i)+1-\vartheta_A(t_i)}{2}\right.$$
$$\left. + \left(1 - \left|\frac{\mu_A(t_i)+1-\vartheta_A(t_i)}{2}\right|\right)^\beta - 1\right] = 1$$

For each $t_i \in T$, we get $\frac{\mu_A(t_i)+1-\vartheta_A(t_i)}{2} = \frac{1}{2} \Rightarrow \mu_A(t_i) = \vartheta_A(t_i)$.

P_3: Let $f(x,y) = \frac{1}{n(2^{1-\beta}-1)}\sum_{i=1}^{n}\left[\left|\frac{x+1-y}{2}\right|^\beta + \left(1 - \left|\frac{x+1-y}{2}\right|\right)^\beta - 1\right]$; where x, y belong to $[0, 1]$.

Assume $\frac{\partial f(x,y)}{\partial x} = 0$ and $\frac{\partial f(x,y)}{\partial y} = 0$. we find $x_{cp} = y$.

Then through Eq. (5) and critical point $x_{cp} = y$, $\frac{\partial f(x,y)}{\partial x} \geq 0$; for $x \leq y$ & $\frac{\partial f(x,y)}{\partial x} \leq 0$; for $x \geq y$.

Thus, $f(x, y)$ be an increases function for x when $x \leq y$ and decreases when $x \geq y$. Likewise, we obtain $\frac{\partial f(x,y)}{\partial y} \leq 0$, for $x \leq y$ and $\frac{\partial f(x,y)}{\partial y} \geq 0$, for $x \geq y$.

Conversely, Consider A and B be two IFSs then.

(i) $\mu_A(t_i) \leq \mu_B(t_i)$ and $\vartheta_A(t_i) \geq \vartheta_B(t_i)$ for $\mu_B(t_i) \leq \vartheta_B(t_i)$ for every $t_i \in T$ Implies that $\mu_A(t_i) \leq \mu_B(t_i) \leq \vartheta_B(t_i) \leq \vartheta_A(t_i)$ then

$$max(\mu_A(t_i), \mu_B(t_i)) = \mu_B(t_i); min(\vartheta_A(t_i), \vartheta_B(t_i)) = \vartheta_B(t_i)$$

i.e. $A \cup B = (\mu_B(t_i), \vartheta_B(t_i)) = B$

$$\Rightarrow E(A \cup B) = E(\mu_B(t_i), \vartheta_B(t_i)) = E(B)$$

Also, $\forall t_i \in T$; $min(\mu_A(t_i), \mu_B(t_i)) = \mu_A(t_i); max(\vartheta_A(t_i), \vartheta_B(t_i)) = \vartheta_A(t_i)$, It shows $A \cap B = (\mu_A(t_i), \vartheta_A(t_i)) = A$;

$$\Rightarrow E(A \cap B) = E(\mu_A(t_i), \vartheta_A(t_i)) = E(A)$$

That implies $E(A) \leq E(B)$.

(ii) $\mu_A(t_i) \geq \mu_B(t_i)$ and $\vartheta_A(t_i) \leq \vartheta_B(t_i)$ for $\mu_B(t_i) \geq \vartheta_B(t_i)$ for every $t_i \in T$, Implies that $\vartheta_A(t_i) \leq \vartheta_B(t_i) \leq \mu_B(t_i) \leq \mu_A(t_i)$ then

$$max(\mu_A(t_i), \mu_B(t_i)) = \mu_A(t_i); min(\vartheta_A(t_i), \vartheta_B(t_i)) = \vartheta_A(t_i)$$

i.e. $A \cup B = (\mu_A(t_i), \vartheta_A(t_i)) = A$;

$$\Rightarrow E(A \cup B) = E(\mu_A(t_i), \vartheta_A(t_i)) = E(A)$$

Again, $\forall t_i \in T$; $min(\mu_A(t_i), \mu_B(t_i)) = \mu_B(t_i); max(\vartheta_A(t_i), \vartheta_B(t_i)) = \vartheta_B(t_i)$ then $A \cap B = (\mu_B(t_i), \vartheta_B(t_i)) = B$;

$$\Rightarrow E(A \cap B) = E(\mu_B(t_i), \vartheta_B(t_i)) = E(B)$$

We obtain that, $E(B) \leq E(A)$.

P$_4$: It's apparent that (5) for $A^c = \{(t_i, \vartheta_A(t_i), \mu_A(t_i)); t_i \in T\}$; it yields $E(A) = E(A^c)$.

Hence, the generalized entropy measure E(A) is a valid entropy.

4   Extended Intuitionistic Fuzzy MARCOS Technique Based on Generalized Entropy Measure for MCDM

The MARCOS technique studies the association of alternatives and context values (ideal (AI) and anti-ideal (AAI) alternatives) Stankovic et al. [17]. The decision-making preferences is studied on the bases of utility function. The utility function is a location of an alternative regarding the AI and AAI answers Stevic et al. [18]. The best alternative is the one that is nearby to the ideal and far apart from the anti-ideal solution. A new intuitionistic fuzzy linguistic variable has been established with quantization in IFNs. Besides, an IF-MARCOS (Intuitionistic-Fuzzy Measurement Alternative and Ranking According to Compromise Solution [17]) is presented in the paper to calculate mobile facilities distributor, the optimal selection to define the grade of peril on them. The MARCOS technique is executed as:

Step1: Built initial decision matrix (DM) of IFS.

In the MCDM procedure, let $P = \{P_1, P_2, \ldots, P_n\}$ is the set of parameters and $S = \{S_1, S_2, \ldots, S_m\}$ be the set of alternatives, and DE's $E = \{E_1, E_2, E_3\}$ to achieve the best alternatives.

Step2: Calculate an initial decision experts'(DEs) weights and decision-making matrix. Consider that l DEs weight $\sum_{\tau=1}^{l} w_\tau = 1$. Based on Born et al. [21].

$$w_\tau = \frac{\mu_\tau + \pi_\tau \left(\frac{\mu_\tau}{\mu_\tau + \vartheta_\tau} \right)}{\sum_{\tau=1}^{l} \left(\mu_\tau + \pi_\tau \left(\frac{\mu_\tau}{\mu_\tau + \vartheta_\tau} \right) \right)}, \tau = 1(1)l \tag{6}$$

Step3: Compute the Aggregation DM corresponding to weights.

$$\varepsilon_{\tau j} = IFWA_w(\xi_1, \xi_2, \ldots, \xi_l) = \left(1 - \prod_{\tau=1}^{l} (1 - \mu_\tau)^{w_\tau}, \prod_{l-1}^{l} (\vartheta_\tau)^{w_\tau} \right) \tag{7}$$

Step4: Calculate the weight on entropy-based formula:

$$\omega_j = \frac{1 - e_j}{\sum_{j=1}^{n} (1 - e_j)} \tag{8}$$

where e_j represents the entropy of the criterion $P_j (j = 1, 2 \ldots, n)$, calculated by Eq. (5).

Step5: Calculate the weight matrix. Magnification is done by finding the product of normalized matrix values and corresponding weights.

$$v_{ij} = \varepsilon_{\tau j} \times \omega_j \tag{9}$$

Step6: Compute an expanded initial intuitionistic-fuzzy DM by evaluating the fuzzy AI and fuzzy AAI solution.

$$AI = \max_j v_{ij} \ if \ j \in B_b \ \& \ AI = \min_j v_{ij} \ if \ j \in C_c \tag{10}$$

$$AAI = \min_j v_{ij} \ if \ j \in B_b \ \& \ AAI = \max_j v_{ij} \ if \ j \in C_c \tag{11}$$

here B_b is the maximum criteria, and C_c is the minimum criteria.

Step7: The evaluate of utility degree of the alternatives κ_i as follows:

$$\kappa_i^+ = \frac{S_i}{Sai} \tag{12}$$

$$\kappa_i^- = \frac{S_i}{Saai} \tag{13}$$

where $S_i = \sum_{i=1}^{n} v_{ij}$ is total sum of weighted matrix.

Step8: Evaluate the utility function of the alternatives $f(\kappa_i); f(\kappa_i^+)$ and $f(\kappa_i^-)$ AI and AAI solution as:

$$f(k_i) = \frac{\kappa_i^+ + \kappa_i^-}{1 + \frac{1 - f(\kappa_i^+)}{f(\kappa_i^+)} + \frac{1 - f(\kappa_i^-)}{f(\kappa_i^-)}} \tag{14}$$

$$f\left(\kappa_i^+\right) = \frac{\kappa_i^-}{\kappa_i^+ + \kappa_i^-} \qquad (15)$$

$$f\left(\kappa_i^-\right) = \frac{\kappa_i^+}{\kappa_i^+ + \kappa_i^-} \qquad (16)$$

Step 9: Rank the alternatives.

Analysis: (Adopted from Mishra [20]) These operative parameters are considered as the criteria for the appraisal of mobile facility distributors. All these constraints are significant for the presentation of facilities, but the changes in the behavior of different operators generates difficulty in analysis. The MFDs valuation process is as follows (Table 1):

Table 1. Operative parameters of MFDs in Madhya Pradesh and Chhattisgarh circle.

Parameters	Benchmark	Audit period (2017)	Airtel	BSNL	Idea	Vodafone	Jio
Network accessibility BTS downtime	≤ 2%	Jan	0.17	1.53	0.07	0.02	0.08
		Feb	0.17	1.69	0.08	0.02	0.06
		Mar	0.10	1.63	0.14	0.03	0.11
Connection availability Call success rate	≥ 95%	Jan	99.87	98.28	99.67	99.61	99.94
		Feb	99.85	96.68	99.75	99.64	99.92
		Mar	99.87	98.35	99.76	99.70	99.95
RRC crowding	≤ 1%	Jan	0.01	0.55	0.05	0.02	0.05
		Feb	0.01	0.062	0.05	0.02	0.04
		Mar	0.01	0.62	0.01	0.02	0.03
RAB crowding	≤ 2%	Jan	0.02	0.31	0.10	0.13	0.00
		Feb	0.02	0.31	0.03	0.16	0.00
		Mar	0.02	0.40	0.01	0.11	0.00
Connection retainability Call drop rate	≤ 2%	Jan	0.25	0.85	0.49	0.33	0.04
		Feb	0.26	0.93	0.47	0.30	0.04
		Mar	0.28	0.70	0.47	0.23	0.03
Connection with better sound quality	≥ 95%	Jan	99.37	NA	99.15	99.85	99.89
		Feb	99.39	NA	99.12	99.84	99.90
		Mar	99.42	NA	98.58	99.85	99.90

Source TRAI Rouyendegh [19] Report for Madhya Pradesh incorporating Chhattisgarh by Bhopal office

Step 1: Creation of an initial decision matrix of IFS.

Step 2: Calculate DE's weights Using Eq. (6) and the (Table 2), the weights of the DEs are evaluated and presented in (Table 4) (Tables 3 and 5).

Table 2. Linguistic variables of criteria and DEs.

Linguistic variables	IFNs
Very significant (VS)	(0.90, 0.10)
Significant (S)	(0.75, 0.20)
Moderate (M)	(0.50, 0.40)
Insignificant (IS)	(0.40, 0.60)
Very insignificant (VI)	(0.10, 0.00)

Table 3. Linguistic variables for rating the alternatives.

Linguistic variables	IFNs
Excessively high (EH)	(1.00, 0.00)
Very high (VH)	(0.90, 0.10)
High (HG)	(0.70, 0.20)
Average (AV)	(0.60, 0.30)
Low (LW)	(0.40, 0.50)
Very low (VL)	(0.20, 0.70)
Excessively low (EL)	(0.10, 0.80)

Table 4. The significance of DM and given weights

	E_1	E_2	E_3
Linguistic variables	Very significant	Significant	Moderate
IFNs	(0.90, 0.1)	(0.75, 0.20)	(0.50, 0.40)
Weight	0.3981	0.3492	0.2527

Table 5. Linguistic variables for rating the MFDs.

Alternative	Experts	P_1	P_2	P_3	P_4	P_5	P_6
S_1	E_1	HG	VH	VH	HG	VH	AV
	E_2	VH	HG	VH	AV	HG	HG
	E_3	AV	VH	VH	VH	VH	AV

(continued)

Table 5. (*continued*)

Alternative	Experts	P₁	P₂	P₃	P₄	P₅	P₆
S₂	E₁	HG	VH	HG	VH	VH	HG
	E₂	HG	AV	AV	HG	HG	HG
	E₃	VH	VH	VH	VH	EH	HG
S₃	E₁	AV	AV	AV	HG	AV	AV
	E₂	HG	AV	AV	AV	HG	HG
	E₃	HG	AV	HG	HG	HG	HG
S₄	E₁	AV	HG	HG	HG	HG	AV
	E₂	HG	HG	HG	HG	HG	HG
	E₃	VH	VH	VH	VH	VH	HG
S₅	E₁	AV	HG	AV	AV	HG	AV
	E₂	HG	HG	AV	AV	AV	AV
	E₃	AV	LW	AV	LW	AV	AV

Table 6. Decision matrix of MFDs in Aggregated intuitionistic fuzzy.

	P₁	P₂	P₃	P₄	P₅	P₆
S1	0.7802, 0.1739	0.8532, 0.1274	0.9000, 0.1000	0.7487, 0.1934	0.8532, 0.1274	0.6382, 0.2604
S2	0.7720, 0.1679	0.8377, 0.1468	0.7487, 0.1934	0.8532, 0.1274	1.000, 0.0000	0.7000, 0.2000
S3	0.6636, 0.2350	0.6000, 0.3000	0.6280, 0.2709	0.6683, 0.2304	0.6636, 0.2350	0.6636, 0.2350
S4	0.7451, 0.1973	0.7727, 0.1679	0.7727, 0.1679	0.7727, 0.1679	0.7727, 0.1679	0.6636, 0.2350
S5	0.6382, 0.2603	0.6426, 0.2521	0.6000, 0.3000	0.5568, 0.3413	0.6433, 0.2553	0.6000, 0.3000

Step 3: Evaluate aggregate DM as shown in (Table 6).

Step 4: Compute criterion weights on generalized entropy measure by Eqs. (5) and (8), is given as:

$$\omega = (0.1309, 0.1638, 0.1582, 0.1437, 0.3251, 0.0783)^T$$

Step 5: Calculate the weight matrix. Aggregation is performed by product of normalized of the matrix Eq. (9) using by IF-MARCOS technique shown below (Table 7).

Table 7. The weighted normalized decision matrix.

ν	P_1	P_2	P_3	P_4	P_5	P_6
S_1	0.1021, 0.0228	0.1398, 0.0209	0.1424, 0.0158	0.1076, 0.0278	0.2774, 0.0414	0.0499, 0.0204
S_2	0.1011, 0.0219	0.1372, 0.0240	0.1184, 0.0306	0.1226, 0.0183	0.3251, 0.0000	0.0548, 0.0157
S_3	0.0869, 0.0308	0.0983, 0.0491	0.0993, 0.0429	0.0960, 0.0331	0.2157, 0.0764	0.0519, 0.0184
S_4	0.0975, 0.0258	0.1266, 0.0275	0.1222, 0.0266	0.1110, 0.0241	0.2512, 0.0546	0.0519, 0.0184
S_5	0.0835, 0.0341	0.1053, 0.0413	0.0949, 0.0475	0.0800, 0.0490	0.2091, 0.0829	0.0469, 0.0235

Sum of the alternatives of each criterion.
{(0.8192, 0.1491), (0.8592, 0.1105), (0.6481, 0.2507), (0.7604, 0.1770), (0.6197, 0.2783)}.
Step 6: Table 8 represents the AI and AAI solution.

Table 8. Entropy measure from each performance to the AI and AAI solution.

Ideal	0.1021, 0.0228	0.1398, 0.0209	0.1424, 0.0158	0.1226, 0.0183	0.3251, 0.0000	0.0548, 0.0157
Anti-ideal	0.0835, 0.0341	0.0983, 0.0491	0.0949, 0.0475	0.0800, 0.0490	0.2091, 0.0829	0.0469, 0.0235

Sum (AI) = [0.8868, 0.0935] and Sum (AAI) = [0.6127, 0.2861].
Step7: The utility degree $f\left(\kappa_i^+\right)$ and $f\left(\kappa_i^-\right)$ of the alternatives are evaluated from the Eqs. (12) and (13), using IF-MARCOS technique on Eq. (14) $f(\kappa_i)$ is computed as shown in the (Table 9).

Table 9. Results of IF-MARCOS technique.

	κ_i^+	κ_i^-	$f\left(\kappa_i^+\right)$	$f\left(\kappa_i^-\right)$	$f(\kappa_i)$	Ranking
S_1	0.9878	1.0773	0.52167	0.47833	0.68664	2
S_2	0.9892	1.0789	0.52169	0.47831	0.68764	1
S_3	0.9169	1.0000	0.52168	0.47832	0.63737	4
S_4	0.9562	1.0429	0.52169	0.47832	0.66471	3
S_5	0.9160	0.9991	0.52170	0.47830	0.63677	5

Step8: Determine the ranking of the MFDs. As can see from (Table 9), S_2 is the highest degree of in these entropies our calculation by IF-MARCOS method. Also, a comparative analysis in the above (Table 10).

Table 10. The comparative study with existing approaches.

Technique	Benchmark	Expert's weight		Criterion weight		Ranking	Optimal MFDs
		Considered	Calculated	Considered	Calculated		
Kumar [22]	AHP & TOPSIS	Yes	–	Yes	–	$S_2 \succ S_1 \succ S_3 \succ S_5 \succ S_4$	S_2
Kumar et al. [23]	fuzzy AHP/DEA	Yes	–	Yes	Yes	$S_1 \succ S_3 \succ S_2 \succ S_4 \succ S_5$	S_1
Kumar et al. [24]	Fuzzy AHP/ ELECTRE	Yes	–	Yes	–	$S_1 \succ S_2 \succ S_3 \approx S_5 \succ S_4$	S_1,S_2
Rouyendegh [19]	AHP & IF-ELECTRE	Yes	Yes	Yes	–	$S_1 \approx S_2 \succ S_3 \approx S_4 \succ S_5$	S_1
Mishra et. al. [20]	IF-ELECTRE	Yes	Yes	Yes	Yes	$S_1 \succ S_2 \succ S_3 \succ S_4 \succ S_5$	S_1
Generalized method	IF- MARCOS	Yes	Yes	Yes	Yes	$S_2 \succ S_1 \succ S_4 \succ S_3 \succ S_5$	S_2

5 Conclusions

Here, in this paper generalized entropy on intuitionistic-fuzzy sets is developed. Here in this paper MFDs problem taken belike a multi-criteria decision making to attain competitiveness. In multi-criteria decision making, vagueness plays a major role in decision forming. IFSs are able to characteristically handling the inherent uncertainties of the day-to-day life problems. Therefore, an outstanding technique for intuitionistic fuzzy sets, as IF-MARCOS, the operations on intuitionistic fuzzy numbers to handle with multi-criteria decision-making problems have been discussed, based on entropy measure. Some alterations have executed in computing the compliance, dominance values and rumpus of this technique. To validate the technique comparison between the generalized technique also existing ones has been discussed. The primary advantage of the evolved technique is easiness of calculation in IFSs.

References

1. Zadeh, L.A.: Fuzzy sets, Inform. Control **8**, 338–353 (1965)
2. Atanassov, K.T.: Intuitionistic fuzzy sets. Fuzzy Sets Syst. **20**, 87–96 (1986)
3. Atanassov, K.T.: New operations defined over intuitionistic fuzzy sets. Fuzzy Sets Syst. **61**, 137–142 (1994)
4. De, S.K., Biswas, R., Roy, A.R.: Some operations on intuitionistic fuzzy sets. Fuzzy Sets Syst. **114**, 477–484 (2000)
5. Ejegwa, P.A., Akowe, S.O., Otene, P.M.: An overview on intuitionistic fuzzy sets. Int. J. Sci. Technol. Res. **3**, 142–145 (2014)
6. Xu, Z.S.: Methods for aggregating interval-valued intuitionistic fuzzy information and their application to decision making. J. Control. Decis. **22**, 215–219 (2007)
7. Xu, Z., Yager, R.R.: Some geometric aggregation operators based on intuitionistic fuzzy sets. Int. J. Gen. Syst. **35**, 417–433 (2006)

8. Liu, P., Tang, G.: some intuitionistic fuzzy prioritized interactive Einstein Choquet operators and their application in decision making. IEEE Access **6**, 72357–72371 (2018)
9. Cali, S., Balaman, S.Y.: A novel outranking based multi criteria group decision making methodology integrating ELECTRE and VIKOR under intuitionistic fuzzy environment. Expert Syst. Appl. **119**, 36–50 (2019)
10. Mishra, A.R., Rani, P., Pandey, K.: Novel multi-criteria intuitionistic fuzzy SWARA-COPRAS approach for sustainability evaluation of the bioenergy production process. Sustainability **12**, 4155 (2020)
11. Mishra, A.R., Rani, P.: Information measures based TOPSIS method for multicriteria decision making problem in intuitionistic fuzzy environment. Iran. J. Fuzzy Syst. **14**, 41–63 (2017)
12. Kozae, A.M., Shokry, M., Omran, M.: Intuitionistic fuzzy set and its application in corona covid-19. Appl. Comput. Math. **9**, 146–154 (2020)
13. Szmidt, E., Kacprzyk, J.: Entropy of intuitionistic fuzzy sets. Fuzzy Sets Syst. **118**, 467–477 (2001)
14. Xu, G.L., Wan, S.P., Xie, X.L.: A selection method based on MAGDM with interval-valued intuitionistic fuzzy sets. Math. Probl. Eng. 1–13 (2015)
15. Havrda, J., Charvát, F.: Quantification method of classification processes. Concept of structural $ a $-entropy. Kybernetika **3**, 30–35(1967)
16. Kapur, J.N.: Measures of fuzzy information. Mathematical Sciences Trust Society (1997)
17. Stankovic, M., Das, D.K., Stevic, Z.: A new fuzzy MARCOS method for road traffic risk analysis. Mathematics **8**, 457 (2020)
18. Stevic, Z., Pamucar, D., Puska, A.: Sustainable supplier selection in healthcare industries using a new MCDM method: measurement of alternatives and ranking according to Compromise solution (MARCOS). Comput. Ind. Eng. **140**, 106231 (2020)
19. Rouyendegh, B.D.: The intuitionistic fuzzy ELECTRE model. Int. J. Manag. Sci. Eng. Manag. **13**, 139–145 (2018)
20. Mishra, A.R., Singh, R.K., Motwani, D.: Intuitionistic fuzzy divergence measure based ELECTRE method for performance of cellular mobile telephone service providers. Neural. Comput. Appl. **32**, 3901–3921 (2020)
21. Born, F.E., Genc, S., Kurt, M., Akay, D.: A multi-criteria intuitionistic fuzzy group decision making for supplier selection with TOPSIS method. Expert Syst. Appl. **36**, 11363–11368 (2009)
22. Kumar, S., Kumar, Y.S.: Evaluation of comparative performance of telecom service providers in India using TOPSIS and AHP. Int. J. Bus. Excell. **6**, 192–213 (2013)
23. Kumar, A., Shankar, R., Debnath, R.M.: Analyzing customer preference and measuring relative efficiency in telecom sector: A hybrid fuzzy AHP/DEA study. Telemat. Inform. **3**, 447–462 (2015)
24. Kumar, P., Singh, R.K., Kharab, K.: A comparative analysis of operational performance of Cellular Mobile Telephone Service Providers in the Delhi working area using an approach of fuzzy ELECTRE. App. Soft Comput. **59**, 438–447 (2017)

Feature Selection in Machine Learning by Hybrid Sine Cosine Metaheuristics

Nebojsa Bacanin[✉][iD], Aleksandar Petrovic[iD], Miodrag Zivkovic[iD], Timea Bezdan[iD], and Milos Antonijevic[iD]

Singidunum University, Danijelova 32, 11000 Belgrade, Serbia
{nbacanin,mzivkovic,tbezdan,mantonijevic}@singidunum.ac.rs
aleksandar.petrovic.17@singimail.rs

Abstract. Feature selection problem from the domain of machine learning refers to selecting only those features from the high dimensional datasets, that have prominent influence on dependent variable(s). In this way, dataset dimensionallity is reduced and only the riches data is kept, training process of machine learning model becomes more efficient and accuracy is increased. This manuscript proposes a new hybridized version of the sine cosine algorithm adjusting for solving feature selection problem. Hybridization is relatively novel approach for combing and improving metaheuristics optimizer. Notwithstanding that the basic sine cosine algorithm establishes good performance for solving NP hard challenges, based on simulation results, it was concluded that there is still space for improvement in its exploitation process. Original sine cosine algorithm and proposed hybridized implementation were tested on a well-known 21 machine learning datasets retrieved from the UCL repository. Comparative analysis between hybrid sine cosine and original one, as well as with 10 other state-of-the-art metaheuristics was conducted. Established results in terms of classification accuracy and fitness prove the robustness and efficiency of proposed method for solving this type of NP hard challenge.

Keywords: Machine learning · Swarm intelligence · NP hardness · Feature selection · Sine and cosine algorithm · Optimization

1 Introduction

In the real-world, machine learning models are used to find useful patterns in datasets with large dimensions. However, most of dimensions (features) do not have significant influence on the dependent variable(s). That is why the dimensionallty reduction process, that shrinks the dimensions of a datasets is very important in a data pre-processing task in machine learning.

Based on the literature, there are two types of dimension reduction: feature extraction and feature selection [12]. Some of the most well-known methods used in feature extraction include principal component analysis (PCA) and linear

© Springer Nature Switzerland AG 2021
M. Singh et al. (Eds.): ICACDS 2021, CCIS 1440, pp. 604–616, 2021.
https://doi.org/10.1007/978-3-030-81462-5_53

discriminant analysis (LDA). At the other side, feature selection does not tweak datasets, but select only those features for the training that are the most relevant. In a search space with large number of dimensions, feature selection is NP hard challenge. It is known that metaheuristics methods are very successful in tackling such problems and many hybrid approaches between metaheuristics and machine learning models for feature selection can be found in the literature survey. Some authors use term for such hybrid approaches learnheuristics, such as L. Calvet et al. [10].

The goal of feature selection is to make a decision which features in high-dimensional datasets to keep for the training process. By using feature selection in the data pre-processing phase, increased accuracy of machine learning models are noted due to the quality of data. The richness of the data, can be measured by the usability of the data in question. The size of the data chunk, as well as the time elapsed on its processing, are all taken into consideration when calculating the quality of data

According to the modern computer science literature, there are three ways of performing feature selection: wrapper method, filter method, and the embedded method, as stated by G. Chandrashekar et al. [11]. The wrapper methods employ a predictive model that assigns scores to features subsets. The scores of that subset are given by the count of the numbers of mistakes made on a hold-out set which is used to be tested on by each new subset used to train the model. Even though being the most expensive to execute, these wrapper methods proved the best performing.

Unlike the wrapper method, the filter method does not use errors to measure the score of the feature subset. Instead, it employs a fast to compute measure without losing out on the quality. The common examples of these measures are the point-wise mutual information, including the mutual information, relief-based algorithms, Pearson product-moment correlation coefficient, and inter/intra class distance or the scores of significance tests for each class/feature combination. In comparison to the previous method, filter method is less hardware demanding, but on the other hand, the production of a feature set cannot be fine-tuned. The consequences of these facts are that this method is more general when put to the previous one and that means it often lags, performance-wise, against the previous method.

Lastly, the embedded methods use feature selection as a part of the model construction process, which makes them the most general one. For example, the LASSO method constructs a linear model. This model uses the penalization of the regression coefficients. Most of them equal to zero after the L1 penalty has been applied. Every non-zero feature gets selected. There is a way to employ the LASSO to the nonlinear problems by extending it with AEFS. In either case regarding the linearity of the problems, the embedded methods are between the filters and wrappers considering the computational complexity.

In the research proposed in this manuscript, relatively new swarm intelligence metaheuristics, sine cosine algorithm (SCA) was improved by hybridization and adopted for solving feature selection problem in machine learning as a wrapper method. Exploitation of the basic SCA is improved by incorporating efficient intensification search procedure from the firefly algorithm (FA) method.

Basic goal of proposed research is to try to further enhance solving feature selection issue in machine learning with hybridized SCA metaheuristics by obtaining better performance metrics than other state-of-the-art methods that were tested on the same datasets.

The paper is organized in the following way. Section 2 gives a short overview of swarm intelligence and its application in different domains. Section 3 provides insight into the basic sine cosine algorithm and the proposed enhanced variant. Section 4 describes the conducted experiments and presents the results. Finally, Sect. 5 gives final remarks and conclude the paper.

2   Related Work

Regarding the metaheuristic approach to the problems of optimization, swarm intelligence, as a field is, in which the solutions are modeled on natural processes. Moreover, group thinking has been observed among the collectives of animals that is greater than the intelligence of each individual. The most common causes of this type of behavior are observed in the groups of animals while looking for food, fleeing from predators, or mating rituals in some species. With this approach, the scientists were able to utilize the social behavior of groups of bees as stated by Karaboga et al. [13], small bats as given in Yang [24].

Applications of swarm intelligence encompass numerous implementations for solving real-world problems that are very diverse and range from energy-saving process, clustering and localization in wireless sensor networks as given by M. Zivkovic et al. [3,25,27], through cloud systems scheduling stated by T. Bezdan, N. Bacanin and I. Strumberger [2,7,18], optimization of artificial neural networks as proposed by N. Bacanin et al. [1,5,16,19], machine learning used by M. Zivkovic et al. in [26], to solving complex problems in MRI classification in the medicine T. Bezdan et al. [6], as well as many others [8,20].

In the literature survey can be seen that the swarm algorithms have successfully been applied for feature selection in machine learning, such as L. Brezocnik et al. [9]. For example, in B. Xue et al. [22], an improved particle swarm optimization (PSO) with novel initialization mechanism and solutions' updated process was validated against 20 datasets and obtained promising achievements. One more representative example include cooperative swarm intelligence approach based on the FA and PSO metaheuristics that also established outstanding results on a well-known UCL datasets, as stated in D. Zouache et al. [28].

3   Proposed Hybrid SCA Metaheuristics

The SCA s based around the mathematical model of sine and cosine functions. The way that the algorithm exploits them is by constantly keeping them fluctuating towards and outwards the best solution. Multiple initial random candidate solutions are generated as the first steps of the algorithm. In addition, several other random and adaptive variables are used in the algorithm. Their role is to emphasize exploration and exploitation during different milestones. This method is introduced in 2016 by Mirjalli [15] and it has been proven that SCA is capable of being highly effective in the field of constrained and unknown search space problems.

Considering that the SCA generates a set of random variables, it is natural that it is beneficial for it to have high exploration and to avoid local optima. These processes are controlled by the returns of the since and cosine function which are between −1 and +1, which allows this algorithm to transit from one end to the other. The best solution is stored in the destination point throughout the execution of the algorithm.

The position updating equations are as follows:

$$X_j^{t+1} = X_j^t + r_1 \cdot \sin(r_2) \cdot |r_3 P_j^t - X_j^t| \tag{1}$$

$$X_i^{t+1} = X_j^t + r_1 \cdot \cos(r_2) \cdot |r_3 P_j^t - X_j^t| \tag{2}$$

The symbols bear meaning as follows: X_j^t - the position of the current solution in the j-th dimension at t-th iteration, r_1, r_2 and r_3 are random numbers from the interval $[0, 1]$, and the P_j - the position of the destination point in the j-th dimension.

These two equations are used in combination as:

$$X_j^{t+1} = \begin{cases} X_j^{t+1} = X_j^t + r_1 \cdot \sin(r_2) \cdot |r_3 P_j^t - X_j^t|, & r_4 < 0.5 \\ X_i^{t+1} = X_j^t + r_1 \cdot \cos(r_2) \cdot |r_3 P_j^t - X_j^t|, & r_4 \geq 0.5, \end{cases} \tag{3}$$

where r_4 is one more psuedo-random number between 0 and 1.

As can be seen from the presented experssions, the algorithm is controlled by random parameters and all four of them are regarding the destination or the position. In this way, the search process is able to balance between the solutions for efficient convergence towards the global optimum. To achieve this, the range of the based functions is changed ad-hoc. The guarantee of the exploitation comes from the ability to re-position around the solution thanks to the cyclic patterns of the main functions.

For successfully carrying out this process it is needed to be allowed for the algorithm to search outside the space between their corresponding destinations. This is accomplished by changing the sine and cosine functions' ranges. The SCA needs for the solution to have its position adjusted outside the overlapping spaces with other solutions. The random number r_2 is responsible for providing more randomness and its value is between 0 and 2Π, which guarantees the exploration. In order for the algorithm to have balance and control between the exploration and exploitation, an equation is used:

$$r_1 = a - t\frac{a}{T}, \tag{4}$$

where T represents the maximum number of iterations in a run and a is constant.

For the purpose of this research, simulations with the original SCA on standard bound-constrained benchmarks, taken from the Congress on Evolutionary Computation 2006 (CEC 2006) benchmark suite [14], were performed and it was concluded that the algorithm establishes good exploration with relatively fast convergence speed. However, the convergence speed that depends on the efficiency of intensification can be further enhanced.

At the other side, founded upon previous research, it was concluded that the search procedure of FA metaheuristics exhibits good convergence [4,21]. Logic assumption is that by employing the FA search procedure, the convergence speed of the SCA can be enhanced.

The FA search procedure is given in Eq. (5). It exhibits strong exploitation around the current solutions by using parameters α, β_0 and γ. More info regarding the FA can be found in [23].

$$x_i^{t+1} = x_i^t + \beta_0 \cdot e^{-\gamma r_{i,j}^2}(x_j^t - x_i^t) + \alpha^t(\kappa - 0.5) \qquad (5)$$

where pseudo-random number drawn from uniform or Gaussian distribution is denoted as $kappa$, distance between solutions x_i^t and x_i^j at iteration t is represented as $r_{i,j}$ and x_i^{t+1} is new position of solution i for the next iteration $(t+1)$. The distance $r_{i,j}$ between solutions x_i and x_j is obtained based on Cartesian distance [23]. Finally, control parameters α, β_0 and γ are randomization, attractiveness at distance $r = 0$ and light absorption coefficient, respectively [23].

Proposed hybrid SCA method in a relatively simple way incorporates the FA's search equation: when iteration number is even $t\% == 0$, standard SCA search procedure is conduction, and when the iteration number is odd, the FA perform exploitation of the search space.

Proposed hybrid method is named SCA with firefly search (SCA-FS) and workings of this method are summarized in Algorithm 1.

Algorithm 1. Pseudo-code of proposed SCA-FS

Initialization. Generate random population of N individuals X within the boundaries of the search space and calculate its fitness
  do
      Update position of the best solution so far $(P = X^*)$
      if t is even **then**
          Update r_1, r_1, r_3, and r_4 parameters.
          Update the positions of search agents using Eq. (3).
      else
          Update the positions of search agents with firefly search (Eq. (5))
      end if
  while $(t < T)$
  return the best solution obtained and visualize results

---

4   Experiments and Comparative Analysis

The research proposed in this manuscript is based around the proposition of the new hybrid algorithm bSCA-FS for the use as a wrapper method in search of the most important features for classification. Each individual from the population is encoded a binary array, which length is equal to the number of features in the dataset. If the feature is taken into account for classification process, it is encoded as 1, otherwise it is represented as 0.

Both, original SCA and hybrid SCA-FS versions are implemented for the purpose of proposed research. However, since every component of each individual is either 0 or 1, basic SCA and proposed SCA-FS are adopted for binary problems by employing V-shaped function, that determines probabilities for updating binary encoded solutions' elements from 1 to 0 and vice-versa.

In most approaches, sigmoid function is used to adapt metaheuristics for a binary problems, however according to [17], it was determined that a sigmoid function is monotonically increasing and as such it does not meet condition that the probability of its position changing increases with the increase of the absolute value of covergence speed. By testing bSCA-FS with sigmoid function, it was concluded that the global search ability is over emphasized. Therefore, as proposed in [17], bSCA-FS utilizes type of V-shaped transfer function, which is shown in Eq. 6.

$$X_{i,j}^{t+1} = \begin{cases} (X_{i,j}^t)^{-1} \; if \; S(X_{i,j}^t) > \text{rnd} \\ (X_{i,j}^t), \; otherwise, \end{cases} \tag{6}$$

where S denotes sigmoid function, $X_{i,j}^t$ and $X_{i,j}^{t+1}$ represent j-th parameter of i-th solution in iteration t and $t+1$, respectively, rnd is pseudo-random number drawn from uniform distribution between 0 and 1 and $(X_{i,j}^t)^{-1}$ is complement operator of $X_{i,j}^t$.

In [22], small, large, mixed and random initializers for feature selection are proposed. By conducting simulations, it was observed that the best results are generated with mixed initialization, where two thirds of all features are selected for classification at the beginning of each run.

To validate solutions', the objective of maximizing classification accuracy is used. The fitness function is expressed as a minimization problem by using the following expression:

$$F = \alpha E_R(D) + \beta \frac{|R|}{|C|}, \tag{7}$$

where $E_R(D)$ denotes the classification error-rate, α and β are weight terms, the selected number of attributes are denoted by R, while the total number of the attributes in the entire data set is denoted by C.

Objective function is calculated by performing classification with the k-nearest neighbors (KNN) classifier.

Since the basic SCA was not tested on this problem before, performance of original bSCA and proposed bSCA-FS were evaluated on the 21 well known UCI datasets, which details are provided in Table 1. To make comparative analysis with other methods fair, the same experimental conditions were used [12]. Simulation for each dataset is executed in 20 independent runs, with 70 iterations per run and the population size N of 8.

Parameters of the FA's search were set as: initial value of α to 1.0, γ to 1.0 and β_0 to 1.0. The value of α was initially decreasing during the run, as suggested in [23].

Achieved results have been put into comparison against the original bSCA and 10 other outstanding well-known metaheuristics, validated under the same conditions, which results were retrieved from [12]. The statistical results of the average fitness are presented in the Table 2, while the average classification accuracy of proposed method and opponent methods on 21 test datasets is shown in the Table 3. In both tables, bold style is used to mark best results for each category of datasets.

Table 1. Details of utilized datasets in simulations

Dataset name	Number of features	Number of samples
Breast Cancer	9	699
Tic Tac Toe	9	958
Zoo	16	101
Wine EW	13	178
Spect EW	22 ˋ	267
Sonar EW	60	208
Ionosphere EW	34	351
Heart EW	13	270
Congress EW	16	435
Krvskp EW	36	3196
Waveform EW	40	5000
Exactly	13	1000
Exactly 2	13	1000
M of N	13	1000
Vote	16	300
Breast EW	30	569
Semeion	265	1593
Clean 1	166	476
Clean 2	166	6598
Lymphography	18	148
PenghungEW	325	73

Table 2. Comparative analysis of the average fitness indicator over different approaches

No.	WOA	bWOA-S	bWOA-v	BALO1	BALO2	BALO3	PSO	bGWO	bDA	bFA	bSCA	bSCA-FS
1	0.054	0.052	0.079	0.100	0.099	0.076	0.031	0.035	0.032	0.019	0.033	**0.017**
2	0.220	0.207	0.215	0.245	0.252	0.246	0.204	0.215	0.209	0.219	0.227	**0.156**
3	0.153	0.148	0.120	0.183	0.146	0.141	0.078	0.096	0.071	0.010	0.060	**0.018**
4	0.925	0.928	0.910	0.935	0.938	0.938	0.884	0.903	0.882	0.053	0.098	**0.047**
5	0.313	0.307	0.289	0.319	0.321	0.312	0.242	0.280	0.255	0.135	0.159	**0.099**
6	0.304	0.286	0.254	0.278	0.298	0.285	0.168	0.235	0.194	0.128	**0.084**	0.122
7	0.159	0.158	0.152	0.156	0.169	0.165	0.113	0.141	0.124	0.134	0.123	**0.077**
8	0.328	0.308	0.259	0.319	0.324	0.308	0.158	0.233	0.167	0.228	0.192	**0.137**
9	0.389	0.380	0.372	0.393	0.397	0.384	0.337	0.359	0.341	0.046	0.042	**0.013**
10	0.071	0.074	0.081	0.074	0.072	0.074	0.040	0.061	0.053	0.030	**0.020**	0.033
11	0.193	0.193	0.195	0.198	0.195	0.193	0.182	0.187	0.188	0.168	0.162	**0.160**
12	0.303	0.308	0.301	0.301	0.307	0.308	0.151	0.272	0.226	0.238	0.120	**0.076**
13	0.241	0.244	0.252	0.237	0.244	0.252	0.238	0.244	0.243	0.248	0.224	0.218
14	0.139	0.133	0.155	0.151	0.150	0.136	0.022	0.112	0.072	0.054	0.098	**0.011**
15	0.084	0.084	0.081	0.089	0.090	0.085	0.048	0.069	0.052	0.073	0.046	**0.014**
16	0.081	0.058	0.062	0.086	0.088	0.086	0.033	0.057	0.031	0.073	0.055	**0.012**
17	0.044	0.043	0.037	0.043	0.043	0.044	0.032	0.034	**0.030**	0.119	0.094	0.072
18	0.191	0.187	0.176	0.184	0.192	0.197	0.136	0.158	0.149	0.161	0.128	**0.091**
19	0.052	0.052	0.049	0.051	0.052	0.052	0.041	0.044	0.042	0.033	0.036	**0.030**
20	0.235	0.230	0.223	0.258	0.243	0.237	0.138	0.211	0.160	0.165	0.121	**0.077**
21	0.260	0.244	0.242	0.276	0.262	0.274	**0.149**	0.217	0.180	0.275	0.223	0.196

Table 3. Comparative analysis of the average accuracy indicator over different approaches

No.	WOA	bWOA-S	bWOA-v	BALO1	BALO2	BALO3	PSO	bGWO	bDA	bFA	bSCA	bSCA-FS
1	0.785	0.619	0.628	0.740	0.725	0.726	0.802	0.962	0.789	0.987	0.973	**0.988**
2	0.787	0.799	0.786	0.686	0.681	0.686	0.720	0.764	0.673	0.786	0.777	**0.850**
3	0.841	0.839	0.822	0.656	0.706	0.680	0.789	0.900	0.779	**0.995**	0.944	0.985
4	0.065	0.056	0.053	0.039	0.033	0.031	0.039	0.086	0.031	0.951	0.905	**0.954**
5	0.678	0.670	0.664	0.635	0.623	0.625	0.656	0.707	0.649	0.869	0.843	**0.903**
6	0.698	0.703	0.703	0.645	0.639	0.647	0.721	0.765	0.705	0.875	**0.917**	0.875
7	0.835	0.836	0.831	0.819	0.803	0.802	0.835	0.860	0.827	0.869	0.879	**0.924**
8	0.656	0.654	0.652	0.625	0.621	0.623	0.668	0.751	0.652	0.775	0.811	**0.865**
9	0.598	0.582	0.595	0.573	0.559	0.577	0.589	0.631	0.571	0.958	0.962	**0.988**
10	0.936	0.930	0.918	0.766	0.765	0.757	0.794	0.943	0.754	0.975	**0.985**	0.970
11	0.812	0.808	0.804	0.642	0.649	0.647	0.763	0.816	0.747	0.836	0.841	**0.842**
12	0.687	0.683	0.691	0.644	0.656	0.648	0.664	0.706	0.642	0.765	0.883	**0.923**
13	0.738	0.740	0.735	0.733	0.711	0.703	0.723	0.735	0.712	0.756	0.769	**0.785**
14	0.865	0.865	0.833	0.734	0.732	0.744	0.761	0.883	0.728	0.952	0.906	**0.992**
15	0.915	0.908	0.900	0.829	0.823	0.829	0.884	0.930	0.866	0.931	0.958	**0.989**
16	0.761	0.610	0.615	0.730	0.744	0.727	0.810	0.944	0.769	0.930	0.948	**0.990**
17	0.964	0.964	0.965	0.924	0.939	0.925	0.956	**0.972**	0.959	0.884	0.910	0.928
18	0.815	0.818	0.803	0.729	0.720	0.724	0.806	0.845	0.791	0.842	0.876	**0.907**
19	0.956	0.956	0.955	0.908	0.910	0.911	0.953	0.962	0.952	0.972	0.969	**0.974**
20	0.756	0.755	0.749	0.639	0.672	0.659	0.705	0.786	0.709	0.838	0.882	**0.926**
21	0.744	0.755	0.725	0.553	0.568	0.563	0.765	0.781	0.730	0.727	0.780	**0.806**

From presented tables, it is obvious that the proposed bSCA-FS in almost all datasets for all indicators substantially outscores original bSCA and also other methods from comparative analysis. Only for datasets Sonar EW and Krvskp EW, basic bSCA obtains better results, which means that in average, improvements of proposed method are significant.

Box and whiskers diagrams for all conducted experiments for bSCA and bSCA-FS are given in Fig. 1 to better visualize performance improvements of proposed method over the original one. These diagrams are a standard practice to display the data distribution, based on the five key metrics, including min,

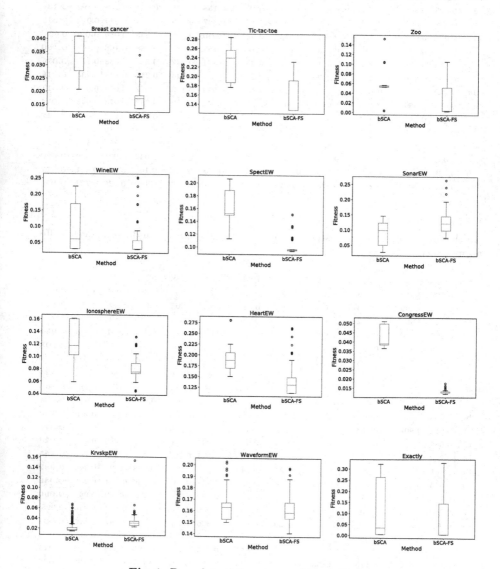

Fig. 1. Box plot of the comparable methods

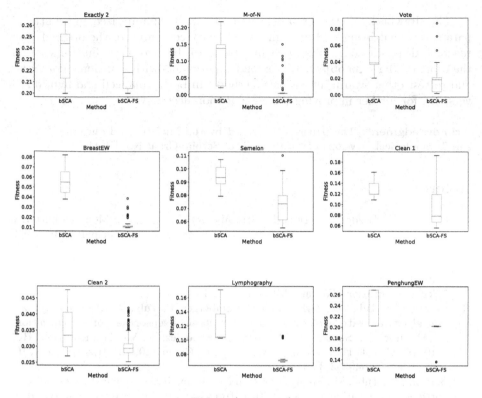

Fig. 1. (*continued*)

first quartile, median value, third quartile and max. These diagrams show how the utilized data is grouped and skewed. From the presented diagrams, it can clearly be seen that the proposed method outperforms the basic bSCA approach on the majority of the datasets included in the research.

5   Conclusion

In this research, a novel method for dimensionality reduction was proposed by applying the swarm intelligence metaheuristics approach. The proposed enhanced SCA method was implemented with a main objective to overcome the drawbacks of the basic SCA metaheuristics, which can be clearly seen from the results of the executed simulations. The proposed method was used as a wrapper method to search for the most important features for the classification process, and later validated on 21 datasets, and finally compared to the results of other supreme swarm intelligence approaches that were tested on the same datasets. The proposed approach achieved the best performances in the conducted comparative analysis.

The drawbacks of the proposed method can be summarized in additional parameters that bring slightly higher complexity compared to the original approach, and require some more time for setting up the algorithm. Future work on the topic will have an objective to extend experiments with additional data sets, and to test other swarm intelligence methods, in both enhanced and hybridized versions, for further improving the classification accuracy.

Acknowledgment. The paper is supported by the Ministry of Education, Science and Technological Development of Republic of Serbia, Grant No. III-44006.

References

1. Bacanin, N., Bezdan, T., Tuba, E., Strumberger, I., Tuba, M.: Monarch butterfly optimization based convolutional neural network design. Mathematics **8**(6), 936 (2020)
2. Bacanin, N., Bezdan, T., Tuba, E., Strumberger, I., Tuba, M., Zivkovic, M.: Task scheduling in cloud computing environment by grey wolf optimizer. In: 2019 27th Telecommunications Forum (TELFOR), pp. 1–4. IEEE (2019)
3. Bacanin, N., Tuba, E., Zivkovic, M., Strumberger, I., Tuba, M.: Whale optimization algorithm with exploratory move for wireless sensor networks localization. In: Abraham, A., Shandilya, S.K., Garcia-Hernandez, L., Varela, M.L. (eds.) HIS 2019. AISC, vol. 1179, pp. 328–338. Springer, Cham (2021). https://doi.org/10.1007/978-3-030-49336-3_33
4. Bacanin, N., Tuba, M.: Firefly algorithm for cardinality constrained mean-variance portfolio optimization problem with entropy diversity constraint. Sci. World J. **2014**, 16 (2014). Special issue Computational Intelligence and Metaheuristic Algorithms with Applications
5. Bezdan, T., Tuba, E., Strumberger, I., Bacanin, N., Tuba, M.: Automatically designing convolutional neural network architecture with artificial flora algorithm. In: Tuba, M., Akashe, S., Joshi, A. (eds.) ICT Systems and Sustainability. AISC, vol. 1077, pp. 371–378. Springer, Singapore (2020). https://doi.org/10.1007/978-981-15-0936-0_39
6. Bezdan, T., Zivkovic, M., Tuba, E., Strumberger, I., Bacanin, N., Tuba, M.: Glioma brain tumor grade classification from MRI using convolutional neural networks designed by modified FA. In: Kahraman, C., Cevik Onar, S., Oztaysi, B., Sari, I.U., Cebi, S., Tolga, A.C. (eds.) INFUS 2020. AISC, vol. 1197, pp. 955–963. Springer, Cham (2021). https://doi.org/10.1007/978-3-030-51156-2_111
7. Bezdan, T., Zivkovic, M., Tuba, E., Strumberger, I., Bacanin, N., Tuba, M.: Multiobjective task scheduling in cloud computing environment by hybridized bat algorithm. In: Kahraman, C., Cevik Onar, S., Oztaysi, B., Sari, I.U., Cebi, S., Tolga, A.C. (eds.) INFUS 2020. AISC, vol. 1197, pp. 718–725. Springer, Cham (2021). https://doi.org/10.1007/978-3-030-51156-2_83
8. Brajevic, I., Tuba, M., Bacanin, N.: Multilevel image thresholding selection based on the cuckoo search algorithm. In: Proceedings of the 5th International Conference on Visualization, Imaging and Simulation (VIS'12), Sliema, Malta, pp. 217–222 (2012)
9. Brezočnik, L., Fister jr, I., Podgorelec, V.: Swarm intelligence algorithms for feature selection: a review. Appl. Sci. **8**, 1521 (2018). https://doi.org/10.3390/app8091521

10. Calvet, L., de Armas, J., Masip, D., Juan, A.A.: Learnheuristics: hybridizing metaheuristics with machine learning for optimization with dynamic inputs. Open Math. **15**(1), 261–280 (2017)
11. Chandrashekar, G., Sahin, F.: A survey on feature selection methods. Comput. Electric. Eng. **40**(1), 16–28 (2014)
12. Hussien, A.G., Oliva, D., Houssein, E.H., Juan, A.A., Yu, X.: Binary whale optimization algorithm for dimensionality reduction. Mathematics **8**(10) (2020). https://doi.org/10.3390/math8101821, https://www.mdpi.com/2227-7390/8/10/1821
13. Karaboga, D., Basturk, B.: Artificial Bee Colony (ABC) optimization algorithm for solving constrained optimization problems. In: Melin, P., Castillo, O., Aguilar, L.T., Kacprzyk, J., Pedrycz, W. (eds.) IFSA 2007. LNCS (LNAI), vol. 4529, pp. 789–798. Springer, Heidelberg (2007). https://doi.org/10.1007/978-3-540-72950-1_77
14. Liang, J., et al.: Problem definitions and evaluation criteria for the cec 2006 special session on constrained real-parameter optimization (2006)
15. Mirjalili, S.: SCA: a sine cosine algorithm for solving optimization problems. Knowl.-Based Syst. **96** (2016). https://doi.org/10.1016/j.knosys.2015.12.022
16. Bacanin, N., Bezdan, T., Tuba, E., Strumberger, I., Tuba, M.: Optimizing convolutional neural network hyperparameters by enhanced swarm intelligence metaheuristics. Algorithms **13**(3), 67 (2020)
17. Rashedi, E., Nezamabadi-Pour, H., Saryazdi, S.: BGSA: binary gravitational search algorithm. Nat. Comput. Int. J. **9**(3), 727–745 (2010)
18. Strumberger, I., Bacanin, N., Tuba, M., Tuba, E.: Resource scheduling in cloud computing based on a hybridized whale optimization algorithm. Appl. Sci. **9**(22), 4893 (2019)
19. Strumberger, I., Tuba, E., Bacanin, N., Zivkovic, M., Beko, M., Tuba, M.: Designing convolutional neural network architecture by the firefly algorithm. In: 2019 International Young Engineers Forum (YEF-ECE), pp. 59–65. IEEE (2019)
20. Tuba, E., Strumberger, I., Zivkovic, D., Bacanin, N., Tuba, M.: Mobile robot path planning by improved brain storm optimization algorithm. In: 2018 IEEE Congress on Evolutionary Computation (CEC), pp. 1–8 (2008)
21. Tuba, M., Bacanin, N.: Improved seeker optimization algorithm hybridized with firefly algorithm for constrained optimization problems. Neurocomputing **143**, 197–207 (2014). https://doi.org/10.1016/j.neucom.2014.06.006
22. Xue, B., Zhang, M., Browne, W.N.: Particle swarm optimisation for feature selection in classification: Novel initialisation and updating mechanisms. Appl. Soft Comput. **18**, 261–276 (2014). https://doi.org/10.1016/j.asoc.2013.09.018, https://www.sciencedirect.com/science/article/pii/S1568494613003128
23. Yang, X.-S.: Firefly algorithms for multimodal optimization. In: Watanabe, O., Zeugmann, T. (eds.) SAGA 2009. LNCS, vol. 5792, pp. 169–178. Springer, Heidelberg (2009). https://doi.org/10.1007/978-3-642-04944-6_14
24. Yang, X.S.: A new metaheuristic bat-inspired algorithm. In: Nature Inspired Cooperative Strategies for Optimization (NICSO 2010), pp. 65–74. Springer, Heidelberg (2010). https://doi.org/10.1007/978-3-642-12538-6_6
25. Zivkovic, M., Bacanin, N., Tuba, E., Strumberger, I., Bezdan, T., Tuba, M.: Wireless sensor networks life time optimization based on the improved firefly algorithm. In: 2020 International Wireless Communications and Mobile Computing (IWCMC), pp. 1176–1181. IEEE (2020)
26. Zivkovic, M., et al.: Covid-19 cases prediction by using hybrid machine learning and beetle antennae search approach. Sustain. Cities Soc. **66**, 102669 (2021)

27. Zivkovic, M., Bacanin, N., Zivkovic, T., Strumberger, I., Tuba, E., Tuba, M.: Enhanced grey wolf algorithm for energy efficient wireless sensor networks. In: 2020 Zooming Innovation in Consumer Technologies Conference (ZINC), pp. 87–92. IEEE (2020)
28. Zouache, D., Ben Abdelaziz, F.: A cooperative swarm intelligence algorithm based on quantum-inspired and rough sets for feature selection. Comput. Ind. Eng. **115**, 26–36 (2018). https://doi.org/10.1016/j.cie.2017.10.025, https://www.sciencedirect.com/science/article/pii/S0360835217305107

Rainfall Prediction Using Logistic Regression and Support Vector Regression Algorithms

Srikantaiah K. C.$^{(\boxtimes)}$ and Meenaxi M. Sanadi

Department of Computer Science and Engineering, SJB Institute of Technology,
Bangalore 560060, India

Abstract. Rainfall Prediction is the use of science and innovation to anticipate the condition of the climate for a given area. Expectation of rainfall is perhaps the most fundamental and requesting undertakings for the climate forecasters. Rainfall expectation assumes a significant part in the field of cultivating and businesses. Exact precipitation forecast is indispensable for recognizing the substantial rainfall and to give the data of alerts with respect to the characteristic disasters. Rainfall expectation includes recording the different boundaries of climate like humidity, wind speed, stickiness, and temperature so on. In this Paper, we distinguish the issues of rainfall prediction and fixing them using the machine learning algorithms like Logistic Regression method and Support Vector Regression. Experimental results show that Logistic Regression algorithm is best suitable for prediction of rainfall with accuracy 96% when compare to the support vector regression algorithm. This prediction results helps in the agriculture work.

Keywords: Logistic regression · Machine learning algorithms · Principal component analysis · Rainfall prediction · Support vector regression

1 Introduction

The agriculture based countries and the accomplishment of horticulture relies upon of rainfall and its financial system is normally based on crop yield and rainfall prediction. Rainfall prediction is helps in agriculture work. Rainfall expectation consists of recording the kind limitations of weather like visibility, humidity, wind speed, rainfall, and temperature and so on.., for breaking down the harvest profitability. Rainfall expectation assumes a sizeable element within side the subject of cultivating and groups depending on that researcher are picking into shape [1]. Machine learning techniques are accompanied to get genuine and proper forecast of rainfall results. In this challenge machine learning algorithms had been investigated which include logistic regression and support vector regression [2].

The accessibility of environment information during the most recent years makes it critical to discover and precise instruments to investigate and extricate the information from this accurate information can assume significant part in understanding the environment fluctuation and environment forecast [3]. Number of techniques and approaches are used by the researchers to predict the rainfall prediction accurately. Climate gauging

© Springer Nature Switzerland AG 2021
M. Singh et al. (Eds.): ICACDS 2021, CCIS 1440, pp. 617–626, 2021.
https://doi.org/10.1007/978-3-030-81462-5_54

assists with climatic problems and in surroundings checking, dry season recognition, some weather expectation, farming and introduction, arranging in power industry, an aeronautics industry, correspondence, infection dispersal [4].

Rainfall Prediction is crucial to exactly determine the weather prediction for effective usage of water assets, crop yield profitability and pre arranging of water structures. Utilizing numerous machine learning techniques it could forecast the rainfall [5]. The maximum methodologies applied for rainfall forecast are relapse. In dynamical technique expectation are created depending on frameworks of rainfall that count on the improvement of the global surroundings frameworks for rainfall prediction with high-quality for expectation [6].

Rainfall prediction is most important fact in agriculture. Here, we get the result of short term rainfall prediction and long term rainfall prediction. In short term prediction, we get few days prediction and in long term prediction we get few months' rainfall prediction in advance [7]. Depending on this rainfall prediction, farmers can start their work. In this paper, first we find the problems of rainfall in agriculture; understand those problems and fixing them. Then, collect the datasets predict the rainfall using the machine learning algorithms like logistic regression and support vector regression.

Rest of the paper is organized as follows. Section 2 includes literature survey. Section 3 includes problem definition, system architecture, and algorithm for rainfall prediction. Section 4 includes experimental setup, experimental results. Section 5 includes conclusion and future work.

2 Literature Survey

Razeef Mohd *et al.* [1] Rainfall prediction system using linear regression algorithm with accuracy 91.84% and errors in the dataset is 8.16%. This model is gives the better performance but it does not contain the larger dataset to prediction. They can also used different type of algorithms in that linear regression is gave the best accuracy prediction of rainfall compare to the other algorithms. Rainfall expectation consists of recording the exclusive obstacles of weather like breeze bearing, wind speed, precipitation, temperature and so forth. This exploration work presentation of few machine learning calculations for foreseeing precipitation utilizing weather facts.

R. Senthil Kumar *et al.* [2] Rainfall prediction system using K-Nearest neighbors algorithm with accuracy of the rainfall is 89% and error is 11%. The main advantage it has a good prediction accuracy rainfall. Data transformation is required and extra computation required for more accurate result and small facts are left for prediction, and characteristic elimination is needed for higher accuracy. India is agriculture of an economy is usually based on rainfall prediction accuracy result. There are essentially approaches to cope with foresee precipitation in India.

Ganesh P *et al.* [3] Rainfall prediction system using linear regression with accuracy of rainfall is 94% and errors 6%. It has a high prediction accuracy and good for analyzing dataset. Small statistics is left for prediction. Do now no longer take care of the nonstop statistics no verification is accomplished here, as opposed to rainfall accurate, an approximated cost is retrieved [8]. Rainfall prediction is the usage of technological know-how and innovation to count on the circumstance of the surroundings for a given area.

Xianggen Gan *et al.* [4] Rainfall prediction system using neural network with accuracy 71.83% and error is 28%. In this prediction model it requires best community is chosen for prediction. Here accuracy varies exceedingly with length of schooling dataset and used many algorithms but that all are not working properly. Based on accuracy prediction neural network is best algorithm [9]. And in proposed work accuracy varies extraordinarily with length of dataset, statistics recorded at abnormal intervals, do now no longer manage nonstop statistics, and attributes normalization is required.

Jharna Majumdar *et al.* [5] Rainfall prediction system using k-nearest neighbors, and helps vector regression with accuracy 85.52% and error is 14.48%. This proposed system is beneficial for dynamic data, good prediction accuracy. Need to combine characteristic choice techniques and dynamic information mining techniques required. Here overall mean value obtained is 2.853241 and standard deviation value is 10.062828 that give the good performance in practical applications [10]. Now no longer correct for long time prediction, cannot are expecting statistics in far flung areas, verification isn't always correct. The huge collection of precipitation estimate strategies are on hand in India.

Andrew Kusiak *et al.* [6] Rainfall prediction system using Regression algorithms like logistic regression multi linear regression with accuracy 82.73% and error is 17.28%. It has acceptable accuracy prediction for rainfall, given by previous year dataset. Disadvantage of this system is not proper for long time prediction, attribute removal required for higher accuracy, and work privacy issues, in mining techniques, security issues in data mining [11]. Precipitation is one in all some huge variables influencing water quality. Studies have validated a stable advantageous dating among precipitation and rainfall prediction.

3 Problem Statement and Architecture

Rainfall prediction is most important fact in agriculture. Here, we get the result of short term rainfall prediction and long term rainfall prediction. Depending on this rainfall prediction, farmers can start their work. In this paper, first we find the problems of rainfall in agriculture; understand those problems and fixing them. Then, collect the datasets predict the rainfall using the machine learning algorithms like logistic regression and support vector regression.

3.1 Problem Definition

India is an agriculture based country and most of the people are dependent on agriculture in their daily life. Rainfall is based on weather forecast, so we need to inquire the weather forecaster. Exact rainfall Prediction is more important in agriculture. The main objective of the paper is to understand the problem of rainfall prediction and algorithms for solving the problem by extensive collection of rainfall datasets and analysis of these datasets using machine learning algorithms like logistic regression, support vector regression. Then, evaluate performance of these two algorithms based on their accuracy.

3.2 System Architecture

The Fig. 1 determines the stream of the system which is developed so as to detect and pre process the data then find the if any missing data is there. Then split the given inputs from the dataset like training datasets and test data sets using the Principal Component Analysis (PCA). Then, apply the suitable algorithm of machine learning like, logistic regression and support vector regression that are affected by rainfall prediction to predict the rainfall accurately. Finally, find which algorithm is more suitable for predicting rainfall.

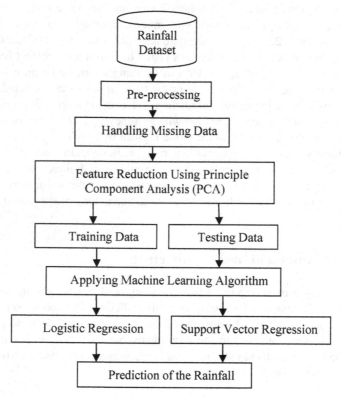

Fig. 1. Rainfall prediction model

3.2.1 Collection of Rainfall Dataset

Dataset is based on the previous year rainfall with 20 attributes as shown in Table 1, which changed into taken from the respective system learning websites together with 24 UCI, https://www.Kaggle.com, towards information technology.

After collecting the required dataset, the next step is to perform statistics cleansing, we carry out the steps concerned in data pre-processing. Preprocessing of data is an insights mining strategy that includes modifying crude data into a reasonable arrangement. Then, handling the missing data, fill it with mean and standard deviation. Feature

Table 1. List of attributes

Name of the attribute	Description
Year	Represents how much rainfall is happened in that year
Month	Represents that the rainfall in different month
Day	Represents the rainfall in the day of the month
TempHigh	Represents the how much temperature in the days
TempAvg	Represents the average of temperature value from weather forecaster
TempLow	Represents the low temperature, it may be happened in the month of July, August and September because of rainy season
DPHigh	Represents DP high in rainfall
DPAvg	Represents the average of DP in the rainfall
DPLow	Represents the low value in DP for rainfall
humidityHigh	Represents the high humidity value per day in rainfall
humidityAvg	Represents the average value of humidity per day
humidityLow	Represents the low value of humidity per day
SPLHigh	Represents the high SPL value in the rainfall dataset
SPLAvg	Represents the average SPL value in the rainfall dataset
SPLLow	Represents the low SPL value in the rainfall dataset
visibilityHigh	Represents the high visibility value per day
visibilityAvg	Represents the average visibility value in the rainfall dataset
visibilityLow	Represents the low visibility value in the rainfall dataset
windAvg	Represents the wind speed in the rainfall
Rainfall	Represents the how much rainfall is happened on that day

reduction using principal component analysis using two steps. Firstly, it making use of a splitting the dataset is more efficient *via* lowering the size of the effective functions. Secondly, increases prediction accuracy by way of getting rid of prediction. In our proposed regression method, PCA targets to create the capabilities and dimension's sequentially relying upon the amount of information it has captured. It addresses two fundamental objectives: Firstly, in general it's miles most perfect to reduce the wide variety of methods that is used to represent the datasets. Secondly it is also suited for set of capabilities to describe massive quantity of facts.

PCA is a well-established method for feature extraction and dimensionality reduction. It is primarily based on the assumption that maximum facts about training are contained within the instructions alongside which the variations are the largest. These instructions are called principal additives. It suggests an instance of principal components in case of two dimensions where first principal aspect is denoted by pc1 30 whereas, 2d one is denoted by way of pc2.

3.2.2 Logistic Regression

Logistic regression analysis is to conduct when the dependent variable i.e., output data is categorical or binary. In other words we will be having categorical response variable, like all regression analysis the logistic regression is a predictive analysis, logistic regression is used to describe data and to explain the relationship between one dependent binary variable and one or more nominal, and ordinal and interval independent variables this is also called as generalized linear regression.

We can use to assess the relationship between one or more predictors variables and a response variables. We can say categorical response as binary when number of categories is two and characteristics are at 2 levels [10]. Data consideration for logistic regression it shows the results are valid or not. The model should provide a good fit to the data if the model does not fit the data then result should be miss leading in that output.

Regression equation of logistic regression is given by,

$$Y = B0 + B1x \qquad (1)$$

Where, the coefficients B0 and B1 are calculated using Eqs. (2) and (3)

$$B1 = sum[(Xi - mean(x) + (Yi - mean(y)]/sum[(Xi - mean(x))^2] \qquad (2)$$

$$B0 = mean(y) - B1 (mean(x)) \qquad (3)$$

Y represents the based variable that wants to be anticipated, X is the incline of the street, N is represents number of objects, Xi and Yi represents the fact that we need to repeat their calculations across values in our dataset and i represents to the ith values of X or Y, Mean(x) and mean(y) represents the average values of X and y.

3.2.3 SVR: Support Vector Regression

SVR is a support vector regression it has big different from SVM. As the name suggest SVR is a regression algorithm so we can use SVR for working with continuous values instead of classification which is SVM. Support vector regression tries to fit as many instances as possible on the street while limiting margin violations.

Kernel, function used to map a lower dimensional data into higher dimensional data. Hyper plane, in SVM this is a basic link the separation line between the data classes all through SVR we are going to define it as the line that will helps us to predict the continuous or target values. Boundary line, this SVM there are two lines other than hyper plan which creates a margin the support vector can be unboundary line or outside the boundary lines separates two classes in SVR also same [11]. Support vectors, these are the data points which are closest to the boundary this distance of the point is minimum or least as shown in Fig. 2.

SVR performs linear regression in higher. We can think of SVR as if each data point in the training represents its dimension. When you evaluate your kernel between a test point and a point in the training set the resulting value gives the coordinate of your test point in that dimension. The vector we get when we evaluate the test point for all points in the training set, k is the representation of the test point in the higher dimensional

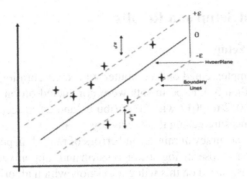

Fig. 2. Support vector regression

space. Once you have that vector you use it to perform a linear regression. hyper plane and its boundary lines are defined as

$$Wx + b = 0 \tag{4}$$

$$Wx + b = +e \tag{5}$$

$$Wx + b = -e \tag{6}$$

SVR is strong to the anomalies. Choice model for support vector regression is effectively refreshed. Can utilize numerous classifiers prepared on the various kinds of information utilizing the likelihood rules. Support vector regression performs lower calculation contrasted with other relapse methods. Its usage is simple compare to other algorithms. The algorithm for predicting rainfall using machine learning technique is shown in Algorithm 1.

Algorithm 1: Rainfall Prediction.

Algorithm Rainfall (D)

//Purpose: To predict the rainfall
//Input: D: Rainfall dataset
//Output: RP- Predicted Rainfall
Step 1: Replace missing values in D with their mean
Step 2: Select the required data from D using Principal Component Analysis
Step 3: Divide the pre-processed data into training and testing data partition
Step 4: Build the model using machine learning algorithm such as logistic regression
         And support vector regression
Step 5: RP= with the model predict the rainfall for test dataset
Step 6: Return RP

4   Experimental Setup and Results

4.1   Experimental Setup

Our algorithm was implemented on a computer with configuration 8GB RAM, 512GB Hard disk, using python 3.0. Experiments were conducted on rainfall dataset of Chittagong, India from 2012 to 2017 with 20 attributes and 2190 records. Performance of the algorithms was measured using their accuracy.

Table 2 shows the accuracy of rainfall and errors of rainfall in percentage, comparing two different algorithms those are logistic regression and support vector regression result in different algorithms. Based on this will get to know which algorithm is more suitable for prediction. Here logistic regression is the best algorithm with accuracy 96.06% and errors is 3.96% compare to the support vector regression.

Formula for find the accuracy,

$$\text{Percent Error} = \frac{\text{Accepted Value - Experimental Value}}{\text{Accepted Value}} \tag{7}$$

Table 2. Accuracy and error of both the algorithm.

Test data size	Logistic regression algorithm results	Support vector regression algorithm result
10	Accuracy Result: 83.28% Error Result: 16.71%	Accuracy Result: 48.15% Error Result: 51.48%
20	Accuracy Result: 75.61% Error Result: 24.38%	Accuracy Result: 59.24% Error Result: 40.75%
25	Accuracy Result: 73.52% Error Result: 26.47%	Accuracy Result: 61.96% Error Result: 38.03%
35	Accuracy Result: 87.39% Error Result: 12.60%	Accuracy Result: 63.97% Error Result: 36.02%
40	Accuracy Result: 84.44% Error Result: 15.55%	Accuracy Result: 72.05% Error Result: 27.94%
50	Accuracy Result: 94.97% Error Result: 5.02%	Accuracy Result: 68.05% Error Result: 31.94%
55	Accuracy Result: 93.74% Error Result: 6.25%	Accuracy Result: 68.84% Error Result: 31.15%
60	Accuracy Result: 96.06% Error Result: 4.67%	Accuracy Result: 71.08% Error Result: 28.91%
65	Accuracy Result: 96.06% Error Result: 3.93%	Accuracy Result: 70.95% Error Result: 29.91%

Table 3 represents the rainfall prediction using logistic regression and support vector regression for every month. Based on this prediction in the month of July, august,

September and October observes above 50% rainfall compare to other months, especially the month of September observes 70% to 80% Rainfall and in the month of February and March observes only 10% to 14%% rainfall in the year of 2021. Based on Table 2 and Table 3 logistic regression algorithm is more suitable for rainfall prediction when compare to support vector regression algorithm. Rainfall prediction result is more helpful to the agriculture, depending upon the prediction result former will decide to start an agriculture work.

Table 3. Rainfall Prediction with LR and SVR for the year 2021

Data and month	Logistic regression prediction	Support vector prediction
January 12	15.61%	12.11%
January 27	14.10%	10.11%
February 13	10.69%	10.26%
February 28	13.10%	11.09%
March 11	13.07%	10.62%
March 25	14.06%	10.89%
April 15	20.10%	17.32%
April 27	21.48%	19.36%
May 14	32.61%	26.89%
May 28	32.68%	28.45%
June 10	47.48%	39.06%
June 22	48.68%	42.32%
July 13	65.01%	58.49%
July 29	68.83%	57.92%
August 03	70.47%	68.56%
August 20	72.79%	68.59%
September 9	77.83%	74.84%
September 21	80.03%	73.86%
October 7	76.81%	67.48%
October 24	76.91%	50.52%
November 14	57.48%	52.62%
November 29	50.03%	48.96%
December 03	48.69%	37.12%
December 26	38.52%	29.13%

5 Conclusion

In agriculture country most of the people are dependent on agriculture in their daily life. Rainfall is based on weather forecast so we need to inquire the weather forecaster due to the fact ages. In this paper, we have measured the performance of linear regression and support vector regression algorithms for rainfall prediction. From experimental results, we conclude that, logistic Regression is best suited for accuracy for prediction with accuracy of 95.32% and shows that, the month of September observes 73% to 80% rainfall and February, March observes only 10% to 14% Rainfall. In future work, we collect the India rainfall datasets. Accuracy for all the algorithms can be increased by making few more modifications to the feature matrix and test application with various parameters and using other machine learning algorithms.

References

1. Razeef, M., Muheet, A.B., Majid, Z.B.: Comparative study of rainfall prediction techniques. J. Comput. Sci. Technol. **7**, 13–19 (2019)
2. Senthil kumar, R., Dr. Ramesh, C.: A study on prediction using data mining technique. Int. Conf. Data Min. **3**, 189–196 (2017)
3. Gaikwad, G.P., Prof. Nikam, V.B.: Rainfall prediction models and general data mining rainfall prediction model. Int. J. Eng. Technol. **2**, 479–508 (2019)
4. Gan, X., Chen, L., Yang, D., Liu, G.: The research of rainfall prediction models. IEEE Paper **4**, 978–985 (2018)
5. Majumdar, J.: Data mining and analysis of big data based on the Hadoop. Int. Conf. Comput. Commun. Control Manage. **8**, 130–134 (2020)
6. Kusiak, A., Wei, X., Verma, A. P., Roz, E.: Modeling and prediction of rainfall using radar reflectivity data: a data-mining approach. IEEE Trans. Geosci. Remote Sens. **51**(4), 2337–2342 (2013). https://doi.org/10.1109/TGRS.2012.2210429
7. Basha, Z., Bhavana, N., Bhavya, P., Sowmya, V.: Rainfall prediction using machine learning and deep leaning techniques. IEEE. **9**, 978–1034 (2019)
8. Bhomia, S., Jaiswal, N., Kishtawal, C.M., Kumar, R.: Multimodel prediction of monsoon rain using dynamical model selection. IEEE Trans. Geosci. Remote Sens. **4**, 0196–2892 (2020)
9. Maruta, R., Tomazowa, A.: Prediction of attenuation by rain. IEEE Trans. Commun. **9**, 457–603 (2017)
10. Kumar, S., Upadhyaes, A., Gola, C.: Rainfall prediction based on 100 years of meteorological data. IEEE **3**, 978–986 (2019)
11. Maitra, A., Chakraborty, R.: Prediction of rain occurrence and accumulation using multi-frequency radiometric observations. IEEE Trans. Geosci. Remote Sens. **56**(5), 2789–2797 (2018). https://doi.org/10.1109/TGRS.2017.2783848

Collaborative Recommender System (CRS) Using Optimized SGD - ALS

Gopal Behera[✉] and Neeta Nain

Department Computer Science and Engineering, Malaviya National Institute
of Technology, Jaipur, India
2019rcp9002@mnit.ac.in

Abstract. Matrix factorization (MF), dimensional reduction techniques
are broadly used in recommender systems (RS) to retrieve the preference of user from explicit ratings. However, the interactions are not
always consistent due to the influence of numerous elements on users
on a product, including friend's recommendation and business publiciz-
ing. In comparison, traditional MF is not able to find consistent ratings.
Find the exact prediction/ratings of a product/item is essential for further improvement of the performance of the collaborative recommender
framework. To find the exact prediction, we propose the parameter optimizing stochastic gradient descent (SGD) and alternate least square
(ALS) over MF. Furthermore, we examine the deviation of prediction
error after setting each parameter over a general parameter distribution
of both techniques (SGD and ALS). To evaluate the performance of the
proposed model, we use two well-known datasets. The exploratory outcomes reveal that our approach gets significant improvement over the
base model.

Keywords: Recommender system · Explicit rating · Matrix
factorization · SGD · ALS

1 Introduction

Nowadays, the recommender framework or system (RS) enormously advances
the improvement of online business. A good quality RS assists clients in finding
items according to his/their choices when they are confronting a massive amount
of products [23], by mitigating the issue of data over-burden [21], also bring more
financial advantages for dealers or sellers [7]. Consequently, to get more exact
results, numerous analysts have proposed different RS techniques. Among all
the techniques, collaborative filtering (CF) is a straightforward and wild used
technique for generating the recommender list to a target client [1]. Perhaps
the most famous techniques for CF is user-based CF (UCF), it expects to discover a few clients who have similar behaviors (for example, browsing history) of
the target client, and afterward, suggest to him those products that are similar
to the target user's choices [6]. Several authors have proposed various similarity estimation strategies to discover similar items and users [10,15,20], yet the

© Springer Nature Switzerland AG 2021
M. Singh et al. (Eds.): ICACDS 2021, CCIS 1440, pp. 627–637, 2021.
https://doi.org/10.1007/978-3-030-81462-5_55

performance is scarcely agreeable on account of exceptionally sparse information [12]. Because of acquiring superb execution in the Netflix Prize rivalry, the model-based CF techniques gain remarkable achievement in RS because of their high exactness and scalability [14,17], and the lattice factorization process is one of them. Matrix factorization (MF) factorized the high dimensional matrix into the item and user factor's low dimensional matrix. Motivated from the MF model, researchers have created various model-based algorithms and improved performance.

Funk [8] introduced the regularized MF to tackle the Netflix challenge and accomplished a decent outcome. Sarwar [22] proposed an incremental MF to make the recommender framework more scalable. Paterek [19] added bias of both user and item to MF for mining the association among them more precisely. Koren [13] incorporated additional data sources in the MF model; however, the time complexity was exceptionally high. The idea of a certain level to quantify client inclinations in an MF has been proposed by Hu [11] on implicit data. He [9] illustrated missing information and described item popularity. Meng [18] proposed weight-based MF and utilized TF-IDF to discover the client's inclinations, yet the technique was just reasonable for text information. Notwithstanding, the above techniques neglect to think about each unequivocal rating's dependability on the client. As a rule, clients have their preferences and conclusions on a product. Even though the client's direct interaction, not all interactions should be given a similar weight [16]. For instance, a few clients like to give high scores, prompting their standard scores a lot higher than the general mean worth. Conversely, different clients rate the most loved things and will, in general, give lower scores on different things [5]. In this circumstance, the inclinations of the two sorts of clients are particularly unique; in the event that they give similar products the same score, the dependability of the two scores should be painstakingly assessed.

In our work, first, we analyze explicit MF without optimizing the model parameters and find the degree of deviation in terms of MSE and MAE. Similarly, on the basis of the degree of deviation of prediction accuracy, we optimized the model parameter through stochastic gradient descent (SGD) and alternate least square (ALS) learning procedure and solved the cost function. Moreover, we used an optimized SGD and ALS to learn the MF model parameters. At last, we conduct the experiments on standard datasets and obtain better performance than baseline MF recommendation algorithms. The rest of the paper is organized as follows. Section 2 describes preliminaries with problem definition and the matrix factorization recommendation algorithm. In Sect. 3 we describe our proposed approach, which defines parameter optimization through SGD and ALS. Section 4, we present the datasets and evaluation metrics, and in Sect. 5 presents experimental results. Finally, we draw the conclusion in Sect. 6.

2   Preliminaries

In this section, we discussed the preliminaries, a brief introduction of traditional MF as well as the problem associated with CF.

2.1 Problem Definition

Let there are M users and N items, then the rating or interaction matrix between user and item is represented as $R \in R^{M \times N}$. Each interaction value $r_{u,i}$ of R represents u^{th} user's rating on i^{th} item, and $\hat{r}_{ui}$: denotes the predicted rating of the user u on item i. Given R, the objective of the RS is to recommend items that an active user might interest in.

2.2 Matrix Factorization

Matrix factorization (MF) techniques have been widely used to get the relationship between user and item. Funk [8] pointed out that the interaction matrix R can be factorized into user-factor and the item-factor according to Eq. 1.

$$R \approx PQ^T \tag{1}$$

where $P \in R^{K \times M}$ and $Q \in R^{K \times N}$ are user features and item features with K latent dimension, whereas, $K \ll min(M, N)$. Therefore, the prediction $\hat{r}_{ui}$ is defined in Eq. 2.

$$\hat{r}_{ui} = p_{.,u} q_{.,i}^T \tag{2}$$

where $p_{.,u}$ represents u^{th} row of P and $q_{.,i}$ represents i^{th} column of Q. Goal is to minimize the regularized squared error loss function L as defined in Eq. 3, to obtain the latent factor of user and item.

$$L = \sum_{(u,i) \in T} (r_{u,i} - \hat{r}_{u,i})^2 + \lambda(||p_{.,u}||^2 + ||q_{.,i}||^2) \tag{3}$$

where λ denotes regularization term, used to avoid over-fitting, $||\cdot||$ is the Frobenius norm and T represents training set.

3 Proposed Method

In this part, we described our proposed technique that is explicit MF with a parameter for optimizing both for stochastic gradient descent (SGD) and alternate least square (ALS) through the grid search (GS) [4] and assume that the optimized parameters of both learning techniques improve the performance of the collaborative recommendation system (CRS). The parameters to be optimized in SGD are λ, η, and $K(number\ of\ factors)$. Similarly, the parameter to be optimized in ALS is λ. We assume that the proposed tuning approach of the factorization technique for CF will further improve the model accuracy and produce a better recommendation.

3.1 Prediction

According to Eq. 2, the prediction is recomputed as $r_{\hat{u},i}$, which is defined in Eq. 4.

$$r_{\hat{u},i} = B_u + B_i + p_{.,u}^T q_{.,i} + \mu \tag{4}$$

where B_u and B_i are the bias of user u item i respectively and global bias denoted as μ. Further we add regularization in the model to avoid over-fitting. The regularized squared error loss L is defined in Eq. 5

$$L = \sum_{u,i}(r_{ui} - r_{\hat{u}i})^2 + \lambda_{pb}\sum_u \|B_u\|^2 + \lambda_{qb}\sum_i \|B_i\|^2 + \lambda_{pf}\sum_u \|\mathbf{P}_u\|^2 + \lambda_{qf}\sum_i \|\mathbf{q}_i\|^2$$
$$\tag{5}$$

where p_{uf} and q_{if} are the user and item factors in the latent dimension f. λ_{pb} and λ_{qb} are the regularization bias for user and item.

3.2 Optimization

The model parameters are optimized through alternate least square (ALS) and stochastic gradient descent (SGD) learning algorithms.

SGD. To minimize the loss defined in Eq. 5, we first use SGD to learn the model parameters because of its high productivity. The procedure of the SGD is shown in Algorithm 1.

Further, each parameter of the model is computed by taking the partial derivative in the gradient's negative direction and updating the parameters until the convergence. The user bias is updated as defined in Eq. 6.

$$b_u \leftarrow b_u - \eta\frac{\delta L}{\delta b_u}$$
$$= b_u - \eta(2(r_{ui} - r_{\hat{u}i}))(-1) + 2\lambda_{u_b}b_u$$
$$= b_u - \eta(-2(err_{ui} + 2\lambda_{u_b}b_u)$$
$$= b_u + \eta((err_{ui} - \lambda_{p_b}b_u) \tag{6}$$

The factor 2 is gets rolled up in the learning rate. Similarly the parameters item bias, item factor and user factor are updated accordingly to the Equation defined from 7 to 9 respectively.

$$b_i \leftarrow b_i + \eta(err_{ui} - \lambda_{qb}b_i) \tag{7}$$

$$p_u \leftarrow p_u + \eta(err_{ui}q_i - \lambda_{pf}p_u) \tag{8}$$

$$q_i \leftarrow q_i + \eta(err_{ui}p_u - \lambda_{qf}y_i) \tag{9}$$

Where η is the learning rate, and λ is the regularization term.

Algorithm 1: Stochastic Gradient Descent (SGD) Algorithm

Input: $R_{N \times M} = r_{ui}$

K: Number of factors

η: Learning rate

λ: Regularization

b_i: item bias

b_u: user bias

Output: latent vector of $U_{N \times K}$ and $V_{K \times M}$

1 Initialize the latent vectors U and V randomly.

2 **for** i *iteration* **do**

3   **for** *each* r_{ui} **do**

4          $b_i \leftarrow \frac{\sum_{u \in R_i}(r_{ui}-\mu)}{\lambda+|R_i|}$

5          $b_u \leftarrow \frac{\sum_{i \in R_u}(r_{ui}-\mu-b_i)}{\lambda+|R_u|}$

6          $r_{ui} = \mu + b_u + b_i + U_u^T \cdot V_i$

7          $e_{ui} = r_{ui} - \hat{r_{ui}}$

8          $e_{ui}' = e_{ui} + b$

9          $U_u = U_u + \eta(e_{ui}'V_i - \lambda U_u)$

10         $V_i = V_i + \eta(e_{ui}'U_u - \lambda V_i)$

11   **end**

12 **end**

13 **return** $V_{n \times k}$ and $U_{k \times m}$

Alternate Least Square (ALS): ALS is a first order minimized procedure where the cost function is minimized by keeping one set of latent vectors as constant and computing derivative with respect to user vectors and vice versa. Then equate the derivative to zero to solve for the non-constant vectors (the user vectors). Similarly, the item vector is solved alternately and these two-step is carried out until convergence reached. Two step derivatives are defined in Eq. 10. Further, the model parameters, such as user and item features are updated according to Eqs. 12 and 13. The Als procedure is defined in Algorithm 2.

$$\frac{\delta L}{\delta p_u} = 0$$

$$\frac{\delta L}{\delta q_i} = 0 \tag{10}$$

$$p_u^T = r_u q(q^T q + \lambda_p I)^{-1} \tag{11}$$

$$p_{uk} = r_{ui}q_{ik}(q_{ki}^T q_{ik} + \lambda_p I_{kk})^{-1} \tag{12}$$

$$q_{ik} = r_{ik}p_{uk}(p^T x + \lambda_q I_{kk})^{-1} \tag{13}$$

Algorithm 2: Algorithm ALS

Input: $R_{N \times M} = r_{ui}$
K: Number of factors
η: Learning rate
λ: Regularization
b_i: *item bias*
b_u: *user bias*
Output: latent vector of $U_{N \times K}$ *and* $V_{K \times M}$
1 Initialize the latent vectors P and Q randomly and $x = 0$.
2 **while** $x \leq$ *iteration* **do**
3   **for** *each rating r_{ui} in R* **do**
4   update p_{uk} *as per* Equation (12);
5   update q_{ik} *as per* Equation (13);
6   **end**
7   i=i+1;
8 **end**

---

4 Experiments

In this part, we present the datasets and evaluation measurements utilized in our tests and afterward investigate the test brings about detail.

4.1 Datasets

MovieLens is one of the predominant datasets in RS. We use MovieLens 100K[1] in our work, where each user has rated at least 20 movies within a range from 1 to 5 and approximately 93% of items do not interact with the user, which signifies that data is more sparse data.

4.2 Evaluation Metrics

To measure the performance of CRS, we use two standard metrics that is mean squared error (MSE) and mean absolute error (MAE) [2,3]. MSE is defined in Eq. 14.

$$MSE = \sum_{(u,i) \in N} \frac{(r_{u,i} - \hat{r_{ui}})^2}{|N|} \qquad (14)$$

N denotes test set pairs of users/items, and the taste set size is denoted as $|N|$. Similarly, MAE is defined in Eq. 15.

$$MAE = \sum_{(u,i) \in N} \frac{|(r_{u,i} - \hat{r_{ui}})|}{|N|} \qquad (15)$$

---

[1] https://grouplens.org/datasets/movielens/100k/.

5   Result and Discussion

This section describes the experimental analysis of our work and compares it with the baseline MF. For the experimental purpose, we split the data into training and testing set with the proportion of 80 : 20. We set the hyper-parameters for the proposed work as shown in Table 1. The optimal hyper-parameters found for the als procedure are K = 20 and λ = 0.01, whereas for the sgd the optimal hyper-parameters are K = 80,λ = 0.01, and η = 0.001. Further, the performance of the proposed approach and the baseline method are shown in Table 2, and it is found that our approach is more suitable than the baseline as it reduced the prediction error in both cases (MSE, MAE).

Table 1. Hyper-parameters.

Hyper-parameters	Values
Learning Rate(η)	[1e−5, 1e−4, 1e−3, 1e−2]
Latent factors (K)	[5, 10, 20, 40, 80, 100]
Regularization(λ)	[0.001, 0.01, 0.01, 1, 10]

Table 2. Performance of the model on movielens 100K dataset.

Techniques	MSE	MAE
ALS without parameter optimization	10.05265	4.4564
Optimization ALS (our approach)	8.06145	2.5704
SGD without parameter optimization	0.9147	0.7696
Optimization SGD (our approach)	0.8847	0.7583

Similarly, Fig. 1(a) shows that there is a reasonable amount of over-fitting that is the test MSE is ≈50% greater than training MSE for als procedure without tunning the parameters. The test MSE bottoms out around five iterations then actually increases after that (even more over-fitting). Simultaneously, after tunning regularization and the number of factors to alleviate some of the over-fitting as shown in Fig. 1(b).

Similarly, Fig. 2(a) shows that there is a reasonable amount of over-fitting that tests MAE is ≈50% greater than training MAE for als procedure without tunning the parameters. The test MAE bottoms out around five iterations then actually increase after that (even more over-fitting). In contrast, after tunning regularization and a number of factors to alleviate the over-fitting as shown in Fig. 2(b).

Figure 3(a) shows that there is a reasonable amount of over-fitting that tests MSE is ≈50% > training MSE for sgd procedure without tunning the parameters. Also, the test MSE reaches as far down as possible around 20 iterations. In contrast, after tunning regularization and a number of factors to alleviate some of the over-fitting as shown in Fig. 3(b).

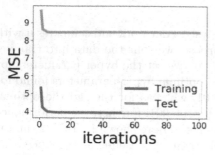
(a) Without optimizing parameters

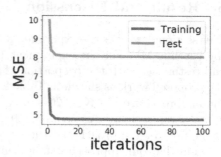
(b) With optimizing the parameters

Fig. 1. Training and testing mean square error using ALS procedure. (a) test MSE is approximately 50% > the training MSE, (b) after tuning the parameters the error is reduced.

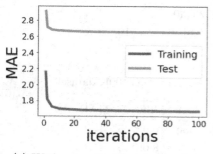
(a) Without optimizing parameters

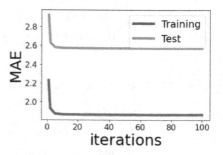
(b) With optimizing the parameters

Fig. 2. Training and testing mean absolute error using ALS procedure. (a) MAE is reasonable amount of overfitting as there is no tunning of regularization. (b) reasonable amount of over-fitting resuced due to tunning of hyperparameters.

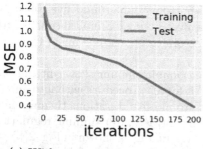
(a) Without optimizing parameters.

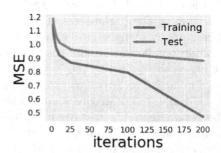
(b) With optimizing parameters.

Fig. 3. Training and testing mean square error using SGD procedure.

Similarly, Fig. 4(a) shows that there is a reasonable amount of over-fitting that is the test MAE is ≈50% > training MAE for als procedure without tunning the parameters. The test MAE reaches as far down as possible around 20 iterations, then actually increases after that (even more over-fitting). Whereas after tunning regularization and the number of factors, it alleviates over-fitting, as shown in Fig. 4(b).

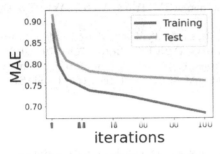
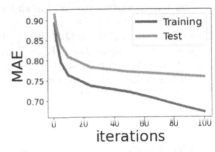

(a) Without optimizing parameters   (b) With optimizing the parameters

Fig. 4. Training and testing mean absolute error using SGD procedure.

6   Conclusion

In our work, we propose an optimized stochastic gradient (SGD) and alternate least squaring (ALS) learning approach for matrix factorization techniques to achieve more accurate prediction for a collaborative recommendation system. In this work, we optimized the hyperparameters such as the number of factors (K), learning rate (η), regularization (λ), and found the optimal values as 80, 0.001, 0.01 for SGD, respectively. Whereas K = 20, λ = 0.01 for ALS.

Our board test results show that we obtain low prediction error compared with baseline matrix factorization (without parameters tunning) algorithms, especially on a sparse dataset. We conclude that our method accomplishes lesser prediction error after optimizing the model parameters than the standard matrix factorization.

References

1. Adomavicius, G., Tuzhilin, A.: Toward the next generation of recommender systems: a survey of the state-of-the-art and possible extensions. IEEE Trans. Knowl. Data Eng. **17**(6), 734–749 (2005)
2. Behera, G., Bhoi, A.K., Bhoi, A.: UHWSF: univariate holt winter's based store sales forecasting. In: Udgata, S.K., Sethi, S., Srirama, S.N. (eds.) Intelligent Systems. LNNS, vol. 185, pp. 283–292. Springer, Singapore (2021). https://doi.org/10.1007/978-981-33-6081-5_25

3. Behera, G., Nain, N.: A comparative study of big mart sales prediction. In: Nain, N., Vipparthi, S.K., Raman, B. (eds.) CVIP 2019. CCIS, vol. 1147, pp. 421–432. Springer, Singapore (2020). https://doi.org/10.1007/978-981-15-4015-8_37
4. Behera, G., Nain, N.: Grid search optimization (GSO) based future sales prediction for big mart. In: 2019 15th International Conference on Signal-Image Technology & Internet-Based Systems (SITIS), pp. 172–178. IEEE (2019)
5. Cacheda, F., Carneiro, V., Fernández, D., Formoso, V.: Comparison of collaborative filtering algorithms: limitations of current techniques and proposals for scalable, high-performance recommender systems. ACM Trans. Web (TWEB) 5(1), 1–33 (2011)
6. Cai, Y., Leung, H.f., Li, Q., Min, H., Tang, J., Li, J.: Typicality-based collaborative filtering recommendation. IEEE Trans. Knowl. Data Eng. 26(3), 766–779 (2013)
7. Chen, L., De Gemmis, M., Felfernig, A., Lops, P., Ricci, F., Semeraro, G.: Human decision making and recommender systems. ACM Trans. Interactive Intell. Syst. (TiiS) 3(3), 1–7 (2013)
8. Funk, S.: Netflix update: Try this at home (2006)
9. He, X., Zhang, H., Kan, M.Y., Chua, T.S.: Fast matrix factorization for online recommendation with implicit feedback. In: Proceedings of the 39th International ACM SIGIR Conference on Research and Development in Information Retrieval, pp. 549–558 (2016)
10. Herlocker, J.L., Konstan, J.A., Terveen, L.G., Riedl, J.T.: Evaluating collaborative filtering recommender systems. ACM Trans. Inf. Syst. (TOIS) 22(1), 5–53 (2004)
11. Hu, Y., Koren, Y., Volinsky, C.: Collaborative filtering for implicit feedback datasets. In: 2008 Eighth IEEE International Conference on Data Mining, pp. 263–272. IEEE (2008)
12. Huang, Z., Chen, H., Zeng, D.: Applying associative retrieval techniques to alleviate the sparsity problem in collaborative filtering. ACM Trans. Inf. Syst. (TOIS) 22(1), 116–142 (2004)
13. Koren, Y.: Factorization meets the neighborhood: a multifaceted collaborative filtering model. In: Proceedings of the 14th ACM SIGKDD International Conference on Knowledge Discovery and Data Mining, pp. 426–434 (2008)
14. Koren, Y., Bell, R., Volinsky, C.: Matrix factorization techniques for recommender systems. Computer 42(8), 30–37 (2009)
15. Li, N., Li, C.: Zero-sum reward and punishment collaborative filtering recommendation algorithm. In: 2009 IEEE/WIC/ACM International Joint Conference on Web Intelligence and Intelligent Agent Technology, vol. 1, pp. 548–551. IEEE (2009)
16. Mavridis, A.: Matrix factorization techniques for recommender systems (2017)
17. Mehta, R., Rana, K.: A review on matrix factorization techniques in recommender systems. In: 2017 2nd International Conference on Communication Systems, Computing and IT Applications (CSCITA), pp. 269–274. IEEE (2017)
18. Meng, J., Zheng, Z., Tao, G., Liu, X.: User-specific rating prediction for mobile applications via weight-based matrix factorization. In: 2016 IEEE International Conference on Web Services (ICWS), pp. 728–731. IEEE (2016)
19. Paterek, A.: Improving regularized singular value decomposition for collaborative filtering. In: Proceedings of KDD cup and Workshop, vol. 2007, pp. 5–8 (2007)
20. Patra, B.K., Launonen, R., Ollikainen, V., Nandi, S.: A new similarity measure using bhattacharyya coefficient for collaborative filtering in sparse data. Knowl.-Based Syst. 82, 163–177 (2015)
21. Ricci, F., Rokach, L., Shapira, B.: Recommender systems: introduction and challenges. In: Recommender Systems Handbook, pp. 1–34. Springer (2015)

22. Sarwar, B., Karypis, G., Konstan, J., Riedl, J.: Incremental singular value decomposition algorithms for highly scalable recommender systems. In: Fifth International Conference on Computer and Information Science, vol. 1, pp. 27–8. Citeseer (2002)
23. Xue, W., Xiao, B., Mu, L.: Intelligent mining on purchase information and recommendation system for e-commerce. In: 2015 IEEE International Conference on Industrial Engineering and Engineering Management (IEEM), pp. 611–615. IEEE (2015)

Violence Detection from CCTV Footage Using Optical Flow and Deep Learning in Inconsistent Weather and Lighting Conditions

R. Madhavan, Utkarsh, and J. V. Vidhya[✉]

SRM Institute of Science And Technology, Chennai, Tamil Nadu, India
{rr4845,un5247,vidhyaj}@srmist.edu.in

Abstract. Physical assault detection in surveillance systems plays an extremely significant role in the safety of our city. The current methods include a primitive approach for video-based pre-processing methods, which includes grayscale conversion, image augmentation, etc. The models include 3D Convolutional Neural Networks (3DCNN) and Support Vector Machines (SVMs). The method which we propose consists of a preprocessing method that involves isolating the pixels into a video (or a group of videos) relevant to the action that is taking place in the original video. In addition, a deep learning model which includes a time-distributed Convolutional Neural Network (CNN) + Long Short Term Memory (LSTM) model extracts the relevant spatial and temporal information or features that can help us predicting violence from an input CCTV footage. Given an input CCTV footage, our model will be able to extract the relevant features to distinguish a violent event from a non-violent one, preferably independent of the lighting. In the current approaches, it was observed that the number of relevant pixels dedicated to the action is low (due to naive cropping). The present system doesn't take into account the inconsistent weather and lighting conditions and its accuracy is expected to drop under those conditions.

Keywords: Violence · CCTV · Physical assault · Deep neural network · Computer vision · CNN · LSTM · Fight · Optical flow

1 Introduction

Violence is defined as the use of physical force with an intention to injure, damage, abuse, and/or destroy. Violence can definitely be preventable to some extent. There is a notable relationship between the extents of violent behaviors and certain factors in a country such as poverty, household income and inequality in genders, harmful alcohol consumption, and the absence of decent relationships between the children and parents. Strategies addressing the underlying causes of violence can be relatively effective in preventing violence, although mental, physical health and their personalities have always been a major factor for the

© Springer Nature Switzerland AG 2021
M. Singh et al. (Eds.): ICACDS 2021, CCIS 1440, pp. 638–647, 2021.
https://doi.org/10.1007/978-3-030-81462-5_56

reason of behaviors like these. Under ideal conditions, the police response for turning up on the violent incident should be as low as possible. Currently, the response time is very high due to a lack of witnesses and a decent form of crime detection from CCTV footage since they may have inaccurate results, or not exist at all.

In primitive approaches such as CNN [13] or 3DCNN and SVM [10,11,16], Convolutional LSTM architecture (ConvLSTM) [15] or even Gaussian Model of Optical Flow (GMOF) [19], though they work well under normal conditions, their performance measure is unclear when it is tested under inconsistent weather and lighting conditions. This might be the case even with approaches involving sound [12] since the sound produced by the result of severe climatic conditions such as loud thunder might mask the sound of the actual violence. Judging the situation by evaluating the facial expression of an individual will work only if the resolution of the video is high enough [3].

Our approach consists of a convolutional layer wrapped in a time distributed layer for spatial feature extraction along with LSTM for the temporal feature extraction. The preprocessing consists of a non action reduction using Optical Flow along with Gaussian blur [14] for removal of noise and increasing padding of the subject. Ideally, our approach to classifying violence should be viable in the above-mentioned inconsistent weather and lighting conditions. The end product is made to be easily deployable to a security system device that is connected to a CCTV camera. We are aiming to detect such types of violence from CCTV cameras in order to minimize the damage caused by that. When the model detects violence in the footage, it is possible to convey that information to higher authorities such as the police. This approach has the advantage of dedicating more pixel space to classification, and the result should be relatively more accurate due to the dynamic elimination variation of lighting. Loss of contextual information is reduced since a time-distributed layer allows the processing of a sequence of images simultaneously in one iteration.

Our paper is structured as follows. In "Related Works", we talked about the different approaches that were used to differentiate violent clips from non-violent clips in videos. It is followed by "Related Works", "Preprocessing Techniques", "Model Architecture" and the "Experimental Setup" sections, where we covered our approach to solve the problem. In conclusion, we have concluded the paper in the "Conclusion" section.

2   Related Works

One notable approach is implementing GMOF [19] in which a novel descriptor, termed as OHOF or Orientation Histogram of Optical Flow was also proposed, which is fed into a linear SVM as input for classification.

Most of the approaches implemented for classifying violence in CCTV footage till now consists of 3DCNN and SVM [10,11,16], but they are not tested on real-time datasets. Also, their approaches yield a lower accuracy in inconsistent weather and lighting conditions. Some approaches make use of the sound information captured from the camera microphone [12], but the lack of microphones in CCTV systems is a major disadvantage. Another approach was classifying based on the face features [3], but that requires a high-resolution video, which is far from what today's average CCTV cameras produce. Pose estimation is also a common approach [18] but the most noticeable disadvantage of the approach is that the model factors the pose of only a single person and it does not consider the motion over time. It is also possible to classify the activity based on multiple camera angles [4] but it doesn't cover the partially overlapping and non-overlapping views to cover more areas in the clips and it requires multiple cameras in the first place.

A novel Grey Centroid algorithm is also used in one approach [17] which was decent for long video clips but was inaccurate in varying weather and lighting conditions. Some approaches implemented transfer learning [1] which led to higher performance levels, but they had the same issue discussed above. Some approaches make use of Bi-LSTM [2] but it is difficult to classify diverse datasets. A few machine learning approaches include detection of HOG features [6] and usage of the ViF descriptors [8] which also fail at classifying in inconsistent weather and lighting conditions. An approach with the Shi-Tomasi corner detection algorithm [7] is fast but the tracking of targets becomes difficult in events of overlaps and occlusion. On top of this, we need to establish the background model in advance. In approaches with the SELayer [9], the accuracy would tend to decrease if there is a low proportion of violent event samples in the data. It is to be noted that, there are quite fewer instances of violence present.

3   Preprocessing

In traditional preprocessing, the size of the frames is reduced by a specific amount before they are passed into the model. This action can potentially remove the pixels corresponding to the action that is taking place in the video. We propose a new preprocessing method that crops the image based on the region where the activity is taking place.

We are implementing Optical Flow, Gaussian Blur, Median Blur, and Boundary calculation. The processed image obtained from the above operations will further undergo resizing, normalization, and augmentation, which is then passed as an input to the model.

3.1   Optical Flow

Video clips consist of multiple frames in a sequence. In most of the videos, especially in CCTV footage, the majority of the pixels do not change over time. It is possible to extract the motion that is taking place in such kinds of videos. The kind of motion that is extracted from the videos can efficiently represent the kind of activity that is taking place, which can in turn help the model to learn.

Optical Flow helps us to extract pixels that are dedicated to the motion (violent or non-violent) that is taking place in the video as shown in Fig. 1.

$$\frac{\partial I}{\partial x}V_x + \frac{\partial I}{\partial y}V_y + \frac{\partial I}{\partial t} = 0 \tag{1}$$

In Eq. 1, $I(x, y, t)$ denotes the pixel value of the video at position (x, y) at time t. V_x and V_y denotes the x and y components of the velocity.

(a) Raw Image   (b) Optical Flow applied

Fig. 1. Optical flow

3.2   Gaussian Blur

Gaussian blur refers to the blurring or smoothing of an image by a Gaussian function. We are using Gaussian blur to temporarily remove the fine details of the frame, which can potentially be noise and not contribute to the final result. This is done for determining the boundaries for cropping as shown in Fig. 2.

$$G(x, y) = \frac{1}{2\pi\sigma^2}e^{-\frac{x^2+y^2}{2\sigma^2}} \tag{2}$$

In Eq. 2, x and y are the horizontal and vertical distances from the origin respectively, and σ is the standard deviation of the Gaussian distribution.

(a) Optical Flow   (b) Gaussian Blur applied

Fig. 2. Gaussian blur

3.3   Median Blur

Median blur is a technique of removing noise from images while preserving the edges in the image. We are making use of median blur to further soften the image without losing the details in or near the edges.

$$x = [24, 56, 36, 41, 69, 2]$$
$$y_1 = median([24, 56, 36]) = median([24, 36, 56]) = 36$$
$$y = [36, 41, 41, 41]$$

$$(3)$$

Equation 3 demonstrates the working of median blur in a hypothetical single-dimensional image. Here, x is the input image and y is the output image.

3.4   Boundary Calculation

The boundaries of the active pixels are calculated, which are then expanded to cover a certain aspect ratio (for example, 1:1). A certain amount of padding is also added to the resulting image to account for extra motion during the next few frames as shown in Fig. 3.

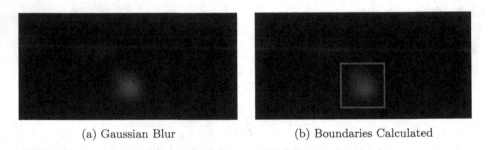

(a) Gaussian Blur   (b) Boundaries Calculated

Fig. 3. Boundary calculation

The original image is then cropped according to the boundaries calculated. In case there are multiple regions where the activity is taking place, multiple boundaries will be calculated and all the regions are extracted into individual frames. The prediction is done for all the frames.

The Gaussian Blur and the boundary calculation steps are performed only once every certain number of frames, to optimize it for real-time usage.

4   Feature Extraction

To differentiate violent video from a non-violent one, the model should be able to process the spatial information in consecutive frames. The information extracted from these consecutive frames points out a certain pattern of action which could be classified as violent or non-violent.

We are using a Deep Neural Network for the recognition of violence in the CCTV clips. A Time Distributed Layer wrapping a Convolutional Neural Network is used to extract the features in every frame of the clip. Those features are then processed simultaneously using a variation of Long Short-Term Memory or LSTM. This combination of LSTM and Time Distributed CNN helps in the analysis of the local motion in the clips by taking the localized spatial and temporal features. As a matter of fact, the differences in the adjacent frames are used as the input to the model, encoding the visual differences in the clips. The architecture diagram is shown in Fig. 4.

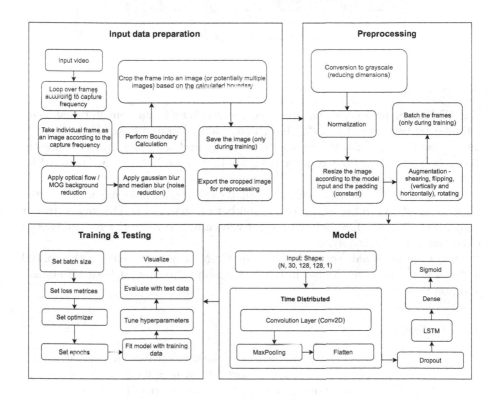

Fig. 4. Architecture diagram

4.1   The Time Distributed Layer

To input a video or a sequence of frames, that are chronologically ordered, we are making use of the Time Distributed Layer. Time Distributed Layer acts as a wrapper that lets us apply a convolutional layer to all the temporal slices which are in this case, the frames of the input video.

In our model, we have wrapped a Conv2D layer, a Max Pooling layer, and a flatten layer, which allows us to perform convolutional operations simultaneously to a large set of frames which are a part of our input video (See Fig. 5).

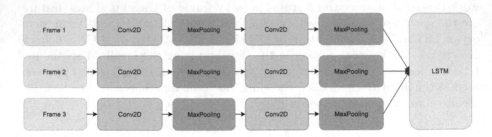

Fig. 5. Time distributed layer

4.2 CNN

The CNN implemented in this approach comprises an input convolutional layer, followed by two sets of Conv2D layer and a Max Pooling layer. Each Conv2D layer is composed of 64, 128 kernels respectively and the filter size is $(3, 3)$ each. The size of the Max Pooling layer is $(2, 2)$. The Conv2D layers are followed by a ReLU activation function.

As input, a CNN takes tensors of shape (h, w, c), where h and w are the dimensions of the image and c is the number of channels in the image. In our model, h and w are 128 and 128 respectively, and c is 3 for the red, green and blue channels respectively.

The Convolutional Neural Network or CNN is used to extract the spatial features from the input frames in order to classify the content between violent, and non-violent.

4.3 LSTM

LSTM or Long Short Term Memory is a variation of RNN (Recurrent Neural Network) which is capable of storing important features for long periods.

The LSTM layer is used to take the processed sequence of frames as the input and gather contextual and temporal information. Since LSTMs can retain information for longer periods, they can process each frame of the video and decipher certain patterns from it. These patterns can be used to determine whether the event is violent or non-violent.

4.4   The Dense Layer

The Dense Layer is the heart of Deep Learning, and it is being used to combine the weights and the input features followed by an activation function, to get the best accuracy over a set amount of epochs. This is also the output layer of the model.

Our model consists of some dense layers, and the final layer consists of 1 neuron. A high value of that neuron (> 0.5) indicates the presence of violence in the video and a low value indicates otherwise.

5   Experimental Setup

5.1   Datasets

We are planning to RWF-2000 [5] as our base dataset since it has a good collection of real-life violent and non-violent examples captured by CCTV cameras. It consists of 1000 violent videos and 1000 non-violent videos. Each video has a duration of 5 s. See Fig. 6.

We are also planning to use the UCF Crime dataset. It is also a real-time dataset. It consists of 13 categories of videos, out of which we have decided to use the videos having the label of 'fights' and 'normal'.

(a) Non Fight Instance   (b) Fight Instance

Fig. 6. RWF-2000 example

5.2   Data Preparation

We are calculating the boundaries every 6 frames and using those measurements for all the 6 frames. After the extraction of the clips in the boundary calculation step, the frames are resized to a shape of 128×128 pixels. We are also batching the input videos by 5. The input to our model will be (N, F, w, h, c), where N is the number of videos per batch, which is 5, F is the number of frames processed in one iteration, which is 6, w and h are the width and height of the frame, which is 128×128, and c is the number of channels, which is 3. Hence, the input shape will be $(5, 6, 128, 128, 3)$.

6 Analysis

In order to implement the approach, we are using the free version of Google Colab which provides one of the following GPUs: Nvidia K80s, Nvidia T4s, Nvidia P4s, and Nvidia P100s. It has an online storage space of 107 GB, a maximum RAM capacity of 12 GB, and an Intel Xeon CPU.

We believe that this approach of classifying violence will yield better accuracy than the state of the art since it retains the significant pixels during the preprocessing using Optical Flow.

Using the pre-processed frames with just the active pixels present helps in a more efficient feature extraction, which could only be possible as a result of using deep learning and optical flow together. The Optical Flow also helps in classification under inconsistent weather and lighting conditions.

7 Conclusion

We have looked at various approaches that have been implemented before for classifying violent and non-violent videos. The major problem we attempted to tackle was the classification in inconsistent weather and lighting conditions. Thanks to our preprocessing, this approach also potentially solves the issue of the dedication of a low amount of pixels for the classification of the video.

References

1. Abdali, A.R., Al-Tuma, R.F.: Robust real-time violence detection in video using CNN and LSTM. In: 2019 2nd Scientific Conference of Computer Sciences (SCCS), pp. 104–108 (2019). https://doi.org/10.1109/SCCS.2019.8852616
2. Akti, Tataroğlu, G.A., Ekenel, H.K.: Vision-based fight detection from surveillance cameras. In: 2019 Ninth International Conference on Image Processing Theory, Tools and Applications (IPTA), pp. 1–6 (2019). https://doi.org/10.1109/IPTA.2019.8936070
3. Ben Ayed, M., Elkosantini, S., Alshaya, S.A., Abid, M.: Suspicious behavior recognition based on face features. IEEE Access 7, 149952–149958 (2019). https://doi.org/10.1109/ACCESS.2019.2947338
4. Bibi, S., Anjum, N., Amjad, T., McRobbie, G., Ramzan, N.: Human interaction anticipation by combining deep features and transformed optical flow components. IEEE Access 8, 137646–137657 (2020). https://doi.org/10.1109/ACCESS.2020.3012557
5. Cheng, M., Cai, K., Li, M.: Rwf-2000: an open large scale video database for violence detection (2020). https://arxiv.org/abs/1911.05913
6. Das, S., Sarker, A., Mahmud, T.: Violence detection from videos using hog features. In: 2019 4th International Conference on Electrical Information and Communication Technology (EICT), pp. 1–5 (2019). https://doi.org/10.1109/EICT48899.2019.9068754
7. Guo, Z., Wu, F., Chen, H., Yuan, J., Cai, C.: Pedestrian violence detection based on optical flow energy characteristics. In: 2017 4th International Conference on Systems and Informatics (ICSAI), pp. 1261–1265 (2017). https://doi.org/10.1109/ICSAI.2017.8248479

8. Hassner, T., Itcher, Y., Kliper-Gross, O.: Violent flows: real-time detection of violent crowd behavior. In: 2012 IEEE Computer Society Conference on Computer Vision and Pattern Recognition Workshops, pp. 1–6 (2012). https://doi.org/10.1109/CVPRW.2012.6239348

9. Jiang, B., Xu, F., Tu, W., Yang, C.: Channel-wise attention in 3d convolutional networks for violence detection. In: 2019 International Conference on Intelligent Computing and its Emerging Applications (ICEA), pp. 59–64 (2019). https://doi.org/10.1109/ICEA.2019.8858306

10. Li, J., Jiang, X., Sun, T., Xu, K.: Efficient violence detection using 3d convolutional neural networks. In: 2019 16th IEEE International Conference on Advanced Video and Signal Based Surveillance (AVSS). pp. 1–8 (2019). https://doi.org/10.1109/AVSS.2019.8909883

11. Perez, M., Kot, A.C., Rocha, A.: Detection of real-world fights in surveillance videos. In: ICASSP 2019–2019 IEEE International Conference on Acoustics, Speech and Signal Processing (ICASSP), pp. 2662–2666 (2019). https://doi.org/10.1109/ICASSP.2019.8683676

12. Ramzan, M., Abid, A., Khan, H.U., Awan, S.M., Ismail, A., Ahmed, M., Ilyas, M., Mahmood, A.: A review on state-of-the-art violence detection techniques. IEEE Access 7, 107560–107575 (2019). https://doi.org/10.1109/ACCESS.2019.2932114

13. Saranya, P., Prabakaran, S.: Automatic detection of non-proliferative diabetic retinopathy in retinal fundus images using convolution neural network. J. Ambient Intell. Humanized Comput., 1–10 (2020). https://doi.org/10.1007/s12652-020-02518-6

14. Saranya, P., Prabakaran, S., Kumar, R., Das, E.: Blood vessel segmentation in retinal fundus images for proliferative diabetic retinopathy screening using deep learning. Visual Comput., 1–16 (2021). https://doi.org/10.1007/s00371-021-02062-0

15. Sharma, M., Baghel, R.: Video surveillance for violence detection using deep learning. In: Borah, S., Emilia Balas, V., Polkowski, Z. (eds.) Advances in Data Science and Management, pp. 411–420. Springer, Singapore (2020)

16. Soliman, M.M., Kamal, M.H., El-Massih Nashed, M.A., Mostafa, Y.M., Chawky, B.S., Khattab, D.: Violence recognition from videos using deep learning techniques. In: 2019 Ninth International Conference on Intelligent Computing and Information Systems (ICICIS), pp. 80–85 (2019). https://doi.org/10.1109/ICICIS46948.2019.9014714

17. Song, W., Zhang, D., Zhao, X., Yu, J., Zheng, R., Wang, A.: A novel violent video detection scheme based on modified 3d convolutional neural networks. IEEE Access 7, 39172–39179 (2019). https://doi.org/10.1109/ACCESS.2019.2906275

18. Teslya, N., Ryabchikov, I., Lipkin, E.: The concept of the deviant behavior detection system via surveillance cameras. In: Misra, S., et al. (eds.) ICCSA 2019. LNCS, vol. 11624, pp. 169–183. Springer, Cham (2019). https://doi.org/10.1007/978-3-030-24311-1_12

19. Zhang, T., Yang, Z., Jia, W., Yang, B., Yang, J., He, X.: A new method for violence detection in surveillance scenes. Multimed. Tools Appl. 75(12), 7327–7349 (2016). https://doi.org/10.1007/s11042-015-2648-8

Speech Based Multiple Emotion Classification Model Using Deep Learning

Shakti Swaroop Patneedi[1]([⊠]) and Nandini Kumari[2]

[1] Department of Information Technology and Engineering, Vellore Institute of Technology
Vellore, Tamilnadu, India
[2] Department of Computer Science and Engineering, Birla Institute of Technology Mesra,
Ranchi, India

Abstract. Speech based Emotion Classification (SEC) can be viewed as a static
or dynamic classification issue, which makes SEC a superb proving ground for
exploring and looking at different Deep learning-based models. We portray a
frame-based feature extraction strategy i.e., MFCC (Mel Frequency Cepstral Coefficient) to extract features from speech signals. These features depend on significant emotion and deep neural network-based strategies has been applied to learn
and demonstrate intra-layer features representation of speech to achieve better classification accuracy. The proposed SEC framework includes layer-based strategies
using Convolutional Neural Network (CNN) and feed-forward neural network
models like Long Short-Term Memory (LSTM) and their hybrid structure. The
final results and analyses directed enlighten the favorable circumstances and constraints of these structures proved to be effective in recognizing multiple emotion
states and achieved better training and testing accuracy with 98.62% and 96.54%
respectively on proposed CNN and 90.28% and 88.19% with CNN + LSTM based
SRC model specifically.

Keywords: Deep neural network · Convolutional neural network · Long
short-term memory · Speech emotion classification · Mel-frequency cepstral
coefficients

1 Introduction

Human speech contains the etymological substance, and it carries the emotion of the
speaker additionally [1]. These emotions may assume as a critical part in various applications such as Automatic Speech Recognition to determine the user's emotions which
could improve the interpretation in entertainment electronics to gather emotional user
behaviors, in smart speech-based home assistant devices commonly available in the market and in text-to-speech frameworks to incorporate genuinely more intuitive speech.
In this manner, in human machine interface applications, it is significant that emotion
states in human vocal speech are completely discerned by computers [2]. Deliberating
the human's emotional condition is an eccentric task and might be utilized as a standard
for any emotion recognition model. The discrete emotion approach is contemplated as

M. Singh et al. (Eds.): ICACDS 2021, CCIS 1440, pp. 648–659, 2021.
https://doi.org/10.1007/978-3-030-81462-5_57

a major methodology among the various classification models utilized for detection of these emotions from speech. It utilizes different emotions, for example, anger, boredom, neutral, surprise, surprise, disgust, joy, happiness, fear, sadness, and shock [3, 4]. Another significant model that incorporates a three-dimensional space with boundaries, for example, arousal, valence, furthermore, intensity. Discrete emotion procedure has been generally utilized for SEC with its benefits of being basic and simple to execute and interpret. As of now, the acoustic features have been broadly applied to consider SEC, and numerous noteworthy accomplishments have been made [5, 6]. These emotional evoked features are broadly utilized for classification of data that is taken from fundamental level features. Spectral based features [7] may be divided into linear spectrum-based features and cepstral based features. The linear spectrum-based features comprise of linear predictor coefficient (LPC) and the log-frequency power coefficient (LFPC). The cepstral based features [8] include the Mel-Frequency Cepstral Coefficient (MFCC) and the linear predictor cepstral coefficient (LPCC). Among them, The MFCC is the extensively preferred spectral feature for SEC because of its attributes being alike the human ear.

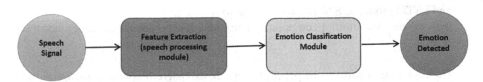

Fig. 1. Speech Emotion Classification (SEC) framework

As appeared in Fig. 1, a Speech Emotion Classification (SEC) framework for the most part consists of following modules: speech procurement module, feature extraction module and emotion classification module [9]. Intuitive speech is acquired using sensors while the speech acquisition module, and afterward sent to the feature extraction module to extricate the speech features. Common speech is extraordinarily influenced by outer components which may prompt a decrease of classification exactness. Hence, before the SEC system can be set up, it is important to gather a speech corpus as indicated by emotion depiction techniques and record a great emotion speech [10]. In the field of speech signal processing, researchers have determined a few features such as source-based excitation features, prosodic features, spectral features, and other hybrid features [1]. At the end, the emotion classification module performs classification procedure dependent on the enthusiastic features separated. This stage incorporates feature classification utilizing linear and nonlinear classifiers. The most ordinarily utilized linear classifiers for emotion detection incorporate Bayesian Networks (BN), Support Vector Machine (SVM) and Maximum Likelihood Principle (MLP). Normally, the speech signal is viewed as nonstationary signal. Henceforth, it is viewed as that nonlinear classifiers work viably for speech emotion detection [11]. There are numerous non-linear classifiers accessible for SEC, including Gaussian Mixture Model (GMM) and Hidden Markov Model (Gee) [12]. Deep Learning based methods for SEC have a few preferences over conventional strategies, including their ability to recognize the perplexing construction and feature learnings with deep layers and weight tuning/adjustments; and capacity to manage and classify data having no prior information.

Deep Learning based methods for SEC have a few preferences over conventional strategies, including their ability to recognize the perplexing construction and feature learnings with deep layers and weight tuning/ adjustments; and capacity to manage and classify data having no prior information. Deep Neural Networks (DNNs) depend on feed-forward structures contained at least one fundamental shrouded layer among initial data. sources and final yields. The feed-forward models for example, DNNs based methods gives competent results to image furthermore, video handling. Then again, repetitive structures for example, Recurrent Neural Networks (RNNs), Convolutional Neural Network (CNNs) and Long Short-Term Memory (LSTM) are a lot of viable in speech and sound based classification, for example, natural language handlining and SEC. The paper has the following accompanying contributions:

1) We Proposed Deep learning-based models to classify different emotional state of human using their speech voice.
2) We trained a Deep Convolutional neural network and combination of convolution and LSTM model on SAVEE (Survey Audio-Visual Expressed Emotion) dataset and then classify the multiple emotional state and evaluate the classification accuracy results.

The following section of the paper presents an extensive literature review, followed by the methodology section including dataset description, the proposed architecture and the implementation of the model. At last, the results are documented followed by the conclusion section.

2   Literature Review

Different classifiers like K-Nearest Neighbour (KNN), Principal Component Examination (PCA) and Decision trees are likewise adapted for emotion classification. Deep Learning has been contemplated as an arising research matter in AI and has acquired consideration in last decades. Traditional techniques for distinguishing the emotion substance of a speech signal have a few difficulties. The primary challenges stem from the way that it is very hard to characterize the literal information regarding emotion and how it tends to be classified based on audio. These techniques for solving this problem are extracting low-level features and training the machine suitably through learning those features. These techniques have been acknowledged as state of art for a long time in AI. However, feature selection is also difficult, and advancement is significantly more troublesome, regularly being essentially tedious in research. Because of this, the traditional trend in audio information retrieval is providing the use of powerful strategies. In recent years, there has been a radical development in the use of neural network utilizing Deep learning to tackle out different classification problems. Deep learning techniques can share low-level representation and normally progress from low-level to significant level structures. Therefore, deep models consequently learn corresponding features by stacking network layers. Feed-forward based network, for example, DNN and CNN have been especially fruitful in image and video classification as well as speech, while recurrent based networks, for example, RNN and LSTM have been viable in speech detection and natural language processing [13, 14]. These architecture and model can be implemented in various ways and have their own points of interest and constraints. For example, ConvNets can manage high-dimensional inputs and learn features that are invariant [15], though LSTMRNNs can manage variable length data sources and model sequential information with long term context [16]. CNN based picture, video, speech, sound and music recognition techniques have been proposed in a few exploration research [17]–[19]. From these investigations, we definitely realize that CNN-based investigation can be applied in one-dimensional signals, for example, speech and sound [20]. Specifically, a CNN based SEC technique has been suggested that learns striking features of SEC utilizing semi-CNNs [21]. The SEC framework utilizing RNNs was proposed in [22], which represents long context-oriented impact in enthusiastic discourse and the vulnerability of enthusiastic names. Moreover, sound/video based multimodal emotion recognition approaches were tested in [23, 24]. SER assumes a successful part in regions of normal human machine connections or interactions, for example, computer instructional exercise applications in which the system reaction relies on emotions of the client [1, 29]. Likewise it can be utilized for in vehicle board framework where data of psychological state of driver might be given to the framework to start wellbeing strategies, whenever required [25]. Moreover, clinical utilization of speech emotion acknowledgment incorporates analytic apparatus for advisors [26]. Further applications incorporate assurance of situational earnestness in emergency call focuses based of human emotion investigation from speech information plays vital role [27]. The fundamental topic of SEC frameworks is to identify specific qualities of speaker's voice in fluctuating passionate conditions. This paper is going to work on improvement of SRC accuracy with less number of layers and parameters utilizing CNN and LSTM models. The accuracy results and comparison are also provided in the results and discussion section to represent the effectiveness of our work.

3 Methodology

3.1 Dataset Description

In this paper, SAVEE dataset (Survey Audio-Visual Expressed Emotion) has been utilized for SEC which is freely available data repository. This is Created by Vlasenko et al. in 2007. This dataset contains the speech recording of four British actors. The 4 native English actors were names as DC, JE, JK, KL. They all were aged between 27 and 31. This dataset consists of 7 different kind of emotions which include happiness, disgust/fear, anger, surprise, sadness, neutral and common. Every sentence which was used to record were selected by experts from TIMIT Corpus. These recordings have 15 phonetically balanced TIMIT sentences and there are also 30 sentences for neutral state. It contains of 480 samples in total which are collected with utmost care and precision. They recorded the data by painting around 60 mark on the face of each actor for the extraction of features.

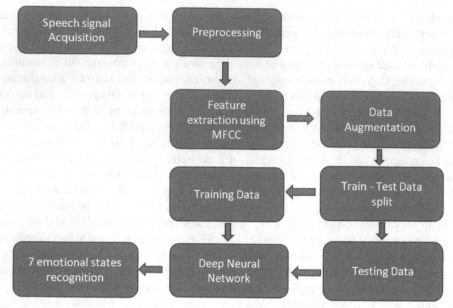

Fig. 2. Overall workflow of proposed Deep neural network-based model for Speech Emotion Classification

3.2 Feature Extraction from Speech Signal Using MFCC (Mel-frequency Cepstral Coefficients).

As described earlier, the MFCC feature extraction has been widely utilized by researchers and it provided the improved results with machine learning algorithms. This feature of speech signals is usually includes windowing signals, applying DFT, to get log of magnitude, and enfolding the frequencies on Mel scale. Ending with applying inverse DCT. The following steps explain the complete MFCC [28].

1) Frame blocking: - The speech signal is a slow time variant signal. For stable acoustic qualities, vocal speech should be analyzed over an adequately brief timeframe. Therefore, speech analysis consistently be completed on short fragments across which the speech signal is assumed to be stationary. Generally, hamming windows with 20ms window size are used. This is achieved to strengthen the harmonics, soften the edges, and decrease the edge influence on the signal when taking the DFT.

2) DFT spectrum: - Each window frame is transformed into a spectrum of magnitude by DFT Application using Eq. (1).

$$X(k) = \sum_{t=0}^{T-1} x(t) e^{\frac{-j2\pi nk}{T}}; 0 \le k \le T-1 \tag{1}$$

Where T is total no of points used for the computation of DFT.

3) Mel spectrum: - The spectrum of Mel is determined by passing the transformed Fourier to a group of bandpass filters called as a Mel-filtering are used to signal

shown in Eq. (2). A Mel is a metric based on the perceived frequency of human ears. It does not comply with the human auditory system, linear to the physical frequency of the tone pitch obviously does not linearly interpret pitch.

$$f_{Mel} = 2595 log_{10}\left(1 + \frac{f}{700}\right)$$ (2)

4) DCT: - The DCT is applied and a collection of cepstral coefficients is generated by transforming Mel frequency coefficients. The Mel spectrum is usually interpreted on a log scale for DCT computation. This, results in a cepstral domain signal with a corresponding frequency peak to the signal pitch. Finally, MFCC has been calculated using Eq. (3).

$$c(t) = \sum_{m=0}^{M-1} log_{10}(s(m)) cos\left(\frac{\pi t(m-0.5)}{M}\right); \qquad t=0,1,2.....C-1$$ (3)

Where $c(t)$ is the cepstrum coefficients and C is number of MFCCs

3.3 Speech Emotion Recognition using Deep Neural Networks

Since the objective is to classify 7 emotion state from speech signals images using 2 proposed Deep learning-based models. To achieve this objective, several experiments were carried out to assess the suitability of proposed method for emotion recognition. Two different set of experiments were performed. A modified CNN based SEC model and a combination of CNN and LSTM based SEC model has been proposed and implemented on SAVEE dataset. Initially the acquired dataset has been preprocessed and 13 MFCC features has been extracted from speech signals. These features are then augmented by using pitch tuning and adding white noise to the speech signals to increase the data size. Finally, these data have been divided into training and testing data and fed to the both proposed Deep neural network based SRC model which classifies the 7 emotional states and provides the overall training and testing classification accuracy. Figure 2 depicts the workflow of overall proposed model architecture. Specifically, many researches has been done using SAVEE dataset, it appears that improved accuracy results has been achieved with traditional as well as machine learning methods.

CNN Based SEC Model.
In this paper, a modified version of CNN model has been implemented and trained on SAVEE dataset and then the model has been tested on 7 emotion states to evaluate the performance of the proposed CNN based SER model. CNN takes 1D MFCC extracted features as input. As the initial data are not sufficient for Deep learning models as they require huge data to train the model. Therefore, data augmentation has been carried out to increase the size of dataset. The proposed CNN model consists of 4 convolutional layers, a MaxPooling layer and a fully connected layer. The overall architecture of proposed CNN based SRC model is shown in Fig. 3 and detailed description with input/output size and other parameter's value are shown in Table 1.

Here, the whole architecture of CNN model has been described significantly:

Table 1. Architecture of convolutional neural network based SEC model

Layer	Input	Operation	Filter size	Strides	Output
1	216 × 1	Conv1D + Relu	256 × 8	1	216 × 256
2	216 × 256	Conv1D + Relu	128 × 8	1	216 × 128
3	216 × 128	Maxpool	8	2	27 × 128
4	27 × 256	Conv1D + Relu	128 × 8	1	27 × 128
5	27 × 128	Conv1D + Relu	128 × 8	1	27 × 128
6		Dropout (0.2)			
7	27 × 128	Dense + Relu	64		221248
8	27 × 128	Dense + Softmax	7		455

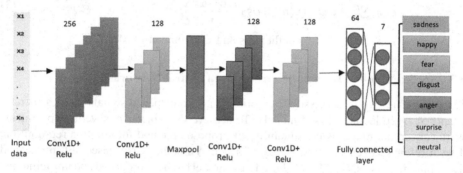

Fig. 3. Proposed CNN based SRC model

1) The first layer of CNN model takes the 1D vector of MFCC feature of size 216 × 1 and applies 256 1D filter with kernel size 8 and outputs 216 × 256 feature map.
2) Then these feature maps are the fed to second Conv1D layer with 128 filter with kernel size 8 and yields 216 × 128 as learned features.
3) The third layer consist of a MaxPooling layer with size 8 and stride rate 2 to reduce the resultant feature size to 27 × 128 and discard the irreverent and redundant information.
4) The next two CNN layer convolve the resultant feature map with same feature size of 128 and kernel size 8 to get the most prominent features.
5) Finally, a dropout of 20% has been applied to reduce over fitting problem and 2 fully connected layer with dense size 64 and 7 are taken. As the last dense layer denotes the 7 emotion states.

CNN + LSTM Based SEC Model.
A combination of CNN and LSTM model has also been implemented, trained on SAVEE dataset, and then evaluated the performance accuracy. CNN takes 1D MFCC extracted features as input. As the initial data are not sufficient for Deep learning models as they

require huge data to train the model. Therefore, data augmentation has been carried out to increase the size of dataset. The proposed CNN model consists of 4 convolutional layers, a MaxPooling layer, a LSTM layer and a fully connected layer. The detailed description of CNN + LSTM based SRC model with input/output size and another parameter's value are shown in Table 2.

1. The first Conv1D layer receives the MFCC feature vector of speech signals taken from SAVEE dataset and the fed to the CNN + LSTM based SER model with 216 × 1 input size and outputs a 126 × 128 feature map by applying 256 and 128 kernels with kernel size 5 and stride 1, followed by a MaxPooling layer of size 5 which suppresses the output to 27 × 128.
2. The third and fourth layer comprises of 128 kernels with 5 kernel size with stride of 1 and yields 27 × 128 feature map.
3. At the end of the convolutional layer, the resultant feature map from the previous convolutional layer is fed to the LSTM layer with 128-dimension size and yields 3456 feature map after fattening the resultant feature map.
4. Finally, a fully connected with a dense layer of output neurons 7 has been applied which derive the probabilities for multiple classification of 7 emotion states, using 20% dropout and SoftMax function.

Table 2. Architecture of CNN + LSTM based SEC model

Layer	Input	Operation	Filter size	Strides	Output
1	216 × 1	Conv1D + Relu	256 × 8	1	216 × 256
2	216 × 256	Conv1D + Relu	128 × 8	1	216 × 128
3	216 × 128	Maxpool	8	2	27 × 128
4	27 × 256	Conv1D + Relu	128 × 8	1	27 × 128
5	27 × 128	Conv1D + Relu	128 × 8	1	27 × 128
6	27 × 128	LSTM	128		138594
7		Dropout (0.2)			
8	27 × 128	Dense + Softmax	7		24199

Additionally, both CNN and CNN + LSTM network was trained with batch size of 32, Categorical CrossEntropy loss function and Adam optimizer was used for 200 epochs. For optimizer, the learning rate was set to 0.0002 and decay to 0.5.

4 Results and Discussion

This paper proposes a Deep learning-based SEC model namely CNN and CNN + LSTM model has been proposed where a modified version of CNN and combination of CNN + LSTM model has been used for multiple classification of speech-based emotions.

The objective of this paper is to classify 7 emotional states from speech signals of subject. In this paper, SAVEE dataset has 480 audio samples from 4 subjects. These audio speeches are firstly gone through preprocessed and then MFCC feature extraction technique has been implemented which extract the prominent and relevant emotion features from speech signals. Although less data may cause overfitting with CNN architecture therefore data augmentation has been used to increase the volume of data for training of proposed methodology. The performance for proposed models is evaluated with the test set and classification accuracy has been recorded. The performance accuracy of proposed CNN based SEC model and CNN + LSTM based SEC model is shown in Table 3. The accuracy and loss graph for both models are shown in Fig. 4 and 5 respectively. It can be seen from accuracy graph that CNN model has been achieved good accuracy as compared to CNN + LSTM model. In general performance accuracy of CNN based SRC model was obviously the best contrasted and other CNN + LSTM based SRC model. The training accuracy of multiple emotion states classification 98.62% and 90.28% and testing accuracy 96.54% and 88.19% has been achieved utilizing CNN and CNN + LSTM based SRC model on 200 epochs.

Table 3. Classification accuracy of proposed CNN and CNN + LSTM based SEC model

Model	Train accuracy (%)	Test accuracy (%)
CNN	98.62	96.54
CNN + LSTM	90.28	88.19

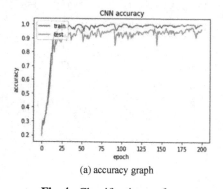

(a) accuracy graph

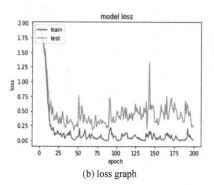

(b) loss graph

Fig. 4. Classification performance graph of proposed CNN based SRC model

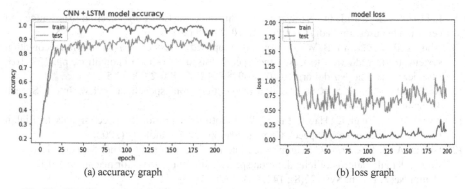

(a) accuracy graph   (b) loss graph

Fig. 5. Classification performance graph of proposed CNN + LSTM based SRC model

5 Conclusion

Diverse Deep learning-based models were investigated on Speech Emotion Classification (SEC) task. Various research work has been illuminating the effectiveness of CNN and LSTM based models in terms of accuracy achieved. CNN showed better discriminative execution contrasted with different designs. the Mel-Recurrence Cepstral Coefficient (MFCC) feature extraction technique has been applied to lead speaker-autonomous tests for multiple class Speech Emotion Classification (SER) framework. To examine the articulation impact of MFCC highlights on audio-based emotions in detail. In this paper, CNN, and CNN + LSTM based SEC model has been proposed for multi emotion classification using SAVEE dataset. The modified CNN model and CNN + LSTM model's performance accuracy are comparable, but CNN worked better than CNN + LSTM. Speech Emotion Recognition (SER) has been widely used in many fields, such as smart home assistants commonly found in the market. In future work, the proposed architecture can be used for such applications and can be also explored for various dataset and real speech samples.

References

1. El Ayadi, M., Kamel, M.S., Karray, F.: Survey on speech emotion recognition: Features, classification schemes, and databases. Pattern Recogn. **44**(3), 572–587 (2011)
2. Cowie, R., et al.: Emotion recognition in human computer interaction. IEEE Sig. Process Mag. **18**(1), 32–80 (2001)
3. Chan, K., Hao, J., Lee, T., Kwon, O.W.: Emotion recognition by speech signals. In: Proceedings of International Conference EUROSPEECH, Citeseer (2003)
4. Picard, R.W.: Affective Computing. MIT press (2000)
5. Zualkernan, I., Aloul, F., Shapsough, S., Hesham, A., El- Khorzaty, Y.: Emotion recognition using mobile phones. Comput. Electric. Eng. **60**, 1–13 (2017)
6. Alonso, J. B., Cabrera, J., Travieso, C. M., López, K., Sánchez-Medina, A.: Continuous tracking of the emotion temperature. Neurocomputing **255**, 17–25 (2017). https://doi.org/10.1016/j.neucom.2016.06.093
7. Jiang, L., Tan, P., Yang, J., Liu, X., Wang, C.: Speech emotion recognition using emotion perception spectral feature. Concurrency and Computation: Practice and Experience, p. e5427

8. Lalitha, S., Geyasruti, D., Narayanan, R., Shravani, M.: Emotion detection using mfcc and cepstrum features. Procedia Comput. Sci. **70**, 29–35 (2015)
9. Abhang, P. A., Gawali, B. W., Mehrotra, S.: proposed eeg/speech-based emotion recognition system. In: Introduction to EEG- and Speech-Based Emotion Recognition, pp. 127–163. Elsevier (2016). https://doi.org/10.1016/B978-0-12-804490-2.00007-5
10. Koolagudi, S.G., Rao, K.S.: Emotion recognition from speech: a review. Int. J. Speech Technol. **15**(2), 99–117 (2012)
11. Kwon, O.-W., Chan, K., Hao, J., Lee, T.-W.: Emotion recognition by speech signals. In: Eighth European Conference on Speech Communication and Technology (2003)
12. Dileep, A.D., Sekhar, C.C.: Gmm-based intermediate matching kernel for classification of varying length patterns of long duration speech using support vector machines. IEEE Trans. Neural Netw. Learn. Syst. **25**(8), 1421–1432 (2013)
13. LeCun, Y., Bengio, Y., Hinton, G.: Deep learning. Nature **521**(7553), 436–444 (2015)
14. Kingma, D.P., Ba, J.: Adam: A method for stochastic optimization. arXiv:1412.6980 (2014)
15. Krizhevsky, A., Sutskever, I., Hinton, G.E.: Imagenet classification with deep convolutional neural networks. In: Advances in Neural Information Processing Systems, pp. 1097–1105 (2012)
16. Graves, A.: Supervised Sequence Labelling with Recurrent Neural Networks [ph. d. dissertation]. Technical University of Munich, Germany (2008)
17. Simonyan, K., Zisserman, A.: Very deep convolutional networks for large-scale image recognition. arXiv:1409.1556 (2014)
18. Abdel-Hamid, O., Mohamed, A.-R., Jiang, H., Deng, L., Penn, G., Yu, D.: Convolutional neural networks for speech recognition. IEEE/ACM Trans. Audio Speech Language Process. **22**(10), 1533–1545 (2014)
19. Muhammad, K., Ahmad, J., Lv, Z., Bellavista, P., Yang, P., Baik, S.W.: Efficient deep cnn-based fire detection and localization in video surveillance applications. IEEE Trans. Syst. Man Cybern.: Syst. **49**(7), 1419–1434 (2018)
20. Nassif, A.B., Shahin, I., Attili, I., Azzeh, M., Shaalan, K.: Speech recognition using deep neural networks: a systematic review. IEEE Access **7**, 19 143–19 165 (2019)
21. Khalil, R.A., Jones, E., Babar, M., Jan, T., Zafar, M., Alhussain, T.: Speech emotion recognition using deep learning techniques: a review. IEEE Access **7**, 117327–117345 (2019). https://doi.org/10.1109/ACCESS.2019.2936124
22. Mirsamadi, S., Barsoum, E., Zhang, C.: Automatic speech emotion recognition using recurrent neural networks with local attention. In: 2017 IEEE International Conference on Acoustics, Speech and Signal Processing (ICASSP), pp. 2227–2231. IEEE (2017)
23. Yoon, S., Byun, S., Jung, K.: Multimodal speech emotion recognition using audio and text. In: IEEE Spoken Language Technology Workshop (SLT), pp. 112–118. IEEE (2018)
24. Zhao, J., Mao, X., Chen, L.: Speech emotion recognition using deep 1d & 2d cnn lstm networks. Biomed. Signal Process. Control **47**, 312–323 (2019)
25. Schuller, B., Rigoll, G., Lang, M.: Speech emotion recognition combining acoustic features and linguistic information in a hybrid support vector machine-belief network architecture. In: 2004 IEEE International Conference on Acoustics, Speech, and Signal Processing, vol. 1, pp. I–577. IEEE (2004)
26. France, D.J., Shiavi, R.G., Silverman, S., Silverman, M., Wilkes, M.: Acoustical properties of speech as indicators of depression and suicidal risk. IEEE Trans. Biomed. Eng. **47**(7), 829–837 (2000)
27. Ahmad, J., Muhammad, K., Kwon, S., Baik, S.W., Rho, S.: Dempster-shafer fusion based gender recognition for speech analysis applications. In: 2016 International Conference on Platform Technology and Service (PlatCon), pp. 1–4. IEEE (2016)

28. Murty, K.S.R., Yegnanarayana, B.: Combining evidence from residual phase and mfcc features for speaker recognition. IEEE Signal Process. Lett. **13**(1), 52–55 (2005)
29. Aouani, H., Ayed, Y.B.: Speech emotion recognition with deep learning. Procedia Comput. Sci. **176**, 251–260 (2020)

A Legal-Relationship Establishment in Smart Contracts: Ontological Semantics for Programming-Language Development

Vimal Dwivedi$^{(\boxtimes)}$ (iD) and Alex Norta (iD)

Tallinn University of Technology, Akadeemia tee 15 a, Tallinn, Estonia
vimal.dwivedi@taltech.ee, alex.norta.phd@ieee.org

Abstract. Machine-readable smart contracts (SC) on blockchains promise drastic enhancements in collaboration efficiency and effectiveness in that cost- and time reductions can be achieved while the quality of services increases. We address existing shortcomings of SCs that are in tendency incomplete for legal recognition especially to smart-contract-enabled funding rounds, not collaborative business-process reflective and are also not aware of their own processing state to justify the claim of smartness. When conflicts occur, tracing the past performance of conventional contract (CC) execution is very slow and expensive while in addition, CCs are challenging to enforce. On the one hand, the legal status of SCs based funding rounds is currently not clarified and the question arises if SCs comprise the necessary legal- concepts and properties. Current SC solutions do not suffice in those regards. To fill this gap, we develop the smart-legal-contract (SCL) ontology to define the legal- and collaborative business concepts and properties in the SCs. Formal methods, such as Colored Petri Nets (CPNs), are suitable to design, develop and analyze processing state of SCs in order to trace the performance of contractual-rights and obligations. In this work, SCL ontology is formalized using Colored Petri Nets resulting in a verifiable CPN model. Furthermore, we conduct a state-space analysis on the resulting CPN model and derive specific model properties. A running case from the automotive supply chain domain demonstrates the utility and validity of our approach.

Keywords: Smart contract · Decentralized autonomous organization · Legal recognition · Blockchain · Ontology · Business process · B2B

1 Introduction

Traditionally the concept of the contract covers a spoken or written agreement governed by court procedures. The first prerequisite to becoming a legally valid contract is that the contracting parties are engaged voluntarily to reach a consensus. In a traditional contract, a service is offered for some form of compensation

© Springer Nature Switzerland AG 2021
M. Singh et al. (Eds.): ICACDS 2021, CCIS 1440, pp. 660–676, 2021.
https://doi.org/10.1007/978-3-030-81462-5_58

(usually money) and some other provisions (e.g., contract terms and conditions, the service delivery dates, liability and compensation for the breach, and so on). Subsequent transactions are based on trust, and contracting parties generally see contracts as a symbol of an existing business deal. Another drawback in traditional way of establishing and managing contracts is that they are often underspecified. More importantly, traditional contracts do not provide sufficient details about the actual process of the transaction and, as a result, frictions between the contracting parties are very likely to occur, e.g., one party assumes a specific product certificate before delivering a partial compensation, and the other party assumes the contrary. The resulting deadlocks result in costly conflict resolution or even the entire contract transaction collapsing. Traditional contract enforcement is also proving to be either too complicated, time-consuming, or impossible, certainly in international circumstances.

Blockchain has established a new type of decentralized autonomous organization (DAO) whose activities run on a peer-to-peer network, involving governance as well as decision making rules. The latter is an organization whose business provisions are written in a programming code, and the necessary business operations are controlled automatically as per the agreeing to the provisions [25]. DAO encourages re-implementing each aspect of traditional organizational governance, replacing voluntary compliance with a business's agreement with actual compliance using pre-agreed smart-contract code. The latter is machine-readable software code that is situated on the protocol layer of a blockchain system to govern transactions between DAOs [2]. Subsequently, Ethereum [1] emerges as the first smart-contract system where the protocol layer is equipped with a Turing-complete programming language. This innovation affects a growing number of application cases, e.g., in logistics [3], e-healthcare [8], cyber-physical systems [28] such as for smart electrical-grid production [10], and so on. In the meantime, various smart-contract systems exist such as Neo[1], Cardano[2], Hyperledger[3], etc. with varying blockchain types, consensus algorithms, and machine-readable languages [15,29].

By punishing opportunistic behavior, contracts and contract law lead to the enforcement of the intentions initially specified in the contract by the acting parties. Contracts are usually defined as legally-binding agreements that stipulate the rights and obligations of the contracting party towards each other [30]. This study [5] shows when code is law, it refers to the idea that, with the advent of digital technology, code has progressively established itself as the predominant way to regulate the behavior of Internet users. Yet, while computer code can enforce rules more efficiently than legal code, there are also limitations, mostly because of the difficulty to transpose the ambiguity and flexibility of legal rules into a formalized language for interpretation by a machine. A number of studies focus on checking the legality of smart contracts. This article [6] considers the

---

[1] https://neo.org/ Neo blockchain—Home Page.
[2] https://cardano.org/ Cardano—Home Page.
[3] https://www.hyperledger.org/ Hyperledger—Home Page.

potential issues with legal and practical enforceability that arise from the use of smart contracts within both civil- and common-law jurisdictions.

This paper [13] shows, the technology of smart contracts neglects the fact that people use contracts as social resources to manage their relations. The inflexibility that they introduce, by design, short-circuit a number of social uses to which law is routinely put. Few studies show that smart contracts are a new form of preemptive self-help that should not be discouraged by the legislatures, or courts [23]. A smart contract gives rise to a novel means of legally enforcing obligations and rights. These issues are treated differently from country to country. Smart contracts that underpin transactions in ICOs (Initial Coin Offerings – e.g., KodakCoin) may be illegal in some jurisdictions, while a smart contract that handles intra-institutional banking and other financial transactions is considered as legal, in the same jurisdiction, or elsewhere [22]. Another study show the necessary requirements and design options for the legality of smart-contract forms and proposes future research directions [4]. The future research direction aims to provide straight-through processing of financial contracts, with highly automated smart contract code to entire semantics of smart contract. This future direction opted by the another study [17]. This research shows the significance of highly automated smart contract so called self-aware smart-contract in the real world financial contract. In [22], another initiative investigates the legal enforceability of smart contracts. In this research, the findings render smart contracts legally enforceable by incorporating crypto primitives such as a digital signature.

Thus, the state of the art shows that CCs cause high transaction costs due to their multiple shortcomings, SCs lack legal relevance and this yields legal uncertainties for users while furthermore, SCs are inflexible code that are not smart. This paper fills the gap by answering the main research question of how to establish legal relevance for smart contracts that have socio-technical utility? To establish a separation of concerns, we deduce the following sub-questions. What conditions need to be fulfilled for SCs to have legal relevance? What are the properties of an ontology that represent these conditions for legal relevance? What enactment mechanisms ensure the legal enforceability for contracts?

The remainder of this paper is structured as follows. Section 2 presents a running case and additional preliminaries. Section 3 comprises the legal problem factors for smart contracts that require legal relevance. Next, Sect. 4 translates these elements into an ontology for SCs and Sect. 5 presents the SLC lifecycle model to monitor contractual rights and obligations. Section 6 demonstrates a feasibility evaluation and discussion that expands the running case of this paper. Finally, Sect. 7 gives conclusions, limitations, open issues and future work.

2   Motivating Example and Preliminaries

In Sect. 2.1 we present the running contract case that stems from real car production supply chain contracts. In Sect. 2.2 we present related background literature that prepare the reader for subsequent section.

2.1 Running Case

To illustrate the approach of the paper, a generic supply chain running case of car production is shown in Fig. 1. The original equipment manufacturer (OEM) actually assembles the delivered car parts. The other parties of the supply chain involve either the supply side, or demand side. E.g., the Supplier A sources and supplies the raw materials to Supplier B, who manufactures the individual car components. The particular car component are then shipped to the OEM, who assembles the final product.

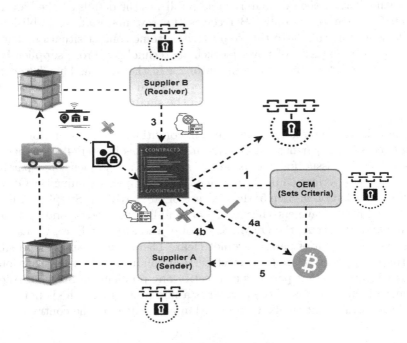

Fig. 1. Supply chain running case.

We deploy the supply chain running case into the blockchain that plays a significant role for checking provenance and tracking of product, which is possible with the integration of smart-contract. The blockchain allows supply-chain partners to access various functions, i.e., partners can access particular function such as checking provenance. For checking the provenance, Supplier A delivers the raw materials to Supplier B and the former publish the records into the blockchain with the integration of sensors. The records include quantity, quality of product and location, time on which the raw product is shipped. This information is immutably stored into the blockchain and can be accessed by supply-chain partners.

For tracking of the product, when Supplier B receives the raw materials, we assume he confirms that the shipment is in order. If the shipment is received on time and at the correct location as per sensor verification, then the payment in the form of digital transaction is executed automatically by a smart-contract.

We can improve supply chain processes efficient by integration of smart-contract into the blockchain. However we introduce various limitation of this technology by considering supply chain running case. In Fig. 1, Step 1 shows the OEM sets the acceptable criteria such as correct location and lead time for each leg of the shipment. The OEM also holds digital currencies while Supplier A is a sender who delivers the raw material and publishes the details on the blockchain in Step 2. When the Supplier B receives the raw material, he publishes the records as well in Step 3. In this step we show some conflict situation that may arise due to immaturity of novel blockchain technology. From supplier B, the data goes to smart contract through sensor for further action. This data can not reliable because it can be altered by external third party. In next phase, when specific criteria meet that are embedded in the smart-contract per Step 4a, then a payment is executed automatically in Step 5. There can be another possibility that the smart contract fails to meet certain conditions. In this situation, an alert is triggered so that partners can rectify any problems in Step 4b. Furthermore, we raise another issue for this paper in asking what happens if a particular obligation is not performed by the partners? For example, assuming the OEM has a payment obligation in Step 5, then in case of late payment, Supplier A has the right to claim late-payment charges. After enabling the corresponding function of rights by Supplier A, the OEM has an obligation to pay. Here, we have seen rights and obligation of parties are not clear. There can be another possibility that the smart contract fails to meet certain conditions. In this situation, an alert is triggered so that partners can rectify any problems in Step 4b. Another challenge of this paper is, if the partner is a non-programmer, he is not capable of understanding what rights and obligation are written in the contract.

2.2   Related Background Literature

In this section, we describe the computation toolkit to understand the solution of the running-case problem. In Sect. 2.1, we propose the situation of the dispute among the parties. To overcome this situation, we develop an ontology that comprises the concept and properties of rights and obligations. The ontology is a formal representation of knowledge by a set of concept and relationship among those concepts [14]. The ontology organizes the class hierarchies of relationships and allows the practitioner to understand the relationships of the particular problem domain. We design the ontology in Protege tool [16] that is open-source ontology editor and comprises the graphical user interface for visualization of the relationship among classes. We employ the HermiT-tool reasoner [7] that checks the correctness of ontology and identify subsumption relationship among classes.

Later, to automate the concept of rights and obligation in smart-contracts, the Protege tool also supports web ontology language[4] (OWL) that express the formal semantics of ontology into machine-readable code. We also employ Coloured Petri Nets (CPN) tools[5] for checking the dynamic behaviour of ontology processing. CPN tool is an software that is useful for simulation and state space analysis of the model. An state space is a directed graph with nodes so-called states and arcs that connect states and transition. The CPN-notation comprises state represented as a circle, transition represented as a rectangle, arcs that connect the states to transitions, and token with color, i.e., attributes with values. Transition fires when all input states hold the tokens and produce condition-adhering tokens into output places.

3   Legal Recognition Factors

In this section, we discuss the legal recognition of the business-to-business (B2B) smart contracts presented in Sect. 2.1. A core principal in contract law is freedom of contract that has two main elements, a) the right of a legal person to freely decide whether to enter into a contract and b) the right to freely decide together with the other contracting party, or parties about the content of the contract [24]. The prevailing view in law [26] concerning a) is that in principal nothing prevents two B2B parties to voluntarily enter into machine-readable sales contracts that are based on a blockchain[6]. To assess how the rules apply regarding the formation of contracts to smart contracts, the analogy of a vending machine has often been used in the literature [27]. In both cases, a legally binding contract is formed due to the consenting actions of the contracting parties. Just as for a vending machine, a smart contract can be independently designed by an offeror and deployed to a blockchain [6]. The design and deployment thereby indicate the offeror's intention to be legally bound according to the terms stipulated in the smart contract. According to this analogy, the offeree agrees to be bound by the smart contract through conclusive conduct. Equivalent to entering a coin into a vending machine he, or she fulfills the triggering requirements of the smart contract. In regard to our running case, this means that an OEM sends cryptocurrencies to Supplier B's smart contract to order intermediate product. By sending the tokens to the smart-contract wallet address, the OEM consents to the contract terms of Supplier B's smart contract and a legally binding agreement is formed [11]. To conclude, from a legal point of view, the right of a legal person to freely decide whether to enter into a contract extends to the right to enter into smart contracts.

---

[4] https://www.w3.org/OWL/.

[5] http://cpntools.org/.

[6] Note that this freedom may be limited in the case of business-to-consumer contracts and special kind of contracts with significant consequences for the contracting parties (e.g., the selling and transfer of real estate) where the law may demand a special form requirements. Still, as these contracts are not part of our running case, we will not discuss the legal problems connected to the operationalization of such contracts as smart contracts any further.

Accordingly, also b) the right to freely decide together with the other contracting party, or parties about the content of the contract, should extend to smart contracts. Still, it is currently unclear how a court, or an arbitrator would interpret smart contacts that do not use legal terminology but are based on programming code [9]. Assuming that for our running case, Supplier B delivers the intermediate product to the OEM who's sensors accurately recognize the delivery of the intermediate product. The cryptocurrency funds are thus released to Supplier B and the goods are stored in the warehouse. Later, the intermediate product is used for production by the OEM and it is discovered that the intermediate products are of inferior quality, most likely due to impurified raw materials delivered by supplier A to Supplier B. The OEM's production process has to stop causing severe consequential damages. The OEM sues supplier A for replacement of the inferior intermediate products and compensation for the lost production. Supplier A claims that the inferior quality of the intermediate products was not caused by impurified raw materials but rather because of inappropriate storage by the OEM. Furthermore, Supplier A argues that the terms and conditions of the smart contract exclude consequential damages and stipulate an immediate notification deadline in case of defective goods that the OEM has missed. The OEM and supplier A also disagree about the responsible dispute resolution forum and the applicable contract law. The current state of smart contracts do not allow a judge to solve these problems. This is because there is currently no complete legal ontology that allows programmers to design smart contracts in the fashion of traditional contracts.

In the next section, we present a comprehensive contract law ontology that allows contracting parties to precisely define the content of a smart contract for machine readability and in case of disputes, also by a human judge, or arbitrator. We argue that only with the help of our contract-law legally binding smart contracts can be designed.

4   Smart-Legal-Contract Ontology

A smart legal contract (SLC) ontology[7] in Fig. 2, Fig. 5 comprises essential concepts and properties to support SC legally enforceability. We develop the SLC ontology in protége tool to render the latter machine-readable and explain the concepts and properties for the running case of the car production business case, as illustrated in Sect. 2.1.

---

[7] Full ontology: https://bit.ly/3c5eYO5.

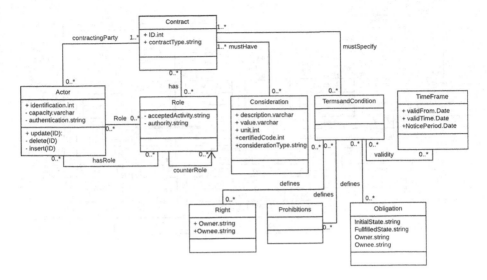

Fig. 2. Outline for upper level smart contract ontology.

The SC ontology defines fundamental concepts such as roles, considerations, rights, and obligations, etc., as presented in Fig. 2. In our running case, Supplier A promises to deliver the raw materials to Supplier B in return a specific amount of money. Supplier B manufactures the car component and promises to deliver to the OEM. To prevent a conflict among the parties the role is defined in the SC. A promise is a statement of commitment to fulfil some act or perform certain deeds and when a promise is given with legal intent of enforcement in court, later becomes legal obligation.

The valid SC must hold two or more contracting parties, such as offeror, offeree, and acceptor etc., [9]. An offeror sends an offer to an offeree for delivering services, and the offeree accepts an offer, offeree becomes acceptor. In our running case, Supplier B acts as an offeror for car components to an OEM that is an offeree. When an offer is accepted by an OEM, then the later transforms into an acceptor. Furthermore, contracting parties must have the necessary competence and capability to perform certain deeds. Supplier B must own the car components as he has promised to deliver the car components, and the OEM must have the capital to pay for it. Further, a valid SC can be a complete-, or incomplete contract. For example, the car manufacturer writes the SC of shipment of cars without giving the deadline, and it considers a valid contract while it is incomplete because there is nothing obligations are specified for the parties (Fig. 3).

Consideration is defined as exchanged value for the trade of which the contracting parties agree to enter into a SC. Car components and raw materials are the asset, or consideration of the contract. The selling of raw material, or car components constitutes the performance that fulfills the promise of the contract. Selling of the car component to an OEM is an obligation of supplier B that is

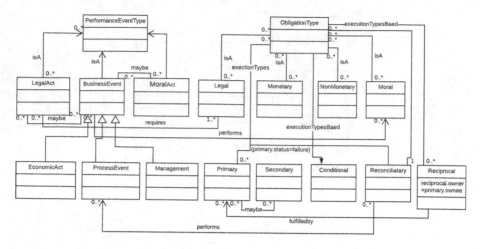

Fig. 3. Rights and obligations.

realized when an actual business act of given car component is performed in return of compensation. An obligation must have obligee who is obliged to perform a particular action and obliger beneficiary who receives the consideration for whom a promise is established. In this case, Supplier B is an obligee who is obliged to deliver car components, and OEM is an obliger who receives the car component.

Also, the SC contains the deadline, or time frame under which the promised performances shall occur. If certain promises are not performed under the deadline, or in unsatisfactory manner, then the state of obligation will change as to unfulfilled. In our running case, if supplier B does not execute the promises as planned and agreed, then the obligation is unfulfilled, resulting in the OEM may seek the pre-agreed rights as compensation. In this case, the OEM may have the right to seek a remedy in the form of a penalty, or can terminate the contract. Also, the OEM can choose not to do anything and settlement occurs by mutual consent.

We explain the simple running case of ontology, where we see an obligation may activate another obligation and rights. Similarly, the rights too may enable new obligations being formed. Therefore, we present the types of rights and obligations in Fig. 5. Also, we describe the change of obligation state under which rights may activate.

We refine an obligation type by adding sub-classes such as monetary-, non-monetary-, moral-, and legal obligation to express a particular remedy. The monetary obligation may create a remedy such as late-payment-charges, penalty, etc. For instance, if the OEM receives a car component from Supplier B and does not pay the money within a deadline, then Supplier B may provide a remedy as to a late-payment-charge. Similarly, if a supplier delivers defective car parts then the OEM may have legal rights to cancel the SC. The OEM may have non-monetary obligations to arrange a carrier to provide the car components but does not have monetary obligation between supplier and OEM. There may be another moral obligation type that is not severely binding but is morally, or ethically an expected obligations. For instance, the OEM has an obligation to pick up the car components from the supplier premises. Still, the OEM may urge for help to arrange the transportation. The supplier may not legally bound to help the OEM, but morally he is bound to assist the OEM (Fig. 4).

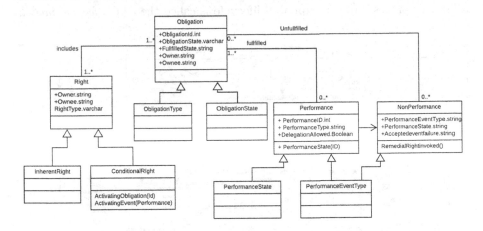

Fig. 4. Common obligation types.

We identify several obligation states to track the SC fulfillment process more systematically, and an individual obligation may exist in more than one category. The proposed obligation states in Fig. 5 are adopted from similar work proposed by LEE [12]. Initially, the SC has an inactive state when the supplier and OEM have signed the SC, but the SC execution does not commence. The SC is said to be active when the performance event is performed. For instance, the OEM demands to deliver the car components to Seller B who receives an order; the supplier obligation to deliver the car components is triggered. The SC obligation state is pending when the supplier has dispatched the car components from the warehouse and is waiting for a carrier for transportation. The obligation state will be pending until the entire essential fulfillment process is completed. When the OEM receives the car components that satisfy the stipulated performance condition, then the obligation state is fulfilled and when the obligation is terminated by the obligee with mutual intent, the obligation state changes to termination.

5   Rights- and Obligations Monitoring

We develop an SLC lifecycle model[8] to monitor contractual rights and obligations defined with the ontology of Sect. 4. We adopt the existing formalized

---

[8] Full download CPN model: shorturl.at/cxBE9.

smart-contracting lifecycle in [18, 20, 21], where the startup phase commences with the configuration of a business network model (BNM). The latter is a cross-enterprise collaboration blueprint and contains the legally valid template contract that inserts the service type with organization roles. The latter enables fast and semi-automatic identification of contracting parties for knowing their identity, services, and reputation. We include the rights and obligations throughout the existing smart-contracting lifecycle to monitor the related contractual fulfillment process, and the updated lifecycle is called the *SLC lifecycle model*.

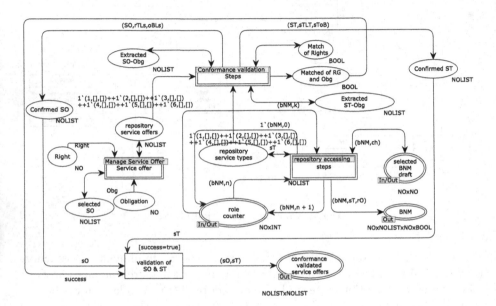

Fig. 5. Rights and obligations selection in BNM.

The SLC lifecycle is divided into two modules, set up- and perform phase. In the *setup module*, the proposal of SC is finalized and negotiated among the contracting parties. Further, the performances of smart- contracting are accomplished in the *perform module*. The rights and obligations are stipulated throughout the entire lifecycle. Still, we present the transaction of rights and obligations in the smart-contracting setup phase and especially in the BNM selection that is an ecosystem for breeding service types with rights, obligations, and roles to become a part of BNM. The latter is divided into several sub-modules namely, *repository accessing*, *manage service offer*, and *manage service type*, as presented in Fig. 7. A *repository accessing* exists in *BNM selection*, where we assume a user inserts the service type with rights and obligations over the time and the same assumptions hold for the repository of service offer in the *manage service offer* module. Finally, the *conformance validation* module is developed to conform to the validation of service offer against the chosen service type in the BNM draft specification.

Next, we provide the details of each module in the subsections below.

5.1   Repository Accessing

We assume a contracting party inserts the rights, roles, and obligation for service types in the repository in Fig. 6.

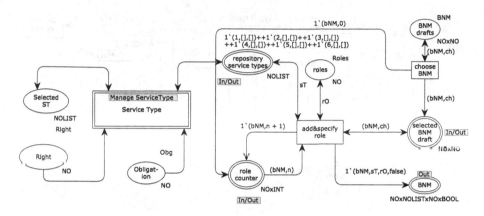

Fig. 6. Repository of service type with rights, roles, and obligations.

As a first step, the contracting party chooses a BNM from stored BNM drafts. Thereafter, the latter inserts the rights and obligations in the *manage service type* module that is stored in the state labeled *repository service types*. Further, the actual BNM selection involves choosing a BNM draft for validating service offers and roles to be filled subsequently with rights and obligations. The rights and obligations to be filled in the *manage service type* module are presented below.

Manage Service Type. The actual repository of service types commences with choosing the rights and obligations in the *Manage service type* module, as presented in Fig. 7. Initially, the list of rights and obligations of service type is empty in the repository. At a first step, the contracting party chooses the service type Id from *choose ST* transition. After that, the latter inserts the rights and obligations simultaneously in *selected ST* by firing the *insert right* and *insert obligation* transitions.

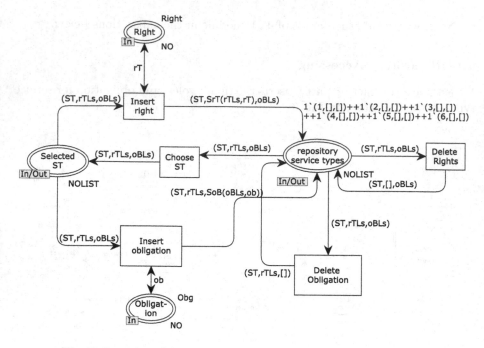

Fig. 7. Insertion, deletion of rights and obligations in service types.

Additionally, the inserted rights and obligations are deleted by firing the transitions *delete right* and *delete obligation*. The same assumption holds for choosing the rights and obligations for service offer in *manage service offer* module.

5.2 Conformance Validation

To be considered as a service offer for finalizing the proto-SC, beforehand, a conformance validation is necessary, as depicted in Fig. 8. The selected service offer and service type from the BNM repository are extracted for the validation. The chosen right and obligation properties inherit the properties of the service type. Thus, we extract the rights and obligations from the state labeled *repository service type* and *repository service offer*. Further, a service offer matched with service types is stored in the state labeled *confirmed SO* and *confirmed ST*.

6   Evaluation

We use the CPN tool to evaluate the model concerning the correctness and performance checking, especially considering aspects that are required for system development. Due to page limitations, we do not present the entire steps that are taken to produce the evaluation results. Several properties, such as reachability, loops, etc., as depicted in Fig. 9, are essential to evaluate in the model. Due to the size of models, computation of states verification through automatic

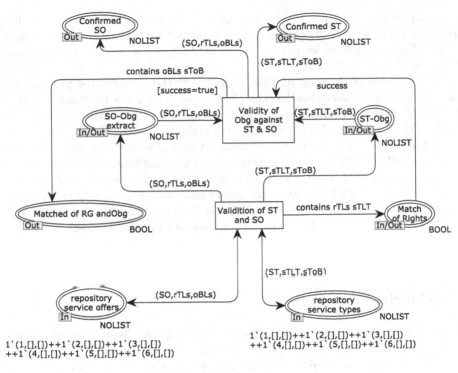

Fig. 8. Conformance validation of service offers and service types.

simulation of token games is challenging. Thus we are focused on the detection of loops to prevent the desired termination reachability. Furthermore, specific attention is required to exit loop-conditions effectively, such as elements of the business-policy control. Performance peaks are calculated during runtime either in designing for sufficient resources or in restricting the load with the business policy control. The utilization property is used to ensure the effectiveness of each model in a specific scenario. Finally, home marking is needed for consistent termination to ensure simple testing of a real system.

We generate the state space on CPN modules where the computation is feasible and present the results in Fig. 9. Loops exist in *manage service type* module. For the Manage service type, a party inserts the rights and obligations.

Module	Loops	Performance peaks	Liveness	Utilization	Home marking	Dead marking
			Module property			
BNM Selection	No	Evenly balanced	ND/NL	Yes	No	Multiple
Repository accessing	No	Manage service type	D*/NL	Yes	No	Multiple
Manage service type	Yes	Insertion of rights and obligations	D*/NL	Yes	No	No
Conformance validation	No	validating service type and service offer	ND/NL	Yes	No	Yes

Fig. 9. Model checking.

Loops exist in *manage service type* module. Managing the service type loop is self-restricting as it only processes the rights and obligations of parties, respectively. The results show for the remaining modules in Fig. 9; they do not contain the loops.

Performance peaks exist in Fig. 9 to represent the places of the SLC lifecycle that are the performance bottlenecks. Peaks exist in each module but not in *BNM selection*. For the repository accessing, the peaks exist for choosing a BNM draft and party's roles and also for inserting the rights and obligations.

There is no home marking, as presented in Fig. 9, and the result for dead marking differ. Multiple dead marking and home marking show test cases are more demanding for practitioners to validate the implementation. D* means a dead marking result that shows the intentional disabling of marking path for the purpose of focusing on a particular module under investigation. Finally, the utilization test in Fig. 9 shows there is no unused sub-module exist.

Due to page limitations, we refer to the reader [19] for more details of evaluation.

7   Conclusion

Smart contracts are machine-readable software code that is situated on the protocol layer of a blockchain system to govern transactions. The later gives rise to a novel means of legally enforcing rights and obligations. State of the art shows that CCs cause high transaction costs due to their multiple shortcomings, SCs lack legal relevance, and this yields legal uncertainties for users while furthermore, SCs are inflexible code that is not smart.

In this paper, we identify a gap between business process management and SC execution, and their fulfillment. Thus, we introduces the ontological concepts of rights and obligations for SC's that are defined in business contracts. For developing the ontology we opt the protege tool that is open source ontology editor and employ the HermiT-tool reasoner that checks the correctness of ontology. Further, we propose the obligation states through which each obligation is passed. For ensuring the legal enforceability of SC's, we present an SLC lifecycle model to monitor contractual rights and obligations. For exploring the SLC lifecycle in a dependable way, we choose CPN Tools that has a modeling notation backed with formal semantics.

The limitation of the paper is that we are only focused on presenting the transaction of rights and obligations in the smart-contracting setup phase. Future work in SC's domain includes analysis and modeling of other types of business legal SC's. Further, resetting of human to the machine for the self-aware SC's are a possible extension to ongoing work.

Acknowledgments. This article is based on research from the Erasmus+ Strategic Partnerships Project - 2018-1-RO01-KA203-049510 "Blockchain for Entrepreneurs - a non-traditional Industry 4.0 curriculum for Higher Education".

References

1. Buterin, V., et al.: Ethereum white paper. Github Repository, pp. 22–23 (2013)
2. Butterin, V.: A next-generation smart contract and decentralized application platform (2014)
3. Casado-Vara, R., González-Briones, A., Prieto, J., Corchado, J.M.: Smart contract for monitoring and control of logistics activities: pharmaceutical utilities case study. In: Graña, M., et al. (eds.) SOCO'18-CISIS'18-ICEUTE'18 2018. AISC, vol. 771, pp. 509–517. Springer, Cham (2019). https://doi.org/10.1007/978-3-319-94120-2_49
4. Clack, C.D., Bakshi, V.A., Braine, L.: Smart contract templates: foundations, design landscape and research directions. arXiv preprint arXiv:1608.00771 (2016)
5. De Filippi, P., Hassan, S.: Blockchain technology as a regulatory technology: from code is law to law is code. arXiv preprint arXiv:1801.02507 (2018)
6. Giancaspro, M.: Is a 'smart contract' really a smart idea? Insights from a legal perspective. Comput. Law Secur. Rev. 33(6), 825–835 (2017)
7. Glimm, B., Horrocks, I., Motik, B., Stoilos, G., Wang, Z.: HermiT: an OWL 2 reasoner. J. Autom. Reason. 53(3), 245–269 (2014)
8. Griggs, K., Ossipova, O., Kohlios, C.P., Baccarini, A., Howson, E., Hayajneh, T.: Healthcare blockchain system using smart contracts for secure automated remote patient monitoring. J. Med. Syst. 42(7), 130 (2018)
9. Idelberger, F., Governatori, G., Riveret, R., Sartor, G.: Evaluation of logic-based smart contracts for blockchain systems. In: Alferes, J.J.J., Bertossi, L., Governatori, G., Fodor, P., Roman, D. (eds.) RuleML 2016. LNCS, vol. 9718, pp. 167–183. Springer, Cham (2016). https://doi.org/10.1007/978-3-319-42019-6_11
10. Imbault, F., Swiatek, M., De Beaufort, R., Plana, R.: The green blockchain: managing decentralized energy production and consumption. In: 2017 IEEE International Conference on Environment and Electrical Engineering and 2017 IEEE Industrial and Commercial Power Systems Europe (EEEIC/I&CPS Europe), pp. 1–5. IEEE (2017)
11. Lauslahti, K., Mattila, J., Seppala, T.: Smart contracts-how will blockchain technology affect contractual practices? ETLA Reports (68) (2017)
12. Lee, R.M., Dewitz, S.D.: Facilitating international contracting: AL extensions to EDI. Int. Inf. Syst. 1(1), 94–123 (1992)
13. Levy, K.E.: Book-smart, not street-smart: blockchain-based smart contracts and the social workings of law. Engag. Sci. Technol. Soc. 3, 1–15 (2017)
14. Maedche, A., Staab, S.: Ontology learning for the semantic web. IEEE Intell. Syst. 16(2), 72–79 (2001)
15. Mohanta, B., Panda, S., Jena, D.: An overview of smart contract and use cases in blockchain technology. In: 2018 9th International Conference on Computing, Communication and Networking Technologies (ICCCNT), pp. 1–4. IEEE (2018)
16. Musen, M.A., et al.: The Protégé project: a look back and a look forward. AI Matters 1(4), 4 (2015)
17. Norta: Self-aware smart contracts with legal relevance. In: 2018 International Joint Conference on Neural Networks (IJCNN), pp. 1–8. IEEE (2018)
18. Norta, A.: Establishing distributed governance infrastructures for enacting cross-organization collaborations. In: Norta, A., Gaaloul, W., Gangadharan, G.R., Dam, H.K. (eds.) ICSOC 2015. LNCS, vol. 9586, pp. 24–35. Springer, Heidelberg (2016). https://doi.org/10.1007/978-3-662-50539-7_3

19. Norta, A., CINCO, C., Computing, I.: Safeguarding trusted ebusiness transactions of lifecycles for cross-enterprise collaboration. Technical report C-2012-1, Department of Computer Science, University of Helsinki, Helsinki, Finland (2012)
20. Norta, A., Othman, A.B., Taveter, K.: Conflict-resolution lifecycles for governed decentralized autonomous organization collaboration (2015). https://doi.org/10. 1145/2846012.2846052
21. Norta, A.: Creation of smart-contracting collaborations for decentralized autonomous organizations. In: Matulevičius, R., Dumas, M. (eds.) BIR 2015. LNBIP, vol. 229, pp. 3–17. Springer, Cham (2015). https://doi.org/10.1007/978-3-319-21915-8_1
22. Patel, D., Shah, K., Shanbhag, S., Mistry, V.: Towards legally enforceable smart contracts. In: Chen, S., Wang, H., Zhang, L.-J. (eds.) ICBC 2018. LNCS, vol. 10974, pp. 153–165. Springer, Cham (2018). https://doi.org/10.1007/978-3-319-94478-4_11
23. Raskin, M.: The law and legality of smart contracts (2016)
24. Schafer, I.: Ott, lehrbuch der okonomischen analyse des zi-vilrechts, 4 (2005)
25. Singh, M., Kim, S.: Chapter four - blockchain technology for decentralized autonomous organizations. In: Kim, S., Deka, G.C., Zhang, P. (eds.) Role of Blockchain Technology in IoT Applications. Advances in Computers, vol. 115, pp. 115–140. Elsevier (2019). https://doi.org/10.1016/bs.adcom.2019.06.001. https://www.sciencedirect.com/science/article/pii/S0065245819300257
26. Smits, J.M.: Contract law: a comparative introduction
27. Szabo, N.: Formalizing and securing relationships on public networks. First Monday 2(9) (1997)
28. Teslya, N.: Industrial socio-cyberphysical system's consumables tokenization for smart contracts in blockchain. In: Abramowicz, W., Paschke, A. (eds.) BIS 2018. LNBIP, vol. 339, pp. 344–355. Springer, Cham (2019). https://doi.org/10.1007/978-3-030-04849-5_31
29. Wang, S., Yuan, Y., Wang, X., Li, J., Qin, R., Wang, F.: An overview of smart contract: architecture, applications, and future trends. In: 2018 IEEE Intelligent Vehicles Symposium (IV), pp. 108–113. IEEE (2018)
30. Wulf, A.J.: Institutional competition of optional codes in European contract law. Eur. J. Law Econ. 38(1), 139–162 (2014)

Aspect Based Sentiment Analysis – An Incremental Model Learning Approach Using LSTM-RNN

Alka Londhe[✉] and P. V. R. D. Prasada Rao

Department of Computer Science and Engineering, Koneru Lakshmaiah Education Foundation,
Vaddeswaram, AP, India
pvrdprasad@kluniversity.in

Abstract. Analysing user opinions have become a necessity in order to understand customer satisfaction and requirements. Sentiment Analysis is the process where framework is capable of analysing and predicting sentiments of a sentence. Several researchers have attempted to perform sentimental analysis with various classification techniques especially machine learning techniques, however still issues faced with respect to analysis of data and decision-making process. Hence, this work attempts to improve the accuracy of classification process by implementing an incremental based novel framework. In this work, sentiments of restaurant reviews are processed and this will help in identifying satisfaction level of customers automatically. Artificial intelligence is used by implementing Recurrent Neural Network and Long Short-Term Memory. The hybrid LSTM-RNN is capable of analysing and predicting polarity of aspects with maximum possible accuracy. Various transformations and pre-processing techniques are applied for cleaning text-data as per required input to classifier. From reviews, lots of different aspects are obtained and trained to the LTSM-RNN classifier. When latest reviews are fed into the classifier, sentiments of each aspects are identified as either positive, negative or neutral aspects. The obtained accuracy is seen to be around 76% which is comparable with existing algorithms. Apart from good accuracy, major outcome is - multiple aspects are correctly extracted from multi-sentence lengthy reviews.

Keywords: Artificial intelligence · Machine learning · Natural language processing · Long Short-Term Memory · Recurrent Neural Networks · Sentiment Analysis

1 Introduction

Aspect Based Sentiment Analysis (ABSA) is a technique where some sets of texts are given as input and the individual aspects of the sentences are obtained as output. The text might be any type of data from a product review to discussions in social media. The sentiment of each sentence can be obtained and also the sentiment of the entire paragraph can be obtained by averaging the existing sentiments. Online sources contain

© Springer Nature Switzerland AG 2021
M. Singh et al. (Eds.): ICACDS 2021, CCIS 1440, pp. 677–689, 2021.
https://doi.org/10.1007/978-3-030-81462-5_59

the different types of feedback by the users which can be used as a strong tool for mining the information about the emotional state of human mind and serves to carry out sentiment analysis [1].

This is done on the basis of customer reviews on various products or the basis of opinion of users on a particular topic. This analysis will aid the vendors, business owners and other involved parties to get valid feedback and analyses them in order to improve their products and services. At time, there might be a rapid decline in the sales of the products or the service users and it is necessary to know the relevant reason for this. The customers' or clients' feedback and the reviews are the only source where the client can get to know the sentiments of the customers. The categories of the review may be divided into positive review, negative review and neutral review. Within these, the reviews might be split into smaller sections based on the severity of the sentiment. Since the number of reviews for a large company maybe high, obtaining and analysing the sentiments is a difficult task [2].

The entire group of reviews may be analysed to get the type of review which may be positive, neutral or negative. In aspect-based sentiment analysis, the review will not analyse only the positivity and negativity in the document, but also the different aspects involved in the reviews. Aspect based analysis is more preferred for large data and especially for reviews. ABSA is the analysis of the obtained data either in part or as a whole for having a clear overview of the aspects [3]. When online data like social media posts are considered for analysis, the aspects of the review of product must be completed. Additionally, the opinion of the user becomes vital in order to have accuracy in the predictions. The accuracy must be high since the data in social media are often informal and do not have traditional structures. Hence, any investigation that relates to sentiments and aspects must focus on enhancing the granularity, which will depend on the extraction of the aspects and classification of sentiments [4].

This analysis can be performed with an efficient algorithm and appropriate classifier and the different aspects like positive, neutral and negative denotes the polarity of the content. The mood or sentiments of the review can be identified depending on the type of polarity. The two most common techniques for mining the data and to identify the data are the Natural Language Processing (NLP) and information mining [5]. To classify them, machine learning techniques are preferred because the high-level features are not required to solve with these techniques [6]. These machine learning techniques may be either supervised or unsupervised or even semi supervised. The analysis will extract the opinion from the texts and then identify the category they belong to and then the polarity is formed for the sentiments [7].

This section has given a brief introduction on the sentimental analysis, its necessity and how it is used. Section 2 studies the related studies related to the different techniques used in the sentimental analysis for identifying the polarity of the aspects. The next section gives a brief description of the LSTM along with its working, followed by the experimental setup and the description of the dataset. The pre-processing techniques for the obtained dataset are discussed and the LSTM-RNN classification for the aspects is discussed. Finally, the work is concluded.

2 Literature Review

A sentimental polarity computational technique for identifying and predicting the senti-ment in online forum has been proposed by Li and Wu [8]. The polarity of each individual text is examined by assigning a value using an algorithm. A machine learning technique has been proposed by combining the SVM and k-means clustering. Text mining tech-nique is used to classify into different clusters. Even though k means clustering is an unsupervised learning technique, it considers spherical distribution of the variables with same variance level. Scaling technique has been used to alter the contents to the required range. This method is however sensitive to over-fitting.

A cross lingual classification of sentiments have been classified in Hajmohammadi et al. [9] to analyse the sentiments of the data. Since most of the available studies have been performed for English, there might be problem with the translation of the data by the machine. Hence, a novel semi supervised model has been proposed to address this issue and to include the unlabelled data from the languages. The training data is enriched by adding labelled instances that has more confidence. The data sets that contained book reviews in three different languages have been considered in this work and show a high classification performance.

A novel model has been proposed in [10, 11] that contains syntactic and lexical features using Condition Random Field (CRF) model. Logistical regression has been performed with the weighted schemes for both positive and negative labels for detecting the polarity. The features that have been extracted contain various features and it is suitable only for linear problems. It suffers through the problem of over-fitting. This model is complicated with respect to training.

A novel technique for classifying the emotions in the text has been developed by Onan et al. [12] by using Latent Dirichlet allocation (LDA). A variable has been intro-duced between the document and variables. Hence, a mixture of different keywords is produced, where different words are applied. This technique has been found to be useful for extracting the aspects at the document level but fails to capture all the aspects at fine grain level.

Opinion extraction has been performed by Zhang et al. [13]. This type of extraction identifies the sentiments in the opinion using different arguments of opinions. Natural language processing has been performed to mine the opinions that have been given previously. Long Short-Term Memory (LSTM) has been used for the encoding process that is bidirectional and has several layers. For the decoding process, it depends on the transitional results and this has been done using global normalization techniques and beam search.

An opinion mining technique has been proposed by Tsirakis et al. [14] for using in real time applications based on the previously available results and based on the customer reviews. The work has been created by using Palapro platform and the results shows that it is efficient. A hybrid algorithm that integrates the k-means clustering and cuckoo search algorithm has been proposed by Chandra Pandey et al. [15]. In this work, tweets are analysed to obtain the sentiments and perspectives of the users. Cluster heads are obtained by using emotional contents of the twitter datasets. A thorough comparison of the iterative techniques like PSO, differential evolution, cuckoo search and n-grams technique has been performed. From the results, it has been seen that the obtained results have good efficiency in the analysis, however, it is necessary to further improve this.

The pre-trained contents are subjected to sentimental analysis by Rezaeinia et al. [16]. The words are converted into vectors in a significant manner. Lots of data is required for the training and to create the appropriate vectors. The suggested approach uses an improved wave vector that utilizes lexicon-based style, by acquiring the part of speech, and positioning the different algorithms for identifying and understanding the words.

Sentiment analysis of images and video have been performed by Song et al. [17]. This has been performed by augmenting and positioning the spatial areas of images. Convolution neural network (CNN) has been used for the classification process and then trained. This technique is seen to be more efficient when the input has more lattice like arrangement with respect to images and videos. The efficiency is higher for those data that has textual data accompanying it. Therefore, a semi supervised technique that can have both labelled and unlabelled data has been used.

Recurrent Neural Network (RNN) contains good structure and functionality. It has functions that can be used as an activation function for future uses and can be used for both structured and unstructured data. The main advantage is that it can work efficiently even when the size of data is less [18]. This is good since for sentiment analysis; there will not be enough data for processing. The problem with this technique as a separate classifier is that the degree of activation reduces with each iteration and this is known as gradient vanishing. Also, this might increase with each iteration in some cases and this is known as gradient explosion. While the vanishing affects the memory by losing data, explosion enables the RNN to receive the future input functions. This problem can be addressed by combining the classifier with Long-Term Short Memory (LSTM) unit [19]. This function standardizes the activation function in the network and also improves the performance of RNN to a great level. With this, the network will not accept any other inputs once the threshold is reached. It will also remember the hidden states whenever required. Even though some of the existing techniques use different frameworks for extracting the opinions, the problem of concept drift has not been focused and the researchers have not mentioned the overfitting problems. Also, the combination of implicit and explicit methods of aspects has not been considered previously. This work has combined both the implicit and the explicit aspects. The incremental learning used in our work will be computationally efficient and accurate so that a genuine prediction can be arrived through streaming data from social media.

LSTM has achieved good success in different NLP tasks like the Target Dependent LSTM (TD-LSTM) and Target Connection LSTM (TC-LSTM) [20, 21]. This considers the target information and achieves a novel performance in classification of the sentiments that depends on the targets. It has obtained the targeted vector by taking the average of the vectors of words, where the phrase is available. Even though these methods are effective, it is still challenging to differentiate the polarities of the different sentiments. Hence, it is necessary to design a novel neural network based sentimental analysis with respect to aspects.

3   Incremental Learning

Incremental learning is performed to learn through multiple iterations, thereby improving through this process. In each iteration, there is more training data and hence the system acquires more vocabulary. It can be used to analyse the opinions of the customer and perform it incrementally. In this work, all the aspects are added as new entities and then the algorithm adapts to each word. In this work, the extracted aspects are given to the classifier where the neural network learns through incrementally. With each iteration, the proposed classifier learns more aspects and hence when all the epochs get completed, the classification will have a high accuracy.

4   LSTM

RNN is a type of neural network that extends the working of traditional feed-forward neural network. The problem with the traditional RNN is that it has gradient vanishing and exploding problems. In order to overcome the issues, Long Short term Memory network (LSTM) has been developed and combined with RNN to achieve better perfor mance. The LSTM framework consists of gates and cell memory. Figure 1 shows the architecture of standard LSTM model.

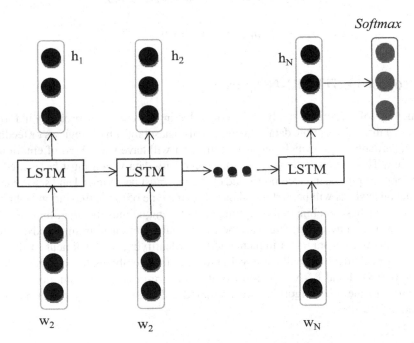

Fig. 1. Architecture of standard LSTM

Here, the series w_1, w_2, ..., w_n represents the word vector in a sentence whose length in N, whereas the hidden layers are represented by h_1, h_2, ..., h_n.

Every cell in LSTM is calculated as per following formulas, where:

- W_i, W_f, W_o are weighted matrices within R^{dx2d}
- b_i, b_f, b_o are biases of LSTM within R^d that are learnt during the training and transformation process.
- The three parameters represent the input, forget and the output.
- σ is the sigmoid function.
- $\odot$ represents the element multiplication.

$$X = \begin{bmatrix} h_{t-1} \\ x_t \end{bmatrix} \tag{1}$$

$$f_t = \sigma(W_f.X + b_f) \tag{2}$$

$$i_t = \sigma(W_i.X + b_i) \tag{3}$$

$$o_t = \sigma(W_o.X + b_o) \tag{4}$$

$$c_t = f_t \odot c_{t-1} + i_t \odot \tanh(W_c.X + b_c) \tag{5}$$

$$h_t = o_t \odot \tanh(ct) \tag{6}$$

5   Proposed LSTM-RNN Approach

The methods of execution usually consist of collecting the data most probably in form of datasets or through real time data collection. This dataset must have reviews or feedback about a particular topic which means that the user will have some kind of emotions in the review. This data will contain lots of unnecessary data which must be cleared and hence, pre-pre-processing techniques and classifiers must be reviewed and selected. The sentimental analysis will be performed based on the type of the optimization technique. Therefore, the best combination of optimization technique must be identified. The files are then validated by comparing with the traditional techniques. In this study, a novel machine learning technique is implemented for identifying the sentiment of the food reviews of restaurants. Initially, a novel algorithm that combines the ensemble classification of LSTM and RNN is performed after the pre-processing techniques. Finally, the metrics of the novel algorithm are evaluated and classified using performance the evaluation measures.

5.1   Dataset

It is necessary to obtain the food reviews from valid datasets since multiple aspects are required in the same sentence. The dataset is selected in such a way that a single review should contain some positive aspects and negative aspects. The data has been obtained from Yelp, that contains reviews of multiple restaurants [22].

5.2   Pre-processing and Feature Extraction

The obtained data must be processed before the classifying process. Hence, it must be subject to pre-processing techniques. Initially, it has to be ensured that the content makes sense grammatically. Hence, lexicon sematic based and rule-based methods are used in this stage for extracting the aspects from the text. A NERSE tool [23] identifies software specific entities from dataset. Same can be trained for other datasets too. SentiWordNet database is used in this stage for all the grammatical processing in the text.

Unnecessary data in the reviews like URLs, slang words, etc. are removed from the text for better classification. The necessary features are obtained from the text. Sentence Grammar checking is not analysed in this phase, but can be done using "Object Knowledge Model" (OKM) proposed in [24]. Further, if there is a large paragraph, only the necessary features must be extracted and not the complete paragraph. However, if the reviews are short, the number of features will be less and hence feature extraction will not be necessary. The dataset is split into a training set and testing set at a ratio of 80:20.

There are limitations in the machine learning techniques since they cannot think naturally like humans, but only can process depending on the training. Hence, to train the algorithm efficiently, some other transformations that are performed on the data. The sentences are transformed into vectors based on the features in form of matrices. Some of them are focused on generating new vectors which keep the data in a more compact way. This is done by ensuring that there is no data loss in the dataset. Case sensitiveness is another feature. Since, the reviews might be a mixture of both upper and lower cases, it is necessary to convert the text to a single type. This is because, the upper case and lower case of a same letter may be read differently by the incremental machine learning algorithm and may be saved differently when they are converted into a numerical value. Hence, in this work, all the text data is transformed into lower case irrespective of the source case. The punctuations are left as such without any changes since their presence is not very relevant. The presence of question mark or an exclamation mark does not have any effect on the text since it might represent both positive and negative sentiment. Hence, they are not processed to used them.

5.3   Classification

In this step, the pre-processed texts are analysed to be classified into positive, negative or neutral. Incremental based Machine learning techniques are used for the classification process which is the sentiment analysis. The nouns in the sentences are highlighted and extracted. These sentiments of these nouns must be identified and classified. A combination of RNN and LSTM is used to extract the sentiments in this work. The labelled training set is used to train the hybrid classifier. Once the classifier is trained, the test set can be given when each of the unlabelled data can be run and its aspects can be obtained. Individual sentences can also be run and its sentiments can be obtained. The sentiments and the polarity are returned as outputs.

5.4 Evaluation

The performance of the system is evaluated on the basis of standard metrics like Recall, Precision and F-Score. The proposed algorithm is compared with other state-of-the-art techniques to prove that the proposed approach is superior. The Flow of the proposed system is given in Fig. 2.

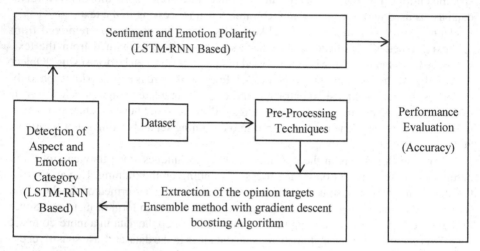

Fig. 2. Implementation flow diagram of proposed method

6 Experimental Setup

When an unlabeled dataset is given, the model is initially trained from the allotted data and then the model which has the best performance is saved with respect to validation of the dataset. The proposed neural model is performed and compared with the traditional algorithm.

The length of the reviews fed into the classifier is not fixed, since every customer is different and hence it might vary from a single sentence to multiple sentences. The maximum number of words in the sentence is set to 80. The dimension for the hidden states in the LTSM is set at 300. For the training, gradient descent is used for effective training. The dropout probability is maintained at 1.0 for both the layers and the learning rate is set at 0.01. The batch size is set to be 25 which means that the maximum reviews in a single batch is 25. The embedding dimension is maintained at 300 and the L2 regularization is kept at 0.001 for a total of 100 epochs. The experiment is validated with respect to the accuracy and compared with the other algorithms.

7 Results and Discussion

The proposed model is run using the techniques on the obtained dataset. The classifier is trained with dataset that contains lots of aspects from different sentences with labelled data. A total of 3699 aspects have been used to train the classifier. The split up of the positive, negative and the neutral data is shown in Table 1.

Table 1. Training Data

Positive	Negative	Neutral
2164	898	637

The training set is used for the training. Individual sentences are fed into the classifier in order to get the aspects. The sentences are split into smaller sentences in such a way that there is one aspect per sentence. These split lines are then subjected to the sentimental analysis in order to get the accurate features. Here, period, comma and colons are used as a separator. These cases have been given in the Figs. 3 and 4.

In the sentence, "Nice biryani. 1 star less for ambience. Service could be a little faster", there are three different aspects. The first one "Biryani" is a positive aspect with respect to word nice. The other two aspects are Service and waiter which are both negative. Even though in the sentence, "Service could be a little faster" the meaning is indirect, the algorithm is able to understand correctly and displays it as negative.

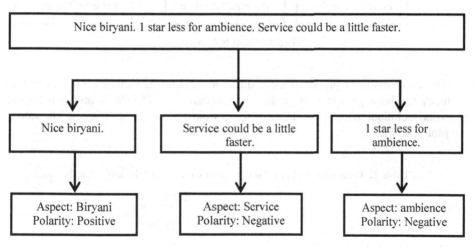

Fig. 3. Case: 3-Aspects

In the sentence, "Food was delicious. Staff was courteous. Nice place for a family outing. Had high chair for my one year. Wasn't pristine clean but the waiter cleaned it up promptly for us Ambience was average", there are six different aspects. The number of aspects in a single review is high and this has been split into six sentences. The algorithm classifies it as four positives and two negatives. In one of the aspects, "Wasn't pristine clean but the waiter cleaned it up promptly for us", there are two aspects; however, the separator or comma was not added by the customer. Hence, the framework has considered them as a single sentence. The case has been shown in Fig. 4.

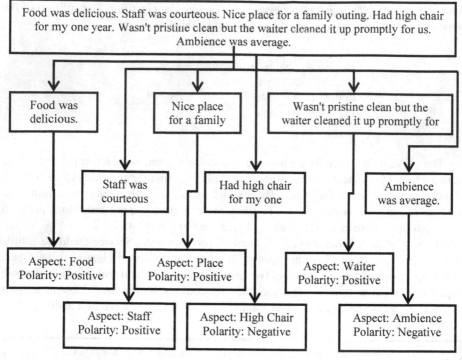

Fig. 4. Case: 6-Aspects

The comparison of proposed technique with traditional methods with respect to accuracy has been presented in Table. 2. An accuracy of 75.89% is obtained for the proposed technique. This is significantly higher than the existing techniques. The values are plotted in Fig. 5.

Table 2. Comparison of proposed method with state of the are methods

Technique	Accuracy
Naïve Bayes	73.65%
Random Forest Method	70.61%
Support Vector Machine	74.36%
Logistic Regression	73.44%
Majority Voting	73.83%
Ankit and Saleena [25]	74.67%
Proposed LSTM-RNN technique	75.89%

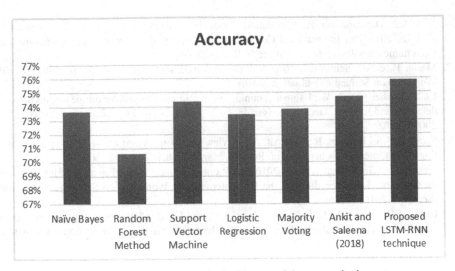

Fig. 5. Comparison of proposed method with state of the are methods w.r.t accuracy

8   Conclusion

Sentimental analysis was performed in this work using artificial intelligence through RNN. The most commonly faced difficulty is the classification of the text in the pre-processing phase. However, this has been overcome. This is because it has an important impact on the number of words that will be learnt by the algorithm. Various transformations and pre-processing techniques are applied for this. From the reviews, lots of different aspects are obtained and trained to the LTSM-RNN classifier. New reviews are fed into the classifier and the sentiments of the individual aspects are obtained with positive, negative and neutral aspects. The obtained accuracy is seen to be around 76% which is higher than the existing algorithms. Apart from good accuracy, major outcome is - multiple aspects are correctly extracted from multi-sentence lengthy reviews. In future, the accuracy is attempted to be improved by further optimizing the proposed system.

References

1. Le, B., Nguyen, H.: Twitter sentiment analysis using machine learning techniques. In: Thi, H. A. L., Nguyen, N. T., Van Do, T. (eds.) Advanced Computational Methods for Knowledge Engineering, pp. 279–289. Springer International Publishing, Cham (2015). https://doi.org/10.1007/978-3-319-17996-4_25
2. Amlaan, B., Joshi, S.: Various Approaches to Aspect-based Sentiment Analysis. arXiv:1805.01984 (2018)
3. Wei, X., Li, T.: Aspect based sentiment analysis with gated convolutional networks. arXiv:1805.07043 (2018)
4. Qiao, L., et al.: Content attention model for aspect-based sentiment analysis. In: Proceedings of the 2018 World Wide Web Conference (2018)
5. Cambria, E., Schuller, B., Xia, Y., Havasi, C.: New avenues in opinion mining and sentiment analysis. IEEE Intell. Syst. **28**(2), 15–21 (2013). https://doi.org/10.1109/MIS.2013.30

6. Jain, A.P., Dandannavar, P.: Application of machine learning techniques to sentiment analysis. In: 2016 2nd International Conference on Applied and Theoretical Computing and Communication Technology (iCATccT). IEEE (2016)

7. Maria, P., et al.: Semeval-2016 task 5: aspect based sentiment analysis. In: International Workshop on Semantic Evaluation (2016)

8. Li, N., Desheng Dash, W.: Using text mining and sentiment analysis for online forums hotspot detection and forecast. Dec. Supp. Syst. **48**(2), 354–368 (2010). https://doi.org/10.1016/j.dss.2009.09.003

9. Sadegh, M.S., Ibrahim, R., Selamat, A.: Bi-view semi-supervised active learning for cross-lingual sentiment classification. Inf. Process. Manage. **50.5**, 718–732 (2014)

10. Hussam, H.: Sentisys at semeval-2016 task 5: opinion target extraction and sentiment polarity detection. In: Proceedings of the 10th International Workshop on Semantic Evaluation (SemEval-2016) (2016)

11. Gopal, P.B., et al.: JU_CSE: A Conditional Random Field (CRF) based approach to aspect based sentiment analysis. In: Proceedings of the 8th International Workshop on Semantic Evaluation (SemEval 2014) (2014)

12. Onan, A., Korukoğlu, S., Bulut, H.: A multiobjective weighted voting ensemble classifier based on differential evolution algorithm for text sentiment classification. Expert Syst. Appl. **62**, 1–16 (2016). https://doi.org/10.1016/j.eswa.2016.06.005

13. Zhang, M., Wang, Q., Guohong, F.: End-to-end neural opinion extraction with a transition-based model. Inf. Syst. **80**, 56–63 (2019). https://doi.org/10.1016/j.is.2018.09.006

14. Tsirakis, N., Poulopoulos, V., Tsantilas, P., Varlamis, I.: Large scale opinion mining for social, news and blog data. J. Syst. Softw. **127**, 237–248 (2017). https://doi.org/10.1016/j.jss.2016.06.012

15. Chandra, A.C., Rajpoot, D.S., Saraswat, M.: Twitter sentiment analysis using hybrid cuckoo search method. Inf. Process. Manage. **53.4**, 764–779 (2017)

16. Rezaeinia, S.M., Rahmani, R., Ghodsi, A., Veisi, H.: Sentiment analysis based on improved pre-trained word embeddings. Expert Syst. Appl. **117**, 139–147 (2019). https://doi.org/10.1016/j.eswa.2018.08.044

17. Song, K., Yao, T., Ling, Q., Mei, T.: Boosting image sentiment analysis with visual attention. Neurocomputing **312**, 218–228 (2018). https://doi.org/10.1016/j.neucom.2018.05.104

18. Anvardh, N., Sherry, L.: Anomaly detection in aircraft data using Recurrent Neural Networks (RNN). In: 2016 Integrated Communications Navigation and Surveillance (ICNS). IEEE (2016)

19. Jiang, H., Yao, L., Jing, X.: Automatic soccer video event detection based on a deep neural network combined CNN and RNN. In: 2016 IEEE 28th International Conference on Tools with Artificial Intelligence (ICTAI). IEEE (2016)

20. Duyu, T., et al.: Effective LSTMs for target-dependent sentiment classification. arXiv:1512.01100 (2015)

21. Wang, Y., et al.: Attention-based LSTM for aspect-level sentiment classification. In: Proceedings of the 2016 Conference on Empirical Methods in NLP (2016)

22. Yelp Dataset- A trove of reviews, businesses, users, tips, and check-in data! https://www.kaggle.com/yelp-dataset/yelp-dataset. Accessed on June 2020

23. Veera Prathap Reddy, M., Prasad, P.V.R.D., Chikkamath, M., Mandadi, S.: NERSE: Named Entity Recognition in Software Engineering as a Service. In: Lam, Ho-Pun., Mistry, Sajib (eds.) ASSRI -2018. LNBIP, vol. 367, pp. 65–80. Springer, Cham (2019). https://doi.org/10.1007/978-3-030-32242-7_6

24. Prabhu, C.S.R., Venkateswara Gandhi, R., Jain, A.K., Lalka, V.S., Thottempudi, S.G., Prasad Rao, P.V.R.D.: A Novel Approach to Extend KM Models with Object Knowledge Model (OKM) and Kafka for Big Data and Semantic Web with Greater Semantics. In: Barolli, L., Hussain, F.K., Ikeda, M. (eds.) Complex, Intelligent, and Software Intensive Systems: Proceedings of the 13th International Conference on Complex, Intelligent, and Software Intensive Systems (CISIS-2019), pp. 544–554. Springer International Publishing, Cham (2020). https://doi.org/10.1007/978-3-030-22354-0_48

25. Ankit, N.S.: An ensemble classification system for twitter sentiment analysis. Procedia Comput. Sci. **132**, 937–946 (2018). https://doi.org/10.1016/j.procs.2018.05.109

Cloud Based Exon Prediction Using Maximum Error Normalized Logarithmic Algorithms

Md. Zia Ur Rahman⬤, Annabathuni Chandra Haneesh⬤,
Bhimireddy Shanmukha Sai Reddy⬤, Sala Surekha$^{(\boxtimes)}$⬤,
and Putluri Srinivasareddy⬤

Department of ECE, Koneru Lakshmaiah Education Foundation, K L University, Vaddeswaram,
Guntur 522002, India

Abstract. Distributed computing gives medical services organizations significant examination and monetary advantages. Cloud administrations guarantee that enormous amounts of such delicate information will be put away and overseen safely. The quality succession labs send crude or gathered data through the Internet to a few arrangement libraries under conventional progression of quality data. Cloud service use will reduce DNA sequencing storage costs to a minimum. In this work, we demonstrated a new genomic bioinformatic system, using Amazon Cloud Services, that stores and processes genomic sequence information. A critical assignment in bio-informatics, which helps in the recognizable proof and plan of sickness drug, is the genuine ID of exon locales in deoxyribonucleic corrosive (DNA) arrangement. All exon recognizable proof procedures depend on three fundamental periodicity (TBP) properties of exons. In contrast with a few different strategies, versatile sign preparing procedures have been promising. This paper utilizes the most extreme blunder standardized least logarithmic outright distinction (MENLLAD) calculation likewise its marked variations to build up various Adaptive Exon Predictors (AEPs) with less computational multifaceted nature. At last, a presentation assessment is performed for various AEPs utilizing different standard quality information successions got from National Biotechnology Information Center (NBI) genomic grouping data set, for example, Sensitivity (Sn), Specificity (Sp) and Precision (Pr) estimations.

Keywords: Bioinformatics · Cloud computing · Computational complexity · Exon predictor · Three base periodicity

1 Introduction

Genomics is an immense field in which areas that code for proteins are identified using AEP based systems presented here [1–7]. Exon areas have a role to play in the assessment of diseases and drug design [8]. To find exon locations accurately, proposed LLAD based AEP's with less number of multiplications can be utilized. In eukaryotes, segments are used for exons it encrypts proteins and fragments are used for non-proteins as introns. Coded region is considered as 3% in eukaryotes of human gene sequence while the extra 97% are non-coded regions. Consequently, it is an important task to detect coded

© Springer Nature Switzerland AG 2021
M. Singh et al. (Eds.): ICACDS 2021, CCIS 1440, pp. 690–700, 2021.
https://doi.org/10.1007/978-3-030-81462-5_60

sections in a DNA sequence. Alongside, precisely finding the coding area is a fundamental task in bio-informatics. By fundamental groupings of DNA, base three periodicity (BTP) property has been appeared, which remains the important property of only exon segments. For a PSD plot, at recurrent point practically identical to f1 = 1/3, a sharp peak is evidently portrayed [9]. Diverse existing methodologies to exon recognition in literature rely on various methods [10–12] for finding exon locations in DNA precisely. To overcome this issue, normalization of the AEP's are utilized to measure exons in DNA sequences [13]. By using this change updated channel wight vector coefficients and their normalization dependent on weight vector standards. Here we think about two sorts of normalization based techniques, as error normalization and maximum error normalization based techniques. They are Error Normalized LLAD (ENLLAD) and Maximum Error Normalized LLAD (MENLLAD) based AEPs [14, 15]. LMS based AEP is also implemented for comparison purpose with all proposed AEPs [16]. The computational complexity is more lessened all around by utilizing proposed maximum normalized based AEPs [17]. Along these lines, most over the top assortments of adaptable tallies are thought of [18, 19].

Disadvantages of LMS are overcome by use of MENLMLS based techniques. To assess this, proposed AEPs are developed to improve exon prediction accuracy and better convergence performance using Maximum Error Normalized LLAD (MENLLAD) based AEPs. Burdens of AEP were overcome by combining all proposed techniques with sign based algorithms and error normalization accordingly [20]. Because of normalization, higher tap length could be confined to one, by applying a methodology named as most uncommon variable standardization paying little mind to tap length [21, 22]. In error normalized LMS based AEPs, relationship among goof and information reference signal is standardized by a worth like squared standard [23]. Best close by standardized strategy combine faster stood apart from standard LMS system and besides beat inclination rattle application issue [24, 25]. Henceforward botch in unsurprising state and speed of relationship of MENLLAD are unavoidable separated and LMS [26]. By using standard NCBI gene database [27], performance measures are evaluated for proposed AEPs. Sensitivity, precision, specificity, computational complexity and convergence features are calculated for estimating the performance of various AEPs. In results section, findings of AEPs, adaptive theory methods and efficiency of AEPs are discussed.

2   Algorithm for the Accurate Prediction of Exon Location

The essential stage is to research quality progression taken as of NCBI game plan informational index subject to dimer nucleotide densities in like manner this gathering was changed as numerical documentation in proposed AEP. It stays a central occupation of genomic preparing on the grounds that solitary computerized or discrete signs can be utilized for signal handling. DNA grouping here is meant paired data portraying four double streams with twofold planning. Advanced change is an urgent errand to deal with quality successions since strategies of sign preparing must be utilized on such signals. Along these lines, 1 demonstrates the presence of a nucleotide and 0 is its nonattendance. The subsequent arrangement presently stays fitting for the versatile calculation as an info. AEP remain thought that is produced utilizing versatile sign handling techniques.

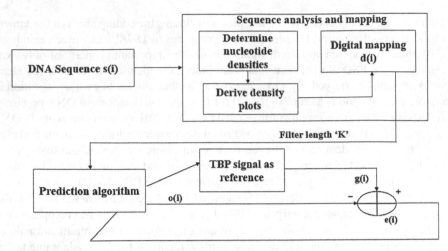

Fig. 1. Square chart of proposed AEP

Let, d(i) as advanced succession that is planned, s(i) as DNA arrangement, g(i) as quality grouping that submits to TBP, o(i) demonstrates result gained by utilization of versatile procedure likewise e(i) address sign of input for modifying weight coefficients produced inside criticism circle. Length in LMS procedure remains viewed as 'K'. In light of present tap coefficient step size boundary 'B', the following tap coefficient was assessed in this calculation with current tap vector as t(i), additionally input paired planned succession showed as d(i) right now. The LMS calculation remains numerically communicated and broke down in [12]. Figure 1 presents delegate block outline for an AEP.

Weight articulation of LMS versatile procedure is composed as

$$t(i + 1) = t(i) + Bs(i) e(i) \tag{1}$$

Adaptable estimations ought to experience least computational difficulty in applications identifying with exon affirmation to pull in nano based bioinformatics applications. In [18], it was discussed with strategies therefore. In these methods, three stamped structures are peddled.

Signum functions is represented as

$$C\{s(i)\} = \left\{ \begin{array}{l} 1 : s(i) > 0 \\ 0 : s(i) = 0 \\ -1 : s(i) < 0 \end{array} \right\} \tag{2}$$

These variations like to decrease the intricacy of LMS calculation. LMS has more prominent computational trouble contrasted with these variations. Information Clipped LMS (DCLMS) strategy can be communicated as LMS recursion by changing information tap vector. For the present circumstance, the mean assessments of vector C[s(i)] remains fill in for s(i), here sign limit C was used for s(i) on reason of part by fragment.

Mass update verbalizations for DCLMS method is implied as

$$t(i + 1) = t(i) + BC\{s(i)\} e(i) \tag{3}$$

The weight connection for ECLMS strategy remains acquired through evolving e(i) by marked structure as

$$t(i + 1) = t(i) + Bs(i) \, C\{e(i)\} \tag{4}$$

Likewise, weight connection for DECLMS is coming about through replacement of s(i), e(i) with utilization of marked structures as

$$t(i + 1) = t(i) + BC\{s(i)\} \, C\{e(i)\} \tag{5}$$

Due to its robustness as well as simplicity, the standard adaptive LMS technique is suitable for exon forecast. In order to choose parameter of step size for the convergence as well as stability, understanding of preceding input power level rate is required for LMS filter. As one of the statistical unknown levels is generally the input power level, it will normally be assessed thru information prior start of adaptation process. The vector of the input information is proportionate to weight update process. Other one being its step size is fixed. Both these remain two setbacks of LMS. An algorithm must be designed so that weak as well as strong signals can be handled in real time. Therefore, the tap coefficients must be adapted accordingly on the basis of filter changes in input as well as output. Thus, for a major information vector, LMS procedure is influenced by inclination clamor intensification impediment.

Normalized LLAD algorithm is considered as a unique LMS algorithm application that takes into consideration signal level variation at the filter output also chooses a logarithmic normalized cost function which leads to a faster converging as well as stable adaptation algorithm. MENLLAD algorithm overwhelms LMS limitations also, improves speed of combination just as exon following capacity. Here, we have used MENLLAD and its adaptive algorithm based on SRA to enhance AEP efficiency. The MENLLAD algorithm overcomes the LMS disadvantages and increases the ability of exon identification and quicker convergence when error is high. This also reduces the surplus EMSE in the exon identification process. These MENLMLS adaptive algorithms are used for developing AEPs in order to cope with computing difficulty of an AEP in practical applications.

Also, the sign capacity is utilized to diminish the unpredictability. With utilization of sign capacity to MENLLAD, three disentangled marked variants MENSRLLAD, MENSLLAD, and MENSSLLAD calculations are determined.

The weight relations for MENSRLLAD, MENSLLAD, and MENSSLLAD techniques becomes

$$t(i + 1) = t(i) + \frac{\beta'}{\varepsilon + max(||e(i)||)^2} C[s(i)] \left[\frac{\alpha(e(i))}{1 + \alpha(|e(i)|)} \right] \tag{6}$$

$$t(i + 1) = t(i) + \frac{\beta'}{\varepsilon + max(||e(i)||)^2} s(i) C \left[e(i) \left[\frac{\alpha(e(i))}{1 + \alpha(|e(i)|)} \right] \right] \tag{7}$$

$$t(i + 1) = t(i) + \frac{\beta'}{\varepsilon + max(||s(i)||)^2} C[s(i)] C \left[e(i) \left[\frac{\alpha(e(i))}{1 + \alpha(|e(i)|)} \right] \right] \tag{8}$$

By using these calculations, we at long last created four AEP's then stood out their outcomes from AEP utilizing LMS. Execution appraisal utilizing sensitivity, specificity and precision so that MENSRLLAD is only underneath to the non-signum variation. Subsequently, MENSRLLAD is superior to its marked forms, alluding to measurements for execution additionally figuring trouble, among different procedures picked for application.

3 Complexity and Convergence Characteristics

The measure of augmentations required is resolved as a measurement to gauge and analyze calculation multifaceted nature specifically. The emphasis isn't on exact investigation for intricacy to perform calculations however on the assessment of unmistakable MENLLAD-based versatile strategies. Furthermore, those methods subject as far as possible are without duplicate figuring's they needed for exon obvious confirmation applications. Let us consider, LMS needs K+1 increase figuring's near to one turn of events, when the mass update condition is enrolled. While to figure 'S. e(n)', simply 2T+1 increment is required for MENSRLLAD based adaptable variety. In the event of two other marked MENLLAD calculations, 2T+1 duplicate calculation stay required. With MENSRLLAD, it needs less duplications with less processing trouble contrasted with other ENLLAD based methods. Computational complexities for LMS and MENLLAD-based assortments is shown in Table 1. The proposed AEPs subject to MENLLAD give offers less intricacy to perform calculations for the area of the ideal quality situation in a genomic input arrangement and can be used in nano gadgets [5].

Table 1. LMS and various ENLLAD based AEP's computations

S. no.	Method	Number of multiplications	Number of additions
1	LMS	K+1	K+1
2	MENLLAD	K+4	K+2
3	MENSRLLAD	4	K+2
4	MENSLLAD	K+3	K+2
5	MENSSLLAD	3	2

Assembly highlights of the proposed MENLLAD and their marked variations are portrayed in Fig. 2. All proposed MENLLAD-based movable calculations are clearly more quickly combining than AEP dependent on LMS. Therefore, the MENSRLLAD adaptive algorithm is considered better, based on computing difficulty as well as convergence efficiency in contrast to LMS and its other signed algorithms, among the algorithms considered for AEP implementation. It was obvious that MENSRLLAD converges quicker compared to MENSLLAD and MENSSLLAD based AEPs from convergence features.

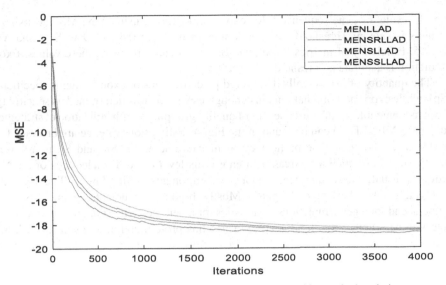

Fig. 2. Assembly curves of MENLLAD algorithm and its marked variations

4 Experimental Results and Analysis

In this section performance comparison of various AEPs are discussed. Block diagram of AEP is shown in Fig. 1. For sign based ENLMS algorithm multiple AEPs are derived. For comparison, LMS based AEP algorithm is created. For evaluating performance, ten genomic datasets are considered from NCBI database [19]. Sensitivity, specificity and precision are considered for performance evaluation of proposed algorithm and theory and expression are explained in [13]. In Table 2, outputs for various algorithms are considered.

As part of the determination of exon segments by using DSP methods, there are few measures based on changes in the threshold level in the output spectrum used for comparison. Nucleotide's amount situated like introns in exon locating phase represented as true negative (TN), whereas exon areas if correctly identified is stated for instance as true positive (TP). In addition, complete amount of exon areas positioned as intron areas is indicated to be false negative (FN), compared to introns amount really anticipated like areas of exons to be false positive (FP). To identify the efficiency of proposed algorithm ten gene datasets of NCBI are considered. The accession for these sequences remains X59065.1, E15270.1, U01317.1, X77471.1, AF009962, X92412.1, AB035346.2, AJ223321.1, AJ225085.1, and X51502.1 respectively.

Expressions for performance metrics are.

$Pr = (TP + TN)/(TP + FP + TN + FN)$
$Sp = TP/(TP + FP)$
$Sn = TP/(TP + FN)$

The number of exons actually found part of sections of exons remains specificity (Sp), whereas exons quantity if exactly forecasted than sensitivity (Sn) is calculated. In

Fig. 4, results of gene sequence 5 for exon identification using MNLMS and its sign variants is shown. Threshold value is chosen in between 0.4 to 0.9 with an interval of 0.05. Efficiency of metrics Pr, Sn, and Sp is evaluated by using these values. Exon prediction is accurate at threshold value of 0.8.

The quantity of exons really discovered piece of areas of exons remains specificity (Sp), while exons amount that remains enough estimated was determined as sensitivity. Exon recognizable proof consequences of quality grouping 5 with utilization of strategies subject to MENLLAD can be found in the Fig. 4. Estimations of edge are browsed 0.4 to 0.9 at a time span. The proficiency of measurements Pr, Sn, and Sp is assessed by utilizing these qualities. Areas with enormous level of A+T nucleotides of a DNA sequence usually show intergenic sequence components, whilst low A+T and greater G+C nucleotides show potential genes. Mostly, high CG dinucleotide content is often found ahead for a gene. Functions of statistics for a gene sequence remains beneficial for determining whether the input gene sequence has protein-coding segments. For DNA sequence having accession AF009962, Fig. 4 shows a standard nucleotide density plot. Its dimer appropriation is shown in a bar delineation utilizing MATLAB programming. It has been appeared from Fig. 3 that T-T base pair dimers are more in this quality grouping 5 when differentiated to every one of its dimers. It comprises of 680 T-T base pair dimers. This quality succession additionally contains 527 A-T and 70 G-C dimers. The G+C content is demonstrated to be more modest than the A+T dimer content which shows that it has less quality check.

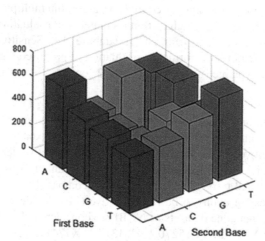

Fig. 3. Plot of gene sequence 5 for Nucleotide density of dimers with accession AF009962

With Accession number AF009962 for all sign based, the assessments of movement 5 is considered. From Table 2, execution extents of MENLLAD based AEP are just average contrasted with MENSRLLAD based AEP with less emphases in light of low multifaceted nature and exon finding limit by finding authoritatively at 3934–4581 with stunning force and a sharp top in PSD plot is observed. The signum function present in all signed versions of MENLLAD reduces the computational complexity and thus all signed versions predict the exon locations more accurately. Of all these algorithms, MENSR-LLAD based AEP is effective in terms of accurate exon prediction when compared to LMS, MENLLAD and its other signed variants with Specificity Sp, 0.7890 (78.90%), Sensitivity Sn 0.7789 (77.89%), also Precision, Pr 0.7806 (78.06%) respectively. At 0.8 threshold value, the exon prediction appears to be better for MENSRLLAD based AEP. PSD plots for MENLLAD and its marked variations are appeared in Fig. 4 (b), (c), and (d) separately. Finally, all proposed MENLLAD based AEPs are more effective to discover exon areas in genomic sequences compared with the prevailing LMS technique. With Accession number AF009962 for all sign based, the assessments of movement 5 is considered. From Table 2, execution extents of MENLLAD based AEP are just average contrasted with MENSRLLAD based AEP with less emphases in light of low multifaceted nature and exon finding limit by finding authoritatively at 3934–4581 with stunning force and a sharp top in PSD plot is observed. The signum function present in all signed versions of MENLLAD reduces the computational complexity and thus all signed versions predict the exon locations more accurately. Of all these algorithms, MENSR-LLAD based AEP is effective in terms of accurate exon prediction when compared to LMS, MENLLAD and its other signed variants with Specificity Sp, 0.7890 (78.90%), Sensitivity Sn 0.7789 (77.89%), also Precision, Pr 0.7806 (78.06%) respectively. At 0.8 threshold value, the exon prediction appears to be better for MENSRLLAD based AEP. These PSD plots for MENLLAD and its marked variations are appeared in Fig. 4 (b), (c), and (d) separately. Finally, all proposed MENLLAD based AEPs are more effective to discover exon areas in genomic sequences compared with the prevailing LMS technique.

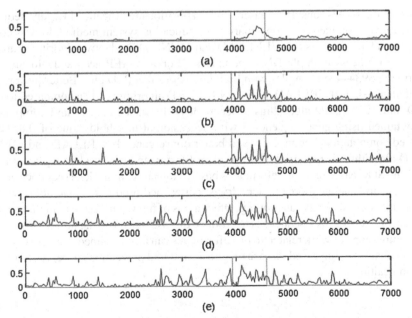

Fig. 4. PSD plots with position of exon (3934–4581) located for genomic data with number AF009962 thru numerous AEP's, (a). Output due to LMS, (b) Output due to MENLLAD, (c) Output due to MENSRLLAD, (d) Output due to MENSLLAD, (e) Output due to MENSSLLAD. *(on x-axis exon Location and PSD on y-axis are shown)*

Table 2. Comparison of various metrics of the considered adaptive algorithms in exon locating experiments.

Algorithm	Metric	Gene Sequence Serial Number									
		1	2	3	4	5	6	7	8	9	10
LMS	Sn	0.6286	0.6384	0.6457	0.6273	0.6481	0.6162	0.6193	0.6241	0.6268	0.6202
	Sp	0.6435	0.6628	0.6587	0.6405	0.6518	0.6324	0.6529	0.6289	0.6452	0.5965
	Pr	0.5922	0.5894	0.5934	0.5858	0.5904	0.5786	0.5896	0.5856	0.5814	0.5761
MENLLAD	Sn	0.8129	0.8119	0.8148	0.8173	0.8188	0.8013	0.8026	0.8125	0.8149	0.8156
	Sp	0.8143	0.8137	0.8166	0.8159	0.8124	0.8024	0.8124	0.8168	0.8126	0.8184
	Pr	0.8231	0.8252	0.8233	0.8237	0.8241	0.8035	0.8150	0.8274	0.8277	0.8265
MENSRLLAD	Sn	0.7834	0.7845	0.7914	0.7826	0.7812	0.7836	0.7845	0.7824	0.7902	0.7858
	Sp	0.7782	0.7816	0.7841	0.7765	0.7836	0.7821	0.7763	0.7816	0.7871	0.7787
	Pr	0.7879	0.7873	0.7857	0.7879	0.7867	0.7795	0.7854	0.7857	0.7829	0.7839
MENSLLAD	Sn	0.7565	0.7642	0.7568	0.7522	0.7585	0.7565	0.7571	0.7567	0.7565	0.7549
	Sp	0.7583	0.7564	0.7544	0.7539	0.7568	0.7674	0.7675	0.7559	0.7558	0.7574
	Pr	0.7670	0.7678	0.7673	0.7651	0.7671	0.7568	0.7549	0.7672	0.7673	0.7671
MENSSLLAD	Sn	0.7450	0.7465	0.7479	0.7484	0.7442	0.7456	0.7437	0.7452	0.7454	0.7454
	Sp	0.7463	0.7476	0.7433	0.7467	0.7455	0.7449	0.7560	0.7476	0.7438	0.7463
	Pr	0.7439	0.7442	0.7459	0.7478	0.7560	0.7457	0.7671	0.7534	0.7582	0.7577

5   Conclusion

In this paper, issue is considered based on exon position identification in a gene sequence. For adaptive exon identification, smart communication system methodology is considered. For addressing this issue, ENLLAD and MENLLAD adaptive algorithms are considered to process multiple DNA sequences. Proposed AEP accurately locates exon location at 3934–4581 samples with high intensities in a PSD plot. Better performance is obtained using ENSRLLAD and MENSRLLAD algorithms when compared to ENLLAD and MENLLAD algorithms in terms of performance metrics and computations are obtained using gene sequence 5 with accession at threshold value of 0.8. Due to reduced computational complexity and better convergence ENSRLLAD and MENSRLLAD algorithms is preferred for exon identification and it is used in applications like LOC and SOC based nano bioinformatics based exon prediction. Gene sequence analysis enables us to generate more efficient drugs, better food products also to attain valuable insight into our own body functioning. Other areas of future interest may include extending our research on DNA sequences to produce disease free crops, healthier farm animals etc. Future scope of work related to bioinformatics can also be aimed at discovery of disease mechanisms, improved disease diagnosis and find prevention strategies to improve human health.

References

1. Kathleen, Ch., Nocle, P., Maria Kmoppers, B.: The adoption of cloud computing in the field of genomics research: the influence of ethical and legal issues. PLoS ONE **11**, 1–33 (2016)
2. Nic Lincoln Stein, D.: The case for cloud computing in genome informatics. Genome Biol. **11**, 1–7 (2010)
3. Ning, L.W., Lin, H., Ding, H., Huang, J., Rao, N., Guo, F.B.: Predicting bacterial essential genes using on sequence composition information. Genet. Mol. Res. **13**, 4564–4572 (2014)
4. Dickerson, J.E., Zhu, A., Robertson, D.L., Hentges, K.E.: Defining the role of essential genes in human disease. PloS ONE **6**, 1–10 (2011)
5. Singh, A.K., Kumar Srivastava, V.: The three base periodicity of protein coding sequences and its application in exon prediction. In: 2020 7th International Conference on Signal Processing and Integrated Networks (SPIN), Noida, India, pp. 1089–1094 (2020). https://doi.org/10.1109/SPIN48934.2020.9071068
6. Ahmad, M., Jung, L.T., Bhuiyan, A.: From DNA to protein: why genetic code context of nucleotides for DNA signal processing? A review. Biomed. Signal Process. Control **34**, 44–63 (2017)
7. Sun, T.M., Wang, Y.C., Wang, F., Du, J.Z., Mao, C.Q.: Cancer stem cell therapy using doxorubicin conjugated to gold nanoparticles via hydrazone bonds. Biomaterials **35**, 836–845 (2014)
8. Massadeh, S., et al.: Nano-materials for gene therapy: an efficient way in overcoming challenges of gene delivery. J. Biosens. Bioelectron. **7**(1), 1–12 (2016)
9. Li, M., Li, Q., Gamage Upeksha, G., Wang, J., Wu, F., Pan, Y.: Prioritization of orphan disease-causing genes using topological feature and go similarity between proteins in interaction networks. Sci. China Life Sci. **57**, 1064–1071 (2014)
10. Scalzitti, N., Jeannin-Girardon, A., Collet, P., et al.: A benchmark study of ab initio gene prediction methods in diverse eukaryotic organisms. BMC Genomics **21**, 293 (2020). https://doi.org/10.1186/s12864-020-6707-9
11. Maji, S., Garg, D.: Progress in gene prediction: principles and challenges. Curr. Bioinform. **8**, 226–243 (2013)
12. Saberkari, H., Shamsi, M., Heravi, H., Sedaaghi, M.H.: A novel fast algorithm for exon prediction in eukaryotes genes using linear predictive coding model and goertzel algorithm based on the Z-curve. Int. J. Comput. Appl. **67**, 25–38 (2013)
13. Tiwari, S., Ramachandran, S., Bhattacharya, A., Bhattacharya, S., Ramaswamy, R.: Prediction of probable genes by Fourier analysis of genomic sequences. Comput. Appl. Biosci. **13**(3), 263–270 (1997)
14. Ismail, Md.A., Ye, Y., Tang, H.: Gene finding in metatranscriptomic sequences. BMC Bioinform. **15**, 01–08 (2014)
15. Voss, R.F.: Evolution of long-range fractal correlations and 1/f noise in DNA base sequences. Phys. Rev. Lett. **68**(25), 3805–3808 (1992)
16. Liu, G., Luan, Y.: Identification of protein coding regions in the eukaryotic DNA sequences based on Marple algorithm and wavelet packets transform. Abstr. Appl. Anal. **2014**, 1–14 (2014)
17. Mahin, G., Hamed, K.: Bioinformatics approaches for gene finding. Int. J. Sci. Res. Sci. Technol. **1**, 12–15 (2015)
18. Putluri, S.R., Rahman, Md.Z.U.: Identification of protein coding region in DNA sequence using novel adaptive exon predictor. J. Sci. Ind. Res. **77**, 1–5 (2018)
19. Azuma, Y., Onami, S.: Automatic cell identification in the unique system of invariant embryogenesis in caenorhabditis elegans. Biomed. Eng. Lett. **4**. 328–337 (2014)

20. Putluri, S., Rahman, Md.Z.U.: Computer based genomic sequences analysis using least mean forth adaptive algorithms. J. Theor. Appl. Inf. Technol. **95**(9), 2006–2014 (2017)
21. Putluri, S., Rahman, Md.Z.U.: New adaptive exon predictors for identifying protein coding regions in DNA sequence. ARPN J. Eng. Appl. Sci. **11**, 13540–13549 (2016)
22. Putluri, S., Rahman, Md.Z.U., Fathima, S.Y.: Cloud based adaptive exon prediction for DNA analysis. IET Healthc. Technol. Lett. **5**(1), 1–6 (2018)
23. Rahman, Md.Z.U., Karthik, G.V.K.S., Fathima, S.Y., L-Ekukaille, A.: An efficient cardiac signal enhancement using time-frequency realization of leaky adaptive noise cancelers for remote health monitoring systems. Measurements **46**, 3815–3835 (2013)
24. Rahman, Md.Z.U., Ahmed Shaik, R., Rama Koti Reddy, D.V.: Efficient and simplified adaptive noise cancelers for ECG sensor based remote health monitoring. IEEE Sens. J. **91**(3), 566–573 (2012)
25. Haykin, S.O.: Adaptive Filter Theory, 5th edn. Pearson Education Ltd., London (2014)
26. Sayin, Md.O., Denizcan Vanli, N. , Serdar Kozat, S.: A novel family of adaptive filtering algorithms based on the logarithmic cost. IEEE Trans. Signal Process. **62**(17), 4411–4424 (2014)
27. National Center for Biotechnology Information. www.ncbi.nlm.nih.gov/

Unsupervised Learning of Visual Representations via Rotation and Future Frame Prediction for Video Retrieval

Vidit Kumar[✉], Vikas Tripathi, and Bhaskar Pant

Graphic Era Deemed to be University, Dehradun, India

Abstract. Due to rapid technological advancements, the growth of videos uploaded to the internet has increased exponentially. Most of these videos are free of semantic tags, which makes indexing and retrieval a challenging task, and requires much-needed effective content-based analysis techniques to deal with. On the other hand, supervised representation learning from large-scale labeled dataset demonstrated great success in the image domain. However, creating such a large scale labeled database for videos is expensive and time consuming. To this end, we propose an unsupervised visual representation learning framework, which aims to learn spatiotemporal features by exploiting two pretext tasks i.e. rotation prediction and future frame prediction. The performance of the learned features is analyzed by the nearest neighbor task (video retrieval). For this, we choose the UCF-101 dataset to experiment with. The experimental results shows the competitive performance achieve by our method.

Keywords: Content based search · Deep learning · Self-supervised learning · Unsupervised learning · Video retrieval

1 Introduction

Video data exists everywhere from social sites to CCTV storage systems. Due to the advancement in technology, this data is increasing day by day. In particular, many videos are uploaded to social sites every minute, which often do not have proper semantic tags. In addition, CCTV recordings (such as surveillance footage), which are streaming video data, are heavily uploaded to storage devices on a daily basis. This huge increase in video data triggered the need for an effective as well as efficient method for analyzing, indexing and retrieval. More importantly, content-based analysis techniques are needed because these videos rarely contain semantic tags or labels. And, therefore, a robust spatio-temporal feature extraction method is required for video representation. With content-based video retrieval (CBVR) as a technology, it provides solutions for applications such as video navigation, video surveillance, content filtering, in-video advertising and many more.

© Springer Nature Switzerland AG 2021
M. Singh et al. (Eds.): ICACDS 2021, CCIS 1440, pp. 701–710, 2021.
https://doi.org/10.1007/978-3-030-81462-5_61

Recently, deep learning based methods have shown tremendous performance by utilizing large scale labeled data in a supervised manner. Also, learning spatio-temporal features requires large-scale labeled video dataset. However, it can be very expensive to annotate such a massive dataset as compared to image dataset. In this regard, unsupervised learning can be a cheap solution to tackle it. Prior works in this regard, learned features through self-supervision task which is usually based on single pretext task. To further improve the feature representation capability, in this paper, we learn the features by joint optimization of two tasks using 3d Convolutional neural network (CNN) as a backbone.

Our main contributions in this paper are summarized below:

- We proposed an unsupervised video representation learning method by joint learning of rotation prediction and future frame prediction.
- We predict future frames with convolutional auto-encoder network and demonstrate its effectiveness in improving the representation learning.
- We used small variant of C3D (refer Sect. 3.2) as our backbone network and tested our method on UCF101 dataset and validate the learned features via nearest neighbor retrieval and action recognition experiments with state-of-the-arts.

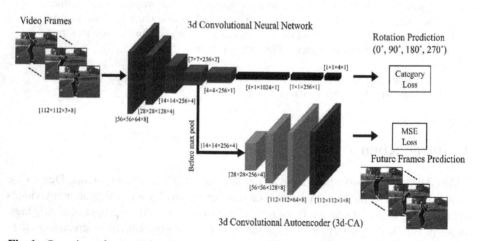

Fig. 1. Overview of proposed approach. The cuboid in green represents convolutional layer and blue color represents transpose convolutional layer. Here, maxpooling layer (in case of C3D, for more info. refer Sect. 3.2) is omitted due to space constraint and we depicted its output as ([.]). (Color figure online)

2 Literature Review

Most of the prior works in video retrieval domain rely on hand crafted features. For instance, Jiang et al. [1] exploits bag-of word model for video representation, Asha et al.

[2] used SURF features for video retrieval. Further, in [3], an improved method for large-scale related videos retrieval using temporal-concentration SIFT features was proposed. The SIFT features were encoded through tracking to generate temporal concentration SIFT, which are local compactness features that help to reduce visual redundancy. Frame-level analysis is used in these techniques to find matching video contents. Moreover, Brindha et al. [4] exploits low level features based on color, texture and shape with motion as a high level feature to bridge the gap between them. In [5], Integrated Histogram Bin Matching (IHBM) similarity measure is proposed for video frames matching, where the frames were represented by HOG descriptor. The main limitation is that these features work well at the static frame level and not at the video level to represent the entire landscape of a scene or an entire video.

The success of deep learning led to research in both image and video domain. For instance, Babenko et al. [6] explore the deep features for image retrieval problem. Kumar et al. [7] present the framework of fine-grained image retrieval based on CNN. In addition, Lou et al. [8] and Podlesnaya et al. [9] exploit CNN features for video retrieval. Furthermore, the problem of movie scene retrieval was addressed by Kumar et al. [10] using CNN and LSTM. In [11], CNN with LSTM was used to generate compact hash code for video retrieval. Few more works like [12] and [13] also used CNN as back bone for retrieval problem. In the context of action recognition, methods like [14] and [15] exploited 2d CNN for video representation, which was further improved by using 3d CNN [16, 17]. However, all the aforementioned methods are built on models that have been extensively pre-trained on a large-scale labelled image or video dataset. And for even better results, we would need to collect even more labelled data or carry out extensive feature engineering to keep the dimensionality low. But creating a large labeled dataset of videos can be very expensive compared to image dataset. So, the economical solution is to train a deep network from freely available web videos, which has recently become a major research spot in the computer vision community.

Most of the works in this direction focuses on self-supervised learning, where a pseudo-task is generated to learn the representations. For instance, Misra et al. [18] proposed sequence order verification task on top of CNN backbone. Lee et al. [19] designed the sequence sorting pretext task. Fernando et. al [20] proposed odd-one-out task. Further, Buchler et al. [21] design the sampling permutations policy and used deep reinforcement learning for order prediction tasks. All these methods typically deploy the 2d CNN, which seems to lack in motion capturing ability. To tackle this, 3d CNN can be a best option along with pretext task. In this regard, Jing et. al [22] proposed a feature learning method using video rotation prediction inspired from the method initially proposed for the image domain [23]. Further, Kim et. al [24] proposed Cubik puzzle solving task with 3d CNN. Also, Xu et. al [25] extends the frame order to clip order prediction task using 3d CNN. Another work focuses on playback speed modelling [26] uses 3d CNN. In contrast, in this paper, we propose to leverage multi self-supervised signals to improve the video representation capability using 3d-CNN as a backbone.

3 Proposed Approach

3.1 Methodology

Overview

Given a set of n training videos $V = [V_1, V_2, ..., V_n]$, where $V \in R^{M \times N \times C \times F}$, M is width, N is height, C is channel and F is frame. Our goal is to learn low dimensional representation $f_i \in R^d$ of video V_i. To achieve this we train a deep 3d convolutional network by jointly optimizing two-tasks as shown in Fig. 1. In next we discuss the used tasks in brief.

Rotation Prediction

Inspired by [23], the author of [22] proposed the rotation prediction task for video representation learning. The goal is to train a 3d-CNN model $f(.)$ to estimate the rotations applied to the input video clip. Let $R(V_i, r)$ be the geometric transformation function applied to input video V_i with $r = \{0°, 90°, 180°, 270°\}$. We implement the video rotation prediction as a classification task and termed the model as Rot3d. The cross entropy loss can be given as

$$L_R = -\frac{1}{|r|} \sum_r \log(f^r(R(V_i, r)|\theta))$$ (1)

where, $f^r(R(V_i, r)|\theta)$ is the predicted likelihood with rotation r applied to input video V_i, and θ are the learnable parameters of $f(.)$.

Future Frames Prediction

To further enhance the representation power and regularize the learning process, we propose to leverage future frames prediction task as an additional objective. To train $f(.)$ to predict future frames we design the 3d convolutional autoencoder (3d-CA) which consist of encoder and decoder as follows:

Encoder

The encoder part consist of initial four convolutional layers and 3 max pooling layers of C3D i.e. Input, conv3d_1, maxpool3d_1 conv3d_2, maxpool3d_2, conv3d_3, maxpool3d_3, conv3d_4. All the parameters are same as of C3D used in this paper (refer Sect. 3.2).

Decoder

To construct future frames, we design the decoder network as: Tconv3d_1 (3,3,3/256), Tconv3d_2 (3,3,3/128), Tconv3d_3 (3,3,3/64), Conv3d_F (3,3,3/3). Where, Tconv3d_1 is a 3d transpose convolutional layer with kernel size of $3 \times 3 \times 3$, stride of $[1, 2]$ and 256 kernels. All the above layers perform transpose convolutions with appropriate padding and followed by batch normalization and relu activation. Overall parameter details are reported in Table 2.

We train our encoder-decoder network with back-propagation with mean square error objective as follows:

$$L_F = \frac{1}{M \times N \times C \times F} \sum_{f=1}^{f=F} \sum_{c=1}^{c=C} \sum_{n=1}^{n=N} \sum_{m=1}^{m=M} \left((V_i)_{fcnm} - (V_i')_{fcnm} \right)^2 \qquad (2)$$

where, $V_i' \in R^{M \times N \times C \times F}$ is reconstructed version.

Overall loss is defined over K minibatch size as:

$$L = \frac{1}{K} \sum_{k=1}^{K} (L_R + L_F) \qquad (3)$$

3.2 3D CNN Architecture

Space-time learning can be done with 3D CNNs such as C3D, since they can model temporal dynamics and well-suited for spatiotemporal learning [16]. This can be considered an extension of 2D CNNs for videos. The C3D network architecture includes 8 convolution layers stacked one on top of the other, with 5 pooling layers interleaved in between, and two fully connected layers that are terminally connected. Each convolutional layers followed by batch normalization and relu activation. The convolution kernels all have dimensions of 3 × 3 × 3. We use the smaller variant of C3D as discussed in [16] with some modification (see Table 1), which consists of 5 convolutions, 5 max pooling and 3 fully connected layers.

Table 1. Architecture and parameters of C3D network

Layer	{Kernel size}/[stride]	Output size
Input	–	112 × 112 × 3 × 8
Conv3d_1	{3,3,3}/[1,1,1]	112 × 112 × 64 × 8
MaxPool3d_1	{2,2,1}/[2,2,1]	56 × 56 × 64 × 8
Conv3d_2	{3,3,3}/[1,1,1]	56 × 56 × 128 × 8
MaxPool3d_2	{2,2,2}/[2,2,2]	28 × 28 × 128 × 4
Conv3d_3	{3,3,3}/[1,1,1]	28 × 28 × 256 × 4
MaxPool3d_3	{2,2,1}/[2,2,1]	14 × 14 × 256 × 4
Conv3d_4	{3,3,3}/[1,1,1]	14 × 14 × 256 × 4
MaxPool3d_4	{2,2,2}/[2,2,2]	7 × 7 × 256 × 2
Conv3d_5	{3,3,3}/[1,1,1]	7 × 7 × 256 × 2
MaxPool3d_5	{2,2,2}/[2,2,2]	4 × 4 × 256 × 1
Fc1	1024	1 × 1 × 1024
Fc2	256	1 × 1 × 256
Fc3	4 (rotation class)	1 × 1 × 4

Table 2. 3d-convolutional Auto-encoder for future frame prediction.

	Layer	{Kernel size}/[stride]	Output size
Encoder	Input	–	$112 \times 112 \times 3 \times 8$
	Conv3d_1	{3,3,3}/[1,1,1]	$112 \times 112 \times 64 \times 8$
	MaxPool3d_1	{2,2,1}/[2,2,1]	$56 \times 56 \times 64 \times 8$
	Conv3d_2	{3,3,3}/[1,1,1]	$56 \times 56 \times 128 \times 8$
	MaxPool3d_2	{2,2,2}/[2,2,2]	$28 \times 28 \times 128 \times 4$
	Conv3d_3	{3,3,3}/[1,1,1]	$28 \times 28 \times 256 \times 4$
	MaxPool3d_3	{2,2,1}/[2,2,1]	$14 \times 14 \times 256 \times 4$
	Conv3d_4	{3,3,3}/[1,1,1]	$14 \times 14 \times 256 \times 4$
Decoder	Tconv3d_1	{3,3,3}/[2,2,1]	$28 \times 28 \times 256 \times 4$
	Tconv3d_2	{3,3,3}/[2,2,2]	$56 \times 56 \times 128 \times 8$
	Tconv3d_3	{3,3,3}/[2,2,1]	$112 \times 112 \times 64 \times 8$
	Conv3d_F	{3,3,3}/[1,1,1]	$112 \times 112 \times 3 \times 8$

Implementation

We choose the smaller variant of C3D with some modification as discussed above. Input size is set to $112 \times 112 \times 3 \times 8$, which is randomly sampled from $128 \times 170 \times 3 \times 8$ sized clip. The minibatch size is set to 44 clips inclusion of rotation transformation. Temporal length of clip is considered of 8 frames with temporal stride of 2. Learning rate is set to 1e-3 initially for 5 epochs and thereupon decreasing with rate of 0.1 every 15th epoch. It is necessary to take care of any pretext task as it can bypass learning by finding low-level cues. In this regard, we utilize channel splitting, random temporal clipping, and random horizontal flipping as data augmentation.

4   Experimental Settings

We evaluate the proposed approach on UCF-101 dataset [27], which consists of approx. thirteen thousand video clips of various human actions. Clips from the training split 1 considered for training our network without using labels. For testing, following the retrieval protocol [25], 10 clips are sampled from each video and their activations corresponds to pool5 are taken as features. All the clips in the testing set (split 1) considered as queries, and retrieval is done on the training set (split 1). We matched videos using cosine distance. Top-k accuracy is used to evaluate retrieval performance such that if the query video's label is matched within top-k retrieved labels, then the query video is said to be predicted correctly. Matlab with Nvidia tesla k40c gpu is used to conduct the experiments.

5 Results

In this section we report the results obtained from the experiments conducted on UCF101, where we first report the effect of future frame prediction task on retrieval accuracy. Then we compare our results with state-of-the-arts (SOTA) and, at last we conduct an ablation study in context of action classification.

5.1 Impact of Future Prediction in Our Approach on Retrieval Performance

Here, we want to see the joint effect of our approach. So, we conduct the experiment with features learn with Rot3d and Rot3d + 3D-CA, and the results are reported in Table 3. We also reported the results obtained through random initialization and considered as a baseline. Training with Rot3d task, model learns fair representative features which is reflected in the retrieval accuracy. When jointly trained with two tasks, the retrieval performance gradually improves with increase in k, which shows the importance of future prediction.

Table 3. Influence of joint learning on retrieval performance (Clip Level)

Methods	k = 1	k = 5	k = 10	k = 20	k = 50
Random	20.23	27.26	31.21	36.17	44.46
Rot3d	26.15	35.23	40.31	46.07	57.86
Rot3d + 3D-CA	28.17	37.92	43.24	51.41	62.93

5.2 Comparison to SOTA

Now, we compare our approach with other SOTA in Table 4, where we clearly outperform all the previous methods. Although our method achieves only 62.93% vs. 66.9% [25] at top-50, but it is able to outperform all methods in top-k accuracy for all k < 50. We achieve 28.17% (top-1), 37.92% (top-5), 43.24% (top-10) and 51.41% (top-20) retrieval accuracy, best compared to rest.

Table 4. (Clip Level) Top-k retrieval accuracy (%) on UCF-101.

Network	Methods	k = 1	k = 5	k = 10	k = 20	k = 50
Alexnet	Jigsaw [28]	19.7	28.5	33.5	40.0	49.4
	OPN [19]	19.9	28.7	34.0	40.6	51.6
	Buchler et al. [21]	25.7	36.2	42.2	49.2	59.5
S3D-G	SpeedNet [26]	13.0	28.1	37.5	49.5	65.0
C3D	Clip Order [25]	12.5	29.0	39.0	50.6	**66.9**
	Ours	**28.17**	**37.92**	**43.24**	**51.41**	62.93

5.3 Ablation Study in Context of Action Recognition

Now, we also test our approach with action recognition task. We first initialize the network with the proposed approach and then fine tune it with labeled dataset. Action recognition training and testing setting protocol is followed as in [25]. The results are summarized in Table 5 with other methods, where we consistently achieve better results than other methods. Note that our method uses only 8 input frames compared to others.

Table 5. Average action recognition accuracy (%) over 3 splits on UCF-101

Method	Network	Accuracy
Jigsaw [28]	Alexnet	51.5
Shuffle&Learn [18]	Alexnet	50.9
OPN [19]	Alexnet	56.3
Buchler et al. [21]	Alexnet	58.6
CubicPuzzle [24]	3D-ResNet18	65.8
3D RotNet [22]	3D-ResNet18	66.0
Speednet [26]	I3D	66.7
Clip Order [25]	C3D	65.6
Ours	C3D	66.8

6 Conclusion

This paper present the unsupervised video representation learning technique by jointly optimizing two pretext task. In this regard, we exploit rotation prediction and future frame prediction. For rotation task, we use four rotations: 0°, 90°, 180° and 270°. For future frame prediction task, we exploit 3d convolutional auto-encoder architecture. We validate our work on UCF-101 dataset on two downstream tasks. For both tasks, our approach performs superiorly to state-of-the-arts. Future work includes to incorporate contrastive learning and other self-supervised approaches.

References

1. Jiang, Y.-G., Ngo, C.-W., Yang, J.: Towards optimal bag-of-features for object categorization and semantic video retrieval. In: Proceedings of the 6th ACM International Conference on Image and Video Retrieval - CIVR 2007. ACM Press (2007)
2. Asha, S., Sreeraj, M.: Content based video retrieval using SURF descriptor. In: 2013 Third International Conference on Advances in Computing and Communications. IEEE (2013)
3. Zhu, Y., Huang, X., Huang, Q., Tian, Q.: Large-scale video copy retrieval with temporal-concentration SIFT. Neurocomputing **187**, 83–91 (2016)
4. Brindha, N., Visalakshi, P.: Bridging semantic gap between high-level and low-level features in content-based video retrieval using multi-stage ESN–SVM classifier. Sādhanā **42**(1), 1–10 (2016). https://doi.org/10.1007/s12046-016-0574-8
5. Ram, R.S., Prakash, S.A., Balaanand, M., Sivaparthipan, C.B.: Colour and orientation of pixel based video retrieval using IHBM similarity measure. Multimedia Tools Appl. **79**(15–16), 10199–10214 (2019). https://doi.org/10.1007/s11042-019-07805-9
6. Babenko, A., Slesarev, A., Chigorin, A., Lempitsky, V.: Neural codes for image retrieval. In: Fleet, D., Pajdla, T., Schiele, B., Tuytelaars, T. (eds.) ECCV 2014. LNCS, vol. 8689, pp. 584–599. Springer, Cham (2014), https://doi.org/10.1007/978-3-319-10590-1_38
7. Kumar, V., Tripathi, V., Pant, B.: Content based underground image retrieval using convolutional neural network. In: 2020 7th International Conference on Signal Processing and Integrated Networks (SPIN). IEEE (2020)
8. Lou, Y., et al.: Compact deep invariant descriptors for video retrieval. In: 2017 Data Compression Conference (DCC). IEEE (2017)
9. Podlesnaya, A., Podlesnyy, S.: Deep learning based semantic video indexing and retrieval. In: Bi, Y., Kapoor, S., Bhatia, R. (eds.) IntelliSys 2016. LNNS, vol. 16, pp. 359–372. Springer, Cham (2017). https://doi.org/10.1007/978-3-319-56991-8_27
10. Kumar, V., Tripathi, V., Pant, B.: Content based movie scene retrieval using spatio-temporal features. IJEAT **9**, 1492–1496 (2019)
11. Kumar, V., Tripathi, V., Pant, B.: Learning compact spatio-temporal features for fast content based video retrieval. IJITEE **9**, 2404–2409 (2019)
12. Mühling, M., et al.: Deep learning for content-based video retrieval in film and television production. Multimedia Tools Appl. **76**, 22169–22194 (2017)
13. Mühling, M., et al.: Content-based video retrieval in historical collections of the German broadcasting archive. Int. J. Digit. Libr. **20**(2), 167–183 (2018). https://doi.org/10.1007/s00799-018-0236-z
14. Simonyan, K., Zisserman, A.: Two-stream convolutional networks for action recognition in videos. In: Proceedings of the 27th International Conference on Neural Information Processing Systems. ACM Press (2014)
15. Karpathy, A., Toderici, G., Shetty, S., Leung, T., Sukthankar, R., Fei-Fei, L.: Large-scale video classification with convolutional neural networks. In: 2014 IEEE Conference on Computer Vision and Pattern Recognition. IEEE (2014)
16. Tran, D., Bourdev, L., Fergus, R., Torresani, L., Paluri, M.: Learning spatiotemporal features with 3D convolutional networks. In: 2015 IEEE International Conference on Computer Vision (ICCV). IEEE (2015)
17. Tran, D., Wang, H., Torresani, L., Ray, J., LeCun, Y., Paluri, M.: A closer look at spatiotemporal convolutions for action recognition. In: 2018 IEEE/CVF Conference on Computer Vision and Pattern Recognition. IEEE (2018)
18. Misra, I., Zitnick, C.L., Hebert, M.: Shuffle and learn: unsupervised learning using temporal order verification. In: Leibe, B., Matas, J., Sebe, N., Welling, M. (eds.) ECCV 2016. LNCS, vol. 9905, pp. 527–544. Springer, Cham (2016). https://doi.org/10.1007/978-3-319-46448-0_32

19. Lee, H.-Y., Huang, J.-B., Singh, M., Yang, M.-H.: Unsupervised representation learning by sorting sequences. In: 2017 IEEE International Conference on Computer Vision (ICCV). IEEE (2017)

20. Fernando, B., Bilen, H., Gavves, E., Gould, S.: Self-supervised video representation learning with odd-one-out networks. In: 2017 IEEE Conference on Computer Vision and Pattern Recognition (CVPR). IEEE (2017)

21. Büchler, U., Brattoli, B., Ommer, B.: Improving spatiotemporal self-supervision by deep reinforcement learning. In: Ferrari, V., Hebert, M., Sminchisescu, C., Weiss, Y. (eds.) ECCV 2018. LNCS, vol. 11219, pp. 797–814. Springer, Cham (2018). https://doi.org/10.1007/978-3-030-01267-0_47

22. Jing, L., Yang, X., Liu, J., Tian, Y.: Self-supervised spatiotemporal feature learning via video rotation prediction. arXiv preprint arXiv:1811.11387 (2018)

23. Gidaris, S., Singh, P., Komodakis, N.: Unsupervised representation learning by predicting image rotations. In: International Conference on Learning Representations (2018)

24. Kim, D., Cho, D., Kweon, I.S.: Self-supervised video representation learning with space-time cubic puzzles. In: Proceedings of the AAAI Conference on Artificial Intelligence (2019)

25. Xu, D., Xiao, J., Zhao, Z., Shao, J., Xie, D., Zhuang, Y.: Self-supervised spatiotemporal learning via video clip order prediction. In: 2019 IEEE/CVF Conference on Computer Vision and Pattern Recognition (CVPR). IEEE (2019)

26. Benaim, S., et al.: SpeedNet: learning the speediness in videos. In: 2020 IEEE/CVF Conference on Computer Vision and Pattern Recognition (CVPR). IEEE (2020)

27. Soomro, K., Zamir, AR., Shah, M.: UCF101: a dataset of 101 human actions classes from videos in the wild. arXiv preprint arXiv:1212.0402 (2012)

28. Noroozi, M., Favaro, P.: Unsupervised learning of visual representations by solving Jigsaw puzzles. In: Leibe, B., Matas, J., Sebe, N., Welling, M. (eds.) ECCV 2016. LNCS, vol. 9910, pp. 69–84. Springer, Cham (2016). https://doi.org/10.1007/978-3-319-46466-4_5

Application of Deep Learning
in Classification of Encrypted Images

Geetansh Saxena[1], Girish Mishra[2(✉)], and Noopur Shrotriya[2]

[1] University of Delhi, Delhi, India
saxenageetansh1@ducic.ac.in
[2] Defence R and D Organisation, Delhi, India
noopurshrotriya@sag.drdo.in

Abstract. In 2020, Citizen Lab in University of Toronto found that the popular video conferencing application, Zoom, was not only ignoring the use of AES-256, but was instead using AES-128 in the Electronic Code Book (ECB) mode. ECB has never been a secure encryption strategy as it preserves the patterns present in the original input. This paper shows a simple deep learning based attack on images encrypted using four different encryption schemes (AES-128, GIFT-64, PICO and PRESENT) under the ECB mode of operation. We build a deep learning model that learns to classify encrypted images with 100% accuracy. We also explore the reasons behind why small size images might not look intelligible to the human eye but are still easily classifiable using a neural network. We also express how this attack can be scaled and further used to exploit encrypted data.

Keywords: Block ciphers · Cryptanalysis · Deep learning · AES · PICO · GIFT-64 · PRESENT · ECB mode of encryption

1 Introduction

Cryptography is concerned with techniques for securing the information transferred over an insecure channel like internet where millions of third parties called adversaries are putting their best efforts round the clock to extract the valuable information. In short, it is a method for secure communication over an insecure channel (like the internet). Though, even after using the most secure encryption techniques, and somehow generating the random keys, confidential information may in certain cases, still be deduced from the encrypted data. This is not by breaking the cipher or getting hold of the secret keys, but by mere observation of data.

In 2020, Citizen Labs released its findings [14] about the popular video conferencing application, Zoom, using weak encryption. In its documentation, Zoom had claimed to use AES-256 for the meetings wherever possible. But in reality, Zoom was using AES-128 in the ECB mode.

© Springer Nature Switzerland AG 2021
M. Singh et al. (Eds.): ICACDS 2021, CCIS 1440, pp. 711–719, 2021.
https://doi.org/10.1007/978-3-030-81462-5_62

David et al. [5] concluded that 83% of the systems have security bugs because of bad implementation/misuse of cryptographic libraries and 17% of the bugs are in the cryptographic library itself. AES-128 is a strong encryption technique but the way it is used can make the achieved security weak. This is not the first time an application's security has been proved to be weak even after using strong encryption techniques. For any given cipher, different modes of encryption exist. Few popular modes of encryption are Electronic Code Book mode (ECB), Cipher Block Chaining (CBC) mode, Cipher Feedback (CFB) mode, Output Feedback (OFB) mode, etc. It is well known in the cryptography community and also well taught that using ECB mode for encryption is not advisable. Under ECB mode of operation, identical data blocks generate identical ciphertext blocks. Hence, the data patterns in the input, get preserved in the encrypted output.

In the case of images, if ECB encryption mode is used, the patterns in the original image, get preserved in the encrypted image. This can be clearly seen in Fig. 2. But this is the case when the image size is large. In the case of small size images, the patterns seem to get diffused and the image looks unintelligible. A clear example can be seen in Fig. 3. After encryption using PICO cipher under ECB mode of operation, the image seems to look randomised to the human eye. Though a neural network can still detect the patterns, and use those patterns to classify the image based on the cipher that was used to encrypt them. This is further explained in Sect. 4.3 later in the paper.

In this paper we build a deep learning model and train it to classify images based on the cipher used. In this experiment, we use small size images, which as explained earlier, look unintelligible to the human eye after encryption under the ECB mode. The training dataset contains images, which were encrypted using any one of the four ciphers, namely AES-128, PICO, GIFT and PRESENT. We explore if given a large dataset, can a neural network learn to examine and classify the images based on the cipher that was used to encrypt.

The paper is arranged as follows: Sect. 2 explores the previous work done along similar lines. Section 3 describes the ECB mode of operation, its advantages and disadvantages. It also describes why image encryption with ECB mode is semantically insecure and preserves patterns of the original image. In Sect. 4, the experiment is described, with the dataset, the classification model and the results. Section 4.3 talks about the analysis of the achieved results. Finally, the paper is concluded in Sect. 5.

2   Previous Work

In 2018, Linus [9] in his masters thesis, performed a traditional image recognition task to encrypted MNIST handwritten images [7] using convolutional neural networks. In a 10 class problem he achieved 10% accuracy for CBC and 42% accuracy for ECB encrypted images, hence, establishing ECB as the weakest mode of encryption. De Mello et al. [10] used a corpora of plaintexts from 7 different languages. These plaintexts were then encrypted by 7 different algorithms in ECB and CBC modes. As expected, the classification accuracy for ECB mode was high. In 2017, Wang et al. [11] performed an image classification task and designed a multi-layer extreme learning machine classifier capable of classifying the encrypted images. The images used were letter databases and handwritten digits. The classification accuracy was high enough. In 2002, Maheshwari [6] and Chandra [8] tried to separateby different algorithms. Nagireddy [12] developed methods to distinguish DES, AES, Blowfish, 3DES and RC5. Saxana [13] used machine learning techniques to classify ciphertexts generated using different algorithms. It should be noted that all of the above distinguishing attacks were operated on ciphertexts generated using ECB mode of encryption. In our paper, we have worked along similar lines for 4 different ciphers and validated the work done by previous researchers. In our experiment, we encrypted the MNIST Handwritten Digits database in ECB mode using 4 different ciphers, and used deep learning techniques to classify them. We explored how ECB encryption in small scale size images does not give out visible patterns or apparent information but is still insecure.

3   ECB Mode of Operation

The ECB mode of encryption is the simplest mode of encryption where one takes a block of data, encrypts it, and moves on to the next block. Hence, each block of data is encrypted independently. Thereafter, the ciphertext blocks are concatenated one after other to form a complete ciphertext.

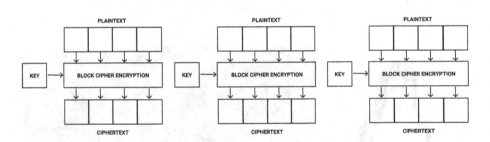

Fig. 1. ECB mode of operation

3.1   Advantages and Disadvantages of ECB

Although ECB mode (shown in Fig. 1) is the simplest mode of encryption in block ciphers, but it has its own advantages and disadvantages.

Advantages of ECB

1. Multiple blocks can be encrypted/ decrypted in parallel.

2. Block synchronisation is not necessary. This means that even if receiver of a data receives less blocks than the sender had sent, he still will be able to decrypt all the received blocks and read the information partly. Also, decryption can start even if the blocks are not in order and aligned later.
3. Bit errors caused by noisy transmissions only affects the corresponding blocks and not all succeeding blocks.

Disadvantages of ECB

1. ECB encrypts deterministically. This means that identical plaintext blocks result in identical ciphertexts (as long as the same key is used).
2. Just by looking at the ciphertexts, an attacker might recognize that the same message has been sent twice. This way the information can be disclosed by observing the metadata of the traffic in a very simple manner.
3. Plaintext blocks are encrypted independently of previous blocks. If an attacker reorders the ciphertext blocks, this may result in valid decrypted data blocks and the reordering might not be detected.

3.2 Image Encryption Using ECB

The main reason behind not using ECB mode is that it is not semantically secure. This means that just by a simple observation, ECB encrypted images can leak the information. This specifically applies to large images. A good demonstration can be found in Fig. 2.

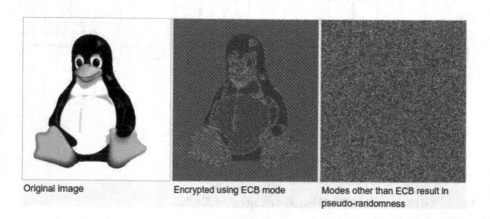

Original image   Encrypted using ECB mode   Modes other than ECB result in
                                                          pseudo-randomness

Fig. 2. ECB image [15]

It is apparent that the image of penguin is clearly identifiable in the ECB mode of encryption. This is the case of a large image where the number of pixels in a row of image is far more than the block size of the cipher and the data in each row does not change much resulting in same crypts. In the case of small images, it is not very apparent though. For example, consider the image of MNIST Digit 1 given in Fig. 3a. Figure 3b is the encrypted version of the same image being encrypted with PICO cipher.

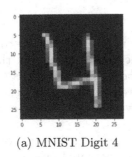

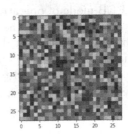

(a) MNIST Digit 1

(b) MNIST Digit 1 - Image encrypted using PICO cipher in ECB Mode

Fig. 3. Pair of original image and corresponding encrypted image using PICO cipher

As it is visible, there is no way a naked eye can deduce that the encrypted image was initially an image of the digit 1. Similar observation can be made when the image is encrypted with AES-128 cipher as shown in Fig. 4a and Fig. 4b. The reason for the encrypted image being unintelligible is discussed in Sect. 4.

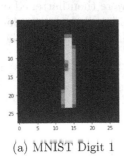

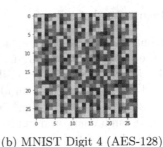

(a) MNIST Digit 4

(b) MNIST Digit 4 (AES-128)

Fig. 4. Pair of original and encrypted image (AES-128 encrypted)

4 Classification of Encrypted MNIST Images

The MNIST Database of Digits consists of 60,000 sample images of handwritten digits. The size of each image is 28 × 28 pixels. We conduct a small experiment over the MNIST Digit database and encrypt the images in ECB operation mode using 4 different block ciphers, namely PICO Cipher [2], GIFT-64 Cipher [1], PRESENT Cipher [3] and Advanced Encryption Standard (AES) Cipher [4].

4.1   Dataset Generation

40,000 out of the available 60,000 images were randomly chosen from the MNIST Digits database and each of them was encrypted using the above mentioned ciphers under ECB mode of operation. These images were then flattened out to form an array of 784 elements. Thereafter, the data is labelled and the label is appended with 0, 1, 2, 3 based on the cipher used for encryption out of the four above mentioned ciphers.

Hence, the dataset contained 1,60,000 rows of encrypted images with a label indicating the cipher used.

4.2   Classification Model

Our classification model consists of 3 layers. The first one being a convolutional layer with 32 filters and a kernel of size 3×3 coupled with max pooling. Flattening of the pooled feature map is then performed. This is done to pass the feature map to an artificial neural network layer. The second is a dense layer of 100 neurons and the last one being the output layer with 4 neurons. The activation functions used in these layers are relu, relu and softmax respectively. We have used the stochastic gradient optimiser with learning rate of 0.01 and a momentum of 0.9.

The dataset is then divided into the training and testing dataset. 80% of the data is used for training and the remaining 20% is used for testing.

4.3   Results and Analysis

When given an encrypted image as an input, the model was able to correctly identify the cipher that was originally used for encrypting the image, that too with a 100% accuracy rate. As visible in Fig. 5, the cross entropy loss reduced to zero and the model became an ideal machine to check and classify encrypted images. It is observed that 100% accuracy is easy to achieve given a dataset containing enough encrypted images.

As seen in Fig. 2, the image preserves its nature and characteristics even after the encryption under ECB mode of operation. The encrypted image is though very unclear but still recognisable. This is because of the way ECB mode works. Repetitive/identical segments of the image give repetitive/identical encrypted results. Hence, the large sections of the image remains preserved. It is quite visible if we notice the feet, belly and the black color on the body. The boundaries on the other hand, constitute a very small section of the image and hence get diffused with the neighbouring pixels. It is clearly visible on the beak and eyes of the penguin. This happens when the input block formed by the combination of pixels is almost equal to the small segments (here, beak and eyes) of an image.

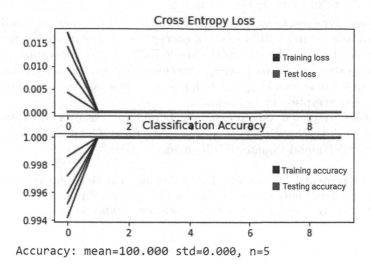

Fig. 5. Classification results

On the other hand, a large section (most preferably with common pixel values) becomes a series of repetitive blocks of encrypted data that the human brain can recognize.

The handwritten digit images are a set of small images of size 28×28. The seemingly large sections of this image are still small when compared to the block/input size of any of the 4 ciphers. After encryption, these small images hence appear to be fully diffused. This is apparent in Fig. 3b or Fig. 4b. It is to be noted that ECB mode is still repetitive in nature, but it is difficult for the human eye to catch those patterns explicitly. Though a deep learning model with adequate number of image samples can easily recognize and classify the same.

The main reason behind achieving the perfect accuracy is the presence of large number of identical input blocks (formed with pixel values) available for encryption in the image. This causes the encryption to give the identical blocks in the encrypted image. Hence, the model can easily map those input blocks to a fixed encrypted block. Hence, the machine learning model learns the underlying mapping easily to classify the encrypted images.

5 Conclusion

In this paper we have tried to classify the encrypted images based on the ciphers that were used for encryption of the images. The images were encrypted using ECB mode of operation. The primary focus of this paper has been to apply deep learning on the small size encrypted images (ECB mode) which are unintelligible and random to the human eyes and establish their deterministic nature for correct classification by the deep learning models.

Through our experiments, we achieved the perfect accuracy in classifying the encrypted images under ECB mode of encryption. This experiment hence gives further validation to why encryption under ECB mode of operation is insecure even in the case of small size images. Given enough data, the adversary can make a model that can accurately predict and get information about what cipher was used for the encryption of an image.

In future, we plan to do our experiments and analysis over encrypted images for other modes of encryption such as Cipher Block Chaining (CBC), Output Feedback (OFB), and Counter (CTR) modes.

Acknowledgements. The authors thank Scientific Analysis Group, DRDO, India for providing the opportunity to work in the area of Cryptography and Cluster Innovation Centre, University of Delhi, India for their constant encouragement.

References

1. Banik, S., Kumar Pandey, S., Peyrin, T., Sasaki, Y., Meng Sim, S., Todo, Y.: GIFT: A Small Present Towards Reaching the Limit of Lightweight Encryption (Full version) (n.d.). https://eprint.iacr.org/2017/622.pdf
2. Bansod, G., Pisharoty, N., Patil, A.: PICO: an ultra lightweight and low power encryption design for ubiquitous computing & electronic engineering. **18**(3), 317–331 (2017). https://doi.org/10.1631/FITEE.1500415
3. Bogdanov, A., et al.: PRESENT: an ultra-lightweight block cipher. In: Paillier, P., Verbauwhede, I. (eds.) CHES 2007. LNCS, vol. 4727, pp. 450–466. Springer, Heidelberg (2007). https://doi.org/10.1007/978-3-540-74735-2_31
4. Daemen, J., Rijmen, V.: Note on naming Rijndael Note on naming (n.d.). https://csrc.nist.gov/CSRC/media/Projects/Cryptographic-Standards-and-Guidelines/documents/aes-development/Rijndael-ammended.pdf
5. Lazar, D., Chen, H., Wang, X., Zeldovich, N.: Why does cryptographic software fail? A case study and open problems (n.d.). http://lib.21h.io/library/3NAZDAR9/download/R6RTZ7MZ/Why_does_cryptographic_software_fail_-_A_case_study_and_open_problems_7p_.pdf
6. Maheshwari, P.: Classification of Ciphers (1999). Www.semanticscholar.org https://www.semanticscholar.org/paper/Classification-of-Ciphers-Maheshwari/18d41ac27b62962d2a008973564841f042028724
7. LeCun, Y., Cortes, C., Burges, C.: MNIST handwritten digit database (n.d.). Yann.lecun.com from http://yann.lecun.com/exdb/mnist
8. Girish, C.: Classification of modern ciphers. Masters thesis (2002)
9. Lagerhjelm, L., Prorok, K.: Extracting Information from Encrypted Data using Deep Neural Networks (2019). http://www.diva-portal.org/smash/get/diva2:1284274/FULLTEXT01.pdf
10. de Mello, F.L., Xexéo, J.A.M.: Identifying encryption algorithms in ECB and CBC modes using computational intelligence. J. Univ. Comput. Sci. (n.d.). Www.jucs.org http://www.jucs.org/jucs_24_1/identifying_encryption_algorithm_in
11. Wang, W., Vong, C.-M., Yang, Y., Wong, P.-K.: Encrypted image classification based on multilayer extreme learning machine. Multidim. Syst. Sign. Process. **28**(3), 851–865 (2016). https://doi.org/10.1007/s11045-016-0408-1

12. Nagireddy, S., Murthy, H.A., Kant, S.: Identification of encryption method for block ciphers using histogram method. J. Discrete Math. Sci. Cryptogr. **13**(4), 319–328 (2010). https://doi.org/10.1080/09720529.2010.10698297
13. Saxena, G.: Classification of ciphers using machine learning. M. Tech thesis, Department of Computer Science and Engineering, Indian Institute of Technology, Kanpur, July 2008
14. Move Fast and Roll Your Own Crypto: A Quick Look at the Confidentiality of Zoom Meetings. The Citizen Lab. https://citizenlab.ca/2020/04/move-fast-roll-your-own-crypto-a-quick-look-at-the-confidentiality-of-zoom-meetings/. Accessed 15 Feb 2021
15. Lunkwill: Block cipher mode of operation. Wikipedia; Wikimedia Foundation, 2 December 2019. https://en.wikipedia.org/wiki/Block_cipher_mode_of_operation

Ear Recognition Using Pretrained Convolutional Neural Networks

K. R. Resmi[1](✉) and G. Raju[2]

[1] School of Computer Sciences, M G University, Kerala, India
[2] Department of Computer Science and Engineering, Christ (Deemed to be University), Bengaluru, India

Abstract. Ear biometrics, which involves the identification of a person from an ear image, is challenging under unconstrained image capturing scenarios. Studies in Ear biometrics reported that the Convolutional Neural Network is a better alternative to classical machine learning with handcrafted features. Two major concerns in CNN are the requirement of enormous computing resources and large datasets for training. The pretrained network concept helps to use CNN with smaller datasets and is less demanding on hardware. In this paper, three pre-trained CNN models, AlexNet, VGG16, and ResNet50 are used for ear recognition. The fully connected classification layers of the nets are trained with AWE, an unconstrained ear dataset. Alternatively, the CNN layers' output (the CNN features) are extracted, and an SVM classification model is built. To improve the classification accuracy, the training dataset size is increased through data augmentation. Data augmentation improved the classification accuracy drastically. The results show that ResNet50, with the fully connected classification layer, results in higher accuracy.

Keywords: Deep learning · Convolutional neural network · Pretrained networks · Ear recognition

1 Introduction

Ear biometrics is gaining more popularity among computer vision experts, especially in developing multi-modal biometric systems. The acceptance of Ear biometrics is primarily due to its "uniqueness, universality, and collectability". Classical 2D ear recognition techniques based on feature extraction and machine learning are of four groups - holistic, local, hybrid, and geometric. The challenging tasks in the classical approaches are the selection of the most discriminating feature set and an appropriate machine learning model for the classification. Despite several studies, the classical models' performance is low when the ear data set is unconstrained. In the last decade, the field of computer vision has been dramatically changed due to the widespread use of Deep Learning (DL). DL models generally perform recognition tasks directly from raw data [1]. Compared to other compute vision tasks, the number of works in Ear biometrics using DL models is less.

Most of the works in Ear biometrics use constrained data sets, and researchers reported very high accuracy. But those algorithms failed to give any impression when

© Springer Nature Switzerland AG 2021
M. Singh et al. (Eds.): ICACDS 2021, CCIS 1440, pp. 720–728, 2021.
https://doi.org/10.1007/978-3-030-81462-5_63

experimented with unconstrained data sets like AWE. To address this, researchers built DL models, which showed a drastic improvement in the performance. Recently several DL models for Ear detection and recognition are proposed. A survey on the recent models highlights the need for better models with high accuracy.

Building a DL model from scratch requires large data sets and high-performance computing facilities. This requirement can be minimized if pretrained networks are used for creating new models. Pretrained nets are developed on the transfer learning concept. This study focuses on Convolutional Neural Networks (CNN), the most common and powerful deep learning architecture in image processing. Several pretrained CNN networks are available for building new models. These pre-trained networks can be retrained for the new application in several ways. CNN extracts the input image features (often raw) and submits them to classification layers. Once the convolutional layers are trained, it is possible to take the features directly from the output of convolutional layers. The features can then be used for building different (or an ensemble of) classifiers.

In this work, three pretrained CNN models are selected [2, 3, 4]. The networks' classification layers are trained using the ear dataset, keeping the parameters of the convolutional layers intact. In a second approach, the final convolutional layer's output is extracted as a feature vector, and an SVM classification model is created. The models Alexnet, VGG-16, and ResNet are popular CNN models due to their superior performance in classifying the ImageNet data set. These models have given a boost to CNN, and hundreds of new variants are now available to choose from. To keep the study simple, we have considered only these three basic models.

Pretrained networks do not require large training data if the new problem/data training is limited to the shallow classification layers. In this work, the parameters of the convolutional layer are kept intact (no training). In spite of this, we have augmented the training dataset using standard image augmentation methods. This resulted in a relatively sizeable training data set, and experiments show considerable improvement in the classification accuracy. Also, classification using Fully connected layers with softmax output layer shows better performance compared to the SVM model used. The work illustrates how a simple and effective CNN model for ear recognition can be built with transfer learning.

2 Related Works

Transfer learning is a process where a model designed for a problem is reused for another problem. It is very costly and time-consuming to build deep models from scratch. Once trained for an application, the same parameter (like weight) can create another model for the same or a different application. In this section, some of the contemporary works in CNN models for ear recognition are presented.

Application of CNN with a relatively small ear data set is reported in [5]. The authors used "AlexNet", "VGG-16", and "SqueezeNet", with the full model and the selective model learning. The dataset used is a combination of the AWE, the CVLED, and 500 images of 50 persons taken from the internet. Selective learning with sqeezeNet achieved the best results, with a rank 1 accuracy of 62%. Hansley et al. employed a method for ear recognition by fusing features derived with a CNN model built for face recognition and handcrafted features [6]. The handcrafted features are – "LBP", "LPQ", "POEM", "HOG", "DSIFT", and "Gabor". The datasets "IIT Delhi", "WPUTE", "AWE", "ITWE" and "UERC". The fusion of CNN features with HOG resulted in the highest Rank1 recognition rate of 75.6% on the AWE dataset. Ear recognition by transfer learning using the pretrained AlexNet, VGG16, VGG 19, ResNet18, and ResNet 50 is presented in [7]. A shallow classifier is used for the recognition. The authors reported Rank1 accuracy of 56.35%, 94.08%, 80.05% on the AWE, CVLE, and AWE+CVLE datasets.

A new database named USTB-Helloear was created by Zhang et al. [8], which contains 612661 images of 1570 subjects. The database includes images with variations in illumination, pose, and images with different levels of occlusion. They fine-tuned and modified a few existing deep models like "GoogLeNet, AlexNet, VGG-Face, VGG-16, and ResNet101". Multilevel features are obtained by replacing the last layers of all the models with spatial pyramid pooling (SPP) layers. The results obtained highlight the superiority of the VGG-Face model. Omara et al. [9] used VGG-M net for extracting deep ear features. The authors used PCA for dimensionality reduction and pairwise SVM for classification. On the USTB I and II datasets, recognition accuracy of 98.3% and 92.0% are reported, respectively.

El-Naggar and Bourlai studied different deep learning models for effective recognition with various ear image distortions, and different yaw poses [10]. The study considered multiple CNN architectures and learning strategies like "transfer learning, data augmentation, and domain adaptation" to suggest the most efficient CNN model for ear recognition. Deep architectures used include AlexNet, MobileNetV2, GoogLeNet, and SqueezeNet. MobileNetV2 gave the highest rank1 recognition rate of 95.67% on the WVU dataset.

Eyiokur et al. [11] proposed the use of "domain adaptation", by creating a new ear dataset from the Multi-PIE face database called the Multi-PIE ear dataset. Domain adaptation improves the accuracy of the various pretrained models in recognition. The study focused on the impact of lighting, image quality, aspect ratio, and "data set bias" on the performance of the different models. The Multi-PIE ear and the UERC dataset is considered for experimentation. The fusion of VGG-16 and GoogLeNet architecture obtained the best recognition performance of 67.5% on the Multi-PIE dataset.

Ensemble learning by combining various deep models' predictions is used to increase the recognition performance. Ear recognition based on ensembles of VGG architectures VGG-11, VGG-13, VGG-16, and VGG-19 is proposed in [12]. The experiments conducted with AMI, AMIC, and WPUT ear datasets resulted in a rank-1 accuracy of 97.50%, 93.21%, and 79.08%. The same authors proposed transfer learning on VGG Net, AlexNet, ResNet, Inception, and ResNeXt 101 models [13]. Experiment on EarVN1.0 [14] achieved the highest rank1 recognition rate of 93.45% with the ResNeXt 101 model. Rank1 recognition rate is increased by 2% by using ensembles of fine-tuned ResNeXt 101 models.

The survey highlights the variety of DL methods proposed for ear recognition and indicates the need for better models. In the following section, a brief overview of the different pretrained models used is given.

3   CNN Pretrained Models

Pretrained models are reused to create a new model using a new dataset from computer vision tasks. In this study, three CNN architectures – AlexNet [2], VGG16 [3], and ResNet 50 [4] are selected. No specific criteria is applied for their selection. These networks are pretrained on the famous ImageNet dataset.

3.1   AlexNet

AlexNet is the first state of art CNN model proposed by Alex Krizhvesky [2] by winning the ImageNet [15] challenge. AlexNet architecture contains 23 layers, which include five convolutional layers with max-pooling, three fully connected layers, and a softmax layer. Earlier layers are used as feature extractors. AlexNet uses ReLu as the activation function instead of the tanh function. To avoid overfitting problems, it employed data augmentation and dropout techniques.

3.2   Vgg-16

VGG-16 is a deep CNN with twelve convolutional layers and four fully connected layers [3]. It uses different 3×3 convolutions filters. It is one of the successful architectures for feature extraction from images.

3.3   ResNet

Deeper networks learn more from the data but give less accuracy during testing. To solve this problem, ResNet uses residual blocks for learning [4]. The main idea of ResNet is to use identity shortcut connection by skipping one or more layers. ResNet is available in different variants, such as ResNet-18 and ResNet-50. In this work, ResNet-50, a 50 layer deep CNN architecture, is used.

4   Methodology

In this work, the pretrained networks (PrNet) are used in two different ways. In the first approach, the PrNets are applied for feature extraction. The images from the data set are preprocessed as per the requirement of the respective PrNet. The PrNets are modified by removing the fully connected classification layers. The images are given as input to the PrNet. The output of the convolutional layers is transformed as the feature vector. The feature vector is subjected to PCA for reducing the dimensionality. These features are used to build a classification model using SVM.

In the second approach, the pretrained architectures are modified by reducing the number of neurons in the last layer (softmax) to 100 from 1000 (AWE, the data set chosen for the work has only 100 classes).. The data set is applied to train the PrNets. Only the parameters associated with the fully connected classification layers are modified during the training process, keeping the convolutional layer parameters intact. The model so build is then tested with the test samples.The two approaches are illustrated in Fig. 1.

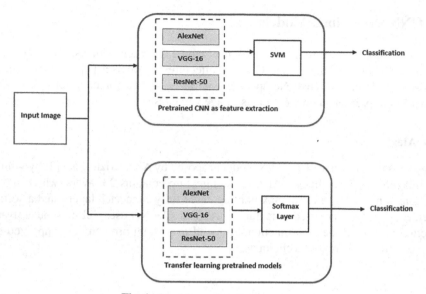

Fig. 1. Models used in the experiment

5 Experiments

The model building is carried out in the Python environment using Tensorflow (Keras). An Intel Core i5 system with 4GB RAM is used for the experimentation. The details of the dataset used are described below.

5.1 Data Set

The unconstrained ear dataset AWE [16] is used for the experiments. It consists of 10 images of varying sizes, each of 100 subjects (total 1000). The database includes both the right and left ears. Figure 2 shows sample images of a single person from the AWE dataset. In CNN, data augmentation is essential if the data set is small. But the deep convolution layers are already trained using a huge dataset, though not ear images. Hence, the task is limited to training the classification layers (or the SVM model training). For this, data augmentation is not mandatory. But data augmentation is incorporated as part of the work to improve the classification performance. The details of data augmentation are given below.

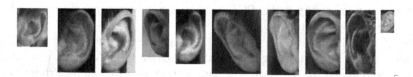

Fig. 2. Sample Images from AWE dataset

5.2 Data Augmentation

More training data improves the model efficiency by lowering overfitting. Augmentation generates a vast, diverse dataset from smaller data sets [17]. Data augmentation is performed using standard augmentation methods. "Imgaug tool", a python library, is used for the augmentation [18]. Augmentation includes changing the brightness in a predefined interval, inversion of the color channel, rotation, and shear within predefined ranges, Gaussian blur, and cropping. From the AWE data set, 80000 samples are generated for model building. Figure 3 shows data augmentation results using standard augmentation techniques on sample images from the AWE database.

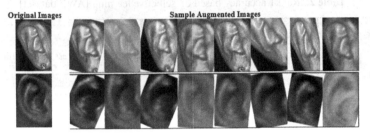

Fig. 3. Data Augmentation results on Sample images from AWE database

6 Results and Discussion

6.1 CNN Features and SVM

Images from the dataset are given as input to the pretrained networks. For AlexNet, resize all the input images to 227 × 227. For VGG-16 and ResNet-50, resize the input images to size 224 × 224. For AlexNet, VGG-16, and ResNet50, the output feature vectors are of length 9216, 25088, and 100352, respectively. PCA is applied to these features independently, and the feature size is reduced to 2720, 4096, and 50176, respectively. The training-test ratio is fixed as 80:20. SVM with a linear kernel is trained using these features. Table 1 shows the Rank1 accuracy with and without data augmentation.

Table 1. Rank1 accuracy – CNN features and SVM [AWE dataset]

Network	Rank-1 recognition accuracy [%]	
	Features with SVM	Features with SVM (augmented data)
AlexNet	30.56%	41.42%
VGG16	31.54%	42.32%
ResNet50	36.25%	58.20%

6.2   Training the Classification Layer of the CNNs

The pretrained networks are modified to suit the number of classes in the AWE. The output layer is reduced to 100 from 1000 in the three models. The weights other than the FC layers are frozen. The network is then trained with the preprocessed and resized image samples, with a training-test split of 80:20. A mini-batch size of 20 is selected. We have experimented by changing the number of epochs. The best performance is observed with 100 epochs. Table 2 shows the Rank1 accuracy of the CNN models (without and without data augmentation).

Table 2. Rank1 accuracy based on selective learning [AWE dataset]

Network	Rank-1 recognition accuracy [%]	
	Deep features and FC layers	Deep features and FC layers (augmented data)
AlexNet	34.67%	46.75%
VGG16	36.25%	49.29%
ResNet50	40.75%	63.00%

From Tables 1 and 2, CNN with classification layer outperforms the SVM classification model with CNN features. Also, data augmentation improved the accuracy drastically. It follows that an augmented data set always enhances the accuracy, whether you are training the deep layers or only the classification layer.

AWE is one of the benchmarking unconstrained ear datasets. For comparison purposes, we have chosen four DL-based works, which used the AWE dataset. Table 3 compares the ResNet50 model's Rank1 recognition rate with those methods. For uniformity, ensemble methods are not included.

Table 3. Comparison of rank1 recognition accuracy -AWE Dataset

Reference	Architectures used	Highest accuracy AWE (rank-1)
Emersic et al., FERA 2017 [5]	AlexNet VGG-16 Squeezenet	62% (Squeezenet)
Hansley et al., IET biom2018 [6]	CNN+Handcrafted features	75.6% (CNN+HOG)
Dodge et al., IET biom2018 [7]	AlexNet VGG-16 VGG-19 ResNet-18 ResNet-50	56.35% (ResNet-18)
Emersic et al., Neurocomputing 2017 [16]	Handcrafted features	49.6% (POEM)
Proposed approach	ResNet-50	63.00%

The scope of the work is limited to only the AWE dataset. Several new datasets with a large number of samples are available. Also, The work considered only three basic CNN pretrained models. Extensive experiments employing different datasets and CNN models are required to select the best models. Also, the efficacy of other machine learning algorithms in conjunction with the CNN features needs to be investigated.

7 Conclusion

Ear Biometrics is still a challenging task, especially when unconstrained data sets are considered. Deep Neural architectures are now widely used in computer vision applications, including ear biometrics. The results reported in the literature indicate the need for better models. Convolutional Neural Networks are one class of Deep learning architectures suitable for ear image analysis. When resources and datasets are limited, pretrained networks are an option to work with. This paper presents a study on ear recognition using three pretrained CNN models - AlexNet, VGG16, and ResNet50.

New models are built following two approaches. In the first, the output layer alone is changed to suit the number of classes, and the fully connected layer is trained with the preprocessed images from AWE. In the second approach, the feature vector derived from the layer prior to the fully connected layers is fed to an SVM for classification. Data augmentation is carried out to improve the accuracy. The experiments show that ResNet50 is a better choice among the three models considered. Also, classification with the Fully Connected shallow network is more effective than the SVM model considered. The proposed approaches are promising and pave the way for further research in ear biometrics.

The study has considered only three basic networks. It shall be extended by incorporating more CNN models. Training can be extended to a few CNN layers also to improve the classification accuracy. Classification with other machine learning algorithms and hybrid classifier models are other possible improvements to the proposed architecture. Above all, the work needs to be extended with more data sets.

References

1. Lecun, Y., Bengio, Y., Hinton, G.: Deep learning. Nature (2015). https://doi.org/10.1038/nature14539
2. Krizhevsky, A., Sutskever, I., Hinton, G.E.: Imagenet classification with deep convolutional neural networks. In: Advances in Neural Information Processing Systems, pp. 1097–1105 (2012)
3. Simonyan, K., Zisserman, A.: Very deep convolutional networks for largescale image recognition. In: International Conference on Learning Representations (2014)
4. He, K., Zhang, X., Ren, S., et al.: Deep residual learning for image recognition. In: IEEE Conference on Computer Vision and Pattern Recognition, pp. 770–778 (2016)
5. Emersic, Z., Stepec, D., Struc, V., Peer, P: Training convolutional neural networks with limited training data for ear recognition in the wild. In: Proceedings - 12th IEEE International Conference on Automatic Face and Gesture Recognition, FERA (2017). https://doi.org/10.1109/FG.2017.123

6. Hansley, E.E., Segundo, M.P., Sarkar, S.: Employing fusion of learned and handcrafted features for unconstrained ear recognition. IET Biom. (2018). https://doi.org/10.1049/iet-bmt.2017.0210
7. Dodge, S., Mounsef, J., Karam, L.: Unconstrained ear recognition using deep neural networks. IET Biom. (2018). https://doi.org/10.1049/iet-bmt.2017.0208
8. Zhang, Y., Mu, Z., Yuan, L., Yu, C.: Ear verification under uncontrolled conditions with convolutional neural networks. IET Biom. (2018). https://doi.org/10.1049/iet-bmt.2017.0176
9. Omara, I., Wu, X., Zhang, H., Du, Y., Zuo, W.: Learning pairwise SVM on hierarchical deep features for ear recognition. IET Biom. (2018). https://doi.org/10.1049/iet-bmt.2017.0087
10. El-Naggar, S., Bourlai, T.: Evaluation of deep learning models for ear recognition against image distortions. In: Proceedings of the 2019 European Intelligence and Security Informatics Conference, EISIC 2019 (2019). https://doi.org/10.1109/EISIC49498.2019.9108870
11. Eyiokur, F.I., Yaman, D., Ekenel, H.K.: Domain adaptation for ear recognition using deep convolutional neural networks. In arXiv (2018)
12. Alshazly, H., Linse, C., Barth, E., Martinetz, T.: Ensembles of deep learning models and transfer learning for ear recognition. Sensors **19**, 4139 (2019)
13. Alshazly, H., Linse, C., Barth, E., Martinetz, T.: Deep convolutional neural networks for unconstrained ear recognition. IEEE Access **8**, 170295–170310 (2020). https://doi.org/10.1109/ACCESS.2020.3024116
14. Hoang, V.T.: EarVN1.0: a new large-scale ear images dataset in the wild. Data Brief **27** (2019)
15. Russakovsky, O., et al.: ImageNet large scale visual recognition challenge. Int. J. Comput. Vis. **115**(3), 211–252 (2015). https://doi.org/10.1007/s11263-015-0816-y
16. Emeršič, Ž., Štruc, V., Peer, P.: Ear recognition: more than a survey. Neuro-computing (2017). https://doi.org/10.1016/j.neucom.2016.08.139
17. Wang, J., Perez, L.: The effectiveness of data augmentation in image classification using deep learning. ArXiv (2017)
18. https://github.com/aleju/imgaug. Accessed 20 Jan 2021

An Adaptive Service Placement Framework in Fog Computing Environment

Pankaj Sharma[ID] and P. K. Gupta[✉][ID]

Department of Computer Science and Engineering, Jaypee University of Information
Technology, Solan 173 234, HP, India
pkgupta@ieee.org

Abstract. In the present scenario, the world is poignant towards smart
devices, particularly after the Internet of Things (IoT). The IoT devices
usually accumulate the data from the sensing environment. It has inhib-
ited the capabilities of computation and storage. This leads to an increase
in IoT integration with cloud computing operations. Fog computing is
the extension of cloud computing environment and enhances the performance of the cloud. The main concern in fog computing is basically
the reliability of the fog nodes that communicates with the various IoT
devices and further with the cloud. In this work, we have proposed a
fault detection framework with service placement to efficient fog node.
This model implements an Adaptive Quality of Service (QoS) conscious
technique with the amalgamation of two methods i.e. Checkpoints and
Replication(CR) and utilize a novel Bee mutation(BM) algorithm with
improved features for best possible service placement to fog node. In the
proposed technique, the performance of the fog nodes is monitored using
a fog service monitor. We have also evaluated the proposed framework
with various metrics for its performance. The proposed framework is also
compared with the existing algorithm based framework. The total execution cost and usage of network of the proposed model are about 84023
USD and 618950 Mbps, respectively.

Keywords: IoT · Fog nodes · Fault tolerance · Service placement ·
Quality of Service

1 Introduction

In this scientific era people are more liable to use smart devices broadly in their
day-to-day life. With the initiation of IoT technology has transform the practice
of smart devices with valuable data congregation [6]. The increase sensors in wide
range of applications are apparent as it supports automation and improves the
day-to-day lives of humans. Numerous IoT-based frameworks have been established that monitors the environment using these sensor devices [14]. However,
the utilization of IoT devices in monitoring of an automated fog environment, is

© Springer Nature Switzerland AG 2021
M. Singh et al. (Eds.): ICACDS 2021, CCIS 1440, pp. 729–738, 2021.
https://doi.org/10.1007/978-3-030-81462-5_64

a composite system and requires a module for accurate detection of fault in the fully mechanized system. This effectual novel fault diagnosing framework may put aside the system and individuals from any disastrous actions [10]. Among a variety of applications of IoT technology, a lot of industrial organizations are now establishing IoT technology for the advancement of their business functions. The major use of IoT technology in industrialization is to monitor fault detection and diagnosis (MFDD) procedures to ensure their proper working [22].

In this work, we have projected a framework that implements a checkpoint and restart mechanism that checks for failed tasks and then restart them from the point of breakdown in the IoT-Fog system. Proposed replication approach generates the variety of job duplication on diverse resources unless or until the complete task gets crash. Further, service allocation in the fog computing framework is optimized using the hybrid Bee Mutation algorithm. This work also consider performance metrics for service placement to a efficient fog node. The proposed technique is the modified version of technique as discussed in [16].

The major sections of this paper are arranged in the subsequent series as Sect. 2 encloses the similar improvements on fault tolerance and service placements in the fog computing environment. Section 4 describes the proposed design. Section 4 also depicts the performance of the proposed framework for discussions. The final section provides the future scope of the this work with concluding remarks.

2   Related Work

In [3], an optimized model based on genetic algorithm has been proposed for mapping of flow of data towards fog nodes. The proposed model considers the current status of the load on the various fog nodes and the latency among the various fog nodes and sensors. This work also deals with the archetypal scenario of a smart city, in which intelligent devices are dispersed over a large geographic area. In [21], a technique for recuperating the reliability of data transfer of fog nodes has been proposed for health care domain. The proposed policy establishes the various methods of directed diffusion and limited flooding. The proposed mechanism employed an implemented scheme that controls the parameter based on the current scenarios of fog. In [13], a fault-tolerant mechanism has been proposed which is based on a four-step protocol. This protocol saves the state of the system with an integrated message logging and checkpoint approach. The proposed protocol also monitors the surrounding environment and creates failure reports. If any probability or perception of failure, the protocol initiates the required steps and avoids appropriate decision-making. Whereas, if any failure takes place, the protocol commences the redesign of the contingent entity through proper recovery actions. In [12], have proposed another fault-tolerant based technique a tree structure has been used for fog computing environment. The proposed technique is a composition of two technique i.e. replication and non-replication. Replication facilitates analogous requests through various fog nodes and the non-replication method reinstate the faulty one with a new fog node. [11], have projected a proficient assignment dilemma to tackle the allocation issue of the fog nodes to IoT devices through the iterative schemes that curtail the energy utilization in delay controlled mobile computing environment. Authors have also utilize numerous meta-heuristic algorithms like Ant Colony Optimization, Genetic Algorithm and Particle Swarm Optimization in the systems that smooth the progress for the existing system in terms of distributed and dynamic computing. Quality of Experience is used to compute the accomplishment of the user agreements in terms of QoS. In [2], Quality of Experience monitor stakeholder privileges and sensitivity in fastidious situation. QoE cognizant technique can improve user trustworthiness and reduce service renounce rate. In [9,19,20], authors have discussed QoE-aware based techniques that have been used for optimizing service coverage. QoE can also be employed for resource estimation [4]. Apart from the service provisioning it can also reduce the data processing time, better resource utilization and quality of network. A multi-tier service placement architecture based on context aware task provisioning has been proposed [18], The work projected in this paper was basically exploit available virtual resources on the network edges to get better the recitation of the system in terms of energy efficiency, average response time and less cost. Energy aware technique for placement of IoT based application has been devised [7]. This divided the service placement into two folds one for critical service and the other for normal one, critical services aims to reduced response time and the objective of the normal one is to reduce the consumption of energy. A fault tolerant aware scheduling method is mainly proposed to reduce the overheads of service, latency and to increase the reliability In [1], authors have proposed a method for classifying the devices that can send requests into three classes depends upon the type of service required. These classes are time-sensitive, time-tolerant and core. In [16], QoS aware technique which was also based on combination of Checkpoints and Replication (CR) faults diagnosis technique for proficient allocation of jobs based on Bee Mutation algorithm have been projected in this work. Checkpoints, Replication and Bee Mutation module interact with the service monitor to make sure that the projected framework is fault-tolerant. Further, the IoT-Fog based framework has been appraised on many performance parameters.

3   Motivation

The process of mapping requests from various IOT devices to a set of fog nodes without infringing the service level agreement(SLA) and without hindering the QoS parameter in the environment is referred to as placement of Service problem in the environment of fog computing. In order to obtain the maximum advantage from a proposed framework, we need to implement the efficient service placement policy that can achieve various goals: i) Response Time Minimization ii) Minimization of Execution Cost iii)Minimization of Network usage analysis

4   Proposed Methodology

The main objective of the proposed framework is to determines the fault in IoT-Fog environment and places the services as shown in Fig. 1. The proposed framework includes various IoT devices, cloud and fog layer. In this environment IoT devices collects the data from the sensing environment and transferred it to the fog nodes to perform computations on this data. The main focus of this framework is to ensure the scalability and accessibility of data through efficient load balancing, and job scheduling for the allocation of resources. Fog nodes process the request after receiving it, and then communicate it to the user. The upcoming request is processing through efficient node among the available fog nodes. The response is quickly processed without any loss of the data.

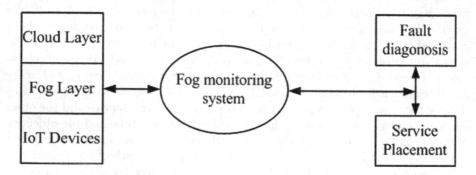

Fig. 1. Proposed IoT- Fog architecture

Further, for efficient functioning of the fog nodes, it should be free from any fault. To achieve better fault tolerance, two techniques have been proposed in the framework. The first step is to generate the checkpoints, and the second step is to buildup different replicas of the fog nodes. The service placement usually depends on the QoS parameter like resource utilization, i.e. CPU, RAM, execution time, processing speed, response time, and delay. In the proposed framework, Bee-Mutation algorithm has been used for effective placement of Fog nodes, and for optimization of the various QoS parameters. The proposed model is a composition of Cloud layer, fog layer, IoT devices fog monitoring system, fault tolerance and service placement module. The functions of the proposed model is as under: i. IOT devices collect data from the real time environment and send it to fog layer for further processing and computation. Fog layer consist of fog devices. ii. After processing, data is available for gratifying client requests and fog node perform reallocation with respect to the demand of clients. iii. Fog nodes receive the request and perform computation on it and resent the response to the client. The processing of the client requests is carried out through the efficient node among all the fog nodes that are utilized for the IoT services. The response is swiftly processed without any loss of data via efficient node. iv. As per the perspective of fault tolerance module checkpoints are placed at the diverse intervals during the processing of the fog nodes and also examining the current status of the fog nodes. The replication mechanism along with checkpoint to defend the node from behaving differently.

4.1   Fault Diagnosing

The proposed framework implements the Checkpoints- Replications mechanism for diagnosing of faults which is carried out in two steps [16]. First of all , checkpoints method are positioned at a different periods of time during the execution of client requests at various fog nodes.Checkpoints can further study the eminence of every fog node and keep the system in operative state irrespective of any failure occurs in the system. The checkpoints technique can also access resource server and scheduler. From the scheduler, it gather the information of fog nodes that are necessary for responding. Secondly, the replication technique uses a passive replica for diagnosing the fault. Here, as the fog nodes receive the processing request, they forms a primary node that processes the particular request. In this technique, the primary fog nodes, and the secondary or back up fog nodes works concurrently in the background. In the event of failure of a primary node, the secondary node takeover the control and ensures that the request is executed effectively in the IoT-Fog environment. This replication and checkpoint technique ensures the protection of the nodes from behaving differently after recovering from any fault.

4.2   Service Placement to Fog Node

The imperative contemplation in this framework is the allocation of incoming service requests to the efficient fog node. For proficient service allocation, the bee mutation technique with improved features has been implemented [5][8]. We have considered the following three steps:

- Performance monitoring
- Decision making
- Service Placement

The Quality of Service metric of the fog nodes like memory and CPU usage, network bandwidth, average response time (ART), cost, and processing speed have been considered by the Bee-Mutation algorithm. The input parameters are fed into the Bee-Mutation algorithm. The proposed algorithm predict the performance of the each Fog node with respect to the QoS parameters. These small set of metric perform an imperative role in increasing the overall performance of the proposed framework, it can also calculate the average response time (ART) for each fog node by using Eq. 1 [15]:

$$ART = \frac{1}{n} \sum Ft - At + TDelay \tag{1}$$

At and Ft is the is the arrival time and finish time of the incoming requests and communication delay can also be calculated using Eq. 2:

$$TDl = Tl + Ttf \tag{2}$$

TDl is the communication delay, Tl is the network latency and Ttf is the time taken to transfer the size of single job request from source to destination location in Eq. 3.

$$Ttransfer = \frac{D}{Bwperuser} \tag{3}$$

$$Bwperuser = \frac{Bwtotal}{Nr} \tag{4}$$

Where, $Bwtotal$ can be consider as available bandwidth in the projected framework and Nr is the number of client requests currently in communication as shown in Eq. 4. The cost of the fog node may vary from node to node and can be calculated using Eq. 5.

$$Cost(D) = ProcessingTime * NumberofFogNode \tag{5}$$

The cross over function of the well known genetic algorithm is put into service over the acquired probabilities, Performance has been monitored based on the requirement of the service, Once an anomaly has been detected, the system begins to select the fog node from the collection of fog nodes and a decision is taken for the allocation of service to a efficient fog node based on its performance. A small subset of parameters has been considered while allocation of requests to efficient node. In real, there can be a dynamic change in IoT devices and there can also be an enhancement of computing cost. Therefore, a monitoring and analyzing mechanism is required that solves the problem of service placement.

5   Results and Discussion

The anticipated framework is simulated with the well known tool iFogSim and implemented on different Window-based platforms. The projected framework has been appraised for its recital by using different metrics. In this work, five fog devices (Fd) have been considered, and their performance is shown in Table 1. The performance of all the five fog devices(Fd) is depicted in Table 1.

The most significant parameter is throughput for effective data communication, mainly in the IoT Fog network which is measured by number of packets transmitted per unit time by the device. The average throughput amongs the five fog devices is about 85.2. Response time was the total amount of time utilized by a particular fog nodes to respond to the received request. Fundamentally response time is the total accretion of time for makespan, node deployment and communication time for the incoming request. Response time for the projected framework was evaluated as 78.8 ms. In this system, the aptitude to process the number of clients request without distressing the ability of fault tolerance is known as scalability. The overall average scalability of the five devices is 84 numbers of client requests processed. The overall effectiveness of the proposed system is computed as 86.4%. For the existing or proposed system, the primary requirement is the availability of fog nodes/devices to receive the request

Table 1. Performance evaluation of fog devices (Fd).

Metric	Fd 1	Fd 2	Fd 3	Fd 4	Fd 5	Average
Throughput	95	81	87	68	95	85.2
Response time	87	78	90	70	69	78.8
Scalability	69	93	69	94	95	84
Availability	96	88	62	77	67	78
Usability	73	83	94	67	85	80.4
Reliability	87	99	78	64	79	81.4
Cost effectiveness	89	68	73	72	73	75

and execute it. In the anticipated Iot Fog system, the overall availability is calculated as 78.2%. Usability can be defined as the mechanism which describes the number of exhausted resources to carry out the process efficient. The reliability is basically a probability of the particular system that premeditated its efficiency to provide precise outcomes. The reliability and usability of the anticipated IoT-Fog framework was evaluated as 80.6% and 79%, respectively. For any efficient system, it is mandatory to comprise cost and other overheads. So, the system be obliged to cost-effective with lower overheads. The overall average overhead of this IoT-fog system was intended as 76.6%, and its corresponding cost-effectiveness was calculated as 73.4%.

5.1 Comparative Analysis

The proposed IoT-Fog has also been compared with the existing CRBM based genetic algorithm based on two parameters i.e. network usage and overall execution cost. The total execution cost of the proposed IoT-Fog algorithm is 84023($) which is less than the cost as obtained from existing algorithm i.e. 85042.20($) as depicted in Fig. 2. The total network usage of the existing GA based IoT-Fog is 635550 (Mbps), which is higher than the proposed IoT-Fog with usage of network is of 618950 (Mbps) and shown in Fig. 3. The comparison of existing Genetic algorithm and proposed algorithms in the IOT-Fog Environment as depicted in Table 2.

Table 2. Comparison between the proposed Qos aware Algorithm and existing algorithms

Parameter	Existing genetic algorithm [17]	Novel bee mutation
Execution cost ($)	85042	84023
Total network usage (Mbps)	635550	618950

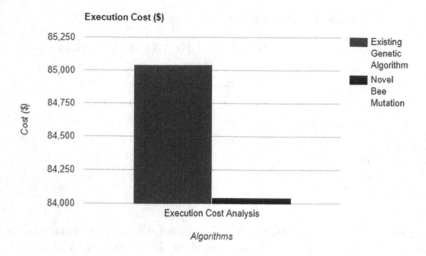

Fig. 2. Execution cost

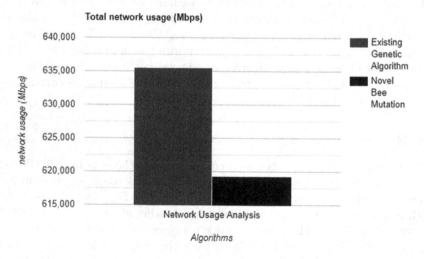

Fig. 3. Network usage analysis

6   Conclusion

The QoS based fault tolerant framework for efficient service allocation has been anticipated. This framework employ the two well-known fault diagnosing techniques recognized as checkpoints and replication by using fog monitoring system. The checkpoint techniques commence the procedure from the establishment of pre-defined checkpoints at the occurrence of failure. The replication process avoid the delay in the checkpoint mechanism. This makes proposed framework fault-tolerant, and ensures the allocation of service requests effectively among the efficient fog nodes by using proactive method for detection of average response time, and execution cost. In order to make efficient service allocation, a hybrid bee-mutation method has been used. Performance of the proposed IoT-Fog based system's has been appraised based on diverse recital metric like throughout, response time, cost effectiveness, and several other metrics. The proposed IoT-Fog model has also been compared with exiting GA based algorithm over the total execution cost and usage of network. In the obtained results it is found that the proposed model surpass the existing GA model for network usage which is 618950 (Mbps) and execution cost which is 84023 ($).

References

1. Alarifi, A., Abdelsamie, F., Amoon, M.: A fault-tolerant aware scheduling method for fog-cloud environments. PloS one **14**(10), e0223902 (2019)
2. Baranwal, G., Yadav, R., Vidyarthi, D.P.: QoE aware IoT application placement in fog computing using modified-topsis. Mob. Netw. Appl. **25**(5), 1816–1832 (2020)
3. Canali, C., Lancellotti, R.: Gasp: genetic algorithms for service placement in fog computing systems. Algorithms **12**(10), 201 (2019)
4. Crespi, N., Molina, B., Palau, C., et al.: QoE aware service delivery in distributed environment. In: 2011 IEEE Workshops of International Conference on Advanced Information Networking and Applications, pp. 837–842. IEEE (2011)
5. Espling, D., Larsson, L., Li, W., Tordsson, J., Elmroth, E.: Modeling and placement of cloud services with internal structure. IEEE Trans. Cloud Comput. **4**(4), 429–439 (2014)
6. Gokhale, P., Bhat, O., Bhat, S.: Introduction to IoT. Int. Adv. Res. J. Sci. Eng. Technol **5**(1), 41–44 (2018)
7. Hassan, H.O., Azizi, S., Shojafar, M.: Priority, network and energy-aware placement of IoT-based application services in fog-cloud environments. IET Commun. **14**(13), 2117–2129 (2020)
8. Kayal, P., Liebeherr, J.: Autonomic service placement in fog computing. In: 2019 IEEE 20th International Symposium on "A World of Wireless, Mobile and Multimedia Networks" (WoWMoM), pp. 1–9. IEEE (2019)
9. Lin, Y., Shen, H.: Cloud fog: towards high quality of experience in cloud gaming. In: 2015 44th International Conference on Parallel Processing, pp. 500–509. IEEE (2015)
10. Lo, N.G., Flaus, J.M., Adrot, O.: Review of machine learning approaches in fault diagnosis applied to iot systems. In: 2019 International Conference on Control, Automation and Diagnosis (ICCAD), pp. 1–6. IEEE (2019)
11. Mebrek, A., Merghem-Boulahia, L., Esseghir, M.: Efficient green solution for a balanced energy consumption and delay in the IoT-Fog-cloud computing. In: 2017 IEEE 16th International Symposium on Network Computing and Applications (NCA), pp. 1–4. IEEE (2017)
12. Oma, R., Nakamura, S., Duolikun, D., Enokido, T., Takizawa, M.: Fault-tolerant fog computing models in the IoT. In: Xhafa, F., Leu, F.-Y., Ficco, M., Yang, C.-T. (eds.) 3PGCIC 2018. LNDECT, vol. 24, pp. 14–25. Springer, Cham (2019). https://doi.org/10.1007/978-3-030-02607-3_2
13. Ozeer, U., Etchevers, X., Letondeur, L., Ottogalli, F.G., Salaün, G., Vincent, J.M.: Resilience of stateful IoT applications in a dynamic fog environment. In: Proceedings of the 15th EAI International Conference on Mobile and Ubiquitous Systems: Computing, Networking and Services, pp. 332–341 (2018)

14. Sethi, P., Sarangi, S.R.: Internet of things: architectures, protocols, and applications. J. Electric. Comput. Eng. **2017** (2017)
15. Sharma, M., Sharma, P.: Performance evaluation of adaptive virtual machine load balancing algorithm. Perf. Eval. **3**(2), 86–88 (2012)
16. Sharma, P., Gupta, P.: QoS-aware CR-BM-based hybrid framework to improve the fault tolerance of fog devices. J. Appl. Res. Technol. **19**(1), 66–76 (2021)
17. Skarlat, O., Nardelli, M., Schulte, S., Borkowski, M., Leitner, P.: Optimized IoT service placement in the fog. Serv. Orient. Comput. Appl. **11**(4), 427–443 (2017)
18. Tran, M.Q., Nguyen, D.T., Le, V.A., Nguyen, D.H., Pham, T.V.: Task placement on fog computing made efficient for IoT application provision. Wirel. Commun. Mob. Comput. **2019** (2019)
19. Varshney, S., Sandhu, R., Gupta, P.K.: QoS based resource provisioning in cloud computing environment: a technical survey. In: Singh, M., Gupta, P.K., Tyagi, V., Flusser, J., Ören, T., Kashyap, R. (eds.) ICACDS 2019. CCIS, vol. 1046, pp. 711–723. Springer, Singapore (2019). https://doi.org/10.1007/978-981-13-9942-8_66
20. Varshney, S., Sandhu, R., Gupta, P.: QoE-based multi-criteria decision making for resource provisioning in fog computing using AHP technique. Int. J. Knowl. Syst. Sci. (IJKSS) **11**(4), 17–30 (2020)
21. Wang, K., Shao, Y., Xie, L., Wu, J., Guo, S.: Adaptive and fault-tolerant data processing in healthcare IoT based on fog computing. IEEE Trans. Netw. Sci. Eng. **7**(1), 263–273 (2018)
22. Yen, I.L., Zhang, S., Bastani, F., Zhang, Y.: A framework for IoT-based monitoring and diagnosis of manufacturing systems. In: 2017 IEEE Symposium on Service-Oriented System Engineering (SOSE), pp. 1–8. IEEE (2017)

Efficient Ink Mismatch Detection Using Supervised Approach

Garima Jaiswal[✉], Arun Sharma, and Sumit Kumar Yadav

Indira Gandhi Delhi Technical University for Women, Delhi, India
{garima002phd18,arunsharma,sumit}@igdtuw.ac.in

Abstract. Hyperspectral imaging captures and analyzes the images from hundreds to thousands of spectral bands, which empowers it to examine the distinctive features that conventional imaging cannot detect. Hyperspectral imaging is a noted technology for the remote sensing domain. But its capacity for ink mismatch detection is still evolving. In this study, we implemented three variants of KNN-fine (F), coarse (CO), and cosine (COS), bagging and boosting ensemble methods, three variants of decision trees-fine (F), Medium (M), and Coarse (CO), and CNN to showcase the impact of varying ink mixing ratios to detect ink mismatch. A comparative analysis with the previous work using the same dataset is illustrated, which indicates that the proposed approach surpasses previous results for blue inks mixed in unequal ratios.

Keywords: Hyperspectral imaging · Ink mismatch detection · KNN · Decision trees · Ensemble methods · CNN

1 Introduction

Analyzing inks plays a crucial role in answering many significant concerns related to the questioned document. Investigating the inks' chemical and physical properties in the questioned documents reveals worthy information concerning their fidelity, which forms the basis for detecting forgery, fraud, backdating, and ink age. To achieve this, one of the main tasks is to differentiate between the visually similar inks, which the naked eye cannot do. The two prime approaches implemented to determine inks are destructive and non-destructive analysis [1]. The non-destructive analysis is preferred over destructive analysis to rescue the document post to an investigation. Hyperspectral imaging is an emerging, effective and non-destructive approach that can reveal the underlying information to discriminate the inks.

Each pixel for the grayscale images requires a single byte, capturing a pixel's brightness in a single band. In comparison, the colored images need three bytes for capturing the same at three bands-red, green, and blue. By extending the number of bands, more meaningful information can be retrieved. Hyperspectral images capture information from hundreds to thousands of bands. For objects under analysis, imaging yields spatial and temporal information while spectroscopy yields spectral information [2]. Hyperspectral

© Springer Nature Switzerland AG 2021
M. Singh et al. (Eds.): ICACDS 2021, CCIS 1440, pp. 739–746, 2021.
https://doi.org/10.1007/978-3-030-81462-5_65

imaging is the concatenation of imaging and spectroscopy techniques. It delivers spectral and spatial information which can't be retrieved by imaging and spectroscopy. The third "spectrum" dimension is appended to the two-dimensional image data resulting in a three-dimensional data cube used for analyzing the hyperspectral images. The spatial dimension disseminates the shape, size, appearance, and texture, whereas the spectral dimension discovers the chemical composition for an object under analysis [3].

Hyperspectral imaging started its root from remote sensing and is now emerging in biomedicine, food quality, document forgery, agriculture, ecology, mining, defense, environment analysis, medical diagnosis, ocean analysis, mineral mapping, history, and archaeology conservation [3–8].

This work aims to demonstrate the potential of hyperspectral imaging for ink mismatch detection. We mixed two, three, four, and five different inks in varied proportions using the UWA Writing Ink Hyperspectral Images (WIHSI) dataset to address the ink mismatch detection. Eight supervised machine learning classifiers and CNN is experimented to showcase the impact of varying ink mixing ratios. A comparative study of the proposed work with previous work on the same dataset is elaborated.

2 Related Work

Khan et al. [8] implemented a k-means algorithm to detect the ink mismatch in hyperspectral document images by mixing only two inks in equal proportions. To work on selected bands, a forward feature selection approach using ten-fold cross-validation has been experimented. The limitation of mixing inks in equal ratios and prior knowledge about a number of inks was addressed by Luo et al. [9]. The authors implemented anomaly detection with unsupervised clustering and mixed two inks in unequal proportions.

The joint sparse band selection and spectral features were explored for automatic ink mismatch detection in [10]. The authors implemented PCA for dimensionality reduction and feature selection. Unsatisfactory results were achieved in the case of varying ink mixing ratios. Abas et al. [11] forecasted disproportionate ink mixing using the concept of hyperspectral unmixing. The authors identified the end members and their abundances by utilizing spectral signatures. Their work had had a limitation that required postprocessing for manually discarding the non-coherent abundance maps.

Spectral features of multispectral images were utilized for illustrating the ink mismatch for two inks mixed in equal proportions in [12]. Khan et al. used a fuzzy C-Means algorithm instead of hard clustering algorithms. The results did not suffice for disproportionate ink combinations.

Khan et al. [13] implemented CNN for ink mismatch detection. By utilizing the spectral features of six authors' for training purposes and one author for testing purposes. The results depicted average accuracy for black inks, and their approach required prior knowledge about ink mixing ratios. Authors in [14] captured spectral and spatial features for implementing CNN to demonstrate ink mismatch detection. The results indicated good accuracy for both blue and black inks. Islam et al. [15] explored spatial and spectral features through CNN for illustrating writer identification problem. They mixed inks of different authors in varying proportions and achieved an accuracy of 71.28%.

Devassy and George [16] provided a comparative analysis of CNN's results and two traditional hyperspectral classification approaches-Spectral Angle Mapper and Spectral Information Divergence. The outcomes exemplified the utility of CNN for ink mismatch detection. The authors in [17] addressed the processing and storing issue due to the presence of a large number of bands in hyperspectral images. They explored the t-Distributed Stochastic Neighbor Embedding (t-SNE) algorithm for dimensionality reduction. Devassay et al. [18] utilized the same approach as in [17] to classify different types of paper data.

3 Proposed Approach

The block diagram for the proposed approach is illustrated in Fig. 1. The dataset is initially exposed to preprocessing. Post to which supervised machine learning classifiers and CNN is implemented followed the inks' classification in five types.

3.1 UWA Writing Ink Hyperspectral Images Dataset

In the study, we utilized UWA Writing Ink Hyperspectral Images Dataset (WIHSI). The dataset comprised seven subjects. Each subject wrote two documents-blue ink and black ink. Each document included five lines written by five different types of ink, respectively. All subjects wrote the same line, "The quick brown fox jumps over the lazy dog," in both documents with five different types of inks. Each document was captured from 400 nm to 720 nm at steps of 10 nm, resulting in fourteen images of size 752 * 480 pixels.

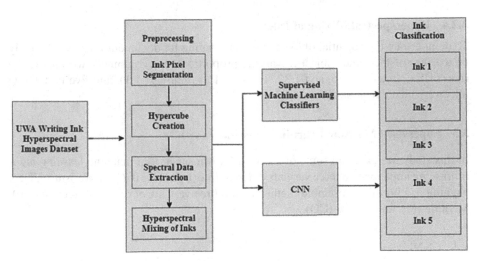

Fig. 1. Block Diagram for the Proposed Approach.

3.2 Preprocessing

The preprocessing phase consists of several steps to extract the ink pixels from the hyperspectral document images. This section elaborates on the preprocessing steps.

3.2.1 Ink Pixel Extraction

The primary task was to segregate the foreground and background pixels. To achieve this task, the local thresholding approach Sauvola was explored. The thresholding was applied on a single selected band (25th band) [8]. Global thresholding was not implemented due to an uneven illumination problem.

3.2.2 Hypercube Creation

Each document is segmented into five lines indicating each ink type after separating the foreground and background pixels and followed by capturing the binary mask for all 33 bands. This phase's outcome is 70 images comprising a single line of the document, each with dimensions of 81 * 627 * 33. The resultant structure is known as a hypercube.

3.2.3 Spectral Data Extraction

The foreground ink pixel values were used for spectral data extraction. The corresponding foreground pixel values across all 33 bands were captured. The resulting spectral vectors of six subjects were utilized for training purposes and used one subject for testing purposes. Prior to implementation, the spectral vectors were exposed to normalization.

3.2.4 Hyperspectral Mixing of Inks

To demonstrate the potential of hyperspectral imaging for document forgery. Manually mixed inks of the same subject in varying proportions. We combined two inks in the ratio of 1:1, 1:4, 1:8, 1:16, and 1:32. Three (1:1:1), four (1:1:1:1), and five (1:1:1:1:1) inks were mixed in equal ratios.

3.3 Supervised Machine Learning Classifiers

In this study, we explored some of the well-known supervised machine learning algorithms. We implemented three variants of KNN-fine (F), coarse (CO), and cosine (COS), bagging and boosting ensemble methods, and three variants of decision trees-fine (F), Medium (M), and Coarse (CO).

3.4 CNN

Convolutional Neural Network is a contemporary deep learning approach used extensively for image classification, object localization, and segmentation [13]. The 1 * 33 spectral response vectors captured during preprocessing phase were padded with three zeros to convert it into a matrix of size 6 * 6. The reshaped matrix was inputted to a CNN. The details of CNN are illustrated in Fig. 2.

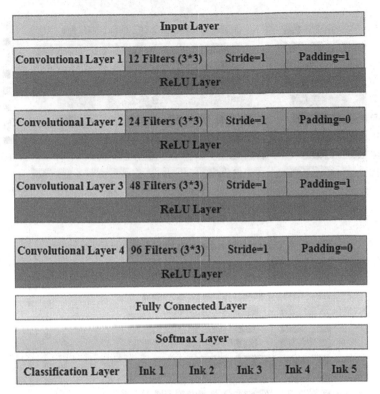

Fig. 2. Details of CNN Architecture.

4   Results

MATLAB's Statistics and Machine Learning Toolbox R202020b were used for the experimentation purpose. Accuracy is the count of rightly classified predictions to the total sum of all predictions made by a classifier. The accuracy attained at varying ink mixing ratios for blue and black inks is illustrated in Fig. 3 and Fig. 4. CNN achieves the highest accuracy among all classifiers for both blue and black ink, as shown in Fig. 5. As black inks are compacted towards a narrow range of reflectance compared to blue inks, huge accuracy variation was noted. A comparative analysis with the previous work using the same dataset is elaborated in Table 1, which indicates that the proposed approach surpasses previous results for blue inks mixed in unequal ratios.

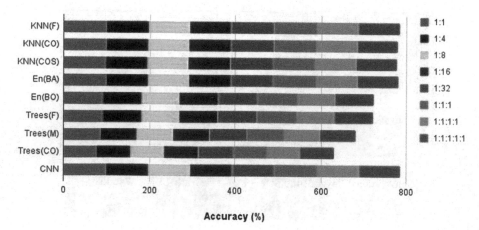

Fig. 3. Accuracy attained at varying blue ink mixing ratios.

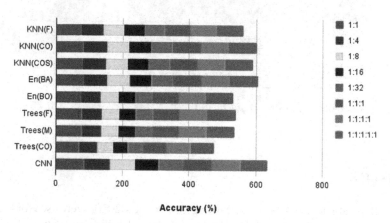

Fig. 4. Accuracy attained at varying black ink mixing ratio

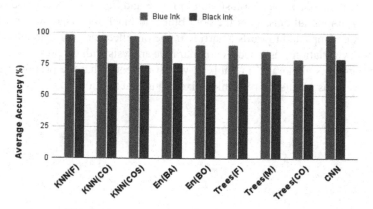

Fig. 5. Average accuracy for blue and black inks.

Table 1. Comparative analysis of the results achieved by proposed approach and previous techniques.

Technique	Average accuracy (%) blue	Average accuracy (%) black	No. of inks mixed	Mixing ratio
Abbas et al. [11]	86.2	83.4	2–4	Equal and unequal
Luo et al. [9]	89.0	82.3	2	Equal and unequal
Khan et al. [8]	85.6	81.4	2	Equal
KNN(F)	97.99	70.11	2–5	Equal and unequal
KNN(CO)	97.53	75.28	2–5	Equal and unequal
KNN(COS)	97.18	73.75	2–5	Equal and unequal
En(BA)	97.63	75.83	2–5	Equal and unequal
En(BO)	90.32	66.42	2–5	Equal and unequal
Trees(F)	90.25	67.34	2–5	Equal and unequal
CNN	98.34	79.32	2–5	Equal and unequal

5   Conclusion and Future Work

In this work, the potential of hyperspectral imaging for ink mismatch detection was examined. For this purpose, two, three, four, and five different inks were mixed in varied proportions. We implemented eight supervised machine learning classifiers and CNN to showcase the impact of varying ink mixing ratios. CNN achieves the highest accuracy among all classifiers for both blue and black ink. A comparative analysis with the previous works demonstrated that the proposed approach surpasses previous results for blue inks mixed in unequal ratios. As black inks are compacted towards a narrow range of reflectance compared to blue inks, huge accuracy variation was noted. In future work, a new approach can be explored to deal with black inks' compactness issue. An unsupervised approach can be examined for automatic ink mismatch detection to obliterate the need for initial information regarding the dataset and trained classifier.

References

1. Deep Kaur, C., Kanwal, N.: An analysis of image forgery detection techniques. Stat. Optim. Inf. Comput. **7**(2), 486–500 (2019)

2. Lu, G., Fei, B.: Medical hyperspectral imaging: a review. J. Biomed. Opt. **19**(1), 010901 (2014)
3. Siche, R., Vejarano, R., Aredo, V., Velasquez, L., Saldana, E., Quevedo, R.: Evaluation of food quality and safety with hyperspectral imaging (HSI). Food Eng. Rev. **8**(3), 306–322 (2016)
4. Lowe, A., Harrison, N., French, A.P.: Hyperspectral image analysis techniques for the detection and classification of the early onset of plant disease and stress. Plant Methods **13**(1), 1–12 (2017)
5. Kamilaris, A., Prenafeta-Boldu, F.X.: Deep learning in agriculture: a survey. Comput. Electron. Agric. **147**, 70–90 (2018)
6. Wang, K., Franklin, S.E., Guo, X., Cattet, M.: Remote sensing of ecology, biodiversity and conservation: a review from the perspective of remote sensing specialists. Sensors **10**(11), 9647–9667 (2010)
7. Van der Meer, F.D., et al.: Multi-and hyperspectral geologic remote sensing: a review. Int. J. Appl. Earth Obs. Geoinf. **14**(1), 112–128 (2012)
8. Khan, Z., Shafait, F., Mian, A.: Hyperspectral imaging for ink mismatch detection. In: 2013 12th International Conference on Document Analysis and Recognition, pp. 877–881. IEEE, August 2013
9. Luo, Z., Shafait, F., Mian, A.: Localized forgery detection in hyperspectral document images. In: 2015 13th International Conference on Document Analysis and Recognition (ICDAR), pp. 496–500. IEEE, August 2015
10. Khan, Z., Shafait, F., Mian, A.: Automatic ink mismatch detection for forensic document analysis. Pattern Recogn. **48**(11), 3615–3626 (2015)
11. Abbas, A., Khurshid, K., Shafait, F.: Towards automated ink mismatch detection in hyperspectral document images. In: 2017 14th IAPR International Conference on Document Analysis and Recognition (ICDAR), vol. 1, pp. 1229–1236. IEEE, November 2017
12. Khan, M.J., Yousaf, A., Khurshid, K., Abbas, A., Shafait, F.: Automated forgery detection in multispectral document images using fuzzy clustering. In: 2018 13th IAPR International Workshop on Document Analysis Systems (DAS), pp. 393–398. IEEE April 2018
13. Khan, M.J., Yousaf, A., Abbas, A., Khurshid, K.: Deep learning for automated forgery detection in hyperspectral document images. J. Electron. Imaging **27**(5), 053001 (2018)
14. Khan, M.J., Khurshid, K., Shafait, F.: A spatio-spectral hybrid convolutional architecture for hyperspectral document authentication. In: 2019 International Conference on Document Analysis and Recognition (ICDAR), pp. 1097–1102. IEEE, September 2019
15. Islam, A.U., Khan, M.J., Khurshid, K., Shafait, F.: Hyperspectral image analysis for writer identification using deep learning. In: 2019 Digital Image Computing: Techniques and Applications (DICTA), pp. 1–7. IEEE, December 2019
16. Devassy, B.M., George, S., Hardeberg, J.Y.: Comparison of ink classification capabilities of classic hyperspectral similarity features. In: 2019 International Conference on Document Analysis and Recognition Workshops (ICDARW), vol. 8, pp. 25–30. IEEE, September 2019
17. Devassy, B.M., George, S.: Ink classification using convolutional neural network. NISK J., 12 (2019)
18. Devassy, B.M., George, S.: Dimensionality reduction and visualisation of hyperspectral ink data using t-SNE. Forensic Sci. Int. **311**, 110194 (2020)

Author Index

Printed in the United States
by Baker & Taylor Publisher Services